COSMOLOGICAL PHYSICS

This textbook provides a comprehensive introduction to modern cosmology, at a level suitable for advanced undergraduates and graduate students. The essential concepts and key equations used by professional researchers in both theoretical and observational cosmology are derived and explained from first principles.

The first third of the book carefully develops the necessary background in general relativity and quantum fields. The remainder of the volume then provides self-contained accounts of the principal topics in contemporary cosmology, including inflation, topological defects, gravitational lensing, the distance scale, large-scale structure and galaxy formation.

Throughout, the emphasis is on helping students to attain a physical and intuitive understanding of the subject. The book is therefore extensively illustrated, and outline solutions to more than 90 problems are included. All necessary astronomical jargon is clearly explained, ensuring that the book is self-contained for students with a background in undergraduate physics.

John Peacock studied physics at the University of Cambridge, obtaining his Ph.D. in 1981. He subsequently worked as a research astronomer at the Royal Observatory, Edinburgh, where he was appointed Head of Research in 1992. In 1998, he moved to become Professor of Cosmology at the University of Edinburgh. His research interests are a mixture of theoretical and observational cosmology: galaxy formation and clustering; high redshift galaxies and quasars; and gravitational lensing. He is married with three children. Time not devoted to science or family is shared between playing classical clarinet and exploring Scottish mountains.

COSMOLOGICAL PHYSICS

J. A. PEACOCK

PUBLISHED BY THE PRESS SYNDICATE OF THE UNIVERSITY OF CAMBRIDGE
The Pitt Building, Trumpington Street, Cambridge CB2 1RP, United Kingdom

CAMBRIDGE UNIVERSITY PRESS
The Edinburgh Building, Cambridge CB2 2RU, UK http://www.cup.cam.ac.uk
40 West 20th Street, New York, NY 10011-4211, USA http://www.cup.org
10 Stamford Road, Oakleigh, Melbourne 3166, Australia

First published 1999

Printed in the United Kingdom at the University Press, Cambridge

Typeset in Monotype Times by the author using TEX

A catalogue record of this book is available from the British Library

Library of Congress Cataloguing in Publication data

Peacock, John A.
Cosmological physics / J.A. Peacock.
p. cm.
Includes bibliographical references and index.
ISBN 0 521 41072 X (hardcover)
1. Cosmology. 2. Astrophysics. I. Title.
QB981.P37 1999
523.1–dc21 98-29460 CIP

ISBN 0 521 41072 X hardback
ISBN 0 521 42270 1 paperback

Contents

Part 4: The early universe

Part 5: Observational cosmology

Preface

This is a textbook on cosmology – a subject that has the modest aim of understanding the entire universe and all its contents. While it can hardly be claimed that this task is complete, it is a fact that recent years have seen astonishing progress towards answering many of the most fundamental questions about the constitution of the universe. The intention of this book is to make these developments accessible to someone who has studied an undergraduate course in physics. I hope that the book will be useful in preparing new Ph.D. students to grapple with the research literature and with more challenging graduate-level texts. I also hope that a good deal of the material will be suitable for use in advanced undergraduate courses.

Cosmology is a demanding subject, not only because of the vast scales with which it deals, but also because of the range of knowledge required on the part of a researcher. The subject draws on just about every branch of physics, which makes it a uniquely stimulating discipline. However, this breadth is undeniably intimidating for the beginner in the subject. As a fresh Ph.D. student, 20 years ago, I was dismayed to discover that even a good undergraduate training had covered only a fraction of the areas of physics that were important in cosmology. Worse still, I learned that cosmologists need a familiarity with astronomy, with all its peculiar historical baggage of arcane terminology. In the past two decades, cosmological knowledge and understanding have advanced almost beyond recognition, and yet undergraduate physics courses have changed only a little. As a result, there is now a yawning gap between the professional literature on cosmology and the knowledge base of a typical physics undergraduate. What I have tried to do in this book is to bridge that gap, by discussing modern cosmology in language that should be familiar to an advanced undergraduate, going back to first principles wherever possible.

The material here is therefore of two kinds: relevant pieces of physics and astronomy that are often not found in undergraduate courses, and applications of these methods to more recent research results. The former category is dominated by general relativity and quantum fields. Relativistic gravitation has always been important in the large-scale issues of cosmology, but the application of modern particle physics to the very early universe is a more recent development. Many excellent texts exist on these subjects, but I wanted to focus on those aspects that are particularly important in cosmology. At times, I have digressed into topics that are strictly 'unnecessary', but which were just too interesting to ignore. These are, after all, the crown jewels of twentieth century physics, and I firmly believe that their main features should be a standard part of an undergraduate education in physics. Despite this selective approach, which aims to concentrate of the essential core of the subject, the book covers a wide range of topics. My original plan was in fact focused more specifically on matters to do with particle physics, the early universe, and structure formation. However, as I wrote, the subject imposed its own logic: it became clear that additional topics simply had to be added in order to tell a consistent story. This tendency for different parts of cosmology to reveal unexpected connections is one of the joys of the subject, and is also a mark of the maturity of the field.

Partly because of the variety of material treated here, my emphasis throughout has been on making the exposition as simple as possible. Nevertheless, I wanted the final result to be useful at a professional level, so I have tried to concentrate on explaining the techniques and formulae that are used in practice in the research literature. In places, I was unable to reduce the full treatment to an argument of tolerable length and complexity, and I then settled for an approximate analysis. In such cases, I have outlined the steps needed to obtain the exact answer and contrasted the approximate analysis with the full result. Wherever possible, I have tried to motivate the more detailed calculations with physical and order-of-magnitude arguments. This is particularly important for the novice, who will be much more willing to accept lengthy and complicated reasoning if it is clear where the argument is going. Because of this desire to expose the logic of the subject, as well as the need to keep the total length manageable, the treatment is not always simple in the sense of giving every last step of a calculation. Readers will have to be prepared to put in some work at various stages to prove to themselves that equation C really does result from combining equations A and B. Usually this sort of manipulation is straightforward, and it can be skipped on a first reading. Where the derivation is a little more challenging, I have left it as a problem, indicated by [problem] in the text. Other problems are collected at the end of each chapter, and include some basic exercises to test comprehension of the material, as well as more discursive and advanced points that would have disrupted the main presentation. There is a set of solutions to all the problems, although these are often quite schematic, since the best way to develop real understanding is to work through the calculations.

A good deal of the book is based on lectures given to final-year undergraduates and early postgraduates, and I have tried to keep the same style in the written version. The advantage of lectures is that they steer a clear course through a subject, whereas textbooks can make a subject that is in essence quite simple seem impenetrable by burdening it with too complete a treatment. What I have attempted to do is to cover the basics thoroughly, so that a student can strike out with confidence towards more specialized topics. I have also tried to make the presentation as locally self-contained as possible, even at the cost of a little duplication. I do not expect that many readers will be masochistic enough to want to read the book in order from cover to cover, and I have tried to ensure that individual chapters and even sections make sense in isolation. In this way, I hope the book can be useful for a variety of courses. It is a very distant ideal to expect that this material in its entirety could be taught to undergraduates. However, some topics are inevitably more straightforward than others, and are suitable for more specialized undergraduate courses.

Many people deserve thanks for having helped this project reach its present stage. First, Malcolm Longair for the initial opportunity to try out my ideas on what could sensibly be lectured to undergraduates. Second, generations of patient students, many of whom at least pretended to enjoy the experience. Third, Malcolm Longair again for the initial suggestion that I should write up my notes – even though he probably realized just how much work this would involve. Fourth, the production of this tome would have been unthinkable without the marvellous electronic tools that are now widely available. The creators of TEX, PostScript, xfig and pgplot all have my humble thanks. Fifth, many scientific colleagues have had a great influence on the contents, both by helpful comments on drafts and particularly through the innumerable conversations and arguments that have formed my understanding of cosmology; I shall not attempt a list, but I owe each one a debt.

Thanks are due especially to all those who generously contributed data for the figures. I have produced many illustrations that involve recent research results, not in some futile attempt to be up to date, but to emphasize that cosmology is above all a subject built on foundations of observational data. Many of the latest results have clarified long-contentious issues, and this reduction of uncertainty has been an unintended benefit of the delay in completing the book. In recent months, I have often felt like a small child in a sweetshop as astronomers all round the world have sent me the most mouthwatering new data. Those who contributed in this way, or gave permission for their results to be used, are as follows. Figure 4.5, T. Muxlow; figure 4.6, W. Sutherland; figure 4.7, R. Ellis; figure 5.1, R. Jimenez; figure 5.2, E. Feigelson; figure 5.3, N. Tanvir; figure 5.4, A. Riess and S. Perlmutter; figure 9.2, M. Smith; figure 12.1, T. Ressell and M. Turner; figure 12.2, C. Steidel and M. Pettini; figure 12.4, T. van Albada; figure 12.5, R. Kolb; figure 13.2, A. Dressler; figure 13.4, S. Driver; figure 13.6, W. Couch; figure 13.7, N. Metcalfe and K. Glazebrook; figure 14.1, P. Hewett and C. Foltz; figure 14.2, C. Carilli; figure 14.6, Y. Dabrowski; figure 15.1, J. Huchra; figure 15.5, J. Colberg; figure 16.9, V. Icke; figure 16.11, M. Davis; figure 18.2, M. White. Figure 13.1 was made using data from the Digitized Sky Survey, which was produced at the Space Telescope Science Institute under US Government grant NAG W-2166. The DSS images are based on photographic data obtained using the Oschin Schmidt Telescope on Palomar Mountain and the UK Schmidt Telescope. I also thank Jörg Colberg, Hugh Couchman, George Efstathiou, Carlos Frenk, Adrian Jenkins, Alistair Nelson, Frazer Pearce, Peter Thomas and Simon White, my colleagues in the Virgo Consortium, for permission to reproduce the cover image. The Virgo Consortium uses computers based at the Computing Centre of the Max-Planck Society in Garching and at the Edinburgh Parallel Computing Centre.

To finish, some apologies. My professional colleagues will doubtless disagree in places with my opinions and emphases. I have tried to make it clear when I am giving a personal view, and have balanced my own simplifications of the issues with frequent references to the literature, so that readers can explore the details of the arguments for themselves. However, the literature is so vast that inevitably many outstanding papers have been omitted. I apologize to those whose work I could have cited, but did not. This is not intended to be a comprehensive review, nor does it try to give a full history of the development of ideas. In places, I have cited classic papers that everyone should read, but in the main papers are cited where I happen to know that they contain further discussion of relevant points. One consequence of this is that the number of references to my own research output exaggerates any contribution I may have made to cosmology.

Lastly, I now appreciate why so many authors acknowledge their families: the conflicting demands for any spare time inevitably make it difficult for families and books to co-exist peacefully. I hope my loved ones will forgive me for the times I neglected them, and I thank the publishers for their patience.

John Peacock

PART 1 GRAVITATION AND RELATIVITY

1 Essentials of general relativity

1.1 The concepts of general relativity

SPECIAL RELATIVITY To understand the issues involved in general relativity, it is helpful to begin with a brief summary of the way space and time are treated in special relativity. The latter theory is an elaboration of the intuitive point of view that the properties of empty space should be the same throughout the universe. This is just a generalization of everyday experience: the world in our vicinity looks much the same whether we are stationary or in motion (leaving aside the inertial forces experienced by accelerated observers, to which we will return shortly).

The immediate consequence of this assumption is that any process that depends only on the properties of empty space must appear the same to all observers: the velocity of light or gravitational radiation should be a constant. The development of special relativity can of course proceed from the experimental constancy of c, as revealed by the Michelson-Morley experiment, but it is worth noting that Einstein considered the result of this experiment to be inevitable on intuitive grounds (see Pais 1982 for a detailed account of the conceptual development of relativity). Despite the mathematical complexity that can result, general relativity is at heart a highly intuitive theory; the way in which our everyday experience can be generalized to deduce the large-scale structure of the universe is one of the most magical parts of physics. The most important concepts of the theory can be dealt with without requiring much mathematical sophistication, and we begin with these physical fundamentals.

4-VECTORS From the constancy of c, it is simple to show that the only possible linear transformation relating the coordinates measured by different observers is the Lorentz transformation:

$$\begin{aligned} dx' &= \gamma\left(dx - \frac{v}{c}c\,dt\right) \\ c\,dt' &= \gamma\left(c\,dt - \frac{v}{c}dx\right). \end{aligned} \tag{1.1}$$

Note that this is written in a form that makes it explicit that x and ct are treated in the same way. To reflect this interchangeability of space and time, and the absence of any preferred frame, we say that special relativity requires all true physical relations to be written in terms of **4-vectors**. An equation valid for one observer will then apply to all others because the quantities on either side of the equation will transform in the same way. We ensure that this is so by constructing physical 4-vectors out of the fundamental interval

$$dx^\mu = (c\,dt, dx, dy, dz) \quad \mu = 0, 1, 2, 3, \tag{1.2}$$

by manipulations with relativistic invariants such as rest mass m and proper time $d\tau$, where

$$(c\,d\tau)^2 = (c\,dt)^2 - (dx^2 + dy^2 + dz^2). \tag{1.3}$$

Thus, defining the 4-momentum $P^\mu = m\,dx^\mu/d\tau$ allows an immediate relativistic generalization of conservation of mass and momentum, since the equation $\Delta P^\mu = 0$ reduces to these laws for an observer who sees a set of slowly moving particles. This is a very powerful principle, as it allows us to reject 'obviously wrong' physical laws at sight. For example, Newton's second law $\mathbf{F} = m\,d\mathbf{u}/dt$ is not a relation between the spatial components of two 4-vectors. The obvious way to define 4-force is $F^\mu = dP^\mu/d\tau$, but where does the 3-force $\mathbf{F}$ sit in F^μ? Force will still be defined as rate of change of momentum, $\mathbf{F} = d\mathbf{P}/dt$; the required components of F^μ are $\gamma(\dot{E}, \mathbf{F})$, and the correct relativistic force–acceleration relation is

$$\mathbf{F} = m\frac{d}{dt}(\gamma\mathbf{u}). \tag{1.4}$$

Note again that the symbol m denotes the **rest mass** of the particle, which is one of the invariant scalar quantities of special relativity. The whole ethos of special relativity is that, in the frame in which a particle is at rest, its intrinsic properties such as mass are always the same, independently of how fast it is moving. The general way in which quantities are calculated in relativity is to evaluate them in the rest frame where things are simple, and then to transform out into the lab frame.

GENERAL RELATIVITY Nothing that has been said so far seems to depend on whether or not observers move at constant velocity. We have in fact already dealt with the main principle of general relativity, which states that the only valid physical laws are those that equate two quantities that transform in the same way under any arbitrary change of coordinates.

Before getting too pleased with ourselves, we should ask how we are going to construct general analogues of 4-vectors. Consider how the components of dx^μ transform under the adoption of a new set of coordinates x'^μ, which are functions of x^ν:

$$\boxed{dx'^\mu = \frac{\partial x'^\mu}{\partial x^\nu}dx^\nu.} \tag{1.5}$$

This apparently trivial equation (which assumes, as usual, the summation convention on repeated indices) may be divided by $d\tau$ on either side to obtain a similar transformation law for 4-velocity, U^μ; so U^μ is a general 4-vector. Things unfortunately go wrong at the next level, when we try to differentiate this new equation to form the 4-acceleration $A^\mu = dU^\mu/d\tau$:

$$A'^\mu = \frac{\partial x'^\mu}{\partial x^\nu}A^\nu + \frac{\partial^2 x'^\mu}{\partial\tau\,\partial x^\nu}U^\nu. \tag{1.6}$$

The second term on the right-hand side (rhs) is zero only when the transformation coefficients are constants. This is so for the Lorentz transformation, but not in general. The conclusion is therefore that $F^\mu = dP^\mu/d\tau$ cannot be a general law of physics, since $dP^\mu/d\tau$ is not a general 4-vector.

INERTIAL FRAMES AND MACH'S PRINCIPLE We have just deduced in a rather cumbersome fashion the familiar fact that $\mathbf{F} = m\mathbf{a}$ only applies in inertial frames of reference. What exactly are these? There is a well-known circularity in Newtonian mechanics, in that inertial frames are effectively defined as being those sets of observers for whom $\mathbf{F} = m\mathbf{a}$ applies. The circularity is only broken by supplying some independent information about $\mathbf{F}$ – for example, the Lorentz force $\mathbf{F} = e(\mathbf{E} + \mathbf{v}\wedge\mathbf{B})$ in the case of a charged particle. This leaves us in a rather unsatisfactory situation: $\mathbf{F} = m\mathbf{a}$ is really only a statement about cause and effect, so the existence of non-inertial frames comes down to saying that there can be a motion with no apparent cause. Now, it is well known that $\mathbf{F} = m\mathbf{a}$ can be made to apply in all frames if certain 'fictitious' forces are allowed to operate. In respectively uniformly accelerating and rotating frames, we would write

$$\begin{aligned} \mathbf{F} &= m\mathbf{a} + m\mathbf{g} \\ \mathbf{F} &= m\mathbf{a} + m\boldsymbol{\Omega}\wedge(\boldsymbol{\Omega}\wedge\mathbf{r}) - 2m(\mathbf{v}\wedge\boldsymbol{\Omega}) + \dot{\boldsymbol{\Omega}}\wedge\mathbf{r}. \end{aligned} \tag{1.7}$$

The fact that these 'forces' have simple expressions is tantalizing: it suggests that they should have a direct explanation, rather than taking the Newtonian view that they arise from an incorrect choice of reference frame. The relativist's attitude will be that if our physical laws are correct, they should account for what observers see from any arbitrary point of view – however perverse.

The mystery of inertial frames is deepened by a fact of which Newton was well aware, but did not explain: an inertial frame is one in which the bulk of matter in the universe is at rest. This observation was taken up in 1872 by Ernst Mach. He argued that since the acceleration of particles can only be measured relative to other matter in the universe, the existence of inertia for a particle must depend on the existence of other matter. This idea has become known as **Mach's principle**, and was a strong influence on Einstein in formulating general relativity. In fact, Mach's ideas ended up very much in conflict with Einstein's eventual theory – most crucially, the rest mass of a particle is a relativistic invariant, independent of the gravitational environment in which a particle finds itself. However, controversy still arises in debating whether general relativity is truly a 'Machian' theory – i.e. one in which the rest frame of the large-scale matter distribution is inevitably an inertial frame (e.g. Raine & Heller 1981).

A hint at the answer to this question comes by returning to the expressions for the inertial forces. The most satisfactory outcome would be to dispose of the notion of inertial frames altogether, and to find a direct physical mechanism for generating 'fictitious' forces. Following this route in fact leads us to conclude that Newtonian gravitation cannot be correct, and that the inertial forces can be effectively attributed to gravitational radiation. Since we cannot at this stage give a correct relativistic argument, consider the analogy with electromagnetism. At large distances, an accelerating charge produces an electric field given by

$$\mathbf{E} = \frac{e}{4\pi\epsilon_0 rc^2}(\hat{\mathbf{r}}\wedge[\mathbf{a}])\wedge\hat{\mathbf{r}}, \tag{1.8}$$

i.e. with components parallel to the retarded acceleration $[\mathbf{a}]$ and perpendicular to the acceleration axis. A charge distribution symmetric about a given point will then generate a net force on a particle at that point in the direction of **a**. It is highly plausible that something similar goes on in the generation of inertial forces via gravity, and we can guess the magnitude by letting $e/(4\pi\epsilon_0) \to Gm$. This argument was proposed by Dennis Sciama, and is known as **inertial induction**. Integrating such a force over all mass in a spherically symmetric universe, we get a total of

$$\frac{F_{\rm tot}}{m} = 2\pi \frac{Ga}{c^2} \int_0^{c/H_0} \int_0^{\pi} \rho\, r \sin^3\theta \, d\theta \, dr = a \, \frac{\pi^2 G\rho}{2H_0^2}. \tag{1.9}$$

This calculation is rough in many respects. The main deficiency is the failure to include the expansion of the universe: objects at a vector distance **r** appear to recede from us at a velocity $\mathbf{v} = H_0\mathbf{r}$, where H_0 is known as Hubble's constant (and is not constant at all, as will become apparent later). This law is only strictly valid at small distances, of course, but it does tell us that objects with $r \simeq c/H_0$ recede at a speed approaching that of light. This is why it seems reasonable to use this as an upper cutoff in the radial part of the above integral. Having done this, we obtain a total acceleration induced by gravitational radiation that is roughly equal to the acceleration we first thought of (the dimensionless factor on the rhs of the above equation is known experimentally to be unity to within a factor 10 or so). Thus, it does seem qualitatively valid to think of inertial forces as arising from gravitational radiation. Apart from being a startlingly different view of what is going on in non-inertial frames, this argument also sheds light on Mach's principle: for a symmetric universe, inertial forces clearly vanish in the average rest frame of the matter distribution. Frames in constant relative motion are allowed because (in this analogy) a uniformly moving charge does not radiate.

It is not worth trying to make this calculation more precise, as the approach is not really even close to being a correct relativistic treatment. Nevertheless, it does illustrate very well the prime characteristic of relativistic thought: we must be able to explain what we see from any point of view.

THE EQUIVALENCE PRINCIPLE In the previous subsection, we were trying to understand the non-inertial effects that are seen in accelerating reference frames as being gravitational in origin. In fact, it is more conventional to state this equivalence the other way around, saying that gravitational effects are identical in nature to those arising through acceleration. The seed for this idea goes back to the observation by Galileo that bodies fall at a rate independent of mass. In Newtonian terms, the acceleration of a body in a gravitational field **g** is

$$m_{\rm I}\, \mathbf{a} = m_{\rm G}\, \mathbf{g}, \tag{1.10}$$

and no experiment has ever been able to detect a difference between the inertial and gravitational masses $m_{\rm I}$ and $m_{\rm G}$ (the equality holds to better than 1 part in 10^{11}: Will 1993). This equality is trivially obvious in the case of inertial forces, and the apparent gravitational acceleration **g** becomes simply the acceleration of the frame **a**. These considerations led Einstein to suggest that inertial and gravitational forces were indeed one and the same. Formally, this leads us to the equivalence principle, which comes in two forms.

The **weak equivalence principle** is a statement only about space and time. It says that in any gravitational field, however strong, a freely falling observer will experience no gravitational effects – with the important exception of tidal forces in non-uniform fields. The spacetime will be that of special relativity (known as **Minkowski spacetime**).

The **strong equivalence principle** takes this a stage further and asserts that not only is the spacetime as in special relativity, but all the laws of physics take the same form in the freely falling frame as they would in the absence of gravity. This form of the equivalence principle is crucial in that it will allow us to deduce the generally valid laws governing physics once the special-relativistic forms are known. Note however that it is less easy to design experiments that can *test* the strong equivalence principle (see chapter 8 of Will 1993).

It may seem that we have actually returned to something like the Newtonian viewpoint: gravitation is merely an artifact of looking at things from the 'wrong' point of view. This is not really so; rather, the important aspects of gravitation are not so much to do with first-order effects as second-order tidal forces: these cannot be transformed away and are the true signature of gravitating mass. However, it is certainly true in one sense to say that gravity is *not* a real force: the gravitational acceleration is not derived from a 4-force F^μ and transforms differently.

GRAVITATIONAL TIME DILATION Many of the important features of general relativity can be obtained via rather simple arguments that use the equivalence principle. The most famous of these is the thought experiment that leads to gravitational time dilation, illustrated in figure 1.1. Consider an accelerating frame, which is conventionally a rocket of height h, with a clock mounted on the roof that regularly disgorges photons towards the floor. If the rocket accelerates upwards at g, the floor acquires a speed $v = gh/c$ in the time taken for a photon to travel from roof to floor. There will thus be a blueshift in the frequency of received photons, given by $\Delta\nu/\nu = gh/c^2$, and it is easy to see that the rate of reception of photons will increase by the same factor.

Now, since the rocket can be kept accelerating for as long as we like, and since photons cannot be stockpiled anywhere, the conclusion of an observer on the floor of the rocket is that in a real sense the clock on the roof is running fast. When the rocket stops accelerating, the clock on the roof will have gained a time Δt by comparison with an identical clock kept on the floor. Finally, the equivalence principle can be brought in to conclude that gravity must cause the same effect. Noting that $\Delta\phi = gh$ is the difference in potential between roof and floor, it is simple to generalize this to

$$\boxed{\frac{\Delta t}{t} = \frac{\Delta\phi}{c^2}.} \tag{1.11}$$

The same thought experiment can also be used to show that light must be deflected in a gravitational field: consider a ray that crosses the rocket cabin horizontally when stationary. This track will appear curved when the rocket accelerates.

The experimental demonstration of the gravitational redshift by Pound & Rebka (1960) was one of the main pieces of evidence for the essential correctness of the above reasoning, and provides a test (although not the most powerful one) of the equivalence principle.

THE TWIN PARADOX One of the neatest illustrations of gravitational time dilation is in resolving the twin paradox. This involves twins A and B, each equipped with a clock.

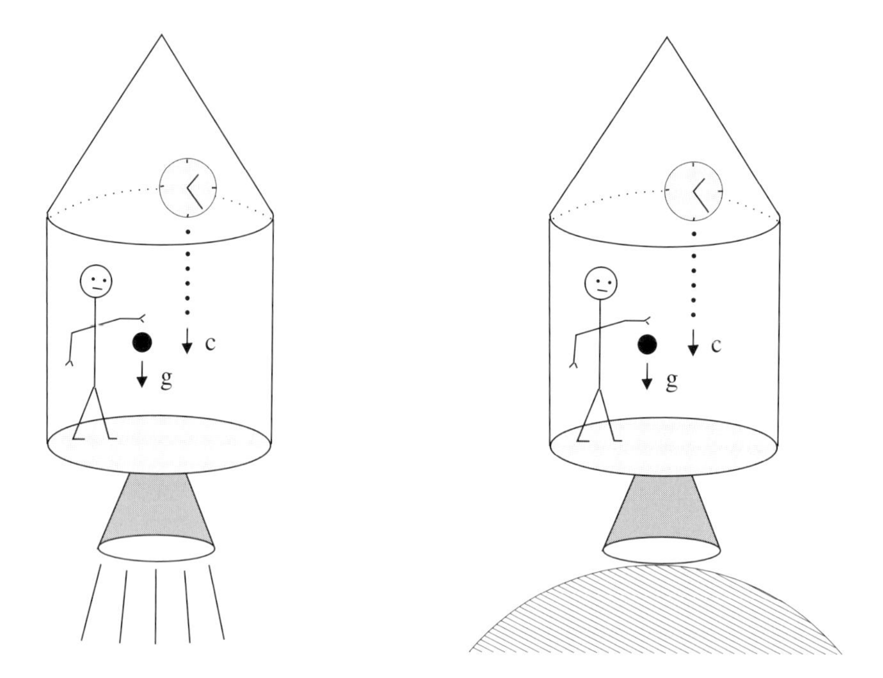

Figure 1.1. Imagine you are in a box in free space far from any source of gravitation. If the box is made to accelerate 'upwards' and has a clock that emits a photon every second mounted on its roof, it is easy to see that you will receive photons more rapidly once the box accelerates (imagine yourself running into the line of oncoming photons). Now, according to the equivalence principle, the situation is exactly equivalent to the second picture in which the box sits at rest on the surface of the Earth. Since there is nowhere for the excess photons to accumulate, the conclusion has to be that clocks above us in a gravitational field run fast.

A remains on Earth, while B travels a distance d on a rocket at velocity v, fires the engines briefly to reverse the rocket's velocity, and returns. The standard analysis of this situation in special relativity concludes, correctly, that A's clock will indicate a longer time for the journey than B's:

$$t_{\rm A} = \gamma\, t_{\rm B}. \tag{1.12}$$

The so-called paradox lies in the broken symmetry between the twins. There are various resolutions of this puzzle, but these generally refuse to meet the problem head-on by analysing things from B's point of view. However, at least for small v, it is easy to do this using the equivalence principle. There are three stages to consider:

(1) Outward trip. According to B, in special relativity A's clock runs slow: $t_{\rm A} = \gamma^{-1} t_{\rm B} \simeq [1 - v^2/(2c^2)](d/v)$.

(2) Return trip. Similarly, A's clock runs slow, resulting in a total lag with respect to B's of $(v^2/c^2)(d/v) = vd/c^2$.

(3) In between comes the crucial phase of turning. During this time, B's frame is non-inertial; there is an apparent gravitational field causing A to halt and start to return to B (at least, what else is B to conclude? There is obviously a force acting on the Earth, but the Earth is clearly not equipped with rockets). If an acceleration g operates for a time $t_{\rm turn}$, then A's clock will run fast by a fractional amount gd/c^2, leading to a total time step of $gdt_{\rm turn}/c^2 = 2vd/c^2$ (since $gt_{\rm turn} = 2v$).

Thus, in total, B returns to find A's clock in advance of B's by an amount

$$t_{\rm A} - t_{\rm B} = -\frac{vd}{c^2} + \frac{2vd}{c^2} \simeq (\gamma - 1)t_{\rm B}, \tag{1.13}$$

exactly (for small v) in accordance with A's entirely special relativity calculation.

1.2 The equation of motion

It was mentioned above that the equivalence principle allows us to bootstrap our way from physics in Minkowski spacetime to general laws. We can in fact obtain the full equations of general relativity in this way, in an approach pioneered by Weinberg (1972). In what follows, note the following conventions: Greek indices run from 0 to 4 (spacetime), Roman from 1 to 3 (spatial). The summation convention on repeated indices of either type is assumed.

Consider freely falling observers, who erect a special-relativity coordinate frame ξ^μ in their neighbourhood. The equation of motion for nearby particles is simple:

$$\frac{d^2\xi^\mu}{d\tau^2} = 0; \qquad \xi^\mu = (ct, x, y, z), \tag{1.14}$$

i.e. they have zero acceleration, and we have Minkowski spacetime

$$c^2 d\tau^2 = \eta_{\alpha\beta}\, d\xi^\alpha d\xi^\beta, \tag{1.15}$$

where $\eta_{\alpha\beta}$ is just a diagonal matrix $\eta_{\alpha\beta} = \text{diag}(1, -1, -1, -1)$. Now suppose the observers make a transformation to some other set of coordinates x^μ. What results is the perfectly general relation

$$d\xi^\mu = \frac{\partial \xi^\mu}{\partial x^\nu} dx^\nu, \tag{1.16}$$

which on substitution leads to the two principal equations of dynamics in general relativity:

$$\boxed{\begin{aligned} \frac{d^2 x^\mu}{d\tau^2} + \Gamma^\mu_{\alpha\beta}\frac{dx^\alpha}{d\tau}\frac{dx^\beta}{d\tau} &= 0 \\ c^2 d\tau^2 &= g_{\alpha\beta}\, dx^\alpha\, dx^\beta. \end{aligned}} \tag{1.17}$$

At this stage, the new quantities appearing in these equations are defined only in terms of our transformation coefficients:

$$\begin{aligned} \Gamma^\mu_{\alpha\beta} &= \frac{\partial x^\mu}{\partial \xi^\nu}\frac{\partial^2 \xi^\nu}{\partial x^\alpha \partial x^\beta} \\ g_{\mu\nu} &= \frac{\partial \xi^\alpha}{\partial x^\mu}\frac{\partial \xi^\beta}{\partial x^\nu}\eta_{\alpha\beta}. \end{aligned} \tag{1.18}$$

COORDINATE TRANSFORMATIONS What is the physical meaning of this analysis? We have taken the special relativity equations for motion and the structure of spacetime and looked at the effects of a general coordinate transformation. One example of such a transformation is a Lorentz boost to some other inertial frame. However, this is not very interesting since we know in advance that the equations retain their form in this case (it is easy to show that $\Gamma^\mu_{\alpha\beta} = 0$ and $g_{\mu\nu} = \eta_{\mu\nu}$). A more general transformation could be one to the frame of an accelerating observer, but the transformation might have *no* direct physical interpretation at all. It is important to realize that general relativity makes no distinction between coordinate transformations associated with motion of the observer and a simple change of variable. For example, we might decide that henceforth we will write down coordinates in the order (x, y, z, ct) rather than (ct, x, y, z) (as is indeed the case in some formalisms). General relativity can cope with these changes automatically. Indeed, this flexibility of the theory is something of a problem: it can sometimes be hard to see when some feature of a problem is 'real', or just an artifact of the coordinates adopted. People attempt to distinguish this second type of coordinate change by distinguishing between 'active' and 'passive' Lorentz transformations; a more common term for the latter class is **gauge transformation**. The term gauge will occur often throughout this book: it always refers to some freedom within a theory that has no observable consequence (e.g. the arbitrary value of $\boldsymbol{\nabla} \cdot \mathbf{A}$, where $\mathbf{A}$ is the vector potential in electrodynamics).

METRIC AND CONNECTION The matrix $g_{\mu\nu}$ is known as the **metric tensor**. It expresses (in the sense of special relativity) a notion of distance between spacetime points. Although this is a feature of many spaces commonly used in physics, it is easy to think of cases where such a measure does not exist (for example, in a plot of particle masses against charges, there is no physical meaning to the distance between points). The fact that spacetime *is* endowed with a metric is in fact something that has been *deduced*, as a consequence of special relativity and the equivalence principle. Given a metric, Minkowski spacetime appears as an inevitable special case: if the matrix $g_{\mu\nu}$ is symmetric, we know that there must exist a coordinate transformation that makes the matrix diagonal:

$$\tilde{\mathbf{\Lambda}} \mathbf{g} \mathbf{\Lambda} = \mathrm{diag}(\lambda_0, \ldots, \lambda_3), \tag{1.19}$$

where $\mathbf{\Lambda}$ is the matrix of transformation coefficients, and λ_i are the eigenvalues of this matrix.

The object $g_{\mu\nu}$ is called a **tensor**, since it occurs in an equation $c^2 d\tau^2 = g_{\mu\nu} dx^\mu dx^\nu$ that must be valid in all frames. In order for this to be so, the components of the matrix $\mathbf{g}$ must obey certain transformation relations under a change of coordinates. This is one way of defining a tensor, an issue that is discussed in detail below.

So much for the metric tensor, what is the meaning of the coefficients $\Gamma^\mu_{\alpha\beta}$? These are known as components of the **affine connection** or as **Christoffel symbols** (and are sometimes written in the alternative notation $\left\{ {\mu \atop \alpha\beta} \right\}$). These quantities obviously correspond roughly to the gravitational force – but what determines whether such a force exists? The answer is that gravitational acceleration depends on spatial change in the metric. For a simple example, consider gravitational time dilation in a weak field: for events at the same spatial position, there must be a separation in proper time of

$$d\tau \simeq dt \left(1 + \frac{\Delta\phi}{c^2} \right). \tag{1.20}$$

This suggests that the gravitational acceleration should be obtained via

$$\mathbf{a} = -\frac{c^2}{2}\boldsymbol{\nabla} g_{00}. \tag{1.21}$$

More generally, we can differentiate the equation for $g_{\mu\nu}$ to get

$$\frac{dg_{\mu\nu}}{dx^\lambda} = \Gamma^\alpha_{\lambda\mu} g_{\alpha\nu} + \Gamma^\beta_{\lambda\nu} g_{\beta\mu}. \tag{1.22}$$

Using the symmetry of the Γ's in their lower indices, and defining $g^{\mu\nu}$ to be the matrix inverse to $g_{\mu\nu}$, we can find an equation for the Γ's directly in terms of the metric tensor:

$$\boxed{\Gamma^\alpha_{\lambda\mu} = \tfrac{1}{2} g^{\alpha\nu} \left(\frac{\partial g_{\mu\nu}}{\partial x^\lambda} + \frac{\partial g_{\lambda\nu}}{\partial x^\mu} - \frac{\partial g_{\mu\lambda}}{\partial x^\nu} \right).} \tag{1.23}$$

Thus, the metric tensor is the crucial object in general relativity: given it, we know both the structure of spacetime and how particles will move.

1.3 Tensors and relativity

Before proceeding further, the above rather intuitive treatment should be set on a slightly firmer mathematical foundation. There are a variety of possible approaches one can take, which differ sufficiently that general relativity texts for physicists and mathematicians sometimes scarcely seem to refer to the same subject. For now, we stick with a rather old-fashioned approach, which has the virtue that it is likely to be familiar. Amends will be made later.

COVARIANT AND CONTRAVARIANT COMPONENTS So far, tensors have been met in their role as quantities that provide generally valid relations between different 4-vectors. If such relations are to be physically useful, they must apply in different frames of reference, and so the components of tensors have to change to compensate for the fact that the components of 4-vectors alter under a coordinate transformation. The transformation law for tensors is obtained from that for 4-vectors. For example, consider $c^2 d\tau^2 = g_{\alpha\beta} dx^\alpha dx^\beta$: substitute for dx^μ in terms of dx'^α and require that the resulting equation must have the form $c^2 d\tau^2 = g'_{\alpha\beta} dx'^\alpha dx'^\beta$. We then deduce the tensor transformation law

$$\boxed{g'_{\alpha\beta} = \frac{\partial x^\mu}{\partial x'^\alpha} \frac{\partial x^\nu}{\partial x'^\beta}\, g_{\mu\nu},} \tag{1.24}$$

of which law our above definition of $g_{\mu\nu}$ in terms of $\eta_{\alpha\beta}$ is an example.

Note that this transformation law looks rather like a generalization of that for a single 4-vector (with one transformation coefficient per index), but with the important difference that the coefficients are upside down in the tensor relation. For Cartesian coordinates, this would make no difference:

$$\frac{\partial x^\mu}{\partial x'^\alpha} = \frac{\partial x'^\alpha}{\partial x^\mu} = \cos\theta, \tag{1.25}$$

where θ is the angle of rotation between the two coordinate axes. In general, though, the transformations are not the same. To illustrate this, consider a set of non-orthogonal basis vectors $\mathbf{e}_i$: there are two ways to define the components of a vector $\mathbf{a}$:

$$\begin{aligned} &(1) \quad \mathbf{a} = \sum a^i \mathbf{e}_i \\ &(2) \quad a_i = \mathbf{a} \cdot \mathbf{e}_i. \end{aligned} \tag{1.26}$$

These clearly differ in general if $\mathbf{e}_i \cdot \mathbf{e}_j \neq 0$, and they are distinguished by writing an index 'upstairs' on one and 'downstairs' on the other.

Something very similar goes on in general relativity. If we define a new vector

$$dx_\nu \equiv g_{\alpha\nu} dx^\alpha, \tag{1.27}$$

then our metric is given by

$$c^2 d\tau^2 = dx_\nu dx^\nu. \tag{1.28}$$

In special relativity, this would yield just $x_\mu = (ct, -x, -y, -z)$. In general relativity, it defines the relations between the **contravariant components** of a 4-vector A^μ and the **covariant components** A_μ. These names reflect that the components transform either in the *same* way as basis vectors (covariant) or oppositely (contravariant). The relevant transformation laws are

$$\boxed{\begin{aligned} A'^\mu &= \frac{\partial x'^\mu}{\partial x^\nu} A^\nu \\ A'_\mu &= \frac{\partial x^\nu}{\partial x'^\mu} A_\nu. \end{aligned}} \tag{1.29}$$

This generalizes to any tensor, by multiplying by the appropriate factor for each index.

INVARIANTS To summarize the above arguments, one can only construct an invariant quantity in general relativity (i.e. one that is the same for all observers) by **contracting** vector or tensor indices in pairs: $A^\mu A_\mu$ is the invariant 'size' or **norm** of the vector A^μ. However, $A^\mu A^\mu$ would not be a constant, since the effects of arbitrary coordinate changes do not cancel out unless upstairs and downstairs indices contract with each other.

This sounds like a tedious complication, but it can be turned to advantage. Suppose we are given an equation such as $A^\mu B_\mu = 1$, and that A^μ is known to be a 4-vector. Clearly, the right-hand side of the equation is invariant, and so the only way in which this can happen in general is if B_μ is also a 4-vector. This trick of deducing the nature of quantities in a relativistic equation is called the principle of **manifest covariance**. As an example, consider the coordinate derivative, for which there exists the common shorthand

$$\boxed{\partial_\mu \equiv \frac{\partial}{\partial x^\mu}.} \tag{1.30}$$

The index must be 'downstairs' since the derivative operates only on one coordinate:

$$\partial_\mu x^\nu = \delta^\nu_\mu = \mathrm{diag}(1, 1, 1, 1). \tag{1.31}$$

Thus, $\partial_\mu x^\mu = 4$ is an invariant, justifying the use of a downstairs index for ∂_μ. More generally, the tensor δ^ν_μ must be an **isotropic tensor** (meaning one whose components are the same in all frames), since it must exist in order to define the inverse matrix to some tensor $T^{\mu\nu}$. Indeed, we have already met an example of this in writing the inverse matrix to $g_{\mu\nu}$ as a contravariant tensor:

$$\boxed{g^{\mu\alpha} g_{\mu\beta} = \delta^\alpha_\beta.} \tag{1.32}$$

The metric tensor is therefore the tool that is used to raise and lower indices, so that

$$\boxed{A_\mu \equiv g_{\mu\nu} A^\nu.} \tag{1.33}$$

For example, in special relativity, the 4-derivatives are therefore

$$\begin{aligned} \partial_\mu &= \left(\frac{\partial}{\partial ct}, \boldsymbol{\nabla}\right) \\ \partial^\mu &= \left(\frac{\partial}{\partial ct}, -\boldsymbol{\nabla}\right). \end{aligned} \tag{1.34}$$

Manifest covariance allows quantities like the **4-current** $J^\mu = (c\rho, \mathbf{j})$ to be recognized as 4-vectors, since they allow the conservation law to be written relativistically: $\partial^\mu J_\mu = 0$.

To summarize, tensor equations with indices in the same relative positions on either side of the expression must be generally valid. An unfortunate term is used for this: the equations are said to be **generally covariant** – i.e. to have the same form for all observers. This has nothing to do with the usage of the term when referring to covariant vectors; it is a historical accident with which one simply has to live.

PSEUDOTENSORS AND TENSOR DENSITIES If we regard a second-rank tensor as a matrix, there is another familiar way of forming a number. In addition to tensor contraction, we can also take the determinant:

$$\boxed{g \equiv -\det g_{\mu\nu}.} \tag{1.35}$$

This is *not* an invariant scalar; thinking of tensor transformations in matrix terms ($\mathbf{g}' = \tilde{\mathbf{\Lambda}}\mathbf{g}\mathbf{\Lambda}$) shows that g' depends on the Jacobian of the coordinate transformation:

$$g' = \left|\frac{\partial x'^\mu}{\partial x^\nu}\right|^{-2} g. \tag{1.36}$$

The reason why this quantity arises in relativity comes from volume elements: under a general coordinate transformation, the hypervolume element behaves as

$$d^4x'^\mu = \left\|\frac{\partial x'^\alpha}{\partial x^\beta}\right\| d^4x^\mu, \tag{1.37}$$

so that an invariant normalization of some scalar ρ can only be constructed via

$$\int \sqrt{-g}\, \rho\, d^4x^\mu = \text{constant}. \tag{1.38}$$

The quantity $\sqrt{-g}\,\rho$ is referred to as a **scalar density**. More generally, an object formed from a tensor and n powers of $\sqrt{-g}$ is called a **tensor density** of weight n.

PROPER AND IMPROPER TRANSFORMATIONS One important consequence of the existence of tensor densities arises when considering coordinate transformations that involve a spatial reflection. It is usual to distinguish between different classes of Lorentz transformations according to the sign of their corresponding Jacobians: **proper Lorentz transformations** have $J > 0$, whereas those with negative Jacobians are termed **improper**. In special relativity, where $g = -1$ always, there are two possibilities: $J = \pm 1$. Thus, a tensor density will in special relativity transform like a tensor if we restrict ourselves to proper transformations. However, on spatial inversion, densities of odd weight will change sign. Such quantities are referred to as **pseudotensors** (or, in special cases pseudovectors or pseudoscalars). The most famous example of this is the totally antisymmetric **Levi–Civita pseudotensor** $\epsilon^{\alpha\beta\gamma\delta}$, which has components $+1$ when $\alpha\beta\gamma\delta$ is an even permutation of 0123, -1 for odd permutations and zero otherwise. One can show explicitly that this frame-independent component definition produces a tensor density of weight -1 by applying the transformation law for such a quantity and verifying the invariance of the components (not too hard since ϵ enters into the definition of the determinant). Lowering indices with the metric tensor produces a covariant density of weight -1:

$$\epsilon_{\alpha\beta\gamma\delta} = g\epsilon^{\alpha\beta\gamma\delta}. \tag{1.39}$$

In special relativity, $\epsilon^{\alpha\beta\gamma\delta}$ is therefore of opposite sign to $\epsilon_{\alpha\beta\gamma\delta}$.

PHYSICS IN GENERAL RELATIVITY So far, we have dealt with how to generalize gravitational dynamics, but how are other parts of physics incorporated into general relativity? A hint at the answer is obtained by looking again at the equation of motion $d^2x^\mu/d\tau^2 + \Gamma^\mu_{\alpha\beta}(dx^\alpha/d\tau)(dx^\beta/d\tau) = 0$. Remembering that $d^2x^\mu/d\tau^2$ is not a general 4-vector, this equation must have added two non-vectors in such a way that the 'errors' in their transformation properties have cancelled to yield a covariant answer. We may say that the addition of the term containing the affine connection has made the equation **gauge invariant**. The term 'gauge' means that there are hidden degrees of freedom (coordinate transformations in this case) that do not affect physical observables.

In fact, we have been dealing with a special case of

$$\boxed{DA^\mu \equiv dA^\mu + \Gamma^\mu_{\alpha\beta} A^\alpha dx^\beta,} \tag{1.40}$$

which is known as the **covariant derivative**. The equation of motion under gravity is then most simply expressed by saying that the covariant derivative of 4-velocity vanishes: $DU^\mu/d\tau = 0$. One can show directly from the definition of the affine connection that the covariant derivative transforms as a 4-vector, but this is a rather messy exercise. It is simpler to use manifest covariance: the form of DU^μ was deduced by transforming the relation $dU^\mu/d\tau = 0$ from the local freely falling frame to a general frame. If $DU^\mu/d\tau$ vanishes in all frames, it must be a general 4-vector. It is immediately clear how to generalize other equations: simply replace ordinary derivatives by covariant ones. Thus, in the presence of non-gravitational forces, the equation of motion for a particle would become

$$m\frac{DU^\mu}{d\tau} = F^\mu. \tag{1.41}$$

There is a frequently used notation to simplify such substitutions. Coordinate partial derivatives may be represented by indices following a comma, and covariant derivatives by a semicolon:

$$\boxed{\begin{aligned} V^\mu{}_{,\nu} &\equiv \frac{\partial V^\mu}{\partial x^\nu} \equiv \partial_\nu V^\mu \\ V^\mu{}_{;\nu} &\equiv \frac{DV^\mu}{\partial x^\nu} \equiv V^\mu{}_{,\nu} + \Gamma^\mu_{\alpha\nu} V^\alpha, \end{aligned}} \tag{1.42}$$

and the notation extends in an obvious way: $V^\mu{}_{;\beta}^{\;\alpha} \equiv D^2 V^\mu / \partial x_\alpha \partial x^\beta$. General relativity thus introduces a simple 'comma goes to semicolon' rule for the bootstrapping of special relativity laws to generally covariant ones. The meaning and origin of the covariant derivative is discussed further below, where two generalizations are proved. The analogous result for the derivatives of covariant vectors is

$$DA_\mu \equiv dA_\mu - \Gamma^\alpha_{\mu\beta} A_\alpha dx^\beta \tag{1.43}$$

(note the opposite sign and different index arrangements).

This procedure is not completely foolproof, unfortunately. A manifestly covariant equation containing only covariant derivatives is clearly a possible generalization of the corresponding special relativity law, but it may not be unique. For example, since the Ricci tensor discussed below vanishes in special relativity, it can be introduced into almost any equation without destroying the special relativity limit: $T^{\mu\nu}_{;\nu} = R^{\mu\nu}_{;\nu}$ has the same limit as $T^{\mu\nu}_{;\nu} = 0$. The best to be said here is that the more complex law is inelegant and lacking in physical motivation; such more complex alternatives can only be ruled out with certainty by experiment. According to the **principle of minimal coupling** we leave out any such extra terms.

A more difficult case arises with equations containing more than one derivative: partial derivatives commute, but covariant derivatives do not:

$$V^\mu_{;\alpha\beta} - V^\mu_{;\beta\alpha} = R^\mu_{\nu\beta\alpha} V^\nu, \tag{1.44}$$

where $R^\mu_{\nu\beta\alpha}$ is the Riemann curvature tensor discussed below. There is no general solution to this problem, which is analogous to the ambiguity of operator orderings in quantum mechanics encountered when going from a classical equation to a quantum one. Sometimes the difficulty can be resolved by dealing with the derivatives one at a time. For example, the natural generalization of Maxwell's equations is clear in terms of the field tensor:

$$\begin{aligned} F^{\mu\nu}_{;\nu} &= -\mu_0 J^\mu \\ F_{\mu\nu} &= A_{\mu;\nu} - A_{\nu;\mu}, \end{aligned} \tag{1.45}$$

even though the combined equation for A^μ is ambiguous.

GEODESICS There is an important way of visualizing the meaning of the general relativity equation of motion. In special relativity a free particle travels along a straight line in space; in general relativity an analogous statement applies to the paths of particles in spacetime. These are *geodesics*: paths whose 'length' in spacetime is stationary with respect to small variations about them in the same way as small perturbations about a straight line in space will increase its total length. This is familiar from special relativity, where particles travel along paths of maximum proper time, as is easily shown by considering a particle that propagates from event A to event B at constant velocity. This motion is most simply viewed from the point of view of the rest frame of this particle. Now consider another path, which corresponds to motion in the frame of the first particle. Suppose the second particle reaches position x at time t_C, and then returns to the origin, both legs of the journey being at constant speed. The proper time for this path is the sum of the proper times for each leg:

$$\tau' = \sqrt{(t_C - t_A)^2 - x^2} + \sqrt{(t_B - t_C)^2 - x^2} < t_B - t_A. \tag{1.46}$$

An arbitrary path can be made out of excursions of this sort, so the unaccelerated path has the maximum proper time. This is the special relativity 'solution' of the twin paradox, although it is actually an evasion, since it refuses to analyse things from the point of view of the accelerated observer. In any case, returning to general relativity, the equivalence principle says that a general path is locally a trajectory in Minkowski spacetime, so it is not surprising that the general path is also one in which the proper time is stationary with respect to variations of the path.

A slightly more formal approach is to express the particle dynamics in terms of an **action principle** and use the calculus of variations. Consider the equation

$$\delta \int L \, dp = 0, \tag{1.47}$$

where p is any parameter describing the path of the particle (t, τ etc.), and δ represents the effects of small variations of the path $x^\mu(p)$. The function L is the **Lagrangian** and Newtonian mechanics can be represented in this form with $L = T - V$, i.e. the difference of kinetic and potential energies for the particle. The principal result of variational calculus is obtained by assuming some arbitrary perturbation $\Delta x^\mu(p)$, expanding L in a Taylor series and integrating by parts for a path with fixed endpoints. This produces the **Euler equation**:

$$\frac{d}{dp}\left(\frac{\partial L}{\partial \dot{x}^\mu}\right) - \frac{\partial L}{\partial x^\mu} = 0, \tag{1.48}$$

where $\dot{x}^\mu$ denotes dx^μ/dp. In special relativity, this equation clearly yields the correct equation of motion for $p = t$, $L = \eta^{\mu\nu} U_\mu U_\nu$, which suggests a manifestly covariant generalization of the action principle:

$$\boxed{\delta \int g^{\mu\nu} U_\mu U_\nu \, d\tau = 0.} \tag{1.49}$$

If this intuitive derivation does not appeal, one can show directly, as an exercise in the calculus of variations, that this yields the correct equation of motion. Now, this formulation of the equations of motion in terms of an action principle has a rather peculiar aspect: the integrand is the Lagrangian, but it must be a constant because it is the norm of the 4-velocity, which is always c^2. Hence the geodesic equation says that the total proper time for a particle to move under gravity between two points in spacetime must be stationary (not necessarily a global maximum, because there may be several possible paths). In short, particles in curved spacetimes do their best to travel in straight lines locally, and it is only the global curvature of spacetime that produces the appearance of a gravitational force.

CONFORMAL TRANSFORMATIONS A related concept is that of making a **conformal transformation** of the metric,

$$g^{\alpha\beta} \to f(x^\mu) g^{\alpha\beta}. \tag{1.50}$$

The name 'conformal' arises by analogy with complex number theory, since such a transformation clearly preserves the 'angle' between two 4-vectors, $A^\mu B_\mu / \sqrt{A^\mu A_\mu B^\mu B_\mu}$. The importance of conformal transformations lies in the structure of null geodesics: clearly the condition $d\tau = 0$ is not affected by a conformal transformation. The paths of light rays are thus independent of these transformations; this result will be useful later in discussing cosmological light propagation.

1.4 The energy–momentum tensor

The only ingredient now missing from a classical theory of relativistic gravitation is a field equation: the presence of mass must determine the gravitational field. To obtain some insight into how this can be achieved, it is helpful to consider first the weak-field limit and the analogy with electromagnetism. The simplest limit of the theory is that of a stationary particle in a stationary (i.e. time-independent) weak field. To first order in the field we can replace τ by t, and the spatial part of the equation of motion is then

$$\ddot{x}^i + c^2 \Gamma^i_{00} = 0, \tag{1.51}$$

where $\Gamma^i_{00} = g^{\nu i}(0 + 0 - \partial g_{00}/\partial x^\nu)/2$. So, as we guessed before via a rough argument from time-dilation considerations, the equation of motion in this limit is

$$\boxed{\ddot{\mathbf{x}} = -\frac{c^2}{2}\boldsymbol{\nabla} g_{00}.} \tag{1.52}$$

THE ELECTROMAGNETIC ANALOGY For a moving particle, it is clear there will be velocity-dependent forces. Before dealing with these in detail, suppose we guess that the weak-field form of gravitation will look like electromagnetism, i.e. that we will end up working with both a scalar potential ϕ and a vector potential $\mathbf{A}$ that together give a velocity-dependent acceleration $\mathbf{a} = -\boldsymbol{\nabla}\phi - \dot{\mathbf{A}} + \mathbf{v}\wedge(\boldsymbol{\nabla}\wedge\mathbf{A})$. Making the usual $e/4\pi\epsilon_0 \to Gm$ substitution would suggest the field equation

$$\partial^\nu \partial_\nu A^\mu \equiv \Box A^\mu = \frac{4\pi G}{c^2} J^\mu, \tag{1.53}$$

where $\Box$ is the **d'Alembertian wave operator**, $A^\mu = (\phi/c, \mathbf{A})$ is the 4-potential and $J^\mu = (\rho c, \mathbf{j})$ is a quantity that resembles a 4-current, whose components are a mass density and mass flux density. The solution to this equation is well known:

$$A^\mu(\mathbf{r}) = \frac{G}{c^2} \int \frac{[J^\mu(\mathbf{x})]}{|\mathbf{r} - \mathbf{x}|}\, d^3x, \tag{1.54}$$

where the square brackets denote retarded values.

Now, in fact this analogy can be discarded immediately as a theory of gravitation in the weak-field limit without any knowledge whatsoever of general relativity. The problem lies in the vector J^μ: what would the meaning of such a quantity be? In electromagnetism, it describes conservation of charge via

$$\partial_\mu J^\mu = \dot{\rho} + \nabla \cdot \mathbf{j} = 0 \tag{1.55}$$

(notice how neatly such a conservation law can be expressed in 4-vector form). When dealing with mechanics, however, we have not one conserved quantity, but *four*: energy and vector momentum. So, although J^μ is a perfectly good 4-vector mathematically, it is not *physically* relevant for describing conservation laws involving mass. For example, conservation laws involving J^μ predict that density will change under Lorentz transformations as $\rho \to \gamma\rho$, whereas the correct law is clearly $\rho \to \gamma^2\rho$ (one power of γ for change in number density, one for relativistic mass increase).

The electromagnetic analogy is nevertheless useful, as it suggests that the source of gravitation might still be mass and momentum: what we need first is to find the object that will correctly express conservation of 4-momentum. Informally, what is needed is a way of writing four conservation laws for each component of P^μ. We can clearly write four equations of the above type in matrix form:

$$\boxed{\partial_\nu T^{\mu\nu} = 0.} \tag{1.56}$$

Now, if this equation is to be covariant, $T^{\mu\nu}$ must be a tensor and is known as the **energy–momentum tensor** (or sometimes as the stress–energy tensor). The meanings of its components in words are: $T^{00} = c^2 \times$ (mass density) $=$ energy density; $T^{12} = x$-component of current of y-momentum etc. From these definitions, the tensor is readily seen to be symmetric. Both momentum density and energy flux density are the product of a mass density and a net velocity, so $T^{0\mu} = T^{\mu 0}$. The spatial stress tensor T^{ij} is also symmetric because any small volume element would otherwise suffer infinite angular acceleration: any asymmetric stress acting on a cube of side L gives a couple $\propto L^3$, whereas the moment of inertia is $\propto L^5$.

For example, a cold fluid with density ρ_0 in its rest frame only has one non-zero component for the energy–momentum tensor: $T^{00} = c^2\rho_0$. Carrying out the Lorentz transformation, we conclude that the tensor's components in another frame are

$$T^{\mu\nu} = c^2\rho_0 \begin{pmatrix} \gamma^2 & -\gamma^2\beta & 0 & 0 \\ -\gamma^2\beta & \gamma^2\beta^2 & 0 & 0 \\ 0 & 0 & 0 & 0 \\ 0 & 0 & 0 & 0 \end{pmatrix}. \tag{1.57}$$

Thus, we obtain in one step quantities such as momentum density $= \gamma^2\rho_0 v$, that would be derived at a more basic level by transforming mass and number density separately.

PERFECT FLUID This line of argument can be taken a little further to obtain a very important result: the energy–momentum tensor for a perfect fluid. In matrix form, the rest-frame $T^{\mu\nu}$ is given by just $\mathrm{diag}(c^2\rho, p, p, p)$ (using the fact that the meaning of the pressure p is just the flux density of x-momentum in the x-direction etc.). We can bypass the step of carrying out an explicit Lorentz transformation (which would be rather cumbersome in this case) by the powerful technique of manifest covariance. The following expression is clearly a tensor and reduces to the above rest-frame answer in special relativity:

$$\boxed{T^{\mu\nu} = (\rho + p/c^2)U^\mu U^\nu - pg^{\mu\nu};} \tag{1.58}$$

thus it must be the general expression for the energy–momentum tensor of a perfect fluid. This is all that is needed in order to derive all the equations of relativistic fluid mechanics (see chapter 2).

1.5 The field equations

Armed with the energy–momentum tensor, we can now return to the search for the relativistic field equations. These cannot be derived in any rigorous sense; all that can be done is to follow Einstein and start by thinking about the simplest form such an equation might take. Our experience with an attempted electromagnetic analogy is again helpful. Consider Maxwell's equations: $\Box A^\mu = \mu_0 J^\mu$. The way in which each component of the 4-current acts as a source in a wave equation for one component of the potential can be generalized to matter if (at least in the weak-field limit) we are dealing with a tensor potential $\phi^{\mu\nu}$:

$$\Box \phi^{\mu\nu} = \kappa T^{\mu\nu}, \tag{1.59}$$

where κ is some constant. The (0, 0) component of this equation will just be Poisson's equation for the Newtonian potential in the case of a stationary field. Since we have already shown that $-c^2 g_{00}/2$ may be identified with this potential, there is a strong suspicion that $\phi^{\mu\nu}$ will be closely related to the metric tensor.

In general, we are looking for a covariant equation that reduces to the above in special relativity. There are many possibilities, but the starting point is to find the simplest alternative that does the job. The reasoning so far suggests that we are looking for a tensor that contains second derivatives of the metric, so why not consider $\partial^2 g_{\mu\nu}/\partial x_\alpha \partial x_\beta$? There are six such creatures, corresponding to the distinct ways in which four indices can be split into two pairs. As usual, such derivatives are not general tensors, and it is not so easy to cure this problem. Normally, one would replace ordinary derivatives with covariant ones and consider $g_{\mu\nu\,;\alpha\beta}$, but the covariant derivatives of the metric vanish identically (see below), so this is no good. It is now not obvious that a tensor can be constructed from second derivatives at all, but there is in fact one combination of the six second-derivative matrices that does work (with the addition of appropriate Γ terms). It is possible (although immensely tedious – see p. 133 of Weinberg 1972) to prove that this is the unique choice for a tensor that is *linear* in second derivatives of the metric.

The tensor in question is the **Riemann tensor**:

$$\boxed{R^\mu{}_{\alpha\beta\gamma} = \frac{d\Gamma^\mu_{\alpha\gamma}}{dx^\beta} - \frac{d\Gamma^\mu_{\alpha\beta}}{dx^\gamma} + \Gamma^\mu_{\sigma\beta}\Gamma^\sigma_{\gamma\alpha} - \Gamma^\mu_{\sigma\gamma}\Gamma^\sigma_{\beta\alpha}.} \tag{1.60}$$

A discussion of the full significance of this tensor will be postponed briefly, but we should note that it is reasonable that some such tensor must exist. The existence of a general metric says that spacetime is curved in a way that is revealed by non-zero second derivatives of $g^{\mu\nu}$. There has to be some covariant description of this curvature, and this is exactly what the Riemann tensor provides.

The Riemann tensor is fourth order, but may be contracted to the **Ricci tensor** $R^{\mu\nu}$, or further to the **curvature scalar** R:

$$R_{\alpha\beta} = R^{\mu}{}_{\alpha\beta\mu}, \quad R = R_{\mu}{}^{\mu} = g^{\mu\nu} R_{\mu\nu}. \tag{1.61}$$

Unfortunately, these definitions are not universally agreed, and different signs can arise in the final equations according to which convention is adopted (see below). All authors, however, agree on the definition of the **Einstein tensor** $G^{\mu\nu}$:

$$G^{\mu\nu} = R^{\mu\nu} - \tfrac{1}{2} g^{\mu\nu} R. \tag{1.62}$$

This tensor is what is needed, because it has zero covariant divergence [problem 1.6]:

$$G^{\mu\nu}_{;\nu} = \frac{DG^{\mu\nu}}{dx^{\nu}} = \frac{dG^{\mu\nu}}{dx^{\nu}} + \Gamma^{\mu}_{\alpha\nu} G^{\alpha\nu} + \Gamma^{\nu}_{\alpha\nu} G^{\mu\alpha} = 0. \tag{1.63}$$

Since we know that $T^{\mu\nu}$ also has zero covariant divergence by virtue of the conservation laws it expresses, it therefore seems reasonable to guess that the two are proportional:

$$\boxed{G^{\mu\nu} = -\frac{8\pi G}{c^4} T^{\mu\nu}.} \tag{1.64}$$

These are Einstein's gravitational field equations, where the correct constant of proportionality has been inserted. This is obtained below by considering the weak-field limit, where Einstein's theory must go over to Newtonian gravity.

PARALLEL TRANSPORT AND THE RIEMANN TENSOR First, however, we ought to take a closer look at the meaning of the crucial Riemann tensor. This has considerable significance in that it describes the degree of curvature of a space: it is the fact that the Riemann tensor is non-zero in general that produces the general relativity interpretation of gravitation as concerned with curved spacetime.

How do we tell whether a space is curved in general? Even the simple case of a surface embedded in 3D space can be tricky. Most people would agree that the surface of a sphere is a curved 2D space, but what about a cylinder? In the sense we are concerned with here, the surface of a cylinder is *not* curved: it can be obtained from a flat plane by bending the plane without folding or distorting it. In other words, the geodesics on a cylinder are exactly those that would apply if the cylinder were unrolled to make a plane; this is a hint of how to proceed in making a less intuitive assessment of curvature.

Gauss was the first to realize that curvature can be measured without the aid of a higher-dimensional being, by making use of the intrinsic properties of a surface. This is a familiar idea: the curvature of a sphere can be measured by examining a (small) triangle whose sides are great circles, and using the relation

$$\text{sum of interior angles} = \pi + 4\pi \frac{\text{area of triangle}}{\text{area of sphere}}. \tag{1.65}$$

Very small triangles have a sum of angles equal to π, but triangles of size comparable to

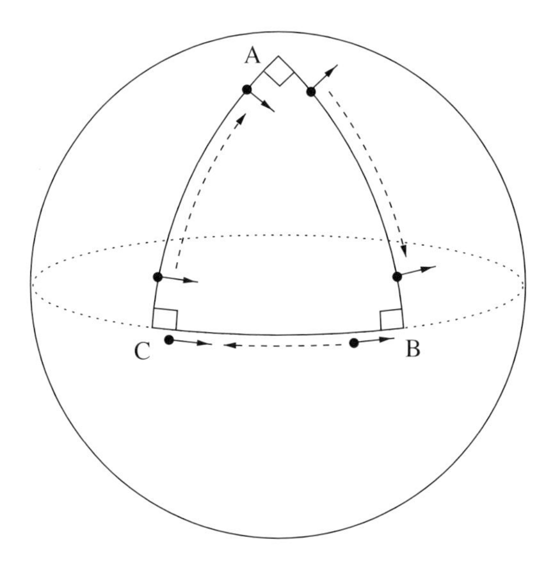

Figure 1.2. This figure illustrates the parallel transport of a vector around the closed loop ABC on the surface of a sphere. For the case of the spherical triangle with all angles equal to 90°, the vector rotates by 90° in one loop. This failure of vectors to realign under parallel transport is the fundamental signature of spatial curvature, and is used to define the affine connection and the Riemann tensor.

the radius of the sphere sample the curvature of the space, and the angular sum starts to differ from the Euclidean value.

To generalize this process, the concept of **parallel transport** is introduced. Here, imagine an observer travelling along some path, carrying with them some vector that is maintained parallel to itself as the observer moves. This is easy to imagine for small displacements, where a locally flat tangent frame can be used to apply the Euclidean concept of parallelism without difficulty. This is clearly a reversible process: a vector can be carried a large distance and back again along the same path, and will return to its original state. However, this need not be true in the case of a loop where the observer returns to the starting point along a different path, as illustrated in figure 1.2. In general, parallel transport around a loop will cause a change in a vector, and it is this that is the intrinsic signature of a curved space. We can think of the effect of parallel transport as producing a change in a vector proportional both to the vector itself (rotation), and to the distance along the loop (to first order), so that the total change in going once round a small loop can be written as

$$\delta V^\mu = -\oint \Gamma^\mu_{\alpha\beta} \, V^\alpha \, dx^\beta. \tag{1.66}$$

Why the minus sign? The reason for this is apparent when we consider the **covariant derivative** of a vector. To differentiate involves taking the limit of $[V^\mu(x+\delta x)-V^\mu(x)]/\delta x$, but the difference of V^μ at two different points is not meaningful in the face of general coordinate transformations. A more sensible procedure is to compare the value of the vector at the new point with the result of parallel-transporting it from the old point.

This is the 'observable' change in the vector, and gives the definition of the covariant derivative as

$$\boxed{V^\mu{}_{;\nu} \equiv V^\mu{}_{,\nu} + \Gamma^\mu_{\alpha\nu} V^\alpha} \tag{1.67}$$

(for a contravariant vector; hence the minus sign above). For a scalar, the covariant and coordinate derivatives are equal. Consider applying this fact to a scalar constructed from two vectors:

$$\begin{aligned} (U^\mu V_\mu)_{;\alpha} &= V_\mu U^\mu{}_{;\alpha} + U^\mu V_{\mu;\alpha} = V_\mu U^\mu{}_{,\alpha} + U^\mu V_{\mu,\alpha} \\ \Rightarrow \quad U^\mu V_{\mu;\alpha} &= U^\mu V_{\mu,\alpha} - V_\mu \left(U^\mu_{;\alpha} - U^\mu_{,\alpha} \right). \end{aligned} \tag{1.68}$$

Since U^μ is an arbitrary vector, this allows us to deduce the covariant derivative for V_μ (different by a sign and index placement in the second term):

$$\boxed{V_{\mu;\nu} \equiv V_{\mu,\nu} - \Gamma^\alpha_{\mu\nu} V_\alpha} \tag{1.69}$$

The covariant derivative of a tensor may be deduced by considering products of vectors, and requiring that the covariant derivative obeys the **Leibniz rule** for differentiation:

$$T^{\mu\nu}_{;\alpha} = (V^\mu U^\nu)_{;\alpha} = V^\mu U^\nu_{;\alpha} + V^\mu_{;\alpha} U^\nu = T^{\mu\nu}_{,\alpha} + \Gamma^\mu_{\beta\alpha} T^{\beta\nu} + \Gamma^\nu_{\beta\alpha} T^{\mu\beta}. \tag{1.70}$$

This introduces a separate Γ term for each index (the appropriate sign depending on whether the index is up or down).

The Riemann tensor arises in terms of the connection Γ when we consider again the change in a vector under parallel transport around a small loop. Since the connection Γ is a function of position in the loop, it cannot be taken outside the loop. What we can do, though, is to make first-order expansions of V^μ and Γ as functions of total displacement from the starting point, Δx^μ. For Γ, the expansion is just a first-order Taylor expansion; for V^μ, what matters is the first-order change in V^μ due to parallel transport. The expression for the total change in V^μ then contains a second-order contribution

$$\delta V^\mu = \left(\Gamma^\mu_{\nu\beta} \Gamma^\nu_{\lambda\alpha} - \Gamma^\mu_{\lambda\beta,\alpha} \right) V^\lambda \oint \Delta x^\alpha \, dx^\beta \tag{1.71}$$

(there is no first-order term because $\oint dx^\mu = 0$ around a loop). Writing this equation twice and permuting α and β allows us to obtain

$$\boxed{\delta V^\mu = \tfrac{1}{2} R^\mu_{\alpha\lambda\beta} \oint \left(\Delta x^\beta \, dx^\alpha - \Delta x^\alpha \, dx^\beta \right),} \tag{1.72}$$

where the tensor $R^\mu_{\alpha\lambda\beta}$ is the **Riemann tensor** encountered earlier:

$$R^\mu{}_{\alpha\beta\gamma} = \Gamma^\mu_{\alpha\gamma,\beta} - \Gamma^\mu_{\alpha\beta,\gamma} + \Gamma^\mu_{\sigma\beta} \Gamma^\sigma_{\gamma\alpha} - \Gamma^\mu_{\sigma\gamma} \Gamma^\sigma_{\beta\alpha}. \tag{1.73}$$

The antisymmetric integral on the right of the expression for δV^μ is clearly a measure of the loop area, so the change in a vector under parallel transport around a loop is proportional both to the loop area and to the Riemann tensor, which is therefore the desired measure of curvature. To show that it is a tensor, even though the Γ's are not, note that it occurs in the impeccably covariant commutator of covariant derivatives:

$$V^{\mu}{}_{;\alpha\beta} - V^{\mu}{}_{;\beta\alpha} = R^{\mu}{}_{\lambda\beta\alpha} V^{\lambda}. \tag{1.74}$$

The Riemann tensor in N dimensions has N^4 components, but few are independent owing to various symmetries. For example, for $N = 2$ there is only one independent number, the **Gaussian curvature** of the surface. This is $R/2$, where $R = R^{\mu\nu}{}_{\mu\nu}$ is the curvature scalar. Written out in full in terms of the metric and its derivatives, this innocuous expression assumes a formidable appearance that quite obscures its intrinsic simplicity. For general relativity, the symmetry of most importance is the **Bianchi identity**, which says that the Einstein tensor is covariantly conserved [problem 1.6]:

$$\boxed{\left(R^{\mu\nu} - \tfrac{1}{2} g^{\mu\nu} R\right)_{;\nu} = 0.} \tag{1.75}$$

It is this property that makes the Einstein tensor such an immediately compelling candidate for field equations in which it is proportional to the energy–momentum tensor.

To complete the logical structure of this approach, we need to relate the connection and the Riemann tensor to the metric. This can be done in a number of ways, the first of which is found in the earlier discussion of the equivalence principle. In the spirit of the discussion here, the most direct route is to note that a scalar should not be changed by the process of parallel transport: $\delta(g_{\mu\nu} V^{\mu} V^{\nu}) = 0$. Applying the rule for parallel transport to the factors separately gives the equation

$$\left(\Gamma^{\alpha}_{\lambda\mu} g_{\alpha\nu} + \Gamma^{\beta}_{\lambda\nu} g_{\beta\nu} - \frac{dg_{\mu\nu}}{dx^{\lambda}}\right) V^{\mu} V^{\nu} dx^{\lambda} = 0, \tag{1.76}$$

which in general requires the factor in parentheses to vanish. As seen previously, this gives the desired relation between the connection Γ and the metric. There is also a simpler way of writing the relation: the metric tensor has zero covariant derivative,

$$\boxed{g^{\mu\nu}{}_{;\nu} = 0.} \tag{1.77}$$

Indeed, this can be taken as the starting point of Riemannian geometry. It is a very easy equation to prove by working in a local inertial frame: the connection is zero there, and the first derivative of the metric also vanishes. This is just the equivalence principle again, saying that inertial frames are those in which the apparent gravitational force (as expressed via derivatives of the metric) is transformed away. This property of the metric is also extremely convenient mathematically, since it means that the raising and lowering of indices commutes with covariant differentiation.

SIGN CONVENTIONS Regrettably, much of the above varies from book to book; in a situation that makes the difference between cgs and SI electromagnetism seem almost trivial, there are few universal conventions in general relativity. The distinctions that exist were analysed into three signs by Misner, Thorne & Wheeler (1973):

$$\begin{aligned} \eta^{\mu\nu} &= [S1] \times \mathrm{diag}\,(-1, +1, +1, +1) \\ R^{\mu}{}_{\alpha\beta\gamma} &= [S2] \times \left(\Gamma^{\mu}_{\alpha\gamma,\beta} - \Gamma^{\mu}_{\alpha\beta,\gamma} + \Gamma^{\mu}_{\sigma\beta}\Gamma^{\sigma}_{\gamma\alpha} - \Gamma^{\mu}_{\sigma\gamma}\Gamma^{\sigma}_{\beta\alpha} \right) \\ G_{\mu\nu} &= [S3] \times \frac{8\pi G}{c^4}\, T_{\mu\nu}. \end{aligned} \tag{1.78}$$

The third sign above is related to the choice of convention for the Ricci tensor:

$$R_{\mu\nu} = [S2] \times [S3] \times R^{\alpha}_{\mu\alpha\nu}. \tag{1.79}$$

With these definitions, Misner, Thorne & Wheeler (unsurprisingly) classify themselves as $(+++)$, whereas Weinberg (1972) is $(+--)$. Peebles (1980, 1993) and Efstathiou (1990) are $(-++)$. This book is $(-+-)$, as are Rindler (1977), Atwater (1974), Narlikar & Padmanabhan (1986), Collins, Martin & Squires (1989). The justification for any of these conventions lies in ease of memory and use of the formulae, which may depend on the application. For example, whether you think $[S1]$ should be positive or negative depends on whether you deal with separations that are mainly space-like or time-like. The important thing is to state which convention is being used.

NEWTONIAN LIMIT The relation between Einstein's and Newton's descriptions of gravity involves taking the limit of weak gravitational fields ($\phi/c^2 \ll 1$). We also need to consider a classical source of gravity, with $p \ll \rho c^2$, so that the only non-zero component of $T^{\mu\nu}$ is $T^{00} = c^2\rho$. Thus, the spatial parts of $R^{\mu\nu}$ must be given by

$$R^{ij} = \tfrac{1}{2} g^{ij} R. \tag{1.80}$$

Converting this to an equation for R^i_j, it follows that $R = R^{00} + \frac{3}{2}R$ and hence that

$$G^{00} = G_{00} = 2R_{00}. \tag{1.81}$$

Discarding nonlinear terms in the definition of the Riemann tensor leaves

$$R_{\alpha\beta} = \frac{\partial \Gamma^{\mu}_{\alpha\mu}}{\partial x^{\beta}} - \frac{\partial \Gamma^{\mu}_{\alpha\beta}}{\partial x^{\mu}} \quad \Rightarrow \quad R_{00} = -\Gamma^{i}_{00,i} \tag{1.82}$$

for the case of a stationary field. We have already seen that $c^2\Gamma^i_{00}$ plays the role of the Newtonian acceleration, so the required limiting expression for G^{00} is

$$G^{00} = -\frac{2}{c^2}\nabla^2\phi, \tag{1.83}$$

and comparison with Poisson's equation gives us the above constant of proportionality in the field equations.

PRESSURE AS A SOURCE OF GRAVITY Newtonian gravitation is modified in the case of a relativistic fluid (i.e. where we cannot assume $p \ll \rho c^2$). It helps to begin by recasting the field equations (this would also have simplified the above discussion). Contract the equation using $g^{\mu}_{\mu} = 4$ to obtain $R = (8\pi G/c^4)\; T$. This allows us to write an equation for $R^{\mu\nu}$ directly:

$$R^{\mu\nu} = -\frac{8\pi G}{c^4}(T^{\mu\nu} - \tfrac{1}{2} g^{\mu\nu} T). \tag{1.84}$$

Since $T = c^2\rho - 3p$, we get a modified Poisson equation:

$$\nabla^2\phi = 4\pi G(\rho + 3p/c^2). \tag{1.85}$$

What does this mean? For a gas of particles all moving at the same speed u, the effective gravitational mass density is $\rho(1 + u^2/c^2)$; thus a radiation-dominated fluid generates a gravitational attraction twice as strong as one would expect from Newtonian arguments. In fact, as will be seen later, this factor applies also to individual particles and leads to an interesting consequence. One can turn the argument round by going to the rest frame of the gravitating mass. We will then conclude that a passing test particle will exhibit an acceleration transverse to its path greater by a factor $(1 + u^2/c^2)$ than that of a slowly moving particle. This appears to be a violation of the equivalence principle: so it is, from the crude 'all objects fall equally fast' point of view, but not in terms of freely falling observers experiencing special relativity conditions.

ENERGY DENSITY OF THE VACUUM One consequence of the gravitational effects of pressure that may seem of mathematical interest only is that a negative-pressure equation of state that achieved $\rho^2 + 3p < 0$ would produce gravitational *repulsion*. Although such a possibility may seem physically nonsensical, it is in fact one of the most important concepts in contemporary cosmology. The origin of the idea goes back to the time when Einstein was first thinking about the cosmological consequences of general relativity. At that time, the universe was believed to be static – although this was simply a prejudice, rather than being founded on any observational facts. It should be obvious, even in the context of Newtonian gravity, that such a universe is not stable: the mutual attraction of all particles would cause the distribution of mass to undergo a global contraction. This could be prevented only by either postulating an expanding universe (which would in retrospect have ranked as perhaps the greatest of all scientific predictions), or by interfering with the long-range properties of gravity. How could this be done? Einstein was loath to complicate the beautiful simplicity of the field equations, and there seemed only one way out. Although we have argued that $G^{\mu\nu}$ is unique as a tensor linear in second derivatives of the metric, this is not so if we consider also terms of *zero* order in these derivatives. Since the metric tensor has zero covariant divergence, it is possible to write a modified set of field equations that are also consistent with the conservation laws:

$$G^{\mu\nu} + \Lambda g^{\mu\nu} = -\frac{8\pi G}{c^4}T^{\mu\nu}. \tag{1.86}$$

Λ is a new constant of nature: the **cosmological constant**. We will see that this modification allows the construction of a static universe. However, it is *ad hoc*; when the expansion of the universe was established, the Λ term was repudiated by Einstein.

Despite this condemnation from its creator, the cosmological constant has refused to die and has assumed a central role in modern cosmology. What is its physical meaning? In the form in which it is written above, it represents the curvature of empty space. A more helpful way of looking at things can be achieved by the apparently trivial act of moving the Λ term to the right-hand side of the field equations. It now looks like the

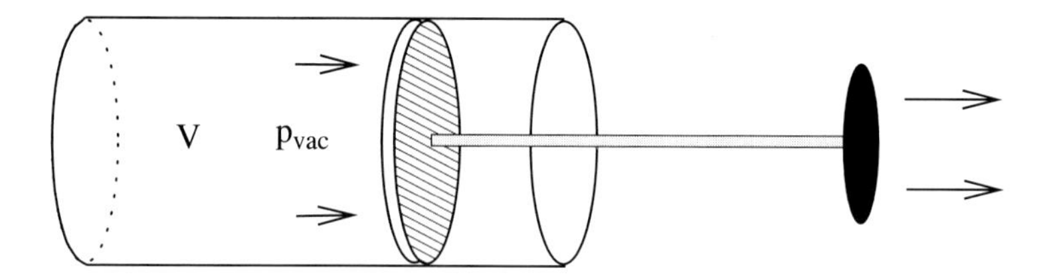

Figure 1.3. A thought experiment to illustrate the application of conservation of energy to the vacuum. If the vacuum density is ρ_{vac} then the energy created by withdrawing the piston by a volume dV is $\rho_{vac}c^2\,dV$. This must be supplied by work done by the vacuum pressure $p_{vac}dV$, and so $p_{vac} = -\rho_{vac}c^2$, as required.

energy–momentum tensor of the vacuum:

$$T^{\mu\nu}_{\rm vac} = \frac{\Lambda c^4}{8\pi G} g^{\mu\nu}. \tag{1.87}$$

How can a vacuum have a non-zero energy density and pressure? Surely these are zero by definition in a vacuum? Later chapters will show just how hard it is to predict what the properties of the vacuum should be. However, we can be sure of one thing: the absence of a preferred frame means that $T^{\mu\nu}_{\rm vac}$ must be the same for all observers in special relativity. Now, apart from zero, there is only one isotropic tensor of rank 2 and it is a very familiar one: $\eta^{\mu\nu}$. Thus, in order for $T^{\mu\nu}_{\rm vac}$ to be unaltered by Lorentz transformations, the only requirement we can have is that it must be proportional to the metric tensor. Therefore, it is inevitable that the vacuum (at least in special relativity) will have a negative-pressure equation of state:

$$\boxed{p_{\rm vac} = -\rho_{\rm vac}\, c^2.} \tag{1.88}$$

In this case, $\rho c^2 + 3p$ is indeed negative: a positive Λ will act to cause a large-scale repulsion. The vacuum energy density can thus play a crucial part in the dynamics of the early universe. Since the vacuum energy is a constant, independent of time, there might seem to be a problem with conservation of energy in an expanding universe: surely the energy density should decline as a given volume expands? (this is certainly what happens to ordinary matter such as galaxies and intergalactic gas). What saves us is the peculiar equation of state of the vacuum: the work done by the pressure is just sufficient to maintain the energy density constant (see figure 1.3). In effect, the vacuum acts as a reservoir of unlimited energy, which can supply as much as is required to inflate a given region to any required size at constant energy density. This supply of energy is what is used in 'inflationary' theories of cosmology to create the whole universe out of almost nothing.

1.6 Alternative theories of gravity

How certain can we be that Einstein's theory of gravitation is correct? Since it does not contain Planck's constant, it is a purely classical apparatus, and so will break down on very small scales. Apart from this restriction, there are no obvious areas of incompleteness: general relativity has plugged the gaps in Newtonian theory that were known to exist in the regime of strong fields and high velocities. Nevertheless, although general relativity has fared better than Newtonian theory in matching observations, it is possible that more accurate experiments will reveal discrepancies. Over the years, this possibility has motivated many suggestions of alternatives to general relativity. Sometimes these ideas can be rather subtle, but three main approaches can be distinguished.

(1) The most radical line is to reject the idea of curved spacetime. Can gravity be described as an extra field that operates in Minkowski spacetime?

(2) An intermediate approach is to retain the idea of curvature, but to add some extra feature to go beyond the simplest implications of the principle of equivalence.

(3) The most straightforward route (and the hardest to rule out) retains general relativity in the sense of the whole apparatus of general covariance, plus the recognition of the metric tensor as the gravitational field. However, the field equations themselves are logically distinct from the rest of the theory, and can be replaced by something different.

The first of these alternatives can be dismissed quite readily. Relativistic theories of gravitation with a scalar, vector or tensor potential can be written down (e.g. chapter 7 of Misner, Thorne & Wheeler 1973), and meet with variable success. The first two cases predict no light deflection, whereas the third makes the same prediction as general relativity (although the precession of planetary orbits is too large by a factor of 4/3). The greatest defect with all such flat-space theories, however, is that they have no place for gravitational time dilation. This is not surprising, since we have seen that both this phenomenon and spacetime curvature are consequences of the equivalence principle.

The main example of a theory that goes beyond these flat spacetime examples, but without becoming something that differs from Einstein only in the field equations, was originated by Cartan (see e.g. Hehl *et al.* 1976). This is a theory with non-zero **torsion**, in which the connection is not symmetric, defining a tensor S:

$$S^{\mu}_{\alpha\beta} = \Gamma^{\mu}_{\alpha\beta} - \Gamma^{\mu}_{\beta\alpha}. \tag{1.89}$$

Now, we saw earlier that the application of the equivalence principle generated a connection that was automatically symmetric. In a local inertial frame, the equivalence principle asserts that $\Gamma = 0$ and the torsion therefore vanishes in this frame; a tensor that vanishes in one frame is zero in all, as may be seen by transforming components to a different frame. However, care is needed with this argument, as may be seen by looking at the equation of motion:

$$\frac{DU^{\mu}}{d\tau} + \Gamma^{\mu}_{\alpha\beta} U^{\alpha} U^{\beta} = 0. \tag{1.90}$$

The symmetric form of the second term means that any antisymmetric contribution to the connection has no effect on the equation of motion. Non-zero torsion is thus not so easy to rule out in practice, and plays a central role in the gravitational theory of Cartan. Like many features of general relativity, its neglect is justified mainly on the grounds of keeping the theory as simple as possible.

We now come to the class of gravitation theories that provides the widest range of plausible alternatives to general relativity, where the sole alteration is in the form of the gravitational field equations. This is most easily viewed from the point of view of the Lagrangian density for the gravitational field (see chapter 7 on quantum fields for a description of how Lagrangian densities generate field equations). In general relativity, the Lagrangian density is just proportional to the Ricci scalar, as derived by Hilbert:

$$\mathscr{L} = \frac{R}{16\pi G}. \tag{1.91}$$

Field equations that are valid from the point of view of general covariance require only that the Lagrangian is a scalar, so that higher powers of R could be added, as well as powers of more complicated scalars that can be formed from the Ricci tensor ($R^{\mu\nu} R_{\mu\nu}$). Any such terms must not interfere with the Newtonian limit of the theory, but this can always be arranged by adding in the extra terms with a small enough coefficient. In the absence of proper strong-field tests of the theory, it is hard to be sure that we have the correct gravitational Lagrangian. Indeed, there is a discouraging analogy in the quantum theory of the weak interaction. There exists a Lagrangian due to Fermi that describes correctly many features of the interaction, but which is nevertheless fundamentally incorrect; Fermi's theory is an effective Lagrangian that applies only at low energies. In view of the divergences encountered in attempts at a quantum theory of gravitation, it is possible that Einstein's theory may have the same status.

A final possibility that has received much attention is to combine gravity with other fields, as is done in **Brans–Dicke gravity**. The motivation for this theory lay in making Einstein gravity more Machian, by allowing the distribution of distant matter to influence the dynamics at a point. This was achieved by making the gravitational constant G into a dynamical field ϕ. This in turn was assumed to be coupled to the trace of the energy–momentum tensor (since we need a scalar quantity to match a scalar field), giving the joint field equations

$$\begin{aligned} G^{\mu\nu} &= -\frac{8\pi}{c^4}\,\phi^{-1}\,T^{\mu\nu} \\ \Box\phi = g^{\mu\nu}\phi_{;\mu\nu} &= \frac{8\pi}{3+2\omega}\,T^{\mu}_{\mu}(\text{matter}). \end{aligned} \tag{1.92}$$

For obvious reasons, this theory is also known as **scalar–tensor gravity**. One subtlety to note is that ϕ is coupled to the matter component of $T^{\mu\nu}$ only, whereas gravity responds to the total, including the contribution of ϕ. The coupling constant for ϕ is written in the peculiar form involving ω for historical reasons. In the limit $\omega \to \infty$, the theory goes over to Einstein gravity: ϕ is then not affected at all by the fluctuating matter density, and so can be set equal to a constant, $\phi = 1/G$. Solar-system tests of gravity have shown that ω must be greater than 600 (Will 1993), so the theory is unlikely to be relevant to the present. However, scalar-tensor theories do arise in some high-energy unified theories of the fundamental interactions (see chapter 8 and chapter 13 of Collins, Martin & Squires 1989).

1.7 Relativity and differential geometry

Before ending this brief survey of general relativity, we should return to the mathematical culture gap that exists between the old-fashioned tensor calculus familiar to physicists and the differential geometry approach preferred by mathematicians. This book is written from the point of view of physicists, but it is important to understand how the traditional approach relates to the more modern notation. This section can do no more than scratch the surface of differential geometry. A good introduction to the subject without excessive formality is Schutz (1980). Many texts in general relativity adopt solely this approach (e.g. Misner, Thorne & Wheeler 1973).

The diversity of approaches centres around the issue of coordinates. Modern practice emphasizes that the quantities one deals with in tensor calculus have a geometrical significance greater than can be easily revealed by discussions of components and their behaviour under coordinate transformation. This is a familiar approach in three-dimensional vector calculus, where both the notation and the very concept of a vector as a 'directed number' possessing magnitude and direction are explicitly coordinate independent. Differential geometry provides a way of realizing these virtues in more general situations.

MANIFOLDS The general arena for the mathematics of relativity is the **differentiable manifold**. Informally, a manifold is a space where the points can be parameterized continuously as an n-dimensional set of real numbers, x^i, which are the coordinates of the point. This means that points can be connected by a curve governed by some parameter along the curve λ, and that the derivatives of the coordinates $dx^i/d\lambda$ exist. Obviously, Euclidean space is an example of a manifold, but the definition also encompasses more abstract spaces such as the thermodynamic space of a fluid, where the coordinates might be (P, V, T).

The difference between Euclidean space and other manifolds is that a manifold need not possess a metric structure, i.e. any notion of distance between points. A weaker but more general concept is that of **affine geometry**. Here, the idea of distance can be introduced at least for vectors that are parallel. Consider expanding a vector in some set of **basis vectors**, $\mathbf{v} = \sum v_i \mathbf{e}_i$, and let the general analogue of coordinate transformations be the linear affine transformation $v_i \to A_{ij} v_j$, where the matrix A_{ij} is unconstrained. If we represent the basis vectors by vectors in Euclidean space, then the transformation looks like a mixture of rotation and shear: any initial 'distance' defined between two points is changed after the transformation. Nevertheless, the transformation keeps parallel lines parallel, so that one can think of distance along a line and compare distances along different lines that are parallel. A good example of an affine space arises in colour vision: the basis vectors are the different colours and the v_i are the intensities in each of red, green and blue; one can compare overall intensities when the colour balance is unaltered, but not for different colour mixtures. The importance of affine geometry is that any geometry that deals with infinitesimals must be linear, so that differential geometry is inevitably affine in nature.

Given a manifold, how is the general analogue of Euclidean vector algebra generated? We are used to writing $\mathbf{v} = \sum_i a^i \mathbf{e}_i$ for the decomposition of a vector as a sum of basis vectors $\mathbf{e}_i$ with components a^i. This allows simple linear vector algebra, in the sense that the sum of two vectors is another vector. In a manifold, points have passing through them parameterized curves, and this can be used to generate **tangent vectors** $d/d\lambda$, where

$$\boxed{\frac{d}{d\lambda} = \sum_i \frac{dx^i}{d\lambda} \frac{\partial}{\partial x^i}.} \tag{1.93}$$

This is an operator relation that applies to any function $f(x^i)$, and is just obtained from the chain rule. What it says, however, is that $d/d\lambda$ is a vector in the sense just described, where $\partial/\partial x^i$ is the basis. A sum of two vectors $a(d/d\lambda) + b(d/d\mu)$ (where a and b are scalars) is just a sum over the basis with a different set of coefficients, and so looks like d/dv, where v is the parameter of another curve passing through the point of interest. By contrast, we have not so far been too explicit concerning the basis vectors used in relativity. Implicitly, these have just been unit vectors in the coordinate directions $\hat{x}$, $\hat{t}$ etc.; however, we have never worried about how such vectors might be constructed in the absence of a metric, and this is the strength of the concept of the tangent vector.

DIFFERENTIAL FORMS Having progressed as far as being able to generate vectors on a manifold, how do we go beyond the simple linear algebra involving addition of vectors? The first concept needed is some analogue of the magnitude of a vector – that is, something that will act as a function of a vector to produce a scalar. Such an entity is called a **1-form**, and is defined by the way it acts on vectors:

$$\tilde{\omega}(\mathbf{v}) = \text{scalar}. \tag{1.94}$$

The function is linear with respect to addition of vectors and multiplication by scalars:

$$\boxed{\begin{aligned} \tilde{\omega}(a\mathbf{v} + b\mathbf{w}) &= a\tilde{\omega}(\mathbf{v}) + b\tilde{\omega}(\mathbf{w}) \\ (a\tilde{\omega} + b\tilde{\sigma})(\mathbf{v}) &= a\tilde{\omega}(\mathbf{v}) + b\tilde{\sigma}(\mathbf{v}). \end{aligned}} \tag{1.95}$$

This latter pair of definitions is important in that it suggests a symmetry: the vector $\mathbf{v}$ can equally well be regarded as operating on the 1-form $\tilde{\omega}$. We say that the objects are **dual**:

$$\tilde{\omega}(\mathbf{v}) = \mathbf{v}(\tilde{\omega}) \equiv \langle \tilde{\omega} | \mathbf{v} \rangle. \tag{1.96}$$

The resemblance to **Dirac notation** from quantum mechanics is no accident: in our present terms, Dirac's 'kets' are vectors and the 'bras' are 1-forms. The concept of vectors as machines that 'eat' a 1-form to produce a number, or vice versa, can be immediately generalized to tensors that may require several 'meals' of each type in order to yield an invariant. An (n, m) tensor T is an object that operates on n 1-forms and m vectors in order to produce a number:

$$T(\tilde{\omega}_1, \tilde{\omega}_2, \ldots, \tilde{\omega}_n, \mathbf{v}_1, \mathbf{v}_2, \ldots, \mathbf{v}_m) = \text{scalar}. \tag{1.97}$$

The components of T are derived by inserting n basis 1-forms and m basis vectors in the appropriate slots of T, and clearly should be written with n upstairs indices and m downstairs indices.

So far, we may seem to have simply invented a new name for a covariant vector; this is true in that the components of 1-forms obey the transformation rule for covariant components. However, the conventional discussion of covariant and contravariant vectors in relativity relies on converting one into the other via the metric tensor, whereas a metric structure may not exist on a general differentiable manifold. Clearly, the idea (which may seem surprising) that a dual structure of vectors and 1-forms can still be constructed in this restricted case is an important insight, which goes beyond the ideas of traditional tensor calculus. Let us now see how this feat is accomplished.

COMPONENTS OF VECTORS AND 1-FORMS If we expand a vector in (contravariant) components, $\mathbf{v} = a^i \mathbf{e}_i$, then the linearity of the 1-form function can be used to write $\tilde{\omega}(\mathbf{v}) = a^i b_i$, where we expand the 1-form in the same way as for the vector: $\tilde{\omega} = b_i \tilde{\mathbf{e}}^i$, defining a **dual basis**

$$\tilde{\mathbf{e}}^i(\mathbf{e}_j) = \delta^i_j. \tag{1.98}$$

To construct such a basis, note the important fact that the gradient of a function is a 1-form, symbolized by $\tilde{\mathbf{d}}f$:

$$\boxed{\tilde{\mathbf{d}}f\left(\frac{d}{d\lambda}\right) \equiv \frac{df}{d\lambda}.} \tag{1.99}$$

Given a set of coordinates x^i, we can write down immediately the components of both a vector (the components of $d/d\lambda$ are $dx^i/d\lambda$) and a 1-form (the components of $\tilde{\mathbf{d}}f$ are df/dx^i). Thus, the required dual basis is

$$\begin{aligned} \mathbf{e}_i &= \frac{d}{dx^i} \\ \tilde{\mathbf{e}}^i &= \tilde{\mathbf{d}}x^i. \end{aligned} \tag{1.100}$$

It is also possible to conceive of sets of basis vectors not derived from a set of coordinates (although these are clearly harder to construct); the above basis is therefore called a **coordinate basis**.

BASIS TRANSFORMATIONS Only at this stage is it necessary to introduce the concept of coordinate transformations that is so central to the traditional formulation of tensor calculus. If we are not using a coordinate basis, we need to consider the behaviour of basis vectors under a basis transformation. A common notation for this transformation is the following:

$$\mathbf{e}'_j = \Lambda^i{}_{j'} \mathbf{e}_i. \tag{1.101}$$

An unfortunate aspect of this notation is that it can start to seem rather clumsy when thinking about taking specific values for i and j: the prime really should refer more clearly to a property of the matrix $\mathbf{\Lambda}$ itself, rather than being associated with an index. Things are easier in the case of a coordinate basis, where we have already met the use of direct partial-derivative notation for the basis transformation coefficients. From the above definition and the dual property of vector and 1-form bases, it is easy to deduce the transformation properties of the basis 1-forms:

$$\tilde{\mathbf{e}}'^i = \Lambda^{i'}_k \tilde{\mathbf{e}}^k, \tag{1.102}$$

from which the transformation properties of the components of a vector are obtained directly:

$$v'^j = \tilde{\mathbf{e}}'^j(\mathbf{v}) = \Lambda^{j'}{}_k \tilde{\mathbf{e}}^k(\mathbf{v}) = \Lambda^{j'}{}_k v^k, \tag{1.103}$$

which is opposite to the transformation law for basis vectors and justifies the name *contra*variant component.

AREAS AND ANTISYMMETRIC TENSORS The 1-form is in fact merely a special case of an important class of tensors, called differential forms, which are antisymmetric in all their arguments. The reason for wanting to introduce such objects is primarily *geometrical*: it is often necessary to be able to deal algebraically with the area (i.e. the parallelogram) defined by two vectors. This must clearly be represented by a (0, 2) tensor, since the magnitude of the area is defined as a bilinear function of two vectors: area($\mathbf{a}, \mathbf{b}$). The antisymmetry comes about by the reasonable requirement that area($\mathbf{a}, \mathbf{a}$) $= 0$; write $\mathbf{a} = \mathbf{b} + \mathbf{c}$ and bilinearity then requires

$$\text{area}(\mathbf{b}, \mathbf{c}) = -\text{area}(\mathbf{c}, \mathbf{b}). \tag{1.104}$$

This naturally generalizes to volumes of parallelepipeds, leading to the introduction of n-forms of arbitrary order. Symmetric terminology then leads us to define an n-vector as a totally antisymmetric $(n, 0)$ tensor.

In index terms, the components of an n-form in m dimensions, $\omega_{ijk\ldots}$, must have a set of indices $ijk\ldots$ that form a distinct combination drawn from $1, \ldots m$. Thus, there can be only ${}^m C_n$ independent components and there exists only one m-form in m dimensions. This number of components is in general different to the number of dimensions; since ${}^3C_2 = 3$, however, the area defined by two vectors can be associated with a vector in three dimensions. The vector cross product is a piece of machinery constructed to exploit this special case, and cannot be defined in other numbers of dimensions.

WEDGE PRODUCTS A mechanism for generating forms of higher order is provided by the **wedge product**:

$$\boxed{\tilde{\mathbf{a}} \wedge \tilde{\mathbf{b}} \equiv \tilde{\mathbf{a}} \otimes \tilde{\mathbf{b}} - \tilde{\mathbf{b}} \otimes \tilde{\mathbf{a}},} \tag{1.105}$$

where the **direct product** is defined by

$$\tilde{\mathbf{a}} \otimes \tilde{\mathbf{b}}(\mathbf{c}, \mathbf{d}) \equiv \tilde{\mathbf{a}}(\mathbf{c})\, \tilde{\mathbf{b}}(\mathbf{d}). \tag{1.106}$$

This generalizes to larger wedge products $\tilde{\mathbf{a}} \wedge \tilde{\mathbf{b}} \wedge \tilde{\mathbf{c}} \cdots$ in an obvious way. The algebra of forms under the anticommutative wedge product is called a **Grassmann algebra**.

Basis 1-forms $\tilde{\boldsymbol{\omega}}_i$ therefore form a basis for n-forms, but care is needed with factors of $n!$: $\tilde{\boldsymbol{\sigma}} = (1/2!)\sigma_{ij}\tilde{\boldsymbol{\omega}}_i \wedge \tilde{\boldsymbol{\omega}}_j$. To see the need for the $1/2!$ factor, evaluate $\tilde{\boldsymbol{\sigma}}(\mathbf{e}_i, \mathbf{e}_j)$, which gives $\sigma_{ij} - \sigma_{ji}$. Since the σ_{ij} coefficients are antisymmetric, this equals $2\sigma_{ij}$, and so we need the normalizing factor in order to keep the sensible definition of components of an n-form as being the result of operating on the basis vectors. Some books avoid this potential problem by incorporating the factor of $n!$ into the definition of the wedge product.

CONNECTIONS, COVARIANT DERIVATIVES AND METRICS In a general manifold, we have seen that tangent vectors are generated by the derivative along some curve, $d/d\lambda$. This means that there is no general way to compare vectors at two different points. Consider the two-dimensional case where points are specified by the numbers (λ, μ), and think about two points A and B separated by some increment $\Delta\lambda$ at constant μ. The vectors $\mathbf{v} = d/d\lambda$ and $\mathbf{u} = d/d\mu$ can be constructed at A and B, but comparing $\mathbf{u}$ at A and B can only be done in a way that depends on the particular set of curves (the **congruence**) chosen. For example, suppose μ is replaced by a new parameter $\mu' = \mu \times f(\lambda)$; the curves at constant λ passing through A and B are unaltered, but the tangent vectors along them have changed. What is required is some extra structure: a rule for parallel-transporting vectors. If this is available, it can be used to define the **covariant derivative**:

$$\nabla_{\mathbf{v}} \mathbf{u} = \lim_{\epsilon \to 0} [\mathbf{u}_{\parallel}(\lambda + \epsilon) - \mathbf{u}(\lambda)]/\epsilon, \tag{1.107}$$

denoting the parallel-transported vector by $\mathbf{u}_{\parallel}$. The obvious analogous definition for scalars (which are trivial to transport) is just $\nabla_{\mathbf{v}} f = df/d\lambda$. Note that the covariant derivative should not be confused with the **Lie derivative**, which is a derivative of a vector that depends on the chosen congruence, as discussed above, and is just the commutator of the two tangent vectors involved:

$$\mathscr{L}_{\mathbf{v}} \mathbf{u} = [\mathbf{v}, \mathbf{u}] = \frac{d}{d\lambda}\frac{d}{d\mu} - \frac{d}{d\mu}\frac{d}{d\lambda}. \tag{1.108}$$

Components of the covariant derivative are introduced in terms of derivatives involving basis vectors via

$$\nabla_{\mathbf{e}_i} \mathbf{e}_j \equiv \Gamma^k_{ji}\, \mathbf{e}_k. \tag{1.109}$$

Using this, it is possible to show that the following definition of a geodesic is equivalent to the equation we had earlier [problem 1.7]:

$$\nabla_{\mathbf{v}} \mathbf{v} = 0; \tag{1.110}$$

i.e. a geodesic curve is one that parallel-transports its own tangent vector.

The Riemann tensor may be defined geometrically in terms of commutators of covariant derivatives:

$$R(\mathbf{v}, \mathbf{u}, \ ,\ ) = [\nabla_{\mathbf{v}}, \nabla_{\mathbf{u}}] - \nabla_{[\mathbf{v},\mathbf{u}]}, \tag{1.111}$$

again denoting commutators by square brackets. There are two empty slots here, one for a vector, one for a 1-form; the covariant derivative $\nabla_{\mathbf{u}}$ maps a vector onto another vector, which then needs to 'eat' a 1-form in order to make an invariant. The Riemann tensor is thus a $(1, 3)$ tensor in its simplest form.

This very nearly completes a basic translation of traditional tensor calculus into the language of differential geometry. It should be a surprise that it has been possible to put off for so long the idea of a metric, given that this is so central to general relativity; but so far there has been no need to introduce this distinct piece of structure. The metric is defined as a symmetric 2-form

$$g(\mathbf{v}, \mathbf{u}) = g(\mathbf{u}, \mathbf{v}), \tag{1.112}$$

which therefore provides a way of associating a 1-form $g(\mathbf{v},\ )$ with a vector, and defines the notion of an **inner product** between vectors. Since such a number should be invariant under parallel transport, it is required that the covariant derivative of the metric be zero with respect to any given vector $\mathbf{v}$:

$$\nabla_{\mathbf{v}}\, g = 0. \tag{1.113}$$

Last of all, this relation in component form gives us the standard relation between g and the components of the connection [problem 1.7]. This therefore completes our aim of looking very briefly at the main results of general relativity in terms of differential geometry.

Problems

(1.1) Use the idea of gravitational time dilation to explain why stars suffer no time dilation due to their apparently high transverse velocities as viewed from the frame of the rotating Earth.

(1.2) Verify that the following trajectory corresponds to that of an observer undergoing constant proper acceleration:

$$\begin{aligned} x &= c^2 g^{-1}[\cosh(g\tau/c) - 1] \\ ct &= c^2 g^{-1} \sinh(g\tau/c). \end{aligned} \tag{1.114}$$

Show that the observer travels less far than would be predicted by a Newtonian distance–time relation in terms of of $x(t)$, but further in terms of $x(\tau)$. How far can an astronaut accelerating at one gravity travel in a proper time of 1 yr, assuming equal acceleration and deceleration phases?

(1.3) A rocket that accelerates at g clearly defines some form of 'accelerated reference frame'. Show that the length contraction of the rocket means that this is not the same frame as that defined by a set of observers who all experience proper acceleration g. Show that the coordinates

$$\begin{aligned} x &= (g^{-1} + x') \cosh gt' - g^{-1} \\ t &= (g^{-1} + x') \sinh gt', \end{aligned} \tag{1.115}$$

with $c = 1$, define time and space coordinates t' and x' that are good candidates for a rocket-based reference frame (i.e. events at constant separation in x' display length contraction as viewed in the lab tx-frame, but 'gravitational' time dilation as viewed in the rocket frame).

(1.4) Consider the path taken by an aeroplane in travelling between two points on the surface of the Earth, which should be an arc of a great circle for optimum fuel efficiency. Attempt to prove this by writing down the equation describing a geodesic on the surface of a sphere.

(1.5) Show that the covariant equivalents of grad and curl ($A_{;\mu}$ and $B_{\mu;\nu} - B_{\nu;\mu}$) are the same as in special relativity, but that the covariant divergence $C^\mu{}_{;\mu}$ is more complicated:

$$C^\mu{}_{;\mu} = \frac{1}{\sqrt{-g}} \frac{\partial}{\partial x^\mu} (\sqrt{-g}\, C^\mu). \tag{1.116}$$

What is the covariant generalization of the d'Alembertian operator $\Box = \partial^\mu \partial_\mu$?

(1.6) By working in a local inertial frame, prove the Bianchi identity

$$\left(R^{\mu\nu} - g^{\mu\nu} R/2\right)_{;\nu} = 0. \tag{1.117}$$

(1.7) Verify that, in terms of components, the geometric equations for a geodesic and for zero covariant derivative of the metric are identical to those derived via traditional tensor calculus.

2 Astrophysical relativity

The idea of this chapter is to go into more detail on a few selected topics in relativity, concentrating on those that are of the most direct interest to the astrophysicist: fluid mechanics, weak fields and orbits therein, gravitational radiation and black holes. A word of warning: so far, all equations have been explicitly dimensional, and contain the correct powers of c; from now on, this will not always be the case. It is common in the literature on relativity to simplify formulae by choosing units where $c = 1$. Since this can be confusing for the beginner, c will be retained where it does not make the algebra too cumbersome. However, sometimes it will disappear from a derivation temporarily. This should encourage the good habit of always being aware of the dimensional balance of any given equation.

2.1 Relativistic fluid mechanics

One of the attractive features of relativity is the economical form that many of its fundamental equations can take. The price paid for this is that the quantities of interest are not always immediately available; the process of 'unpacking' some of these expressions can become rather painful and reveal considerable buried complexity. It is worth illustrating this with the example of fluid mechanics, not just for its own sake, but because we will end up with some results that are rather useful in astrophysics and cosmology. To keep things simple, the treatment will be restricted to special relativity and perfect fluids, where we have already seen that the energy–momentum tensor is

$$T^{\mu\nu} = (\rho + p/c^2)U^\mu U^\nu - pg^{\mu\nu}. \tag{2.1}$$

To use this expression, all that needs to be done is to insert the specific components $U^\mu = \gamma(c, \mathbf{v})$ into the fundamental conservation laws defined by setting the 4-divergence of the energy–momentum tensor to zero: $\partial T^{\mu\nu}/\partial x^\nu = 0$. The manipulation of the resulting equations is a straightforward exercise. Note that it is immediately clear that the results will involve the total or **convective derivative**:

$$\boxed{\frac{d}{dt} \equiv \frac{\partial}{\partial t} + \mathbf{v}\cdot\boldsymbol{\nabla} = \gamma^{-1}U^\mu\partial_\mu.} \tag{2.2}$$

The idea here is that the changes experienced by an observer moving with the fluid are inevitably a mixture of temporal and spatial changes. If I start to feel rain as I cycle towards my destination, it might be a good idea to cycle harder in the hope of arriving before the downpour really stars, but it could also be that it is raining near my destination, and I should stop and wait for the local shower to finish. With only local information on $d[\text{rain}]/dt$ there is no way to tell which is the right strategy. Notice that this two-part derivative arises automatically in the relativistic formulation through the 4-vector dot product $U^\mu \partial_\mu$, which arises from the 4-divergence of an energy–momentum tensor containing a term $\propto U^\mu U^\nu$.

The equations that result from unpacking $T^{\mu\nu}{}_{,\nu} = 0$ in this way have a familiar physical interpretation. The $\mu = 1, 2, 3$ components of $T^{\mu\nu}_{,\nu} = 0$ give the relativistic generalization of **Euler's equation** for momentum conservation in fluid mechanics (not to be confused with Euler's equation in variational calculus):

$$\frac{d}{dt}\mathbf{v} = -\frac{1}{\gamma^2(\rho + p/c^2)}\,(\boldsymbol{\nabla} p + \dot{p}\mathbf{v}/c^2), \tag{2.3}$$

and the $\mu = 0$ component gives a generalization of conservation of energy:

$$\frac{d}{dt}\left[\gamma^2(\rho + p/c^2)\right] = \dot{p}/c^2 - \gamma^2(\rho + p/c^2)\boldsymbol{\nabla}\cdot\mathbf{v}, \tag{2.4}$$

where $\dot{p} \equiv \partial p/\partial t$. The meaning of this equation may be made clearer by introducing one further conservation law: particle number. This is governed by a 4-current having zero 4-divergence:

$$\frac{d}{dx^\mu}J^\mu = 0, \qquad J^\mu \equiv nU^\mu = \gamma n(c, \mathbf{v}). \tag{2.5}$$

If we now introduce the **relativistic enthalpy** $w = \rho + p/c^2$, then energy conservation becomes

$$\frac{d}{dt}\left(\frac{\gamma w}{n}\right) = \frac{\dot{p}}{\gamma n c^2}. \tag{2.6}$$

Thus, in steady flow, $\gamma\times$(enthalpy per particle) is constant.

NONRELATIVISTIC LIMIT For low velocities and normal temperatures ($p \ll \rho c^2$), these equations reduce to more familiar forms. Letting $\gamma \to 1$ and $w \to \rho$, the right-hand sides of the momentum and energy equations tend respectively to $-(\boldsymbol{\nabla} p + \dot{p}\mathbf{v})/\rho$ and $\dot{p}/c^2 - \rho\boldsymbol{\nabla}\cdot\mathbf{v}$. We can also neglect $\dot{p}$ in both cases, via the following dimensional argument. Let $|\boldsymbol{\nabla} p| \sim p/L$ and $|\dot{p}| \sim P/T$, where L and T are characteristic length and time scales for change in the fluid; the ratio defines a characteristic internal velocity $V = L/T$. The order of magnitude of the two terms is thus $|\dot{p}\mathbf{v}/c^2|/|\boldsymbol{\nabla} p| \sim (v/c)(V/c)$, and so the $\dot{p}$ term is negligible if the fluid is both internally and externally nonrelativistic. The first two equations thus reduce to

$$\begin{aligned} \frac{d}{dt}\mathbf{v} &= -\frac{1}{\rho}\boldsymbol{\nabla} p \\ \frac{d}{dt}\rho &= -\rho\boldsymbol{\nabla}\cdot\mathbf{v}, \end{aligned} \tag{2.7}$$

which represent just conservation of momentum and of mass. The energy equation in the version involving enthalpy takes an illuminating form if we define $\rho = \rho_0(1 + \epsilon/c^2)$, where the rest-mass density is $\rho_0 = nm$, and ϵ is thus the internal energy per unit mass. The nonrelativistic limit of $\gamma w/\rho_0$ is $1 + \epsilon/c^2 + p/(\rho_0 c^2) + v^2/(2c^2)$, so that the limiting nonrelativistic form of the conservation equation is $(d/dt)[\epsilon + p/\rho_0 + v^2/2] = \dot{p}/\rho_0$. This does not look particularly familiar, but it can be manipulated into

$$\frac{\partial}{\partial t}\left(n\epsilon + nv^2/2\right) = -\boldsymbol{\nabla}\cdot\left[n\mathbf{v}(\epsilon + v^2/2) + p\mathbf{v}\right], \tag{2.8}$$

which has an immediate interpretation in terms of variation of kinetic energy density responding to a kinetic energy flux density and a rate of working.

EQUATION OF STATE It is common practice to close the system of fluid equations by means of an equation of state. This is done by defining the rest-mass density, ρ_0 to be $\Sigma m_i n_i$, where the sum is taken over all particle species:

$$\frac{p}{c^2} = (\Gamma - 1)(\rho - \rho_0). \tag{2.9}$$

For a cold gas, Γ would be the ratio of specific heats ($= 5/3$ for a monatomic gas). In general, this interpretation of Γ is not quite correct, although it does also hold for an ultrarelativistic, **radiation-dominated fluid** ($p = \rho c^2/3 \Rightarrow \Gamma = 4/3$). The meaning of Γ becomes more obvious if we consider an adiabatic change: deform a fluid element in such a way that $\Delta E = -p\Delta V$. Writing $E = \rho c^2 V$ and using $\Delta V/V = -\Delta\rho_0/\rho_0$ implies the generalization of the familiar adiabatic relation

$$(\rho - \rho_0) \propto p \propto \rho_0^\Gamma. \tag{2.10}$$

Inserting this in the relativistic version of Bernoulli's equation ($\gamma w/\rho_0 =$ constant) leads us to the relation $\gamma \propto p^{-(\Gamma-1)/\Gamma}$ for the steady flow of a radiation-dominated fluid. This gives an idea of how the relativistic jets seen in active galaxies may be produced: a drop in pressure of even a moderate factor is inevitable in moving to the outer regions of a galaxy, and so a relativistic plasma will convert its internal energy into bulk relativistic motion (Blandford & Rees 1974; see also Hughes 1991 and chapter 14).

SOUND WAVES A very useful general procedure can be illustrated by **linearizing** the fluid equations. Consider a small perturbation about each quantity ($\rho \to \rho + \delta\rho$ etc.) and subtract the unperturbed equations to yield equations for the perturbations valid to first order. This means that any higher-order term such as $\delta\mathbf{v}\cdot\boldsymbol{\nabla}\,\delta\rho$ is set equal to zero. If we take the initial state to have constant density and pressure and zero velocity, then the resulting equations are simple:

$$\begin{aligned} \frac{\partial}{\partial t}\delta\mathbf{v} &= -\frac{1}{\rho + p/c^2}\,\boldsymbol{\nabla}\delta p \\ \frac{\partial}{\partial t}\delta\rho &= -(\rho + p/c^2)\,\boldsymbol{\nabla}\cdot\delta\mathbf{v}. \end{aligned} \tag{2.11}$$

Now eliminate the perturbed velocity (via the divergence of the first of these equations minus the time derivative of the second) to yield the wave equation:

$$\boxed{\nabla^2\delta p - \left(\frac{\partial\rho}{\partial p}\right)\frac{\partial^2\delta p}{\partial t^2} = 0.} \tag{2.12}$$

This defines the **speed of sound** to be $c_s^2 = \partial p/\partial\rho$. Notice that, by a fortunate coincidence, this is exactly the same as is derived from the nonrelativistic equations, although we could not have relied upon this in advance. Thus, the speed of sound in a **radiation-dominated fluid** is just $c/\sqrt{3}$.

2.2 Weak fields

Moving from special to general relativity, we should now consider the application of relativistic gravitation to the normal conditions of (astro)physics – in which the Newtonian approximation is very nearly valid. Studying the subject of weak-field general relativity provides proficiency in handling non-Newtonian effects such as gravitational radiation, and also a deeper insight into the meaning of gauge transformations in general relativity.

The obvious way to attack the weak-field problem in general relativity is by way of perturbation theory, assuming that the metric is nearly that of Minkowski spacetime:

$$g^{\mu\nu} = \eta^{\mu\nu} + h^{\mu\nu}, \qquad h^{\mu\nu} \ll 1. \tag{2.13}$$

In first-order perturbation theory, which is all that will be considered here, matters are often simplified by the fact that we can replace the full metric tensor by its special relativity form whenever we want to perform index manipulation on $h^{\mu\nu}$. Thus, for example, the scalar perturbation h may be written

$$h = g_{\mu\nu} h^{\mu\nu} = h^{00} - (h^{11} + h^{22} + h^{33}). \tag{2.14}$$

Similarly, we now have for the Ricci tensor and the connection

$$\begin{aligned} R_{\mu\nu} &= \frac{\partial}{\partial x^\nu}\Gamma^\beta_{\mu\beta} - \frac{\partial}{\partial x^\beta}\Gamma^\beta_{\mu\nu} \\ \Gamma^\alpha_{\lambda\mu} &= \tfrac{1}{2}\eta^{\alpha\nu}\left(\frac{\partial h_{\mu\nu}}{\partial x^\lambda} + \frac{\partial h_{\lambda\nu}}{\partial x^\mu} - \frac{\partial h_{\mu\lambda}}{\partial x^\nu}\right). \end{aligned} \tag{2.15}$$

If we now define a new field via

$$\bar{h}^{\mu\nu} \equiv h^{\mu\nu} - \tfrac{1}{2}\eta^{\mu\nu} h, \tag{2.16}$$

then it is just a matter of straightforward manipulation (plus one vital simplification: see below) to obtain the field equations in the linearized form

$$\boxed{\Box \bar{h}^{\mu\nu} = -\frac{16\pi G}{c^4} T^{\mu\nu}.} \tag{2.17}$$

This is a wave equation, and it shows that gravitational waves exist and propagate at the speed of light, as can be argued on very general grounds (see below). In the absence of matter, simple plane waves $\bar{h}^{\mu\nu} \propto \exp(i\mathbf{k}\cdot\mathbf{x} - kct)$ are a solution to the equation; with matter as a source of gravitation, the solution is analogous to electromagnetism:

$$\bar{h}^{\mu\nu} = \frac{4G}{c^4}\int \frac{[T^{\mu\nu}]}{|\mathbf{r}-\mathbf{r}'|}\, d^3r', \tag{2.18}$$

in terms of the retarded source, indicated by the square brackets. The same Green function is used as appears in the solution of Maxwell's equations even though the wave equation now applies to a tensor rather than a vector field; this just corresponds to solving the wave equation separately for each independent component.

As in electromagnetism, this simple form of the equations is achieved only by means of a **gauge condition**, which is the analogue of the Lorentz gauge and takes a similar form:

$$\boxed{\frac{\partial}{\partial x^\mu}\bar{h}^{\mu\nu} = 0.} \tag{2.19}$$

The fact that general relativity contains this freedom should be immediately obvious: if a solution of Einstein's equations for the tensor gravitational potential is known, we only need make a coordinate transformation in order to obtain a new potential. There are four degrees of freedom introduced by the transformation, which are fixed by the four equations in the above gauge condition. The concept of gauge is discussed in more detail below, where the gauge condition is shown to have an important consequence: it allows us to see clearly that the waves are transverse.

The weak-field equations can be further simplified if we now make a restriction to slowly moving sources and work to first order only in the source velocity: for a classical source $p \ll \rho c^2$, so that only T^{00} is significantly non-zero, and all diagonal components of $\bar{h}^{\mu\nu}$ except the (0, 0) component may therefore be treated as zero. Since $h^{\mu\nu} = \bar{h}^{\mu\nu} - \frac{1}{2}\eta^{\mu\nu}\bar{h}$ ($\bar{h} = -h$ as $\eta^\mu_\mu = 4$), we see that all diagonal components of $h^{\mu\nu}$ are identical and equal to $\bar{h}^{00}/2$, and the off-diagonal components are zero except for $h^{0i} = \bar{h}^{0i}$. These components can be regrouped into scalar and vector 'potentials':

$$\begin{aligned} \Phi = -\frac{c^2}{4}\bar{h}^{00} &= -G\int \frac{[\rho]}{|\mathbf{r}-\mathbf{r}'|}\, d^3r' \\ A^i = -c\bar{h}^{0i} &= -\frac{4G}{c^2}\int \frac{[\rho u^i]}{|\mathbf{r}-\mathbf{r}'|}\, d^3r'. \end{aligned} \tag{2.20}$$

Apart from the factor 4 in the second equation, these look identical to the solutions for the scalar and vector potentials in electromagnetism. The potential **A** could of course be redefined to absorb the factor 4, but the equation of motion shows that this would still not make the analogy perfect. The spatial parts of the equation of motion are given by

$$\begin{aligned} \frac{d^2x^i}{dt^2} &= \frac{d\tau}{dt}\frac{d}{d\tau}\left(\frac{d\tau}{dt}\frac{dx^i}{d\tau}\right) \\ &= \left(\frac{d\tau}{dt}\right)^2\frac{d^2x^i}{d\tau^2} - \left(\frac{d\tau}{dt}\right)^3\frac{d^2t}{d\tau^2}\frac{dx^i}{d\tau} \\ &= -\Gamma^i_{\alpha\beta}\frac{dx^\alpha}{dt}\frac{dx^\beta}{dt} + \Gamma^0_{\alpha\beta}\frac{dx^\alpha}{dt}\frac{dx^\beta}{dt}\frac{dx^i}{dt} \\ &\simeq -c^2\Gamma^i_{00} - 2c\Gamma^i_{0j}v^j + c^2\Gamma^0_{00}v^i, \end{aligned} \tag{2.21}$$

where the second step is $[(t')^{-1}]' = -[t']^{-2}t''$, using primes to denote $d/d\tau$, and the last step involves expanding to first order in $\mathbf{v}$. Using the above weak-field forms for the connection coefficients in terms of $h^{\mu\nu}$ produces the following equation in terms of our definition of the gravitational scalar and vector potentials:

$$\boxed{\ddot{\mathbf{x}} = -\boldsymbol{\nabla}\Phi - \dot{\mathbf{A}} + \mathbf{v}\wedge(\boldsymbol{\nabla}\wedge\mathbf{A}) + 3\mathbf{v}\dot{\Phi}.} \tag{2.22}$$

In many circumstances (e.g. the gravitational field of a rotating body having cylindrical symmetry) the last term on the right-hand side would vanish, leaving an equation of motion identical to that of electromagnetism. The tensor nature of the force still cannot be completely hidden, however, and manifests itself as the unexpected factor 4 in the equation for the vector potential. The analogy between gravity and electromagnetism cannot be made to work at any level of approximation save that of Newtonian gravity.

MOTION OF RELATIVISTIC PARTICLES If we do not take the low-velocity limit of the equation of motion, a simple result can still be obtained by considering the case of a constant Newtonian field, i.e. $h_{\mu\nu} = h_{00}\,\mathrm{diag}(1,1,1,1)$, that acts perpendicular to the particle's path (or rather, by considering only the effect of the perpendicular component of the field). Let the particle propagate in the x-direction and experience an acceleration in the y-direction. Since the result of all derivatives except d/dy is zero, the only surviving terms in the equation of motion are

$$\ddot{y} = -c^2\Gamma^2_{00} - v^2\Gamma^2_{11} - 2vc\Gamma^2_{01}. \tag{2.23}$$

The last term vanishes if there are no off-diagonal terms in $h_{\mu\nu}$, and what remains is the famous result

$$\boxed{\frac{d^2y}{dt^2} = -\left(1 + \frac{v^2}{c^2}\right)\frac{d\Phi}{dy}.} \tag{2.24}$$

Thus, photons 'fall sideways' twice as fast as slowly moving particles. This is one case where the prediction of general relativity is not simply a small perturbation away from the Newtonian prediction, and its verification in the Solar eclipse of 1919 led to the rapid acceptance of the theory. This gravitational deflection of light is of great cosmological importance in gravitational lensing.

GAUGE FREEDOM IN GENERAL RELATIVITY We have already seen the necessity for distinguishing between 'active' and 'passive' Lorentz transformations in relativity. A more common term for the latter is **gauge transformation**, meaning some mathematical degree of freedom within a theory that has no observable consequence (e.g. the arbitrary value of $\boldsymbol{\nabla} \cdot \mathbf{A}$, where $\mathbf{A}$ is the vector potential in electrodynamics).

From a mathematical point of view, the gravitational field equations contain a gauge ambiguity through a counting of degrees of freedom. In the field equations, both $G^{\mu\nu}$ and $T^{\mu\nu}$ are symmetric, thus defining 10 independent equations for the components of the metric $g^{\mu\nu}$, which also has 10 independent components. Things seem to be proceeding well, but of course the 10 field equations are not all independent: they are related by the fact that the 4-divergence of both the Einstein and energy–momentum tensors is zero, i.e. by the Bianchi identity $G^{\mu\nu}_{\;;\nu} = 0$. The fact that we are then left with four degrees of freedom is of course obvious in hindsight: given a solution of the field equations, we can readily generate another by a change of the coordinate system: $x^\mu \to x^\mu + \epsilon^\mu$.

This situation is closely reminiscent of that in electromagnetism, where there is the freedom to add the 4-gradient of some scalar function to the 4-potential, without affecting Maxwell's equations: $A^\mu \to A^\mu + \partial^\mu\psi$. As will be seen in later chapters, this freedom exists because the photon is massless: a massive particle has more quantum-mechanical degrees of freedom available than a massless particle. The vanishing of the photon mass causes these extra degrees of freedom (which would be associated with longitudinal waves, rather than transverse ones) to appear in the theory as gauge degrees of freedom. Since the graviton is also massless, we are faced with a similar problem.

As in electromagnetism, the solution is to add some condition that fixes the unphysical degrees of freedom. What is needed is the analogue of the Lorentz gauge $\partial_\mu A^\mu = 0$. Since this corresponds to fixing a frame of reference, any such condition must be explicitly non-covariant. As was mentioned above, a convenient way of dealing with weak fields is to impose the **harmonic or Lorentz gauge**

$$g^{\mu\nu}\Gamma^\lambda_{\mu\nu} = 0, \tag{2.25}$$

where it is easy to use the linear form of the connection to show that this is just an alternative form of the Lorentz condition, written above as $\bar{h}^\mu_{\nu,\mu} = 0$. It allows the weak-field equations to be cast into a simple form, and plays very much the same algebraic role as the Lorentz gauge in obtaining Maxwell's equations in 4-vector form.

Although the gauge freedom can be exploited in the case of electromagnetism to impose $\partial_\mu A^\mu = 0$, it is well known that this does not remove the gauge freedom entirely: applying the transformation $A^\mu \rightarrow A^\mu + \partial^\mu \psi$ does not change the electric and magnetic fields, and yet any transformation that satisfies $\Box\psi = 0$ will leave the Lorentz condition unchanged. The same situation exists in general relativity, where the condition $\bar{h}^{\mu\nu}_{,\mu} = 0$ is preserved under Lorentz transformations. A large range of possible specific gauge choices therefore exist and are used in the literature; some of the most common are as follows.

THE SYNCHRONOUS GAUGE This confines any perturbations away from Minkowski spacetime to the spatial part of the metric:

$$g^{0\mu} = (1, 0, 0, 0). \tag{2.26}$$

This gauge is commonly used in the study of cosmological perturbations.

THE NEWTONIAN GAUGE Here one expresses the deviations of the metric in terms of a function that looks like the Newtonian potential, Φ. The metric perturbation is purely diagonal, $h^{\mu\nu} = h\,\mathrm{diag}(1, 1, 1, 1)$:

$$c^2 d\tau^2 = \left(1 + \frac{2\Phi}{c^2}\right) c^2 dt^2 - \left(1 - \frac{2\Phi}{c^2}\right)(dx^2 + dy^2 + dz^2). \tag{2.27}$$

The Newtonian gauge does not allow gravitational waves, but it was shown earlier that this is the correct choice of metric for weak gravitational fields generated by static mass distributions. This is therefore a convenient gauge for problems such as planetary dynamics. More generally, the relation of the metric perturbations to the familiar gravitational potential helps the use of physical intuition in understanding the meaning of weak-field calculations, particularly in cosmology.

THE TRANSVERSE TRACELESS GAUGE This is the gravitational analogue of the Coulomb gauge, which allows electromagnetic radiation in free space to be described in terms of a vector potential only. In gravity, the corresponding gauge definition is

$$h_{\mu 0} = 0, \quad h^\mu_\mu = 0. \tag{2.28}$$

This name comes from application of the wave equation and the Lorentz condition to a wave

$$\bar{h}^{\mu\nu} = A^{\mu\nu} e^{ik^\mu x_\mu}, \tag{2.29}$$

which gives respectively

$$\begin{aligned} k^\mu k_\mu &= 0 \\ k^\mu A_{\mu\nu} &= 0. \end{aligned} \tag{2.30}$$

The first equation just says that the wave travels at the speed of light (k^μ is null); the second equation says that the wave is spatially transverse, since $A_{00} = 0$ in this gauge. Note that, since $h = 0$, there is no distinction in this gauge between $h_{\mu\nu}$ and $\bar{h}_{\mu\nu}$.

2.3 Gravitational radiation

GEODESIC DEVIATION AND TIDES The observational manifestation of gravity is tidal forces: the relative acceleration of different particles that cannot be removed by change of frame. In Newtonian language, the relative acceleration of two test particles separated by **X** is

$$\ddot{X}_i = -\frac{\partial^2 \Phi}{\partial x_i \partial x_j} X_j; \tag{2.31}$$

what is the relativistic analogue of this expression? We need to consider two particles on neighbouring geodesics, each of which obeys the equation of motion

$$\frac{d^2 x^\mu}{d\tau^2} = -\Gamma^\mu_{\alpha\beta}(x) \frac{dx^\alpha}{d\tau} \frac{dx^\beta}{d\tau}. \tag{2.32}$$

The next step is expand this as a Taylor series in X^μ to obtain the equation of motion at the neighbouring particle: $x^\mu \to x^\mu + X^\mu$. To first order, this generates an expression for $d^2 X^\mu / d\tau^2$ that involves Γ and its first derivatives. These terms can be collected in terms of the Riemann tensor to obtain the equation of **geodesic deviation**:

$$\boxed{\frac{D^2}{d\tau^2} X^\mu = R^\mu_{\alpha\beta\gamma} \frac{dx^\alpha}{d\tau} \frac{dx^\beta}{d\tau} X^\gamma} \tag{2.33}$$

(see e.g. section 6.10 of Weinberg 1972 or section 6.6 of Narlikar & Padmanabhan 1986 for details of this transition).

This now gives the desired generalization of the Newtonian tidal acceleration; in a local inertial frame, the left-hand side is just the relative 4-acceleration of the two particles. The distinction between ordinary and covariant derivatives may seem to be a problem in making this transition, since the double covariant derivative will generate terms involving $d\Gamma/d\tau$, and the derivatives of the connection are non-zero in general (the signature of curvature). However, by using a coordinate system tied to a freely falling observer, the local inertial frame carried by the observer remains inertial at all τ: Γ always vanishes along the world-line of the observer, and so $d\Gamma/d\tau = 0$ also.

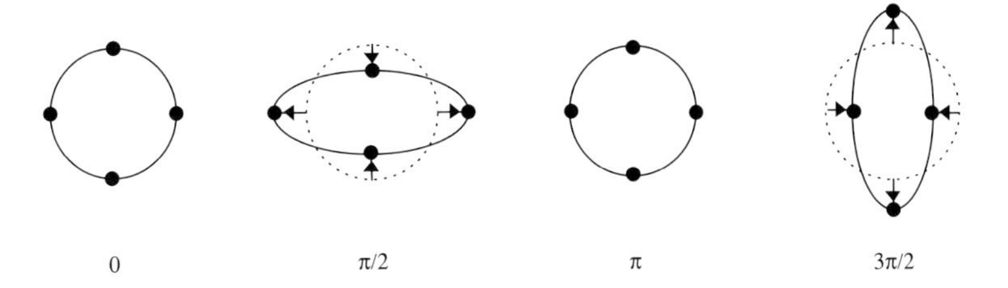

Figure 2.1. This figure illustrates the displacement of test particles caused by the passage of a plane polarized gravitational wave, with wave-vector normal to the page. The quadrupolar nature of the wave shows itself by the production of shear in the particle distribution, and there is zero average translational motion.

LABORATORY EXPERIMENTS Observers thus experience the passage of a gravitational wave in the form of a non-zero **tidal tensor**, whose spatial parts are the generalization of Φ_{ij}:

$$\Delta_{\mu\nu} \equiv R_{\mu\alpha\beta\nu} U^\alpha U^\beta, \tag{2.34}$$

where U^μ is the observer's 4-velocity. The Riemann tensor is expressed to first order in terms of the metric perturbation as

$$R_{\mu\nu\alpha\beta} = \tfrac{1}{2}(-\partial_\mu \partial_\alpha h_{\nu\beta} - \partial_\nu \partial_\beta h_{\mu\alpha} + \partial_\mu \partial_\beta h_{\nu\alpha} + \partial_\nu \partial_\alpha h_{\mu\beta}). \tag{2.35}$$

In the rest frame, where $U^\mu = (1, 0, 0, 0)$, the tidal tensor reduces to

$$\Delta_{\mu\nu} = \frac{1}{2}\frac{\partial^2}{\partial t^2} h_{\mu\nu}. \tag{2.36}$$

We know several general properties of the tidal tensor: from the symmetries of the Riemann tensor, Δ must be symmetric. Because the Ricci tensor vanishes outside matter, Δ must be traceless. In the TT gauge $\Delta_{\mu 0} = 0$; as we saw above, the Lorentz condition is equivalent to $\Delta_{\mu\nu} k^\nu = 0$, which implies that the waves are spatially transverse in this case. If we take the simple case of a wave propagating in the $\hat{z}$-direction and work in the rest frame of the observer, the general form of the wave can thus depend on only two independent parameters, Δ_+ and $\Delta_\times$:

$$\Delta_{\mu\nu} = \begin{pmatrix} 0 & 0 & 0 & 0 \\ 0 & \Delta_+ & \Delta_\times & 0 \\ 0 & \Delta_\times & -\Delta_+ & 0 \\ 0 & 0 & 0 & 0 \end{pmatrix} \exp -ik^\mu x_\mu. \tag{2.37}$$

The displacement pattern produced by one of these modes is shown in figure 2.1; the other is related by a rotation of 45°. Consider just the Δ_+ mode, which produces a displacement field

$$\delta \mathbf{X} = -\Delta_+ \omega^{-2} (X, -Y, 0) e^{i\omega t}. \tag{2.38}$$

All this makes perfect sense on Newtonian grounds: the tidal field must be divergenceless, and so any tendency for test particles to approach each other along one direction means that they will be separated along the other direction.

These modes can be combined: $\Delta_\times = \pm i\Delta_+$ gives circularly polarized radiation where the motion of test particles is circular. It is easy to verify in that case that the wave changes as

$$\Delta_{\mu\nu} \to e^{\pm 2i\theta}\Delta_{\mu\nu} \tag{2.39}$$

under a rotation of θ about the $\hat{z}$-axis (transform using the rotation matrix, R, so that $\Delta' = R^T \Delta R$). In other words, the gravitational 'wave function' has a helicity of 2, which is the classical basis for expecting the quantum of the field (the graviton) to have spin 2.

ENERGY OF GRAVITY WAVES Gravitational waves are able to cause the motion of bodies at some distance from the source, so they clearly carry energy. There ought to exist an energy–momentum tensor that describes the effect of the waves, but where do we find it in Einstein's equations? The only explicit energy–momentum tensor is that of the matter, which vanishes for the case of vacuum radiation, so that the field equations are then just $G^{\mu\nu} = 0$. The trick that must be used is to divide the Einstein tensor into two parts, one of which represents the large-scale properties of spacetime on a scale much greater than the wavelength of the gravity waves; the remaining part of $G^{\mu\nu}$ can be interpreted as the energy–momentum tensor of the waves. This is similar to the expansion that was made previously in obtaining the linear field equations:

$$g_{\mu\nu} = g^{(\mathrm{B})}_{\mu\nu} + h_{\mu\nu}, \tag{2.40}$$

where (B) denotes the background. The difference now is that the background metric is not regarded as something static, but as responding to the waves: if $G^{\mu\nu} = 0$, then the field equation for the background is

$$G^{(\mathrm{B})}_{\mu\nu} = -G^{(h)}_{\mu\nu}, \tag{2.41}$$

where the rhs is the contribution of the waves to the Einstein tensor. This cannot be accomplished within linearized theory, since we have previously seen that the linear Ricci tensor vanishes in the absence of matter. The Einstein tensor is then, to lowest order in h,

$$G_{\mu\nu} \simeq G^{(\mathrm{B})}_{\mu\nu} + G^{(2)}_{\mu\nu}, \tag{2.42}$$

where the (2) superscript denotes the second-order contribution. The effective source term for the gravity waves is then

$$T^{\mathrm{GW}}_{\mu\nu} = \frac{c^4}{8\pi G} G^{(2)}_{\mu\nu}. \tag{2.43}$$

This means that it is necessary to expand the Einstein tensor to second order in h, which is an algebraic exercise that would consume too much space here. The resulting expression for the higher-order corrections to the Riemann tensor is simplified if it is averaged over several wavelengths or periods of the waves, to obtain the mean flux of energy (averaging is denoted below by angle brackets), so that terms of odd order in field derivatives vanish (see chapter 35 of Misner, Thorne & Wheeler 1973). Physically, this means that the energy transport by the waves cannot be seen clearly on small scales; the same applies in electromagnetism, where the Poynting vector for a plane wave is not constant, reflecting the oscillating energy density. The final result is the **Landau–Lifshitz pseudotensor**

$$T^{\mathrm{GW}}_{\mu\nu} = \frac{c^4}{32\pi G}\left\langle \bar{h}_{\alpha\beta,\mu}\bar{h}^{\alpha\beta}{}_{,\nu} - \tfrac{1}{2}\bar{h}_{,\mu}\bar{h}_{,\nu} \right\rangle. \tag{2.44}$$

In the TT gauge, things are simpler still, because the second term inside the angle brackets vanishes.

Accepting this result, the above linear solutions to the wave equation can be used to obtain the energy radiated by a given configuration of matter, $T^{\mu\nu}$. The answer can be made simple in a general way first by assuming that the observer is in the far field, so that factors of $1/r$ can be taken out of integrals. Also, the fact that the source satisfies the usual conservation law can be used to recast the source term in terms of the quadrupole moment:

$$\begin{aligned} & T^{\mu\nu}{}_{,\nu} = 0 \quad \Rightarrow \quad T^{00}{}_{,0} = -T^{0i}{}_{,i} \\ \Rightarrow \quad & T^{00}{}_{,00} = -T^{0i}{}_{,i0} = -(T^{i0}_{,0})_{,i} \\ \Rightarrow \quad & T^{00}{}_{,00} = +T^{ij}{}_{,ij}, \end{aligned} \tag{2.45}$$

where roman letters denote spatial components, as usual, and the last step uses $T^{ij}_{,j} + T^{i0}_{,0} = 0$. Now $(T^{ij}x^i x^j)_{,ij}$ equals $2T^{ij} + T^{ij}{}_{,ij}x^i x^j$ + derivative terms that integrate to zero (because the source can be made to vanish on the boundary). Using the above result shows finally that the desired integral over the source is related to the second time derivative of the inertia tensor:

$$\int T^{ij}\, d^3x = \frac{1}{2}\frac{d^2}{dt^2}\int T^{00} x^i x^j\, d^3x. \tag{2.46}$$

QUADRUPOLE FORMULA This assembles all the items needed in order to calculate the energy loss rate through gravitational radiation of an oscillating body. The linear solution gives the field in terms of $T^{\mu\nu}$ for the source, which we have just shown to be related to the time derivatives of the matter quadrupole moment. Integrating the resulting $T^{\rm GW}_{\mu\nu}$ gives the total energy loss rate. Unfortunately, the amount of algebra needed to finish the calculation is considerable; see chapter 36 of Misner, Thorne & Wheeler (1973). The final product is the **quadrupole formula**, which is the general expression for the energy lost by an oscillating nonrelativistic source via gravitational radiation:

$$\boxed{-\dot{E} = \frac{G}{5c^5}\left\langle \dddot{I}_{ij}\dddot{I}^{ij}\right\rangle,} \tag{2.47}$$

where the 3-tensor I^{ij} is the reduced quadrupole-moment tensor of the source,

$$I_{ij} = \int \rho\,(x_i x_j - r^2\delta_{ij}/3)\, d^3x, \tag{2.48}$$

$\dddot{I}$ denotes the third time derivative, and the average is over a period of oscillation.

It should be no surprise that the radiation depends on the quadrupole moment; the gravitational analogue of dipole radiation is impossible owing to conservation of momentum. Whenever a mass is accelerated in astronomy, another mass or group of masses suffers an equal and opposite change in momentum, and so any attempt by an individual mass to produce dipole radiation is automatically cancelled. It is a pity that such a simple key equation cannot be derived more simply. The best that can be said is that it can be understood by means of the electromagnetic analogy, which gets the energy loss rate wrong by a factor 4 [problem 2.3].

ASTRONOMICAL SIGNALS AND SENSITIVITIES It is instructive to calculate some characteristic numbers for terrestrial and astronomical gravity waves. As an application of the quadrupole formula, consider a rod of length L that rotates about its midpoint. The rate of radiation of gravitational-wave energy is

$$\dot{E}_{\rm GW} = \frac{2G}{45c^5} M^2 L^4 \omega^6 \simeq 1.2 \times 10^{-54}\ {\rm W}\,(M/{\rm kg})^2\,(L/{\rm m})^4\,(\omega/{\rm s}^{-1})^6, \tag{2.49}$$

which is extremely small on a laboratory scale. Astronomical systems can do a good deal better, but only up to a limit. The typical frequency expected from orbiting systems of size L is the virial $\omega^2 \sim GM/L^3$, so that the power expected from the $M^2L^4\omega^6$ scaling is

$$\dot{E}_{\rm GW} \sim \frac{c^5}{G}\left(\frac{GM}{c^2L}\right)^5. \tag{2.50}$$

Inside the parentheses is the ratio of the Schwarzschild radius to the scale of the system, which will never exceed unity. The maximum gravity-wave power that any system can emit is therefore independent of mass:

$$\dot{E}_{\rm GW} \lesssim \frac{c^5}{G} = 10^{52.6}\ {\rm W}. \tag{2.51}$$

The sort of astrophysical phenomenon that might yield this optimistic output level is coalescing binary black holes, where the orbital decay would be caused precisely by the emission of gravitational radiation (cf. the binary pulsar below). Normal systems are not in danger of being affected significantly by gravitational-wave energy loss: the Earth emits around 200 W of gravitational radiation by virtue of its orbit around the sun, and the corresponding orbital decay time is approximately 10^{23} years [problem 2.4].

To see that the maximal rate is indeed optimistic, consider the total energy emitted by a binary radiating at the limit. Most of the energy comes out in the last orbital period, so $E \sim \dot{E}/\omega \sim Mc^2$ (for an object with $GM/c^2r \sim 1$). The maximal emission is thus attained only by an object that is capable of transforming all its rest-mass energy into gravitational radiation. As with many astrophysical processes involving the release of gravitationally generated energy, it is more plausible to introduce an efficiency $\epsilon < 1$:

$$\Delta E = \epsilon M c^2, \tag{2.52}$$

and estimates of possible realistic ϵ values range from $\sim 10^{-3}$ down (e.g. Thorne 1987).

To complete this thumbnail sketch of the simplest astrophysical emitters of gravitational waves, the characteristic frequency for an object with $GM/c^2r \sim 1$ is

$$\omega \sim \frac{c^3}{GM} = \left(\frac{M}{10^{5.3} M_\odot}\right)^{-1}\ {\rm Hz}. \tag{2.53}$$

The most common circumstance in which something close to such a situation might occur is during the formation of a neutron star in the course of a type II supernova explosion. This is an event that occurs roughly once per century in a given galaxy, producing an object of around Solar mass and $GM/c^2r \sim 0.1$. Thus, one might expect to find bursts of radiation with $\dot{E} \lesssim 10^{-5}c^5/G$ at roughly kHz frequencies.

What kind of signal would gravitational waves of this sort produce on Earth? The fundamental effect is a **gravitational strain**: a fractional distortion in the length of objects induced by the fluctuating gravitational acceleration. Two main classes of detector have been proposed to search for this signal: resonant bars and interferometers. The former have been the dominant class of detector, but as yet have registered no signal. In the coming years, however, great improvements are expected from laser-based interferometers; these are also somewhat simpler to analyse, so what follows is restricted to this case. Consider the equation of motion for a test particle subjected to a plane-polarized wave of frequency ω:

$$\ddot{x}_i = \Delta_{ij} x_j. \tag{2.54}$$

To obtain the **spatial strain** corresponding to this motion, suppose only one mode is present, $\bar{h}_{\mu\nu} \propto \mathrm{diag}(0, h_+, -h_+, 0)$, which implies that the tidal tensor is

$$\Delta_{\mu\nu} = \mathrm{diag}(0, \ddot{h}_+/2, -\ddot{h}_+/2, 0) \equiv \mathrm{diag}(0, \Delta_+, -\Delta_+, 0). \tag{2.55}$$

Now integrate the equation of motion twice to find the amplitude of the spatial oscillation Δx_i, obtaining

$$\left|\frac{\Delta x}{x}\right| = \frac{\Delta_+}{\omega^2} = \frac{h_+}{2}, \tag{2.56}$$

with a similar definition for $h_\times$, if this mode is present. So, h is (twice the) spatial strain produced by the wave; in general, we would define the total strain as

$$h \equiv \sqrt{h_+^2 + h_\times^2}. \tag{2.57}$$

Now, the energy flux density of the wave is

$$T_{01} = \frac{1}{32\pi G}\left\langle \bar{h}_{\alpha\beta,0}\bar{h}^{\alpha\beta}{}_{,1}\right\rangle = \frac{\omega^2}{32\pi G} \times 2h^2 \times \tfrac{1}{2} = \frac{\omega^2}{32\pi G} h^2 \tag{2.58}$$

(remembering the factor 1/2 from taking the time average of the square of an oscillating quantity; h is a wave amplitude, not an rms). Inserting the dimensions gives

$$h = \left(\frac{32\pi G\, T_{01}}{c^3\omega^2}\right)^{1/2}. \tag{2.59}$$

Assuming a maximally emitting source, $T_{01} = (c^5/G)/(4\pi r^2)$, and $h = \sqrt{8}c/\omega r$. At 1 kHz, this gives $h \sim 10^{-14}$ at 1 kpc (a galactic source), or 10^{-17} at 1 Mpc (a nearby galaxy). As discussed above, a more realistic treatment of supernova-induced formation of a neutron star would give an energy output at least 10^5 times lower, and thus a strain perhaps as small as 10^{-20} at 1 Mpc. This is an absurdly small number, approximately one atomic dimension in a baseline of 1 a.u., and yet it is a remarkable fact that such a signal is expected to be within the reach of a new generation of interferometer-based 'gravity telescopes': the **LIGOs** (laser interferometer gravity-wave observatories).

The most promising idea for the detection of gravitational waves seems to be the construction of large Michelson interferometers, in which light is passed along two perpendicular paths, reflected at the end of each arm, and recombined to create fringes. Movement of the fringes measures a relative change in length of the two arms. The length of arm should be as long as possible to increase the fractional strain that can be measured, but the signal will become incoherent if the light travel time is comparable to the period of the gravitational waves. Since the typical frequency of interest may be in the kHz region, the maximum arm length may need to be a few hundred km. A practical means of attaining this is to have a smaller baseline (currently $\sim$ 10 m, likely to increase 100-fold within a few years) with multiple reflections, using a Fabry–Pérot etalon.

What are the fundamental limits on such a device? Assuming that isolation from seismic, thermal and other external perturbations can be arranged, the ultimate limit is set by the uncertainty principle. For the present application, this states that the end mirror of one of the arms is only observed over some integration time τ, so there is a restriction on how precisely its position can be determined. The joint rms uncertainties in position and conjugate momentum of the end mirror (of mass m) satisfy $\Delta x \Delta p \gtrsim \hbar$, as usual. After integration for a time τ, the momentum uncertainty produces an additional position uncertainty of $\tau \Delta p/m$, added in quadrature to the initial Δx, producing a minimal position uncertainty of $\sqrt{\tau\hbar/m}$. The minimum strain detectable is then fundamentally limited at

$$\boxed{h_{\rm min} \sim \frac{\sqrt{\tau\hbar/m}}{L_{\rm arm}}.} \tag{2.60}$$

Note that it is the arm length that enters, rather than the effective total path after multiple reflections; this is an argument for having as long an arm as possible. For an idealized future instrument with $L_{\rm arm} = 1$ km, $m = 1$ tonne and $\tau = 1$ ms, this gives a limit of 10^{-23}. The detection of plausible astronomical signals will thus require an instrument operating very close to the quantum limit, and this will only be possible if the power output of the laser is set at an optimal value. Too few photons mean that the sensitivity to fringe shifts is reduced; too many, and the fluctuating radiation pressure overcomes the gravity-wave perturbation. Suppose N photons are counted during one integration period, and let n be the effective number of multiple reflections in the interferometer arm. The momentum uncertainty through shot-noise fluctuations in the number of photons is $\Delta p \sim (\hbar\omega/c)\, n\sqrt{N}$ (note that the uncertainty scales linearly with the number of passes through the arm). The radiation-pressure position uncertainty then translates to a strain uncertainty of

$$\Delta h \sim (\hbar\omega/c)\,(n\tau/mL)\,\sqrt{N}. \tag{2.61}$$

Now consider the measuring error arising from finite photon number. If the path difference between two arms is Δx, the phase difference is $\phi = 2\Delta x n\omega/c$, and the number of photons detected at the output of the interferometer is $O = N\cos^2(\phi/2)$, where N is now the total number of photons passing down the interferometer. The shot-noise limit on the phase difference that can be detected is then

$$\Delta\phi = \frac{d\phi}{dO}\sqrt{O} = \left[\sqrt{N}\sin(\phi/2)\right]^{-1}. \tag{2.62}$$

This is minimized at a dark fringe, $\phi = \pi$, and the corresponding strain uncertainty is

$$\Delta h \sim \frac{c}{\omega n L\sqrt{N}}. \tag{2.63}$$

The optimum laser output is that for which these two sources of uncertainty are equal, and at this point the total uncertainty reaches the quantum limit. The power required to attain this happy state is

$$\text{power} = \frac{N\hbar\omega}{\tau} = \frac{c^2 m}{\omega n^2 \tau^2}, \tag{2.64}$$

which is about 3000 kW for the above parameters, with a wavelength of 600 nm and taking $n = 100$ to obtain an effective armlength of 100 km. This is well above any likely practical power output (10–100 W), and so future interferometers seem likely to be shot-noise limited. Nevertheless, current interferometers have already achieved sensitivities of order $h \sim 10^{-19}(\tau/\mathrm{s})^{-1/2}$, so astronomically interesting performance may be expected from scaling up to km baselines with a moderate increase in laser power. It will be an outstanding triumph for experimental physics if the new-generation LIGOs are successful. For a detailed survey of the field, consult Blair (1991).

2.4 The binary pulsar

It seems inevitable that gravitational radiation of some sort must exist. Causality dictates that the gravitational effects of moving bodies must propagate at the speed of light, and J.J. Thomson's argument gives a reasonable idea of the sort of field configurations that must result from accelerated bodies. This argument consists of imagining a body or bodies that make a brief accelerated transition between different initial and final static configurations. Requiring the lines of force to join continuously over the sphere at $r = ct$ shows that there is a transverse pulse of radiation, and allows the correct field strength to be calculated in the case of electric dipole radiation. Thus, the detection of gravitational radiation of some sort would only be a rather weak test of the basic principles of general covariance. Nevertheless, the detailed properties of the radiation do depend on the field equations and their weak-field limit, so in principle a test of Einstein's theory is possible.

Laboratory experiments remain some way from even the detection of gravitational waves, never mind the detailed categorization needed to distinguish competing theories. Nevertheless, most physicists agree that such waves have been detected astrophysically, and that they have the properties expected in Einstein's theory. These conclusions follow from the study of a remarkable object in our galaxy: the **binary pulsar** 1913+16. This was discovered in 1975 by Joseph Taylor and Russell Hulse, who received the 1993 Nobel prize for Physics in recognition of their subsequent use of this source as a test-bed for relativity theories.

Despite its name, the binary pulsar is one pulsar rather than two; but it does occur in a binary system with a non-pulsating star as a companion. Pulsars are neutron stars that rotate with periods of typically 10^{-2} to 10^{-1} s, yielding regular pulses of radiation with a stability that makes them highly accurate clocks. The binary pulsar system is nature's realization of one of Einstein's many thought experiments involving moving clocks, and it allows relativistic effects to be probed in great detail. The period of the binary pulsar is $P = 0.059$ s, spinning down at a fractional rate $\dot{P} \simeq 10^{-17}$, which gives some idea of the clock precision that can be achieved by counting a large number of pulses.

The analysis of the binary pulsar is similar to that for a spectroscopic binary in stellar astronomy. There, one has a star with a time-varying Doppler redshift caused by its orbital motion about a companion. For the binary pulsar, the pulses act as a low-frequency analogue of the spectral lines used in the stellar case; it was the Doppler-induced variations in the pulsar's period that led to its being identified as a member of a binary system. What information can be deduced about a system for which the radial velocity curve as a function of time is known? It is a standard exercise in stellar astronomy to fit such a curve to the parameters of a Keplerian orbit. The system is characterized by the following parameters: the masses m_1 and m_2 of the bodies and the eccentricity e and semi-major axis a of the orbit. In addition, there are two angles that govern the system as viewed from the Earth: the inclination of the orbit to the line of sight i, and the position angle in the orbital plane of periastron ω ($i = 0$ corresponds to

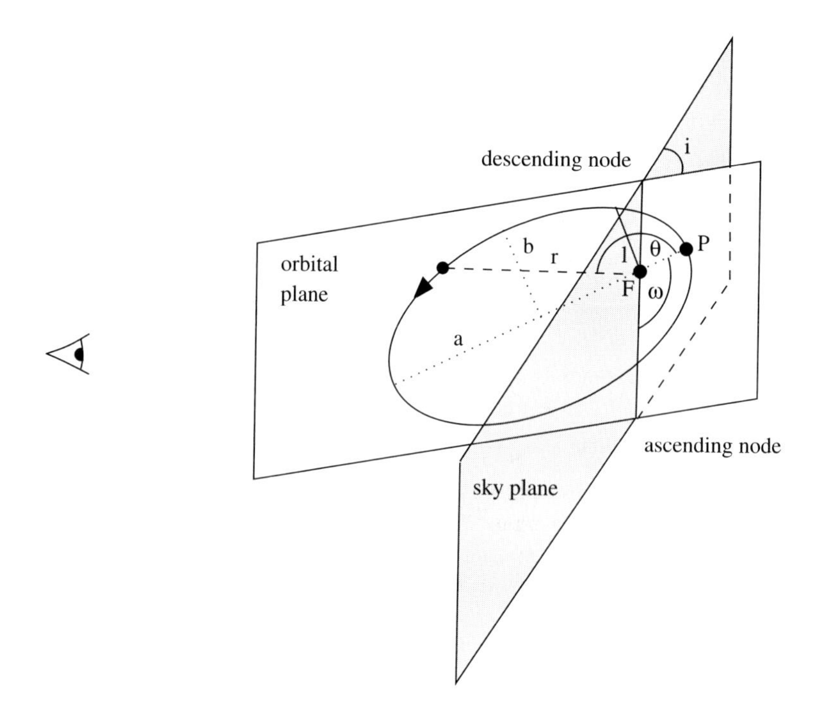

Figure 2.2. The orbital elements for an inclined elliptical orbit. The radius r is measured from the focus (F), and the angle θ is measured from periastron (P). The equation of the elliptical orbit is then $r^{-1} = l^{-1}(1 + e\cos\theta)$, in terms of which the semi-minor axis $b = l/(1-e^2)^{1/2}$, and the semi-major axis is $a = l/(1-e^2)$. The angles i and ω define the orientation of the orbit relative to the sky plane.

an orbit in the plane of the sky); see figure 2.2. Of these six parameters, two (e and ω) can be found exactly, but the other four can be constrained only by two combinations, the **projected semi-major axis** and the **mass function**:

$$
\begin{aligned}
a_\perp &= a\,\sin i \\
f(m) &= \frac{(m_2 \sin i)^3}{(m_1 + m_2)^2},
\end{aligned}
\tag{2.65}
$$

where m_1 denotes the pulsar and m_2 its companion. For a purely Newtonian system, this is all that can be measured; however, relativistic effects allow this degeneracy to be broken.

The first effect to come to mind is advance of the periastron. The short orbital period of 7.8 hours suggests a very tight orbit, where the advance rate will be significant, and $\dot{\omega}$ is indeed large: $4.2\,\mathrm{deg\,yr^{-1}}$. A second piece of information is the extra contributions to the pulsar redshift: higher-order corrections to the Doppler shift and gravitational redshifts in the field of the companion. These two pieces of data alone suffice to determine the parameters of the system, implying that both pulsar and companion have nearly equal masses $m \simeq 1.4\ M_\odot$, and that the inclination is $i \simeq 45°$. See e.g. chapter 12 of Will (1993) for an analysis of these constraints.

Impressive as such a feat of applied relativity is, the most significant aspect of the system is that the orbital period is observed to be decreasing:

$$\dot{P}_{\rm orbit} = -2.1 \pm 0.4 \times 10^{-12}. \qquad (2.66)$$

Of course, there are a variety of mechanisms that could in principle cause such a loss of orbital energy, but the most intriguing is the possibility of gravitational radiation – especially as the perfectly specified system gives a prediction for the loss rate with no free parameters. The quadrupole formula for the energy lost by a binary system can be manipulated into the form

$$-\dot{E} = \frac{32}{5} F(e) \frac{m_1^2 m_2^2 (m_1 + m_2)}{a^5}, \qquad (2.67)$$

where $F(e) = (1 + 73e^2/24 + 37e^4/96)(1 - e^2)^{-7/2} \simeq 11.8$ for the present case of $e = 0.617$.

From simple orbital dynamics, the rate of change of energy is related to the period change via $\dot{P}/P = -3\dot{E}/(2E)$, and so there is a direct prediction for the rate of change of the orbital period. Within the observational uncertainties, the actual change is consistent with the prediction of the quadrupole formula. Many, including the Nobel prize committee, have taken this as constituting effectively an experimental detection of gravitational radiation. Although there are possible alternative explanations for a period change (acceleration by a massive companion and tidal dissipation in the primary being the most plausible), these can generally be argued as likely to be smaller effects. In any case, for these to yield a result equivalent to the general relativity prediction requires something of a conspiracy – in much the same way that it is implausible that the anomalous precession of Mercury could be produced by a large Solar quadrupole or some other unaccounted-for effect. Physicists will still persist with attempts at the laboratory detection of gravitational waves; they have the motivation of not only the sheer technical challenge, but also the possibility of doing a radically new kind of astronomy. However, the binary pulsar provides them with the strongest reassurance that their quarry exists and has very much the properties expected in general relativity.

2.5 Black holes

Moving away from weak fields, we now consider the analogue in general relativity of one of the simplest problems in Newtonian gravitation: the field of a point mass. Exact solutions of the field equations are rare, owing to the nonlinear nature of the equations, but one of the most important solutions was found within a year of Einstein's publication of his full theory, by Karl Schwarzschild in 1916.

THE SCHWARZSCHILD METRIC Schwarzschild's solution for the metric was

$$\boxed{c^2 d\tau^2 = \left(1 - \tfrac{2GM}{c^2 r}\right) c^2 dt^2 - \frac{dr^2}{\left(1 - \tfrac{2GM}{c^2 r}\right)} - r^2 d\psi^2.} \qquad (2.68)$$

At large r this expression becomes identical to the metric for weak Newtonian fields discussed above, with $\Phi = -GM/r$. It is therefore immediately plausible that this is a candidate for the relativistic generalization of the gravitational field from an object of mass M. Proving that the Schwarzschild solution does in fact satisfy the field equations exactly is left as an exercise [problem 2.5]. The derivation of the metric uses spherical symmetry arguments to reduce the form of the metric to two unknown functions, which are obtained from the vacuum field equations. This argument requires no other information about the distribution of the matter giving rise to the gravitational field: it only needs to be spherically symmetric and of zero density at the radius of interest. A derivation of the Schwarzschild metric thus also proves **Birkhoff's theorem**, which states that the solution to the vacuum Einstein equations for a spherically symmetric source always looks like the Schwarzschild metric. Two of the most important features of Newtonian gravity therefore also apply to general relativity: the gravity of a spherical body appears to act from a central point mass; the gravitational field inside a spherical shell vanishes.

Features to note about this solution include the non-Euclidean nature of the spatial part of the metric, as expected from our above discussion of weak fields, and the apparent singularity at the **Schwarzschild radius**, $r_S = 2GM/c^2$. This latter feature will be covered in more detail after a discussion of some aspects of gravitational dynamics in the Schwarzschild field. However, we can immediately note that something strange happens to the speed of light at r_S, since $dr/dt = \pm c(1 - r_S/r)$ for a radial null trajectory. It will therefore take an infinite time (as measured by t, which is the proper time for an observer at large radii) for a photon to propagate from $r = r_S$ to larger radii, or to reach the Schwarzschild radius from smaller r. The Schwarzschild radius is thus also termed the **event horizon**, since events occurring inside it cannot propagate light signals to infinity. Any body that is small enough to exist within its own event horizon is therefore in some sense disconnected from the rest of the universe: it manifests itself only through its relativistically deep gravitational potential well, leading to the adoption of the evocative term **black hole** during the late 1960s.

ORBITS IN SCHWARZSCHILD SPACETIME As with the Newtonian two-body problem, the symmetry of the situation tells us that it will be sufficient to consider orbits in the equatorial plane ($\theta = \pi/2$ in spherical polars). We have to deal with the following non-zero 4-velocity components:

$$U^\mu \equiv \frac{\partial}{\partial \tau} x^\mu = (ct', r', 0, \phi'). \tag{2.69}$$

It is now a question of substitution into the equation of motion. The equations for t and ϕ are simple since the metric is diagonal and depends on neither of these variables; $\Gamma^\mu_{\alpha\beta}$ depends only on derivatives of g^{00} or g^{33} for $\mu = 0$ and 3 respectively. The equations of motion that follow integrate to:

$$\boxed{\begin{aligned} t'\left(1 - \frac{2GM}{c^2 r}\right) &= \text{constant} \\ r^2 \phi' &= \text{constant} \equiv L. \end{aligned}} \tag{2.70}$$

The latter of these corresponds to conservation of angular momentum. The former looks like gravitational time dilation, but the power of $(1 - r_S/r)$ is different; this is a freely falling body, which suffers both gravitational and kinematic time dilation. The connection coefficients needed for the radial equation are more complicated, but this difficulty may be avoided by the powerful trick of using the invariant $U^\mu U_\mu = c^2$. Taking this equation and eliminating ϕ' in terms of the angular momentum $L = r^2\phi'$ yields an equation that looks like conservation of energy:

$$\frac{1}{2}(r')^2 + \Phi(r) = \text{constant} \equiv E. \tag{2.71}$$

The effective potential here is similar to the usual Newtonian sum of the gravitational and centrifugal potentials:

$$\Phi(r) = -\frac{GM}{r} + \frac{L^2}{2r^2} - \frac{L^2 GM}{c^2 r^3}. \tag{2.72}$$

However, there is an additional term on the rhs. This depends on r in the same way as a quadrupole potential, and is responsible for the 'anomalous' precession of planetary orbits (it is distinguished from a true quadrupole by the L^2 coefficient, and so has an effect that falls off much more rapidly with radius).

PRECESSION OF ORBITS The analysis of orbital precession is easiest for orbits that are close to circular. Differentiate the generalized conservation equation, remembering that $(d/dr)\,r'^2/2 = r''$, to get the equation of motion for the radial 'acceleration':

$$r'' = -\frac{GM}{r^2} + \frac{L^2}{r^3} - \frac{3L^2 GM}{c^2 r^4}, \tag{2.73}$$

which vanishes for a circular orbit. Expanding the rhs to first order in a radial perturbation and using the equation for a circular orbit gives the perturbed equation

$$\Delta r'' = \left(-\frac{L^2}{r^4} + \frac{6L^2 GM}{c^2 r^5}\right) \Delta r. \tag{2.74}$$

This describes simple harmonic motion in the radial direction, with angular frequency

$$\omega \simeq \frac{L}{r^2}\left(1 - \frac{3GM}{c^2 r}\right). \tag{2.75}$$

Since the orbital period is $\tau = 2\pi/\phi' = 2\pi r^2/L$, this just says that the orbit is almost an ellipse. If $\omega\tau = 2\pi$, the radial oscillations are in phase with the orbit, which would repeat itself as a non-precessing ellipse. The small perturbation away from $\omega\tau = 2\pi$ thus implies a precession, of magnitude

$$\Delta\phi = \frac{6\pi GM}{c^2 r} \text{ radians per orbit.} \tag{2.76}$$

For an orbit that is significantly non-circular, we have to work a bit harder [problem 2.6]. The answer is that the general formula for the perihelion advance is $\Delta\phi = 6\pi GM/(c^2\ell)$, where $r^{-1} = \ell^{-1}(1 + e\cos\phi)$ is the equation of the elliptical orbit.

The application of this result to the orbit of Mercury was one of the classic early tests of general relativity. The theoretical precession is 43 arcsec per century, and is in accord with observation to about 1%. This is even more impressive than it sounds, since the 43 arcsec is the unexplained residual left from 532 arcsec due to tidal perturbations from other planets. It is a tribute to the skill of early generations of professionals in celestial mechanics that the small discrepancy was believed to be real – and led to the search for bodies interior to Mercury's orbit, the name 'Vulcan' being reserved for the hypothetical new planet. One of Einstein's achievements was thus to remove a planet from the Solar system (only for it to reappear many years later in popular culture).

ORBITAL STABILITY It is interesting to look at orbits rather closer to the black hole than the weak-field regime inhabited by planets. The equation derived earlier for radial perturbations about a circular orbit is exact and applies in the strong-field limit. From it, we see an interesting behaviour that is completely absent from Newtonian gravitation: circular orbits close to the hole are unstable. The angular frequency of radial oscillations is imaginary and deviations from a circular orbit grow exponentially, provided

$$\boxed{r < \frac{6GM}{c^2} = 3r_{\rm S}.} \tag{2.77}$$

Circular orbits are possible within the **marginally stable orbit** at $r = 3r_{\rm S}$, but would not exist in practice. Any particles inserted into this zone would spiral inwards and be captured by the black hole within a few orbital times. Even if we are prepared to ignore the instability, however, there is a limit to the radius of circular orbits, which can be derived by considering the orbits of photons. The equation for such an orbit is a special case of that obtained above, but in the limit that the velocity tends to c. As this happens, time dilation increases without limit and the 'angular momentum' $L = r^2\phi'$ diverges, since it is defined in terms of proper time. That means that terms in the orbit equation not involving L can be neglected, leading to

$$\frac{1}{2r^4}\left(\frac{dr}{d\phi}\right)^2 + \left(\frac{1}{2r^2} - \frac{GM}{c^2r^3}\right) = \frac{E}{L^2} = \text{constant}. \tag{2.78}$$

As before, $[(r')^2]' = 2r''$ whether differentiation is with respect to proper time or ϕ, which gives an equation for $d^2r/d\phi^2$. Setting this and the first derivative equal to zero gives the radius of circular photon orbits:

$$r_{\rm photon} = \frac{3GM}{c^2}. \tag{2.79}$$

Within this radius, photons must move away from the hole to avoid capture, with the required motion becoming more precisely radial as the horizon is approached.

Similar behaviour will apply for other orbits: objects that approach too close to the hole will in general be captured. As a last example, consider a particle with velocity v_0 at infinity and impact parameter b. The constants describing the orbit can be evaluated easily at infinity: $L = \gamma v_0 b$ and $E = \gamma^2 v_0^2/2$, so that the equation for the point of closest approach, with $r' = 0$, is

$$1 - \frac{2GM}{c^2r} - \frac{2GMr}{\gamma^2 v_0^2 b^2} = \frac{r^2}{v_0^2 b^2}. \tag{2.80}$$

For photons, the last term on the left can be neglected since $\gamma \to \infty$. Putting $r_{\rm min} = 3GM/c^2$ for a circular orbit gives $b = \sqrt{27}r_{\rm S}$ as the condition for the hole to capture a photon into a bound orbit.

THE NATURE OF THE EVENT HORIZON We now return to the most unexpected feature of the Schwarzschild geometry: the singularity at $r = 2GM/c^2$, where the gravitational time dilation becomes infinite. From a Newtonian analogy, one might have expected that the general-relativity gravitational field of a point mass would increase without limit as $r \to 0$, and that time dilation would become infinite there. It is therefore a major surprise to find that the time-dilation factor diverges at the finite **Schwarzschild radius** $r_S = 2GM/c^2$.

However, this singularity is less grave than it first appears, and owes much to the particular coordinate system being used. For example, curvature invariants such as R^μ_μ and $R^{\mu\nu}R_{\mu\nu}$ are non-singular there. More directly, it is perfectly possible for a body to fall into this potential well and cross the Schwarzschild radius. For purely radial orbits, the equation of motion is exactly of the Newtonian form

$$\frac{(r')^2}{2} = \frac{GM}{r} + \text{constant}, \tag{2.81}$$

where the constant vanishes for a particle that has zero velocity at infinity. The solution is

$$\tau = \pm\frac{2}{3}\frac{r^{3/2}}{\sqrt{2GM}}, \tag{2.82}$$

where the origin of proper time is chosen to be $\tau = 0$ at $r = 0$. Nothing strange happens on crossing the event horizon, confirming the artificial nature of the singularity there, and an observer who has fallen from infinity reaches the centre in a proper time $\tau = 2r_S/c$ after crossing the horizon. Equally, there is apparently nothing to prevent outgoing trajectories that are the time-reversed versions of the ingoing ones.

How does this behaviour appear to an external observer? To calculate in terms of coordinate time (i.e. the proper time measured by clocks at large distances), we can use the equation of motion for t derived earlier: $t'(1 - r_S/r)^{-1} = 1$ (the constant must be unity for any orbits that reach to infinity at nonrelativistic speeds). This equation can also be derived directly by substituting the expression for $r(\tau)$ into the equation for a radial geodesic. The coordinate time is therefore

$$\begin{aligned} t &= \int \frac{d\tau}{1 - r_S/r} = -\frac{r_S}{c}\int \frac{x^{3/2}\,dx}{x-1} \\ &= \frac{r_S}{c}\left(\ln\left|\frac{\sqrt{x}+1}{\sqrt{x}-1}\right| - 2\sqrt{x} - \frac{2x^{3/2}}{3}\right), \end{aligned} \tag{2.83}$$

where $x \equiv r/r_S$. The coordinate time taken for the freely falling body to reach the event horizon is therefore infinite, as measured by an observer at large radii. Note that this is not simply an artefact of the light travel time, which has not been included yet. In a real sense, an external observer can say that the body never falls into the black hole, but is condemned to hover for ever infinitesimally outside the event horizon. In principle, we could verify that it is still there by programming it to emit a photon radially outwards at some very large time in the future. Of course, this is not practical, since the body crosses the event horizon after a finite *proper* time. If it releases photons at anything like a constant proper rate, there will be a last photon emitted prior to crossing, and external observers will receive no further information.

If we cannot receive signals that the 'hovering' body emits, can we verify its presence directly by sending a rocket to retrieve the body? The answer is no: even travelling at the speed of light, it is impossible to catch up with a falling body once it has been left to fall for a critical time. Here we need to solve for a radial null geodesic:

$$\begin{aligned} ct &= ct_0 + \int_r^{r_0} \frac{dr}{(1 - r_S/r)} \\ &= ct_0 + r_0 - r - r_S \ln\left(\frac{r - r_S}{r_0 - r_S}\right). \end{aligned} \tag{2.84}$$

It can be shown from this that a particle dropped from r_0 into a black hole can be contacted by a light signal only if the signal is sent within a coordinate time similar to the proper time taken for the particle to reach the central singularity [problem 2.8]. After this, the particle cannot be contacted. This is not surprising: it is only the time dilation that freezes the infall close to the horizon. As the rescue mission approaches the horizon, this relative time dilation is lessened, unfreezing the infalling body.

GRAVITATIONAL COLLAPSE AND BLACK HOLE FORMATION How might black holes be formed in practice? What is required is a mass density reasonably close to the critical value for a body to lie within its own horizon. This is a function of mass, since the critical density scales as M^{-2}, being roughly the density of water for black holes of mass $10^8 M_\odot$. The direct formation of very massive black holes may thus be easier than the production of small ones. The centre of a dense massive star cluster or galaxy may evolve via stellar collisions so that it is close to the critical state (Shapiro & Teukolsky 1983).

For the formation of smaller black holes, the main idea that has been pursued is the end state of stellar evolution. If a star ceases to generate energy, it is liable to undergo gravitational collapse, which may increase its density beyond the critical point. The only known way to save the star from this fate is to use degeneracy pressure, either from electrons in **white dwarfs**, or in their relativistic analogues, the **neutron stars**. The maximum mass for a white dwarf (the **Chandrasekhar mass**) is about 1.4 $M_\odot$, and this arises roughly as follows. Consider the energy density for zero temperature in the limit that the electrons are all highly relativistic (i.e. the Fermi momentum is $p_F \gg m_e c$, which will be true at high enough density, since the typical spacing for degenerate electrons is $\sim \hbar/p_F$). The energy density is

$$u_e = \tfrac{3}{4}(3\pi^2)^{1/3}\, \hbar c\, n_e^{4/3}. \tag{2.85}$$

In the opposite limit of highly nonrelativistic electrons, the energy density is now

$$u_e = \frac{\hbar^2}{5m_e}\,(3\pi^2)^{2/3}\, n_e^{5/3}. \tag{2.86}$$

Suppose the star is in the relativistic regime, so that the total kinetic energy of the electrons is $E_K \propto u_e V \propto n_e^{4/3} V \propto M^{4/3}/r$. The gravitational energy is proportional to $-M^2/r$, so that the total energy can be written as

$$E_{\rm tot} = (AM^{4/3} - BM^2)/r. \tag{2.87}$$

There thus exists a critical mass where the two terms in the bracket are equal. If the mass is smaller, then the total energy is positive and will be reduced by making the star expand until the electrons reach the mildly relativistic regime and the star can exist as a stable white dwarf. If the mass exceeds the critical value, the binding energy increases without limit as the star shrinks: gravitational collapse has become unstoppable.

To find the exact limiting mass, we need to find the coefficients A and B above, where the argument has implicitly assumed the star to be of constant density. In this approximation, the kinetic and potential energies are

$$\begin{aligned} E_K &= \left(\frac{243\pi}{256}\right)^{1/3} \frac{\hbar c}{r} \left(\frac{M}{\mu m_p}\right)^{4/3} \\ E_V &= -\frac{5GM^2}{3r}, \end{aligned} \tag{2.88}$$

where the mass per electron is μm_p. This estimate of the critical mass is

$$M_{\rm crit} = \frac{0.80}{\mu^2} \left(\frac{\hbar c}{G}\right)^{3/2} m_p^{-2} \simeq \frac{1.5}{\mu^2} M_\odot. \tag{2.89}$$

Now, since the fluid of relativistic electrons will be a $p = u_e/3 \propto n_e^{4/3} \propto \rho^{4/3}$ **polytrope**, the density will not in fact be constant. To do the calculation properly, we need to solve the equation of hydrostatic equilibrium: $dp/dr = -\rho[GM(< r)/r^2]$, with $p = K\rho^{4/3}$. Unfortunately this equation is nonlinear and must be solved numerically (see e.g. section 11.3 of Weinberg 1972). Nevertheless, some nice scaling properties of the solutions are apparent. Suppose we solve the equation for $\rho(r)$ and then try to find a new solution with ρ increased everywhere by a factor α. Since $M(< r) = \int 4\pi r^2 \rho\, dr$, it is immediately clear that the equation is unchanged provided r is decreased by a factor $\alpha^{-1/3}$: there is a whole family of solutions characterized by increasing mean density and decreasing size, but with density profiles of the same shape. In detail, such a relativistic polytrope turns out numerically to have a well-defined outer edge where the density becomes zero, and a central density that is approximately 54 times the mean density inside this radius. For our present purpose, however, it is most important to note that the mass of a stable configuration is independent of the density: $M \propto \rho R^3$, with $R \propto \langle\rho\rangle^{-1/3}$. This unique mass depends only on the constant K and is of course the same as the mass found earlier to within a numerical factor, which is slightly larger than the constant-density estimate:

$$M_{\rm crit} = \frac{3.10}{\mu^2} \left(\frac{\hbar c}{G}\right)^{3/2} m_p^{-2} \simeq \frac{5.9}{\mu^2} M_\odot. \tag{2.90}$$

For a star near the end of the evolutionary track, much of the initial fuel has been burned to iron, so that $\mu \simeq 2$ and the critical mass is about 1.4 $M_\odot$.

For a star slightly more massive than this, a last-gasp strategy for evading total collapse would be to combine its electrons and protons into neutrons, which becomes possible once the density increases sufficiently that the Fermi energy rises above $m_e c^2$. The above analysis would then be applied with neutrons as the fermions and $\mu \simeq 1$. This apparently gives a limiting mass 4 times larger, but is the Newtonian analysis valid for these **neutron stars**?

To find the approximate size of such degenerate stars, recall that stability requires a star on the relativistic borderline, i.e. $p_{\rm F} \sim m_e c$ for the Fermi momentum. For a degenerate configuration, the number density is thus $n_e \sim (m_e c/\hbar)^3$, i.e. a particle spacing of about one de Broglie wavelength. The number density is also $n_e \sim (M/\mu m_p)/R^3$, which gives R in terms of M. At the critical mass, this gives a gravitational potential at the star's surface of

$$\frac{\Phi}{c^2} \sim \frac{m_e}{\mu m_p}. \tag{2.91}$$

Newtonian theory is therefore adequate for white dwarfs, but fails entirely for neutron stars where $m_e \to m_n$ and $\mu \to 1$. Detailed calculations of the limiting mass for neutron stars gives a mass of about 0.7 $M_\odot$ and a radius of about 10 km – only nine times the Schwarzschild radius, in the case of an ideal neutron gas (Shapiro & Teukolsky 1983). For more realistic equations of state, the number becomes more uncertain, perhaps up to 3 $M_\odot$; for comparison, there is a real example in the binary pulsar of $M = 1.4\ M_\odot$.

It is therefore commonly believed that any detection of a relativistically deep potential well with a mass well above 1 $M_\odot$ effectively constitutes the detection of a black hole. It is worth noting that this reasoning is circular: the assumption that stable exotic stars of greater mass cannot exist depends on the correctness of general relativity, in which case observations are not required to prove that black holes exist. Nevertheless, it is clearly of interest to astronomers to look for candidate black holes, and a number have been found as part of **X-ray binaries** where a compact object in orbit about a normal star reveals itself by accreting material from the companion and thereby emitting X-rays (e.g. Cowley 1992).

How exactly does a black hole form once a body has become unable to support itself against its own gravity? The main features of the problem may be understood by studying the simplest possible situation: the collapse of a star that is taken to be a uniform pressureless sphere. The symmetry of the situation simplifies things considerably, as does Birkhoff's theorem, which tells us that any vacuum solution of the field equations for a spherically symmetric mass distribution is just the Schwarzschild solution, so that the field inside a spherical cavity vanishes. The metric outside the surface of the collapsing star is thus the Schwarzschild form

$$d\tau^2 = (1 - 2GM/r)\, dt^2 - (1 - 2GM/r)^{-1}\, dr^2 - r^2\, d\psi^2, \tag{2.92}$$

where $d\psi$ is the element of angular distance, as usual. What is needed is a metric that is spatially isotropic, and where the mass density is spatially homogeneous. The answer is the Robertson–Walker metric for a space of constant positive spatial curvature (such as the surface of a sphere in 2D; see chapter 3 for further discussion):

$$d\tau^2 = dt^2 - R^2(t)(dr^2 + \sin^2 r\, d\psi^2). \tag{2.93}$$

The interior of the star here behaves exactly as a small portion of a closed universe, and undergoes uniform gravitational collapse, described by a decrease in the scale factor $R(t)$ with time. Notice that the star does not 'know' about its edge: the metric remains as the Robertson–Walker form everywhere within the body of the star, and would not be altered if more mass were added to extend the star. This arises because the requirements of uniform collapse and isotropy reduce the freedom in the metric to a single radial function. Since the field equations are local, this function can be constructed by integration from the centre without reference to boundary conditions at the surface of the star. Of course, boundary conditions at the surface do exist, but they simply govern the matching between the interior and exterior solutions. Both the metric and its first derivative must be continuous over a boundary where the density is non-singular (in Newtonian terms, the gravitational force must not change on crossing the boundary). Some aspects of the matching are obvious; for example, matching transverse length elements requires $R \sin(r_{\rm in}) = r_{\rm out}$. However, establishing the relation between the interior and exterior time and radius coordinates is a messy exercise, which is performed in section 11.9 of Weinberg (1972); we shall not need it here.

The collapse of the sphere from its initial state of rest at scale factor $R(t = 0) = R_0$

is easily deduced in a nearly Newtonian fashion: the collapse is uniform, and so we need only consider the dynamics of the central part, where spacetime curvature may be neglected. The solution is the cycloid, where time and radius are parameterized in terms of an angle θ:

$$\begin{aligned} R &= \frac{R_0}{2}(1+\cos\theta) \\ t &= \frac{t_c}{\pi}(\theta+\sin\theta). \end{aligned} \tag{2.94}$$

Note that the time evolution is independent of the size of the star. The scale factor vanishes when $\theta = \pi$, or $t = t_c$, the **collapse time**. This is related to the density as follows. To satisfy the equation of motion, the above form requires $R_0^3 = 8GM\, t_c^2/\pi^2$, and so the proper time taken for collapse from rest to infinite density is just determined by the initial density:

$$\boxed{t_c = \sqrt{\frac{3\pi}{32\, G\, \rho_0}}.} \tag{2.95}$$

Although the collapse of a pressure-free spherical body to a singularity therefore takes only a finite proper time, things will appear very different to an external observer. By Birkhoff's theorem, the outermost shell of mass marking the stellar surface falls as if into a pre-existing black hole, and so the previous analysis of an infalling observer can be invoked to show that the formation of a black hole takes an infinite coordinate time. Does this mean that it is impossible in practice to form black holes? It may seem that the answer is yes, since coordinate time is certainly the one that matters to external observers. Indeed, the term 'frozen star' was once used to describe the end-point of gravitational collapse. Only when it was realized that the frame of the infalling observer would still show the formation of a horizon and a singularity was the black hole point of view accepted as a more complete description. In any case, we have shown that material undergoing free collapse cannot be recovered at late times; neither can it send measurable light signals if the proper rate of emission of light is to remain finite at the horizon. Therefore, it is effectively impossible to derive an experiment that would distinguish in practice between a Schwarzschild black hole and a 'frozen star'.

ROTATING BLACK HOLES The solution to the field equations that looks like the gravitational field due to a rotating mass took much longer to arrive than the Schwarzschild metric, being discovered only in 1963 by R.P. Kerr. The metric takes its simplest form in **Boyer–Lindquist coordinates**, which use the normal spherical polar coordinates θ and ϕ, plus a number of functions of both radius and angle. First define a length proportional to the angular momentum per unit mass,

$$a \equiv \frac{Jc}{M}, \tag{2.96}$$

and then the following powers of lengths:

$$\begin{aligned} \Delta &\equiv r^2 + a^2 - \frac{2GMr}{c^2}; \\ \rho^2 &\equiv r^2 + a^2\cos^2\theta; \\ \Sigma^2 &\equiv (r^2+a^2)^2 - a^2\,\Delta\,\sin^2\theta. \end{aligned} \tag{2.97}$$

In these terms, the metric takes the form

$$c^2 d\tau^2 = \frac{\rho^2 \Delta}{\Sigma^2} c^2 dt^2 - \frac{\rho^2}{\Delta} dr^2 - \rho^2 d\theta^2 - \frac{\Sigma^2}{\rho^2} \sin^2\theta (d\phi - \omega dt)^2, \qquad (2.98)$$

where the 'angular velocity' ω is

$$\omega = \frac{2GMra}{c\Sigma^2}. \qquad (2.99)$$

The way in which this equation is written anticipates an important result in this spacetime: the rotation of the hole tends to drag observers round with it. The Kerr metric looks rather like the spacetime of a rotating frame.

As in the Schwarzschild metric, something important happens at a critical radius: $g_{rr} \to \infty$ at $\Delta = 0$. At this point, the **lapse function** (which is the square root of the term multiplying $c^2 dt^2$ in the metric, and differs from $\sqrt{g_{00}}$ in this case because of the 'rotational' term involving ωdt) also goes to zero, implying infinite gravitational redshift. In fact, there are *two* such radii:

$$r_{\rm H} = \frac{GM}{c^2} \pm \left[\left(\frac{GM}{c^2} \right)^2 - a^2 \right]^{1/2}. \qquad (2.100)$$

These clearly merge at $r = GM/c^2$ (half the Schwarzschild value) for a sufficiently large angular momentum $a = GM/c^2$, giving an **extreme Kerr black hole**.

The gravitational field in the Kerr geometry is thus fixed completely by the mass and spin of the hole. This can be taken a step further to black holes with charge (see chapter 33 of Misner, Thorne & Wheeler 1973). A general black hole is characterized only by these three parameters, leading to the statement 'a black hole has no hair'. This means that all other properties of the collapsing material (e.g. baryon or lepton number) are lost when the hole forms.

2.6 Accretion onto black holes

Black holes have an importance in astrophysics beyond that of a diverting mathematical puzzle because they are highly efficient generators of energy: matter that accretes onto a hole in the correct way can liberate a substantial fraction of its rest mass as radiation. To analyse how this happens, we need to discuss the concept of orbital energy in general relativity. The energy of a particle ought to be be included in the 4-momentum P^μ; in special relativity it would be given by simply $E = P^0 = P_0$ – but what about in general? The answer is that the conserved quantities are given by the *covariant* 4-momentum. As an example, far from a black hole $P_0 = mt'$ and $P_3 = -mr^2 \sin^2\theta \phi'$ ($t' \equiv \partial t/\partial\tau$ etc.) gives us the energy (γm) and (minus the) angular momentum (the factor $-r^2 \sin^2\theta$ comes directly from the metric). To see that this is the correct identification in general, we need the Lagrangian $L = (g_{\mu\nu} U^\mu U^\nu)^{1/2}$. As usual, the fact that this has no explicit dependence on time or ϕ tells us that there must be conserved quantities

$$\begin{aligned} \epsilon &= \frac{\partial L}{\partial t'} \\ h &= \frac{\partial L}{\partial \phi'}. \end{aligned} \qquad (2.101)$$

Using $L = 1$ shows that the energy and angular momentum per unit mass are indeed given in general by the t and ϕ components of U_μ. So, for example, the energy per unit mass of a particle orbiting a Kerr black hole is

$$\epsilon = \left(1 - \frac{2rGM/c^2}{r^2 + a^2\cos^2\theta}\right) t' + \frac{2\sin^2\theta\, arGM/c^2}{r^2 + a^2\cos^2\theta}\, \phi'. \tag{2.102}$$

Combining this with the corresponding equation for h and with $L = 1$ yields an orbit equation, which may be presented in the equatorial plane for simplicity ($c = 1$):

$$\begin{aligned} &\tfrac{1}{2}r'^2 - \Phi = \tfrac{1}{2}(\epsilon^2 - 1) \\ &\Phi = \left(\frac{GM}{r} + \frac{h^2}{2r^2} + \frac{(h^2 + a^2)GM}{r^3}\right) - \frac{2GMa\epsilon h}{r^3} + \tfrac{1}{2}\left(\frac{a^2}{r^2} + \frac{2GMa^2}{r^3}\right). \end{aligned} \tag{2.103}$$

For circular orbits, we have the conditions $\Phi = (1 - \epsilon^2)/2$ and $\partial\Phi/\partial r = 0$, which can be manipulated to give eventually (see Lynden-Bell 1978) the energy for circular orbits as a function of radius:

$$\epsilon = \frac{|1 - 2\xi \pm \alpha\xi^{3/2}|}{|1 - 3\xi \pm 2\alpha\xi^{3/2}|^{1/2}}, \tag{2.104}$$

where $\xi = GM/(c^2 r)$ and $\alpha = ac^2/(GM)$. There is therefore a special radius called the **marginally bound orbit** at which $\epsilon = 1$. For a Schwarzschild hole, this occurs at $r = 4Gm/c^2$, i.e. twice the radius of the event horizon; for an extreme Kerr hole, it lies at $(GM/c^2)(\sqrt{2} \pm 1)^{-2}$ (+ for prograde orbits, − for retrograde).

ACCRETION EFFICIENCY For a particle to reach $\epsilon = 1$ implies radiating away all its rest-mass energy, but in practice the marginally bound orbit would not be reached by slow dissipative orbital decay. Imagine the situation where material orbits the hole in circular orbits that slowly spiral inwards through quasi-viscous interaction of material in an **accretion disk**, giving out radiation through the gravitational heating of the disk. This process must end where $\epsilon(r)$ has a minimum; at this point the orbit ceases to be stable and we expect that material will thereafter rapidly spiral into the hole without radiating away any more energy. The accretion efficiency of the hole is then given by working out $1 - \epsilon$ at the **marginally stable orbit**

$$\frac{\partial^2}{\partial r^2}\Phi = 0. \tag{2.105}$$

This works out to give $r = 6GM/c^2$ for a Schwarzschild hole and $r = (5 \mp 4)GM/c^2$ in the extreme Kerr case. Thus, the accretion efficiencies become

$$1 - \epsilon(r_{\rm ms}) = \begin{cases} 1 - \frac{2}{3}\sqrt{2} & \simeq 6\% \quad \text{(Schwarzschild)} \\ 1 - \sqrt{\frac{1}{3}} & \simeq 42\% \quad \text{(extreme Kerr; prograde)} \\ 1 - \frac{5}{3}\sqrt{\frac{1}{3}} & \simeq 4\% \quad \text{(extreme Kerr; retrograde).} \end{cases} \tag{2.106}$$

For the Kerr hole, prograde rotation is more probable (i) because the hole may well attain its angular momentum from the same source as the accreted material, and (ii) because of frame-dragging effects. Thus, we see that accretion onto black holes may plausibly release $\gtrsim 10\%$ of the rest-mass energy of the accreted material. This is the reason for the popularity of black holes in theories for the operation of extremely active objects; no other astrophysical mechanism can approach the efficiency of gravity in energy generation.

Problems

(2.1) The energy–momentum tensor $T^{\mu\nu} = (\rho + p/c^2)U^\mu U^\nu - pg^{\mu\nu}$ tells us that the energy flux density is $cT^{01} = \gamma^2(\rho c^2 + p)v$ and the momentum flux density is $T^{11} = \gamma^2(\rho + p/c^2)v^2 + p$, for a fluid moving at relativistic speeds. Use these relations to obtain the jump conditions for a relativistic shock and solve these in the cases of (a) a strong shock propagating into cold fluid; (b) a radiation-dominated fluid ($p = \rho c^2/3$).

(2.2) Use linearized gravity to obtain the metric inside a thin uniform shell of mass M and radius R, which rotates with angular velocity $\omega_{\rm shell}$. Show that the metric looks like a frame of reference that rotates at angular velocity

$$\frac{\omega}{\omega_{\rm shell}} = \frac{4GM}{3c^2R}. \tag{2.107}$$

Note the moral for Mach's principle: in a loose sense, $GM/(c^2R) \simeq 1$ for the universe as a whole and so local inertial frames are tied to the global mass distribution.

(2.3) Show that the electromagnetic analogy of using electric quadrupole radiation to predict the gravitational radiation quadrupole formula yields an energy loss rate too low by a factor 4.

(2.4) Show that the gravitational power emitted by a binary system in a circular orbit is

$$-\dot{E} = \frac{32c^5}{5G}\left(\frac{\mu}{M}\right)^2\left(\frac{GM}{c^2a}\right)^5, \tag{2.108}$$

where M is the total mass, μ the reduced mass and a the semi-major axis. Deduce the orbital decay time. What is a typical atomic transition rate for graviton emission?

(2.5) Show that the Schwarzschild metric is indeed a solution of the Einstein field equations in a vacuum. Argue that the metric must have the same form in the zero-density regions of any spherically symmetric mass distribution, and thus prove Birkhoff's theorem.

(2.6) Show that the perihelion advance for an elliptical orbit is $\Delta\phi = 6\pi GM/(c^2\ell)$ per orbit, where $\ell = L^2/(GM)$ and L is the orbital angular momentum.

(2.7) Consider the modified Newtonian potential

$$\Phi = -\frac{Gm}{r - a}, \tag{2.109}$$

and show that the stability properties of the orbits in this potential are the same as those of the Schwarzschild black hole for a suitable choice of a.

(2.8) A particle is dropped from rest at a distance $r_0 \gg r_{\rm s}$ from a black hole, and a light signal is sent after it at a time Δt later. What is the maximum value of Δt that will allow the particle to be contacted before it crosses the event horizon?

PART 2 CLASSICAL COSMOLOGY

3 The isotropic universe

3.1 The Robertson–Walker metric

To the relativist, cosmology is the task of finding solutions to Einstein's field equations that are consistent with the large-scale matter distribution in the universe. Modern observational cosmology has demonstrated that the real universe is highly symmetric in its large-scale properties, but the evidence for this was not known at the time when Friedmann and Lemaître began their pioneering investigations. Just as Einstein aimed to write down the simplest possible relativistic generalization of the laws of gravity, so cosmological investigation began by considering the simplest possible mass distribution: one whose properties are **homogeneous** (constant density) and **isotropic** (the same in all directions).

ISOTROPY IMPLIES HOMOGENEITY At first sight, one might think that these two terms mean the same thing, but it is possible to construct universes that are homogeneous but anisotropic; the reverse, however, is not possible. Consider an observer who is surrounded by a matter distribution that is perceived to be isotropic; this means not only that the mass density is a function of radius only, but that there can be no preferred axis for other physical attributes such as the velocity field. This has an important consequence if we take the velocity strain tensor $\partial v_i / \partial x_j$ and decompose it into symmetric and antisymmetric parts. The antisymmetric part corresponds to a rotation, so the velocity field can be written to first order in r as

$$\mathbf{v} = \mathbf{\Sigma} \cdot \mathbf{r} + \mathbf{\Omega} \wedge \mathbf{r}. \tag{3.1}$$

For isotropy, not only must $\mathbf{\Omega}$ vanish, but all the principal values of the shear tensor $\mathbf{\Sigma}$ must be equal. Thus, the only allowed velocity field on a local scale is expansion (or contraction) with velocity proportional to distance:

$$\boxed{\mathbf{v} = H\mathbf{r}.} \tag{3.2}$$

Now, as was discovered by Slipher in the 1920s, we do in fact see such an isotropic recession about us; what does it tell us about the universe? It is easy to produce an isotropic velocity field: just set off an explosion that scatters debris in all directions with a range of speeds v. If the amount of matter involved were too small for self-gravity to be important, then the distance travelled by a piece of debris would be just vt and an observer left in the centre would inevitably see an isotropic distribution of objects with a

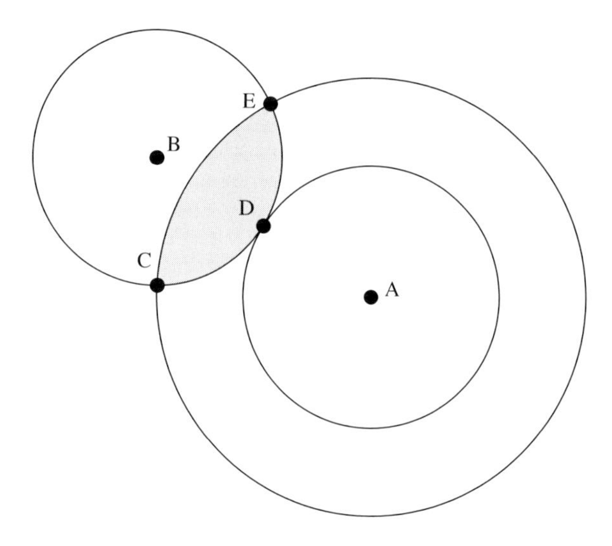

Figure 3.1. Isotropy about two points A and B shows that the universe is homogeneous. From isotropy about B, the density is the same at each of C,D,E. By constructing spheres of different radii about A, the shaded zone is swept out and shown to be homogeneous. By using large enough shells, this argument extends to the entire universe.

recessional velocity proportional to their distance. However, most scientists believe that it is not reasonable to adopt a cosmological model in which the universe is simply a joke played for the benefit of mankind. This attitude is encapsulated in the **Copernican principle**, which states that humans are not privileged observers. Although obviously false on a local scale (a point selected at random would be most unlikely to lie so close to a star), there is no reason why our existence should depend on special features of the large-scale appearance of the universe. It is therefore a reasonable supposition that, if the universe appears isotropic about our position, it would also appear isotropic to observers in other galaxies; the term 'isotropic' is therefore often employed in cosmology as a shorthand for 'isotropic about all locations'.

This is a crucial shift in meaning, for the properties of such a universe are highly restricted. In fact, given only two points (A and B) from which conditions appear isotropic, we can prove that they must be homogeneous everywhere. Construct a sphere about A, on which the density has a constant value; considering the intersection of this sphere with similar ones centred on B shows that the density must be constant everywhere within the sphere, which can then be made as large as we desire (figure 3.1). Observationally, this may seem rather abstract, since it is not easy to verify isotropy from another's viewpoint. However, a universe which is isotropic about one point and which is homogeneous has to be isotropic in the general sense. Consider the expansion: since $\boldsymbol{\nabla} \cdot \mathbf{v}$ must be a (homogeneous) constant, $\mathbf{v} \propto \mathbf{r}$; the expansion is therefore isotropic about every point, since $\mathbf{v}_1 - \mathbf{v}_2 \propto \mathbf{r}_1 - \mathbf{r}_2$. We now see how it is possible to imagine universes that are homogeneous but anisotropic. For example, a model that expands at different rates along three principal axes would preserve a homogeneous density, but the distance–redshift relation could be a function of direction on the sky.

Having chosen a model mass distribution, the next step is to solve the field equations to find the corresponding metric. Since our model is a particularly symmetric one, it is perhaps not too surprising that many of the features of the metric can be deduced from symmetry alone – and indeed will apply even if Einstein's equations are replaced by something more complicated. These general arguments were put forward independently by H.P. Robertson and A.G. Walker in 1936.

COSMOLOGICAL TIME The first point to note is that something suspiciously like a universal time exists in an isotropic universe. Consider a set of observers in different locations, all of whom are at rest with respect to the matter in their vicinity (these characters are usually termed **fundamental observers**). We can envisage them as each sitting on a different galaxy, and so receding from each other with the general expansion (although real galaxies have in addition random velocities of order $100\,\mathrm{km\,s^{-1}}$ and so are not strictly fundamental observers). We can define a global time coordinate t, which is the time measured by the clocks of these observers – i.e. t is the proper time measured by an observer at rest with respect to the local matter distribution. The coordinate is useful globally rather than locally because the clocks can be synchronized by the exchange of light signals between observers, who agree to set their clocks to a standard time when e.g. the universal homogeneous density reaches some given value. Using this time coordinate plus isotropy, we already have enough information to conclude that the metric must take the following form:

$$c^2 d\tau^2 = c^2 dt^2 - R^2(t)\left[f^2(r)\,dr^2 + g^2(r)\,d\psi^2\right]. \tag{3.3}$$

Here, we have used the equivalence principle to say that the proper time interval between two distant events would look locally like special relativity to a fundamental observer on the spot: for them, $c^2 d\tau^2 = c^2 dt^2 - dx^2 - dy^2 - dz^2$. Since we use the same time coordinate as they do, our only difficulty is in the spatial part of the metric: relating their dx etc. to spatial coordinates centred on us.

Because of spherical symmetry, the spatial part of the metric can be decomposed into a radial and a transverse part (in spherical polars, $d\psi^2 = d\theta^2 + \sin^2\theta\, d\phi^2$). Distances have been decomposed into a product of a time-dependent **scale factor** $R(t)$ and a time-independent **comoving coordinate** r. The functions f and g are arbitrary; however, we can choose our radial coordinate such that either $f = 1$ or $g = r^2$, to make things look as much like Euclidean space as possible. Furthermore, we can determine the form of the remaining function from symmetry arguments.

To get some feeling for the general answer, it should help to think first about a simpler case: the metric on the surface of a sphere. A balloon being inflated is a common popular analogy for the expanding universe, and it will serve as a two-dimensional example of a space of constant curvature. If we call the polar angle in spherical polars r instead of the more usual θ, then the element of length on the surface of a sphere of radius R is

$$d\sigma^2 = R^2\left(dr^2 + \sin^2 r\, d\phi^2\right). \tag{3.4}$$

It is possible to convert this to the metric for a 2-space of constant **negative curvature** by the device of considering an imaginary radius of curvature, $R \to iR$. If we simultaneously let $r \to ir$, we obtain

$$d\sigma^2 = R^2\left(dr^2 + \sinh^2 r\, d\phi^2\right). \tag{3.5}$$

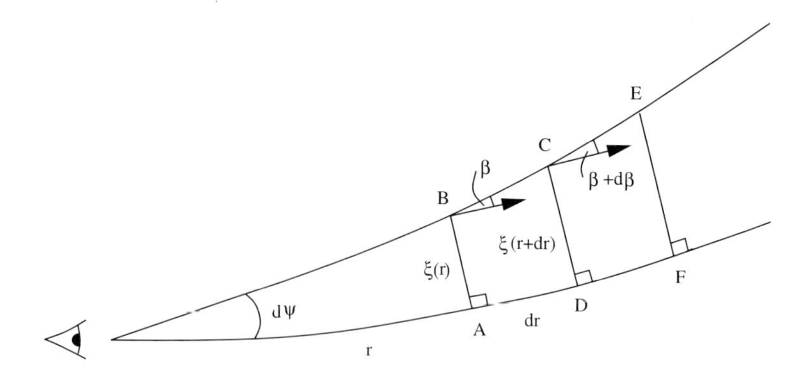

Figure 3.2. The geometry of two light rays in a curved space. The angle between them is $d\psi$ and the transverse separation is ξ. The test for curvature is when a vector that makes an angle β with one of the rays is parallel-transported around the loop ABCD; if the angle changes to $\beta + d\beta$, then $d\beta$ is a measure of the degree of curvature.

These two forms can be combined by defining a new radial coordinate that makes the transverse part of the metric look Euclidean:

$$d\sigma^2 = R^2 \left(\frac{dr^2}{1 - kr^2} + r^2 \, d\phi^2 \right), \tag{3.6}$$

where $k = +1$ for positive curvature and $k = -1$ for negative curvature. This argument is hardly a proof that spaces of constant negative curvature exist and can be justified only by a fuller treatment. However, it does give a feeling for the meaning of the final result. Indeed, the equation we have just written is in fact the exact form of the spatial part of the Robertson–Walker metric, which applies in any number of dimensions. The fact that we can obtain the correct answer using a 2-space is not surprising in retrospect, because of isotropy: we only need consider the non-Euclidean nature of the geometry in a plane to determine the functions f and g.

It is clear that this metric nicely incorporates the idea of a uniformly expanding model with no centre. For small separations, where space is Euclidean, we have a simple scaling-up of vector separations: $\mathbf{x}(t) \propto R(t)\mathbf{x}(t_0)$. The same law applies irrespective of the origin we choose: $\mathbf{x}_1(t) - \mathbf{x}_2(t) \propto R(t)[\mathbf{x}_1(t_0) - \mathbf{x}_2(t_0)]$. Every fundamental observer with a sufficiently large ego will conclude that the universe is centred on them.

AN INFORMAL DERIVATION There exists an illuminating half-way house between the above hand-waving argument and a totally formal approach. This is an argument due to P.A.G. Scheuer (given in Longair 1986). Consider the geometry shown in figure 3.2: ξ is the comoving separation between two radial light rays and the radial increment of comoving distance is dr, i.e. we are working with $f = 1$ and trying to find ξ, which is equal to $g(r)d\psi$. Consider also a vector that makes an angle β with one of the light rays. The effect of non-Euclidean geometry will be to cause this angle to change by some amount $\delta\beta$ on parallel-transport around the loop. For our present purposes the crucial step in the argument is to see that isotropy and homogeneity require that this change must be proportional to the *area* of the loop. For example, going around the loop ABCD twice clearly produces twice the $\delta\beta$ obtained from one circuit, but from homogeneity this should be the same as going round two identical loops, ABCD followed by DCEF (the multiple traversing of the line CD has no effect). Similarly, one can argue that the effect of any loop should be the sum of the effects of the sub-loops into which it may be divided. Given this, it is straightforward to derive ξ, since $\beta = \partial\xi/\partial r$; $\delta\beta \propto$ area therefore becomes

$$\frac{\partial^2 \xi}{\partial r^2}\, dr = \pm A^2 \xi\, dr, \tag{3.7}$$

where the constant of proportionality could have either sign. Apart from the Euclidean case, $A = 0$, there are thus two possible geometries:

$$\xi \propto \sin Ar \quad \text{or} \quad \xi \propto \sinh Ar. \tag{3.8}$$

If we impose the boundary condition $\xi \rightarrow r\, d\psi$ as $r \rightarrow 0$, then the constants of proportionality may be found. We end up with the same metric as we obtained from the 2-sphere. It is possible to give derivations of considerably greater rigour (consult Weinberg 1972 or Misner, Thorne & Wheeler 1973 for examples). However, this will suffice here.

The conclusion of this argument is that an isotropic universe has the same form for the comoving spatial part of its metric as the surface of a sphere. This is no accident, since it it possible to define the equivalent of a sphere in higher numbers of dimensions, and the form of the metric is always the same. For example, a **3-sphere** embedded in four-dimensional Euclidean space would be defined by the coordinate relation $x^2 + y^2 + z^2 + w^2 = R^2$. Now define the equivalent of spherical polars and write $w = R\cos\alpha$, $z = R\sin\alpha\cos\beta$, $y = R\sin\alpha\sin\beta\cos\gamma$, $x = R\sin\alpha\sin\beta\sin\gamma$, where α, β and γ are three arbitrary angles. Differentiating with respect to the angles gives a four-dimensional vector (dx, dy, dz, dw), and it is a straightforward exercise to show that the squared length of this vector is

$$|(dx, dy, dz, dw)|^2 = R^2 \left[d\alpha^2 + \sin^2\alpha\,(d\beta^2 + \sin^2\beta\; d\gamma^2)\right], \tag{3.9}$$

which is the Robertson–Walker metric for the case of positive spatial curvature. This $k = +1$ metric describes a **closed universe**, in which a traveller who sets off along a trajectory of fixed β and γ will eventually return to their starting point (when $\alpha = 2\pi$). In this respect, the positively curved 3D universe is identical to the case of the surface of a sphere: it is finite, but unbounded. By contrast, the $k = -1$ metric describes an **open universe** of infinite extent; as before, changing to negative spatial curvature replaces $\sin\alpha$ with $\sinh\alpha$, and α can be made as large as we please without returning to the starting point. The $k = 0$ model describes a **flat universe**, which is also infinite in extent. This can be thought of as a limit of either of the $k = \pm 1$ cases in which the curvature scale R tends to infinity.

NOTATION AND CONVENTIONS The Robertson–Walker metric (which we shall often write in the shorthand **RW metric**) may be written in a number of different ways. The most compact forms are those where the comoving coordinates are *dimensionless*. Define the very useful function

$$S_k(r) = \begin{cases} \sin r & (k=1) \\ \sinh r & (k=-1) \\ r & (k=0), \end{cases} \tag{3.10}$$

and its cosine-like analogue, which will be useful later,

$$C_k(r) \equiv \sqrt{1 - kS_k^2(r)} = \begin{cases} \cos r & (k=1) \\ \cosh r & (k=-1) \\ 1 & (k=0). \end{cases} \tag{3.11}$$

The metric can now be written in the preferred form that we shall use throughout:

$$c^2 d\tau^2 = c^2 dt^2 - R^2(t)\left[dr^2 + S_k^2(r)\, d\psi^2\right]. \tag{3.12}$$

The most common alternative is to use a different definition of comoving distance, $S_k(r) \to r$, so that the metric becomes

$$c^2 d\tau^2 = c^2 dt^2 - R^2(t)\left(\frac{dr^2}{1-kr^2} + r^2 d\psi^2\right). \tag{3.13}$$

There should of course be two different symbols for the different comoving radii, but each is often called r in the literature, so we have to learn to live with this ambiguity; the presence of terms like $S_k(r)$ or $1-kr^2$ will usually indicate which convention is being used. Alternatively, one can make the scale factor dimensionless, defining

$$a(t) \equiv \frac{R(t)}{R_0}, \tag{3.14}$$

so that $a=1$ at the present. In this case, the metric becomes

$$\begin{aligned} c^2 d\tau^2 &= c^2 dt^2 - a^2(t)\left(dr^2 + \frac{S_k^2(Ar)}{A^2} d\psi^2\right) \\ c^2 d\tau^2 &= c^2 dt^2 - a^2(t)\left(\frac{dr^2}{1-k(Ar)^2} + r^2 d\psi^2\right), \end{aligned} \tag{3.15}$$

where A is the constant of proportionality from Scheuer's argument, which is equal to R_0, the current value of $R(t)$. We shall generally use the metric in the form with dimensionless comoving distances. However, it is important to note that cosmologists tend to use the term 'distance' as meaning comoving distance unless otherwise specified, usually in units of Mpc: this means that one generally deals with the combination $R_0 r$.

Other forms of the metric are encountered less commonly; in particular, the rarely used **isotropic** form is

$$c^2 d\tau^2 = c^2 dt^2 - \frac{R^2(t)}{(1+kr^2/4)^2}\left(dr^2 + r^2 d\psi^2\right). \tag{3.16}$$

There is, however, one other modification of some importance; this is to define the **conformal time**

$$\eta = \int_0^t \frac{c\, dt'}{R(t')}, \tag{3.17}$$

which allows a factor of R^2 to be taken out of the metric. This is a special case of a **conformal transformation** in general relativity, such transformations correspond to $g^{\mu\nu} \rightarrow f g^{\mu\nu}$, where f is some arbitrary spacetime function. The universe with $k = 0$, although certainly possessing spacetime curvature, is obviously directly related to Minkowski spacetime via a conformal transformation and so tends to be loosely known as the 'flat' model. In fact, the Robertson–Walker metric turns out to be conformally flat for all values of k – i.e. a coordinate transformation can always be found that casts the metric in the form of the Minkowski metric multiplied by some spacetime function.

COMOVING SEPARATION Having erected a comoving coordinate system at the site of a given observer, we can calculate the comoving separation of two events (or astronomical 'sources') that lie close to each other. How does this generalize to the case when dr and $d\psi$ are not small? The analogy with a 2-sphere is invaluable in this case, as the answer is obviously just the spherical cosine rule (appropriately generalized for positive and negative curvature):

$$C_k(r_{12}) = C_k(r_1)C_k(r_2) + kS_k(r_1)S_k(r_2)\cos\psi. \tag{3.18}$$

This can be recast in terms of the more usual $S_k(r)$ via $S_k^2 = (1 - C_k^2)/k$:

$$\begin{aligned} S_k^2(r_{12}) = {} & S_k^2(r_1)C_k^2(r_2) + S_k^2(r_2)C_k^2(r_1) + kS_k^2(r_1)S_k^2(r_2)\sin^2\psi \\ & - 2S_k(r_1)S_k(r_2)C_k(r_1)C_k(r_2)\cos\psi. \end{aligned} \tag{3.19}$$

For a flat universe ($k = 0$), this goes over to the cosine rule, as expected. To see how complicated a non-intuitive derivation of this result can become, see Osmer (1981).

THE REDSHIFT It is time to re-establish contact with the expanding universe in terms of Hubble's law, $v = Hr$. Our discussion so far has been in terms of comoving coordinates, which are time independent. At small separations, where things are Euclidean, it is easy enough to see what happens: the proper separation of two fundamental observers is just $R(t)\, dr$, so that we obtain Hubble's law with

$$H = \frac{\dot{R}}{R}. \tag{3.20}$$

At large separations where spatial curvature becomes important, the concept of radial velocity becomes a little more slippery – but in any case how could one measure it directly in practice? At small separations, the recessional velocity gives the Doppler shift

$$\frac{\nu_{\rm emit}}{\nu_{\rm obs}} \equiv 1 + z \simeq 1 + \frac{v}{c}. \tag{3.21}$$

This defines the **redshift** z in terms of the shift of spectral lines. What is the equivalent

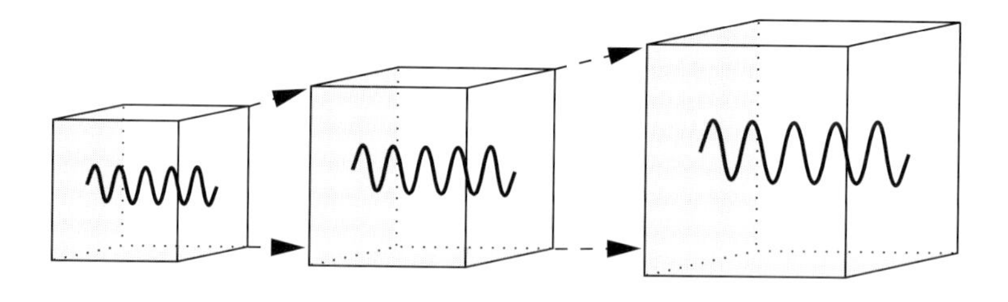

Figure 3.3. Suppose we trap some radiation inside a box with silvered sides that expands with the universe. At least for an isotropic radiation field, the photons trapped in the box are statistically equivalent to any that would pass into this space from outside. Since the walls expand at $v \ll c$ for a small box, it is easily shown that the Doppler shift maintains an adiabatic invariant, which is the ratio of wavelength to box side, and so radiation wavelengths increase as the universe expands. This argument also applies to quantum-mechanical standing waves: their momentum declines as $a(t)^{-1}$.

of this relation at larger distances? Since photons travel on null geodesics of zero proper time, we see directly from the metric that

$$r = \int \frac{c\,dt}{R(t)}. \tag{3.22}$$

The comoving distance is constant, whereas the domain of integration in time extends from $t_{\rm emit}$ to $t_{\rm obs}$; these are the times of emission and reception of a photon. Photons that are emitted at later times will be received at later times, but these changes in $t_{\rm emit}$ and $t_{\rm obs}$ cannot alter the integral, since r is a comoving quantity. This requires the condition $dt_{\rm emit}/dt_{\rm obs} = R(t_{\rm emit})/R(t_{\rm obs})$, which means that events on distant galaxies time-dilate according to how much the universe has expanded since the photons we see now were emitted. Clearly (think of events separated by one period), this dilation also applies to frequency, and we therefore get

$$\boxed{\frac{\nu_{\rm emit}}{\nu_{\rm obs}} \equiv 1 + z = \frac{R(t_{\rm obs})}{R(t_{\rm emit})}.} \tag{3.23}$$

In terms of the normalized scale factor $a(t)$ we have simply $a(t) = (1+z)^{-1}$. Photon wavelengths therefore stretch with the universe, as is intuitively reasonable; see figure 3.3.

This is the only correct interpretation of the redshift at large distances; it is common but misleading to convert a large redshift to a recession velocity using the special-relativistic formula $1 + z = [(1 + v/c)/(1 - v/c)]^{1/2}$. Any such temptation should be avoided – it is only valid in a universe with zero matter density [problem 3.4].

3.2 Dynamics of the expansion

EXPANSION AND GEOMETRY The equation of motion for the scale factor can be obtained in a quasi-Newtonian fashion. Consider a sphere about some arbitrary point, and let the radius be $R(t)r$, where r is arbitrary. The motion of a point at the edge of the sphere will, in Newtonian gravity, be influenced only by the interior mass. We can therefore write down immediately a differential equation (**Friedmann's equation**) that expresses conservation of energy: $(\dot{R}r)^2/2 - GM/(Rr) = \text{constant}$. In fact, to get this far, we do require general relativity: the gravitation from mass shells at large distances is not Newtonian, and so we cannot employ the usual argument about their effect being zero. In fact, the result that the gravitational field inside a uniform shell is zero does hold in general relativity, and is known as **Birkhoff's theorem** (see chapter 2). General relativity becomes even more vital in giving us the constant of integration in Friedmann's equation [problem 3.1]

$$\boxed{\dot{R}^2 - \frac{8\pi G}{3}\rho R^2 = -kc^2.} \tag{3.24}$$

Note that this equation covers all contributions to ρ, i.e. those from matter, radiation and vacuum; it is independent of the equation of state. A common shorthand for relativistic cosmological models, which are described by the Robertson–Walker metric and which obey the Friedmann equation, is **FRW models**.

The Friedmann equation reveals the astonishing fact that there is a direct connection between the density of the universe and its global geometry. For a given rate of expansion, there is a **critical density** that will yield $k = 0$, making the comoving part of the metric look Euclidean:

$$\boxed{\rho_c = \frac{3H^2}{8\pi G}.} \tag{3.25}$$

A universe with density above this critical value will be **spatially closed**, whereas a lower-density universe will be **spatially open**. If we give a Newtonian interpretation to the Friedmann equation, the total 'energy' on the rhs thus changes with the geometry: closed universes are bound and open universes are unbound. This suggests that the future of the expansion should depend on density: just as the Earth's gravity defines an escape velocity for projectiles, so a universe that expands sufficiently fast should continue to expand forever. Conversely, gravity should eventually halt the expansion if the density is above the critical value. We give the detailed solutions of the Friedmann equation below and show that this intuitive reasoning is almost correct: provided vacuum energy is negligible, the universe either expands forever or eventually recollapses, depending on its geometry.

The proof of the Friedmann equation is 'only' a question of inserting the Robertson–Walker metric into the field equations [problem 3.1], but the question inevitably arises of whether there is a quasi-Newtonian way of seeing that the result must be true; the answer is 'almost'. First note that any open model will evolve towards undecelerated expansion provided its equation of state is such that ρR^2 is a declining function of R – the potential energy becomes negligible by comparison with the total and $\dot{R}$ tends to a constant. In this mass-free limit, there can be no spatial curvature and the open RW metric must be just a coordinate transformation of Minkowski spacetime. We will exhibit this transformation later in this chapter and show that it implies $R = ct$ for this model, proving the $k = -1$ case.

An alternative line of attack is to rewrite the Friedmann equation in terms of the Hubble parameter:

$$H^2 - \frac{8\pi G}{3}\rho = \frac{\text{constant}}{R^2}. \tag{3.26}$$

Now consider holding the local observables H and ρ fixed but increasing R without limit. Clearly, in the RW metric this corresponds to going to the $k = 0$ form: the scale of spatial curvature goes to infinity and the comoving separation for any given proper separation goes to zero, so that the comoving geometry becomes indistinguishable from the Euclidean form. This case also has potential and kinetic energy much greater than total energy, so that the rhs of the Friedmann equation is effectively zero. This establishes the $k = 0$ case, leaving the closed universe as the only stubborn holdout against Newtonian arguments.

It is sometimes convenient to work with the time derivative of the Friedmann equation, because acceleration arguments in dynamics can often be more transparent than energy ones. Differentiating with respect to time requires a knowledge of $\dot{\rho}$, but this can be eliminated by means of conservation of energy: $d(\rho c^2 R^3) = -p d(R^3)$. We then obtain

$$\ddot{R} = -4\pi G R(\rho c^2 + 3p)/3. \tag{3.27}$$

Both this equation and the Friedmann equation in fact arise as independent equations from different components of Einstein's equations for the RW metric [problem 3.1].

DENSITY PARAMETERS ETC. The 'flat' universe with $k = 0$ arises for a particular **critical density**. We are therefore led to define a **density parameter** as the ratio of density to critical density:

$$\boxed{\Omega \equiv \frac{\rho}{\rho_c} = \frac{8\pi G\rho}{3H^2}.} \tag{3.28}$$

Since ρ and H change with time, this defines an epoch-dependent density parameter. The current value of the parameter should strictly be denoted by Ω_0. Because this is such a common symbol, it is normal to keep the formulae uncluttered by dropping the subscript; the density parameter at other epochs will be denoted by $\Omega(z)$. The critical density therefore just depends on the rate at which the universe is expanding. If we now also define a dimensionless (current) Hubble parameter as

$$\boxed{h \equiv \frac{H_0}{100\,\text{km}\,\text{s}^{-1}\text{Mpc}^{-1}},} \tag{3.29}$$

then the current density of the universe may be expressed as

$$\begin{aligned}\rho_0 &= 1.88 \times 10^{-26}\Omega h^2\ \text{kg}\,\text{m}^{-3} \\ &= 2.78 \times 10^{11}\Omega h^2\ M_\odot\,\text{Mpc}^{-3}.\end{aligned} \tag{3.30}$$

A powerful approximate model for the energy content of the universe is to divide it into pressureless matter ($\rho \propto R^{-3}$), radiation ($\rho \propto R^{-4}$) and vacuum energy (ρ constant). The first two relations just say that the number density of particles is diluted by the expansion, with photons also having their energy reduced by the redshift; the third relation applies for Einstein's **cosmological constant**. In terms of observables, this means that the density is written as

$$\frac{8\pi G\rho}{3} = H_0^2(\Omega_v + \Omega_m a^{-3} + \Omega_r a^{-4}) \tag{3.31}$$

(introducing the normalized scale factor $a = R/R_0$). For some purposes, this separation is unnecessary, since the Friedmann equation treats all contributions to the density parameter equally:

$$\boxed{\frac{kc^2}{H^2R^2} = \Omega_m(a) + \Omega_r(a) + \Omega_v(a) - 1.} \tag{3.32}$$

Thus, a flat $k = 0$ universe requires $\sum \Omega_i = 1$ at all times, whatever the form of the contributions to the density, even if the equation of state cannot be decomposed in this simple way. In this chapter, Ω without a subscript denotes the total density parameter. In the later chapters, and in the research literature, Ω generally stands for the matter contribution alone. The reason is that the only models normally considered in practice are either matter-dominated models with $\Omega_v = 0$, or $k = 0$ models with $\Omega_m + \Omega_v = 1$. In either case, Ω_m is the only free parameter.

In terms of the **deceleration parameter**,

$$\boxed{q \equiv -\frac{\ddot{R}R}{\dot{R}^2},} \tag{3.33}$$

the $\ddot{R}$ form of the Friedmann equation says that

$$q = \Omega_m/2 + \Omega_r - \Omega_v, \tag{3.34}$$

which implies $q = 3\Omega_m/2 + 2\Omega_r - 1$ for a flat universe. One of the classical problems of cosmology is to test this relation experimentally.

Lastly, it is often necessary to know the present value of the scale factor, which may be read directly from the Friedmann equation:

$$\boxed{R_0 = \frac{c}{H_0}\left[\frac{(\Omega - 1)}{k}\right]^{-1/2}.} \tag{3.35}$$

The Hubble constant thus sets the **curvature length**, which becomes infinitely large as Ω approaches unity from either direction. Only in the limit of zero density does this length become equal to the other common measure of the size of the universe – the **Hubble length**, c/H_0.

SOLUTIONS TO THE FRIEDMANN EQUATION The Friedmann equation is so named because Friedmann was the first to appreciate, in 1922, that Einstein's equations admitted cosmological solutions containing matter only (although it was Lemaître who in 1927 both obtained the solution and appreciated that it led to a linear distance–redshift relation). The term **Friedmann model** is therefore often used to indicate a matter-only cosmology, even though his equation includes contributions from all equations of state.

The Friedmann equation may be solved most simply in 'parametric' form, by recasting it in terms of the conformal time $d\eta = c\,dt/R$ (denoting derivatives with respect to η by primes):

$$R'^2 = \frac{8\pi G}{3c^2}\rho R^4 - kR^2. \tag{3.36}$$

Because $H_0^2 R_0^2 = kc^2/(\Omega - 1)$, the Friedmann equation becomes

$$a'^2 = \frac{k}{(\Omega - 1)}\left[\Omega_r + \Omega_m a - (\Omega - 1)a^2 + \Omega_v a^4\right], \tag{3.37}$$

which is straightforward to integrate provided $\Omega_v = 0$. Solving the Friedmann equation for $R(t)$ in this way is important for determining global quantities such as the present age of the universe, and explicit solutions for particular cases are considered below. However, from the point of view of observations, and in particular the distance–redshift relation, it is not necessary to proceed by the direct route of determining $R(t)$.

To the observer, the evolution of the scale factor is most directly characterized by the change with redshift of the Hubble parameter and the density parameter; the evolution of $H(z)$ and $\Omega(z)$ is given immediately by the Friedmann equation in the form $H^2 = 8\pi G\rho/3 - kc^2/R^2$. Inserting the above dependence of ρ on a gives

$$\boxed{H^2(a) = H_0^2\left[\Omega_v + \Omega_m a^{-3} + \Omega_r a^{-4} - (\Omega - 1)a^{-2}\right].} \tag{3.38}$$

This is a crucial equation, which can be used to obtain the relation between redshift and comoving distance. The radial equation of motion for a photon is $R\,dr = c\,dt = c\,dR/\dot{R} = c\,dR/(RH)$. With $R = R_0/(1+z)$, this gives

$$\boxed{\begin{aligned} R_0 dr &= \frac{c}{H(z)}\,dz \\ &= \frac{c}{H_0}\left[(1-\Omega)(1+z)^2 + \Omega_v + \Omega_m(1+z)^3 + \Omega_r(1+z)^4\right]^{-1/2} dz. \end{aligned}} \tag{3.39}$$

This relation is arguably the single most important equation in cosmology, since it shows how to relate comoving distance to redshift, Hubble constant and density parameter. The comoving distance determines the apparent brightness of distant objects, and the comoving volume element determines the numbers of objects that are observed. These aspect of observational cosmology are discussed in more detail below in section 3.4.

Lastly, using the expression for $H(z)$ with $\Omega(a) - 1 = kc^2/(H^2R^2)$ gives the redshift dependence of the total density parameter:

$$\boxed{\Omega(a) - 1 = \frac{\Omega - 1}{1 - \Omega + \Omega_v a^2 + \Omega_m a^{-1} + \Omega_r a^{-2}}.} \tag{3.40}$$

This last equation is very important. It tells us that, at high redshift, all model universes apart from those with only vacuum energy will tend to look like the $\Omega = 1$ model. This is not surprising given the form of the Friedmann equation: provided $\rho R^2 \to \infty$ as $R \to 0$, the $-kc^2$ curvature term will become negligible at early times. If $\Omega \neq 1$, then in the distant past $\Omega(z)$ must have differed from unity by a tiny amount: the density and rate of expansion needed to have been finely balanced for the universe to expand to the present. This tuning of the initial conditions is called the **flatness problem** and is one of the motivations for the applications of quantum theory to the early universe that are discussed in later chapters.

MATTER-DOMINATED UNIVERSE From the observed temperature of the microwave background (2.73 K) and the assumption of three species of neutrino at a slightly lower temperature (see later chapters), we deduce that the total relativistic density parameter is $\Omega_r h^2 \simeq 4.2 \times 10^{-5}$, so at present it should be a good approximation to ignore radiation. However, the different redshift dependences of matter and radiation densities mean that this assumption fails at early times: $\rho_m/\rho_r \propto (1+z)^{-1}$. One of the critical epochs in cosmology is therefore the point at which these contributions were equal: the redshift of **matter–radiation equality**, given by

$$\boxed{1 + z_{\rm eq} \simeq 23\,900 \Omega h^2.} \tag{3.41}$$

At redshifts higher than this, the universal dynamics were dominated by the relativistic-particle content. By a coincidence discussed below, this epoch is close to another important event in cosmological history: **recombination**. Once the temperature falls below $\simeq 10^4$ K, ionized material can form neutral hydrogen. Observational astronomy is only possible from this point on, since Thomson scattering from electrons in ionized material prevents photon propagation. In practice, this limits the maximum redshift of observational interest to about 1000 (as discussed in detail in chapter 9); unless Ω is very low or vacuum energy is important, a matter-dominated model is therefore a good approximation to reality.

By conserving matter, we can introduce a characteristic mass M_*, and from this a characteristic radius R_*:

$$\frac{4\pi G}{3c^2}\rho R^3 = \frac{c}{H_0}\frac{\Omega[k(\Omega-1)]^{3/2}}{2} \equiv \frac{GM_*}{c^2} \equiv R_* \tag{3.42}$$

where we have used the expression for R_0 in the first step. When only matter is present, the conformal-time version of the Friedmann equation is simple to integrate for $R(\eta)$, and integration of $dt = d\eta/R$ gives $t(\eta)$:

$$\boxed{\begin{aligned} R &= kR_*[1 - C_k(\eta)] \\ ct &= kR_*[\eta - S_k(\eta)]. \end{aligned}} \tag{3.43}$$

This **cycloid solution** is a special case of the general solution for the evolution of a spherical mass distribution: $R = A[1 - C_k(\eta)]$, $t = B[\eta - S_k(\eta)]$, where $A^3 = GMB^2$ and the mass M need not be the mass of the universe. In the general case, the variable η is known as the **development angle**; it is only equal to the conformal time in the special case of the solution to the Friedmann equation. We will later use this solution to study the evolution of density inhomogeneities. The evolution of $R(t)$ in this solution is plotted in figure 3.4. A particular point to note is that the behaviour at early times is always the same: potential and kinetic energies greatly exceed total energy and we always have the $k = 0$ form $R \propto t^{2/3}$.

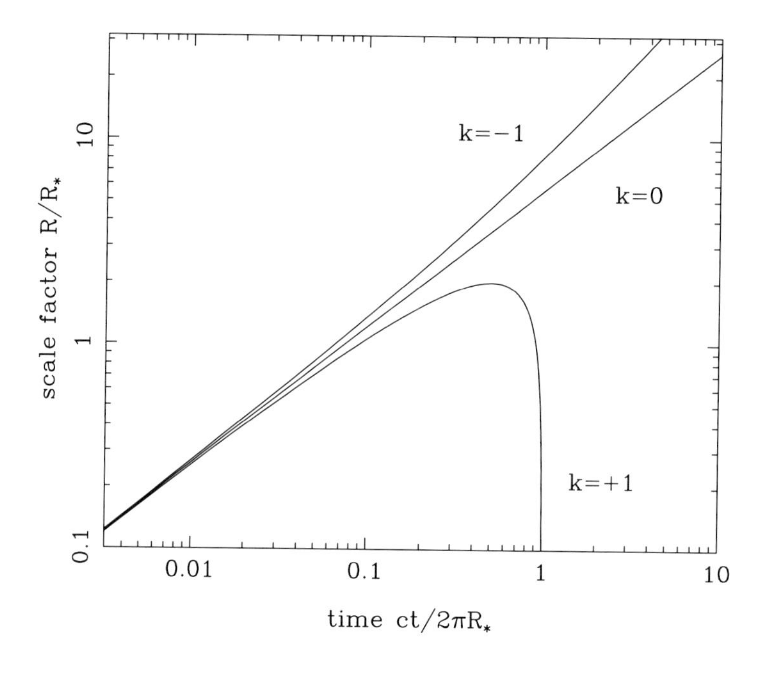

Figure 3.4. The time dependence of the scale factor for open, closed and critical matter-dominated cosmological models. The upper line corresponds to $k = -1$, the middle line to the flat $k = 0$ model, and the lowest line to the recollapsing closed $k = +1$ universe. The log scale is designed to bring out the early-time behaviour, although it obscures the fact that the closed model is a symmetric cycloid on a linear plot of R against t.

At this point, we have reproduced one of the great conclusions of relativistic cosmology: the universe is of finite age, and had its origin in a mathematical **singularity** at which the scale factor went to zero, leading to a divergent spacetime curvature. Since zero scale factor also implies infinite density (and temperature), the inferred picture of the early universe is one of unimaginable violence. The term **big bang** was coined by Fred Hoyle to describe this beginning, although it was intended somewhat critically. The problem with the singularity is that it marks the breakdown of the laws of physics; we cannot extrapolate the solution for $R(t)$ to $t < 0$, and so the origin of the expansion becomes an unexplained boundary condition. It was only after about 1980 that a consistent set of ideas became available for ways of avoiding this barrier, as discussed in chapter 11.

The parametric solution cannot be rearranged to give $R(t)$, but it is clearly possible to solve for $t(R)$. This is most simply expressed in terms of the density parameter and the age of the universe at a given stage of its development:

$$\boxed{t(z) = H(z)^{-1}\left\{[1-\Omega(z)]^{-1} + \frac{k\Omega(z)}{2[k(\Omega(z)-1)]^{3/2}}C_k^{-1}\left[\frac{2-\Omega(z)}{\Omega(z)}\right]\right\}.} \tag{3.44}$$

When we insert the redshift dependences of $H(z)$ and $\Omega(z)$,

$$\begin{aligned} H(z) &= H_0(1+z)\sqrt{1+\Omega z} \\ \Omega(z) &= \frac{\Omega(1+z)}{1+\Omega z}, \end{aligned} \tag{3.45}$$

this gives us the time–redshift relation. An alternative route to this result would have been to use the general differential expression for comoving distance dr/dz; since $c\,dt = [R_0/(1+z)]\,dr$, this gives the age of the universe as an integral over z.

An accurate and very useful approximation to the above exact result is

$$t(z) \simeq H(z)^{-1}\left[1 + \tfrac{1}{2}\Omega(z)^{0.6}\right]^{-1}, \tag{3.46}$$

which interpolates between the exact ages H^{-1} for an empty universe and $\frac{2}{3}H^{-1}$ for a critical-density $\Omega = 1$ model.

MATTER PLUS RADIATION BACKGROUND The parametric solution can be extended in an elegant way for a universe containing a mixture of matter and radiation. Suppose we write the mass inside R as

$$\frac{4\pi G\rho}{3}R^3 = M_m + M_r a^{-1}, \tag{3.47}$$

reflecting the R^{-3} and R^{-4} dependences of matter and radiation densities respectively. Now define dimensionless masses of the form $y \equiv GM/(c^2 R_0)$, which reduce to $y_{m,r} = k\Omega_{m,r}/[2(\Omega - 1)]$. The parametric solutions then become

$$\boxed{\begin{aligned} a &= \sqrt{2y_r}\,S_k(\eta) - k y_m[C_k(\eta) - 1] \\ \frac{ct}{R_0} &= k\sqrt{2y_r}\,[1 - C_k(\eta)] - k y_m[S_k(\eta) - \eta]. \end{aligned}} \tag{3.48}$$

MODELS WITH VACUUM ENERGY The solution of the Friedmann equation becomes more complicated if we allow a significant contribution from vacuum energy – i.e. a non-zero cosmological constant. Detailed discussions of the problem are given by Felten & Isaacman (1986) and Carroll, Press & Turner (1992); the most important features are outlined below.

The Friedmann equation itself is independent of the equation of state, and just says $H^2R^2 = kc^2/(\Omega - 1)$, whatever the form of the contributions to Ω. In terms of the cosmological constant itself, we have

$$\Omega_{\rm v} = \frac{8\pi G\rho_{\rm v}}{3H^2} = \frac{\Lambda c^2}{3H^2}. \tag{3.49}$$

STATIC UNIVERSE The reason that the cosmological constant was first introduced by Einstein was not simply because there was no general reason to expect empty space to be of zero density but also because it allows a non-expanding cosmology to be constructed. This is perhaps not so obvious from some forms of the Friedmann equation, since now $H = 0$ and $\Omega = \infty$; if we cast the equation in its original form without defining these parameters, then zero expansion implies

$$\rho = \frac{3kc^2}{8\pi G R^2}. \tag{3.50}$$

Since Λ can have either sign, this appears not to constrain k. However, we also want to have zero acceleration for this model, and so need the time derivative of the Friedmann equation: $\ddot{R} = -4\pi G R(\rho + 3p)/3$. A further condition for a static model is therefore that

$$\rho = -3p. \tag{3.51}$$

Since $\rho = -p$ for vacuum energy, and this is the only source of pressure if we ignore radiation, this tells us that $\rho = 3\rho_{\rm v}$ and hence that the mass density is twice the vacuum density. The total density is hence positive and $k = 1$; we have a closed model.

Notice that what this says is that a positive vacuum energy acts in a repulsive way, balancing the attraction of normal matter. This is related to the idea of $\rho + 3p$ as the effective source density for gravity. This insight alone should make one appreciate that the static model cannot be stable: if we perturb the scale factor by a small positive amount, the vacuum repulsion is unchanged whereas the 'normal' gravitational attraction is reduced, so that the model will tend to expand further (or contract, if the initial perturbation was negative). Thinking along these lines, a tidy history of science would have required Einstein to predict the expanding universe in advance of its observation. However, it is perhaps not so surprising that this prediction was never clearly made, despite the fact that expanding models were studied by Lemaître and by Friedmann in the years prior to Hubble's work. In those days, the idea of a quasi-Newtonian approach to cosmology was not developed; the common difficulty of obtaining a clear physical interpretation of solutions to Einstein's equations obscured the meaning of the expanding universe even for its creators.

DE SITTER SPACE Before going on to the general case, it is worth looking at the endpoint of an outwards perturbation of Einstein's static model, first studied by de Sitter and named after him. This universe is completely dominated by vacuum energy, and is clearly the limit of the unstable expansion, since the density of matter redshifts to zero while the vacuum energy remains constant. Consider again the Friedmann equation in its general form $\dot{R}^2 - 8\pi G\rho R^2/3 = -kc^2$: since the density is constant and R will increase without limit, the two terms on the lhs must eventually become almost exactly equal and the curvature term on the rhs will be negligible. Thus, even if $k \neq 0$, the universe will have a density that differs only infinitesimally from the critical, so that we can solve the equation by setting $k = 0$, in which case

$$R \propto \exp Ht, \quad H = \sqrt{\frac{8\pi G\rho_{\rm v}}{3}} = \sqrt{\frac{\Lambda c^2}{3}}. \tag{3.52}$$

An interesting interpretation of this behaviour was promoted in the early days of cosmology by Eddington: the cosmological constant is what *caused* the expansion. In models without Λ, the expansion is merely an initial condition: anyone who asks why the universe expands at a given epoch is given the unsatisfactory reply that it does so because it was expanding at some earlier time, and this chain of reasoning comes up against a barrier at $t = 0$. It would be more satisfying to have some mechanism that set the expansion into motion, and this is what is provided by vacuum repulsion. This tendency of models with positive Λ to end up undergoing an exponential phase of expansion (and moreover one with $\Omega = 1$) is exactly what is used in inflationary cosmology to generate the initial conditions for the big bang.

THE STEADY-STATE MODEL The behaviour of de Sitter space is in some ways reminiscent of the **steady-state universe**, which was popular in the 1960s. This theory drew its motivation from the philosophical problems of big-bang models – which begin in a singularity at $t = 0$, and for which earlier times have no meaning. Instead, Hoyle, Bondi and Gold suggested the **perfect cosmological principle** in which the universe is homogeneous not only in space, but also in time: apart from local fluctuations, the universe appears the same to all observers at all times. This tells us that the Hubble constant really is constant, and so the model necessarily has exponential expansion, $R \propto \exp(Ht)$, exactly as for de Sitter space. Furthermore, it is necessary that $k = 0$, as may be seen by considering the transverse part of the Robertson–Walker metric: $d\sigma^2 = [R(t)S_k(r)\, d\psi]^2$. This has the convention that r is a dimensionless comoving coordinate; if we divide by R_0 and change to physical radius r', the metric becomes $d\sigma^2 = [a(t)\, R_0\, S_k(r'/R_0)\, d\psi]^2$. The current scale factor R_0 now plays the role of a curvature length, determining the distance over which the model is spatially Euclidean. However, any such curvature radius must be constant in the steady-state model, so the only possibility is that it is infinite and that $k = 0$. We thus see that de Sitter space is a steady-state universe: it contains a constant vacuum energy density, and has an infinite age, lacking any big-bang singularity. In this sense, some aspects of the steady-state model have been resurrected in inflationary cosmology. However, de Sitter space is a rather uninteresting model because it contains no matter. Introducing matter into a steady-state universe violates energy conservation, since matter does not have the $p = -\rho c^2$ equation of state that allows the density to remain constant. This is the most radical aspect of steady-state models: they require **continuous creation** of matter. The energy to accomplish this has to come from somewhere, and Einstein's equations are modified by adding some 'creation' or 'C-field' term to the energy–momentum tensor:

$$T'^{\mu\nu} = T^{\mu\nu} + C^{\mu\nu}, \quad T'^{\mu\nu}_{\ \ ;\nu} = 0. \tag{3.53}$$

The effect of this extra term must be to cancel the matter density and pressure, leaving just the overall effective form of the vacuum tensor, which is required to produce de Sitter space and the exponential expansion. This *ad hoc* field and the lack of any physical motivation for it beyond the cosmological problem it was designed to solve was always the most unsatisfactory feature of the steady-state model, and may account for the strong reactions generated by the theory. Certainly, the debate between steady-state supporters and protagonists of the big bang produced some memorable displays of vitriol in the 1960s. At the start of the decade, the point at issue was whether the proper density of active galaxies was constant as predicted by the steady-state model. Since the radio-source count data were in a somewhat primitive state at that time, the debate remained inconclusive until the detection of the microwave background in 1965. For many, this spelled the end of the steady-state universe, but doubts lingered on about whether the radiation might originate in interstellar dust. These were perhaps only finally laid to rest in 1990, with the demonstration that the radiation was almost exactly Planckian in form (see chapter 9).

BOUNCING AND LOITERING MODELS Returning to the general case of models with a mixture of energy in the vacuum and normal components, we have to distinguish three cases. For models that start from a big bang (in which case radiation dominates completely at the earliest times), the universe will either recollapse or expand forever. The latter outcome becomes more likely for low densities of matter and radiation, but high vacuum density. It is however also possible to have models in which there is no big bang: the universe was collapsing in the distant past, but was slowed by the repulsion of a positive Λ term and underwent a 'bounce' to reach its present state of expansion. Working out the conditions for these different events is a matter of integrating the Friedmann equation. For the addition of Λ, this can only in general be done numerically. However, we can find the conditions for the different behaviours described above analytically, at least if we simplify things by ignoring radiation. The equation in the form of the time-dependent Hubble parameter looks like

$$\frac{H^2}{H_0^2} = \Omega_v(1 - a^{-2}) + \Omega_m(a^{-3} - a^{-2}) + a^{-2}, \tag{3.54}$$

and we are interested in the conditions under which the lhs vanishes, defining a turning point in the expansion. Setting the rhs to zero yields a cubic equation, and it is possible to give the conditions under which this has a solution (see Felten & Isaacman 1986), which are as follows.

(1) First, negative Λ always implies recollapse, which is intuitively reasonable (either the mass causes recollapse before Λ dominates, or the density is low enough that Λ comes to dominate, which cannot lead to infinite expansion unless Λ is positive.

(2) If Λ is positive and $\Omega_m < 1$, the model always expands to infinity.

(3) If $\Omega_m > 1$, recollapse is only avoided if Ω_v exceeds a critical value

$$\Omega_v > 4\Omega_m \left[\cos\left(\tfrac{1}{3}\cos^{-1}(\Omega_m^{-1} - 1) + \tfrac{4}{3}\pi\right)\right]^3. \tag{3.55}$$

(4) If Λ is large enough, the stationary point of the expansion is at $a < 1$ and we have a bounce cosmology. This critical value is

$$\Omega_v > 4\Omega_m \left[f\left(\tfrac{1}{3}f^{-1}(\Omega_m^{-1} - 1)\right)\right]^3, \tag{3.56}$$

where the function f is similar in spirit to C_k: cosh if $\Omega_m < 0.5$, otherwise cosine. If the universe lies exactly on the critical line, the bounce is at infinitely early times and we have a solution that is the result of a perturbation of the Einstein static model. Models that almost satisfy the critical $\Omega_v(\Omega_m)$ relation are known as **loitering models**, since they spend a long time close to constant scale factor. Such models were briefly popular in the early 1970s, when there seemed to be a sharp peak in the quasar redshift distribution at $z \simeq 2$. However, this no longer seems a reasonable explanation for what is undoubtedly a mixture of evolution and observational selection.

In fact, bounce models can be ruled out quite strongly. The same cubic equations that define the critical conditions for a bounce also give an inequality for the maximum redshift possible (that of the bounce):

$$1 + z_{\rm B} \leq 2f\left(\tfrac{1}{3}f^{-1}(\Omega_m^{-1} - 1)\right) \tag{3.57}$$

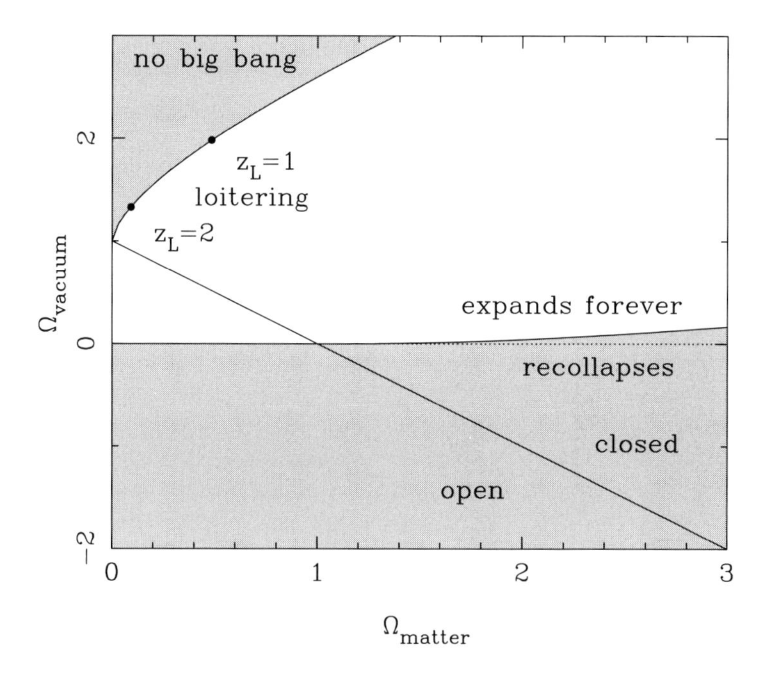

Figure 3.5. This plot shows the different possibilities for the cosmological expansion as a function of matter density and vacuum energy. Models with total $\Omega > 1$ are always spatially closed (open for $\Omega < 1$), although closed models can still expand to infinity if $\Omega_v \neq 0$. If the cosmological constant is negative, recollapse always occurs; recollapse is also possible with a positive Ω_v if $\Omega_m \gg \Omega_v$. If $\Omega_v > 1$ and Ω_m is small, there is the possibility of a 'loitering' solution with some maximum redshift and infinite age (top left); for even larger values of vacuum energy, there is no big bang singularity.

A reasonable lower limit for Ω_m of 0.1 then rules out a bounce once objects are seen at $z > 2$.

The main results of this section are summed up in figure 3.5. Since the radiation density is very small today, the main task of relativistic cosmology is to work out where on the $\Omega_{matter}\Omega_{vacuum}$-plane the real universe lies. The existence of high-redshift objects rules out the bounce models, so that the idea of a hot big bang cannot be evaded. As subsequent chapters will show, the data favour a position somewhere near the point (1,0), which is the worst possible situation: it means that the issues of recollapse and closure are very difficult to resolve.

FLAT UNIVERSE The most important model in cosmological research is that with $k = 0$, which implies $\Omega_{total} = 1$; when dominated by matter, this is often termed the **Einstein–de Sitter** model. Paradoxically, this importance arises because it is an unstable state: as we have seen earlier, the universe will evolve away from $\Omega = 1$, given a slight perturbation. For the universe to have expanded by so many **e-foldings** (factors-of-e

expansion) and yet still have $\Omega \sim 1$ implies that it was very close to being spatially flat at early times. Many workers have conjectured that it would be contrived if this flatness was other than perfect – a prejudice raised to the status of a prediction in most models of inflation.

Although it is a mathematically distinct case, in practice the properties of a flat model can usually be obtained by taking the limit $\Omega \to 1$ for either open or closed universes with $k = \pm 1$. Nevertheless, it is usually easier to start again from the $k = 0$ Friedmann equation, $\dot{R}^2 = 8\pi G\rho R^2/(3c^2)$. Since both sides are quadratic in R, this makes it clear that the value of R_0 is arbitrary, unlike models with $\Omega \neq 1$: the comoving geometry is Euclidean, and there is no natural curvature scale.

It now makes more sense to work throughout in terms of the normalized scale factor $a(t)$, so that the Friedmann equation for a matter–radiation mix is

$$\dot{a}^2 = H_0^2 \left(\Omega_m a^{-1} + \Omega_r a^{-2}\right), \tag{3.58}$$

which may be integrated to give the time as a function of scale factor:

$$H_0 t = \frac{2}{3\Omega_m^2} \left[\sqrt{\Omega_r + \Omega_m a}\,(\Omega_m a - 2\Omega_r) + 2\Omega_r^{3/2}\right]; \tag{3.59}$$

this goes to $\frac{2}{3}a^{3/2}$ for a matter-only model, and to $\frac{1}{2}a^2$ for radiation only.

One further way of presenting the model's dependence on time is via the density. Following the above, it is easy to show that

$$\begin{aligned} t &= \sqrt{\frac{1}{6\pi G\rho}} \quad &\text{(matter domination)} \\ t &= \sqrt{\frac{3}{32\pi G\rho}} \quad &\text{(radiation domination).} \end{aligned} \tag{3.60}$$

The whole universe thus always obeys a rule-of-thumb for the collapse from rest of a gravitating body: the collapse time $\simeq 1/\sqrt{G\rho}$.

Because Ω_r is so small, the deviations from a matter-only model are unimportant for $z \lesssim 1000$, and so the distance–redshift relation for the $k = 0$ matter plus radiation model is effectively just that of the $\Omega_m = 1$ Einstein–de Sitter model. An alternative $k = 0$ model of greater observational interest has a significant cosmological constant, so that $\Omega_m + \Omega_v = 1$ (radiation being neglected for simplicity). This may seem contrived, but once $k = 0$ has been established, it cannot change: individual contributions to Ω must adjust to keep in balance. The advantage of this model is that it is the only way of retaining the theoretical attractiveness of $k = 0$ while changing the age of the universe from the relation $H_0 t_0 = 2/3$, which characterizes the Einstein–de Sitter model. Since much observational evidence indicates that $H_0 t_0 \simeq 1$ (see chapter 5), this model has received a good deal of interest in recent years. To keep things simple we shall neglect radiation, so that the Friedmann equation is

$$\dot{a}^2 = H_0^2[\Omega_m a^{-1} + (1 - \Omega_m)a^2], \tag{3.61}$$

and the $t(a)$ relation is

$$H_0 t(a) = \int_0^a \frac{x\, dx}{\sqrt{\Omega_m x + (1 - \Omega_m)x^4}}. \tag{3.62}$$

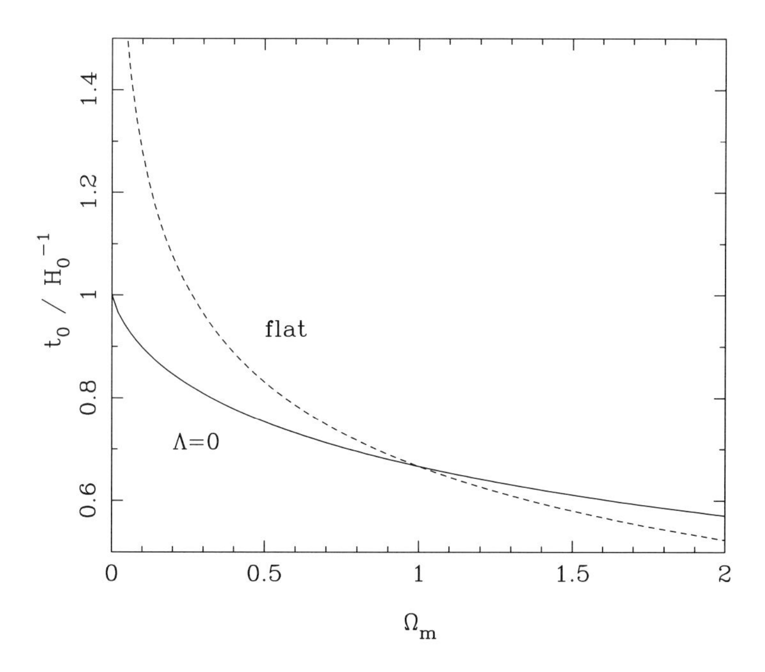

Figure 3.6. The age of the universe versus the matter density parameter. The solid line shows open models; the broken line shows flat models with $\Omega_m + \Omega_v = 1$. For values of a few tenths for Ω_m, the flat models allow a significantly greater age for a given H_0, and so a non-zero Λ can be inferred if t_0 and H_0 can be measured sufficiently accurately.

The x^4 in the denominator looks troublesome, but it can be rendered tractable by the substitution $y = \sqrt{x^3|\Omega_m - 1|/\Omega_m}$, which turns the integral into

$$H_0 t(a) = \frac{2}{3}\,\frac{S_k^{-1}\left(\sqrt{a^3|\Omega_m - 1|/\Omega_m}\right)}{\sqrt{|\Omega_m - 1|}}. \tag{3.63}$$

Here, k in S_k is used to mean sine if $\Omega_m > 1$, otherwise sinh; these are still $k = 0$ models. This $t(a)$ relation is compared to models without vacuum energy in figure 3.6. Since there is nothing special about the current era, we can clearly also rewrite this expression as

$$H(a)\,t(a) = \frac{2}{3}\,\frac{S_k^{-1}\left(\sqrt{|\Omega_m(a) - 1|/\Omega_m(a)}\right)}{\sqrt{|\Omega_m(a) - 1|}} \simeq \frac{2}{3}\,\Omega_m(a)^{-0.3}, \tag{3.64}$$

where we include a simple approximation that is accurate to a few per cent over the region of interest ($\Omega_m \gtrsim 0.1$). In the general case of significant Λ but $k \neq 0$, this expression still gives a very good approximation to the exact result, provided Ω_m is replaced by $0.7\Omega_m - 0.3\Omega_v + 0.3$ (Carroll, Press & Turner 1992).

HORIZONS For photons, the radial equation of motion is just $c\,dt = R\,dr$. How far can a photon get in a given time? The answer is clearly

$$\Delta r = \int_{t_0}^{t_1} \frac{c\,dt}{R(t)} = \Delta\eta, \tag{3.65}$$

i.e. just the interval of conformal time. What happens as $t_0 \to 0$ in this expression? We can replace dt by $dR/\dot{R}$, which the Friedmann equation says is $\propto dR/\sqrt{\rho R^2}$ at early times. Thus, this integral converges if $\rho R^2 \to \infty$ as $t_0 \to 0$, otherwise it diverges. Provided the equation of state is such that ρ changes faster than R^{-2}, light signals can only propagate a finite distance between the big bang and the present; there is then said to be a **particle horizon**. Such a horizon therefore exists in conventional big bang models, which are dominated by radiation at early times.

A particle horizon is not at all the same thing as an event horizon: for the latter, we ask whether Δr diverges as $t \to \infty$. If it does, then seeing a given event is just a question of waiting long enough. Clearly, an event horizon requires $R(t)$ to increase more quickly than t, so that distant parts of the universe recede 'faster than light'. This does not occur unless the universe is dominated by vacuum energy at late times, as discussed above. Despite this distinction, cosmologists usually say **the horizon** when they mean the particle horizon.

There are some unique aspects to the idea of a horizon in a closed universe, where you can in principle return to your starting point by continuing for long enough in the same direction. However, the related possibility of viewing the back of your head (albeit with some time delay) turns out to be more difficult once dynamics are taken into account. For a matter-only model, it is easy to show that the horizon only just reaches the stage of allowing a photon to circumnavigate the universe at the point of recollapse – the 'big crunch'. A photon that starts at $r = 0$ at $t = 0$ will return to its initial position when $r = 2\pi$, at which point the conformal time $\eta = 2\pi$ also (from above) and the model has recollapsed. Since we live in an expanding universe, it is not even possible to see the same object in two different directions, at radii r and $2\pi - r$. This requires a horizon size larger than π; but conformal time $\eta = \pi$ is attained only at maximum expansion, so antipodal pairs of high-redshift objects are visible only in the collapse phase. These constraints do not apply if the universe has a significant cosmological constant; loitering models should allow one to see antipodal pairs at approximately the same redshift. This effect has been sought, but without success.

3.3 Common big bang misconceptions

Although the concept of an isotropically expanding universe is straightforward enough to understand locally, there are a number of conceptual traps lurking in wait for those who try to make global sense of the expansion. The most common sources of error in thinking about the basics of cosmology are discussed below. Some of these may seem rather elementary when written down, but they are all fallacies to be encountered quite commonly in the minds of even the most able students.

THE INITIAL SINGULARITY Beginning at the beginning, a common question asked by laymen and some physicists is what the universe expands *into*. The very terminology of the 'big bang' suggests an explosion, which flings debris out into some void. Such a picture is strongly suggested by many semi-popular descriptions, which commonly include a description of the initial instant as one 'where all the matter in the universe is gathered at a single point', or something to that effect. This phrase can probably be traced back to Lemaître's unfortunate term 'the primaeval atom'. Describing the origin of the expansion as an explosion is probably not a good idea in any case; it suggests some input of energy that moves matter from an initial state of rest. Classically, this is false: the expansion merely appears as an initial condition. This might reasonably seem to be evading the point, and it is one of the advantages of inflationary cosmology that it supplies an explicit mechanism for starting the expansion: the repulsive effect of vacuum energy. However, if the big bang is to be thought of explosively, then it is really many explosions that happen everywhere at once; it is not possible to be outside the explosion, since it fills all of space. Because the density rises without limit as $t \to 0$, the mass within any sphere today (even the size of our present horizon) was once packed into an arbitrarily small volume. Nevertheless, this does not justify the 'primaeval atom' terminology unless the universe is closed. The mass of an open universe is infinite; however far back we run the clock, there is infinitely more mass outside a given volume than inside it.

THE ORIGIN OF THE REDSHIFT For small redshifts, the interpretation of the redshift as a Doppler shift ($z = v/c$) is quite clear. What is not so clear is what to do when the redshift becomes large. A common but incorrect approach is to use the special-relativistic Doppler formula and write

$$1 + z = \sqrt{\frac{1 + v/c}{1 - v/c}}. \tag{3.66}$$

This would be appropriate in the case of a model with $\Omega = 0$ (see below), but is wrong in general. Nevertheless, it is all too common to read of the latest high-redshift quasar as 'receding at 95 per cent of the speed of light'. The reason the redshift cannot be interpreted in this way is that a non-zero mass density must cause gravitational redshifts [problem 3.4]; even this interpretation is hard to apply rigorously when redshifts of order unity are attained. One way of looking at this issue is to take the rigid point of view that $1 + z$ tells us nothing more than how much the universe has expanded since the emission of the photons we now receive. Perhaps more illuminating, however, is to realize that, although the redshift cannot be thought of as a *global* Doppler shift, it is correct to think of the effect as an accumulation of the infinitesimal Doppler shifts caused by photons passing between fundamental observers separated by a small distance:

$$\frac{\delta z}{1 + z} = H(z)\, \delta\ell(z) \tag{3.67}$$

(where $\delta\ell$ is a radial increment of proper distance). This expression may be verified by substitution of the standard expressions for $H(z)$ and $d\ell/dz$. The nice thing about this way of looking at the result is that it emphasizes that it is momentum that gets redshifted; particle de Broglie wavelengths thus scale with the expansion, a result that is independent of whether their rest mass is non-zero. To see this last point, replace $d\ell$ by $v\, dt$ in general, and consider the effect of an infinitesimal Lorentz transformation: $\delta p = (E/c^2)\delta\beta$.

THE NATURE OF THE EXPANSION An inability to see that the expansion is locally just kinematical also lies at the root of perhaps the worst misconception about the big bang. Many semi-popular accounts of cosmology contain statements to the effect that 'space itself is swelling up' in causing the galaxies to separate. This seems to imply that all objects are being stretched by some mysterious force: are we to infer that humans who survived for a Hubble time would find themselves to be roughly four metres tall?

Certainly not. Apart from anything else, this would be a profoundly anti-relativistic notion, since relativity teaches us that properties of objects in local inertial frames are independent of the global properties of spacetime. If we understand that objects separate now only because they have done so in the past, there need be no confusion. A pair of massless objects set up at rest with respect to each other in a uniform model will show no tendency to separate (in fact, the gravitational force of the mass lying between them will cause an *inward* relative acceleration). In the common elementary demonstration of the expansion by means of inflating a balloon, galaxies should be represented by glued-on coins, not ink drawings (which will spuriously expand with the universe).

THE EMPTY UNIVERSE A very helpful tool for regaining one's bearings in cases of confusion is a universe containing no matter at all, but only non-gravitating test particles in uniform expansion. This is often called the **Milne model** after E. Milne, who attempted to establish a *kinematical* theory of cosmology (e.g. Milne 1948). We no longer believe that his model is relevant to reality, but it has excellent pedagogical value. The reason is that, since the mass density is zero, there is no spatial curvature and the metric has to be the familiar Minkowski spacetime of special relativity. Our normal intuition concerning such matters as length and time can then be brought into play in cases of confusion.

We therefore start with the normal Minkowski metric,

$$d\tau^2 = dt^2 - \left(dr^2 + r^2 d\psi^2\right), \tag{3.68}$$

and consider how this is viewed by a set of fundamental observers, in the form of particles that are ejected from $r = 0$ at $t = 0$, and which proceed with constant velocity. The velocity of particles seen at radius r at time t is therefore a function of radius: $v = r/t$ ($t = H_0^{-1}$, as required); particles do not exist beyond the radius $r = ct$, at which point they are receding from the origin at the speed of light. If all clocks are synchronized at $t = 0$, then the cosmological time t' is just related to the background special-relativity time via time dilation:

$$t' = t/\gamma = t\sqrt{1 - r^2/c^2t^2}. \tag{3.69}$$

If we also define $d\ell$ to be the radial separation between events measured by fundamental observers at fixed t', the metric can be rewritten as

$$c^2 d\tau^2 = c^2 dt'^2 - d\ell^2 - r^2 d\psi^2. \tag{3.70}$$

To complete the transition from Minkowski to fundamental-observer coordinates, we need to eliminate r. To do this, define the velocity variable ω:

$$v/c = \tanh\omega \quad \Rightarrow \quad \gamma = \cosh\omega. \tag{3.71}$$

Now, the time-dilation equation gives r in terms of t and t' as

$$r = ct'\sqrt{t^2 - t'^2} = ct'\sinh\omega, \tag{3.72}$$

and the radial increment of proper length is related to dr via length contraction (since $d\ell$ is at constant t'):

$$d\ell = dr/\gamma = ct'\, d\omega. \tag{3.73}$$

The metric therefore becomes

$$d\tau^2 = dt'^2 - t'^2 \left(d\omega^2 + \sinh^2\omega \, d\psi^2 \right). \tag{3.74}$$

This is the $k = -1$ Robertson–Walker form, and it demonstrates several important points. First, since any model with $\Omega < 1$ will eventually go over to the empty model as $R \to \infty$, this allows us to demonstrate that the constant of integration in the Friedmann equation really does correspond to $-kc^2$ (see earlier) in the $k = -1$ case without using Einstein's field equations. Second, it shows that it is more useful to think of the comoving distance $S_k(\omega)$ as representing the distance to a galaxy than ω itself. In the empty model, the reason for this is that the proper length elements $d\omega$ become progressively length-contracted at high redshift, giving a spuriously low value for the distance.

3.4 Observations in cosmology

The various distances that we have discussed are of course not directly observable; all that we know about a distant object is its redshift. Observers therefore place heavy reliance on formulae for expressing distance in terms of redshift. For high-redshift objects such as quasars, this has led to a history of controversy over whether a component of the redshift could be of non-cosmological origin:

$$1 + z_{\rm obs} = (1 + z_{\rm cos})(1 + z_{\rm extra}). \tag{3.75}$$

For now, we assume that the cosmological contribution to the redshift can always be identified.

DISTANCE–REDSHIFT RELATION The general relation between comoving distance and redshift was given earlier as

$$\begin{aligned} R_0 dr &= \frac{c}{H(z)}\, dz \\ &= \frac{c}{H_0} \left[(1-\Omega)(1+z)^2 + \Omega_v + \Omega_m (1+z)^3 + \Omega_r (1+z)^4 \right]^{-1/2} dz. \end{aligned} \tag{3.76}$$

For a matter-dominated Friedmann model, this means that the distance of an object from which we receive photons today is

$$R_0 r = \frac{c}{H_0} \int_0^z \frac{dz'}{(1+z')\sqrt{1+\Omega z'}}. \tag{3.77}$$

Integrals of this form often arise when manipulating Friedmann models; they can usually be tackled by the substitution $u^2 = k(\Omega - 1)/[(\Omega(1 + z)]$. This substitution produces **Mattig's formula** (1958), which is one of the single most useful equations in cosmology as far as observers are concerned:

$$R_0 S_k(r) = \frac{2c}{H_0} \frac{\Omega z + (\Omega - 2)[\sqrt{1+\Omega z} - 1]}{\Omega^2 (1+z)}. \tag{3.78}$$

A more direct derivation could use the parametric solution in terms of conformal time, plus $r = \eta_{\rm now} - \eta_{\rm emit}$. The generalization of the sine rule is required in this method: $S_k(a-b) = S_k(a)C_k(b) - C_k(a)S_k(b)$.

Although the above is the standard form for Mattig's formula, it is not always the most computationally convenient, being ill defined for small Ω. A better version of the relation for low-density universes is

$$R_0 S_k(r) = \frac{c}{H_0} \frac{z}{(1+z)} \frac{1+\sqrt{1+\Omega z}+z}{1+\sqrt{1+\Omega z}+\Omega z/2}. \tag{3.79}$$

It is often useful in calculations to be able to convert this formula into the corresponding one for $C_k(r)$. The result is

$$C_k(r) = \frac{(2-\Omega)(2-\Omega+\Omega z)+4(\Omega-1)\sqrt{1+\Omega z}}{\Omega^2(1+z)}, \tag{3.80}$$

remembering that $R_0 = (c/H_0)[(\Omega-1)/k]^{-1/2}$.

It is possible to extend Mattig's formula to the case of contributions from pressureless matter (Ω_m) and radiation (Ω_r):

$$\boxed{R_0 S_k(r) = \frac{2c}{H_0} \frac{\Omega_m z + (\Omega_m + 2\Omega_r - 2)\left[\sqrt{1+\Omega_m z + \Omega_r(z^2+2z)} - 1\right]}{[\Omega_m^2 + 4\Omega_r(\Omega_r + \Omega_m - 1)](1+z)}.} \tag{3.81}$$

There is no such compact expression if one wishes to allow for vacuum energy as well (Dąbrowski & Stelmach 1987). The comoving distance has to be obtained by numerical integration of the fundamental dr/dz, even in the $k=0$ case. However, for all forms of contribution to the energy content of the universe, the second-order distance–redshift relation is identical, and depends only on the deceleration parameter:

$$R_0 S_k(r) \simeq \frac{c}{H_0}\left(z - \frac{1+q_0}{2}z^2\right) \tag{3.82}$$

[problem 3.4]. The sizes and flux densities of distant objects therefore determine the geometry of the universe only once an equation of state is assumed, so that q_0 and Ω_0 can be related.

CHANGE IN REDSHIFT Although we have talked as if the redshift were a fixed parameter of an object, having a status analogous to its comoving distance, this is not correct. Since $1+z$ is the ratio of scale factors now and at emission, the redshift will change with time. To calculate how, differentiate the definition of redshift and use the Friedmann equation. For a matter-dominated model, the result is [problem 3.2]

$$\dot{z} = H_0\,(1+z)\,[1-\sqrt{1+\Omega z}] \tag{3.83}$$

(e.g. Lake 1981; Phillipps 1982). The redshift is thus expected to change by perhaps 1 part in 10^8 over a human lifetime. In principle, this sort of accuracy is not completely out of reach of technology. However, in practice these cosmological changes will be swamped if the object changes its peculiar velocity by more than 3 m s^{-1} over this period. Since peculiar velocities of up to 1000 km s^{-1} are built up over the Hubble time, the cosmological and intrinsic redshift changes are clearly of the same order, so that separating them would be very difficult.

AN OBSERVATIONAL TOOLKIT We can now assemble some essential formulae for interpreting cosmological observations. Since we will mainly be considering the post-recombination epoch, these apply for a matter-dominated model only. Our observables are redshift, z, and angular difference between two points on the sky, $d\psi$. We write the metric in the form

$$c^2 d\tau^2 = c^2 dt^2 - R^2(t) \left[dr^2 + S_k^2(r)\, d\psi^2 \right], \tag{3.84}$$

so that the *comoving* volume element is

$$\boxed{dV = 4\pi [R_0 S_k(r)]^2\, R_0 dr.} \tag{3.85}$$

The *proper* transverse size of an object seen by us is its comoving size $d\psi\, S_k(r)$ times the scale factor at the time of emission:

$$\boxed{d\ell = d\psi\ R_0 S_k(r)/(1+z).} \tag{3.86}$$

Probably the most important relation for observational cosmology is that between monochromatic flux density and luminosity. Start by assuming isotropic emission, so that the photons emitted by the source pass with a uniform flux density through any sphere surrounding the source. We can now make a shift of origin, and consider the RW metric as being centred on the source; however, because of homogeneity, the comoving distance between the source and the observer is the same as we would calculate when we place the origin at our location. The photons from the source are therefore passing through a sphere, on which we sit, of proper surface area $4\pi [R_0 S_k(r)]^2$. But redshift still affects the flux density in four further ways: photon energies and arrival rates are redshifted, reducing the flux density by a factor $(1+z)^2$; opposing this, the bandwidth $d\nu$ is reduced by a factor $1+z$, so the energy flux per unit bandwidth goes down by one power of $1+z$; finally, the observed photons at frequency ν_0 were emitted at frequency $\nu_0(1+z)$, so the flux density is the luminosity at this frequency, divided by the total area, divided by $1+z$:

$$\boxed{S_\nu(\nu_0) = \frac{L_\nu([1+z]\nu_0)}{4\pi R_0^2 S_k^2(r)(1+z)}.} \tag{3.87}$$

A word about units: L_ν in this equation would be measured in units of $\mathrm{W\,Hz^{-1}}$. Recognizing that emission is often not isotropic, it is common to consider instead the luminosity emitted into unit solid angle – in which case there would be no factor of 4π, and the units of L_ν would be $\mathrm{W\,Hz^{-1}\,sr^{-1}}$.

The flux density received by a given observer can be expressed by definition as the product of the **specific intensity** I_ν (the flux density received from unit solid angle of the sky) and the solid angle subtended by the source: $S_\nu = I_\nu\, d\Omega$. Combining the angular size and flux-density relations thus gives the relativistic version of surface-brightness conservation. This is independent of cosmology (and a more general derivation is given in chapter 4):

$$I_\nu(\nu_0) = \frac{B_\nu([1+z]\nu_0)}{(1+z)^3}, \tag{3.88}$$

where B_ν is the **surface brightness** (the luminosity emitted into unit solid angle per unit area of source). We can integrate over ν_0 to obtain the corresponding total or **bolometric** formulae, which are needed e.g. for spectral-line emission:

$$S_{\rm tot} = \frac{L_{\rm tot}}{4\pi R_0^2 S_k^2(r)(1+z)^2}; \tag{3.89}$$

$$I_{\rm tot} = \frac{B_{\rm tot}}{(1+z)^4}. \tag{3.90}$$

The form of the above relations lead to the following definitions for particular kinds of distances:

$$\begin{array}{rl} \textbf{angular-diameter distance} & D_{\rm A} = (1+z)^{-1} R_0 S_k(r) \\ \textbf{luminosity distance} & D_{\rm L} = (1+z)\; R_0 S_k(r). \end{array} \tag{3.91}$$

At least the meaning of the terms is unambiguous enough, which is not something that can be said for the term **effective distance**, sometimes used to denote $R_0 S_k(r)$. Angular-diameter distance versus redshift is illustrated in figure 3.7.

The last element needed for the analysis of observations is a relation between redshift and age for the object being studied. This brings in our earlier relation between time and comoving radius (consider a null geodesic traversed by a photon that arrives at the present):

$$c\, dt = R_0\, dr/(1+z). \tag{3.92}$$

So far, all this is completely general; to complete the toolkit, we need the crucial input of relativistic dynamics, which is to give the distance–redshift relation. For almost all observational work, it is usual to assume the matter-dominated Friedmann models with differential relation

$$R_0\, dr = \frac{c}{H_0} \frac{dz}{(1+z)\sqrt{1+\Omega z}}, \tag{3.93}$$

which integrates to

$$R_0 S_k(r) = \frac{c}{H_0} \frac{2}{\Omega^2(1+z)} \left[\Omega z + (\Omega - 2)\left(\sqrt{1+\Omega z} - 1\right)\right]. \tag{3.94}$$

PREDICTING BACKGROUNDS The above formalism can be given an illustrative application to obtain one of the fundamental formulae used for calculations involving background radiation. Suppose we know the emissivity j_ν for some process as a function of frequency and epoch, and want to predict the current background seen at frequency ν_0. The total spectral energy density created during time dt is $j_\nu([1+z]\nu_0, z)\, dt$; this reaches the present reduced by the volume expansion factor $(1+z)^{-3}$. The redshifting of frequency does not matter: ν scales as $1+z$, but so does $d\nu$. The reduction in energy density caused by the lowering of photon energies is then exactly compensated by a reduced bandwidth, leaving the spectral energy density altered only by the change in proper photon number density. Inserting the redshift–time relation, and multiplying by

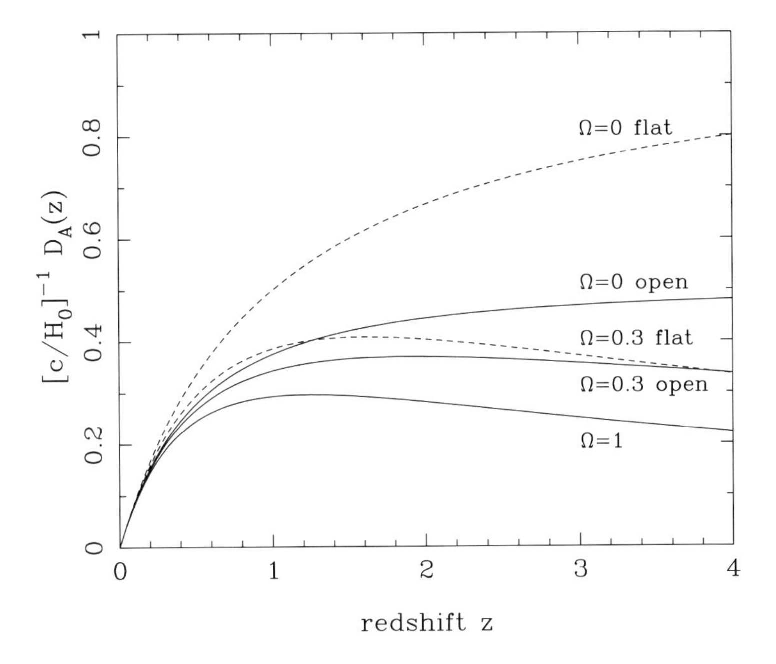

Figure 3.7. A plot of dimensionless angular-diameter distance versus redshift for various cosmologies. The solid curves show models with zero vacuum energy; the broken curves show flat models with $\Omega_m + \Omega_v = 1$. In both cases, the curves for $\Omega_m = 1, 0.3, 0$ are shown; higher density results in lower distance at high z, due to gravitational focusing of light rays (see chapter 4).

$c/4\pi$ to obtain the specific intensity, we get

$$\boxed{I_\nu = \frac{1}{4\pi} \frac{c}{H_0} \int j_\nu([1+z]\nu_0, z) \frac{dz}{(1+z)^5 \sqrt{1+\Omega z}}.} \tag{3.95}$$

This may seem a bit like cheating: we have considered how the energy density evolves at a point in space, even though we know that the photons we see today originated at some large distance. This approach works well enough because of large-scale homogeneity, but it may be clearer to obtain the result for the background directly. Consider a solid angle element $d\Omega$: at redshift z, this area on the sky for some radial increment dr defines a proper volume

$$V = [R_0 S_k(r)]^2 \, R_0 \, dr \, d\Omega \, (1+z)^{-3}. \tag{3.96}$$

This volume produces a luminosity $V j_\nu$, from which we can calculate the observed flux density $S_\nu = L_\nu / [4\pi (R_0 S_k)^2 (1+z)]$. Since surface brightness is just flux density per unity solid angle, this gives

$$I_\nu = \frac{1}{4\pi} \int j_\nu \, \frac{R_0 \, dr}{(1+z)^4}, \qquad (3.97)$$

which is the same result as the one obtained above.

3.5 The anthropic principle

Our discussion of cosmology so far has leaned heavily on the **Copernican principle** that we are not privileged observers of the universe. However, there are some aspects of the subject for which this principle has to be discarded; as in quantum mechanics, we have to ask whether the mere presence of an observer influences the system under study. At first sight, it may seem absurd to think of humanity influencing in any way the global properties of the universe, but the simple fact of our existence does turn out to have profound consequences. These ideas are given the label 'anthropic', and we speak of the **anthropic principle** when discussing the universe in the context of the existence of observers. However, this is an area that generates controversy, and not everyone will agree on the degree of importance of anthropic ideas. See Barrow & Tipler (1986) for a thorough review of the subject. Like religious schisms, the anthropic dogma comes in varying forms and degrees of elaboration. We can distinguish perhaps three sets of attitudes to anthropic reasoning: trivial, weak and strong anthropic principles.

THE TRIVIAL ANTHROPIC PRINCIPLE This is not strictly part of the anthropic principle proper, but consists of anthropic ideas on which almost everyone can agree – i.e. that we can use observations of humanity as cosmological information in the same way as we use data from telescopes. For example, we can deduce that the universe is $\gtrsim 1$ Gyr old merely by noting that carbon-based life has formed, and that there needed to be time for typical stars to go through their life cycle and distribute heavy elements in order for this to have happened. The existence of humanity thus gives us a bound on H_0. That is about as far as such trivial anthropic arguments go in cosmology; they are in a sense unnecessary, as we have direct dating of the Earth to set a much more precise limit on H_0, a constraint that was astronomically important in the early days of the distance scale, when values of $H_0 \simeq 500 \, \mathrm{km\,s^{-1}Mpc^{-1}}$ were suggested.

However, anthropic arguments of this type have an honourable history from the nineteenth century, when the Earth could not be dated directly. At that time, Lord Kelvin was advocating an age for the Earth of only $\sim 10^7$ years, based on its cooling time. Evolutionary biologists were able to argue that this was an inadequate time to allow the development of species, a conclusion that was vindicated by the discovery of radioactivity (which allowed both the dating of the Earth and showed the flaw in Kelvin's argument). Here was an excellent example of important astronomical conclusions being drawn from observations of life on Earth.

THE WEAK ANTHROPIC PRINCIPLE Moving on from the astronomical consequences of observations of their planet by human beings, are there cosmological inferences to be drawn from the existence of humanity itself? This question arises because the very possibility of carbon-based life seems to depend on a series of striking coincidences, as first pointed out in a seminal article by Dicke (1961).

The most striking of Dicke's arguments is based upon the mechanism for the production of the elements. As we shall see, it is now thought virtually certain that the abundances of the light elements up to ^{7}Li were determined by the progress of nuclear reactions in the early stages of the big bang. Heavier elements were produced at a much later stage by fusion in stars. Incidentally, this division of labour represents an ironic end to a historically important debate concerning the origin of the universe, which dominated cosmology in the 1960s. The epochal paper that became universally known just as B^{2}FH (Burbidge, Burbidge, Fowler & Hoyle 1957) was concerned with showing how the elements could be built up by nuclear reactions in stars. Although this was not the motivation for the work, these mechanisms provided a vital defence for the steady-state model (which never passes through a hot phase) against the belief of Gamow and coworkers that all elements could be synthesized in the big bang (Gamow 1946; Alpher, Bethe & Gamow 1948). Although the steady-state model passed away, the arguments of B^{2}FH have become part of current orthodoxy, leaving only the lightest elements adhering to Gamow's vision.

Now, the fascinating aspect of all this is that synthesis of the higher elements is rather difficult, owing to the non-existence of stable elements with atomic weights $A = 5$ or $A = 8$. This makes it hard to build up heavy nuclei by collisions of ^{1}H, ^{2}D, ^{3}He and ^{4}He nuclei. The only reason that heavier elements are produced at all is the reaction

$$3\,{}^4\mathrm{He} \to {}^{12}\mathrm{C}. \tag{3.98}$$

A three-body process like this will only proceed at a reasonable rate if the cross-section for the process is **resonant**: i.e. if there is an excited energy level of the carbon nucleus that matches the typical energy of three alpha particles in a stellar interior. The lack of such a level would lead to negligible production of heavy elements – and no carbon-based life. Using these arguments, Hoyle made a breathtaking leap of the imagination to predict that carbon would display such a resonance, which was duly found to be the case (Hoyle *et al.* 1953).

In a sense, this is just trivial anthropic reasoning: we see carbon on Earth, and nuclear physics gives us an inevitable conclusion to draw. And yet, one is struck by the *coincidence*: if the energy levels of carbon had been only slightly different, then it is reasonable to assume that the development of life anywhere in the universe would never have occurred. Does this mean that some controlling agent designed nuclear physics specifically to produce life? Perhaps, but it is not so easy to prove one way or the other, as we can see from the following arguments.

Suppose we imagine some process that produces an ensemble of a large number of universes with widely varying properties or even physical laws. What the **weak anthropic principle** points out is that only those members of the ensemble that are constructed so as to admit the production of intelligent life at some stage in their evolution will ever be the subject of cosmological enquiry. The fact that we are observers means that, if the production of life is at all a rare event, any universe that we can see is virtually guaranteed to display bizarre and apparently improbable coincidences. The ^{12}C energy level is one such; another goes under the heading of **Dirac's large-number hypothesis**. Dirac noted that very large dimensionless numbers often arise in both particle physics and cosmology. The ratio of the electrostatic and gravitational forces experienced by an electron in a hydrogen atom is

$$\frac{e^2}{4\pi\epsilon_0 G m_e m_p} \simeq 10^{39.4}; \qquad (3.99)$$

one of the problems of unifying gravity and other forces is understanding how such a vast dimensionless number can be generated naturally. In a cosmological context, the weakness of gravity manifests itself in the fact that the Hubble radius is enormously greater than the Planck length:

$$\frac{c/H_0}{\sqrt{\hbar G/c^3}} \simeq 10^{61}. \qquad (3.100)$$

This number is very nearly the 1.5 power of the previous large number: it is as if these large numbers were quantized in steps of 10^{20}. Dirac proposed that the coincidence must indicate a causal relation; requiring the proportionality to hold at all times then yields the radical consequence

$$G \propto t^{-1} \qquad (3.101)$$

(because H_0 declines as $\sim t^{-1}$ as the universe ages).

Geological evidence shows that this prediction is not upheld in practice, since the sun would have been hotter in the past. A less radical explanation of the large-number coincidence uses the anthropic idea that life presumably requires the existence of elements other than hydrogen and helium. The universe must therefore be old enough to allow typical stars to go through their life cycle and produce ‘metals’. This condition can be expressed in terms of fundamental constants as follows. Stars have masses of the order of the **Chandrasekhar mass**

$$M_{\rm Chandra} \equiv \left(\frac{\hbar c}{G m_p^2}\right)^{3/2} m_p, \qquad (3.102)$$

where m_p is the proton mass (see chapter 2). The luminosity of a star dominated by electron-scattering opacity is [problem 5.1]

$$L \sim \frac{G^4 m_p^5 M^3}{\hbar^3 c^2 \sigma_{\rm T}}. \qquad (3.103)$$

The characteristic lifetime of a star, $M_{\rm Chandra} c^2/L$, can thus also be expressed in fundamental constants, and is

$$t_* \sim \frac{c\,\sigma_{\rm T}}{G\,m_p} = 5.7 \times 10^9 \text{ years}. \qquad (3.104)$$

Comparing with the above, we see that the large-number coincidence is just $t_* \sim H_0^{-1}$; i.e. the universe must be old enough for the stars to age. The fact that the universe is not very much older than this may tell us that we are privileged to be in the first generation of intelligence to arise after the big bang: civilizations arising in $\gg 10^{10}$ years time will probably not spend their time in cosmological enquiry, as they will be in contact with experienced older races who know the answers already. Lastly, the coincidence can be used to argue for the fundamental correctness of the big bang as against competitors such as the steady-state theory; if the universe is in reality very much older than t_*, there is no explanation for the coincidence between t_* and H_0^{-1}.

In short, the **weak anthropic principle** states that, because intelligent life is necessary for cosmological enquiry to take place, this already imposes strong selection effects on cosmological observations. Note that, despite the name, there is no requirement for the life to be human, or even carbon based. All we say is that certain conditions can be ruled out if they do not lead to observers, and this is one of the weaknesses of the principle: are we really sure that life based on elements less massive than carbon is impossible? The whole point of these arguments is that it only has to happen once.

It is nevertheless at least plausible that the 'anthropic' term may not be a complete misnomer. It has been argued (see Carter 1983) that intelligent life is intrinsically an extremely unlikely phenomenon, where the mean time for development could be very long:

$$\bar{t}_{\rm Intelligence} \gg t_*. \tag{3.105}$$

If the inequality was sufficiently great, it would be surprising to find even one intelligent system within the current horizon. The anthropic principle provides a means for understanding why the number is non-zero, even when the expectation is small, but there would be no reason to expect a second system; humanity would then probably be alone in the universe. On the other hand, $\bar{t}_{\rm Intelligence}$ may be shorter, and then other intelligences would be common. Either possibility is equally consistent with the selection effect imposed by our existence, although the fact that life on Earth took billions of years to develop is consistent with the former view; if life is a rare event, it would not appear early in the allowed span of time.

OTHER WEAK ANTHROPIC DEDUCTIONS What other features of the universe might be amenable to anthropic arguments? The striking aspects to explain are that we live in a universe with roughly critical density in matter, dominating a radiation background of 2.73 K by roughly four powers of 10. Our starting point is the age argument given above, plus the assumption that life will only arise (a) after recombination (so that we are not cooked); (b) after matter domination (so that matter can self-gravitate into stars):

$$t_0 \sim t_* > t_r,\ t_{\rm eq}. \tag{3.106}$$

A first coincidence to consider is that recombination and matter–radiation equality occur at relatively similar times; why is this? Recombination requires a temperature of roughly the ionization potential of hydrogen:

$$kT_r \sim \frac{m_e c^2}{137^2}. \tag{3.107}$$

The mass ratio of baryons and photons at recombination is therefore $137^2 m_p/m_e \sim 10^{7.5}$, and the universe will be matter dominated at recombination unless the ratio of photon and proton number densities exceeds this value. At high temperatures, $kT > m_p c^2$, baryon, antibaryon and photon numbers will be comparable; it is thought that the present situation arises from a particle-physics asymmetry between matter and antimatter, so that baryons and antibaryons do not annihilate perfectly (see chapter 9). Matter–radiation equality thus arises anywhere after a redshift $10^{7.5}$ times larger than that of last scattering, and could be infinitely delayed if the matter–antimatter asymmetry were small enough. The approximate coincidence in epochs says that the size of the particle–antiparticle asymmetry is indeed roughly $137^2 m_p/m_e$, and it is not implausible that such a relation might arise from a complete particle physics model. At any rate, it seems to have no bearing on anthropic issues.

The anthropic argument for the age of the universe lets us work out the time of recombination, if we accept for the moment that $\Omega \sim 1$ (see below). The age for an Einstein–de Sitter model would tell us both H_0 and ρ_0, and hence the current number density of photons, if we could obtain the photon-to-baryon ratio from fundamental arguments. Since $n_\gamma \propto T_\gamma^3$, that would give the present photon temperature, and hence the redshift of recombination. The fact that this was relatively recent (only at $z \sim 10^3$) reflects the fact that the photon-to-baryon ratio is $\sim 10^7$ rather than a much larger number.

Finally, what do anthropic arguments have to say about the cosmic density parameter, Ω? We have already mentioned the instability in Ω:

$$\left|\Omega^{-1}(z) - 1\right| = (1+z)^{-1}\left|\Omega_0^{-1} - 1\right|, \tag{3.108}$$

so that $\Omega(z)$ has to be fine-tuned at early epochs to achieve $\Omega \simeq 1$ now. We can go some way towards explaining this by appealing to some results that will be proved in later chapters. The formation of galaxies can only occur at a redshift z_f such that $\Omega(z_f) \sim 1$, otherwise the growth of density perturbations switches off; Ω_0 must be unity to within a factor $\sim (1+z_f)$. The redshift z_f must occur after matter–radiation equality, otherwise radiation pressure would prevent galaxy-scale systems from collapsing. Any universe with $\Omega \ll 10^{-3}$ at $t = t_*$ would then have great difficulty in generating the nonlinear systems needed for life. Similar arguments can be made concerning the contribution of a cosmological constant to Ω (Weinberg 1989; Efstathiou 1995): structures cannot form after the time at which the universe becomes dominated by any cosmological term. However, because formation is constrained to occur between a redshift of order unity and 10^3, the vacuum density could presently exceed the density in matter by a factor of up to 10^9 without preventing structure formation. There is no anthropic reason for it to be any smaller.

These arguments are impressive, but frustratingly limited. Although a rough allowed area of parameter space can be delineated, proper physical mechanisms are required to explain why a particular point in the parameter space was selected by reality. Given such a mechanism, why do we need anthropic reasoning at all? Worse still, anthropic reasoning tends to predict a universe at the extremes of the allowed range. Consider a generator of an ensemble of universes creating them with various values of $\Omega(t_*)$. Owing to the fine-tuning required, we expect that the probability of a given result is a strongly increasing function of $|\Omega - 1|$. Multiply this by a probability of generating intelligence that declines with $|\Omega - 1|$ slowly at first, and we should expect a life-selected universe to have a peak probability near to the very largest or smallest possible values of $|\Omega - 1|$. This is not the observed situation: anthropic arguments may therefore not be relevant to determining Ω, even if their importance in explaining the large-number ratios cannot be challenged.

THE STRONG ANTHROPIC PRINCIPLE The ultimate form of anthropic reasoning is to assert that the coincidences we have remarked on are more than that: that the universe *must* be such as to admit the production of intelligent life at some time. This idea is known as the **strong anthropic principle**. Is such an idea a part of testable science? The whole basis of the weak anthropic principle is the argument that life-free universes cannot be observed, and observations of these counter-examples would be required in order to falsify the strong anthropic principle. However, this extension of anthropic ideas does have some attractions. Much weak anthropic reasoning invokes the generation of an ensemble of universes, but we have no idea whether such a concept is valid. It does apply to certain forms of inflationary cosmology, but it is equally possible that there is only one universe. In this case of a unique event, the arguments of statistical selection effects that lie at the core of weak anthropic reasoning are less satisfying; was it inevitable that life should arise on the one occasion that it has a chance? A possible position here is that the universe was designed for life, but this does not constitute a mechanism within the bounds of physics for enforcing a strong anthropic principle. It is consistent with the facts, but is in no way a proof for the existence of a creator. More profitable ideas are to be found in the area of the interpretation of quantum mechanics, which is a topic discussed to some extent in chapter 6 (chapter 7 of Barrow & Tipler 1986 gives a full discussion of the relation between quantum and anthropic ideas). Here, the role of the observer is critical in determining how the universe evolves. In the (almost) standard 'Copenhagen' interpretation, the critical events in time are the moments of wave-function collapse when the act of observation singles out a concrete state from undetermined possibilities (e.g. spin up or down?). In this sense, the act of the observer is necessary in order to bring the universe into being at all.

Problems

(3.1) Show that the Robertson–Walker metric satisfies Einstein's equations and obtain the Friedmann equation(s).

(3.2) An object is observed at redshift z in a Friedmann universe with density parameter Ω. Calculate the observed rate of change of redshift for the object. What fractional precision in observed frequency would be needed to detect cosmological deceleration in a decade?

(3.3) An object at comoving radius r is observed at two periods six months apart, allowing the parallax angle θ to be observed for a baseline of D perpendicular to the line of sight. Show that the **parallax distance** D/θ is $R_0 S_k(r)/C_k(r)$ and give an expression for this valid to second order in redshift.

(3.4) Consider the second-order corrections to the relation between redshift and angular-diameter distance $(c/H_0)D(z)$:

$$D(z) \simeq z - \frac{3+q_0}{2} z^2 \quad \Rightarrow \quad z \simeq D(z) + \frac{3+q_0}{2} D(z)^2, \tag{3.109}$$

to second order. Attempt to account for this relation with a Newtonian analysis.

(3.5) The steady-state universe has $R \propto e^{Ht}$, zero curvature of its comoving coordinates ($k = 0$), and a *proper* density of all objects that is constant in time. Show that the comoving volume out to redshift z is $V(z) = 4\pi(cz/H)^3/3$, and hence that the number-count slope for objects at typical redshift z becomes $\beta = [(3+\alpha)\ln z]^{-1}$ for $z \gg 1$, where α is the spectral index of the objects.

4 Gravitational lensing

The previous chapter showed how to understand the propagation of light in a homogeneous universe, allowing the intrinsic properties of cosmologically distant objects to be inferred from the observations we make today. The real universe is of course not perfectly homogeneous, and the propagation of light is influenced by the gravitational fields of collapsed objects. This phenomenon of **gravitational lensing** (known in French literature by the perhaps more appropriate term **gravitational mirages**) was first studied by Einstein in 1912 (see Renn, Sauer & Stachel 1997), well before the 1919 eclipse verification of Einstein's formula for light deflection. The Sun can deflect light by angles of order 1 arcsec; over a large enough baseline, this deflection can focus light rays, so that the Sun acts as a gravitational telescope. Over cosmological baselines in particular, the distortions produced by this form of gravitational imaging can be large and significant. In 1979, the discipline of gravitational lensing leapt from theoretical speculation to being an observational tool with the discovery of 0957+561, a quasar split into two images 6 arcsec apart by the gravitation of an intervening galaxy (Walsh, Carswell & Weymann 1979). The subject of gravitational lensing has since developed into one of the major tools of cosmology. It is an invaluable way of probing mass distributions directly, irrespective of whether the matter is dark or visible. For (much) more detail, consult the textbook by Schneider, Ehlers & Falco (1992), or the review by Blandford & Narayan (1992).

4.1 Basics of light deflection

The deflection geometry for a single gravitational lens is shown in figure 4.1. Source, lens and observer are shown as lying in the same plane, which is always possible even for a non-symmetric lens. A plot of this type is known as an **embedding diagram**: a curved space is drawn by treating a set of coordinates that describe the space as if they were Euclidean. The parameters $D_{\rm LS}$ etc. are angular-diameter distances. The **bend angle**, α is given by the line integral of the gravitational acceleration perpendicular to the path, $a_\perp$:

$$\alpha = \frac{2}{c^2}\int a_\perp \, d\ell. \tag{4.1}$$

This simply says that photons are deflected twice as much as would be expected from a naive Newtonian argument (chapter 1). When the light bending takes place within a small enough distance that the deflection can be taken as sudden we have a **geometrically thin lens**; this is normally a good assumption provided light is only deflected by one lens. **Thick lenses** are a good deal more complicated to analyse.

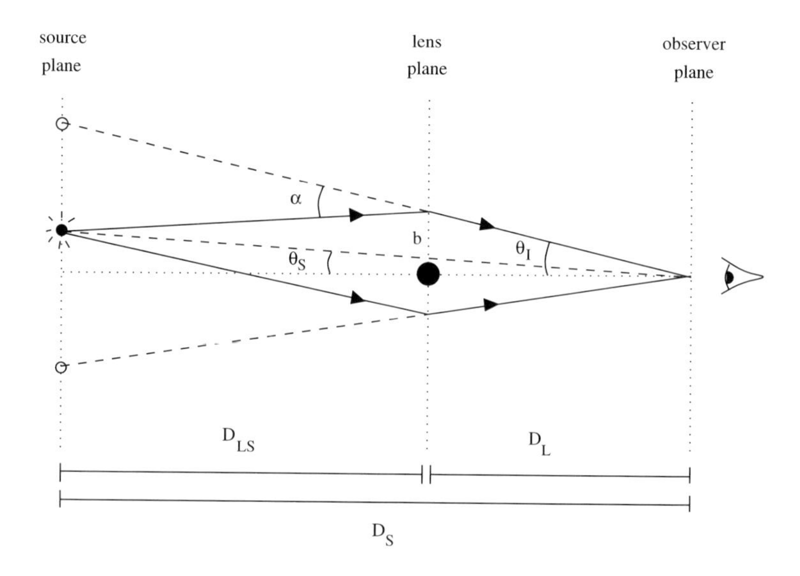

Figure 4.1. The geometry of a gravitational lensing event. For a thin lens, deflection through the small bend angle α may be taken as instantaneous. The angles $\theta_{\rm I}$ and $\theta_{\rm S}$ (two-dimensional vectors in general) specify respectively the observed and intrinsic positions of the source on the sky.

Adding up distances at the left-hand side of figure 4.1 in two different ways yields the fundamental lensing equation

$$\boxed{\alpha(D_{\rm L}\theta_{\rm I}) = \frac{D_{\rm S}}{D_{\rm LS}}(\theta_{\rm I} - \theta_{\rm S}),} \tag{4.2}$$

where the angles $\theta_{\rm I}, \theta_{\rm S}$ and α are in general two-dimensional vectors on the sky. In principle, this equation can be solved to find the mapping between the **object plane** and the **lens plane** or **image plane** – i.e. $\theta_{\rm I}(\theta_{\rm S})$ – and hence positions of the images. If the lensing deflection is small, this mapping is just a one-to-one and invertible distortion of the coordinates: this is **weak lensing**. For larger deflections, however, a unique inverse mapping from source to lens plane may not exist: this is the regime of **strong lensing**, associated with multiple imaging.

The only other thing we need in order to find the appearance of the images is the fact that gravitational lensing does not alter surface brightness. Formally, this comes about from the relativistic invariant I_ν/ν^3, which is proportional to the photon phase-space density and is conserved along light rays through Liouville's theorem (Kristian & Sachs 1966, Sachs 1973). If that seems too compact an argument, then recall that surface brightness is well known to be conserved in Euclidean space [problem 4.1]. It is also easy to show that a Doppler shift causes a change in surface brightness governed by $I_\nu \propto \nu^3$, so that I_ν/ν^3 is an invariant in special relativity. Now, by the equivalence principle, light can be thought of as propagating along a path that is locally Euclidean, but for which there is a gravitational redshift between the start and end of any given segment of the ray; the total change in surface brightness therefore just depends on the total frequency change along the path. In cosmology, this is usually almost identical to the mean redshift for the object's distance (extra blueshifts caused by falling into the lens potential well are cancelled by redshifts on leaving). Hence, the **amplification** of image flux densities is given simply by a ratio of image areas, using the Jacobian of the transformation:

$$A = \left| \frac{\partial(\theta_{\rm I})}{\partial(\theta_{\rm S})} \right| . \tag{4.3}$$

In the circumstances, magnification might be a more accurate term than amplification; however, the former is usually reserved for discussions of changes of apparent linear sizes. Some images will have $A > 1$, some $A < 1$; the total flux received, summing over all images, is always increased by lensing, and at least one image has $A > 1$ – although we will not prove this here (Schneider 1984).

It is worth noting that we are dealing here with the relativistic equivalent of geometrical optics; this approximation is discussed in detail by Misner, Thorne & Wheeler (1973). Wave effects are only important for low-mass lenses, where the first Fresnel zone becomes comparable in size to the lens impact parameter ($M \lesssim \lambda c^2/G$); this limit is of little importance in practice.

ANGULAR-DIAMETER DISTANCES In a Friedmann universe, the quantity $D_{\rm LS}$ is given by an extension of the usual distance–redshift formula:

$$D_{\rm LS} = \frac{R_0 S_k(r_{\rm S} - r_{\rm L})}{1 + z_{\rm S}}, \tag{4.4}$$

which yields (for a matter-only model)

$$D_{\rm LS} = \frac{2c}{H_0} \frac{\sqrt{1 + \Omega z_{\rm L}}\,(2 - \Omega + \Omega z_{\rm S}) - \sqrt{1 + \Omega z_{\rm S}}\,(2 - \Omega + \Omega z_{\rm L})}{\Omega^2 (1 + z_{\rm S})^2 (1 + z_{\rm L})} \tag{4.5}$$

(Refsdal 1966); $D_{\rm L}$ and $D_{\rm S}$ are special cases of this formula.

The above formula is strictly correct only when dealing with a single lens embedded in a background Friedmann universe. The relevant distance measure in a universe populated with many lenses should really be found self-consistently by solving the thick gravitational lens. Although this is virtually impossible in general, quite a bit of progress has been made with variants of the so-called **empty light cone** model. Here, one considers a light ray that passes far from all clumps, experiencing a density that is a fraction α of the mean. It is more natural to have $\alpha < 1$, but in principle one can also consider uniformly *over*dense lines of sight. In general, the propagation of such a ray is governed by the general-relativistic **optical scalar equations** (Sachs 1961; Dyer & Roeder 1973), which describe how a beam of light rays is focused and sheared by gravitational fields. In the present case, the beam is shear free and the full apparatus is not needed. Consider the light beam as a cylinder of proper diameter D: the relative gravitational acceleration of the two sides is $a_\perp = 2\pi G\rho D$. This tells us the rate of change of the relative momentum of a pair of photons that travel along opposite sides of the tube:

$$\dot{p}_\perp = \frac{2a_\perp}{c} p_\parallel. \tag{4.6}$$

Now, the proper rate of change of D is just the cone angle: $\partial D/\partial r_{\rm prop} = p_\perp/p_\parallel$. Differentiating and using the redshifting of $p_\parallel$, $\delta p/p = H(z)\delta r_{\rm prop}/c$, we obtain a second-order equation for D. This is most simply expressed in terms of the relativistic **affine distance**, symbolized by λ, which is a distance variable that scales along light rays in such a way as to make the photon momentum expressed in affine units a constant: $\delta\lambda = (H_0/c)\delta r_{\rm prop}/(1+z)$. In these terms, the equation for D becomes

$$\frac{\partial^2 D}{\partial \lambda^2} = -\frac{3\alpha}{2}\Omega_0(1+z)^5 D, \tag{4.7}$$

for which several useful solutions exist:

$$D = \begin{cases} \dfrac{c}{H_0}\displaystyle\int_0^z \frac{dz'}{(1+z')^3\sqrt{1+\Omega z'}} = \frac{c\lambda}{H_0} & (\alpha = 0) \\ \dfrac{c}{H_0}\dfrac{2}{\beta}(1+z)^{(\beta-5)/4}\,[1-(1+z)^{-\beta/2}] & (\Omega = 1,\ \alpha < \frac{25}{24}) \\ \dfrac{c}{H_0}\dfrac{4}{\beta}(1+z)^{-5/4}\sin\left[\dfrac{\beta}{4}\ln(1+z)\right] & (\Omega = 1,\ \alpha > \frac{25}{24}), \end{cases} \tag{4.8}$$

where $\beta \equiv |25 - 24\alpha|^{1/2}$. For an empty beam ($\alpha = 0$), the angular-diameter distance is just the affine distance. For non-zero α, D is reduced by the gravitational focusing of matter in the beam; the universe as a whole acts as a thick gravitational lens to magnify distant objects. This interpretation applies to the homogeneous $\alpha = 1$ case, which is why the normal $D(z)$ relation has a maximum.

Although it is important for the insight it gives into light propagation in an inhomogeneous universe, the empty-cone model should not be used very often in practice, because gravitational lensing must obey **flux conservation**. Imagine a sphere centred on a source of radiation; the energy leaving that sphere is independent of whether the matter inside is smooth or in lumps. In the latter case, the flux amplification caused by discrete lenses must compensate for the reduction in flux along under-dense lines of sight. In other words, the standard distance-redshift relation for a Friedmann model will be correct on average; unless a detailed multiple-lensing calculation is being performed, this is still the quantity to use.

THE LENSING POTENTIAL There is an alternative way of presenting the lensing equation for thin lenses that is more useful in general situations, and this is to work in terms of a two-dimensional **lensing potential**. The bend angle depends on the gravitational potential Φ as follows:

$$\boldsymbol{\alpha} = \frac{2}{c^2}\int \boldsymbol{\nabla}_2\Phi\, d\ell = \boldsymbol{\nabla}_2 \frac{2}{c^2}\int \Phi\, d\ell, \tag{4.9}$$

where $\boldsymbol{\nabla}_2$ is the two-dimensional gradient operator in the lens plane (i.e. the components of $\boldsymbol{\nabla}$ perpendicular to the photon path). Thus, $\boldsymbol{\alpha}$ is proportional to the gradient of some 2D function ψ, which can be chosen to be dimensionless, so that the lensing equation becomes:

$$\boldsymbol{\theta}_{\rm I} - \boldsymbol{\theta}_{\rm S} = \frac{D_{\rm LS}}{D_{\rm S}}\boldsymbol{\alpha} \equiv \boldsymbol{\nabla}_\theta \psi(\boldsymbol{\theta}_{\rm I}) \tag{4.10}$$

(introducing the angular gradient operator). This is a quite remarkable result: for thin lenses, any two systems with the same surface density have the same lens effect, which is independent of the details of their 3D density structure.

Using Poisson's equation for the 3D gravitational potential and neglecting a zero boundary term, it is easy to see [problem 4.2] that ψ obeys Poisson's equation in a form involving the surface density $\Sigma = \int \rho \, d\ell$:

$$\boxed{\nabla_\theta^2 \psi = \frac{D_\mathrm{L} D_\mathrm{LS}}{D_\mathrm{S}} \frac{8\pi G}{c^2} \Sigma \equiv 2 \frac{\Sigma}{\Sigma_\mathrm{C}}.} \tag{4.11}$$

This defines Σ_C, the critical surface density (see below). This equation can be solved to find ψ knowing Σ just as in 3D: we just need the Green function for ∇_θ^2, which is $\ln|\theta - \theta'|/2\pi$ (a point mass in 2D gives a $1/r$ force as compared to $1/r^2$ in 3D). The lens potential is then

$$\psi(\boldsymbol{\theta}) = \frac{1}{\pi \Sigma_\mathrm{C}} \int \Sigma(\boldsymbol{\theta}') \, \ln|\boldsymbol{\theta} - \boldsymbol{\theta}'| \, d^2\theta'. \tag{4.12}$$

In practical calculations, one would not use this integral to evaluate the lens potential; rather, the efficient alternative is to solve the 2D Poisson equation in Fourier space (so that $\nabla^2 \to -k^2$), by gridding the surface density and Fourier transforming with an FFT (fast Fourier transform) algorithm (see e.g. chapter 12 of Press *et al.* 1992). The deflection field can then be obtained either by numerical differentiation or by a further transform. The only disadvantage of this method is that the resolution of the potential and deflection fields is then limited by the grid. This is a similar limitation to that encountered in cosmological N-body gravitational dynamics, and some of the methods discussed under that heading in chapter 16 may be used to extend the resolution.

4.2 Simple lens models

CIRCULARLY SYMMETRIC LENSES The lenses that have received the most attention are those where the mass distribution appears circularly symmetric on the sky. For these, there is the nice result [problem 4.3] that

$$\alpha = \frac{4G}{c^2} \frac{M(<b)}{b}, \tag{4.13}$$

where $b = D_\mathrm{L}\theta_\mathrm{I}$ is the distance of closest approach (*not* the impact parameter) and $M(<b)$ is the mass seen *in projection* within a radius b. The two most common implementations of the above formula are (i) the point mass and (ii) the singular isothermal sphere. This latter provides a convenient semi-realistic model for a galactic 'heavy halo', satisfying $\rho = \sigma_v^2/(2\pi G r^2)$, $\Sigma = \sigma_v^2/(2Gr)$; the corresponding bend angle is constant at $\alpha = 4\pi\sigma_v^2/c^2$, where σ_v is the line-of-sight velocity dispersion.

The solution of the lensing equation for a symmetrical lens may be visualized graphically as the intersection of a straight line with the bend-angle curve. This is illustrated in figure 4.2, which also shows the two-dimensional image structure; different portions of the object are displaced along radial lines, leading to crescent-shaped images. Note that, in this simple case, image C has parity opposite to that of A, B and the original. For cases of close alignment (i.e. $\theta_\mathrm{S} \to 0$) the principal outer pair of images has

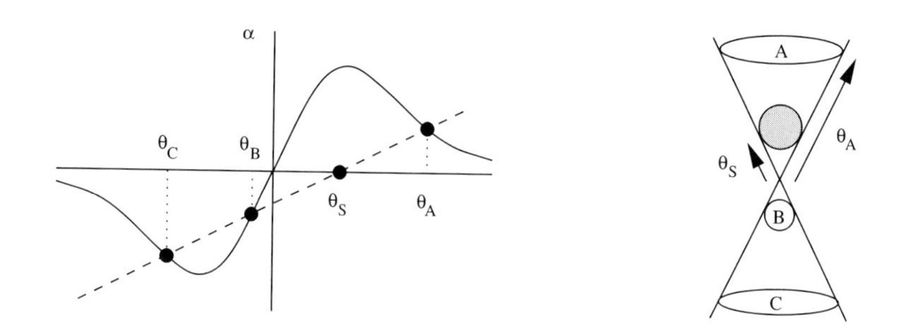

Figure 4.2. The graphical solution (left) and image geometry (right) for a spherical gravitational lens. The shaded circle on the right shows the image position in the absence of the lens. Images A, B and C are produced where there is an intersection between the bend-angle curve $\alpha(\theta)$ and a straight line that crosses the θ-axis at the source radius θ_S. Because the acceleration is radially inwards, different portions of the images are displaced along different radial lines, which often leads to crescent-shaped images.

an angular separation very close to the diameter of the **Einstein ring** formed by perfect alignment. The radius of this ring is the most important characteristic property of the lens, and is known as the **Einstein radius**:

$$
\begin{aligned}
\text{point mass} \quad \theta_E &= \left(\frac{4GM}{c^2}\,\frac{D_{LS}}{D_L D_S}\right)^{1/2} \\
&= \left(\frac{M}{10^{11.09} M_\odot}\right)^{1/2} \left(\frac{D_L D_S / D_{LS}}{\text{Gpc}}\right)^{-1/2} \text{arcsec} \\
\text{isothermal sphere} \quad \theta_E &= \left(\frac{4\pi\sigma_v^2}{c^2}\right) \frac{D_{LS}}{D_S} \\
&= \left(\frac{\sigma_v}{186\,\text{km s}^{-1}}\right)^2 \frac{D_{LS}}{D_S} \quad \text{arcsec.}
\end{aligned} \tag{4.14}
$$

The term 'Einstein radius' usually refers to the case of a point mass. The angle provides the natural way of constructing dimensionless lensing variables. For example, the total amplification for the two images produced when a point-mass lens images a point source that lies at an undeviated angle θ_S from the lens is [problem 4.4]

$$
\boxed{A = \frac{1 + x^2/2}{x\sqrt{1 + x^2/4}},} \tag{4.15}
$$

where $x \equiv \theta_S/\theta_E$. This formula is one of the most useful in gravitational lensing, and is the basis for many practical calculations.

CROSS-SECTIONS It is often of practical interest to calculate the probability of a lensing event. For this, it is useful to have the **lensing cross-section** – i.e. the proper area around a given lens through which the undeviated light ray would need to pass to cause an amplification greater than A of some point source ($= \pi[D_L\theta_S(A)]^2$). The cross-section for a point mass is

$$\sigma(>A) = \frac{8\pi GM}{c^2} \frac{D_L D_{LS}}{D_S} \frac{1}{(A^2-1)+A\sqrt{A^2-1}}. \tag{4.16}$$

For an isothermal sphere, the corresponding result is

$$\begin{aligned} \sigma(>A) &= \left(\frac{4\pi\sigma_v^2}{c^2}\right)^2 \left(\frac{D_L D_{LS}}{D_S}\right)^2 \frac{4\pi}{A^2} \qquad (A>2) \\ \sigma(>A) &= \left(\frac{4\pi\sigma_v^2}{c^2}\right)^2 \left(\frac{D_L D_{LS}}{D_S}\right)^2 \frac{\pi}{(A-1)^2} \qquad (A<2). \end{aligned} \tag{4.17}$$

Note in both cases the factor $D_L D_{LS}/D_S$, which peaks when $D_L \simeq D_{LS}$; the lenses are most effective when midway between source and observer.

We can now work out the probability of obtaining a lens event along a given line of sight to high redshift: $P(>A) = \int n(z)\,\sigma(z,A)\,dr_{\rm prop}(z)$. In the case of both point and isothermal lenses, the high-A dependence is $P \propto 1/A^2$, and the **lensing optical depth**, τ, can be defined as follows:

$$\boxed{P(>A) \rightarrow \frac{\tau}{A^2} \quad \text{as} \quad A \rightarrow \infty.} \tag{4.18}$$

Another way of looking at this is that τ gives the fraction of the sky covered by the Einstein rings of the lenses. For uniformly distributed point masses that make a contribution Ω_L to the density parameter, the optical depth is

$$\tau = \frac{3\Omega_L}{2} \int_0^{z_S} \frac{d_L d_{LS}}{d_S} \frac{1+z_L}{\sqrt{1+\Omega z_L}}\, dz_L, \tag{4.19}$$

where dimensionless angular-diameter distances have been defined by factoring out the Hubble constant: $D_L(z) = (c/H_0)\,d_L(z)$ etc. At high redshift, this shows that $\tau \sim \Omega_L$, and that interesting lensing effects would be common in a highly inhomogeneous universe (Press & Gunn 1973). On the other hand, the constant of proportionality is not very large until we reach very high redshifts: for $\Omega = 1$, $\tau(z=1) \simeq 0.12\,\Omega_L$, rising to $\tau(z=3) \simeq 0.5\,\Omega_L$ (see figure 4.3).

Since the visible cores of galaxies produce $\Omega \sim 10^{-3}$, we might expect roughly one in 10^3 to 10^4 high-z quasars to be strongly lensed. However, because galaxies have extended haloes of dark matter (see chapter 12) it is really necessary to integrate over the galaxy luminosity function with a luminosity–velocity dispersion relation to find the total optical depth on the assumption of isothermal-sphere lenses. The result is

$$\tau \simeq 0.75 \int_0^{z_S} \frac{d_L^2 d_{LS}^2}{d_S^2} \frac{1+z_L}{\sqrt{1+\Omega z_L}}\, dz_L \tag{4.20}$$

(Peacock 1982; Turner, Ostriker & Gott 1984). Again, this suggests that the order of magnitude of the optical depth will be $\tau \sim 1$ at high redshift, but numerical values are in practice somewhat lower (see figure 4.3); for $\Omega = 1$, $\tau(z=1) \simeq 0.01$ and $\tau(z=3) \simeq 0.03$; the effective Ω_L for galaxies is thus approximately 0.07. This higher value reflects the fact that the Einstein rings for galaxies sample the mass distribution at 10–100 kpc radii, where the dark-matter contribution is becoming significant.

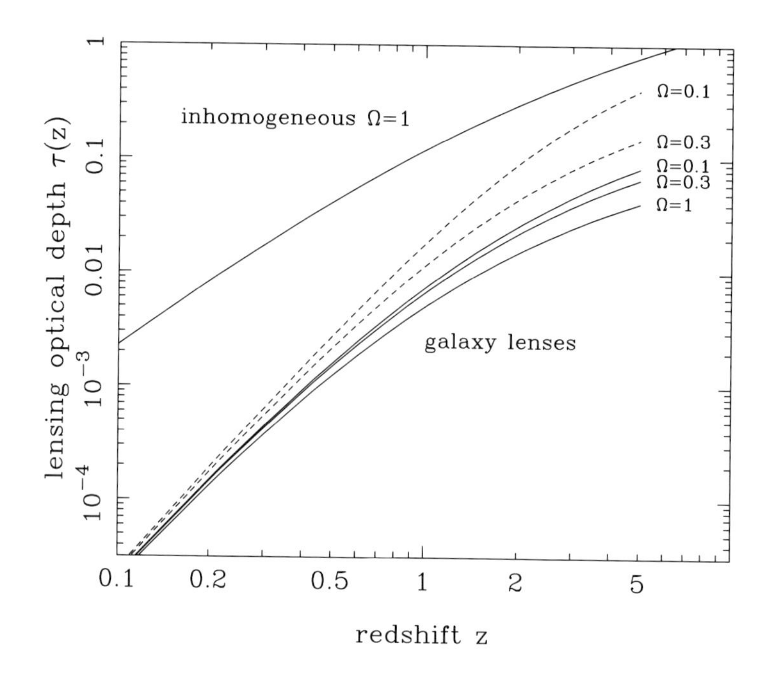

Figure 4.3. The optical depth to gravitational lensing, τ, in various models. The lower set of curves deals with gravitational lensing by the observed galaxy population only; results are shown for zero vacuum energy with $\Omega = 1$, 0.3, and 0.1 (solid lines) and flat vacuum-dominated models with $\Omega = 0.3$ and 0.1 (broken lines). By contrast, the upper solid line shows a totally inhomogeneous universe (all dark matter as point-mass lenses) with $\Omega = 1$. For universes containing a smaller fraction of point-mass lenses, the optical depth scales as $\tau \propto \Omega_{\rm L}$, with non-zero Λ having a similar relative effect on τ as in the case of galaxy lensing. The increased lensing probability for Λ-dominated models is potentially an important signature of vacuum energy.

ELLIPTICAL QUASI-ISOTHERMAL POTENTIALS The circularly symmetric distributions discussed above are special cases, and it is often of interest in practice to consider mass distributions with lower symmetry. A simple way of analysing mass distributions with elliptical symmetry is to write an expression for the two-dimensional effective lensing potential directly, rather than attempting to analyse the effect of some density distribution (Blandford & Kochanek 1988, Kochanek & Blandford 1988). Blandford and Kochanek's potential has the form

$$\psi = \frac{(\Sigma_0/\Sigma_{\rm c})\, s^2}{2\alpha} \left\{ \left[1 + (1-\epsilon)\left(\frac{x}{s}\right)^2 + (1+\epsilon)\left(\frac{y}{s}\right)^2 \right]^\alpha - 1 \right\}. \tag{4.21}$$

This corresponds to a projected mass density with a central value Σ_0, an (angular) core size s, a parameter α describing the degree of concentration of the potential and a flattening parameter ϵ. These four parameters are usually satisfactory for describing most of the restricted lens data likely to be encountered in astronomical reality, especially when a number of components of such form, but with different centres, are superimposed. To see why the coefficient of proportionality in the potential has the quoted form, remember that the potential satisfies the 2D Poisson equation, and that the surface density can be found just by differentiating (this is why it is better to specify the potential than to give an analytical form for Σ).

The conversion from the 2D surface density ($\nabla^2\psi/2$) to the corresponding 3D surface density is sometimes of interest, and this deprojection can be accomplished, for lenses with spherical or transverse cylindrical symmetry, by using a form of the **Abel integral equation** (see e.g. Binney & Tremaine 1987).

$$\begin{aligned}\Sigma(R) &= 2\int_R^\infty \rho(r)\,\frac{r\,dr}{\sqrt{r^2-R^2}}\\ \rho(r) &= -\frac{1}{\pi}\int_r^\infty \frac{d\Sigma}{dR}\,\frac{dR}{\sqrt{R^2-r^2}}.\end{aligned} \tag{4.22}$$

Once the density is known, the gravitational acceleration g can be calculated and related to the isotropic velocity dispersion via hydrostatic equilibrium:

$$\frac{d}{dr}\left(\rho\sigma_v^2\right) = -\rho\, g. \tag{4.23}$$

The forms with $\alpha = 1/2$ are of particular interest, as they are nearly isothermal. For $\epsilon = 0$, and putting $r \equiv (x^2+y^2)^{1/2}/s$, we have the three-dimensional density

$$\frac{\rho}{\rho_0} = \frac{1}{3}\left[\frac{1}{(1+r^2)} + \frac{2}{(1+r^2)^2}\right], \tag{4.24}$$

where the central density and projected density are related by $\Sigma_0 = 2\pi\rho_0 s/3$ (implicitly converting s from angular to linear units). From the density, the corresponding one-dimensional isotropic velocity dispersion is found to be

$$\sigma_v^2 = \sigma_\infty^2\,\frac{2+r^2}{3+r^2} \tag{4.25}$$

and the 'structural length' s equals $(3\sigma_\infty^2/2\pi G\rho_0)^{1/2}$. In the opposite limit, $\epsilon = 1$, the model reduces to the isothermal cylinder (Ostriker 1965), where the velocity dispersion is exactly constant (unlike the 3D case, where the only analytical isothermal model is the singular isothermal sphere with zero core radius):

$$\rho/\rho_0 = (1+2r^2)^{-2}, \tag{4.26}$$

where in this case $\Sigma_0 = \pi\rho_0 s/(2\sqrt{2})$ and $s = [4\sigma_\infty^2/(\pi G\rho_0)]^{1/2}$.

4.3 General properties of thin lenses

TIME DELAYS A potentially important aspect of gravitational lensing is that the lens alters the time taken for light to reach the observer, producing a time lag between any multiple images. There are two effects at work: a geometrical delay caused by the second-order increase in the path length,

$$c\Delta t_g = (1+z_{\rm L})\,\frac{D_{\rm L}D_{\rm LS}}{D_{\rm S}}\,\frac{\alpha^2}{2}, \tag{4.27}$$

where the factor $(1 + z_L)$ converts path increase at the lens to observed time delay (consider figure 4.1). For some time it was thought that this geometrical term alone governed the time delay, but in fact there is a second term due to the gravitational potential (Cooke & Kantowski 1975). The origin of this may be seen by writing the general static weak-field metric in the form

$$c^2 d\tau^2 = \left(1 + \frac{2\Phi}{c^2}\right) c^2 dt^2 - \left(1 - \frac{2\Phi}{c^2}\right)(dx^2 + dy^2 + dz^2), \tag{4.28}$$

where Φ is the Newtonian potential. The reduced coordinate speed of light leads to a potential-delay term:

$$c\Delta t_p = -\int (1 + z_L)\frac{2\Phi}{c^2}\, d\ell, \tag{4.29}$$

where $d\ell$ is the element of proper length. Sadly, although this second term is rather more model-dependent than the geometrical term, it is usually of the same order of magnitude. Were this not the case, it would be easy to estimate H_0 quite accurately.

FERMAT'S PRINCIPLE The error in omitting the potential-delay term becomes most obvious when considering the formulation of gravitational lensing in terms of Fermat's principle (Schneider 1985, Blandford & Narayan 1986), which relies on the fact that images must form where the time delay is stationary (constructive interference). The geometrical delay is stationary only for $\alpha = 0$, in which case there is no lensing. The formulation of lensing in terms of time delays begins by writing the total time delay in terms of the dimensionless 2D lensing potential:

$$\boxed{\tau \equiv \frac{D_{LS}}{(1 + z_L) D_L D_S}\, c\Delta t = \tfrac{1}{2}(\theta_I - \theta_S)^2 - \psi(\theta_I);} \tag{4.30}$$

differentiation with respect to θ_I clearly recovers the basic lens equation, demonstrating the equivalence of Fermat's principle to the geometrical methods used so far.

CRITICAL SURFACE DENSITY Differentiate the lensing equation $\theta_I - \theta_S = \nabla_\theta \psi$ to find the Jacobian:

$$\left(\frac{\partial \theta_I}{\partial \theta_S}\right)_{ij} = \delta_{ij} + \frac{\partial^2 \psi}{\partial \theta_i \partial \theta_j}. \tag{4.31}$$

Because this is symmetric, the answer can always be expressed in the principal coordinate system in terms of the isotropic and trace-free parts (Young 1981):

$$\left(\frac{\partial \theta_I}{\partial \theta_S}\right)_{ij} = \begin{pmatrix} 1 + \gamma - \kappa & 0 \\ 0 & 1 - \gamma - \kappa \end{pmatrix}. \tag{4.32}$$

The Poisson equation for ψ gives us the trace of the Jacobian, yielding

$$\kappa = \frac{4\pi G}{c^2} \int \frac{D_L D_{LS}}{D_S}\, \rho\, d\ell \equiv \frac{\Sigma}{\Sigma_C}. \tag{4.33}$$

The **convergence**, κ, thus measures the ratio of the observed surface density to the **critical surface density**, which is the density that can just produce refocusing of the light beam in the case of zero **shear** ($\gamma = 0$); this is clearly also the condition for multiple imaging. This critical surface density is given by

$$\Sigma_{\rm C} = \frac{c^2 D_{\rm LS}}{4\pi G\, D_{\rm L} D_{\rm S}} \simeq 3.5 \,\frac{(D_{\rm LS}/{\rm Gpc})}{(D_{\rm L}/{\rm Gpc})\,(D_{\rm S}/{\rm Gpc})}\; {\rm kg\,m^{-2}}. \tag{4.34}$$

One reason that gravitational lensing is interesting for cosmology is that this surface density is indeed of the order of that found in clusters of galaxies and the central parts of galaxies.

Of course, the shear is usually non-zero (the surface density has to be circularly symmetric about the light ray for γ to vanish). For a circularly symmetric lens, a necessary and sufficient condition for multiple imaging is that there exists a point with

$$\kappa > \tfrac{1}{2} + \bar{\kappa}, \tag{4.35}$$

where $\bar{\kappa}$ is the mean value of κ over radii internal to the critical point. For a general lens, there is in fact no lower limit to the surface density required to produce multiple imaging; such conditions arise only for lenses constrained by symmetry (Subrahmanian & Cowling 1986). Nevertheless, in practice the criterion $\Sigma_{\rm C}$ does very nearly distinguish those lenses that will produce non-trivial imaging.

The convergence has a further interpretation if one considers a lens formed of a large number of point masses (a galaxy of stars, for example). It is easy to see from our previous discussion of cross-sections that κ also gives the probability for strong lensing by one of the masses in the screen: $\tau \simeq \kappa/A^2$. Hence, a lens that is capable of splitting the image of a background object will also be likely to cause **microlensing**, whereby the exact flux densities of the macroscopic images will be influenced by whether there is a point mass close to the beam. This can produce variability on short timescales as microlenses move past the beam, and allows a powerful probe of the fine structure of the lensing object.

THE ODD-NUMBER THEOREM It is clear from the graphical lensing solution for a symmetric lens that the number of images formed will be *odd*, provided that the lens is (i) **bounded** so that α does not diverge at infinity; (ii) **non-singular** in density so that α is continuous. In fact, this result is independent of the symmetry of the lens. The general proof is rather involved, but can be made plausible by thinking about the wavefront emitted by a point source. This becomes distorted from a sphere by gravitational potential irregularities, until it becomes a convoluted surface. As this surface sweeps out from the source, it will hit an observer several times, with each hit corresponding to a different image. There must be an odd number of hits: one last hit plus two for every fold of the surface that passes through the observer. For a more formal proof, see Burke (1981). It is also clear why singular mass distributions are able to evade the theorem: the infinite depth of their potential wells is able to trap part of the wavefront for an indefinite time, and point mass lenses therefore produce only two images.

CAUSTICS AND CATASTROPHES We have seen above a number of important features of symmetric lens systems: the fact that high-amplification cross-sections behave as $\sigma(>A) \propto A^{-2}$; the fact that new multiple images are created in pairs with infinite amplification. It is natural to ask if such behaviour is *generic* or associated only with

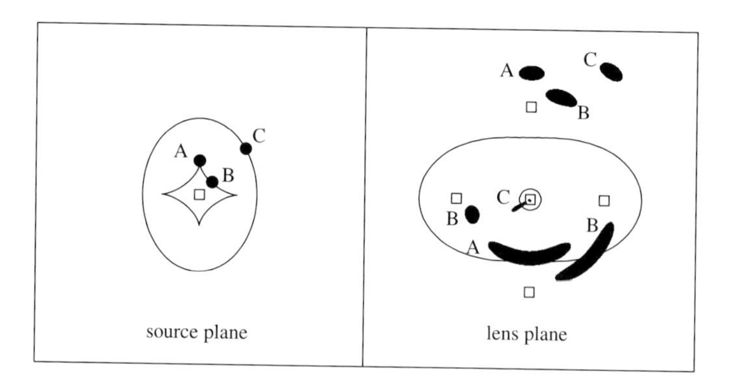

Figure 4.4. The caustics (left panel) and critical lines (right panel) for an example of an elliptical, supercritical isothermal gravitational lens, using the Blandford–Kochanek potential (with $\Sigma_0/\Sigma_C = 20$, $\epsilon = 0.2$, $\alpha = 1/2$; both panels are $60s$ on a side, where s is the core size in the Blandford–Kochanek potential). The outer caustic maps onto the inner critical line, and vice versa. The different circular sources (A, B, C) on the source plane and their mapping onto the image plane illustrate the images produced at different locations for the source. The open squares show the five image positions for an on-axis source, but image area does not indicate amplification in this case.

some special symmetric models. This question was investigated by Blandford & Narayan (1986). From a mathematician's point of view, gravitational lensing is merely an example of a mapping between one plane and another. The properties of such mappings are governed by the area of mathematics known as **catastrophe theory**, and some rather deep insights into the operation of lensing can be obtained in this way.

Consider the mapping from the image plane to the object plane: this is one-to-one, and a unique value $\theta_S(\theta_I)$ exists for every ray back-projected from the observer past the lens; the inverse map, however, is in general many-to-one. Most points on the object plane will yield a finite amplification despite the multiple imaging, but at some locations the Jacobian $\partial\theta_I/\partial\theta_S$ of the transformation will diverge. These **singularities of the mapping** constitute the **caustics**: as sources approach these loci of infinite amplification (which map onto **critical lines** in the image plane), the observer perceives two or more images blending together.

As the sources cross caustics on the source plane, in general a new pair of images will appear at a different location in the image plane, on one of the critical lines. This behaviour is not seen to best advantage in the symmetric lens models, but is illustrated well in figure 4.4, which shows results for an elliptical potential.

The properties of images near caustics depend on the form of the time-delay surface in the caustic's vicinity; the crucial input of catastrophe theory, which has to be simply quoted here, is that the Taylor expansion of the time delay can take one of only two forms, known as the **cusp catastrophe** and the **fold catastrophe**. Armed with this knowledge, Blandford & Narayan (1986) were able to show that the $1/A^2$ behaviour is indeed general.

4.4 Observations of gravitational lensing

MULTIPLE QUASARS The subject of gravitational lensing became a major part of cosmology with the detection of the first multiply-imaged quasar, 0957+561, by Walsh, Carswell & Weymann (1979). This was found as a result of a radio-source identification programme. Two stellar images were found near the location of the radio source, 6 arcsec apart; a reasonable assumption would have been that one was a quasar and that the other was a galactic star, unrelated to the radio source. To the undying credit of the astronomers involved, however, they did not leave this source when the first of the two images proved to be a quasar, but took a spectrum of the second object for completeness. When both turned out to be quasars, with the identical redshifts and line ratios, the case for a lens was very strong, and was quickly confirmed by the discovery of a galaxy between the images (Young *et al.* 1980).

In subsequent years, $\sim 10^4$ quasars have been examined for multiple imaging and about 10 examples of the genre now exist – confirming our rough estimate of the probability of the event. One reason for the relatively slow rate of progress is that pairs of distinct quasars do exist: confirmation of a lens requires good data to prove close similarity of the spectra, together with tight limits on any relative velocity. Despite all this effort, the **double quasar** 0957+561 remains in many ways the best case. This is so because it is a radio source with extended and compact components (see figure 4.5). The compact core appears to lie close to a caustic, as only the base of the quasar jet is multiply-imaged (see chapter 14 for more on jets in active galaxies and quasars). High-resolution data on the milliarcsecond jets in the compact cores allow the parity and shear of the quasar images to be measured, providing additional constraints on the lensing potential. These constraints are needed because the lensing is produced not just by the central galaxy but also by the surrounding cluster, whose dark-matter content is not easy to measure. It should be clear that this will be the case for many of the larger-separation lenses; the splitting from a galaxy alone would be only around 1 arcsec. The majority of the known cases are rather larger than this, and it appears that this is not simply an observational bias against small separations: clusters and galaxies often work together.

RINGS AND ARCS The other class of objects where gravitational lensing has been observed is where the source does not involve an effectively point-like quasar component, but is smooth and extended: either the diffuse lobe of a radio galaxy, or the optical image of a galaxy. About four cases of near-perfect radio Einstein rings with diameter of order 1 arcsec are known.

In the case where an optical galaxy is imaged, the lensing galaxy would prevent the easy recognition of rings of this diameter, but similar events have been found where the lens is a cluster core, producing spectacular **luminous arcs** several tens of arcsec in length. Such events are confined to atypically compact clusters, since most clusters have central surface densities that are below critical. The cluster arcs are of especial interest, as the source is usually a low-luminosity galaxy of high redshift, and the lensing event provides a 'gravitational telescope' to aid the study of objects that would otherwise be

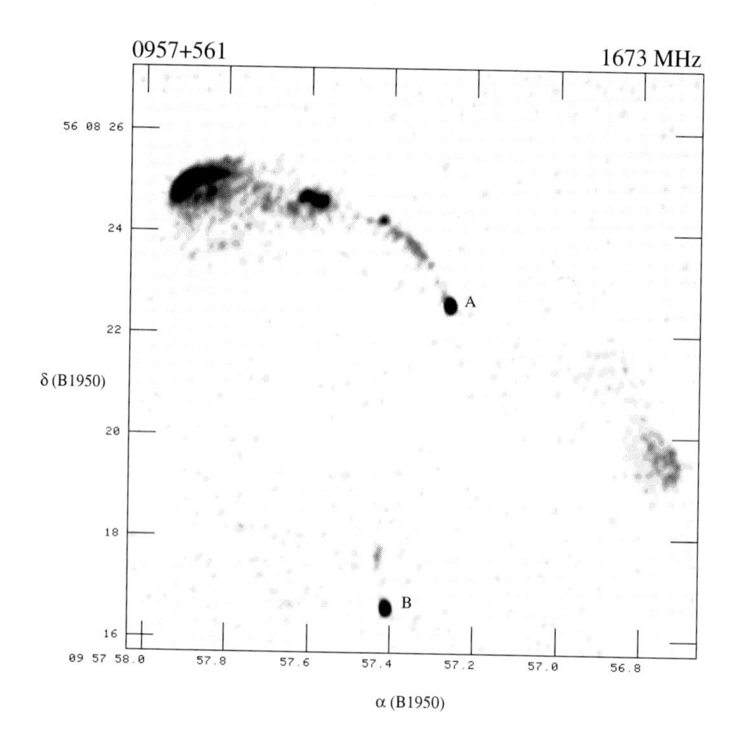

Figure 4.5. The double quasar, 0957+561. The 'A' and 'B' images are images of the same quasar nucleus, whereas the extended structure lies outside the caustic and is imaged only once. Combined with the micro-arcsecond jets seen in the nuclei, this gives a very tight constraint on the lens potential.

barely detectable. Smaller distortions in galaxies further from the cluster centre give further constraints on the cluster potential well. Faint galaxies at high redshift act as a kind of 'cosmic wallpaper', a background screen whose distortions allow us to map the lensing potential over the whole extent of the lens, not just at a few special locations (as with multiply-imaged quasars). Of course, thc background pattern is random, but the surface density of galaxies can become extremely high if very deep images are taken. A statistical knowledge of properties such as the ellipticity distribution of faint galaxies is then sufficient to enable maps of the cluster dark matter to be made (see below).

THE HUBBLE CONSTANT VIA LENSING Gravitational lensing provides a novel route towards two of the major cosmological problems: the distance scale and the weighing of astronomical objects. We have already seen that the time delays between different pairs of images scale as H_0^{-1}, if the imaging geometry and potential structure of the lens are understood. The best case for applying this is 0957+561, where the time delay has been measured at around 415 days (see chapter 5). This system is rather special in a number of ways: both quasars are radio-loud and have milliarcsecond jets for which the position angles and relative sizes can be measured, allowing all components of the magnification matrix to be determined. Also, the system is probably dominated by a single massive galaxy, so that the modelling is made relatively simple. However, the effect of the cluster surrounding the lensing galaxy cannot be neglected (no normal galaxy could produce a splitting of 6 arcsec unaided), and it is the uncertainty in the cluster properties that limits current models.

To illustrate the effect of a cluster of unknown strength, consider the simple model of an isothermal-sphere galaxy of velocity dispersion σ_v surrounded by a screen of uniform surface density and convergence κ. The lens potential and corresponding lens equation are given by

$$\psi = s^2|\theta_{\rm I}| + \kappa\theta_{\rm I}^2/2 \quad \Rightarrow \quad \theta_{\rm I} - \theta_{\rm S} = \pm s^2 + \kappa\theta_{\rm I}, \tag{4.36}$$

where $s^2 = (D_{\rm LS}/D_{\rm S})4\pi\sigma_v^2/c^2$; the $\pm s^2$ term is positive when $\theta_{\rm I}$ has the same sign as $\theta_{\rm S}$, otherwise it is negative. We can now evaluate the time delay, $\tau \propto (\theta_{\rm I} - \theta_{\rm S})^2/2 - \psi$, which may be written as

$$\tau = (s^2 + \kappa\theta)^2/2 - s^2\theta - \kappa\theta^2/2, \tag{4.37}$$

where θ stands for $|\theta_{\rm I}|$. If the two images have measured separations from the central galaxy of θ_1 and θ_2, then the difference in time delay is

$$\delta\tau = s^2(1-\kappa)(\theta_1 - \theta_2) + \kappa(1-\kappa)(\theta_1^2 - \theta_2^2). \tag{4.38}$$

The lens equation gives the image splitting as $\theta_1 + \theta_2 = 2s^2/(1-\kappa)$, allowing κ to be eliminated from the time delay:

$$\tau = s^2(\theta_1 - \theta_2). \tag{4.39}$$

This system contains a degeneracy between σ_v and κ, since the two observables θ_1 and θ_2 cannot determine the three unknowns s^2, κ and $\theta_{\rm S}$. If most of the splitting comes from the galaxy, the time delay is larger than if the identical image configuration is produced with κ being dominant.

The more detailed modelling of the double quasar faces the identical degeneracy; the predicted time delay scales proportional to σ_v^2, and therefore so does the estimated Hubble constant:

$$\boxed{H_0 = (79 \pm 7 \ {\rm km\,s^{-1}Mpc^{-1}}) \left(\frac{\sigma_v}{300 \ {\rm km\,s^{-1}}}\right)^2 \left(\frac{\Delta t}{\rm yr}\right)^{-1}.} \tag{4.40}$$

(Grogin & Narayan 1996). The quoted uncertainty is intended to represent the uncertainty in the theoretical modelling. As an indication of the robustness of this procedure, it is worth comparing an earlier model for this system, which was considered the standard model for a number of years (Falco *et al.* 1991); the value of H_0 quoted there for the same observables was lower by 33%. Observations indicate $\sigma_v \simeq 280\,{\rm km\,s^{-1}}$, implying $h \simeq 0.61$ (see chapter 5). It has been argued that what counts in this formula is the dispersion of the dark matter, which could be a factor $\sqrt{3/2}$ higher (for stars with $\rho \propto r^{-3}$ in a $\rho \propto r^{-2}$ isothermal halo), and this would raise the estimate to $h \simeq 0.72$. In practice, the correction will be smaller than this, since it assumes that the stars are test particles, whereas they are probably gravitationally dominant over the dark matter in the centre of the galaxy. In any case, the total mass in the 6-arcsec tube is well constrained, and all that is uncertain is how this is partitioned between galaxy and cluster. Once the cluster properties can be determined independently (see Kundić *et al.* 1997), lensing should become one of the best methods for determining H_0.

The measurement of masses is in principle rather more precise because, in circular symmetry, only the mass within the Einstein ring contributes to the lensing. Since the total light within the ring is directly observable, we get a measurement of the M/L ratio that is rather robust, and in practice not very sensitive to small departures from symmetry in the lens. The results are again in reasonable agreement with numbers obtained from other techniques: M/L measured in blue light is $\simeq 300h$ Solar units in the cores of luminous-arc clusters (e.g. Squires *et al.* 1996) and about one-twentieth of that for the individual galaxy haloes probed by the radio Einstein rings (e.g. Langston *et al.* 1990). There is however some suggestion that lensing mass estimates may exceed other estimates by a small factor (Miralda-Escudé & Babul 1995); this is discussed further in chapter 12.

4.5 Microlensing

It is commonly the case that lensing bodies have substructure: galaxies are made of stars; the dark matter may consist of elementary particles, and so be effectively a continuum, but it too could consist of discrete bodies. The gravitational-lens properties of a body will be affected by its small-scale structure, and this can in turn be used as a powerful probe of that structure. Consider a quasar observed through the outer parts of a galaxy halo, where the surface density is a small fraction of critical: $\Sigma/\Sigma_c = \kappa \ll 1$. For a smoothly declining density profile, the shear γ will be $\lesssim \kappa$, so that the amplification becomes approximately $A \simeq 1 + 2\kappa$. If the galaxy mass is treated as a continuum, this small amplification of flux is the only effect. If, however, the angular size of the background source is smaller than the Einstein ring of a $1M_\odot$ lens ($\sim 10^{-6}$ arcsec), then it is possible for a close alignment between the source and a star in the galaxy to generate a very large amplification. The corresponding image splitting cannot be observed without micro-arcsecond resolution. The term **microlensing** is thus appropriate, even though it is used irrespectively of the microlens mass to indicate that substructure is important, and that **macrolensing** by the smooth galaxy potential does not give a complete description.

MICROLENSING PROBABILITIES Microlensing is thus a statistical problem: a source seen through a screen of lenses has a probability distribution for its flux, rather than a well-defined value; an interesting question is then how to calculate this probability distribution. A greater challenge is to work out how to verify the prediction, given that the image splittings are unobservably small. This verification is in fact only feasible because very precise alignment is required of a low-mass microlens, if a large amplification is to be obtained. Orbital velocities of source, observer and microlens all alter this alignment, and can lead to a characteristic pattern of apparent variability in the source.

From the earlier results on cross-sections, we can derive the probability distribution for the amplification caused by the microlensing effects of a set of randomly distributed point masses. The number of lenses expected with individual amplification A is

$$N(>A) = \frac{2\tau}{(A^2-1) + A\sqrt{A^2-1}}, \tag{4.41}$$

where the **lensing optical depth** is

$$\tau = \frac{4\pi G}{c^2} \int \frac{D_{\rm L} D_{\rm LS}}{D_{\rm S}} \, \rho \, d\ell. \tag{4.42}$$

This shows that, for a thin screen of microlenses, the optical depth is the same as the convergence $\kappa = \Sigma/\Sigma_{\rm C}$. For large A, the number is small and we just obtain the strong-lensing probability $N \simeq \tau/A^2$, whereas N diverges as A approaches unity. This is quite reasonable, since an analysis based on cross-sections treats each lens independently, and this is an approximation that must break down at large angles from the source, where there are many lenses with roughly the same effect. Since the lens equations are nonlinear, we cannot simply superimpose their effects. The best that can be done analytically is to note that there will always be one lens closest to the line of sight to the source, and to calculate the probability distribution for the amplification due to this object alone. For this, we need Poisson statistics, and obtain

$$\frac{dp}{dA} = \frac{dN}{dA} \exp[-N(>A)], \tag{4.43}$$

which automatically has the correct normalization. For small optical depths $\tau \lesssim 0.1$, this is a reasonable approximation to the observed distribution (Peacock 1986; Pei 1993), but extending the calculation to large optical depths is a challenging numerical problem, since the caustic network becomes extremely complicated for $\tau \simeq 1$, and there are always several lenses strongly affecting every line of sight.

With the above analysis, it is possible to demonstrate explicitly that statistical lensing satisfies flux conservation, at least for low optical depths. Using the nearest-lens distribution, the mean amplification is

$$\langle A \rangle = \int_0^\infty A \, dP(A) = 2\tau \exp(2\tau) K_1(2\tau), \tag{4.44}$$

where K_1 is a modified Bessel function. For small τ, this becomes $\langle A \rangle \simeq 1 + 2\tau$, which can be derived by neglecting the $\exp(-N)$ term in dP/dA altogether (i.e. the spike at $A \simeq 1$ does not give an important contribution to $\langle A \rangle$ in this limit). For a thin lens, we have $A = [(1-\kappa)^2 - \gamma^2]^{-1}$, which just becomes $1 + 2\kappa$ in the linear regime where the second-order quantities κ^2 and γ^2 may be neglected. The statistical effects of microlensing therefore combine to give the correct average amplification. This is also true for a thick lens, but is not so easily demonstrated, even for low optical depths (Weinberg 1976; Peacock 1986).

This statistical distribution of amplifications has important implications for the selection of extragalactic objects, especially quasars. The apparent surface density on the sky of a set of lensed objects is modified to

$$n'(\ln S) = \int_{-\infty}^{\infty} n(\ln(S/A)) \, p(\ln A) \, d\ln A. \tag{4.45}$$

This is just a convolution of the magnitude distribution. If the luminosity function is steep, then the reservoir of intrinsically faint objects available to be lensed up can be very large and it is quite feasible to find $n' \gg n$. Furthermore, such selection effects can mean that the fraction of a flux-density-limited sample which is found to be amplified by a factor $> A$ can greatly exceed the raw probability $P(>A)$ [problem 4.5]. This phenomenon is referred to as **amplification bias**.

A number of workers have investigated the influence of cosmologically distributed lenses on the quasar luminosity function, where the bright-end slope can be as steep as $N(>L) \propto L^{-2.5}$. The main conclusion is that the selection effects of lensing are not important for known (galaxy) lenses. However, if $\Omega \sim 1$ is postulated in lenses massive enough to lens quasar continuum sources coherently ($\gtrsim 10^{-4} M_\odot$), then there are significant effects, both in terms of distortions of the shape of the luminosity function and in the scatter of quasar line-to-continuum ratios. Between them, these arguments limit the possible density in dark compact objects to $\Omega \lesssim 1$ (Canizares 1982). In a related calculation, Canizares (1981) suggested that such effects might be responsible for apparent associations between quasars and low-redshift galaxies, as claimed by Arp (1981). Here, the calculation is much the same, except that we need to introduce an additional factor to allow for the macrolensing of the galaxy, which will dilute the observed number density of background quasars by stretching the separation between them: $n' \to n'/\langle A \rangle$. It remains a matter of some controversy whether this effect has been detected (Webster *et al.* 1988; Narayan 1989).

The gross lensing effects produced by models with $\Omega \sim 1$ in point masses are far greater than the modest effects that are produced by galaxy lenses. Figure 4.3 shows that the probability for strong lensing at $z = 2$ is $\sim 1\%$ – a number that is in practice not greatly boosted by selection effects. An interesting aspect of this calculation is that, for a given galaxy population, the lensing probabilities are about a factor of 2 higher in a vacuum-dominated universe than in an open universe of the same Ω. If the lens population and its evolution can be modelled sufficiently well, this opens the door to using lensing statistics in order to detect a non-zero cosmological constant. Indeed, a number of papers have carried this calculation through in detail, and claimed that the observed number of multiply imaged quasar pairs is too low to be consistent with vacuum-dominated models with $\Omega \simeq 0.2$–0.3 (e.g. Maoz & Rix 1993; Kochanek 1996). However, it has also been claimed that only vacuum-dominated models yield sufficient lenses by comparison with observations (Chiba & Yoshii 1997). The difference in these conclusions stems mainly from observational uncertainty about the lens properties of the local galaxy population, so this method is a promising one for pinning down Λ in the future.

MICROLENSING LIGHT CURVES Let us turn now to the question of microlensing-induced variability. Suppose initially that only the lens is moving, with a transverse velocity v. We can define a dimensionless observed angular velocity that uses the Einstein radius as a length unit and allows for the lens redshift in time-dilating to an observed angular velocity

$$u_{\rm L} = \frac{v_{\rm L}}{(1+z_{\rm L})D_{\rm L}\theta_{\rm E}}. \tag{4.46}$$

If either observer or source has a motion, these also change the lensing geometry; the total u will be the 2D vector sum of all three effects, with the following effective values of u for observer motion and source motion:

$$\begin{aligned} u_{\rm S} &= \frac{v_{\rm S}}{(1+z_{\rm S})D_{\rm L}\theta_{\rm E}} \\ u_{\rm O} &= \frac{v_{\rm O}\, D_{\rm LS}}{(1+z_{\rm L})D_{\rm L}D_{\rm S}\theta_{\rm E}}. \end{aligned} \tag{4.47}$$

The derivation of the factor $(1+z_{\rm L})$ in the last equation is rather subtle. What is needed is the angular-diameter distance of the observer as seen at the lens: this is $R(z = 0)S_k(r_{\rm L})$, whereas $D_{\rm L} = R(z = z_{\rm L})S_k(r_{\rm L})$.

If time is now measured from the point of closest approach, where $\theta_{\rm S} = b\theta_{\rm E}$, then there is a characteristic time–amplification relation, using the previous result for $A(\theta_{\rm S})$:

$$A(t) = \frac{1 + x^2/2}{x\sqrt{1 + x^2/4}}, \quad x(t) = \sqrt{b^2 + u^2t^2}. \tag{4.48}$$

This defines a characteristic symmetric 'bump' in the light curve, described by the two dimensionless parameters b and u. The second of these is the more important, as it allows the mass of the microlens to be estimated, given assumptions about the physical velocities of lens, source and observer, plus knowledge of (or assumptions about) the distances involved ($D_{\rm LS}$ etc.).

There are two rather different sets of circumstances in which microlensing has been detected. The first involves quasars that are strongly macrolensed by a foreground galaxy. Here, microlensing is rather likely, since any image splitting requires a roughly critical surface density; each macroimage therefore has $\tau \sim 1$ and the probability of microlensing is high provided the mass of the galaxy is in lumps rather than being smooth dark matter. The most clear-cut case is the 'cloverleaf lens' 2237+030, where four images are seen through the centre of a bright low-redshift galaxy. Here the mass must be dominated by stars, and so microlensing is inevitable. The signature of microlensing would be that one of the four macro-images varies in a way that is not correlated with variations in the other images (the relative time delay is only a few days in this system). Irwin *et al.* (1989) indeed reported such an isolated event with a timescale of order 100 days, implying a mass $\sim 0.01\, M_\odot$ for the microlens. However, the interpretation of such high-τ configurations is complicated: typically, several microlenses have an effect simultaneously, so that the variability is not simply an isolated flare of the above form.

MICROLENSING AND GALACTIC DARK MATTER A more challenging application is to look for the microlensing of stars by microlenses within the Milky Way. The reason this is interesting is that τ for lines of sight out of the galaxy will be dominated by the contribution of dark matter in the galactic halo. There is therefore the opportunity to determine whether the galactic dark matter occurs in discrete lumps by monitoring large numbers of stars in a convenient external target such as the Large Magellanic Cloud (LMC) and searching for variability with the expected symmetric light curve. Such an experiment was proposed by Paczynski (1986), and a number of groups undertook to carry it out.

It is immediately clear that the optical depth to lensing will be tiny in this case, and that searches for microlensing events need very good rejection against noise spikes and various forms of intrinsic variability. A rough idea of τ may be obtained from the expression for the optical depths through an isothermal sphere at a distance r from the centre:

$$\tau = 2\pi\, \frac{\sigma_v^2}{c^2}\, \frac{D_{\rm L} D_{\rm LS}}{r\, D_{\rm S}}. \tag{4.49}$$

When looking out from within an isothermal sphere to distances a few times r, we clearly need an expression of similar form, so that $\tau \simeq \sigma_v^2/c^2 \sim 10^{-6}$. A properly detailed calculation (see Griest 1991) gives $\tau \simeq 5 \times 10^{-7}$ for the line of sight to the LMC, assuming all the dark matter is capable of microlensing. For the known stars alone, the optical depth is smaller by a factor of about 10. To detect microlensing events, then, it is necessary to monitor several million stars; it is a tribute to the capabilities of modern observational astronomy that this has been done successfully by several groups.

How does the event rate relate to the optical depth? Here we want to classify events by the maximum amplification undergone in a given time, so the calculation is a little different from the previous probabilistic sums, which were concerned with the number of events above a given amplification at a single time. This change is quite easy: suppose we are concerned with events having 'impact parameters' (minimum θ_S/θ_E) within some value b. Now imagine a background star to be moving at velocity u; in time t the star sweeps out a strip of width $2b$ whose area is just $2but$. If the motions of the lens or observer dominate, the variability can still be thought of as arising from the source motion, but with an effective value of u, as described above. Since the fraction of the sky covered by Einstein rings is τ, this implies that the number of lenses in one steradian of the sky is $\tau/(\pi\theta_E^2)$, which is just τ/π per unit area measured in Einstein-ring length units. The microlensing event rate is therefore

$$\boxed{\Gamma = \frac{2bu\tau}{\pi}.} \tag{4.50}$$

Note that this scales linearly with b, unlike the single-time probability, which depends on b^2. Finally, converting as above between b and $A_{\rm max}$,

$$\begin{aligned} b/\sqrt{2} &= \left[(A_{\rm max}^2 - 1) + A_{\rm max}\sqrt{A_{\rm max}^2 - 1}\right]^{-1/2} \\ &= \left[A_{\rm max}/\sqrt{A_{\rm max}^2 - 1} - 1\right]^{1/2}, \end{aligned} \tag{4.51}$$

so that $\Gamma \simeq 2^{3/2}u\tau/(\pi A_{\rm max})$ for large $A_{\rm max}$. This slow decline with $A_{\rm max}$ means that significant numbers of events with very high peak amplifications are expected.

Several groups have attacked the problem of detecting this signature, spawning a variety of lurid acronyms (**MACHO**, massive astrophysical compact halo object; **OGLE**, optical gravitational lensing experiment). Several candidate events corresponding to the above analysis have been seen towards the LMC. The MACHO consortium (Alcock *et al.* 1993) describe an extremely good example with $b = 0.15$, $u^{-1} = 17$ days (see figure 4.6). For typical physical transverse velocities of perhaps $200\,{\rm km\,s^{-1}}$, this u value is as expected for a lens mass of perhaps $0.1 M_\odot$, although this depends on exactly where the lens is positioned along the line of sight. Moreover, a large number of events have been seen by the OGLE collaboration in the direction of the galactic bulge, where the optical depth is much higher: $\tau \simeq 2 \times 10^{-6}$ (Udalski *et al.* 1994). These results demonstrate that such experiments can detect the predicted rate of events with high efficiency down to the masses that can lens background stars over a detectable timescale $\gtrsim 10^{-3} M_\odot$. At the time of writing, the results on the microlensing rate towards the LMC are puzzling, as they favour approximately 50% of the halo mass being in microlenses of mass typically $0.5\ M_\odot$ (Alcock *et al.* 1996). These are certainly not dark stars below the nuclear-burning limit; it is possible that this result represents the detection of an unexpectedly large population of white dwarfs, although exotica such as small black holes are also a logical possibility. The key question is whether the statistics firmly rule

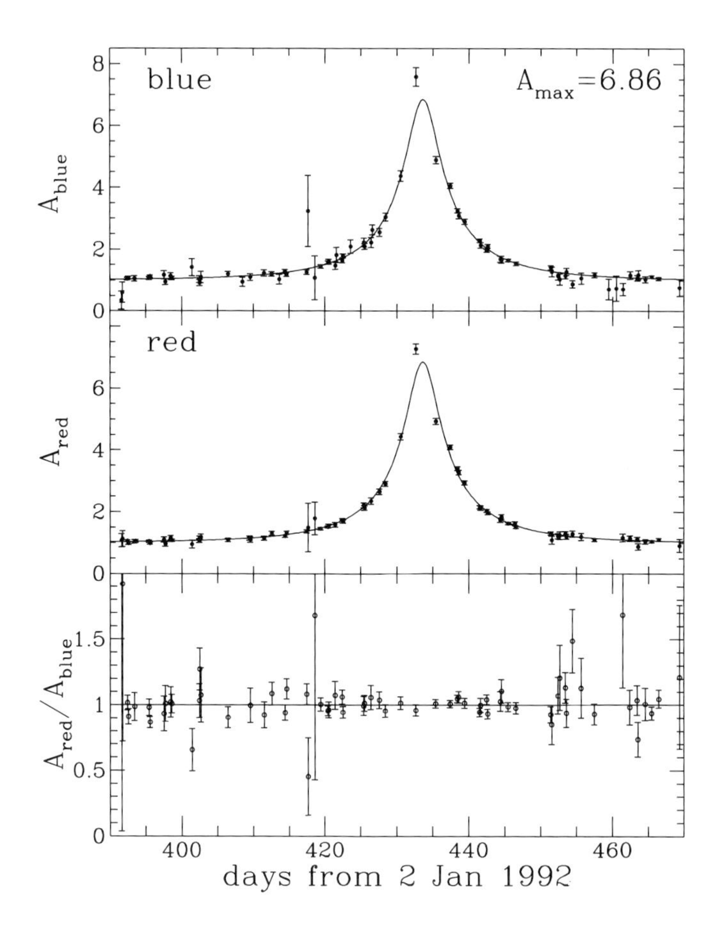

Figure 4.6. A light curve from a star in the Large Magellanic Cloud that is apparently being lensed by a foreground object in the halo of the Milky Way (Alcock *et al.* 1993). Note the very good fit of the (two-parameter) lens model and the high degree of achromaticity.

out the possibility that all the halo mass could consist of MACHOs; if confirmed, this would be the first direct indication that the dark matter may indeed be non-baryonic in form.

4.6 Dark-matter mapping

Perhaps the most exciting application of gravitational lensing in cosmology has been the ability to visualize the dark matter in clusters of galaxies. In the central 10–100 kpc of a cluster, the surface density may be above critical and luminous arcs can be used to measure directly the mass within this central region (see figure 4.7). Nevertheless, in many ways it is the mass profile at larger radii that is of most interest; this is where most of the mass lies, and where we would most like to be able to study the relative

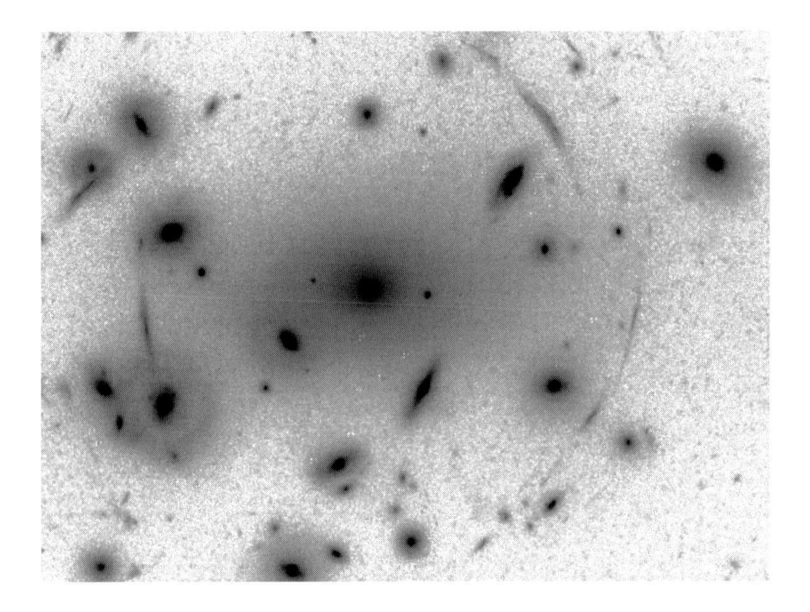

Figure 4.7. An image of the central 1.3 × 1.0 arcmin of the rich galaxy cluster A2218, taken with the Hubble Space Telescope (see Kneib *et al.* 1996). Note the spectacular arcs resulting from strong lensing of background galaxies. At larger radii, the gravitational shear is weak and less obvious to the eye, but the mean 'polarization' of the galaxy field can still be detected by averaging over many galaxies, allowing the dark matter to be mapped to large radii (see Squires *et al.* 1996 for a shear-based map of A2218).

distributions of mass and light. Tyson, Wenk & Valdes (1990) had the great insight that the mass could still be mapped in this weak-lensing regime, by exploiting the high surface density of faint background galaxies. Although the individual images can be rather elliptical, the distribution of their orientations should be random. However, any shear introduced by the cluster potential will produce a weak alignment, which can be detected by averaging over many images. This produces a map of the cluster potential limited in spatial resolution only by the depth of the image being used and the number of background galaxies that result. This insight was carried to completion by Kaiser & Squires (1993), with spectacular and important results.

The main step is to ask how the image of a circular source is distorted by the cluster potential. The surface brightness is only affected by the coordinate transformation of positional translation

$$I'(\boldsymbol{\theta}) = I(\boldsymbol{\theta} - \boldsymbol{\nabla}\psi), \tag{4.52}$$

where ψ is the usual lensing potential. Now define the second derivative matrix $\psi_{ij} = \partial^2\psi/\partial\theta_i\partial\theta_j$; in terms of this, the Jacobian matrix of the coordinate transformation that lensing applies to the sky is

$$\frac{\partial\theta'_i}{\partial\theta_j} = J_{ij} \equiv \delta_{ij} + \psi_{ij}. \tag{4.53}$$

This is symmetric, so the lens mapping contains no rotational component. The change of image area is given by the Jacobian determinant, which to first order is just the trace of **J**,

$$A \simeq 1 + \psi_{11} + \psi_{22} = 1 + \nabla^2\psi = 1 + 2\Sigma/\Sigma_c. \tag{4.54}$$

This overall amplification is not so easily detected, since galaxy number counts scale roughly as $N \propto (\text{flux})^{-1}$, and the increase in numbers through greater depth is compensated by the $1/A$ decrease of the galaxy surface density, which arises because the lens stretches the observed angular distances between galaxies. These effects can, however, be separated if redshift information exists for the background galaxies; the lens induces a systematic distortion of the galaxy luminosity function, which can be used to infer the dark matter density (Broadhurst, Taylor & Peacock 1994).

If we subtract the term corresponding to the trace of **J**, this leaves as an observable effect the **shear distortion**

$$\frac{\partial\theta'_i}{\partial\theta_j} = S_{ij} \equiv \delta_{ij} + \psi_{ij} - \tfrac{1}{2}\delta_{ij}\nabla^2\psi = \begin{pmatrix} 1+A & B \\ B & 1-A \end{pmatrix}, \tag{4.55}$$

involving the two independent components $A \equiv (\psi_{11} - \psi_{22})/2$ and $B \equiv \psi_{12}$. What effect does this shear have on a circular background source? To answer this question, it is easiest to start in the frame where **S** is diagonal and consists only of a stretch along the x-axis and a compression along the y-axis.

$$\mathbf{S} = \begin{pmatrix} 1+\gamma & 0 \\ 0 & 1-\gamma \end{pmatrix}. \tag{4.56}$$

To transform to a general set of axes involves the rotation matrix **R** that describes a rotation by θ, under which S transforms like a tensor

$$\mathbf{S} = \mathbf{R} \begin{pmatrix} 1+\gamma & 0 \\ 0 & 1-\gamma \end{pmatrix} \mathbf{R}^{\mathrm{T}}. \tag{4.57}$$

Evaluating this matrix product tells us that the principal shear and position angle of the ellipse relate to the derivatives of ψ as follows

$$\begin{aligned} A &= \gamma\cos 2\theta = (\psi_{11} - \psi_{22})/2 \\ B &= \gamma\sin 2\theta = \psi_{12}. \end{aligned} \tag{4.58}$$

In practice, the shear components of an elliptical image would be estimated from the quadrupole moments of the intensity (taking moments about the image centroid, so that first-order moments vanish):

$$Q_{ij} = \int I(\boldsymbol{\theta})\, \theta_i\theta_j\, d^2\theta. \tag{4.59}$$

The components of the matrix Q_{ij} again obey the tensor transformation law under rotation, so we can repeat the above idea of rotating from the coordinate frame in which the long axis of the elliptical image is along the x-axis: $Q_{ij} = \text{diag}(a^2, b^2)$ $(a > b)$. If we divide by the invariant Tr $Q/2 = (a^2 + b^2)/2$, then $Q_{ij} = \text{diag}(1+\epsilon, 1-\epsilon)$, where the distortion parameter ϵ is

$$\epsilon \equiv \frac{a^2 - b^2}{a^2 + b^2}. \tag{4.60}$$

Note that this is not the **ellipticity**, which would be $1-(b/a)$, nor the **eccentricity**, which would be $[1-(b/a)^2]^{1/2}$. Rotating by an angle θ as before now gives

$$\mathbf{Q} = \begin{pmatrix} 1+\epsilon\cos 2\theta & \epsilon\sin 2\theta \\ \epsilon\sin 2\theta & 1-\epsilon\cos 2\theta \end{pmatrix}, \tag{4.61}$$

and ϵ and θ are related to the Q_{ij} via a two-component 'polarization' (although it does not transform as a vector)

$$\begin{pmatrix} e_1 \\ e_2 \end{pmatrix} = \begin{pmatrix} \epsilon\cos 2\theta \\ \epsilon\sin 2\theta \end{pmatrix} = \begin{pmatrix} [Q_{11}-Q_{22}]/[Q_{11}+Q_{22}] \\ 2Q_{12}/[Q_{11}+Q_{22}] \end{pmatrix}. \tag{4.62}$$

The physical meaning of the two components e_1 and e_2 is that pure e_1 is an ellipse with position angle 0 or 90°, whereas pure e_2 is an ellipse with position angle 45°.

Now, for the case of a circular image undergoing gravitational shear, $a \propto 1+\gamma$, $b \propto 1-\gamma$, so that $\epsilon = 2\gamma/(1+\gamma^2) \simeq 2\gamma$ in the linear regime. When the shear is weak, there are therefore two observational quantities that relate directly to shear and give the derivatives of the lensing potential:

$$\boxed{\begin{pmatrix} e_1 \\ e_2 \end{pmatrix} = \begin{pmatrix} \psi_{11}-\psi_{22} \\ 2\psi_{12} \end{pmatrix}.} \tag{4.63}$$

These equations can be solved to find the surface density Σ. It is easiest to proceed in 2D Fourier space, denoting angular wavenumber by $\mathbf{k} = (k_1, k_2)$, and Fourier-transformed quantities by a k subscript. The transformed shear equation is then

$$\begin{pmatrix} e_1 \\ e_2 \end{pmatrix}_k = \begin{pmatrix} (k_1^2-k_2^2)/k^2 \\ 2k_1k_2/k^2 \end{pmatrix} \frac{\Sigma_k}{\Sigma_c/2} \equiv \begin{pmatrix} \chi_1 \\ \chi_2 \end{pmatrix} \frac{\Sigma_k}{\Sigma_c/2}, \tag{4.64}$$

where ψ has been eliminated by using the k-space version of the lensing Poisson equation. The two components of this equation yield independent estimates of the density, proportional to e_1/χ_1 and e_2/χ_2. Since the errors on e_1 and e_2 should be equal, we should combine these estimates with the usual reciprocal-variance weighting, to obtain the final result for the estimator of Σ as the dot product of the χ and e 'vectors':

$$\boxed{\hat{\Sigma}_k = \frac{\Sigma_c}{2} \begin{pmatrix} \chi_1 \\ \chi_2 \end{pmatrix} \cdot \begin{pmatrix} e_1 \\ e_2 \end{pmatrix}_k} \tag{4.65}$$

This equation can be multiplied by a window function, to yield (after transforming) a map of the dark matter with any desired resolution. Since the equation is in the form of a k-space product, it is just a convolution in real space, and the reconstruction is presented in this form by Kaiser & Squires (1993). Since the observables are related to the shear only, this technique has the defect that an arbitrary constant may be added to the derived Σ, i.e. we have to impose a sensible criterion such as the vanishing of Σ at the boundary of the data. Images of sufficient size can usually be obtained, so that this is not a problem in practice.

All the above analysis assumes that we have available a network of circular background sources. Real galaxies are of course intrinsically elliptical, but the net polarization can be measured by averaging over a number of galaxies in a cell (whose size sets the resolution of the mass map). At least when the galaxies themselves are weakly elliptical, the polarization parameters e_i add linearly, so $\langle e_i \rangle$ can be estimated accurately. A practical application of the technique has to consider the following complications. The limited resolution of a telescope (the 'seeing') will circularize the images, systematically reducing the polarization; real galaxies are also sometimes highly elliptical. These effects can be simulated, and imply that good ground-based images can yield $\langle e_i \rangle$ values about 70% of ideal. Since $\langle |e_i|^2 \rangle \simeq 0.5$, the fractional error on the reconstructed density is of order $0.5/\sqrt{n}$, when averaging over n galaxies. Attainable surface densities are $\sim$ 10–100 objects per arcmin^2, so an rms error of order one-tenth of the critical density can be obtained with a pixel size of 0.1 arcmin: $\sim 100h^{-1}$ kpc for a lensing cluster at $z = 0.2$. Some impressive results have emerged from this technique, as applied by e.g. Squires *et al.* (1996). The studies to date suggest that the mass distribution at radii of order several hundred Mpc follows the distribution of galaxies; this kind of constraint will be vital in attempting to test models of biased galaxy formation (see chapter 17).

Problems

(4.1) Give a simple proof that surface brightness is conserved in Euclidean space. Now show that volume elements in phase space are Lorentz invariant, and hence that the relativistic expression of surface-brightness conservation is that I_ν/ν^3 is an invariant. Show from this that black body radiation appears thermal to all observers. If this is so, how is it possible to use the microwave background to determine that the Earth has an absolute velocity of $\simeq 370 \text{ km s}^{-1}$?

(4.2) Show that the lensing potential ψ obeys a two-dimensional version of Poisson's equation. Is the use of a lensing potential general, or is it limited to geometrically thin lenses?

(4.3) Derive the gravitational deflection formula $\alpha = 4GM(< r)/(c^2 r)$ for a mass distribution that is circular in projection, where M is the mass within a projected radius r (remember only the projected density matters). Show that this formula yields $\alpha = 4\pi\sigma_v^2/c^2$ for the isothermal sphere.

(4.4) Show that the total amplification of a point source produced by a point lens is

$$A = \frac{1 + x^2/2}{x\sqrt{1 + x^2/4}}, \tag{4.66}$$

where $x \equiv \theta_{\rm S}/\theta_{\rm E}$. What are the amplifications of the individual images?

(4.5) If a population with a power-law luminosity function $N(> L) \propto L^{-\beta}$ is subjected to statistical gravitational lensing by an optical depth τ of point lenses, calculate the factor by which the observed numbers of objects is boosted and the amplification distribution for a given observed L.

5 The age and distance scales

5.1 The distance scale and the age of the universe

The most striking conclusion of relativistic cosmology is that the universe has not existed forever, and that only a finite period of time separates humanity from the extremes of energy and density in which the expanding cosmos began. But just how long ago was this? The age of the universe is largely determined by the rate at which it expands, and the current value of the Hubble 'constant' fixes the **Hubble time**

$$H_0^{-1} = 9.78\, h^{-1}\ \mathrm{Gyr}, \tag{5.1}$$

which is the basic timescale for the universe (recall the definition of the dimensionless Hubble parameter: $h \equiv H_0/100\,\mathrm{km\,s^{-1}Mpc^{-1}}$).

If the expansion did not decelerate, the Hubble time would be exactly the age of the universe: $v = Hd$ and $d = vt$, so $t = 1/H$. Chapter 3 showed that, ignoring radiation, the dynamical result for the age of the universe may be written as

$$H_0 t_0 = \int_0^\infty \frac{dz}{(1+z)\sqrt{(1+z)^2(1+\Omega_m z) - 2(2+z)\Omega_v}}. \tag{5.2}$$

Over the range of interest ($0.1 \lesssim \Omega_m \lesssim 1$, $|\Omega_v| \lesssim 1$), this exact answer may be approximated to a few % accuracy by

$$\boxed{H_0 t_0 \simeq \tfrac{2}{3}\left(0.7\Omega_m + 0.3 - 0.3\Omega_v\right)^{-0.3}.} \tag{5.3}$$

If the density content of the universe is known, then H_0 and t_0 are directly related. Alternatively, if both H_0 and t_0 are known, then we measure a combination of Ω_m and Ω_v.

In practice, all these numbers are frustratingly uncertain. Ages of astronomical objects can be measured by a variety of methods, discussed below, but these give only a lower limit to t_0. The expansion rate H_0 can be measured only if we have absolute distances to objects at significant redshifts. This is the other role of H_0: as well as setting the characteristic age of the universe, it also sets its characteristic size. The distance to an object of known redshift is always of the form

$$D = \frac{c}{H_0}\, z\, f(z, \Omega), \tag{5.4}$$

where f is close to unity provided z is small.

The problem of determining the Hubble constant can be viewed as one of bridging the divide between two extremes. We wish to apply $H_0 = v/r$, and just one object for which both v and r are accurately determined will suffice. Nearby objects may have their distances measured quite easily, but their radial velocities are dominated by deviations from the ideal Hubble flow, which typically have a magnitude of several hundred $\mathrm{km\,s^{-1}}$. Although the Earth–Moon distance is known to high precision, this separation will not increase as the universe expands. However, objects at redshifts $z \gtrsim 0.01$ will have observed recessional velocities that differ from their ideal values by $\lesssim 10\%$, but absolute distances are much harder to supply in this case. The traditional way of presenting the solution to this problem is the construction of the **distance ladder**: an interlocking set of methods for obtaining relative distances between various classes of object, which begins with absolute distances at the 10–100 pc level and terminates with galaxies at significant redshifts. This is a vast subject with a long history, about which whole books have been written (Rowan-Robinson 1985). The intention here is to concentrate on a restricted set of more recent results and techniques, which hold the promise of bringing some peace to this turbulent and controversial topic (see e.g. Jacoby *et al.* 1992; VandenBergh 1996). We shall also invert the common order of presentation and start with the step from distant to nearby galaxies. This divides the problem into two: first transfer the *velocity* scale from large redshifts, so that we have some local galaxies whose ideal Hubble-flow redshifts are known; we then have to find some method of obtaining absolute distances for these. This approach reflects what is done in practice, and also has the merit of emphasizing the close relation between studies of the distance scale and of peculiar velocity fields.

5.2 Methods for age determination

NUCLEAR COSMOCHRONOLOGY The most accurate means of obtaining ages for astronomical objects is based on the natural clocks provided by radioactive decay (see e.g. Fowler 1988; Cowan, Thielemann & Truran 1991). There exist a number of heavy nuclei with decay lifetimes of order 10 Gyr (not the common 'half-life': this is smaller by a factor $\ln 2$). The most useful decay clocks are based on thorium and uranium: $^{232}\mathrm{Th} \to {}^{208}\mathrm{Pb}$ (20.27 Gyr); $^{235}\mathrm{U} \to {}^{207}\mathrm{Pb}$ (1.02 Gyr); $^{238}\mathrm{U} \to {}^{206}\mathrm{Pb}$ (6.45 Gyr). The use of these clocks is complicated by a lack of knowledge of the initial conditions of the decay. Suppose (as is effectively the case in the above examples) that a given 'parent' (P) element decays only to a single 'daughter' (D), and that this daughter is produced by no other reaction. Once a sample of material is isolated from the nuclear reactions that produce the heavy elements (largely in supernovae), the abundances, by mass, say, of the two elements concerned satisfy

$$D = D_0 + P_0[1 - \exp(-t/\tau)] = D_0 + P[\exp(t/\tau) - 1]. \tag{5.5}$$

The quantities D and P are observed and τ is known; however, the age cannot be found unless the initial daughter abundance D_0 is known.

The way round this impasse is to exploit what at first sight seems to be a complication. Once solid bodies condense from a cloud of chemically uniform material, chemical fractionation produces spatial variations in P_0/D_0. Now pick a stable isotope S of D and express all abundances as a ratio to S; at the initial time, D_0/S will be a constant, whereas P_0/S will have a range of values. As a result, both D/S and P/S

will have a range of values at some later time, but the scatter in these variables will be correlated:

$$D/S = D_0/S + (P/S)[\exp(t/\tau) - 1]. \tag{5.6}$$

The age of a sample can thus be determined from the slope of a plot of P/S against D/S. This gives the time at which spatial homogeneity in P/S was first broken, and corresponds to the time of solidification of a given rock. It is only the introduction of a scatter in P/S with no scatter in D/S that makes the age measurement possible at all. When applied to meteorites, these methods give a highly precise age for the Solar system: this formed 4.57 Gyr ago, with an uncertainty of only about 1%. The oldest rocks on Earth are slightly younger: about 3.7 Gyr.

Such studies tell us not only the age of the Solar system, but also the abundances in the pre-Solar material: $^{235}U/^{238}U \simeq 0.33$; $^{232}Th/^{238}U \simeq 2.3$. Can we use these numbers to date the time of production of the heavy elements, and hence obtain an age for the galaxy itself? This can only be done if we know what abundance ratios are produced in the initial supernova explosion, which requires an input from nuclear physics theory.

Heavy elements this close to the end of the periodic table are not synthesized in stars during normal burning, and are thought to be made during a supernova explosion, when the neutron flux is high. As such, they form via the **r process**, in which neutron capture is 'rapid' in the sense that it is more probable that a nucleus will capture a further neutron than undergo decay. The opposite ('slow') possibility, where nuclei only rarely capture a further neutron prior to decay, is the **s process**. By solving the rate equations for the abundances of the massive elements, it is possible to predict the relative abundances of the massive elements. This was first done by Burbidge *et al.* (1957); a modern version of the calculation gives $^{235}U/^{238}U \simeq 1.3$ and $^{232}Th/^{238}U \simeq 1.7$, with an uncertainty in the calculation of order 10% (Fowler 1988). Comparing with the inferred primordial Solar-system abundances tells us something about how old the elements were when the Solar system formed. It does not say everything, because of the unknown time distribution of the star-formation history. We know that this has to be more complex than a single spike of activity at one time, because the uranium and thorium ratios do not give a consistent age: the former implies 1.6 Gyr, whereas the latter gives 2.9 Gyr. A very safe lower limit to the age of the galaxy comes from adding the latter figure to the Solar-system 4.6 Gyr to obtain 7.5 Gyr. However, the uranium figure says, unsurprisingly, that there must have been some later activity. Fowler (1988) solves for a model consisting of an initial spike followed by uniform activity, and obtains an age of 5.4 ± 1.5 Gyr (an exponentially declining rate of element formation would give very similar numbers). The age of the local part of the Milky Way on this basis is 10.0 ± 1.5 Gyr; The main source of uncertainty is in the theoretical production abundances.

The same calculations can be performed using observations of chronometer elements in other stars. Butcher (1987) found an unvarying ratio of Th/Nd, which sets a limit to the maximum age possible for a model of roughly constant element production. His limit was a maximum of 9.6 Gyr at 95% confidence, although it increased by about 2 Gyr for an exponentially declining production rate. This approach agrees reasonably well with that based on Solar-system abundances.

In principle these estimates of the age of the galaxy should be believable, since they are based on the straightforward physics of nuclear decay. The complications of astrophysical messiness cannot be avoided entirely, however, and there are two significant difficulties. The first is that the calculations of the production abundances are uncertain not through theoretical difficulties but through a lack of knowledge about the physical circumstances that apply. Changing the temperature and density by amounts corresponding to the difference between different plausible sites for nucleosynthesis has a non-negligible effect on the abundances. Also, the history of nucleosynthesis is almost certainly more complex than the simple models that are often used. How much all of this matters is somewhat controversial; the above treatment is optimistic in the sense that it assumes that the deviations from simple models will not be too disastrous. For a much more pessimistic view, see Arnould & Takahashi (1990).

AGES FROM STELLAR EVOLUTION The other major means of obtaining cosmological age estimates is based on the theory of stellar evolution. This is a highly developed branch of astrophysics, and only a very brief outline of the arguments can be given here. For further reading, consult e.g. Clayton (1983); Collins (1989); Kippenhahn & Weigert (1990); Tayler (1994). The book by Celnikier (1989) has a discursive and informal approach that provides an attractive contrast to more standard treatments.

Consider figure 5.1, which is a colour–magnitude diagram for a **globular cluster** – a system of roughly 10^5 stars. These systems are found in the outer parts of galaxy haloes, and are believed to be examples of a stellar population at a single age (**coeval**). The colour–magnitude diagram is the practical version of what can be predicted from theory: the **Hertzsprung–Russell diagram** (HR diagram) or plot of luminosity against effective temperature. On this plot, we can see very clearly the main elements of the life-cycle of a star, as follows.

(1) The **main sequence** is the locus extending to low luminosities and red colours. This is a line parameterized by mass, where massive stars are bluer and more luminous. At the time of creation of the cluster, only the diagonal line of the main sequence would have been occupied. The approximate form of the sequence can be deduced [problem 5.1] with no knowledge of the nuclear reactions that power stars, assuming only virial equilibrium and that the main scattering mechanism in stars is Thomson scattering (true for the more massive examples). The luminosity, temperature and size are then expected to scale as follows: $L \propto M^3$; $T \propto M^{1/2}$; $R \propto M^{1/2}$. A star will shine on the main sequence until its fuel is exhausted, which corresponds to having fused a significant fraction of its total mass. The lifetime is then given by $L\tau \propto Mc^2 \Rightarrow \tau \propto L^{-2/3}$. The most luminous stars therefore leave the main sequence soonest.

(2) The giant branch. Once hydrogen is exhausted in the core, it continues to burn in a shell around the core. The helium-rich core contracts until it is supported by electron degeneracy pressure, as in white dwarfs. This phase is associated with an expansion in size of the outer parts of the star by roughly a factor 10 to form a red giant. There is thus a **turnoff point** on the main sequence, corresponding to stars that have just reached this point. In the initial phases of the process, the turnoff stars are subgiants and become somewhat cooler at constant luminosity, before increasing luminosity greatly at roughly constant temperature as they ascend the giant branch proper. It would be nice if there existed a simple reason why giant stars increase in size so much, but one of the embarrassments of the subject is that the behaviour seems more complicated the more closely it is examined. One answer, but not a very satisfying one, is that this is predicted by a solution of the equations of stellar structure when the appropriate boundary conditions

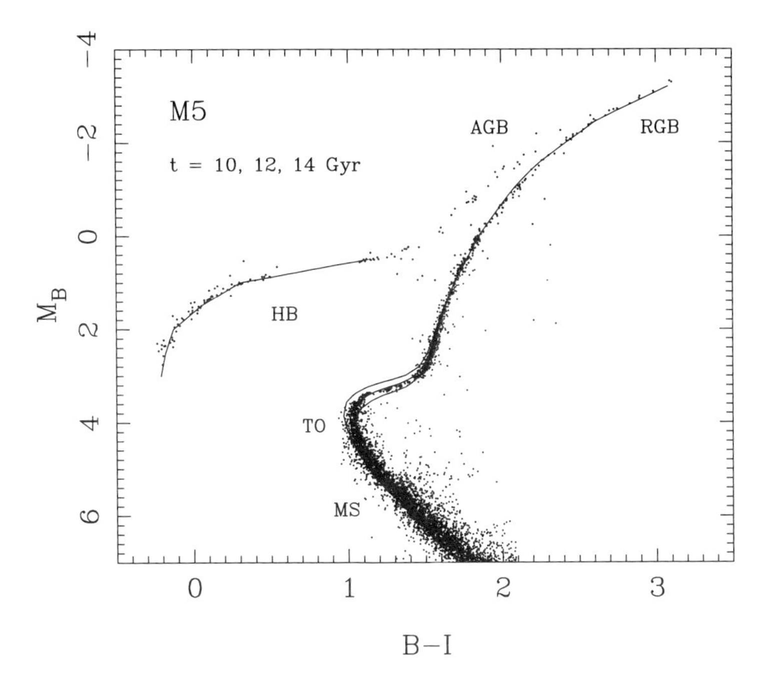

Figure 5.1. A colour–magnitude plot (the observational version of the Hertzsprung–Russell diagram) for stars in the globular cluster M5 (data courtesy of R. Jimenez). This illustrates well the main features of stellar evolution: the main sequence (MS) and its turn-off point (TO), followed by the red giant branch (RGB), horizontal branch (HB) and asymptotic giant branch (AGB). The main-sequence turn-off occurs at about $M_B = 4$, and is well-fitted by a theoretical isochrone of age 12 Gyr.

are applied to match the central core to the outer envelope. A common and deceptively simple argument is that it is all to do with the virial theorem, where the kinetic and potential energies must satisfy $2K + V = 0$ in equilibrium: since the contraction of the core causes V to become more negative, the balance can be restored if the outer parts expand to become less tightly bound. This argument works because the expansion is approximately isothermal, but it misses out the fact that the expansion is associated with a vast increase in nuclear energy output. In any case, the classical red-giant expansion is only found in stars of $\lesssim 3\, M_\odot$.

(3) The horizontal branch. As red giants burn more hydrogen, the mass and temperature of the helium core increases, until it too can fuse. Stars then follow their usual pattern of defying intuition by reducing luminosity somewhat but becoming smaller and hotter. They end up more or less at the top end of the original main sequence, which is reassuring, since we argued that this was largely independent of the details of nuclear energy generation for stars where the energy originated in the core. During this phase, some stars will be vibrationally unstable: the **RR Lyrae stars**.

(4) The AGB. History now repeats itself as helium in the core is exhausted. Shell burning returns, with both a helium-burning shell and an outer hydrogen-burning shell, and the star ascends the giant branch once more, now termed the **asymptotic giant branch**, although its location in the HR diagram is rather similar to the red giant branch.

(5) Late stages of evolution. Stars that come to the end of the AGB stage contain high-density cores enhanced in heavy elements. The normal behaviour is for the remaining outer layers to be lost, in a greatly exaggerated analogue of the Solar wind, leaving behind a **white dwarf**, which simply cools via radiation without generating further nuclear energy. These stars lie below the main sequence, being of very low luminosity for their temperature.

(6) A caveat on high-mass stars. The older star clusters can be misleading, since they contain no stars more massive than around $1\,M_\odot$. For $M \gtrsim 5\,M_\odot$ the story is somewhat different, because the core in such stars is convective. Core fuel exhaustion and simple shell burning therefore do not occur in the same way, although such stars still depart from the main sequence when hydrogen ceases to be the main central fuel. Most importantly for the question of the distance scale, such massive post-main-sequence stars pass through a phase as classical **Cepheid variables** with periods in the range ~ 1–100 days. These are thus in many ways the ideal distance indicator, not only through the period–luminosity relation already discussed, but also because they are among the most extreme supergiants, with luminosities $\sim 10^4\,L_\odot$.

On the basis of the above thumbnail sketch, two ways are clearly open for the use of stellar evolution to estimate ages. One is to identify the turnoff point in star clusters; the other is to exploit the relatively simple cooling behaviour of white dwarfs.

The age of a globular cluster may be obtained via a fit of a single-age evolution model (an **isochrone**) to the data in the colour–magnitude diagram, as illustrated in figure 5.1. The fits are impressively good in their detailed agreement with observation, although some significant difficulties are hidden when only the final plot is inspected. First of all, the distance to the cluster under study is not known, and the conversion from apparent magnitudes to luminosities is correspondingly uncertain. This might not seem a problem: the main sequence defines a colour–luminosity relation, and there will only be one choice of distance at which the model will match. However, both the metallicity of the cluster and the line-of-sight reddening will alter the main-sequence locus, and uncertainties in these parameters alter the best-fit distance. The alternative of using the near-constant luminosity of the horizontal branch to fix the distance is also affected by sensitivity to metallicity, and also because this is a more complicated evolutionary stage than the relatively simple point of turnoff. The result is that the formal uncertainty in the fitted age (ignoring any systematics in the model) is dominated by the uncertainty in the distance. If we assume that the turnoff mass can be fitted correctly, then the inferred age scales inversely with the turnoff luminosity:

$$\delta \ln t \simeq 2\delta \ln D. \tag{5.7}$$

Given the accuracy of distances within the galaxy, we therefore expect a minimum age uncertainty of around 10% rms. The ages determined in this way for the low-metallicity clusters that are plausibly the oldest systems consistently come out in the range 13–17 Gyr (VandenBergh 1991). Given the random errors in the fitting procedure alone, it is plausible that the very largest figures may be upward fluctuations, but a figure of about 15 Gyr may be regarded as a good estimate of the typical age. However, because any uncertainty in the distance scale is systematic, this must carry a minimum error of ± 1.5 Gyr. The 95% confidence limit for the age of the halo of the Milky Way is thus approximately $t > 12$ Gyr.

This ignores any possible systematic uncertainties in the models. The calculation of a full stellar evolution model is a complex business, and this is one of the difficulties facing a non-specialist who wants to make an assessment of the accuracy of the results. Some of the features of the physics are simple in principle, but complex in detail: the nuclear reactions in the core, or the various sources of opacity that determine the structure of the atmosphere and shape the emergent spectrum. More difficult to treat exactly are the effects of convective mixing at intermediate radii. The results produced by computer models of stellar evolution have not altered greatly in recent years, but it is hard for an outsider to know whether this reflects more than an increasingly accurate implementation of the same set of basic assumptions. One way of proceeding is to attempt to fit more detailed features of the data, going beyond the simplest feature of the main-sequence turnoff. For example, Jimenez *et al.* (1996) and Padoan & Jimenez (1997) fit the horizontal branch and the full number distribution along the locus in the colour–magnitude diagram, rather than just its shape. They claim good agreement with models, and favour slightly lower globular-cluster ages of around 13 Gyr.

WHITE DWARF COOLING White dwarfs should in principle be a simpler set of objects to consider. The equation of state for degenerate matter is independent of temperature, so that these endpoints of stellar evolution do not change their internal structure as they cool. Nevertheless, matters are not completely trivial, since the rate at which the stars release energy is controlled by the opacity in the outer, non-degenerate, atmosphere. On the assumption that this is dominated by ions, Mestel (1952) calculated that the luminosity should scale with mass and age as follows:

$$L \propto M t^{-7/5} \tag{5.8}$$

(see also Liebert 1980). Such a relation implies that there should be a sharp cutoff in the luminosity function of white dwarfs, and this is observed rather convincingly in the disk stars of the Milky Way. Fitting to the most detailed available model atmospheres, Winget *et al.* (1987) deduce an age for the disk (including the pre-white dwarf stellar evolution) of

$$t_{\rm disk} = 9.3 \pm 2.0 \ \text{Gyr}. \tag{5.9}$$

Is there a consistent picture for the age of the universe here? We have seen that nuclear decays and white dwarfs imply ages for the galactic disk of respectively 10.0 ± 1.5 and 9.3 ± 2.0 Gyr, giving a formal average of 9.7 ± 1.3 Gyr. When did this system form? The evidence from the abundance of massive neutral hydrogen disks at high redshift (see chapter 13) is that disk formation happened at relatively low redshift, $z \lesssim 3$, but perhaps not at much lower redshift. At $z = 3$ in an $\Omega = 1$ universe, the age is $1/8$ of the present age, and the age of the universe is therefore at least $8/7$ times that of the disk, or 11 ± 1.4 Gyr. A formation redshift of $z = 1$ would raise this to 15 ± 1.9 Gyr. For the globular cluster ages, it is hard to be sure what redshift of formation to assume. If they are associated with the formation of elliptical galaxies, the lack of primaeval galaxies (see chapter 16) means that $z \gtrsim 8$ may be appropriate, introducing an offset of perhaps only a few hundred million years between the ages of the oldest globular clusters and that of the universe itself.

5.3 Large-scale distance measurements

We now turn from ages to the distance scale, starting far away, and working closer to home. At the largest distances, discrete objects within galaxies cannot be resolved (with the exception of supernovae; see below). We therefore have to use methods based on the global properties of galaxies. These effectively rely on using a circular velocity or velocity dispersion to 'weigh' the galaxy and thus correct for the large dispersion in galaxy luminosities that prevents their employment as **standard candles** – i.e. fictitious objects of constant luminosity for which apparent magnitude is directly related to distance.

THE FUNDAMENTAL PLANE FOR ELLIPTICALS The properties of elliptical galaxies are perhaps the most easily understood. These are all very similar objects, with surface-brightness profiles that are well described by a universal function that contains only two parameters: a characteristic radius and surface brightness. A convenient fitting form for these profiles that works surprisingly well is de Vaucouleurs' $\mathbf{r^{1/4}}$ **law**:

$$I(r) = I_0 \exp[-(r/r_0)^{1/4}]. \tag{5.10}$$

This model may be integrated exactly to obtain $L = 20160 I_0 r_0^2$; clearly a dimensionally similar scaling will apply with any two-parameter family. From an observational point of view, it is more common to encounter r_e, the radius that contains half the light ($r_e = 3459 r_0$), and the surface brightness at this radius, I_e. A third observable is the central velocity dispersion σ_v (the Doppler width of spectral lines in the central parts of the galaxy). It is found empirically that these three observables are closely correlated, so that elliptical galaxies lie on a **fundamental plane**:

$$L \propto I_0^x \, \sigma_v^y, \tag{5.11}$$

with $(x, y) \simeq (-0.7, 3)$ (Dressler *et al.* 1987; Djorgovski & Davis 1987). This relation is quite easy to understand: it just says that elliptical galaxies are self-gravitating systems with roughly constant mass-to-light ratios. By the virial theorem (in fact just by dimensions), the two-parameter galaxies must satisfy the Keplerian relation

$$\sigma_v^2 \propto \frac{M}{r_0}. \tag{5.12}$$

Now add a weakly-varying mass-to-light ratio

$$M/L \propto M^a; \tag{5.13}$$

elimination of r_0 via $L \propto I_0 r_0^2$ yields

$$L^{1+a} \propto \sigma_v^{4-4a} I_0^{a-1}, \tag{5.14}$$

which is very close to what is observed, provided $a \simeq 0.25$. The fundamental plane sometimes also goes by the name of the $\mathbf{D_n - \sigma}$ **relation**. Consider measuring the diameter D_n of the aperture within which the mean surface brightness takes some reference value: in general this will have an approximate power-law dependence $D_n \propto r_0^\alpha I_0^\beta$. For the correct choice of reference level, the dependence is such that the fundamental plane may be re-expressed as that for which σ_v is proportional to some power of D_n. The reference brightness used in practice is 20.75 mag arcsec^{-2} in the B waveband, corrected for $(1+z)^{3+\alpha}$ cosmological effects to give a rest-frame value (the strange units mean that the flux from 1 arcsec^2 is equivalent to a point source of magnitude 20.75). This number arises only because it is comparable to the night-sky brightness, and so it is somewhat fortuitous that the $D_n - \sigma$ relation approximates the fundamental plane so well.

There is thus an excellent physical basis for the use of the fundamental plane as a distance indicator: measurement of the distance-independent (when corrected for redshift) observables σ_v and I_0 gives a calibration of the relative luminosity of a given elliptical galaxy. Relative distances then follow directly from the apparent magnitude. In practice, there is inevitably some scatter about the relation, which limits the precision of any given distance estimate to about 20% rms. However, this can be greatly reduced by averaging over many galaxies in a cluster.

FABER–JACKSON RELATION Although we have shown that elliptical galaxies must lie on a fundamental plane, this does not tell us *where* on the plane they must lie. However, it is a fact that only a relatively narrow locus is occupied: in the absence of surface-brightness information, there is still a relation between luminosity and surface brightness:

$$L \propto \sigma_v^\alpha, \tag{5.15}$$

with $\alpha \simeq 3$–4, as shown by Faber & Jackson (1976). Other ways of expressing this relation are clearly possible given the existence of the fundamental plane; it was first discovered in a form involving potential energy by Fish (1964). The most plausible reason for the existence of this relation is that given by Faber (1982): it is the result of collapse in a gravitational hierarchy. Denote the rms density variation in the universe after smoothing by a box containing mass M by σ_0, and let a dimensionless size of perturbation be $\nu \equiv (\delta\rho/\rho)/\sigma_0$. Lastly, for simplicity make a power-law approximation $\sigma_0 \propto M^{-\lambda}$. Gravitational collapse theory tells us that galaxies will form when $\delta\rho/\rho \simeq 1$ (see chapter 15), and that the density contrast evolves as $(1+z)^{-1}$; the collapse redshift therefore scales as

$$1 + z \propto \nu M^{-\lambda}. \tag{5.16}$$

At collapse, the density of the galaxy is some multiple of the background density, so that

$$M/r_0^3 \propto (1+z)^3. \tag{5.17}$$

Eliminating z and using the fundamental-plane equations gives

$$L^{2-3\lambda} \propto \nu^{-3(1+a)} \sigma_v^{6(1+a)}. \tag{5.18}$$

A low value of λ appears to be appropriate for the observed fluctuation spectrum on galaxy scales (see chapter 16); taking $\lambda = 1/6$ gives roughly the observed relation. Finally, as a bonus, the scatter in ν expected for peaks in random density perturbations accounts for the observed ~ 1 magnitude of scatter in the relation (see chapter 17).

TULLY–FISHER RELATION The reason for going through the Faber–Jackson relation in some detail is that a very similar relation applies for spiral galaxies. The relation is now between a measure of luminosity and a measure of the rotational velocity of the spiral. There are several possibilities: the velocity may be measured in the optical or via the 21-cm emission line of neutral hydrogen, and the luminosity may be measured in the optical or infrared or via the 21-cm line. One needs to specify the radius at which the rotation velocity is measured, but this is usually taken to be at a large enough distance that the asymptotic flat rotation curve is reached. In practice, the smallest scatter is achieved in the infrared, although the originators of this method (Tully & Fisher 1977) worked principally with neutral hydrogen. The acronym **IRTF** (infrared Tully–Fisher) is thus a common one. As with the fundamental plane for ellipticals, an rms precision of about 20% in distance can be achieved. Attempts have been made to bring in extra variables to improve the precision of the method – in a way analogous to the improvement of the Faber–Jackson law by using surface brightness. There is so far no general agreement that these have been successful (see the references in Jacoby *et al.* 1992). Indeed, the entire physical basis of the Tully–Fisher method is somewhat more obscure than for the elliptical-based distance indicators [problem 5.2].

THE COMA–VIRGO DISTANCE Given the limited accuracy of these methods, it is desirable to improve things by averaging over many galaxies that can be assumed to lie at the same distance – i.e. a cluster of galaxies. From this point of view, the two most important clusters are those in Virgo and Coma. The Virgo cluster is the closest moderately rich concentration of galaxies; the Coma cluster is one of the closest examples of a cluster of the highest richness. Coma is moreover sufficiently distant that one can assume with confidence that its redshift derives almost entirely from the expansion of the universe: at a heliocentric redshift of $z = 0.0231$ (Faber *et al.* 1989), possible deviations from the mean Hubble flow of several hundred $\mathrm{km\,s^{-1}}$ alter the redshift by only $\lesssim 10\%$. Measuring H_0 therefore 'only' requires an absolute distance to Coma. Alternatively, we can regard Coma as an excellent calibrator for the extragalactic velocity scale, which we can attempt to carry inwards to Virgo. The idea will be to deduce the redshift that Virgo 'ought' to have in the absence of deviations from the Hubble flow.

The $D_n - \sigma$ data in these two clusters are extensive and determine the offset in D_n (and hence in angular-diameter distance) to be a factor 5.82 within about 1% formal random error (Feigelson & Babu 1992; see figure 5.2). If we correct the redshift of Coma to the microwave-background frame (i.e. allowing for the component of the Earth's absolute motion towards Coma of about 270 $\mathrm{km\,s^{-1}}$, see chapter 9), the result is $z = 0.0240$, implying an ideal redshift for Virgo of $z_{\rm Virgo} = 1186\,\mathrm{km\,s^{-1}}$. The Hubble constant can therefore be expressed in terms of the distance to Virgo:

$$h = \frac{11.9\ \mathrm{Mpc}}{D_{\rm Virgo}}. \tag{5.19}$$

The likely amplitude of deviations of Coma's redshift from the ideal Hubble flow means that this estimate has a random error of approximately 5% rms.

Since the Virgo and Coma clusters are not of identical richness, there is always the worry that the $D_n - \sigma$ relation may be systematically different in the two systems. The result needs to be verified by independent methods, and we shall see below that this value is in excellent agreement with independent measurements using supernovae. The Tully–Fisher method is not so useful here, since spirals in Virgo are not very concentrated towards the centre; there is uncertainty about which galaxies are associated with the cluster, and which are background or foreground objects. The Tully–Fisher method may still be used, treating each galaxy individually. The disadvantages of this

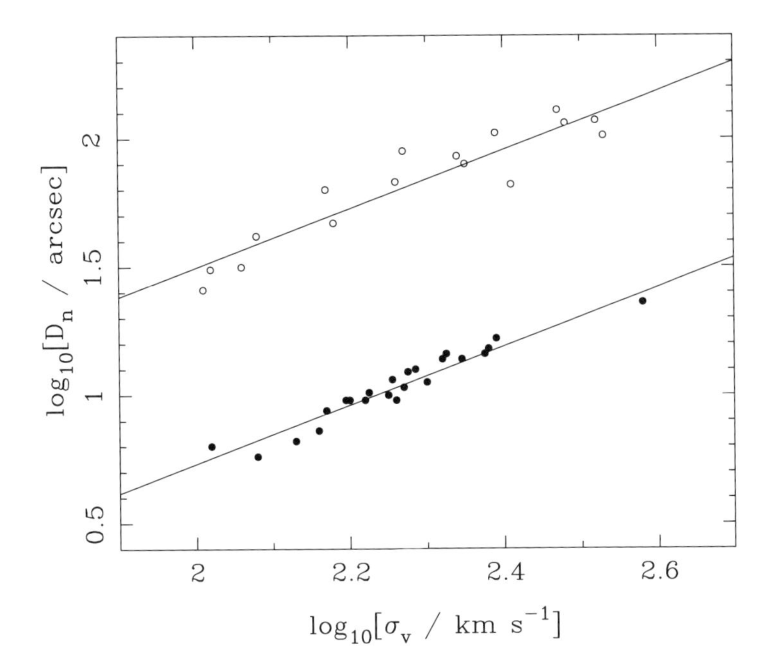

Figure 5.2. A comparison of the $D_n-\sigma$ data for the galaxies in the Virgo (open circles) and Coma (closed circles) clusters. Since the velocity dispersion σ is independent of distance, the offset between these clusters allows the ratio of angular-diameter distances to be determined (adapted from Feigelson & Babu 1992).

approach are twofold. First, dealing with a set of objects in a cluster verifies that the objects under study really do obey the distance-indicator relation being used; this is an important guard against systematics. Second, the individual-galaxy approach gives ideal Hubble velocities for a number of nearby galaxies, each of which then requires an absolute distance estimate. Nevertheless, the Tully–Fisher approach does give results in good agreement with the distance-scale results emphasized here (Jacoby *et al.* 1992).

INTERMEDIATE DISTANCES Bridging the gap between Virgo and more local distance measures was for a long time the most uncertain part of the distance scale. Until the advent of the Hubble Space Telescope, Virgo was too distant for the most reliable distance indicator (Cepheid variable stars; see below) to be used. Although the Tully–Fisher and fundamental-plane methods can be used to provide relative distances between Virgo and more nearby galaxies, the small number of close galaxies means that one cannot use averaging to obtain a very accurate answer. Fortunately, recent years have seen the arrival of two further whole-galaxy methods of superior accuracy. These are harder to apply at large distances, and so do not supplant the techniques discussed previously further than Virgo. However, they do very nicely fill this previously difficult range.

SURFACE-BRIGHTNESS FLUCTUATIONS This method is a statistical version of older methods based on trying to find the brightest supergiants in a galaxy. If the upper limit to stellar mass is universal or has a variation with galaxy type that can be calibrated, then we have a valid indicator. At large distances, though, it may be impossible to resolve individual stars, or to be sure that a single object is truly a supergiant associated with the galaxy. However, it is possible to count the stars in a galaxy without resolving them, which is the basis of a method invented by Tonry & Schneider (1988); see also Tonry (1991). The main idea is very simple, and may be illustrated by supposing that all stars in all galaxies are of the same intrinsic luminosity. Even though a galaxy may be so distant that there are many stars (N, say) in one resolution element, the number can be inferred because there will be statistical $1/\sqrt{N}$ fluctuations in brightness between different resolution elements. Thus, an image of a nearby galaxy does not look smooth: it is mottled by white noise with a coherence length set by the astronomical seeing. The fluctuations let us count the stars; knowing the total apparent brightness of the galaxy then tells us how distant it is [problem 5.3]. With corrections based on apparent colour to account for the different stellar populations in different galaxies, this method is claimed to provide accuracies in distance of around 5% (based on the scatter for galaxies in a single cluster).

PLANETARY NEBULAE This method is based on the apparent constancy of the upper limit to the luminosities of planetary nebulae (shells of gas shed and then illuminated by evolving massive stars). These objects are easily detected through their narrow intense emission-line spectra, and so can be found even against the bright background of the body of a galaxy by using narrow-band filters. Again, a precision of around 5% is claimed (e.g. Jacoby, Ciardullo & Walker 1990; Jacoby, Ciardullo & Ford 1990).

THE M31–VIRGO DISTANCE The two above methods both relate in some way to the limit on maximum stellar mass, but they are sufficiently independent that a comparison should be rather exacting: these are the two techniques of greatest claimed rms precision, so any systematics should be especially obvious. Remarkably, such systematics seem to be lacking over a factor of about 20 in distance (see Tonry 1991). The nearest galaxies that can be used to calibrate the methods are M31 and M32, from which a good relative distance to the Virgo cluster can be established:

$$D_{\rm virgo}/D_{\rm M31} = 20.65 \pm 6\%. \tag{5.20}$$

5.4 The local distance scale

CEPHEID VARIABLES The above discussion of the distance scale has now brought us fairly close to the Milky Way. Distances to the nearest few galaxies may be determined most accurately by the use of Cepheid variable stars. These have a positive correlation between luminosity and period (roughly $L \propto P^{1.3}$) that is very tightly defined. Data in different wavebands can be used to find the relative distance between two groups of Cepheids and also to determine the relative extinctions involved, so this is not a source of uncertainty in the method.

Cepheid variables have many advantages; not only are they among the most luminous stars known, but the physical basis for their variability is relatively simple. All stars have a natural oscillation period, deriving from the crossing time for sound waves, where the speed of sound is $c_S = \sqrt{\gamma p/\rho}$. Since the thermal energy density is also of order p, the condition for virial equilibrium is then

$$p R^3 \sim GM^2/R \sim G\rho^2 R^5 \;\Rightarrow\; R/c_S \sim (G\rho)^{-1/2}. \tag{5.21}$$

The fundamental oscillation period should then be roughly equal to the pressure-free collapse time. This does not of course explain why some stars become unstable to radial oscillations with this period whereas others (such as the Sun) do not. What is needed is in fact a stellar structure where the opacity is sensitive to small changes in density. This arises in Cepheids because of the existence of a transition zone between singly and doubly ionized helium, which is sensitive to the small changes in temperature associated with small radial perturbations. Why this should arise in evolved stars but not main-sequence stars requires detailed modelling (see e.g. chapter 39 of Kippenhahn & Weigert 1990). However, this does not prevent us using the fact of variability if and where it occurs.

The density of a star, as well as its observables of luminosity and colour, all plausibly have some power-law correlation with its mass – although not one that is easy to calculate *a priori* because Cepheids arise in a post-main sequence phase of evolution. Because giant stars have increased their luminosity principally by a vast expansion in size, we expect that the density will be anticorrelated with the luminosity, so that the most luminous stars will have the longest periods. What is found in practice (the famous 1907 discovery of Henrietta Leavitt) is approximately $\tau \propto L^{1.3}$. Because the colours of Cepheids do not correlate perfectly with their luminosities, there have long been suggestions that the use of colour data could reduce still further the approximate 0.2 magnitude scatter in the relation (see section 3.1 of Rowan-Robinson 1985). In practice, this possibility is usually ignored, since the effects of reddening of the colours may introduce error. A safer procedure is to examine the $P-L$ relation in different wavebands and fit an extinction law to the offsets in luminosity at given period (see figure 5.3).

The Cepheid method is limited by the closest point at which there is a large group of Cepheids to calibrate the period–luminosity relation. Such a group is found in the **Large Magellanic Cloud** (LMC), and its relative distance to M31 is found to be (Freedman & Madore 1990)

$$D_{\rm M31}/D_{\rm LMC} = 15.28 \pm 5\%. \tag{5.22}$$

The main concern with using the LMC as a zero point is that this is a dwarf galaxy of low metal content relative to the Sun. It is therefore conceivable that there could be a systematic effect, since Cepheids will tend to be found in more luminous and metal-rich galaxies at high redshift. Some workers have claimed to detect such an effect, albeit at a level comparable with the random errors in the distance (Gould 1994; Sasselov *et al.* 1996).

DISTANCES WITHIN THE GALAXY We have now come to the point where calibration of the extragalactic distance scale requires methods for measuring distances on scales up to a few kpc; this is the domain of the traditional approach to the distance scale. Both because this is covered so well elsewhere (e.g. Rowan-Robinson 1985) and because the SN1987A result mentioned below has arguably short-circuited so much of the classic

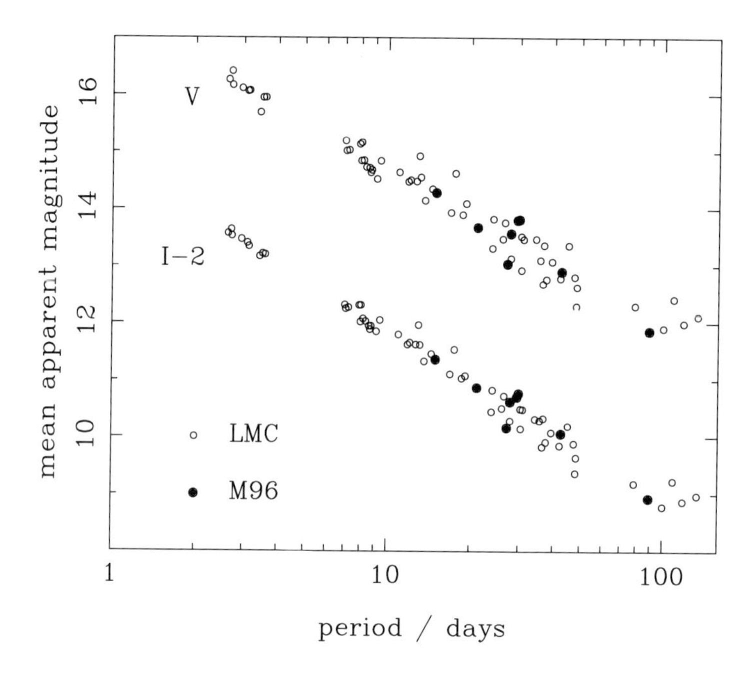

Figure 5.3. A plot of the Cepheid $P-L$ relation for stars in the Large Magellanic Cloud and in M96 (Tanvir *et al.* 1995). The M96 stars have been shifted to overlay the LMC data, and the required shift gives the apparent difference in distance modulus. Examining this offset as a function of wavelength and fitting an extinction model allows the true relative distance to be established.

work in this field (from the narrow point of view of determining H_0, that is) we can be extremely brief here.

The traditional approach to distance measurement begins with trigonometric parallaxes based on the Earth's orbit. Aided by the astrometric **Hipparcos** satellite (see e.g. Lindgren *et al.* 1992; Feast & Catchpole 1997), this will very soon give a distance to the nearest star cluster, the Hyades, although at present a less direct method based on angular motions of the Hyades stars is used. From there, by main-sequence fitting (finding the offset in apparent magnitude at a given colour) distances to further clusters can be found, leading eventually to distances to Cepheid variables. By this route (and others) a standard value for the LMC distance has become established:

$$D_{\rm LMC} = 50.1 \text{ kpc } \pm 6\%. \tag{5.23}$$

SN1987A The most nearby part of the distance scale was thus for many years perhaps the most dubious. However, this has been changed by a marvellously neat direct distance measure for the LMC by Panagia *et al.* (1991). The 1987 supernova in that galaxy was observed to produce a ring of emission that was elliptical to high precision – and therefore almost certainly a circular ring seen inclined. Different parts of the ring were observed being illuminated at different times owing to finite light travel-time effects. Knowing the inclination, plus the observed angular size of the ring, the distance to the supernova follows [problem 5.4]. Note that the original analysis of Panagia *et al.* was flawed, although not by enough to affect their distance estimate (see Dwek & Felten 1992). Making a small correction for the position of the supernova within the LMC, a distance $D_{\rm LMC} = 51.8$ kpc $\pm 6\%$ follows – very close to the traditional value.

SUMMARY OF THE TRADITIONAL DISTANCE LADDER We can now put everything together. There are three main steps in the traditional distance ladder: the LMC distance; the LMC–Virgo distance; the Virgo–Coma distance (alternatively, the corrected redshift for Virgo).

For the LMC, let us adopt here the average of the traditional and the SN1987 distances: 51 kpc $\pm 4\%$. The LMC–Virgo distance ratio obtained above was 315.5 $\pm 8\%$. In an important development, Cepheids in Virgo have been detected both from the ground and with the HST, giving alternative one-step measurements of this distance ratio as 288 $\pm 7\%$ for NGC4571 (Pierce *et al.* 1994) or 340 $\pm 9\%$ for M100 (Freedman *et al.* 1994). These numbers are well consistent with the above estimate, and yield a formal average of 314 $\pm 5\%$. Finally, taking the ideal Hubble-flow velocity for Virgo, we get

$$H_0 = 74\,\mathrm{km\,s^{-1}Mpc^{-1}} \pm 8\%. \tag{5.24}$$

The error estimate here should be an honest reflection of the random uncertainties resulting from the various steps in the argument. Of course, the tortuous history of the subject has been one of unrecognized systematics that exceed the random errors, whereas we have naively combined the available numbers on the assumption that such factors are now negligible. Caution is therefore advisable, but it is worth noting that each of the elements of this calculation has been confirmed by at least two independent methods with claimed accuracies of $\sim 5\%$, so it is tempting to believe that the above calculation cannot be seriously incorrect. If the errors are Gaussian in $\log h$, the quoted rms error then allows a 95% confidence band for H_0 of 63–87 $\mathrm{km\,s^{-1}Mpc^{-1}}$.

5.5 Direct distance determinations

SUPERNOVA HUBBLE DIAGRAM The exception to the above rule that individual objects cannot be seen in distant galaxies is provided by supernovae. These come in two-and-a-bit varieties, SNe Ia, Ib and II, distinguished according to whether they display absorption and emission lines of hydrogen. The SNe II do show hydrogen; they are associated with massive stars at the endpoint of their evolution, and are rather heterogeneous in their behaviour. The former, especially SNe Ia, are much more homogeneous in their properties, and can be used as standard candles. There is a characteristic rise to maximum, followed by a symmetric fall over roughly 30 days, after which the light decay becomes less rapid. Type Ib SNe are a complication to the scheme; they do not have the characteristic light curve, and also lack hydrogen lines.

The simplest use of these supernovae is to note that they empirically have a very small dispersion in luminosity at maximum light ($\lesssim 0.3$ magnitudes). The Hubble diagram of magnitude against redshift can be plotted out to redshifts in excess of 0.1 (figure 5.4), and so direct measurements of distances to some of the nearer SN Ia suffice

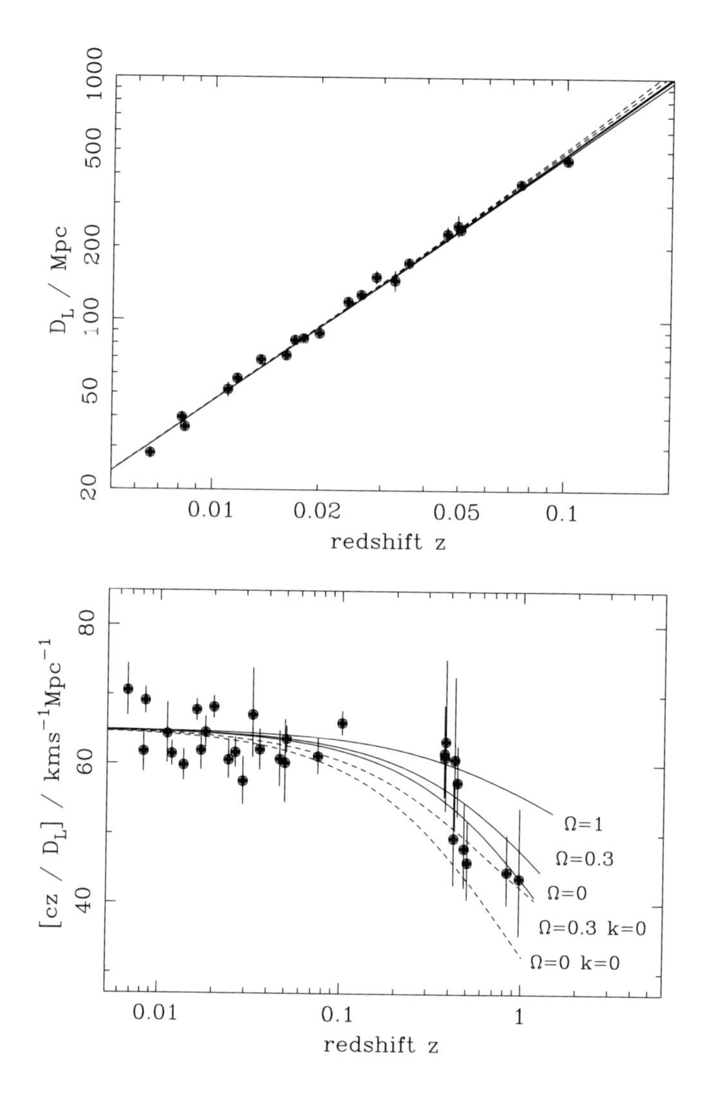

Figure 5.4. The type Ia supernova Hubble diagram. Using a measure of characteristic time as well as peak luminosity for the light curve, relative distances to individual SNe can be measured to 6% rms. Setting the absolute distance scale (D_L is luminosity distance) using local SNe in galaxies with Cepheid distances produces the upper panel (Riess, Press & Kirshner 1995, 1996). This shows that the large-scale Hubble flow is indeed linear and uniform, and gives an estimate of $H_0 = 64 \pm 3\ \mathrm{km\,s^{-1}Mpc^{-1}}$. The lower panel shows an expanded scale, with the linear trend divided out, and the high-z supernovae of Perlmutter *et al.* (1997a, b, c) and Garnavich *et al.* (1997) added. Comparison with the prediction of FRW models appears to favour a vacuum-dominated universe.

to give H_0. As usual, greater power is obtained by averaging over objects in clusters, and an accurate estimate for the relative distance between Virgo and clusters at or beyond the distance of Coma can be obtained. Sandage & Tammann (1993) quote

$$h = \frac{11.4 \text{ Mpc}}{D_{\text{Virgo}}} \tag{5.25}$$

to within an rms of about 5%; this is almost identical to the figure quoted earlier. Again, Virgo is of critical importance because it is the closest large concentration of supernovae. Sandage & Tammann (1993) also attempt to calibrate the Hubble diagram with Cepheid-based distances for local galaxies that have hosted supernovae, obtaining $H_0 = 50 \pm 7 \text{ km s}^{-1}\text{Mpc}^{-1}$.

Although impressive, these supernovae-based results use a completely empirical method, and one might legitimately ask why SNe Ia should be standard candles. After all, presumably the progenitor stars vary in mass, and this should affect the energy output. A more satisfactory treatment of the supernovae distance scale takes this possibility into account by measuring both the height of the light curve (apparent luminosity at maximum light) and the width (time taken to reach maximum light, or other equivalent measures). For SNe where relative distances are known by some other method, these parameters are seen to correlate: the maximum output of SNe scales as roughly the 1.7 power of the characteristic timescale (Phillips 1993). The physical interpretation of this relation is that both the measured parameters depend on *mass*: a more massive star has more fuel and so generates a more energetic explosion, but the resulting fireball has to expand for longer in order for its optical depth to reach unity, allowing the photons to escape. Given good enough data to measure the timescale, it is possible to turn SNe Ia into genuine standard candles, and the accuracy is astonishingly good: a corrected dispersion of 0.12 magnitudes, implying that relative distances to a single SN can be measured to 6% precision.

This method has been applied both to a large set of local SNe to constrain H_0, and also to a smaller set of high-z SNe, allowing the first meaningful test of cosmological geometry to be carried out. Some of these results are shown in figure 5.4. The local results are taken from Riess, Press & Kirshner (1995, 1996). Calibrating with three local SNe with Cepheid distances, they obtain $H_0 = 64 \pm 3 \text{ km s}^{-1}\text{Mpc}^{-1}$. The error is statistical, and assumes that the Cepheid scale is known exactly. Certainly, it is clear (i) that Hubble's law is linear to high precision, and (ii) that H_0 is known, provided we believe distance determinations in the Cepheid regime.

Moving to higher redshifts, the same method has been applied to a number of SNe at $0.35 < z < 1$ (Perlmutter *et al.* 1997a, b, c; Garnavich *et al.* 1997). With this level of precision, it is possible to contemplate a determination of the **deceleration parameter** q_0 with good statistical precision. Better still, because the laws of stellar evolution should be the same at all distances, the results will not be subject to the usual evolutionary uncertainty. These high-z results are compared with the low-z data in the second panel of figure 5.4 (although note that Perlmutter *et al.* in fact use a different set of low-z comparison data). The different lines show the predictions of a variety of FRW models, and the precision of the data is clearly getting good enough to discriminate between them. The Einstein–de Sitter $\Omega = 1$ model is in very strong disagreement with the data, and the highest-redshift SNe appear to require a negative value of q_0, which would imply a non-zero cosmological constant. It is probably too early to take this as in any way definitive evidence for a vacuum-dominated universe, but it is clear that this technique is likely to yield a definitive verdict on Λ quite soon. The observational progress on these projects is rapid, and it is reasonable to expect SNe at $z \gtrsim 1$ to be detected in large numbers within the next few years.

BAADE–WESSELINK METHOD In principle, the neatest use of supernovae is to obtain absolute distances in one step, without calibration via any other distance indicator. This is possible because supernovae are approximately black body radiators: knowing the temperature (which is independent of distance) means that we can deduce the angular radius from the observed flux density:

$$\theta = \left[\frac{S_\nu}{\pi I_\nu(T)} \right]^{1/2}. \tag{5.26}$$

Now, spectroscopic observations tell us not only the temperature but also the blueshift of spectral lines in the photosphere, from which a radius may be estimated: $r = \int v \, dt$. Comparing the angular and true radii gives the distance. The possible flaws are (i) departures from a black body, (ii) spectral lines may originate at a different depth from the continuum; a practical application therefore requires some sophisticated corrections. However, in the best-studied case (SN1987A), the method gave $D = 44$–50 kpc (Schmidt *et al.* 1992), in excellent agreement with other determinations (see above). These authors also consider Baade–Wesselink distances for 19 other type II SNe, with distances out to 120 Mpc, and obtain $H_0 = 60 \pm 10 \ \mathrm{km\,s^{-1} Mpc^{-1}}$. In many ways, this is an impressive result.

RADIOACTIVE DECAY MODELS An alternative absolute method is based on the assumption that SNe Ia produce their luminosity over a good part of their light curves via the decay of ^{56}Ni and ^{56}Co, an assumption that was well verified by observations of SN1987A. Since the progenitors of SN Ia are thought to be white dwarfs that have accreted material until they reach the Chandrasekhar limit, the masses of the elements can be predicted *ab initio* and an estimate of H_0 follows from the SN Ia Hubble diagram. Branch (1992) found $H_0 = 61 \pm 10 \ \mathrm{km\,s^{-1} Mpc^{-1}}$.

GRAVITATIONAL LENSES Chapter 4 dealt with the question of estimating H_0 directly from the time delay in gravitational lenses. For the double quasar, Grogin & Narayan (1996) find

$$H_0 = (79 \pm 7 \,\mathrm{km\,s^{-1} Mpc^{-1}}) \left(\frac{\sigma_v}{300 \,\mathrm{km\,s^{-1}}} \right)^2 \left(\frac{\Delta t}{\mathrm{yr}} \right)^{-1}, \tag{5.27}$$

where the relevant parameters are the galaxy velocity dispersion and the time delay. After some initial uncertainty caused by poor sampling of the observations, the time delay has been measured at around 1.1 years. Schild & Thomson (1997) quote 404 ± 26 days; Kundić *et al.* (1997) quote 417 ± 3 days (see Press *et al.* 1992, Pelt *et al.* 1996 for earlier controversy over this result). The observed velocity dispersion has been quoted as $303 \pm 50 \ \mathrm{km\,s^{-1}}$ (Rhee 1991), but Kundić *et al.* (1997) give $275 \pm 15 \ \mathrm{km\,s^{-1}}$. If we take a delay of 415 days and a dispersion of $280 \ \mathrm{km\,s^{-1}}$, then the dominant source of uncertainty in H_0 probably lies in the theoretical modelling. The resulting estimate is $H_0 = 61 \pm 5 \ \mathrm{km\,s^{-1} Mpc^{-1}}$.

THE SUNYAEV–ZELDOVICH EFFECT Chapter 12 discusses the SZ effect in some detail. Briefly, this is the diminution in surface brightness of the Rayleigh–Jeans portion of the spectrum of the microwave background caused by Compton scattering from electrons in hot intracluster gas. For a given temperature, it is proportional to the line integral of the electron number density through the cluster, $\Delta T/T \propto \int n_e \, d\ell$. However, the X-ray bremsstrahlung surface brightness for a given temperature scales as $I_\nu \propto \int n_e^2 \, d\ell$ (being a two-body process). Since the temperature of the gas can be found from X-ray spectroscopy, the depth of the cluster can be inferred by eliminating n_e. If the cluster is symmetric, the transverse angular size then gives an estimate of the distance. This method therefore has some difficulties, since the radial dependence of T is not well known, and the symmetry assumption may not be exact. Nevertheless, Birkinshaw & Hughes (1994) quote $H_0 = 55 \pm 17 \ \mathrm{km\,s^{-1}Mpc^{-1}}$, which at least shows that such effects cannot be too serious.

5.6 Summary

We have covered the main independent methods for obtaining H_0: the traditional distance ladder; two absolute determinations from supernovae; an absolute determination from gravitational lensing; an absolute determination from the SZ effect. All these methods have formal errors of around 10%–15%, and all are in reasonably good agreement, yielding values of respectively 74, 63, 61, 61 and 55 $\mathrm{km\,s^{-1}Mpc^{-1}}$. The interesting thing here is that the absolute methods all prefer values lower than the central value from the traditional distance ladder. In any individual case, one might well suspect that a systematic error could be afflicting the absolute method, but not when four totally independent methods come to the same conclusion. It therefore seems likely that the higher H_0 values permitted by the traditional distance ladder (74–87) can be ruled out, whereas the lower range (63–74) is consistent with all other data.

Interestingly, these figures are close to the value (65 ± 15) recommended in the book by Rowan-Robinson (1985) as a result of a critical assessment of the available results at that time. Many of the most important techniques discussed above have arisen since 1985; the older methods are, however, in most cases not discredited, but are simply of lower accuracy. There is thus something of a case to be made that the controversy on the distance scale has converged on a true value of H_0 in the range 60–70 $\mathrm{km\,s^{-1}Mpc^{-1}}$. In a subject that has generated much acrimony, however, it is important to remember that this author's personal opinion may not be universally shared.

As discussed earlier, values for t_0 in the range 12–16 Gyr are a reasonable summary of the present estimates from stellar evolution. If the globular-cluster ages are not trusted, however, nuclear decay and white-dwarf cooling do not compel us to believe that the universe is any older than 9–10 Gyr. Since published Hubble constants span $0.4 < h < 0.9$, that allows an extreme range of

$$0.37 < H_0 t_0 \simeq \tfrac{2}{3} \left(0.7\Omega_m + 0.3 - 0.3\Omega_v\right)^{-0.3} < 1.47. \tag{5.28}$$

If $\Omega_m > 0.1$ is accepted as a hard lower bound, then vacuum energy is required on the basis of this formula if $H_0 t_0 > 0.90$, and a total $\Omega_m + \Omega_v > 1$ is required if $H_0 t_0 > 1.33$. A more substantial density of $\Omega_m > 0.3$ would lower these limits to 0.82 and 0.96 respectively. The Einstein–de Sitter model requires $H_0 t_0 = 2/3$.

In view of the tortuous history of this subject, venturing a prediction on the eventual outcome of the timescale debate is risky in the extreme. Nevertheless, the importance of the age argument means that we should try to reach some conclusions. Here are a number of possible positions that could be adopted:

(1) The Einstein–de Sitter model ($\Omega = 1$) must be correct on grounds of simplicity. Since we are certain from nuclear dating that $t_0 > 9$ Gyr, that requires $h < 0.72$. A conservative stellar age of 12 Gyr would extend this to $h < 0.54$. Therefore, either the stellar ages are grossly in error, or $h \lesssim 0.5$ in spite of all the evidence.

(2) The Einstein–de Sitter model is clearly ruled out. Since slightly non-zero cosmological constants are contrived, the universe must be open. Values for Ω_m of a few tenths are indicated by most studies, and so a realistic model would have $H_0 t_0 \simeq 0.85$. This still argues for somewhat low values of either H_0 or t_0, but not unreasonably so, given the errors; $h \simeq 0.60$ and $t_0 \simeq 13.5$ Gyr would be acceptable.

(3) A straightforward reading of the evidence argues for $h \simeq 0.65$ and $t_0 \simeq 14$ Gyr; taken as an average over various techniques, neither of these numbers has changed greatly in a decade, despite individual results that differ greatly. Most arguments against such a data pair have been motivated by a desire to save $\Omega = 1$. With $H_0 t_0 \simeq 0.93$ from these figures, this is in good agreement with a $k = 0$ model with vacuum energy, provided the matter contribution is substantial, $\Omega_m \simeq 0.3$; lower-density $k = 0$ models predict too large a value of $H_0 t_0$.

Probably the only thing that can be predicted with any confidence is that each of these positions will still be supported by significant numbers of astronomers a decade or more hence. A resolution of this fundamental controversy will probably require measurements of H_0 and t_0 to better than 5% rms. Such a development is plausible for H_0, but the complexities of stellar evolution modelling mean that it will not be easy to give a convincing demonstration that the age of the universe is known to this precision. Furthermore, even if $H_0 t_0$ is known accurately, the conclusion that is drawn will depend on the density parameter. If $H_0 t_0 > 1$, then vacuum energy is required; otherwise the value of $H_0 t_0$ determines $0.7\Omega_m - 0.3\Omega_v$. Very small values of Ω_m will thus favour an open universe, whereas more substantial values will argue for $k = 0$. Given the realistic uncertainties in all these parameters, it will be difficult to determine the geometry of the universe from the age argument alone.

Problems

(5.1) Show that, for a star in which the opacity is due to Thomson scattering, the luminosity, temperature and size are expected to scale as follows: $L \propto M^3$, $T \propto M^{1/2}$, $R \propto M^{1/2}$, independently of the detailed mechanism for central energy generation.

(5.2) Show that disk galaxies will display an idealized $L \propto V^4$ Tully–Fisher relation provided they have (i) constant mass-to-light ratios, (ii) constant central surface brightnesses.

(5.3) Show that the method of surface-brightness fluctuations in effect measures a magnitude that corresponds to a given moment of the stellar luminosity function in a galaxy. Given a digital image of a galaxy, what are the steps needed to process it into a form suitable for measuring this magnitude, and what are the possible pitfalls?

(5.4) The ring around SN1987A consists of a perfect circle that lies in a plane inclined at an angle θ to the line of sight. If the ring scatters line radiation from the early phase of the supernova, what is the time dependence of the scattered flux density, and how can this be used to measure the distance to the supernova?

PART 3 BASICS OF QUANTUM FIELDS

6 Quantum mechanics and relativity

It is no exaggeration to say that fundamental physics in the twentieth century has been dominated by the problem of combining relativity with quantum theory. This task is still far from complete, but great progress has been made on many fronts; we may now be close to a complete understanding of the fundamental interactions between particles, with the exception of gravity. This part of the book aims to present the essence of these fundamental developments, whose implications for cosmology will be the subject of later parts. The present chapter covers a few of the critical aspects of relativistic quantum mechanics, to lay the foundation for discussing quantum fields and gauge theories in later chapters.

6.1 Principles of quantum theory

The logic that leads to quantum mechanics starts with the experimental quantization relations for the energy and momentum of photons in terms of their angular frequency and wave vector:

$$E = \hbar\omega, \quad \mathbf{p} = \hbar\mathbf{k}. \tag{6.1}$$

Now suppose that any wave-like behaviour of matter might manifest itself as a plane wave $\psi \propto e^{i\mathbf{k}\cdot\mathbf{x}-i\omega t}$, and guess that the energy–frequency and momentum–wavenumber relations for matter are the same as is observed for photons. This **de Broglie hypothesis** gives the operator equation

$$\boxed{P^\mu = i\hbar\,\partial^\mu.} \tag{6.2}$$

Notice how neatly the relativistic form gives the correct signs: $\partial^\mu \equiv \partial/\partial x_\mu = (\partial/\partial ct, -\boldsymbol{\nabla})$, so $\mathbf{p} = -i\hbar\boldsymbol{\nabla}$ and $E = i\hbar\partial/\partial t$, as required for the conventional choice of signs in the exponent of the plane wave. The de Broglie relations predict a group velocity $\partial\omega/\partial k = pc^2/E$, so this procedure makes physical sense if particles are thought of as wave packets.

Nonrelativistic quantum mechanics follows from inserting the above operator substitutions into the energy–momentum relation, $E = p^2/2m + V$, to generate the **Schrödinger equation**. This must apply to a general wave function, owing to the linearity of the equation and the fact that plane waves form a complete set:

$$\left(\frac{\hbar^2}{2m}\nabla^2 + V\right)\psi = i\hbar\dot{\psi}. \tag{6.3}$$

Note that the wave function *must* be a complex quantity in this equation, because the differential operator involved is a mixture of real spatial derivatives and an imaginary time derivative. Historically, Schrödinger first obtained the time-independent equation, for which the wave function can be real, and only reluctantly moved to a complex ψ in the time-dependent case (see the biography by Moore 1992 for many interesting insights on the development of Schrödinger's ideas). The significance of the entry of complex numbers into quantum mechanics is not so clear at this point, and is discussed more fully towards the end of this chapter. Later, when studying the Dirac equation, we shall see that it is possible in the relativistic case to have a real ψ – provided that the particle involved is uncharged. The true role of a complex ψ is thus bound up with the electromagnetic interaction and the existence of charge.

DIRAC NOTATION For practical quantum-mechanical calculations, the notation derived by Dirac is invaluable. It is worth summarizing the essential elements that will be used here. The wave function is replaced by the more general notion of the **state function**, written $|\psi\rangle$. This encodes all the information about the quantum state, including internal degrees of freedom, such as spin, in addition to spatial variations in the wave function. States of particular interest are the **eigenfunctions** of operators: $A\,|i\rangle = a_i|i\rangle$.

The most common calculation in quantum mechanics is to evaluate some matrix element for an operator, and the Dirac notation for this is particularly simple, emphasizing the similarity to an inner product:

$$\int \psi_1^*(\mathbf{x})\, A\, \psi_2(\mathbf{x})\, d^3x \quad \longrightarrow \quad \langle 1|A|2\rangle. \tag{6.4}$$

The matrix element is always constructed from a **bra** $\langle\psi|$ and a **ket** $|\psi\rangle$ – making a bracket. A crucial definition is the **Hermitian conjugate** of an operator:

$$\boxed{\langle\phi|A|\psi\rangle^* \equiv \langle\psi|A^\dagger|\phi\rangle.} \tag{6.5}$$

Using this relation twice shows that we can think of a bracket as arising either through A operating to the right, or $A^\dagger$ to the left: $\langle\phi|A|\psi\rangle = \langle\phi|A\psi\rangle$, but also $\langle\phi|A|\psi\rangle = \langle\psi|A^\dagger|\phi\rangle^* = \langle A^\dagger\phi|\psi\rangle$.

A Hermitian operator has $A^\dagger = A$, and so $\langle n|A|n\rangle^* = \langle n|A|n\rangle \Rightarrow a_n^* = a_n$: the eigenvalues are real. Similarly,

$$\langle n|A|m\rangle^* = \langle m|A|n\rangle \quad \Rightarrow \quad a_m^*\langle m|n\rangle = a_n\langle m|n\rangle; \tag{6.6}$$

if the eigenvalues are non-degenerate, the eigenfunctions are therefore orthogonal, and can thus be defined to be orthonormal: $\langle i|j\rangle = \delta_{ij}$. Such a set of eigenfunctions is called a **complete set**, since any other state function can be expanded in terms of them:

$$\begin{aligned} |\psi\rangle &= \sum_i c_i|i\rangle \\ \langle\psi| &= \sum_i c_i^*\langle i| \\ c_i &= \langle i|\psi\rangle. \end{aligned} \tag{6.7}$$

The coefficient c_i can be extracted by operating on the first sum with a bra; this proves that the coefficients in the analogous series for $\langle\psi|$ are c_i^*. Now apply this expansion to two states: $|\psi\rangle = \sum_i c_i|i\rangle$ and $|\phi\rangle = \sum_i d_i|i\rangle$. The bracket between them is $\langle\phi|\psi\rangle = \sum_i d_i^* c_i = \sum_i \langle\phi|i\rangle\langle i|\psi\rangle$. Because this equation must be true for all states, we deduce the **completeness relation**

$$\boxed{\sum_i |i\rangle\langle i| = 1,} \tag{6.8}$$

which is a most powerful calculational tool.

Finally, it should be emphasized that the state $|\psi\rangle$ is not the same as the wave function ψ; the latter is a function of position, but the former is an operator. See problem 7.6 for more on this distinction.

OBSERVATIONS IN QUANTUM MECHANICS Schrödinger's equation tells us the probability of finding a particle at a given location. This interpretation is justified by showing that $|\psi|^2$ satisfies a conservation equation and is therefore reasonably identified as a probability density [problem 6.1]. However, the interpretation of quantum mechanics is less straightforward when we deal with experiments designed to measure the properties of just a single one of the particles that are described by the wave function. It is worth going over these problems to emphasize just what a peculiar theory quantum mechanics really is.

Despite the limitations set by the uncertainty principle to our knowledge of the microscopic universe, the teaching of physics often clings to the Newtonian world view. Quantum-mechanical calculations commence by writing down an entirely classical Hamiltonian, and then introduce the wave function to provide a probability distribution over the variables involved in the system. We proceed much as if particles did have well-defined positions and momenta, even if the process of measuring one of these quantities inevitably introduces uncertainty into the other (cf. the Heisenberg microscope; see e.g. chapter 1 of Omnès 1994).

However, quantum uncertainty cannot be thought of as being introduced only by a clumsy observer. There are really two uncertainty principles: one for the intrinsic properties of particles, and one for the errors introduced by measurements. That these are connected is inevitable, because any measuring apparatus also obeys the uncertainty principle, as may be seen by the following thought experiment. Put a particle in an eigenstate of x-momentum; according to the uncertainty principle, its x-coordinate is completely uncertain. Now measure x with some machine of rms accuracy Δx; this sets the uncertainty in x, since we initially had no information on x. The measurement must introduce some uncertainty into p_x, which must be the final uncertainty in p_x since this was initially perfectly determined. If the final uncertainty obeys the uncertainty principle, we deduce the uncertainty relation between the measurement of x and the perturbation to p_x:

$$\Delta x^{\rm meas} \Delta p_x^{\rm perturb} \geq \hbar/2. \tag{6.9}$$

However, this induced uncertainty is a result of assuming the uncertainty principle for the *initial* state – we cannot say that it is all the fault of the measuring apparatus. This of course was the view of Bohr and the **Copenhagen interpretation** of quantum mechanics: it makes no sense to talk of properties that cannot be measured, and it is misleading to believe that a particle in any sense 'knows' in advance the properties that an experimenter may wish to determine.

BELL'S THEOREM The Copenhagen point of view nevertheless seemed to be challenged by a famous thought experiment known as the **EPR paradox** after Einstein, Podolsky & Rosen (1935). They considered a system of two particles, which today is usually illustrated by a pair of fermions emitted by the decay of a spin-0 particle. If we let the particles separate to large distances and then measure (say) the x-component of spin for particle 1, then we know for certain that a subsequent measurement on particle 2 will yield the opposite answer: if we obtain 'up', then particle 2 must be 'down'. We can now interrogate a friend who performed the measurement on particle 2 at a location causally disconnected from ours and find that they did indeed measure 'down'; it seems we can conclude that particle 2 always 'knew' that it had this x-component of spin. Does this mean that each particle carried with it an instruction manual for how it should respond to any given spin component measurement? It may seem quite plausible that one could produce the statistical aspect of quantum mechanics by giving different instructions to each particle pair, so as to yield the correct distribution of results. This was Einstein's view: that the appearance of quantum uncertainty arose because of the existence of **hidden variables**. The idea is that there exists an 'objective reality' that obeys straightforward causal physics, with the probabilistic aspect arising through our ignorance of all the properties of the particles. A plausible analogy might be with the pseudo-random numbers often used in numerical computation. These are produced causally via a simple algorithm, and yet appear truly random to anyone without access to the algorithm.

Remarkably, any 'hidden realism' scheme of this sort turns out not to be tenable in quantum mechanics, as was shown by John Bell (1964). The argument is quite simple, but constitutes one of the most important advances in understanding quantum phenomena since the glory days at the start of the twentieth century.

Begin by writing down the wave function for the two-particle state, as a superposition of eigenstates. Let $\uparrow$ and $\downarrow$ stand for 'spin up' and 'spin down', where the spin is projected onto some arbitrary axis $\hat{\mathbf{a}}$:

$$|\psi\rangle = \frac{1}{\sqrt{2}}\Big[\,|\uparrow\downarrow\rangle - |\downarrow\uparrow\rangle\Big], \tag{6.10}$$

where the possible spin values along $\hat{\mathbf{a}}$ are rescaled to be $S_a = \pm 1$. Now, suppose that instead of measuring particles 1 and 2 along the same axis, we measure the spin of particle 2 along a different axis $\hat{\mathbf{b}}$. The spin operators projected along these two arbitrary axes are just $A \equiv \boldsymbol{\sigma}\cdot\hat{\mathbf{a}}$, $B \equiv \boldsymbol{\sigma}\cdot\hat{\mathbf{b}}$, where $\boldsymbol{\sigma}$ is the vector of Pauli spin matrices, and so the expected spin correlation function for the two axes is

$$C(\mathbf{a},\mathbf{b}) \equiv \langle S_a(1)S_b(2)\rangle = \langle\psi|A(1)B(2)|\psi\rangle = -\hat{\mathbf{a}}\cdot\hat{\mathbf{b}} = -\cos\theta \tag{6.11}$$

(using the **anticommutation relation** $\{\sigma_i, \sigma_j\} = 2\delta_{ij}$ obeyed by the Pauli spin matrices – see below). So, for spin experiments with $\theta = 0$, there is perfect anticorrelation: the EPR 'objective reality'. Thus far we have no test: both quantum mechanics and causal theories make the same prediction in this case. Things become more interesting when the correlation is less than perfect.

For general angles, the EPR viewpoint requires some hidden variable λ, which controls the values of the spin components of the particles. The correlation for one pair is a deterministic function of λ:

$$S_a(1)S_b(2) = S_a(1,\lambda)S_b(2,\lambda). \tag{6.12}$$

So far, we can clearly reach accord with $C(\mathbf{a},\mathbf{b})$ in quantum mechanics, by allowing λ to have some appropriate probability density, $d\rho(\lambda)$. However, the trouble starts when we realize that the properties of the particles must be set before the experimenter decides which axes to measure, and we cannot fool all the experiments all the time. Pick a third axis, $\hat{\mathbf{c}}$, and take the difference of two spin correlations:

$$\begin{aligned} C(\mathbf{a},\mathbf{b}) - C(\mathbf{a},\mathbf{c}) &= \int [S_a(1)S_b(2) - S_a(1)S_c(2)]\, d\rho(\lambda) \\ &= \int S_a(1)S_b(2)[1 + S_b(1)S_c(2)]\, d\rho(\lambda), \end{aligned} \tag{6.13}$$

where the second step follows because $S_b(1)S_b(2) = -1$. Lastly, because $|S| \leq 1$, the term in square brackets in the integrand cannot be negative and we clearly have

$$\boxed{|C(\mathbf{a},\mathbf{b}) - C(\mathbf{a},\mathbf{c})| \leq \int [1 + S_b(1)S_c(2)]\, d\rho(\lambda) = 1 + C(\mathbf{b},\mathbf{c}).} \tag{6.14}$$

This is **Bell's inequality**, and it is easy to find cases where it is violated by the quantum $C = -\cos\theta$ correlation (consider **a**, **b** and **c** to be coplanar and separated by rotations of ϕ: the inequality fails for $\phi > 52^\circ$). This example illustrates how quantum mechanics violates the inequality, which in this case reduces to $1 + C(2\phi) \leq 2[1 + C(\phi)]$. Hidden realism allows only a linear relaxation away from perfect anticorrelation, whereas quantum mechanics generates probabilities quadratically from the linearly changing amplitudes.

This argument says that in a sense particles have no properties until these are brought into being via the act of observation. The only way in which this conclusion may be evaded is if we allow the existence of **non-local communication**: measurements on particle 1 can then affect particle 2 in some fashion not limited by the propagation of causal signals. Not surprisingly, most physicists prefer the conceptual difficulties of quantum mechanics to any such tinkering with the basics of relativity. These rather disturbing conclusions about the nature of elementary particles provide some motivation for moving towards a view in which the prime entity in physics is not the particle, but the quantum field of which particles are a manifestation.

Finally, although the above is presented as a thought experiment, and indeed was enough to convince many even in this form, Bell's ideas have been implemented in a series of experiments performed in the 1980s. These used photons rather than fermions; the only difference is that, instead of the two possible spin components along a given axis, we have two possible polarizations – either parallel or perpendicular to an analyser. The correlation function for quantum mechanics is now $C(\theta) = -\cos(2\theta)$, but otherwise things follow as above. Bell's inequality is clearly violated experimentally (see e.g. Bransden & Joachain 1989).

WAVE-FUNCTION COLLAPSE Perhaps the most puzzling aspect of quantum theory is the way in which observation makes concrete one of the many statistical possibilities that the wave function allows for the outcome of a given experiment. Quantum mechanics is deterministic and time-reversible (the evolution of ψ is described by Schrödinger's equation) except at the moment of observation, when an irreversible change is introduced into the system. The question of the physical mechanism that produces this collapse of the wave function has produced much debate, which continues today with little sign of a resolution (see, for example, Wheeler & Zurek 1983; Omnès 1994). The problems are at their most perplexing when we consider the application of quantum mechanics to macroscopic objects, in particular living beings. **Schrödinger's cat** is a well-known beast that lives in a box with poison gas that will be released if a certain nucleus decays; the cat is formally in a mixed state between alive and dead before the box is opened. Many people feel instinctively that this must be nonsense: the cat surely knows if it has been poisoned, whether or not an observer has opened the box.

The dilemma is sharpened if another observer (often called **Wigner's friend**) opens the box first, but does not tell a second observer what is seen. We know that physics works in a consistent way, so that the two observers will always agree on the health of the cat, but when does the wave function actually collapse? Do we need a wave function for the brain of Wigner's friend also? These questions raise the issue of **solipsism**: how can a given observer be sure that they are not the only conscious being in the universe? The Copenhagen interpretation of quantum mechanics is simply a theory of how information enters the brain of this single observer, and so may seem to deny the self-awareness of other individuals such as Wigner's friend. There is an analogy here with special relativity: we do not say that the aether does not exist – only that its existence makes no difference to what we actually observe and calculate. Similarly, quantum mechanics gets the right answers to questions of observables independently of whether other people are truly self-aware, in effect saying that this is not a question that is accessible to experimental physics.

DECOHERENCE A more recent development in this debate claims to solve some of the deeper problems. First note that we have assumed that a wave function description is always applicable, which is not always true. An isolated system may be described by a wave function, but a sub-part of a single system will not be so described unless the overall wave function factorizes. In other words, the overall system of Schrödinger's cat plus killing apparatus may be described by a wave function, but the cat is not so easily disentangled from its environment.

A more general concept is the **density matrix** (see e.g. section 42 of Schiff 1955). To describe simply what this is, suppose that a given system has eigenstates $|i\rangle$ corresponding to some operator. If this system is described by a wave function, this will be expandable in terms of the eigenstates: $|\psi\rangle = \sum_i a_i |i\rangle$, where $a_i = \langle i|\psi\rangle$. We can therefore introduce the **projection operator** $|i\rangle\langle i|$, which operates on $|\psi\rangle$ to yield its ith 'component': $|i\rangle\langle i|\,|\psi\rangle = a_i|i\rangle$. Now, when a non-isolated system is observed, the result will still be one of the eigenvalues of the operator, and there will be some probability p_i that the system will have been prepared in this state. The density operator is the weighted sum of the projectors over all possible states:

$$\boxed{\rho = \sum_i p_i\, |i\rangle\langle i|.} \tag{6.15}$$

In this language, noting that ρ can be represented by a matrix, the way to find the probability of being in a state $|j\rangle$ is $p_j = \mathrm{Tr}\left(|j\rangle\langle j|\rho \right)$:

$$\begin{aligned}
\mathrm{Tr}\left(|j\rangle\langle j|\rho \right) &= \sum_n \langle n|j\rangle \, \langle j|\rho\, |n\rangle \\
&= \sum_{n,i} p_i \langle n|j\rangle \, \langle j|i\rangle \, \langle i|n\rangle \\
&= \sum_{n,i} p_i \langle i|n\rangle \, \langle n|j\rangle \, \langle j|i\rangle \\
&= \sum_i p_i \langle i|j\rangle \, \langle j|i\rangle = p_j,
\end{aligned} \tag{6.16}$$

where some reordering of terms and completeness on n has been used. Similarly, the density operator gives a general means of obtaining expectation values of operators [problem 6.3]: $\langle O\rangle = \mathrm{Tr}(O\rho)$.

Note that this formalism includes the case of a **pure state**, where the system is described by a single wave function. Consider for simplicity the case where this is a mixture of two other states: $|\psi\rangle = a|1\rangle + b|2\rangle$, where $\langle 1|2\rangle = 0$. The density matrix is

$$\rho = |\psi\rangle\langle\psi| = |a|^2|1\rangle\langle 1| + |b|^2|2\rangle\langle 2| + ab^*|1\rangle\langle 2| + a^*b|2\rangle\langle 1|. \tag{6.17}$$

The indication of the possibility of quantum interference here is the mixed terms $|1\rangle\langle 2|$ and $|2\rangle\langle 1|$. For a 'classical' system, which would definitely be either in state $|1\rangle$ or $|2\rangle$, these off-diagonal terms would be zero, so that

$$\rho = |a|^2|1\rangle\langle 1| + |b|^2|2\rangle\langle 2|. \tag{6.18}$$

Now, the key point is that in both classical and non-classical cases the probability of a measurement finding the system in state $|1\rangle$ is independent of the off-diagonal terms: $\mathrm{Tr}\left(|1\rangle\langle 1|\rho \right) = |a|^2$. This gives a possible solution to our intuitive worries about macroscopic objects possibly being in a mixed state. Suppose we set up a cat in a pure state that is a mixture of the $|\text{alive}\rangle$ and $|\text{dead}\rangle$ states. If the cat is an isolated system, that mixture will of course be preserved, and interference will be possible at some later time. In practice, however, the cat is a complex system of many sub-parts all of which interact with its environment – and these interactions are the very agency that will cause the balance of probabilities to tilt from life to death. In these circumstances, the rapid phase changes of the microscopic components of the cat will mean that the wave function at later times will not stay coherent with its initial value, and the opportunity for macroscopic interference will be lost. The idea of decoherence is thus that these microscopic interactions very rapidly cause the off-diagonal components of the density matrix to decay to zero. When we open the box, we find a classical system in which the cat is definitely dead or alive, not a superposition of these states.

These ideas are discussed at length in, for example, chapter 7 of Omnès 1994. The removal of the need to think of a cat as a superposition is certainly progress, but is it a final solution? The quantum uncertainty remains: we cannot predict whether the cat will be alive or dead. The chance of observing these events is independent of whether the cat can interfere with itself, so in a way we are no further forward. Asking questions about whether the cat was alive or dead before the box was opened is analogous to debating whether the aether exists. If it cannot be observed, there can be no final answer to the question.

6.2 The Dirac equation

Conceptual difficulties aside, the programme of nonrelativistic quantum mechanics as summarized above leaves open two obvious questions.

(1) Why does $E = \hbar\omega$ apply to photons? Since we are proceeding with matter by analogy with light, it seems reasonable to want to understand the starting point.

(2) Surely we should be using the correct relativistic energy–momentum relation?

The first point is harder to deal with and is postponed to the next chapter. The second may seem more straightforward: surely it is only necessary to insert the normal operator substitutions into $E = \sqrt{p^2c^2 + m^2c^4} + V$? For a free particle, this produces the **Klein–Gordon equation**

$$\boxed{\left(\Box + \frac{m^2c^2}{\hbar^2}\right)\psi = 0.} \tag{6.19}$$

This equation was in fact written down by Schrödinger prior to the invention of nonrelativistic quantum mechanics. Impressively, he solved it as far as obtaining the fine structure of the hydrogen atom – but obtained answers that failed to fit reality. Only as a result of this did Schrödinger turned to nonrelativistic quantum mechanics. As we shall see, this 'failure' arose because the Klein–Gordon equation does not fully describe fermions.

The Klein–Gordon equation seems rather puzzling from the standpoint of nonrelativistic experience. First, it is second order in time, so that the initial information needed to specify the time evolution of ψ is completely different in nature. Second, the conserved quantity associated with ψ need not be positive [problem 6.2]. These difficulties can be overcome, but only at the price of ceasing to assume that ψ describes a single particle. Faced with these apparently fatal problems, Dirac's approach was to insist that one should have a wave equation that, while still relativistic, was linear in time; relativity then requires that it should be linear in spatial derivatives also, and we are inevitably led to the general form

$$H\psi = i\hbar\dot{\psi}, \quad H = c(\boldsymbol{\alpha}\cdot\mathbf{p}) + \beta mc^2, \tag{6.20}$$

where $\boldsymbol{\alpha}$ and β are vector and scalar constants. Inserting this Hamiltonian into the relativistic energy–momentum relation gives the following conditions for consistency:

$$\begin{aligned} \beta^2 &= 1, \\ \{\beta, \alpha_i\} &= 0, \\ \{\alpha_i, \alpha_j\} &= 2\delta_{ij}. \end{aligned} \tag{6.21}$$

The fact that these relations involve **anticommutators** $\{a, b\} \equiv ab + ba$ seems bad news from the point of view of obtaining a simple wave equation. Dirac nevertheless persisted in taking these relations seriously: $\boldsymbol{\alpha}$ and β must then be interpreted as *operators* whose existence reveals a new internal degree of freedom.

It is illuminating to look for matrices that represent these operators. A simple counting of the number of equations to be satisfied by four Hermitian matrices shows that the smallest viable candidates must be 4×4. The most common representation is the **Dirac representation**:

$$\beta = \begin{pmatrix} I & 0 \\ 0 & -I \end{pmatrix}, \qquad \alpha_i = \begin{pmatrix} 0 & \sigma_i \\ \sigma_i & 0 \end{pmatrix}, \tag{6.22}$$

where we have used the **Pauli spin matrices**,

$$\sigma_1 = \begin{pmatrix} 0 & 1 \\ 1 & 0 \end{pmatrix}, \quad \sigma_2 = \begin{pmatrix} 0 & -i \\ i & 0 \end{pmatrix}, \quad \sigma_3 = \begin{pmatrix} 1 & 0 \\ 0 & -1 \end{pmatrix}. \tag{6.23}$$

These represent the algebra of the $J = \hbar/2$ angular momentum operator and obey the anticommutation relation $\{\sigma_i, \sigma_j\} = 2\delta_{ij}$.

To use this matrix representation, the wave function must be represented by a column vector, which is known as a **spinor**. The justification for this name will become apparent shortly, but the fact that the Pauli matrices enter into the algebra should already give a strong hint that this will be the physical interpretation of the extra degrees of freedom in ψ.

RELATIVISTIC FORM The Dirac equation can be put in a more compact form by defining a new set of matrices known as the **gamma matrices**, $\gamma^0 \equiv \beta$, $\gamma^i \equiv \beta\alpha_i$. From the anticommutation relations between the α and β matrices, these obey the anticommutation relation

$$\boxed{\{\gamma^\mu, \gamma^\nu\} = 2g^{\mu\nu}.} \tag{6.24}$$

In these terms, and remembering the de Broglie operator relation $P_\mu = i\hbar\partial_\mu$, the Dirac equation becomes

$$\boxed{(\gamma^\mu P_\mu - mc)\psi = 0.} \tag{6.25}$$

The equation in this form includes the relativistic scalar product of a 4-vector with γ^μ; this occurs sufficiently often that there is a special notation for it:

$$\not{a} \equiv \gamma^\mu a_\mu; \tag{6.26}$$

in words, 'a slash'.

CONSERVATION EQUATION One of the motivations for introducing the Dirac equation was to obtain a physically sensible probability density. To achieve this aim, we need some more notation: define the **adjoint spinor** $\overline{\psi}$ by

$$\overline{\psi} \equiv \psi^\dagger \gamma^0; \tag{6.27}$$

as usual, the Hermitian conjugate $\psi^\dagger$ denotes the complex conjugate of the transposed wave function (i.e. taking the usual vector transpose of the column vector ψ). Now write the Dirac equation in adjoint form: $P_\mu(\overline{\psi}\gamma^\mu) + mc\overline{\psi} = 0$ (for this, the relations $\gamma^{0\dagger} = \gamma^0$, $\gamma^{j\dagger} = -\gamma^j$; $j \neq 0$ are needed), and take the combination $\overline{\psi}\,[\text{dirac}] - \overline{[\text{dirac}]}\,\psi$. This gives

$$\boxed{\partial_\mu(\overline{\psi}\gamma^\mu\psi) = 0,} \tag{6.28}$$

which is in the form of the continuity equation, suggesting that $J^\mu = \overline{\psi}\gamma^\mu\psi$ is to be interpreted as the probability 4-current. Certainly, the density J^0 proportional to $\psi^\dagger\psi$ describes a conserved quantity that is always positive, making it plausible that this is the probability density for the particle (exactly similar algebra and reasoning leads to the identification of $|\psi|^2$ as the probability density in nonrelativistic quantum mechanics).

The relativistic form of the Dirac equation looks like a manifestly covariant equation, but in fact is nothing of the sort: the '4-vector' of matrices γ^μ is constant in all frames and the wave function ψ is not a relativistic scalar invariant. Instead, the transformation properties of the Dirac spinors are determined by some matrix S:

$$\psi'(x'^\mu) = S\,\psi(x^\mu), \tag{6.29}$$

where the coefficients of the corresponding Lorentz transformation are

$$x'^\mu = \Lambda^\mu{}_\nu\, x^\nu. \tag{6.30}$$

If ψ satisfies the Dirac equation in the unprimed frame, then is is easy to see that ψ' satisfies the same equation in the primed frame, provided

$$\gamma^\nu = \Lambda^\nu{}_\mu\, S\,\gamma^\mu\, S^{-1}. \tag{6.31}$$

In addition, requiring $\bar{\psi}\psi$ to be an invariant implies the further condition on S

$$S^{-1} = \gamma^0 S^\dagger \gamma^0. \tag{6.32}$$

In principle, these equations can be solved to find the transformation matrix S for a given Lorentz transformation, although it may not seem so obvious that this can always be done; the trick is to consider only infinitesimal transformations $\Lambda^\mu{}_\nu = \delta^\mu{}_\nu + a^\mu{}_\nu$, $S = I + b$, and solve only to first order in the small quantities a and b. Finite transformations may then be produced by repeated application of these generators.

ELECTROMAGNETIC FIELD The meaning of the Dirac equation is illuminated when we consider the equation as modified to allow for the presence of an electromagnetic field. Recall how this affects the Schrödinger equation: we simply write down the classical Hamiltonian for a particle in an electromagnetic field,

$$H = \frac{1}{2m}(\mathbf{p} - e\mathbf{A})^2 + e\phi, \tag{6.33}$$

and make the usual operator substitution for $\mathbf{p}$. What we are doing here is subtracting the field momentum $e\mathbf{A}$ from the total to obtain the particle momentum. Note that the quantity e is the charge of the particle: for an electron, e is a negative number.

The above Hamiltonian is usually justified by showing that it leads to the correct equation of motion (the Lorentz force) in Hamiltonian mechanics. The result is sufficiently central to much of the material in this part that it is worth giving a simpler derivation to gain a better feeling for what is going on. Consider the equation of motion $\dot{\mathbf{p}} = e(\mathbf{E} + \mathbf{v} \wedge \mathbf{B})$, and consider the rate of doing work on the particle, $\dot{E} = \dot{\mathbf{p}} \cdot \mathbf{v} = -e(\dot{\mathbf{A}} \cdot \mathbf{v} + \mathbf{v} \cdot \boldsymbol{\nabla}\phi)$, since the magnetic field does no work. Rearranging this expression gives $(\dot{\mathbf{p}} + e\dot{\mathbf{A}}) \cdot \mathbf{v} = -e\mathbf{v} \cdot \boldsymbol{\nabla}\phi$, which suggests an interpretation in which $\mathbf{p} + e\mathbf{A}$ is a total momentum and the total energy is modified only by the term involving ϕ: $E \to E + e\phi$. So, to obtain the particle energy and momentum, we must *subtract* the field contribution from the total:

$$\boxed{P^\mu \to P^\mu - eA^\mu.} \tag{6.34}$$

This substitution is the general one that is always made to allow for the effects of an electromagnetic field.

This argument is at best only heuristic. To do things properly, we need the Lagrangian that describes particle motion. This can be guessed on relativistic grounds: $\int L\, dt = \int L\gamma\, d\tau$ must be stationary, so that γL must be an invariant that looks like a kinetic energy minus a potential. For a slowly moving particle, this suggests $\gamma L = -mc^2 - e\phi$ (the minus sign on the mass term is so that the non-electromagnetic part of the Lagrangian is $L = -mc^2/\gamma \simeq -mc^2 + mv^2/2$ and the kinetic energy has positive sign). The natural relativistic generalization of this is

$$\begin{aligned} \gamma L &= -mc^2 - eU_\mu A^\mu \\ \Rightarrow\; L &= -mc^2/\gamma - e\phi + e\mathbf{v}\cdot\mathbf{A}, \end{aligned} \tag{6.35}$$

where U_μ is the 4-velocity of the particle. It is not to hard to check that this reasoning has worked, by showing that Euler's equation gives the Lorentz force (remember that $d/dt = \partial/\partial t + \mathbf{v}\cdot\boldsymbol{\nabla}$). The **canonical momentum** is by definition $p_i = \partial L/\partial v_i = \gamma m v_i + eA_i$, as required. In these terms, the Hamiltonian is

$$H \equiv \mathbf{p}\cdot\mathbf{v} - L = \left[m^2c^4 + (\mathbf{p}-e\mathbf{A})^2c^2\right]^{1/2} + e\phi, \tag{6.36}$$

as required for the $P^\mu \to P^\mu - eA^\mu$ substitution to be the correct procedure.

In the case of the Dirac equation in an electromagnetic field, two coupled equations are obtained if the wave function is divided into an upper and lower part, $\psi = \binom{\psi_{\rm U}}{\psi_{\rm D}}$:

$$\begin{aligned} [\boldsymbol{\sigma}\cdot(\mathbf{p}-e\mathbf{A})]\psi_{\rm D} &= \frac{1}{c}(E - e\phi - mc^2)\psi_{\rm U} \\ [\boldsymbol{\sigma}\cdot(\mathbf{p}-e\mathbf{A})]\psi_{\rm U} &= \frac{1}{c}(E - e\phi + mc^2)\psi_{\rm D}. \end{aligned} \tag{6.37}$$

If we consider $E \simeq mc^2$, then to order of magnitude these equations imply $\psi_{\rm D} \sim (v/c)\psi_{\rm U}$; for nonrelativistic particles in weak fields, the lower part of ψ goes to zero.

Leaving aside the interpretation of $\psi_{\rm D}$ for now, solve for $\psi_{\rm U}$. The equation obeyed by this part of the wave function is

$$[\boldsymbol{\sigma}\cdot(\mathbf{p}-e\mathbf{A})]^2\psi_{\rm U} = \frac{1}{c^2}(E - e\phi - mc^2)(E - e\phi + mc^2)\psi_{\rm U}. \tag{6.38}$$

In the nonrelativistic limit with $E - mc^2 = E_{\rm NR}$ and $E - e\phi + mc^2 \to 2mc^2$, this looks very similar to Schrödinger's equation. The complicating factor is the appearance of $\boldsymbol{\sigma}$, the vector of Pauli spin matrices. To convert this to something more illuminating, we can use the relation $\sigma_i\sigma_j = \delta_{ij} + i\sigma_k\epsilon_{ijk}$ (which just expresses the fact that the Pauli matrices are orthogonal and obey the cyclic relations $\sigma_1\sigma_2 = i\sigma_3$ etc.). From this, we conclude that $(\boldsymbol{\sigma}\cdot\mathbf{a})^2 = a^2 + i\boldsymbol{\sigma}\cdot(\mathbf{a}\wedge\mathbf{a})$. Now, classically $\mathbf{a}\wedge\mathbf{a} = \mathbf{0}$, but not if $\mathbf{a}$ is an operator. Something magical happens when $\mathbf{a} = \mathbf{p} - e\mathbf{A}$; because $\mathbf{p} \propto \boldsymbol{\nabla}$, $\mathbf{a}\wedge\mathbf{a} = -e(\mathbf{p}\wedge\mathbf{A}) = ie\hbar\mathbf{B}$. Thus, in the nonrelativistic limit, the Dirac equation yields the familiar form

$$\boxed{\left[\frac{1}{2m}(\mathbf{p}-e\mathbf{A})^2 - \frac{e\hbar}{2m}\boldsymbol{\sigma}\cdot\mathbf{B}\right]\psi = (E_{\rm NR} - e\phi)\psi.} \tag{6.39}$$

This is the Schrödinger equation for a particle interacting with an electromagnetic field and possessing a magnetic moment of $e\hbar/(2m)$ – i.e. 1 Bohr magneton. The Dirac theory thus yields the correct 'g-factor' of $g = 2$ automatically.

There are many comments to make about this marvellous result. First, and most important, it is sometimes stated on the basis of this analysis that spin is in some sense a relativistic effect. This is not so. If we adopt the nonrelativistic Hamiltonian $H = (\boldsymbol{\sigma} \cdot \mathbf{p})^2/(2m) + V$, then we have just shown that the appearance of a magnetic moment arises automatically. This is a consistent choice at the classical level: $(\boldsymbol{\sigma} \cdot \mathbf{a})^2$ may be replaced by a^2 when $\mathbf{a}$ is not an operator – because $\sigma_i \sigma_j + \sigma_j \sigma_i = 2\delta_{ij}$. This provides another instance of the general problem that the transition from a classical system to a quantum one cannot be performed uniquely.

What has happened is that the Dirac equation has made this the *natural* choice for the nonrelativistic Hamiltonian. To obtain a consistent equation following Dirac's line of argument, it was necessary to include an internal degree of freedom, which is now identified with spin.

NEGATIVE ENERGIES AND THE DIRAC SEA So much for ψ_{U}, what about ψ_{D}? To make the meaning of this part of the wave function clearer, it is interesting to consider a particle at rest and without electromagnetic interactions. We then have the relations

$$(E - mc^2)\psi_{\mathrm{U}} = (E + mc^2)\psi_{\mathrm{D}} = 0, \tag{6.40}$$

which is consistent with our earlier statement that ψ_{D} vanishes as $v \to 0$. However, the symmetry of the equations suggests another solution, with $\psi_{\mathrm{U}} = 0$ and $E = -mc^2$. The general solution for the wave function would then be

$$\psi = \left[\begin{pmatrix} a \\ 0 \\ 0 \\ 0 \end{pmatrix} + \begin{pmatrix} 0 \\ b \\ 0 \\ 0 \end{pmatrix} \right] e^{-imc^2 t/\hbar} + \left[\begin{pmatrix} 0 \\ 0 \\ c \\ 0 \end{pmatrix} + \begin{pmatrix} 0 \\ 0 \\ 0 \\ d \end{pmatrix} \right] e^{+imc^2 t/\hbar}. \tag{6.41}$$

This may seem unreasonable; mathematically, the negative energies arise because we have taken a root: $E = \pm[p^2c^2 + m^2c^4]^{1/2}$, but surely they are unphysical? Well, it is impossible to ignore the negative-energy solutions. For non-zero momentum, all four components of the wave function will be non-zero in general. It is possible to superpose states so that at some initial time $\psi_{\mathrm{D}} = 0$, but the different time dependences of the various states will cause ψ_{D} to evolve with time, so that it becomes non-zero.

Trying to take the concept of negative-energy states literally leads to some unpleasant consequences. In particular, if states with $E < -mc^2$ are accessible to an electron, why do all particles not immediately make downward transitions – emitting photons with $\omega \gtrsim mc^2/\hbar$ ($\sim 10^{20}$ Hz) as they go? Dirac's solution to this was direct: the 'sea' of negative-energy states must be filled, so that transitions are forbidden through the exclusion principle. This gives us our first particle-physics estimate of the cosmological constant, although it hardly counts as a very great success:

$$\rho_{\mathrm{vac}} = -\infty. \tag{6.42}$$

This is only the first of several divergent contributions to the energy density of the vacuum that arise with quantum fields.

PLANE WAVES Extending the discussion now to states with non-zero momentum, we need a notation for the two independent states corresponding to the two signs of the energy for each momentum. The conventional choice is

$$\psi = \begin{cases} u_r(\mathbf{p}) \exp\left(-ip^\mu x_\mu/\hbar\right) & (E > 0) \\ v_r(\mathbf{p}) \exp\left(+ip^\mu x_\mu/\hbar\right) & (E < 0), \end{cases} \tag{6.43}$$

where r is an index labelling the two independent solutions. Inserting this into the Dirac equation shows that the time-independent spinors u_r and v_r satisfy respectively $(\not{p} - mc)u_r = 0$ and $(\not{p} + mc)v_r = 0$. As before, these spinors will have an upper part and a lower part, and we can choose the independent states such that the upper parts of u_1 and u_2 are respectively

$$\propto \begin{pmatrix} 1 \\ 0 \end{pmatrix} \text{ and } \propto \begin{pmatrix} 0 \\ 1 \end{pmatrix},$$

and similarly for the lower parts of v_1 and v_2. The momentum-dependent remaining parts of the spinors are then fixed by the Dirac equation. The overall normalization is chosen conventionally as follows:

$$u_r^\dagger(\mathbf{p})u_s(\mathbf{p}) = v_r^\dagger(\mathbf{p})v_s(\mathbf{p}) = \frac{|E|}{mc^2}\,\delta_{rs}. \tag{6.44}$$

HELICITY AND CHIRALITY The different states u_1 and u_2 correspond loosely to spin up and spin down, but the direction of the spin axis has not yet been specified. The natural choice is the momentum vector of the particle, and it is standard in nonrelativistic quantum mechanics to define the **spin projection operator**, which selects the desired spin component:

$$\sigma_p = \frac{\boldsymbol{\sigma} \cdot \mathbf{p}}{|\mathbf{p}|}, \tag{6.45}$$

where $\boldsymbol{\sigma}$ is a vector of Pauli spin matrices. For the 4-spinors occurring in the Dirac equation, this definition needs extending. The obvious generalization is to a vector of 4×4 matrices $\boldsymbol{\sigma} = (\sigma^{23}, \sigma^{31}, \sigma^{12})$, where

$$\sigma^{ij} = \begin{pmatrix} \sigma_k & 0 \\ 0 & \sigma_k \end{pmatrix}, \tag{6.46}$$

and i, j, k are in cyclic order. This expression applies for the Dirac representation, but a more general definition would be in terms of the commutator of Dirac matrices:

$$\sigma^{\mu\nu} = \frac{i}{2}\left[\gamma^\mu, \gamma^\nu\right]. \tag{6.47}$$

The plane wave states are thus constructed to be eigenstates of the spin projection operator, and a fermion will have spin either parallel or antiparallel to the momentum; these are states of definite **helicity**, where the name is suggested by the classical picture of the electron as a spinning ball. We can define **helicity operators** that pick out just one of these states (i.e. have eigenvalues 0 or 1):

$$\Pi^\pm = \tfrac{1}{2}(1 \pm \sigma_p). \tag{6.48}$$

These operators take an important form for massless particles, which satisfy $\not{p}\psi = 0 \Rightarrow \gamma^0|\mathbf{p}|\psi = \gamma^i p^i \psi$. First, we have to define a new Dirac matrix,

$$\boxed{\gamma^5 \equiv i\gamma^0\gamma^1\gamma^2\gamma^3,} \tag{6.49}$$

whose name reflects a lack of agreement about whether 4-vector indices run from 0 to 3 or from 1 to 4. In the Dirac representation,

$$\gamma^5 = \begin{pmatrix} 0 & I \\ I & 0 \end{pmatrix}. \tag{6.50}$$

Now premultiply $\gamma^0|\mathbf{p}|\psi = \gamma^i p^i \psi$ by $\gamma^5\gamma^0$ and divide by $|\mathbf{p}|$:

$$\gamma^5\psi = \frac{\gamma^5\gamma^0\gamma^i p^i}{|\mathbf{p}|}\,\psi = \sigma_p\psi; \tag{6.51}$$

the last step requires the identity $\sigma^{ij} = \gamma^5\gamma^0\gamma^k$ (i, j, k cyclic), which is most easily verified by direct inspection of the matrices involved. So, for zero mass, the helicity operators become identical to the **chirality operators** (from the Greek for 'handedness'),

$$\Pi^{\pm} = \tfrac{1}{2}(1 \pm \gamma^5). \tag{6.52}$$

A state of definite chirality thus produces only particles of one helicity in the massless limit. In general, the fractional mix of different helicities will be biased in one direction, in the ratios 1 to $\sim (mc^2/E)^2$. We shall see later that the weak interaction involves $1 - \gamma^5$ and therefore produces neutrino states of one chirality only, left-handed.

HIGHER SPINS In all this, it might be objected that the appearance of spin in the Dirac equation was fixed by including the Pauli matrices from the start, but this is not correct: any representation of the Dirac gamma matrices would yield a result that looked like spin, whether or not the matrices involved the σ_i explicitly. Furthermore, representations of the gamma matrices that are larger than 4×4 do not correspond to spins higher than $1/2$ – despite statements to this effect in some textbooks. For spin $3/2$, the analogue of the Dirac equation is known as **Rarita–Schwinger theory** (1941). Here the wave function is a '4-vector' of spinor fields ψ_μ, thus containing 4×4 degrees of freedom. As well as satisfying the Dirac equation, the Rarita–Schwinger field obeys two extra conditions

$$\partial^\mu \psi_\mu = \gamma^\mu \psi_\mu = 0. \tag{6.53}$$

This leaves eight degrees of freedom – the correct number for a particle of spin $3/2$ (four helicity states for each of the positive- and negative-energy states). This is the highest spin a fermion can have. In general, quantum mechanics with particles of high spin becomes hard to make consistent in the sense of producing finite answers to quantum field calculations, and the $s = 2$ graviton has the highest spin of any particle that is considered to be fundamental.

6.3 Symmetries

ANTIPARTICLES The concept of a sea of filled states suggests an analogy with 'holes' in solid-state physics: the absence of a particle in a state with momentum $\mathbf{p}$, energy E and charge e may appear to correspond to the existence of a particle with the negative of these properties. Dirac predicted the existence of such counterparts for all particles; the first of these, the **positron**, was duly discovered by Anderson in 1932. To make such a notion mathematically precise, we need to look at the Dirac equation and see how a given negative-energy state may be transformed into something that looks like the wave function for an antiparticle.

Begin with the wave function for a negative-energy electron: $\psi \propto e^{i(\mathbf{p}\cdot\mathbf{x}+|E|t)/\hbar}$. Complex conjugation will reverse the signs of $\mathbf{p}$ and E, which achieves the desired spatial and time dependences. However, two further manipulations must be performed: we should swap $\psi_{\rm U}$ and $\psi_{\rm D}$, so that the negative-energy states map onto the part of the wave function that behaves sensibly at low energies; we also need to achieve a spin flip $\mathbf{S} \to -\mathbf{S}$, which corresponds to swapping the (1, 2) and (3, 4) components. A little thought shows that the γ^2 matrix will achieve exactly this effect. This heuristic reasoning therefore suggests that

$$\psi_{\rm C} \equiv \gamma^2 \psi^* \tag{6.54}$$

should be regarded as the **charge-conjugate** wave function, which defines the operation of charge conjugation: $C\psi = \psi_{\rm C}$. It is easily verified that $\psi_{\rm C}$ obeys the Dirac equation with the appropriate charge for an antiparticle [problem 6.6]:

$$\left[\gamma^\mu(P_\mu + eA_\mu) - mc\right]\psi_{\rm C} = 0. \tag{6.55}$$

OTHER SYMMETRIES The charge-conjugation operator C is one example of a class of **symmetry operators**, the other two important cases being P (**parity** or **space inversion**) and T (**time reversal**). Here, we are trying to find the effect on the wave function of the coordinate transformations $\mathbf{x} \to -\mathbf{x}$ and $t \to -t$. It turns out [problem 6.6] that the corresponding transformations of the wave function are

$$\begin{aligned} P\psi &= \gamma^0\psi \\ T\psi &= \gamma^1\gamma^3\psi^*. \end{aligned} \tag{6.56}$$

What we are doing here mathematically is showing that the Dirac equation displays **form invariance** under the transformations we have been considering. In other words, the symbolic form of the equation is unaltered: if a mathematically valid solution of the old equations is possible, one will also exist for the new variable. For example, the time-reversal operator in the nonrelativistic Schrödinger equation just corresponds to complex conjugation. If we give as initial data for the first-order equation $\psi(t = 0) = \psi_0$, the forward evolution of ψ is determined: $\psi = \psi_0 + f(t)$. The identical evolution also applies to the time-reversed wave function $\psi_{\rm T}$ if we start with $\psi_{\rm T} = \psi_0$, and so the evolution backwards in time would be $\psi_{\rm T} = \psi_0 + f(-t)$. We have used initial conditions that differ in phase, but this is in any case unobservable; the probability density is symmetric in time: $|\psi(t)|^2 = |\psi_{\rm T}(-t)|^2$. What this means in practice is that, if we had a film of some particular process involving particles that obeyed equations invariant under T, P or C, we could respectively run the film backwards, view it in a mirror, or relabel all particles as the corresponding antiparticles, and *still* see a valid physical process that might occur elsewhere.

More straightforward examples of time-reversal invariance are found in classical physics. Newtonian mechanics involves accelerations $\ddot{\mathbf{x}}$, which are clearly time-reversal invariant. An example that looks more difficult is the diffusion equation $D\nabla^2\rho = \dot{\rho}$; the single time derivative means that ρ does not obey the time-reversed equation. However, thinking of the physical significance of the diffusion coefficient D tells us that this should also change sign if t does. The conclusion is then that diffusive behaviour is symmetric: given a Gaussian density concentration at some time, we expect it to be broader in the past and the future – in agreement with the time-symmetric dynamics of the particles that undergo the random walks described macroscopically by diffusion. Both these classical symmetries and the quantum symmetry discussed above should make it clear that the increase of entropy cannot have a microscopic foundation. Although many workers from Boltzmann onwards have struggled to prove an **H theorem** that shows that microscopic processes inevitably cause an increase in disorder, there is a good case to be made (see Gull 1991 or chapter 19 of Waldram 1985) that the appearance of increase of entropy is only an artefact of an observer's incomplete knowledge of the initial state [problem 6.4].

SYMMETRY VIOLATION In the early days of particle physics, it was thought that the three symmetries C, P and T might each apply separately. However, it became apparent in 1956 that this was not the case when the weak nuclear force was involved (**parity violation**). Up to this point, it had been assumed that each particle had an intrinsic parity quantum number, and that these could be multiplied to find the overall (conserved) parity of a state. In particular, the pion has negative parity, so that decays that produce different numbers of pions must apparently involve different particles; if

$$\begin{aligned} \theta &\to \pi^+ + \pi^0 \\ \tau &\to \pi^+ + \pi^+ + \pi^-, \end{aligned} \tag{6.57}$$

then θ and τ must be of respectively even and odd intrinsic parity and hence be different particles if parity is conserved. The problem was that the mass and lifetime of the θ and τ appeared to be identical. Since the slow rate of the above reactions suggested that they proceeded via the weak interaction, this prompted Lee & Yang (1956) to suggest that parity was not conserved under the weak interaction, and that only a single particle was involved (now known as the K^+).

The important mathematical feature for dealing with parity involves the γ^5 Dirac matrix: $\gamma^5 \equiv i\gamma^0\gamma^1\gamma^2\gamma^3$. The combination $\overline{\psi}\gamma^5\psi$ behaves as a pseudoscalar and reverses sign under coordinate inversion (because $\gamma^0\gamma^5\gamma^0 = -\gamma^5$). The appearance of this sort of quantity in a quantum theory indicates that the theory will not obey parity symmetry, as we have seen above in the case of chirality.

In 1963, experimental evidence became available that the combined symmetry CP was also not obeyed in general. However, one of the great formal results of quantum field theory is that any consistent formulation of the theory must obey the combined symmetry CPT. There is no space to prove this fundamental theorem here; see e.g. Sakurai (1964).

MAJORANA PARTICLES It is possible that a particle will be its own antiparticle; this is familiar in the case of the photon. For fermions obeying the Dirac equation, this corresponds to the requirement that $\psi_C = \psi$. In this case, it is advantageous to adopt the **Majorana representation** in which the gamma matrices are purely imaginary:

$$\gamma^\mu_{\rm majorana} = \left(\gamma^1_{\rm dirac}, \gamma^2_{\rm dirac}, i\gamma^0_{\rm dirac}, -i\gamma^5_{\rm dirac}\right). \tag{6.58}$$

At this point, we should recast some of the earlier discussion of the Dirac equation in a way that is representation independent. If we put $\overline{\psi} \equiv \psi^\dagger A$, then the condition for Lorentz invariance of $\overline{\psi}\psi$ is

$$\gamma^{\mu\dagger} = A\gamma^\mu A^{-1}. \tag{6.59}$$

If $\psi_C \equiv B\psi^*$, then the condition for ψ_C to satisfy the charge-conjugate Dirac equation is

$$B^{-1}\gamma^{\mu *}B = -\gamma^\mu. \tag{6.60}$$

In the Majorana representation, this leads simply to $\psi_{\rm C} = \psi^*$. Thus, the equivalence between particles and antiparticles in this representation is achieved by having a *real* wave function, and thus a real wave equation. This makes mathematical and physical sense: the wave function has half the number of degrees of freedom, and the particle must be uncharged because we cannot introduce the electromagnetic interaction via $\partial^\mu \to \partial^\mu + ieA^\mu/\hbar$.

For a case without an obvious label such as charge for distinguishing particles and antiparticles, it is unclear *a priori* whether we are dealing with Majorana particles. Neutrinos are usually taken to be Majorana in nature, but it is possible to imagine that they might not be. In this case, the confusing terminology is to speak of **Dirac neutrinos**, even though both types satisfy the Dirac equation. The distinction between Majorana and Dirac particles vanishes if the neutrino mass is zero, since we have seen above that only left-handed neutrinos are produced in this limit. If the mass is non-zero, however, both left- and right-handed helicities are allowed and the choice between Dirac and Majorana neutrinos muct be faced. This topic is discussed further in the section on massive neutrinos.

6.4 Spinors and complex numbers

So far, we have encountered spinors as simply representing the wave function for a particle with spin. There is much more to spinors than this: just as vectors are more than a straightforward column of numbers, so spinors have fundamental *geometrical* significance. The algebra of tensors in general and Riemannian geometry in particular is an impressive edifice, and it may be hard to believe that not all geometrical concepts can be treated in this way. However, even simple rotations have some more subtle aspects that require the introduction of a new kind of geometrical entity, which is the **classical spinor**. The theory of spinors can often seem rather abstract, but it is well worth trying to explore briefly a few aspects of the subject. A knowledge of spinors leads to a fuller understanding of the nature of spacetime and of why something like spin in quantum mechanics should exist at all. For a (much) more detailed treatment, see Corson (1953) or Penrose & Rindler (1984).

ROTATIONS AND THE NEED FOR SPINORS The main example of the use of spinors with which most physicists are familiar is the treatment of spin-$\frac{1}{2}$ particles in quantum mechanics, where the existence of non-integral angular momentum has some peculiar consequences. For a state with z-component of angular momentum L_z, we would expect to find a wave function looking like

$$\psi \propto e^{iL_z\phi/\hbar} \tag{6.61}$$

(making the usual $i\hbar\partial/\partial\phi$ operator assignment for the azimuthal angular momentum operator). Continuity of the wave function thus appears to demand that L_z is quantized in multiples of $\hbar$. However, we can show from the algebra of ladder operators that half-integral angular momentum can exist in principle [problem 6.7]. This means that ψ would change sign under a rotation of 2π and only return to its original value after a rotation of 4π. This peculiar behaviour is something we tend to accept because of the empirical existence of spin. It shows that, for fermions, the wave function cannot be described by any familiar geometrical object: we must introduce a new quantity, called a spinor.

Motivating the introduction of spinors through quantum mechanics is unsatisfactory in several ways. It gives the impression that spin is purely a quantum phenomenon and does not make clear what it is about space that *requires* the existence of non-integral spin.

THE TOPOLOGY OF ROTATIONS In fact, the paradoxical failure of wave functions to return to their initial state after a rotation of 2π is entirely a classical phenomenon. To analyse this, we need to look in more detail at the properties of rotations, which are described by the Lie group $SO(3)$ (rotations are described by three-dimensional orthogonal matrices with unit determinant); see chapter 8 for more detail on Lie groups. Three spatial dimensions is a rather special number because $SO(3)$ then has just three independent **generators**. The generators are the matrices corresponding to infinitesimal rotations: $\mathbf{R} = \mathbf{I} + \epsilon\mathbf{G}$, and $\mathbf{G}$ has to be antisymmetric and traceless in order for $\mathbf{R}$ to be a rotation matrix. For $n \times n$ $SO(n)$ matrices, there are thus $n(n-1)/2$ independent generators, and so only in three dimensions can a rotation be represented by a vector.

This coincidence is the root of the peculiar **topological properties** of $SO(3)$, because there is an ambiguity in the assignment of rotation axis: a rotation of θ about the axis $\mathbf{x}$ is equivalent to a rotation of $-\theta$ about $-\mathbf{x}$. Suppose we represent the effect of a rotation by a position vector inside a sphere, so that the direction of the vector gives the rotation axis and the length of the vector gives the rotation angle, as illustrated in figure 6.1. Because of the ambiguity in rotations through an angle π, the maximum radius of the sphere is π, and antipodal points on the surface of the sphere are equivalent. Now apply a set of infinitesimal rotations to a body; the final state is always equivalent to one rotation, so the movement of the body can be represented by following a locus inside the sphere. A closed loop inside the sphere would then represent a set of rotations that were topologically equivalent to no rotation at all. We would demonstrate this by showing that the loop could be shrunk to a point by a series of small deformations of the loop. However, suppose we now deal with a rotation by 2π. The line heads out radially along some direction $\hat{\mathbf{x}}$; when it reaches radius π, it reappears on the opposite side of the sphere in the direction $-\hat{\mathbf{x}}$. This line cannot be shrunk to a point, because opposite points of the sphere at $r = \pi$ are identified: a small movement in the 'exit' point will be exactly reflected in the corresponding 'entry'. A rotation of 4π does not share this problem.

So, a rotation of 2π is not equivalent classically to no rotation at all. The proof may seem rather abstract, but receives startling illustration in the party game sometimes dubbed 'Dirac's scissors' (see figure 6.2).

SPINOR ALGEBRA The discussion of the previous section was designed to motivate the existence of spinors, by showing that it would be physically useful if there existed mathematical quantities that transform in such a way that they are left unaltered by a rotation of 4π, rather than one of 2π. This gives no hint about how to construct such a

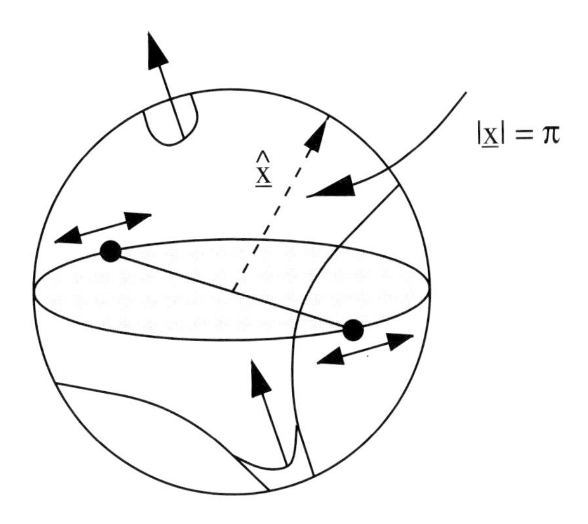

Figure 6.1. This figure illustrates the peculiar topological properties of the rotation group $SO(3)$. The effect of a rotation is represented by a point inside a sphere of radius π (radius = magnitude of rotation). A set of infinitesimal rotations moves a body along a locus inside the sphere. Opposite points on the sphere are identified, so that small deformations of the locus can eliminate pairs of points that cross the boundary. However, a simple rotation by 2π corresponds to a line that crosses a diameter of the sphere, and this can never be removed: attempting to move one of the boundary points of this line simply causes the other to move so as to maintain opposition. Rotation by 2π cannot be transformed continuously into zero rotation.

quantity, but one place to look is inside vectors, since they clearly describe many aspects of space correctly.

A 4-vector can be represented by a 2×2 Hermitian matrix $\mathbf{Q}$, where

$$\mathbf{Q} = \begin{pmatrix} x^0 + x^3 & x^1 - ix^2 \\ x^1 + ix^2 & x^0 - x^3 \end{pmatrix}, \tag{6.62}$$

so that $\det \mathbf{Q}$ is the invariant 4-vector norm. Consider now a null vector, so that the matrix has zero determinant. Such a matrix can always be written as a direct product of two vectors, $Q_{ij} = A_i B_j$, but a Hermitian matrix of zero determinant requires only one vector:

$$Q_{ij} = \chi_i \chi_j^*. \tag{6.63}$$

Thus, a null 4-vector can be constructed out of a two-component complex vector χ_i. This is a hint of the sort of substructure in vectors we were hoping for; the numbers χ_i are the components of a **Weyl spinor**. This name distinguishes the two-component quantities being used here from the four-component spinors that occur in the Dirac equation. Each Dirac spinor is a combination of two Weyl spinors (one each for the 'up' and 'down' parts).

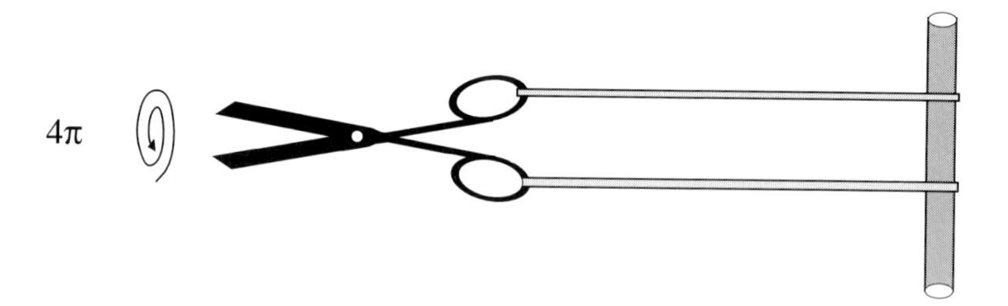

Figure 6.2. Illustrating how rotations of 2π and 4π are not equivalent in a simple classical system. The scissors are connected to a bar by loops of string and the scissors are rotated by 2π about their long axis, twisting the string. The challenge is to untwist the string without reversing the rotation of the scissors. The solution is to rotate 2π *further*, and then pass a loop of string over the end of the scissors. The same thing is going on in the common contortion where a plate held waiter-style is rotated by 2π, thus twisting the arm; a further rotation then frees the arm without having to reverse the rotation of the plate.

For vectors of non-zero norm, we can add two such terms:

$$Q_{ij} = \chi_i \chi_j^* \pm \xi_i \xi_j^* \tag{6.64}$$

(the $\pm$ allows for time-like or space-like 4-vectors). Taking the determinant, we now find $\det \mathbf{Q} = \pm|\chi_1 \xi_2 - \chi_2 \xi_1|^2$ and hence

$$\boxed{(\chi_1\ \chi_2)\begin{pmatrix} 0 & 1 \\ -1 & 0 \end{pmatrix}\begin{pmatrix} \xi_1 \\ \xi_2 \end{pmatrix} = \text{invariant.}} \tag{6.65}$$

The 2×2 antisymmetric Levi–Civita symbol ϵ_{ij} thus plays a similar role to that of the metric tensor in allowing the construction of invariants, and is thus named the **metric spinor**.

SPINOR INDEX NOTATION We should now make the above a little less informal by introducing some of the peculiarities of the index notation for spinors. Suppose that the change in the components of the spinor under coordinate transformation is specified by some matrix $\mathbf{t}$:

$$\psi'^A = t^A{}_B \psi^B. \tag{6.66}$$

ψ_A is a **contravariant spinor**, and the label A can take values 1 or 2; as usual, repeated indices are summed over. As with vectors, we need to have objects with 'downstairs indices' to form invariants:

$$\psi'_A = (t^{-1})_A{}^B \psi_B. \tag{6.67}$$

Where the **covariant spinor** is formed via the metric spinor, which is of second rank (the

rank of a spinor being the number of spinor indices):

$$\boxed{\psi_A = \epsilon_{AB}\psi^B.} \tag{6.68}$$

One annoying aspect of this subject is the existence of conflicting conventions. If we follow the vector–tensor analogy, the matrix ϵ^{AB} will be defined as the inverse to ϵ_{AB}, in which case $\epsilon_{AB} = -\epsilon^{AB}$. This is not particularly appealing, as the choice of whether to have the index up or down is arbitrary, and so there is an alternative convention where both versions of ϵ are the antisymmetric symbol. The price paid for this is a slightly bizarre index-raising formula:

$$\boxed{\psi^A = \epsilon^{BA}\psi_B.} \tag{6.69}$$

This is the relation adopted by Corson (1953); Penrose & Rindler (1984) use the former convention.

Spinors are complex, and so is the matrix **t**, in general. We therefore have to distinguish between spinors, which transform with **t**, and **conjugate spinors**, which transform with $\mathbf{t}^*$. To indicate the latter, it is usual to put a dot over the appropriate index:

$$\psi'^{\dot{A}} = (t^*)^A{}_B\psi^{\dot{B}}. \tag{6.70}$$

This is a little clumsy, as $\dot{A}$ is just a number, and the dot indicates a property of the spinor, not of its indices. Nevertheless, the convention is easy enough to manipulate: complex conjugation of a spinor equation removes dots from dotted indices, and vice versa.

RELATION OF SPINORS AND TENSORS Armed with a smattering of formalism, let us now return to the question of expressing 4-vectors and tensors in terms of spinors. Looking at the earlier expression for a tensor in terms of 2-spinors, $Q_{ij} = \chi_i\chi_j^* \pm \xi_i\xi_j^*$, we now recognize that the indices on the terms with complex conjugate signs should properly be dotted indices. A tensor with a given number of indices can thus in general be associated with a spinor that has twice the number of indices – one ordinary and one dotted spinor index for each tensor index. Every tensor operation therefore has a spinor equivalent that involves operations on pairs of indices. This may seem a retrograde step, which doubles the amount of work needed. However, there will be spinor operations that have no immediate tensor equivalent, so in a sense this gives a deeper insight into the operation of tensor calculus. As an example, consider the relation that defines a tensor to be symmetric, and its spinor equivalent

$$T_{\mu\nu} = T_{\nu\mu} \quad \longrightarrow \quad T_{A\dot{A}B\dot{B}} = T_{B\dot{B}A\dot{A}}. \tag{6.71}$$

Each of the four possible values of the 4-vector index μ maps onto the four possible combinations of the two values of the indices A and $\dot{A}$. Now, a common modification of a symmetric tensor is **trace reversal**: $T_{\mu\nu} \to T_{\mu\nu} - \frac{1}{2}T^\alpha_\alpha g_{\mu\nu}$. The spinor equivalent of this operation is unexpectedly simple:

$$T_{A\dot{A}B\dot{B}} \to T_{A\dot{B}B\dot{A}} = T_{B\dot{A}A\dot{B}}; \tag{6.72}$$

trace reversal amounts to simply exchanging a pair of spinor indices (see chapter 3 of Penrose & Rindler 1984 for a proof).

CLIFFORD ALGEBRAS An illuminating alternative viewpoint on the need for spinors and complex numbers in quantum mechanics comes from a formalism known as **geometric calculus** or **Clifford algebras**. This also provides an alternative way of looking a tensors and differential forms. For more details at an introductory level, see Hestenes (1985). A more advanced treatment is given in Hestenes & Sobczyk (1984). The starting point is to ask why we treat vector algebra differently from the algebra of scalars.

No-one should need convincing that vector algebra is a good thing. Rather than having to write explicitly the Cartesian components of a vector, and so have to specify not only the number of dimensions we are dealing with but also the particular set of axes being used, we can write physical equations in a way that is independent of these details. In short, vector algebra allows us for many purposes to deal with the **directed numbers** that constitute vectors exactly as if they were scalars. We can add vectors in a way that is commutative and associative:

$$\begin{aligned}\mathbf{a}+\mathbf{b}&=\mathbf{b}+\mathbf{a}\\ \mathbf{a}+(\mathbf{b}+\mathbf{c})&=(\mathbf{a}+\mathbf{b})+\mathbf{c}.\end{aligned} \tag{6.73}$$

We can also handle the multiplication of vectors by scalars, which is again commutative and associative:

$$\begin{aligned}\alpha\mathbf{a}&=\mathbf{a}\alpha\\ \alpha(\beta\mathbf{a})&=(\alpha\beta)\mathbf{a}.\end{aligned} \tag{6.74}$$

It is also distributive over addition:

$$\begin{aligned}\alpha(\mathbf{a}+\mathbf{b})&=\alpha\mathbf{a}+\alpha\mathbf{b}\\ (\alpha+\beta)\mathbf{a}&=\alpha\mathbf{a}+\beta\mathbf{a}.\end{aligned} \tag{6.75}$$

All that is missing is some way of multiplying vectors themselves. If this existed, and had the same properties as multiplication by scalars, we could carry out vector algebra exactly as for scalars, and only worry about the geometrical interpretation of our results at the end of the calculation. Of course, we know that this is not possible and that two different sorts of vector multiplication are necessary: scalar $(\mathbf{a}\cdot\mathbf{b})$ and vector $(\mathbf{a}\wedge\mathbf{b})$. Why is this? We are quite familiar with the fact that vectors in two dimensions can be represented by **complex numbers** $z = a + ib$, which can indeed be manipulated exactly like scalars. The fact that this is possible has all sorts of advantages, such as being able to apply conformal transformations to give simple solutions of tricky potential problems. Why should this not be possible in other numbers of dimensions? It is, as the example of **quaternions** shows.

QUATERNIONS Suppose we wanted to extend the idea of complex numbers to embrace non-real numbers other than i. We might try writing a generalized complex number as a sum over a basis of three quantities $(1, i, j)$, where j is to play the role of describing the third spatial axis:

$$z = x + iy + jw. \tag{6.76}$$

Again by analogy with two dimensions, we will want to introduce the **conjugate** z^*, such that

$$zz^* = x^2 + y^2 + z^2. \tag{6.77}$$

Write $z^* = x + i^*y + j^*w$, and multiply out the terms in zz^*. There are three types of condition for obtaining the desired result; the first two are as expected,

$$\begin{aligned} i^* = -i, \quad j^* = -j \\ i^2 = j^2 = -1, \end{aligned} \tag{6.78}$$

but the third is the critical feature:

$$ij + ji = 0. \tag{6.79}$$

Trying to satisfy this is what gives problems when we try to make this number system *closed* – i.e. attempt to ensure that the product $z_1 z_2$ is also a complex number. $z_1 z_2$ will contain non-zero terms in ij, and so we must have $ij = a + bi + cj$ to stay within the system we have set up. This is easily shown to be impossible: multiply by i on the left or j on the right and we conclude successively that $a = b = c = 0$. However, ij must be non-zero: $(ij)i = 0$ if ij vanishes, but $(ij)i = (-ji)i = j \neq 0$. The only way out is to increase the dimensionality of the system, so that

$$ij = k, \tag{6.80}$$

and we end up with the **quaternion algebra**, invented by Hamilton in 1843:

$$\boxed{\begin{aligned} &i^2 = j^2 = k^2 = -1 \\ &\{i, j\} = \{i, k\} = \{k, j\} = 0 \\ &ij = k, \quad ki = j, \quad jk = i. \end{aligned}} \tag{6.81}$$

This seems very puzzling; why is it possible to define complex numbers in four but not three dimensions?

GEOMETRIC CALCULUS To explain these puzzles, we need to introduce another: why should there be a connection between complex numbers and 2D vectors anyway? The answer is to be found in the formalism of **geometric calculus**, which tackles the puzzle about vector products head on, hypothesizing that such products can be defined by writing an equation that treats vectors as if they were numbers: $\mathbf{A} = \mathbf{ab}$. Now, without prior knowledge, we might not want to assume that such a **geometric product** was commutative, and so in general we must distinguish *two* different kinds of product, taking symmetric and antisymmetric parts:

$$\begin{aligned} \mathbf{a} \cdot \mathbf{b} \equiv \tfrac{1}{2}(\mathbf{ab} + \mathbf{ba}) \\ \mathbf{a} \wedge \mathbf{b} \equiv \tfrac{1}{2}(\mathbf{ab} - \mathbf{ba}) \end{aligned} \tag{6.82}$$

(the choice of notation here is not a coincidence). Alternatively, we can say that the geometric product has produced a new kind of quantity, a **bivector**, which is constructed from two distinct parts:

$$\boxed{\mathbf{ab} = \mathbf{a} \cdot \mathbf{b} + \mathbf{a} \wedge \mathbf{b}.} \tag{6.83}$$

At first sight, it may seem odd to add together quantities of two distinct kinds (assuming that the '$\cdot$' and '$\wedge$' symbols will indeed turn out to be related to the familiar scalar and vector products). However, this is *exactly* what we do when we write a complex number $z = a + ib$.

To see how these concepts work in a simple case, consider a two-dimensional basis $(\boldsymbol{\sigma}_1, \boldsymbol{\sigma}_2)$, defined such that $\boldsymbol{\sigma}_1^2 = \boldsymbol{\sigma}_2^2 = 1$, $\boldsymbol{\sigma}_1 \cdot \boldsymbol{\sigma}_2 = 0$ as for orthonormal vectors. What distinct geometrical entities can we construct with this system? The product of two vectors $\mathbf{z} = a\boldsymbol{\sigma}_1 + b\boldsymbol{\sigma}_2$ yields three distinct kinds of objects: scalars, vectors (linear combinations of $\boldsymbol{\sigma}_1$ and $\boldsymbol{\sigma}_2$) and bivectors (proportional to $\boldsymbol{\sigma}_1\boldsymbol{\sigma}_2$). This is all there is, as further multiplication does not yield trivectors (the definitions imply $\boldsymbol{\sigma}_1\boldsymbol{\sigma}_2 = -\boldsymbol{\sigma}_2\boldsymbol{\sigma}_1$, so products of bivectors with vectors yield vectors, and bivectors with bivectors yield scalars).

Now, the important thing to note is that vectors of even kind form a closed system. If we write

$$\mathbf{z} = a + b\,\boldsymbol{\sigma}_1\boldsymbol{\sigma}_2 \tag{6.84}$$

then products of vectors of this form can never generate terms proportional to $\boldsymbol{\sigma}_i$. If we define

$$i \equiv \boldsymbol{\sigma}_1\boldsymbol{\sigma}_2 \tag{6.85}$$

and note that $i^2 = -1$, we see that we have invented complex algebra: it arises inevitably from the structure of geometry in two dimensions.

It should be clear by now what will happen in general. In n dimensions, we can construct the distinct quantities

$$1, \quad \boldsymbol{\sigma}_i, \quad \boldsymbol{\sigma}_i\boldsymbol{\sigma}_j, \quad \ldots, \quad \boldsymbol{\sigma}_1 \cdots \boldsymbol{\sigma}_n \tag{6.86}$$

(n^2 of them for even n, otherwise $n^2 - 1$). The algebra of such quantities is known as a **Clifford algebra** (after the British mathematician); this is a term that generally refers to the set of quantities produced by taking products of a set of elements that contain an anticommutator (for example, the set of matrices generated from products of Dirac matrices is a Clifford algebra). The full algebra divides into odd and even **subalgebras**, of which only the even is closed. This will contain $n^2/2$ (even n) or $(n^2 - 1)/2$ (odd n) quantities, which will look like the analogues of complex numbers. Thus, ordinary complex numbers are followed by quaternions, and the next stop will be the analogue of complex algebra in eight dimensions, although it is actually generated by 4D geometry.

CODA This brief discussion has confirmed that Schrödinger was right to be puzzled about the appearance of i in his wave equation. We have shown that the inclusion of spin and complex numbers in relativistic wave equations reflects many subtle features of the geometry of spacetime. The next chapter turns to the quantum theory of fields, which is generally perceived to be a more difficult area than quantum mechanics. However, these difficulties are at least mainly algebraic, and few conceptual problems will be added to the existing list.

Problems

(6.1) Show that the nonrelativistic Schrödinger equation can be manipulated into the form of a conservation law $\dot{\rho} + \boldsymbol{\nabla} \cdot \mathbf{j} = 0$, where $\rho = |\psi|^2$ is thus to be interpreted as the conserved density of charge, mass, or probability. What is the expression for the current $\mathbf{j}$?

(6.2) Show that the 4-current $J^\mu = i(\phi^* \partial^\mu \phi - \phi \partial^\mu \phi^*)$ is conserved for a field that obeys the Klein–Gordon equation. Show that plane-wave solutions to the Klein–Gordon equation have $E = \pm\sqrt{k^2 + m^2}$ and that the negative-energy solutions have negative J^0. If this is not a probability density, what is being conserved here?

(6.3) Define the density matrix ρ as

$$\rho \equiv \sum_i p_i \, |i\rangle\langle i|, \tag{6.87}$$

where p_i is the probability that the system under study has been prepared in the state $|i\rangle$. Show that the expectation of an operator A is given by $\mathrm{Tr}(A\rho) = \sum_n \langle n|A\rho|n\rangle$, and hence that the density matrix for a system in thermal equilibrium is $\rho \propto \exp(-H/kT)$.

(6.4) Define the entropy of a quantum-mechanical system in terms of the occupation probabilities of its microstates:

$$S = -k \sum_i p_i \, \ln p_i. \tag{6.88}$$

Use the golden rule and first-order perturbation theory to show that this quantity always increases with time. How is this possible, given that the nonrelativistic Schrödinger equation is symmetric under time reversal?

(6.5) Show that electromagnetism is invariant under parity transformation, in the sense that the Lorentz force law applies independently of the transformation $\mathbf{r} \to -\mathbf{r}$.

(6.6) Show that the time-reversal operator for the Dirac equation is $T\psi = \gamma^1\gamma^3\psi^*$ in the Dirac representation. Defining $P\psi = \gamma^0\psi$ and $C\psi = \gamma^2\psi^*$, find the matrix corresponding to CPT and show directly that the Dirac equation is invariant under this operation together with $x^\mu \to -x^\mu$ and $e \to -e$.

(6.7) Construct the operators that represent angular momentum and obtain ladder operators to show that allowed eigenvalues of L_z are separated by $\hbar$. Show that the largest such eigenvalue must be a multiple of $\hbar/2$.

7 Quantum field theory

7.1 Quantum mechanics of light

As emphasized in the previous chapter, quantum mechanics is developed from the observed existence of energy quantization in light, $E = \hbar\omega$, so it is clearly essential to try to understand how this phenomenon arises. We now face the challenge of applying quantum mechanics to continuous fields, a process often called **second quantization**. This is not a very helpful name, since it obscures the pleasant fact that most of the main ideas of standard quantum mechanics are used unchanged in tackling the problem of field quantization. It makes sense to begin with the case of electromagnetism, although many of the points to be made apply to other fields.

COULOMB GAUGE Electromagnetism concerns the six components of the electric and magnetic fields; these all relate to the 4-potential A^μ, but the number of degrees of freedom can be reduced still further by exploiting the **gauge freedom** in electromagnetism. The transformation $A^\mu \to A^\mu + \partial^\mu\psi$, where ψ is any scalar field, has no observable effect, because it does not alter the field tensor $F^{\mu\nu} = \partial^\mu A^\nu - \partial^\nu A^\mu$, whose components are the $\mathbf{E}$ and $\mathbf{B}$ fields. This means that it is always possible to make the scalar potential ϕ vanish (so that, for example, the electrostatic field from a point charge would correspond to a time-varying $\mathbf{A}$). Alternatively, it is possible to choose the **Coulomb gauge**, in which $\nabla \cdot \mathbf{A} = 0$.

Electromagnetism would look particularly simple if we could satisfy both these conditions simultaneously: $\phi = \nabla \cdot \mathbf{A} = 0$. We would then automatically satisfy the **Lorentz condition** $\partial_\mu A^\mu = 0$, which is required in order to write Maxwell's equations as the wave equation $\Box A^\mu = \mu_0 J^\mu$. In general, this simplicity is unattainable; the scalar potential in the Coulomb gauge is independent of time, but does not vanish in all cases. However, it is possible to make the simple choice in a vacuum. This result can be proved in two stages: if the initial form of the scalar potential is ϕ_i, choosing $\dot{\psi} = -\phi_i$ sets ϕ to zero. Now, in free space $\Box\phi_i = 0$, so that $\Box\psi$ is a function of position only. The effect of the gauge transformation on the Lorentz condition is $\partial_\mu A^\mu \to \partial_\mu A^\mu + \Box\psi$; if the Lorentz condition holds initially, then enforcing $\phi = 0$ makes $\nabla \cdot \mathbf{A}$ a function of position only. Finally, a second gauge transformation, where ψ is independent of time, can be used to enforce $\nabla \cdot \mathbf{A} = 0$. This procedure is not covariant, and a different gauge transformation is required in each frame. However, because the four components of A^μ are not independent, it is hard to cast quantum electrodynamics in a fully covariant form. This can be done, but the simpler approach is preferred here.

FIELD EXPANSION Things simplify further if we only try to describe electromagnetism in some finite volume, rather than fields throughout the whole universe. This has the effect of introducing discreteness into the problem, which may be an advantage if we are eventually expecting something like particles to emerge. The field may then be described by a set of Fourier expansion coefficients:

$$\mathbf{A} = \frac{1}{\sqrt{V}} \sum_{\mathbf{k},\alpha} \left[c\boldsymbol{\epsilon}^{(\alpha)} e^{i(\mathbf{k}\cdot\mathbf{x}-\omega t)} + c^* \boldsymbol{\epsilon}^{(\alpha)} e^{-i(\mathbf{k}\cdot\mathbf{x}-\omega t)} \right]. \tag{7.1}$$

The allowed values of wave number are set by choosing **harmonic boundary conditions** for the expansion. This makes the field periodic on the scale of a box of side L: $k_x = 2n\pi/L$ etc. This is a common procedure, which should be familiar from statistical mechanics. Although these conditions may seem artificial, the assumption is that this treatment will give the correct properties for fields in free space if we take the limit $L \to \infty$. The expansion coefficients c are functions of the wavenumber $\mathbf{k}$ and of α; α is an index labelling the polarization, which is described by the two unit vectors $\boldsymbol{\epsilon}^{(1)}$ and $\boldsymbol{\epsilon}^{(2)}$, which are perpendicular to $\mathbf{k}$ and to each other. The normalization factor $1/\sqrt{V}$ (where $V = L^3$ is the box volume) is chosen for later convenience. The Hamiltonian of this system is

$$H = \tfrac{1}{2} \int (B^2 + E^2)\, dV = \tfrac{1}{2} \int \left[(\boldsymbol{\nabla}\wedge\mathbf{A})^2 + \dot{A}^2 \right] dV, \tag{7.2}$$

which is not hard to evaluate. For example, for one Fourier term we have

$$\begin{aligned} (\boldsymbol{\nabla}\wedge\mathbf{A})_{\mathbf{k}} &= ci\mathbf{k}\wedge\boldsymbol{\epsilon}^{(\alpha)} e^{i(\mathbf{k}\cdot\mathbf{x}-\omega t)} - c^* i\mathbf{k}\wedge\boldsymbol{\epsilon}^{(\alpha)} e^{-i(\mathbf{k}\cdot\mathbf{x}-\omega t)} \\ \mathbf{k}\wedge\boldsymbol{\epsilon}^{(1)} &= k\boldsymbol{\epsilon}^{(2)} \quad \text{etc.} \end{aligned} \tag{7.3}$$

From this it is clear that most cross terms integrate to zero, either because of the orthogonality of the polarization vectors or because of the harmonic boundary conditions $k_x = 2n\pi/L$, so that $\int e^{i\Delta\mathbf{k}\cdot\mathbf{x}} dV = 0$ unless $\Delta\mathbf{k} = 0$. We then find that each half of the Hamiltonian gives the same result (as it must for radiation), so that the total is

$$H = \sum_{\mathbf{k},\alpha} \omega^2 (c\,c^* + c^* c). \tag{7.4}$$

Why not equate $c\,c^*$ and c^*c? Although these are equivalent at the classical level, they may not be so in quantum theory and it is safer to keep them distinct. Now, since the amplitudes c and c^* are constants, we have a Hamiltonian that is explicitly independent of time. However, we could equally well have performed the analysis with time-dependent coefficients $c_{\text{new}} = c_{\text{old}} e^{-i\omega t}$, in which case we would obtain the same equation for H, but with the new c. The difference seems trivial, but is vital if we define new variables

$$\begin{aligned} q &= c + c^* \\ p &= \frac{\omega}{i}(c - c^*), \end{aligned} \tag{7.5}$$

in terms of which the Hamiltonian is

$$H = \sum \tfrac{1}{2}(p^2 + \omega^2 q^2). \tag{7.6}$$

This should look familiar: it is the Hamiltonian of a set of simple harmonic oscillators, and in retrospect it is hardly surprising that the electromagnetic field can be decomposed in this way.

To show that the oscillator analogy is exact, note that the variables obey **Hamilton's equations of motion**

$$\begin{aligned} \frac{\partial H}{\partial q} &= -\dot{p} \\ \frac{\partial H}{\partial p} &= \dot{q}. \end{aligned} \tag{7.7}$$

These imply the relations $\dot{p} = -\omega^2 q$ and $p = \dot{q}$, so that p and q look like the momentum and position coordinates for a particle of unit mass in a quadratic potential.

FIELD QUANTIZATION The quantization of an oscillator is a familiar problem. We assume that p and q are to be interpreted as operators acting on some wave function (*not* to be confused with the electromagnetic field **A**), and obeying a commutation relation

$$\boxed{[q, p] = i\hbar.} \tag{7.8}$$

Now we see that it was indeed wise to distinguish cc^* and c^*c, because $[q, p] = -(2\omega/i)[c, c^*]$. If we define the **ladder operator**

$$a = \sqrt{\frac{2\omega}{\hbar}}\, c, \tag{7.9}$$

and its Hermitian conjugate in terms of c^*, then we end up with the familiar algebra

$$\boxed{\begin{aligned} H &= \sum \frac{\hbar\omega}{2} \left(aa^\dagger + a^\dagger a\right) \\ &= \sum \left(N + \tfrac{1}{2}\right) \hbar\omega, \end{aligned}} \tag{7.10}$$

where $[a, a^\dagger] = 1$ and the **number operator** $N \equiv a^\dagger a$.

This is rather disappointing, as it indicates the end of any hopes we had of understanding more deeply *why* photons obey $E = \hbar\omega$. There is a circularity in quantum mechanics: we bootstrap ourselves from the observation of quantization in photons to the concept of non-commuting operators, only to find that these are needed in order to explain the initial observation. Planck's constant must be introduced by definition somewhere, and this seems not a bad place to do it. The existence of the fundamental commutator represents the concept that it may not be possible to measure position and momentum simultaneously for a microscopic system.

This process of splitting a wave field into oscillators and quantizing each via the fundamental commutator is a procedure that also works with other fields and does not apply only to the familiar case of electromagnetism.

LADDER OPERATORS It is worth briefly revising the algebra of ladder operators. The number operator $N \equiv a^\dagger a$ is Hermitian because $(AB)^\dagger = B^\dagger A^\dagger$, so that we can label states by the real eigenvalues of N:

$$N|n\rangle = n|n\rangle. \tag{7.11}$$

Using the commutator of a and $a^\dagger$ gives the commutators

$$\begin{aligned}[a, N] &= a \\ [a^\dagger, N] &= -a^\dagger;\end{aligned} \tag{7.12}$$

this shows that a and $a^\dagger$ are the **annihilation** and **creation** operators, which respectively lower or raise n by one unit:

$$\begin{aligned} Na^\dagger|n\rangle &= (n+1)\, a^\dagger|n\rangle \quad &\Rightarrow \quad a^\dagger|n\rangle &\propto |n+1\rangle \\ Na|n\rangle &= (n-1)\, a|n\rangle \quad &\Rightarrow \quad a|n\rangle &\propto |n-1\rangle. \end{aligned} \tag{7.13}$$

Finally, we use normalization of the wave function, $\langle an|an\rangle = 1 = \langle n|N|n\rangle$, where $|an\rangle$ stands for $a|n\rangle$, and the last step follows from the definition of the Hermitian conjugate, $\langle m|a|n\rangle^* \equiv \langle n|a^\dagger|m\rangle$, to obtain the relations

$$\boxed{\begin{aligned} a|n\rangle &= \sqrt{n}\;|n-1\rangle \\ a^\dagger|n\rangle &= \sqrt{n+1}\;|n+1\rangle. \end{aligned}} \tag{7.14}$$

This shows us that there must be a ground state at $n = 0$: $a|0\rangle$ produces the empty set, and n cannot be lowered further. The state $|0\rangle$ is often referred to as the **vacuum state**. In quantum field theory, the vacuum is often anything but empty, but it does correspond to there being no excitations in any of the normal modes of the field.

REPRESENTATIONS To minimize technical distractions, we have up till now been deliberately a little vague about the time dependence of quantities in the field expansion. In simple quantum mechanics, one usually adopts the **Schrödinger representation**, where the time dependence of the system is carried by the wave function and the operators are time independent. The opposite is the **Heisenberg representation**, which involves the unitary transformation of wave function and operators:

$$\begin{aligned} \psi_{\rm H} &= e^{iHt/\hbar}\psi_{\rm S} \\ O_{\rm H} &= e^{iHt/\hbar} O_{\rm S} e^{-iHt/\hbar}. \end{aligned} \tag{7.15}$$

In the Heisenberg representation, the equation of motion corresponding to the Schrödinger equation is now an equation for the operator: $i\hbar\dot{O}_{\rm H} = [O, H]$ (this is proved by considering the time derivative of the matrix element $\langle i|O|j\rangle$ and using the facts that $O_{\rm S}$ is time independent and $H_{\rm S} = H_{\rm H}$). The Heisenberg representation is what we have implicitly used in the analysis of the free electromagnetic field, where a and $a^\dagger$ were taken as time-dependent quantities prior to quantization, and the wave function was implicitly time independent.

FIELD QUANTA AS PARTICLES Well, we have a quantized wave field, but an excitation of a wave is not yet a particle. The discreteness of the energy content of the box suggests an interpretation in terms of particles, but where are they? All we know is that they must lie inside the box of volume $V = L^3$; by the uncertainty principle, therefore, their momentum should be uncertain by an amount of order $\hbar/L$. This compares favourably with the spacing of the momentum eigenvalues imposed by the harmonic boundary conditions, $p_x = 2\pi n\hbar/L$. If we continue with the assumption that the energy density is due to particles, then the properties of the particles can be found as follows.

(1) Mass. The momentum in the radiation field can be found in the same way as we obtained the total energy, by integrating the Poynting vector:

$$\mathbf{p} = \int (\mathbf{E} \wedge \mathbf{H})\, dV = \sum \left(N + \tfrac{1}{2}\right) \hbar \mathbf{k}. \tag{7.16}$$

Each mode has $E = pc$, which is the signature of a *massless* particle ($E^2 = p^2c^2 + m^2c^4$).

(2) Spin. One might think of obtaining the spin of the particles by integrating up the angular momentum in the same way as above for the momentum, but this approach runs into problems with boundary terms (see e.g. chapter 1 of Baym 1968). Instead, we shall make a rather more general argument connected with the fact that electromagnetism is a vector field, as follows.

The angular momentum carried by an object is defined by how that object transforms under rotations: in an eigenstate of L_z with eigenvalue $m\hbar$, the behaviour of the wave function would be

$$\psi \propto \exp(im\phi). \tag{7.17}$$

For spin s, there are $2s + 1$ eigenvalues of L_z, and so the field must contain this number of 'internal' degrees of freedom. A scalar field must therefore have spin zero, whereas a 3D vector field has $s = 1$ and a 3D second-rank traceless tensor field (such as the gravitational tidal field, $\partial^2\Phi/\partial x_i \partial x_j$) has $s = 2$.

This counting argument seems to go wrong in the case of electromagnetism, where only two independent states of circular polarization exist (electromagnetic waves being transverse). The $m = 0$ state therefore is not possible for photons. This is a general result: massless particles of any spin can have only two **helicity eigenstates**. The helicity $h = \mathbf{s} \cdot \hat{\mathbf{p}}$ can take values $\pm s$ only. This corresponds to the fact that a massless particle must travel at c to have finite energy; a massive particle could be overtaken, changing the sign of $\hat{\mathbf{p}}$ and hence passing through zero helicity, but this is not possible if the mass vanishes. We can see that the result holds in general by counting the degrees of freedom for relativistic fields. The 4-potential A^μ has four degrees of freedom, of which two are lost via the transversality condition $A^\mu k_\mu = 0$ and the gauge freedom $A^\mu \to A^\mu + \partial_\mu \chi$, which allows us to specify the arbitrary function $\chi(x^\mu)$. Similarly, the symmetric field $g_{\mu\nu}$ has 10 degrees of freedom, of which four are removed by the transversality condition $k^\mu \Delta_{\mu\nu} = 0$ and four by gauge degrees of freedom, leaving again the two degrees of freedom corresponding to the two helicity choices.

7.2 Simple quantum electrodynamics

SPONTANEOUS EMISSION One of the most important things we can do with the above basics of field quantization is to give the proper explanation for why an excited atom radiates. Recall how this is tackled in the semiclassical approximation. We write down the Hamiltonian for an electron in an electromagnetic field (nonrelativistic, for simplicity):

$$H = \frac{1}{2m}(\mathbf{p} - e\mathbf{A})^2 + e\phi + V. \tag{7.18}$$

If we adopt the Coulomb gauge, the first-order perturbation is just $H' = (e/m)\mathbf{A}\cdot\mathbf{p}$ (operating $\mathbf{p} = -i\hbar\boldsymbol{\nabla}$ on $\mathbf{A}$ has no effect because $\boldsymbol{\nabla}\cdot\mathbf{A} = \mathbf{0}$ in this gauge). The matrix element needed to calculate transition rates is therefore $\langle i|\mathbf{A}\cdot\mathbf{p}|j\rangle$, which in lowest order is proportional to the dipole term $\langle i|\mathbf{A}\cdot\mathbf{x}|j\rangle$ [problem 7.1]. The usual puzzle is that this apparently leads to equal rates of transition for $i \to j$ and $j \to i$, and a transition rate that vanishes for zero applied field.

The reason why these conclusions are incorrect is that they ignore the quantum nature of the electromagnetic field. Remember the expansion for $\mathbf{A}$:

$$\mathbf{A} = \sum_{\mathbf{k},\alpha} \sqrt{\frac{\hbar}{2\omega V}} \left[a\, \boldsymbol{\epsilon}^{(\alpha)} e^{ik^\mu x_\mu} + a^\dagger\, \boldsymbol{\epsilon}^{(\alpha)} e^{-ik^\mu x_\mu} \right]. \tag{7.19}$$

Now, from $\mathbf{A}$'s point of view, the only important event is the emission or absorption of photons. The required matrix element therefore involves an initial state $|n\rangle$ and a final state $|m\rangle$ (considering the occupation number of just one mode). Because $\mathbf{A}$ is a mixture of creation and annihilation operators, we have

$$H'|n\rangle \propto \sqrt{n}\,|n-1\rangle + \sqrt{n+1}\,|n+1\rangle. \tag{7.20}$$

This constrains the allowed final states with non-zero matrix elements quite severely; the atom can emit or absorb only one photon at a time. Also, squaring the modulus of the matrix element to find transition rates, we get

$$\boxed{r_{ij} \propto \begin{cases} n & \text{(absorption)} \\ n+1 & \text{(emission).} \end{cases}} \tag{7.21}$$

The expression for the rate of absorption agrees with the semiclassical one in being proportional to the incident intensity. This is just good luck: the semiclassical approximation should only work in the classical limit $n \gg 1$. The beauty of the full analysis is that emission is automatically given as the sum of a **spontaneous** term and a **stimulated** one ($\propto n$). Note that the stimulated term only produces emission into the same state that is responsible for the stimulation. Consider just two photon states, with initial occupation numbers n and 0 respectively: the matrix element for transitions in which a photon is emitted into the second state is independent of n and corresponds to spontaneous decay only. Only if the transition adds a photon to the first state do we see the effect of the stimulated term. Lasers thus produce coherent light because every stimulated photon is an identical copy of the initial photon. This is not something that can be easily understood from the semiclassical analysis.

EINSTEIN COEFFICIENTS It is illuminating to reconsider the Einstein A and B coefficients in the light of the above discussion. Recall the definition of these coefficients in terms of the transition rates $r_{ij} = BI_\nu$ for absorption and $r_{ji} = BI_\nu + A$ for emission. Invoking detailed balance plus Boltzmann occupation probabilities for the two electronic levels under consideration gives us

$$A = I_\nu(e^{\hbar\omega/kT} - 1)B = \frac{4\pi\hbar\nu^3}{c^2}B, \tag{7.22}$$

using the Planck expression for I_ν in the case of thermal radiation. The final expression is the usual result to quote, yet it leaves the meaning of A completely obscured. It is much better to note that the photon occupation number is $n = (e^{\hbar\omega/kT} - 1)^{-1}$, and hence that

$$\frac{r_{ji}}{r_{ij}} = \frac{n+1}{n}, \tag{7.23}$$

just as above. An interesting alternative is to *derive* the Planck law from this full quantum expression plus Boltzmann level populations.

ZERO-POINT ENERGIES AND VACUUM DENSITY One surprising aspect of the radiation field is the zero-point energy, $\hbar\omega/2$ per mode. This implies that the energy density of the vacuum (defined as all $n_i = 0$) will be non-zero. Integrating $\hbar\omega/2c^2$ per mode over k-space (with a degeneracy of 2 for polarization) gives a total density of

$$\rho_{\rm vac} = \frac{\hbar}{2\pi^2 c^5} \int \omega^3 \, d\omega, \tag{7.24}$$

which diverges horribly. Is it possible that the upper limit of the integral should be finite? This would be the case if space were a lattice, which is perhaps conceivable on some unobservably small scale. However, even with a cutoff at the hardly microscopic level of $\lambda \sim 1$ mm, $\rho_{\rm vac}$ already exceeds the critical density of the universe ($\sim 10^{-26}\,{\rm kg\,m^{-3}}$); something is badly wrong.

Different physicists have differing reactions to this embarrassing failure. Some state that only relative energies are measurable, which is true as far as the actual field dynamics are concerned (see below), but is a weak standpoint given that gravity measures total energy. A better argument is to note that the expression that led to the troublesome zero-point energy might be re-ordered. We had $H = \sum(aa^\dagger + a^\dagger a)\hbar\omega/2$, which just goes over to $\sum(c\,c^* + c^* c)\omega^2$ in the classical limit. In fact, we deduced the quantum limit from the classical one, and so we are quite at liberty to replace $a^\dagger a$ by $aa^\dagger$, which cures the problem by eliminating the zero-point energy. This is fine so far as it goes, but leaves a nasty taste: which of the many quantum-inequivalent classical expressions should we choose, and why? Note that in solid-state physics, where the lattice does provide a well-defined cutoff frequency, the zero-point energy is indeed $\hbar\omega/2$ per mode and is measurable through X-ray diffraction (see section 2.9 of Ziman 1972).

CASIMIR EFFECT An illuminating insight on the reality or otherwise of zero-point energy in quantum field theory is gained by considering the modification of the fields that arise when conductors are introduced. Specifically, suppose we place ourselves between two infinite conducting sheets, a distance L apart. Clearly, the energy density of the vacuum in our vicinity must have dropped, since modes with $k_x < \pi/L$ are now forbidden owing to the boundary conditions of zero field at the plate (k_x denotes the component of wavenumber in the x-direction, which we take to be the axis along which the plates are separated). The energy density change means in general that there must be some force per unit area between the plates. Suppose this scaled as a power of the separation: $F \propto L^{-\alpha}$. If we denote by ρ the *difference* in energy density between the plates and outside the plates, then conservation of energy implies that ρ is also proportional to $L^{-\alpha}$ ($F\,dL = d[\rho L]$ if ρ is uniform between the plates). The force is therefore attractive if $\alpha > 1$. It can now be argued that dealing with zero-point energy, frequencies and cutoffs at L implies that we need a combination of c, $\hbar$ and L;

$$F \propto \frac{\hbar c}{L^4} \tag{7.25}$$

is the only one with the required dimensions. The constant required on the right-hand side turns out to be $\pi^2/240$ (see section 3.2.4 of Itzykson & Zuber 1980).

This argument begs a number of questions: without a proper resolution of the vacuum-energy puzzle, we have performed the equivalent of obtaining the force by subtracting two infinite numbers. Things are actually a little better than this, since we can **regularize** the divergent zero-point integral by introducing some upper cutoff in frequency, which we can allow to go to ∞ after calculating the change in energy density due to the altered low-frequency modes (this latter change is clearly independent of the cutoff if the cutoff wavelength is $\ll L$). Moreover, the above force can be detected experimentally (the **Casimir effect**). On the one hand, this seems to prove the reality of the zero-point energy, and reinforces the puzzle of the finite vacuum density. On the other hand, perfect conductors do not really exist, and a more prosaic view of the origin of the attractive force might be the Van der Waals forces between the individual atoms in the conducting sheets.

So, far from resolving the conceptual problems about vacuum energy, the Casimir effect merely muddies the waters. It does not not tell us why the global zero-point energy fails to be infinite, but it does show that physically sensible finite results can be obtained by calculating differences in observable quantities that have been regularized to prevent divergences. In this respect, it illustrates well the general philosophy of quantum field theory, which has been to sweep the big conceptual difficulties under the carpet and get on with calculating things.

7.3 Lagrangians and fields

The treatment so far has concentrated on the differential equations obeyed by the field. This is much the same as analysing classical mechanics in terms of forces, and indeed differential equations such as the Dirac or Klein–Gordon equations are often known as **equations of motion** for the field. Given this analogy, it should not be surprising that a neater analysis in field theory can be achieved via energy arguments. Consider the formulation of classical mechanics in terms of an **action principle**. We write the variational equation

$$\delta \int L \, dt = 0, \tag{7.26}$$

where the **Lagrangian** L is the difference of the kinetic and potential energies, $L = T - V$, for some set of particles with coordinates $q_i(t)$. The integral $\int L \, dt$ is the **action**; this equation says that the paths $q_i(t)$ travelled by each particle between given starting and finishing positions are such that the action is extremal (not necessarily a minimum, even though one often speaks of the principle of least action) with respect to small variations in the paths. If we take the 'positions' $q_i(t)$ and 'velocities' $\dot{q}_i(t)$ to be independent variables, then expansion of $L(q, \dot{q})$ in terms of small variations in the paths and integration by parts leads to **Euler's equation** for each particle

$$\frac{d}{dt}\left(\frac{\partial L}{\partial \dot{q}_i}\right) = \frac{\partial L}{\partial q_i}. \tag{7.27}$$

When the coordinates are Cartesian, this clearly gives Newton's law of motion (at least, for the simple case of velocity-independent forces). The power of the Lagrangian approach is that, having deduced the action principle, one can choose **generalized coordinates** to obtain less obvious equations of motion.

LAGRANGIAN DENSITIES A field may be regarded as a dynamical system, but with an infinite number of degrees of freedom. How do we handle this? A hint is provided by electromagnetism, where we are familiar with writing the total energy in terms of a density which, as we are dealing with generalized mechanics, we may formally call the Hamiltonian density:

$$H = \int \mathcal{H} \, dV = \int \left(\frac{\epsilon_0 E^2}{2} + \frac{B^2}{2\mu_0} \right) dV. \tag{7.28}$$

This suggests that we write the Lagrangian in terms of a **Lagrangian density** $\mathcal{L}$: $L = \int \mathcal{L} \, dV$. This quantity is of such central importance in quantum field theory that it is usually referred to (incorrectly) simply as 'the Lagrangian'. In these terms, our action principle now takes the pleasingly relativistic form

$$\delta \int \mathcal{L} \, d^4x^\mu = 0, \tag{7.29}$$

although note that to be correct in general relativity, the 'Lagrangian' $\mathcal{L}$ needs to take the form of a invariant scalar times the Jacobian $\sqrt{-g}$.

We can now apply the variational principle as before, considering $\mathcal{L}$ to be a function of a general coordinate, ϕ, which is the field, and the '4-velocity' $\partial_\mu \phi$. This yields the **Euler–Lagrange equation**

$$\boxed{\frac{\partial}{\partial x^\mu} \left[\frac{\partial \mathcal{L}}{\partial(\partial_\mu \phi)} \right] = \frac{\partial \mathcal{L}}{\partial \phi}.} \tag{7.30}$$

The Lagrangian $\mathcal{L}$ and the field equations are therefore generally equivalent, and any statement about the field derived using one method can in principle also be derived using the other. We shall hereafter stick mainly to the Lagrangian formalism, not only because it is the language of professional field theorists but also because it often in practice makes derivations simpler. In any case, the Lagrangian arguably seems more fundamental: we can obtain the field equations given the Lagrangian, but inverting the process is less straightforward.

EXAMPLES The Euler–Lagrange formalism may seem rather abstract, but it is undeniably a powerful tool, as the following examples illustrate.

(1) Waves on a string. A good classical example of the Euler–Lagrange formalism is provided by waves on a string. If the density per unit length is σ, the tension is T, and we call the transverse displacement of the string y, then the (one-dimensional) Lagrangian density is

$$\mathcal{L} = \tfrac{1}{2}\sigma \dot{y}^2 - \tfrac{1}{2} T y'^2 \tag{7.31}$$

(at least for small displacements). The potential term comes from the work done in stretching the string. Inserting this in the Euler–Lagrange equation yields a familiar result:

$$\sigma \ddot{y} - T y'' = 0. \tag{7.32}$$

This is just the wave equation, and it tells us that the speed of sound in a plucked string is $\sqrt{T/\sigma}$.

When applied to other classical or quantum fields, the Euler–Lagrange equation again gives the appropriate wave equation. The form of the Lagrangians for all these fields are often quite similar: if we want a wave equation linear in second derivatives of the field, then the Lagrangian must contain terms quadratic in first derivatives of the field.

(2) Electromagnetic field. Here we need the **field tensor** $F^{\mu\nu} \equiv \partial^\mu A^\nu - \partial^\nu A^\mu$, in terms of which Maxwell's equations are $\partial_\nu F^{\mu\nu} = -\mu_0 J^\mu$. The Lagrangian that generates this is

$$\mathscr{L} = -\tfrac{1}{4} F^{\mu\nu} F_{\mu\nu} - \mu_0 J^\mu A_\mu. \tag{7.33}$$

(3) Complex scalar field. Here we want a Lagrangian that will yield the Klein–Gordon equation. The required form is

$$\mathscr{L} = (\partial^\mu \phi)(\partial_\mu \phi)^* - \mu^2 \phi \phi^*, \tag{7.34}$$

where the only subtlety is that we treat ϕ and ϕ^* as independent variables (rather than the real and imaginary parts of ϕ). For a real field (which we saw in chapter 6 corresponds to a neutral particle), the Lagrangian becomes

$$\mathscr{L} = \tfrac{1}{2}(\partial^\mu \phi \partial_\mu \phi - \mu^2 \phi^2). \tag{7.35}$$

(4) Dirac field. Here we treat ψ and $\overline{\psi}$ as independent. The Dirac equation then follows from

$$\mathscr{L} = i\overline{\psi}\gamma^\mu \partial_\mu \psi - m\overline{\psi}\psi. \tag{7.36}$$

(5) Gravity. The Lagrangian for gravity in the absence of matter is about as simple as it could be given the constraints of general covariance, and is known as the **Einstein–Hilbert Lagrangian**:

$$\mathscr{L} = -\frac{1}{16\pi G}(R + 2\Lambda)\sqrt{-g}, \tag{7.37}$$

where R is the Ricci scalar. To show that this yields Einstein's equations is not trivial, and the relation $\delta\sqrt{-g} = \frac{1}{2}\sqrt{-g}\, g^{\mu\nu}\delta g_{\mu\nu}$ is needed [problem 7.2]. To obtain the field equations in the presence of matter, we need to add the matter Lagrangian, analogous to the $-J^\mu A_\mu$ term in the electromagnetic Lagrangian. This term is more complicated in the case of gravity, however, owing to the presence of the $\sqrt{-g}$ term. Comparison with Einstein's equations shows that the matter Lagrangian must satisfy $\sqrt{-g}\, T^{\mu\nu} = -2\partial\mathscr{L}_m/\partial g_{\mu\nu}$. Further discussion of this point is deferred until the next chapter, where we will see how the energy–momentum tensors of fields arise from their Lagrangians.

SCALINGS Notice that the Lagrangian has two degrees of freedom: the Euler–Lagrange equation will yield the same result if we multiply $\mathscr{L}$ by a constant or add a constant term (more generally, a total divergence; this contributes to the action integral only on the boundary, whereas the paths in the integral are varied about fixed endpoints and so boundary terms have no effect). The meaning of the additive constant is clear enough:

the idea that the zero point of energy is arbitrary is familiar enough in e.g. potential problems. Of course, this arbitrariness lies at the root of the problem of obtaining a well-defined vacuum energy, since the zero point of energy yields a mass density that is observable through its coupling to gravity. What about the multiplicative degree of freedom? This is clearly directly observable, since doubling the Lagrangian also doubles the Hamiltonian and affects the observable field energy density. The choice of a particular scaling thus comes down to a choice of units for the field.

NATURAL UNITS To simplify the appearance of equations, it is universal practice in quantum field theory to adopt **natural units**, where we take

$$\boxed{\hbar = c = \mu_0 = \epsilon_0 = 1.} \tag{7.38}$$

This convention makes the meaning of equations clearer by reducing the algebraic clutter, and is also useful in the construction of intuitive arguments for the order of magnitude of quantities in field theory. As with the question of setting $c = 1$ in relativity, this streamlining of the formulae can be unhelpful to the novice, which is why it has been resisted so far; however professional practice must eventually be faced.

The adoption of natural units corresponds to fixing the units of charge, mass, length and time relative to each other. This leaves one free unit, usually taken to be energy. Natural units are thus one step short of the Planck system, in which $G = 1$ also, so that all units are fixed and all physical quantities are dimensionless. In natural units, the following dimensional equalities hold:

$$\begin{aligned} [E] &= [m] \\ [L] &= [m]^{-1}. \end{aligned} \tag{7.39}$$

Hence, the dimensions of energy density are

$$[\mathscr{L}] = [m]^4, \tag{7.40}$$

with units usually quoted in GeV^4. Thus, when we deal with quadratic mass terms ($m^2\phi^2$, $m^2 A^\mu A_\mu/2$ etc.), the dimensions of the fields must clearly be

$$[\phi] = [A^\mu] = [m]. \tag{7.41}$$

MASS TERMS In the above quantum examples, the terms that are quadratic in the field are associated with the mass of the particle. This shows us that, if the photon were massive, we should need to add a **mass term** to the Lagrangian $\mathscr{L} \rightarrow \mathscr{L} + \mu^2 A^\mu A_\mu/2$. The resulting field equations are called the **Proca equations**:

$$(\Box + \mu^2)A^\mu = J^\mu. \tag{7.42}$$

Since we can apparently deal with a massive photon so easily, the question of why the photon should have zero mass at all arises. This is one of the principal topics of the next chapter.

INTERACTION TERMS The term $-J^\mu A_\mu$ in the electromagnetic Lagrangian describes the interaction of the field with external currents. Similar terms are to be expected whenever a field is coupled to a source. For a neutral scalar field coupled to a source density ρ, the corresponding interaction Lagrangian would be $g\rho\phi$, leading to the wave equation $(\Box + \mu^2)\phi = g\rho$ (where g is some coupling constant). This approach gives some insight into why forces between particles are sometimes attractive (gravity) but in other cases repulsive (electrostatics). Consider the scalar field case, with a constant delta-function source of unit strength $\rho = \delta(\mathbf{r})$; the field produced is $\phi = g\exp(-\mu r)/4\pi r$. If we now have another contribution to ρ from a further δ-function at r, the change in the Lagrangian due to the effect on this second particle of the field due to the first is $\Delta L = \int \Delta\mathscr{L}\, dV = \int g\phi\rho\, dV$, which is just $\Delta L = g^2 \exp(-\mu r)/4\pi r$. Now, the perturbation to the Hamiltonian is just $\Delta H = -\Delta L$ (see below), and so the interaction is attractive, independently of the sign of g. Similar arguments for vector and tensor fields show that the forces in these cases are respectively repulsive and attractive, thus depending on the spin of the particle involved. The scalar interaction was Yukawa's original model of a nuclear strong interaction governed by the exchange of **mesons** (π^0 particles). The density ρ would be the density of nucleons and ϕ would represent the pion field. The true theory of the strong interaction is, however, considerably more complex, as will be seen in the next chapter.

FIELD EXPANSIONS Given the Lagrangian for a field, it is then straightforward to obtain the quantized version of the field. The normal procedure is to identify the momentum density π corresponding to a field ϕ via $\pi = d\mathscr{L}/d\dot{\phi}$ and apply the **equal-time commutators**

$$\begin{aligned} [\phi(\mathbf{r}), \phi(\mathbf{r}')] &= [\pi(\mathbf{r}), \pi(\mathbf{r}')] = 0 \\ [\phi(\mathbf{r}), \pi(\mathbf{r}')] &= i\hbar\delta^{(3)}(\mathbf{r} - \mathbf{r}'). \end{aligned} \tag{7.43}$$

This is the same as we had in the case of electromagnetism; the delta function is present so that spatial integration gives the usual commutator $[q, p] = i\hbar$. This is of course a non-covariant procedure, since the time coordinate is singled out and treated specially. The question of the commutator of fields at a general spacetime separation is discussed below under the heading of field propagators.

Given the field-momentum commutator, an expansion of the field in the free-particle plane-wave states allows the creation and annihilation operators to be identified, and the form of the field expansion is always of these operators multiplying the plane wave states. For electromagnetism we have seen that this looks like

$$\mathbf{A} = \sum_{\mathbf{k},\alpha} \sqrt{\frac{\hbar c^2}{2\omega V}} \left[a\, \boldsymbol{\epsilon}^{(\alpha)} e^{-ik^\mu x_\mu} + a^\dagger\, \boldsymbol{\epsilon}^{(\alpha)} e^{ik^\mu x_\mu} \right]. \tag{7.44}$$

For scalar fields, the corresponding expression is simpler because there is no polarization to consider:

$$\phi = \sum_{\mathbf{k}} \sqrt{\frac{\hbar c^2}{2\omega V}} \left[a\, e^{-ik^\mu x_\mu} + a^\dagger\, e^{ik^\mu x_\mu} \right]. \tag{7.45}$$

The complex scalar field is more complicated, because ϕ and $\phi^\dagger$ are now distinct. Particles and antiparticles thus both need their own sets of creation and annihilation operators ($a^\dagger$ and a, $b^\dagger$ and b) and the field operator is

$$\phi = \sum_{\mathbf{k}} \sqrt{\frac{\hbar c^2}{2\omega V}} \left[a\, e^{-ik^\mu x_\mu} + b^\dagger\, e^{ik^\mu x_\mu} \right] \tag{7.46}$$

(and similarly for $\phi^\dagger$).

FERMION FIELDS AND ANTICOMMUTATORS For the Dirac field, things are similar, but involve the momentum-dependent spinors u and v:

$$\psi = \sum_{\mathbf{p},r} \sqrt{\frac{mc^2}{EV}} \left[b\, u_r(\mathbf{p})\, e^{-ik^\mu x_\mu} + d^\dagger\, v_r(\mathbf{p})\, e^{ik^\mu x_\mu} \right]. \tag{7.47}$$

Here, $b^\dagger$ and $d^\dagger$ create respectively electrons and positrons.

Of course, Dirac particles are fermions, for which the exclusion principle arises if fields are quantized via anticommutators rather than commutators [problem 8.6]. For the Dirac equation, the momentum conjugate to ψ is $\pi = \bar{\psi} i\gamma^0$, and the usual quantization conditions with anticommutators become

$$\begin{aligned} \{\psi_\alpha(\mathbf{x}), \psi_\beta(\mathbf{x}')\} &= \{\pi_\alpha(\mathbf{x}), \pi_\beta(\mathbf{x}')\} = 0 \\ \{\psi_\alpha(\mathbf{x}), \pi_\beta(\mathbf{x}')\} &= i\hbar\, \delta_{\alpha\beta}\, \delta^{(3)}(\mathbf{x} - \mathbf{x}'). \end{aligned} \tag{7.48}$$

The creation and annihilation operators in the Dirac expansion thus satisfy

$$\{b_r, b_s^\dagger\} = \{d_r, d_s^\dagger\} = \delta_{rs}, \tag{7.49}$$

And the Dirac Hamiltonian comes out to be

$$H = \sum E_k \left(b^\dagger b + d^\dagger d \right) \tag{7.50}$$

(although re-ordering of operators is required, as with electromagnetism). This many-particle interpretation of the Dirac equation is the proper implementation of the idea of positrons as holes, and avoids the need to consider explicitly the negative-energy sea.

Although it would be possible to change the anticommutators to commutators and apply this to the Dirac equation, only anticommutators guarantee a positive Hamiltonian (see e.g. section 4.3 of Mandl & Shaw 1984). The **spin-statistics theorem** (proved by considering in turn all relevant wave equations) says that sensible results are only obtained if integral spin particles are quantized with boson statistics and half-integral spin particles with fermion statistics. It is a pity that no transparent argument exists for such a fundamental result.

Lastly, a conceptual difficulty with the Dirac field is how the classical limit arises for particles that satisfy the Dirac equation. Because fermions are involved, the occupation numbers cannot exceed unity, and the classical $n \gg 1$ condition familiar from the boson case can never be attained. If we stick with anticommutators, pursuing the idea of a classical limit leads to a new kind of number that will anticommute even when we are considering independent modes free from quantum interference: **Grassmann variables**. These have some curious properties, which are explored in problem 7.3.

7.4 Interacting fields

The Lagrangian approach to quantum fields may seem an elegant luxury – just a neater way of getting wave equations. The real power of the approach, however, becomes apparent when several fields are present. In particular, consider a fermionic field plus electromagnetism; we just add the Lagrangians, obtaining

$$\mathcal{L} = \mathcal{L}_\mathrm{f} + \mathcal{L}_\mathrm{em} = i\bar{\psi}\gamma^\mu \partial_\mu \psi - m\bar{\psi}\psi - \tfrac{1}{4} F^{\mu\nu} F_{\mu\nu} - J^\mu A_\mu. \tag{7.51}$$

So far, we have treated the 4-current J^μ as a classical quantity. To be consistent, we should now use the *quantum* current from the fermion field (charge times probability 4-current):

$$J^\mu = e\overline{\psi}\gamma^\mu\psi. \tag{7.52}$$

This introduces a vital term into the Lagrangian: one that *mixes* the two fields. This is always known as an **interaction term**: regarding the fields as sums of creation and annihilation operators means that such mixed terms allow us to change the numbers of the two particle species simultaneously. The resulting theory is known as **quantum electrodynamics (QED)**.

To use the QED Lagrangian in perturbative quantum mechanics to calculate observed quantities such as scattering cross-sections, we need the perturbation to the Hamiltonian. This is quite straightforward to obtain. Consider the definition of the Hamiltonian in terms of some coordinates q_i:

$$H \equiv \sum p_i\dot{q}_i - L(q_i, \dot{q}_i). \tag{7.53}$$

If we make a perturbation to the Lagrangian, $L \to L + \delta L$, where δL is independent of the $\dot{q}_i$, then the corresponding perturbation to the Hamiltonian is just $\delta H = -\delta L$. Thus, in the case of QED, we get the perturbation to the Hamiltonian density

$$\boxed{\delta\mathcal{H} = e\overline{\psi}\gamma^\mu\psi A_\mu.} \tag{7.54}$$

INTERACTION REPRESENTATION We have previously dealt with the choices of representation: whether to put the time dependence in the wave function (Schrödinger) or in the operators (Heisenberg). The intermediate case is used in field theory, where we split the Hamiltonian into the free-field parts plus an interaction term: $H = H_0 + H_{\mathrm{I}}$. The **interaction representation** makes the unitary Heisenberg transformation, but using only H_0. This yields wave functions and operators that are time dependent, obeying the equations of motion

$$\begin{aligned} i\hbar\dot{\psi} &= H_{\mathrm{I}}\psi \\ i\hbar\dot{O} &= [O, H_0]. \end{aligned} \tag{7.55}$$

S-MATRIX The main processes of interest in particle physics are scattering events, where one set of objects approach from infinity, interact, and a different set then recede. The transition process between different states can be described by the operator equation

$$\psi(t) = U(t, t_0)\psi(t_0). \tag{7.56}$$

The asymptotic transition operator, $U(\infty, -\infty) \equiv S$ is known as the **S-matrix**. In general, this important quantity cannot be found exactly, but only as a perturbation series. The U operator satisfies the interaction-representation Schrödinger equation, which is integrated to yield the integral equation

$$U = 1 - \frac{i}{\hbar}\int_{t_0}^{t} H_{\mathrm{I}}\, U\, dt. \tag{7.57}$$

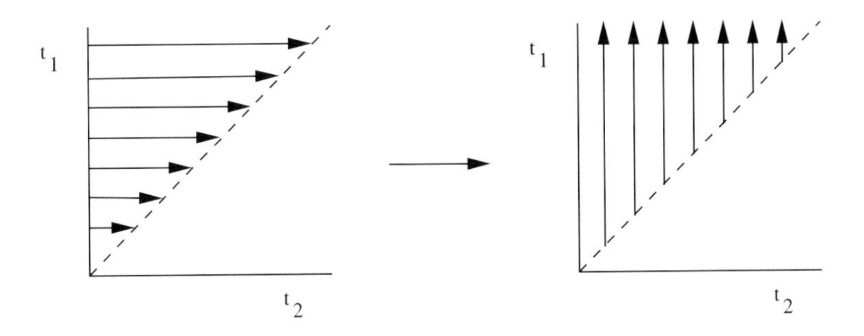

Figure 7.1. Illustrating how the order of time integration in the second-order part of the S-matrix is inverted: t_1 running from $-\infty$ to ∞ and t_2 from $-\infty$ to t_1 is equivalent to t_2 running from $-\infty$ to ∞ and t_1 from t_2 to ∞.

This can be solved iteratively, starting with the zeroth-order approximation $U = 1$, and resubstituting successive approximations inside the time integral. We thus obtain the series

$$\boxed{\begin{aligned} S = 1 - i \int_{-\infty}^{\infty} H_{\rm I}\, dt \\ + (-i)^2 \int_{-\infty}^{\infty} H_{\rm I}(t_1)\, dt_1 \int_{-\infty}^{t_1} H_{\rm I}(t_2)\, dt_2 + \cdots \end{aligned}} \tag{7.58}$$

A mathematician would hardly be satisfied at this point, since we have not proved that the series converges. In fact, it does not. Quantum field theory frequently produces **asymptotic approximations**: series in which successive terms initially decline with increasing order, only to grow again at large orders. Useful results are obtained by taking the first few terms, and the error involved is of the order of the last term included. For more details of these rather magical properties, see e.g. the books by Jeffries (1962) or Copson (1965).

A last important property of the U-matrix is that it is **unitary** ($U^{-1} = U^\dagger$). This is easily proved because U obeys the Schrödinger equation (symbolized $[SE]$): $i\hbar\dot{U} = H_{\rm I}U$, where $H_{\rm I}$ is Hermitian. Taking $U^\dagger[SE] - [SE]^\dagger U$ shows that $U^\dagger U$ is a constant, which must be unity because $U(t_0, t_0) = 1$. This is important because it relates to probability conservation in the scattering process: for $|\psi|^2$ to be unchanged after the scattering, $S^\dagger S = 1$ is required. The requirement of unitarity often allows rather general constraints on quantum field models to be obtained.

THE T-PRODUCT The S-matrix formalism should also be covariant. It certainly is in first order, where we need the integral $\int \mathcal{H}_{\rm I}\, d^4x^\mu$. At higher orders, this is not so clear, since a specific relation of time coordinates is imposed ($t_1 > t_2$ in second order). We can amend this problem by realizing that the double integral can be performed in the opposite order, as illustrated in figure 7.1.

Symmetry suggests that we write the result as the mean of the two possibilities:

$$S^{(2)} = \frac{(-i)^2}{2} \left[\int_{-\infty}^{\infty} dt_1 \int_{\infty}^{t_1} H_I(t_1) H_I(t_2)\, dt_2 + \int_{-\infty}^{\infty} dt_2 \int_{t_2}^{\infty} H_I(t_1) H_I(t_2)\, dt_1 \right]. \tag{7.59}$$

This is simplified by defining the **T-product** or **Wick product**:

$$T[H(t)H(t')] \equiv \begin{cases} H(t)H(t') & (t > t') \\ \pm H(t')H(t) & (t < t') \end{cases} \tag{7.60}$$

where the minus sign applies for fermionic fields (cf. the classical idea of spinors as anticommuting **Grassmann variables**). The T-product of more than two terms simply means writing them in order of decreasing time from left to right and setting commutators to zero for bosons or anticommutators to zero for fermions.

In these terms, the nth term in the perturbation series can be written as the unrestricted time integral

$$S^{(n)} = \frac{(-i)^n}{n!} \int_{-\infty}^{\infty} dt_1 \cdots \int_{-\infty}^{\infty} dt_n\; T[H_I(t_1) \cdots H_I(t_n)] \tag{7.61}$$

and the formal sum of the perturbation series may be written as follows:

$$S = T\left[\exp\left(i \textstyle\int \mathcal{L}_I\, d^4x^\mu\right)\right]. \tag{7.62}$$

What do the terms in the S-matrix expansion mean? The interaction Lagrangian will be a mixture of creation and annihilation operators (e.g. $\mathcal{L}_I = e\bar{\psi}\gamma^\mu\psi A_\mu$ for quantum electrodynamics). The successive higher orders of perturbation theory thus describe a scattering process that proceeds through the production and subsequent annihilation of ever increasing numbers of particles. The meaning of these various terms is illustrated using **Feynman diagrams**, to which we shall turn shortly.

PROPAGATORS The linear form of the S-matrix brings to mind a general concept for evolving wave functions forward in time: the **propagator** in quantum mechanics, which arises because the wave equations are linear and first order in time. The wave function must then be expressible as a linear combination of the values of ψ at different points in space at some earlier time (the quantum equivalent of Huygens' wavelets). This encourages us to write

$$\psi(\mathbf{r}, t) = i \int \psi(\mathbf{x}, t')\, G(\mathbf{r}, t\,;\, \mathbf{x}, t')\, d^3x, \tag{7.63}$$

where G acts as a Green function, giving the response of the quantum system at a given time to the quantum 'disturbance' at some earlier time; the reason for the factor i will be seen shortly.

So far, G is not quite a propagator, since the very linearity of quantum wave equations does not permit an 'arrow of time': there is no guarantee that the above relation has $t > t'$, and indeed the wave function can be specified in terms of the values it will have in the future. To put in by hand commonsense notions of causality, we shall have to split G into the **retarded Green function** G^+, which vanishes unless $t > t'$, and its counterpart the **advanced Green function** G^-. The retarded function is more properly what is called the propagator.

Enforcing causality in this way naturally involves the Heaviside or step function: $\Theta(t-t')$ is zero unless $t > t'$, in which case it is unity, and the propagator now satisfies

$$\Theta(t-t')\,\psi(\mathbf{r},t) = i\int \psi(\mathbf{x},t')\,G^+(\mathbf{r},t\,;\,\mathbf{x},t')\,d^3x. \tag{7.64}$$

Operating on this equation with the Schrödinger operator shows that the propagator satisfies the differential equation

$$\left[i\frac{\partial}{\partial t} - \frac{H}{\hbar}\right] G^+(\mathbf{r},t\,;\,\mathbf{x},t') = \delta(t-t')\,\delta^{(3)}(\mathbf{r}-\mathbf{x}) \tag{7.65}$$

(to verify this, multiply both sides by $\psi(\mathbf{x},t')$ and integrate d^3x). Thus, the propagator is indeed the response of the Schrödinger differential system to an applied delta-function impulse.

The solution for the propagator is easily performed in the absence of an external potential, and is most transparent in Fourier space, where the normal convention for 4D Fourier transforms is

$$\begin{aligned} F(x^\mu) &= (2\pi)^{-4}\int \widetilde{F}(k^\mu)\,\exp(-ik^\mu x_\mu)\,d^4k \\ \widetilde{F}(k^\mu) &= \int F(x^\mu)\,\exp(+ik^\mu x_\mu)\,d^4x. \end{aligned} \tag{7.66}$$

In this form, the spatial and temporal delta functions are just constants, whereas the Schrödinger operator becomes $\omega - k^2/2m$. The Green function is then

$$G^+(\mathbf{r},t\,;\,\mathbf{x},t') = \frac{1}{(2\pi)^4}\int d^3k\,d\omega\,\frac{1}{\omega - k^2/2m}\,e^{i\mathbf{k}\cdot(\mathbf{r}-\mathbf{r}')-i\omega(t-t')}. \tag{7.67}$$

What does this mean? The integration over ω cannot give a well-defined result because the integrand has a pole at $\omega = k^2/2m$. A way of avoiding the problem is to displace the pole from the real ω-axis by adding $i\epsilon$ ($\epsilon > 0$) to the denominator of the propagator:

$$\frac{1}{\omega - k^2/2m} \quad\to\quad \frac{1}{\omega - k^2/2m + i\epsilon}. \tag{7.68}$$

To see how this helps, look at the contours in figure 7.2. If ω contains a positive imaginary part, then $\exp-[i\omega(t'-t)]$ becomes vanishingly small on a semicircle of large radius in the upper part of the Argand diagram when $t' > t$. Since the pole is then outside the contour, the integral along the real axis therefore vanishes, and so the propagator is zero, as required. Similarly, for $t' < t$ the integral along a lower semicircle vanishes, but the pole is now included, giving a non-zero contribution to the propagator.

GREEN FUNCTIONS We now return to quantum fields and address the apparent problem that the use of the T-product is not manifestly covariant, since it singles out the time coordinate for special treatment. To show that this is not a difficulty, we need to consider the **Green functions** that are important in quantum field theory. A Green function is the response of a differential equation to an applied impulse. For example,

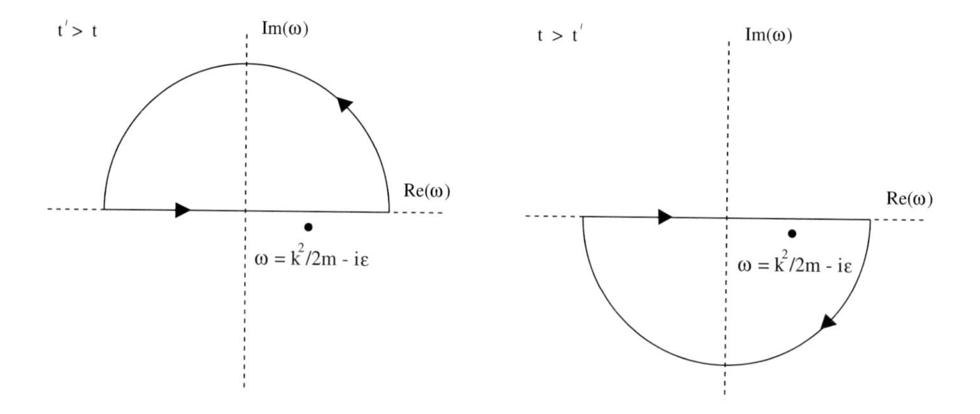

Figure 7.2. Contours for the evaluation of the frequency integral in the Schrödinger propagator. By shifting the pole to $k^2/2m - i\epsilon$, the result is made well defined. Closing the contour in the upper half-plane gives zero when $t < t'$, as required for the retarded propagator.

consider the scalar-field wave equation with a delta-function source:

$$(\Box + m^2)G = -\delta^{(4)}(x^\mu - x'^\mu). \tag{7.69}$$

The Green function $G(x^\mu, x'^\mu)$ that solves this equation is found by going to momentum space, in which

$$\delta^{(4)}(x) = \int \frac{d^4k}{(2\pi)^4}\, e^{-ikx}. \tag{7.70}$$

To simplify the notation, we shall drop the explicit denotion of 4-vectors. This convention is very common in the literature on this subject, and it should be clear from the context what is intended. Thus, henceforth

$$\boxed{q^\mu x_\mu \to qx.} \tag{7.71}$$

Similarly, k^2 denotes $k^\mu k_\mu$, and must be carefully distinguished from the square of the spatial part of the 4-vector, $\mathbf{k}^2$: $k^2 = k_0^2 - \mathbf{k}^2$.

The momentum-space Green function is now given by the 'inverse' of the Klein–Gordon operator:

$$G = -(\Box + m^2)^{-1} = (k^2 - m^2)^{-1}. \tag{7.72}$$

This gives trouble when $k^2 = m^2$: the momentum-space integral has to be interpreted as a contour integral that avoids these poles, and different choices of contour lead to different properties of the Green function. Going to the complex k_0-plane, the poles are at $k_0 = \pm\sqrt{\mathbf{k}^2 + m^2} = \pm\omega$. The arguments in the relativistic case are similar to those for the nonrelativistic Schrödinger equation. For a retarded Green function, G should vanish when $t < t'$; this is achieved if the contour passes above both poles. Similarly, the advanced function requires a contour that passes below both poles. It may seem that there are to be no new features here, but a major surprise is in store: relativistic quantum

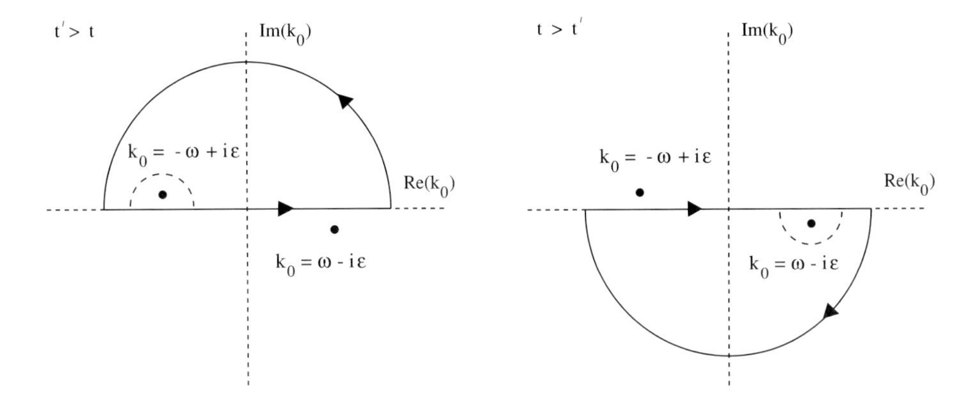

Figure 7.3. The different choices of contour for evaluating the momentum-space integral for the Green function of the Klein–Gordon field. The poles at $k_0 = \pm\omega$ have to be avoided. A contour that passes above them gives a retarded Green function, whereas the advanced function contour passes below. Feynman's contour passes below the pole at $-\omega$ and above the one at $+\omega$.

fields turn out to require a *third* contour, as will be proved shortly. This is shown in figure 7.3; it is a contour that passes above one pole and below the other, being closed in the appropriate half of the plane, depending on the sign of $t - t'$. The same effect can be obtained by subtracting a small imaginary amount from the poles, so that they now lie at $k_0^2 = \omega^2 - i\epsilon$, or $k_0 = \pm(\omega - i\epsilon')$, with a redefinition of ϵ to some other small positive quantity. This then gives the **Feynman propagator**

$$\boxed{G_{\mathrm{F}} = \frac{1}{k^2 - m^2 + i\epsilon},} \tag{7.73}$$

where the small positive ϵ is to be set to zero after performing momentum-space integrations. What does this mean? For positive frequencies it follows the retarded contour, but chooses the advanced contour for negative frequencies. In a loose sense, then, the Feynman function propagates positive frequencies forwards in time, but negative frequencies backwards in time. It is the mathematical basis for Feynman's statement that a positron propagating forwards in time is equivalent to a negative-energy electron propagating backwards in time.

We now show that Feynman's covariant Green function is exactly what is required in order to evaluate the integrals over T-products of fields that arise in perturbation theory. Consider $\langle 0|\, T[\phi(x)\phi(x')]\, |0\rangle$; we can evaluate this as follows. First recall the field expansion in terms of creation and annihilation operators:

$$\phi(x) = \sum \frac{\hbar c^2}{2\omega V}\left[a(\mathbf{k})e^{-ikx} + a^\dagger(\mathbf{k})e^{ikx}\right]. \tag{7.74}$$

Write this twice to produce the product $\phi(x)\phi(x')$ and take the vacuum expectation value. The only non-zero terms will be those where $\phi(x)$ creates a particle that is than annihilated by $\phi(x')$. Taking the limit $\Sigma \to [V/(2\pi)^3] \int d^3k$ then gives

$$\langle 0|\phi(x)\phi(x')|0\rangle = \hbar \int \frac{d^3k}{(2\pi)^3\, 2\omega} e^{ik(x-x')}. \tag{7.75}$$

To involve the T-product brings in the Heaviside function, since

$$T[\phi(x)\phi(x')] = \Theta(t-t')\phi(x)\phi(x') + \Theta(t'-t)\phi(x')\phi(x). \tag{7.76}$$

The Fourier representation of the Heaviside function is

$$\Theta(t) = \lim_{\epsilon\to 0^+} \frac{i}{2\pi} \int_{-\infty}^{\infty} e^{-i\omega t} \frac{d\omega}{\omega + i\epsilon}, \tag{7.77}$$

as may be seen by differentiating and using $\int_{-\infty}^{\infty} e^{-i\omega t}\, d\omega = 2\pi\delta(t)$; a step is clearly the result of integrating the delta function. The only subtlety is to check that the constant of integration is correct, and that $\Theta(t)$ vanishes for $t < 0$. This is easily seen, since in this case the integral can be converted to a contour integral closed by a semicircle in the upper half of the Argand diagram ($e^{-i\omega t}$ is vanishingly small), and the contour contains no poles.

An alternative integral representation is

$$\Theta(t)e^{-i\omega t} = \int \frac{dk_0}{2\pi} \frac{ie^{-ik_0 t}}{k_0 - (\omega - i\epsilon)}, \tag{7.78}$$

where ϵ is small and positive as before. This expression can be used to substitute for the terms like $\Theta(t-t')\exp(i\omega t - t')$ in the propagator, leading to

$$\langle 0|T[\phi(x)\phi(x')]|0\rangle = i \int \frac{d^4k}{(2\pi)^4} \frac{1}{2\omega} \left[\frac{e^{ik(x-x')}}{k_0 - (\omega - i\epsilon)} + \frac{e^{-ik(x-x')}}{k_0 - (\omega - i\epsilon)} \right]. \tag{7.79}$$

To make sense of this expression, remember that $\omega = \sqrt{\mathbf{k}^2 + m^2}$, whereas k_0 is unrestricted; ω is a positive number, but k_0 takes either sign. Finally, the sign of the integration variables can be reversed for the second term in the integral, and a new ϵ can be defined (still small and positive), leading to the form

$$\boxed{\langle 0|T[\phi(x)\phi(x')]|0\rangle = i \int \frac{d^4k}{(2\pi)^4} e^{ik(x-x')} \frac{1}{k^2 - m^2 + i\epsilon}.} \tag{7.80}$$

Thus the integral over the T-product yields a covariant Green function to within a factor i. This expression is also known as the **propagator** of the scalar field, for the reasons given above. It is the Feynman propagator that turns out to be of importance in quantum fields, justifying the contour choice we made earlier.

The factor i above is a source of some annoyance, since it leads to a variety of different conventions for propagators in the literature. Some authors define $G_{\rm F}$ with an extra factor of i, so that $\langle 0|T[\phi(x)\phi(x')]|0\rangle = G_{\rm F}$ in k-space. Also, it is possible to define $G_{\rm F}$ with the opposite sign by changing the sign of the delta-function source in the Klein–Gordon equation. The particular choice made here attempts to maximize the chance of remembering the correct signs, but this is an area where particular care is needed.

To sum up this section, the critical ingredients for perturbation-theory calculations of scattering processes involving quantum fields are the T-products of fields, which are most simply expressed as propagators in Fourier space. The derivations of the propagators for other fields follow the same line as for the scalar field, but we shall not give the details here. For a spinor field, the Fourier-space equivalent of the real-space propagator is

$$\frac{1}{i\hbar}\langle 0|T[\psi_\alpha(x)\bar{\psi}_\beta(x')|0\rangle \rightarrow (\not{k}+m)_{\alpha\beta}\, G_{\rm F} = \frac{(\not{k}+m)_{\alpha\beta}}{k^2-m^2+i\epsilon}. \tag{7.81}$$

For a massive vector field B_μ, it is

$$\frac{1}{i\hbar}\langle 0|T[B_\mu(x)B_\nu(x')|0\rangle \rightarrow \left(-g_{\mu\nu}+\frac{k_\mu k_\nu}{m^2}\right) G_{\rm F} = \frac{(-g_{\mu\nu}+k_\mu k_\nu/m^2)}{k^2-m^2+i\epsilon}. \tag{7.82}$$

For the photon, the result is not quite as expected from the $m \rightarrow 0$ limit of the massive-field expression:

$$\frac{1}{i\hbar}\langle 0|T[A_\mu(x)A_\nu(x')|0\rangle \rightarrow -g_{\mu\nu}\, G_{\rm F} = \frac{-g_{\mu\nu}}{k^2+i\epsilon}. \tag{7.83}$$

This peculiarity is related to the gauge-field nature of the photon, and is discussed in some detail in section 3.2.3 of Itzykson & Zuber (1980).

7.5 Feynman diagrams

A convenient way to keep track of the products of creation and annihilation operators that occur in the perturbation expansions of quantum field theory is via the device of **Feynman diagrams**. Most readers will have met these, but will perhaps not be fully aware of their meaning. These diagrams provide a way of injecting physical intuition and particle concepts into the world of quantum fields and second quantization. Consider as an example the interaction Hamiltonian for QED:

$$\mathscr{H}_{\rm I} = e\,\bar{\psi}\gamma^\mu\psi\, A_\mu. \tag{7.84}$$

It is convenient to decompose this via

$$\begin{aligned} \psi &= \psi^{(+)} + \psi^{(-)} \\ \bar{\psi} &= \bar{\psi}^{(+)} + \bar{\psi}^{(-)}, \end{aligned} \tag{7.85}$$

where

$$\begin{aligned} \psi^{(+)} &= \sum \sqrt{\frac{mc^2}{EV}}\; b\; u(\mathbf{p})\; e^{i(\mathbf{p}\cdot\mathbf{x}-Et)/\hbar} \\ \psi^{(-)} &= \sum \sqrt{\frac{mc^2}{EV}}\; d^\dagger\; v(\mathbf{p})\; e^{i(Et-\mathbf{p}\cdot\mathbf{x})/\hbar}, \end{aligned} \tag{7.86}$$

and the sum is over momentum and spin states. The $(-)$ terms create respectively electrons and positrons, which are annihilated by the $(+)$ terms. At first order, the QED interaction Hamiltonian therefore gives rise to eight possibilities, represented diagramatically in figure 7.4.

These are spacetime diagrams; where did spacetime arise? We have seen how the S-matrix elements involve an integration of propagators over spacetime. Since propagators determine the way in which wave functions at one spacetime point influence wave functions at another, it is perhaps reasonable to think pictorially about the field

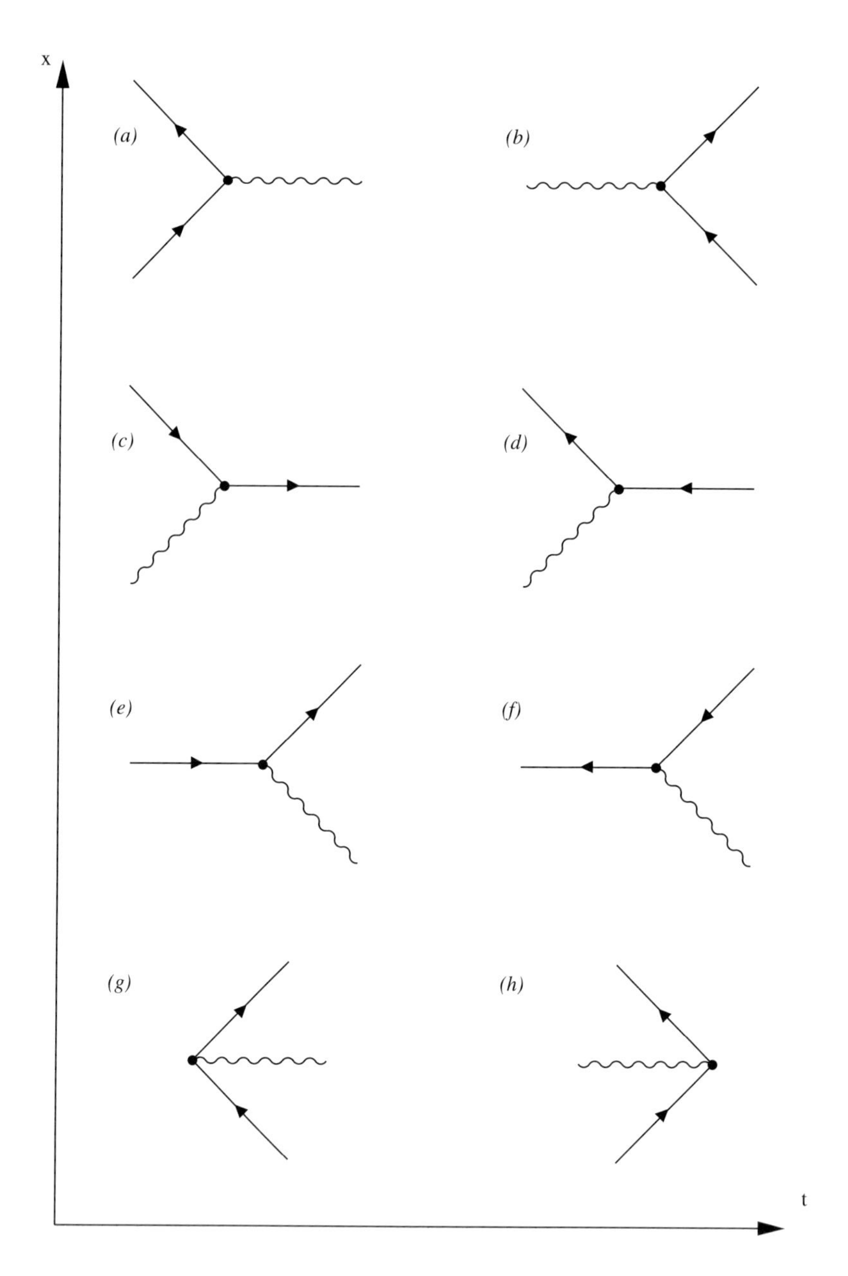

Figure 7.4. The Feynman diagrams for QED. These represent the possible combinations of the creation and annihilation of photons, electrons and positrons allowed at first order by the QED Lagrangian. Lines with an arrow pointing towards increasing time denote electrons, backwards-pointing arrows are positrons, and wavy lines denote photons.

quantum propagating the information between x and x'. Alternatively, the non-zero parts of the propagator matrix element always involve the creation of one particle and the annihilation of another, again suggesting a diagram depicting a particle appearing at one spacetime point and disappearing at another. However, this is not really what is going on. The creation and annihilation operators deal with excitations of plane wave states, rather than particles that are spatially localized. The spacetime coordinates are variables to be integrated over, rather than representing in any sense the location of a particle. The main function of the diagrams is really as a mnemonic that represents in a non-algebraic fashion the mathematical content of the S-matrix element.

The diagrams have been drawn with t as the horizontal axis, and x as the vertical axis. This is not a universal convention, and many books draw diagrams with the opposite orientation. Fortunately, because this is relativistic field theory, the orientation is not so critical. Changing the axis labels alters the physical process being represented, but it will still be a valid physical interaction. For example, consider the first diagram, which represents e^+e^- annihilation leading to a photon as it stands. Rotated by 120 degrees, it would correspond to diagram (c), in which an electron is scattered by absorbing a photon.

It is important to be clear that none of these processes can be realized in nature exactly as shown, because in general they fail to conserve energy–momentum. For example, in process (a), a pair cannot annihilate to yield a single photon: in the rest-frame of the pair the 4-momentum is just $P^\mu = (E, 0)$, which does not correspond to the 4-momentum of any single photon. The particles produced by these processes are **virtual particles**, and are discussed further below. To make an observable process, the failure of energy–momentum conservation in one of these diagrams has to be cancelled by combining it with a second diagram.

FIRST-ORDER SCATTERING The simplest example of quantum field theory in action is in a first-order process, where we treat e.g. electron scattering off a classical potential. The initial state is $|\mathbf{p}\rangle = b^\dagger(\mathbf{p})|0\rangle$, and the final state is $b^\dagger(\mathbf{p}')|0\rangle$ The required S-matrix element is

$$S_{fi} = -ie \left\langle b'^\dagger\ 0 \middle| \int \bar{\psi}\gamma^\nu\psi A_\nu\, d^4x \middle| b^\dagger\ 0 \right\rangle. \tag{7.87}$$

Now, commonsense tells us that only combinations in H_I looking like $b'^\dagger b$ matter (annihilate the incoming electron and create the outgoing one), so that the first-order matrix element becomes

$$S_{fi} = -ie\, \frac{m}{EV}\, \bar{u}'\gamma^\nu u \int A_\nu\, d^4x\, e^{-i(P^\nu - P'^\nu)x_\nu}. \tag{7.88}$$

This is now starting to look similar to the 'classical' Born approximation from nonrelativistic quantum mechanics. Let us sketch how this leads to the well-known Rutherford cross-section if we make the classical potential just the central electrostatic potential from a nucleus,

$$A^\mu = (\phi, 0, 0, 0); \quad \phi = \frac{-Ze}{4\pi\epsilon_0 r}. \tag{7.89}$$

The integral in S_{fi} factorizes into a product:

$$\int A_\nu \, d^4x^\mu \, e^{-i(P^\nu - P'^\nu)x_\nu} = \frac{-Ze}{4\pi\epsilon_0} \int e^{i\Delta E\, t}\, dt \int \frac{1}{r}\, e^{-i\Delta \mathbf{p}\cdot\mathbf{x}}\, d^3x$$
$$= \frac{-Ze}{\epsilon_0\, |\Delta\mathbf{p}|^2} \int e^{i\Delta E\, t}\, dt. \tag{7.90}$$

The time integral is dealt with by the same process as in nonrelativistic perturbation theory: let the interaction be considered to exist for a time T, and let $T \to \infty$:

$$\left| \int e^{i\Delta E\, t}\, dt \right|^2 \to 2\pi T\, \delta(E - E'). \tag{7.91}$$

The cross-section is nearly calculated. Defining the scattering rate, $|S_{fi}|^2/T$, disposes of T. Now divide this rate by the input particle flux density $(V^{-1}|\mathbf{p}|/E)$ and multiply by the density of final states (which brings in a factor of V to make the result independent of V, as it must be). This gives us

$$\frac{d\sigma}{d\Omega} = \frac{Z^2 e^4 m^2}{4\pi^2 \epsilon_0^2} \, \frac{|\bar{u}'\gamma^0 u|^2}{|\Delta\mathbf{p}|^4}. \tag{7.92}$$

This may not look so familiar, but we have $|\Delta\mathbf{p}|^4 = 4|\mathbf{p}|^2 \sin^2\theta/2$, where θ is the scattering angle. The term $|\bar{u}'\gamma^0 u|^2$ looks more formidable, but it is easily evaluated at least in the nonrelativistic limit (everything so far has been completely general), since

$$u \to \begin{pmatrix} 1 \\ 0 \\ 0 \\ 0 \end{pmatrix} \quad \text{or} \quad \begin{pmatrix} 0 \\ 1 \\ 0 \\ 0 \end{pmatrix} \tag{7.93}$$

for the two spin states of the electron. The problematic term on the rhs is thus either 1 or 0, depending on whether there is a spin flip. Averaging over either input spin and summing over all output spins thus leaves us with 1, and the cross-section is the familiar **Rutherford formula**, which may be re-expressed as

$$\boxed{\frac{d\sigma}{d\Omega} = \frac{(Z\alpha)^2}{4m^2 v^4 \sin^4\theta/2}.} \tag{7.94}$$

If the term involving spinors is evaluated without taking the nonrelativistic limit, what results is the same expression, multiplied by the factor $\gamma^{-2}(1 - \beta^2 \sin^2\theta/2)$, which is the cross-section for **Mott scattering**.

This is hardly the most transparent derivation for Rutherford scattering, but it is just about the simplest possible calculation one can attack using the methods of quantum electrodynamics. Things can get considerably more complicated when one considers processes of higher order, or those where several fields are involved (e.g. processes that include the quantization of the electromagnetic field fully, rather than treating it semiclassically as in the above calculation). Without considering interacting fields, some of the most important concepts in this subject never come into play. However, the above example has the merit of illustrating not only that the formalism works, but that it can hide something relatively simple under its surface complexity; it is hoped that this will give some reassurance in other cases if the algebra should appear to be getting out of control. For the purposes of this book, it would in any case not be appropriate to delve too far into detailed calculations. The remaining sections will focus mainly on important general results.

SECOND-ORDER PROCESSES Consider now a second-order process in quantum electrodynamics. The S-matrix element is of the form

$$S_{fi} = (-ie)^2 \int d^4x_1 \int d^4x_2 \, \langle f|F(x_1)F(x_2)|i\rangle, \tag{7.95}$$

where the function $F = \bar{\psi}\gamma^\mu\psi A_\mu$. The matrix element can be factorized into a part for the photons $\langle f|A_\mu A_\nu|i\rangle$ and a term for the remaining fermionic factors (since these independent fields commute).

Both these terms introduce wave-like spacetime variations. We have seen that the fermionic factor requires the introduction of the T-product or Wick product:

$$T[A(x), B(x')] = \begin{cases} AB & (t > t') \\ \pm BA & (t < t') \end{cases} \tag{7.96}$$

where the minus sign applies for fermionic fields. The exact form of these terms depends on the initial and final states; consider the special case of Compton scattering, which is probably one of the cases of greatest practical interest. Initial and final states thus both consist of one electron and one photon, but different in each case. Now, the field $\mathbf{A(x)}$ can both create and annihilate a photon, so that a second-order matrix element $\langle f|A_\mu A_\nu|i\rangle$ is just what is required to annihilate one photon and create another. Recall also that we are working in the Coulomb gauge, so that the 0-component of A^μ vanishes. The photon matrix element is thus

$$\langle f|A_\mu(x_1)A_\nu(x_2)|i\rangle = \left[\frac{1}{\sqrt{2E_1V}}\epsilon^{(1)}_\mu e^{ik_1x_1}\right]\left[\frac{1}{\sqrt{2E_2V}}\epsilon^{(2)}_\nu e^{ik_2x_2}\right] + \text{perm}, \tag{7.97}$$

where 'perm' stands for the same expression with the indices 1 and 2 interchanged.

The fermion part of the matrix element is a little more complicated. There is a product of two ψ and two $\bar{\psi}$ terms to consider; however, most of these give zero, owing to the particular initial and final states under consideration. The matrix element is divided up by inserting the completeness relation $\sum |n\rangle\langle n| = 1$ between each element, and using the fact that only $n = 0$ matters (since the initial and final states are represented by the operation of one creation operator upon the vacuum). Schematically, this process looks like

$$\langle e'|\bar{\psi}\psi\bar{\psi}\psi|e\rangle = \langle e'|\bar{\psi}|0\rangle\langle 0|\bar{\psi}\psi|0\rangle\langle 0|\psi|e\rangle. \tag{7.98}$$

The outer terms may be evaluated, leaving a fermionic matrix element

$$\left(\sqrt{\frac{m}{E'V}}\bar{u}e^{-ip'x_1}\right)\gamma^\mu \, \langle 0|T[\psi(x_1)\bar{\psi}(x_2)]|0\rangle \, \gamma^\nu \left(\sqrt{\frac{m}{EV}}\bar{u}e^{ipx_2}\right). \tag{7.99}$$

To obtain the S-matrix element, we now have to integrate over x_1 and x_2. The integrand has an oscillatory dependence on both x_1 and x_2, and since the regime of integration is infinite, we recognize that the Dirac δ-function will be involved in the answer:

$$\delta^{(4)}(p_1 - p_2) = \frac{1}{(2\pi)^4}\int e^{i(p_1-p_2)x}\, d^4x. \tag{7.100}$$

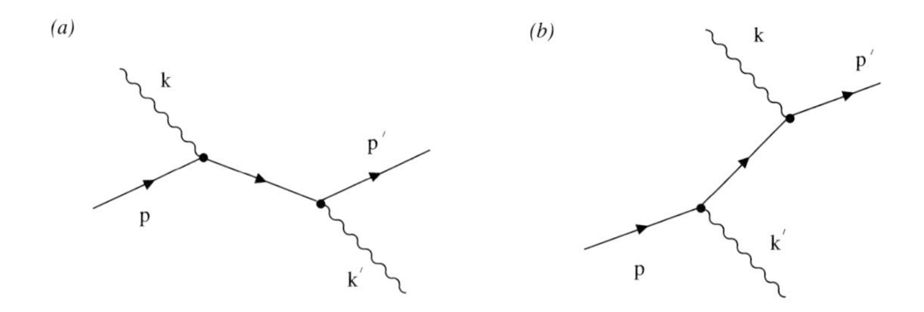

Figure 7.5. The diagrams describing the scattering of an electron with 4-momentum p and a photon with 4-momentum k to a final state with p' and k' respectively. The two distinct possibilities correspond to the final photon being emitted either before or after the initial photon is annihilated.

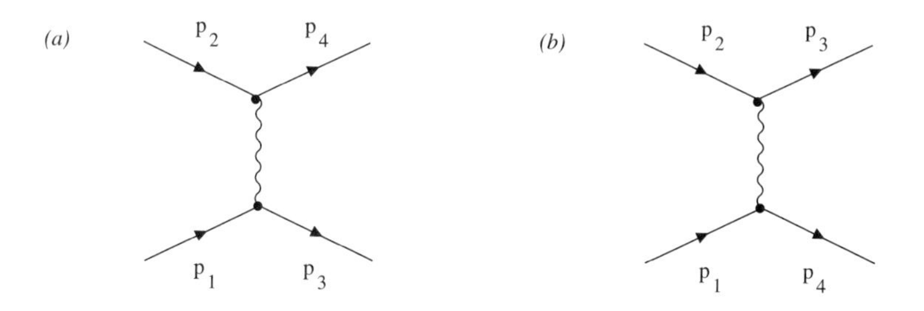

Figure 7.6. The diagrams describing Møller scattering; two electrons change momenta via the exchange of a virtual photon. Because electrons are identical particles, there have to be two diagrams, corresponding to exchanging the labels of a pair of electrons.

This is very satisfying: although we have made no attempt to impose it by hand, the formalism has produced the conservation of energy and momentum as a natural consequence. We can attempt to summarize the above by writing down the Feynman diagrams for this process (figure 7.5).

Although dressed in the language of quantum fields, Compton scattering is a familiar enough process. More illuminating for our present aims, however, is the closely-related second-order process where two electrons scatter off each other. In this case, no photon is present in either the initial or final state, but one is present in the intermediate state. The diagrams for this process, **Møller scattering** are shown in figure 7.6 and make very clear that the function of the photon in this case is to mediate the interaction; the electromagnetic force between the two electrons arises only through this mechanism of photon exchange.

VIRTUAL PARTICLES A Feynman diagram has **external lines**, representing the physical particles that escape to infinity after the scattering process and can be detected. More subtle are the **internal lines**. If we adopt a literal interpretation of the diagram, these are particles that only come into existence in being passed between two of the physical particles. Indeed, processes such as scattering can be thought of as being brought about precisely through the exchange of these **virtual particles**. This is a familiar concept, which goes back to Yukawa's ideas on nuclear forces.

There is a further difference between these lines, which goes beyond their position in the diagram. First note that 4-momentum is conserved at the **vertex** where lines meet. However, this does not imply that the energy and momentum of the various particles always obeys the **mass-shell condition**

$$P^\mu P_\mu = m^2. \tag{7.101}$$

This is true for physical particles, but not for virtual particles. The production of virtual particles is a peculiar trick, which balances the books by carrying away the correct summed energy and momentum from the other particles that meet at the vertex, but in a ratio that is normally incorrect for that type of particle (e.g. a virtual photon will not satisfy $E = pc$). This is yet another illustration of the difficulties that arise in trying to think of quantum particles as Newtonian billiard balls. The field method has no difficulty in dealing with these processes; the problems arise when we try to impose an over-literal meaning to the diagrams.

FEYNMAN RULES Rather than start from scratch every time, the diagrammatic approach to field theory allows scattering problems to be solved relatively quickly, by setting down a recipe known as the **Feynman rules** for the theory. The calculations for many scattering processes proceed along very similar lines, so it is worth defining a covariant scattering matrix element $\mathcal{M}_{fi}$ that factors out the general features of the problem from the S-matrix:

$$S_{fi} = \delta_{fi} + (2\pi)^4\, \delta^{(4)}(P^\mu_{\text{in}} - P^\mu_{\text{out}})\, \sqrt{F(E)}\, \mathcal{M}_{fi}. \tag{7.102}$$

Here, P^μ is the total 4-momentum summed over all particles, and the function $F(E)$ accounts for all the normalization constants in the field expansions:

$$F(E) = \prod_{\text{in}} (2VE)^{-1} \prod_{\text{out}} (2VE)^{-1} \prod_{\text{f}} (2m). \tag{7.103}$$

The last term in the product accounts for the fermions that are present in the initial and final states.

Feynman realized that the invariant amplitude $\mathcal{M}$ can be written down more or less at sight by inspection of the diagrams involved in a given process, by means of the following rules. These will not be given a formal justification in detail; it is best for readers to convince themselves that the procedure gives sensible results by examining a few particular cases, more or less as Feynman did originally. The amplitude $\mathcal{M}$ for nth-order perturbations is obtained by drawing all topologically different diagrams with n vertices that contain the appropriate external lines for the initial and final particles of interest. The diagrams are transcribed into algebraic quantities following the recipe below. These various factors are then multiplied together and integrated over the 4-momenta of the internal lines as $(2\pi)^{-4} \int d^4q$.

(1) For each vertex, write $ie\gamma^\alpha$.

(2) For each internal photon line, of 4-momentum k, write $-ig_{\alpha\beta}/(k^2 + i\epsilon)$.

(3) For each internal fermion line, of 4-momentum q, write $i/(\not{q} - m + i\epsilon)$.

Table 7.1. The Feynman rules for QED

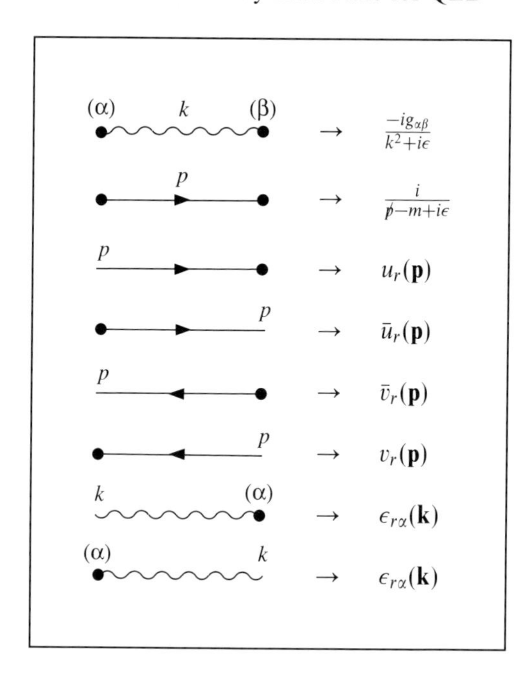

(4) For each external line, write a factor depending on type: $u_r(\mathbf{p})$ for an input electron; $\bar{u}_r(\mathbf{p})$ for an output electron; $\bar{v}_r(\mathbf{p})$ for an input positron; $v_r(\mathbf{p})$ for an output positron; $\epsilon_{r\alpha}(\mathbf{k})$ for an input photon; $\epsilon_{r\alpha}(\mathbf{k})$ for an output photon.

(5) For each closed fermion loop, multiply by -1 and take the trace over spinor indices (which corresponds to summing over spin states of the virtual $e^{\pm}$ pair).

The rules for QED are represented diagramatically in table 7.1.

To this recipe a last complication must be added, owing to the anticommuting nature of fermion fields. The exact ordering of the fermion factors can change the final result by a sign. We shall omit the procedure for getting this aspect right, since it only matters when several diagrams are combined.

THE KLEIN–NISHINA CROSS-SECTION In the case of Compton scattering, which was discussed earlier, there are two relevant diagrams, shown in figure 7.5. It should not be too hard to be convinced that the Feynman rules in this case give the $\mathcal{M}$-matrix element as

$$\begin{aligned}\mathcal{M}_{fi} = &- e^2\bar{u}(\mathbf{p}')\left[\not{\epsilon}(\mathbf{k}')G_{\mathrm{F}}(q=p+k)\not{\epsilon}(\mathbf{k})\right]u(\mathbf{p}) \\ &- e^2\bar{u}(\mathbf{p}')\left[\not{\epsilon}(\mathbf{k})G_{\mathrm{F}}(q=p-k')\not{\epsilon}(\mathbf{k}')\right]u(\mathbf{p}),\end{aligned} \tag{7.104}$$

where the expression for the Feynman propagator is

$$G_{\mathrm{F}}(y) \equiv \frac{\not{y}+m}{y^2-m^2+i\epsilon} = \frac{1}{\not{y}-m+i\epsilon}. \tag{7.105}$$

This gives us everything we need in order to deduce the relativistic generalization of the cross-section for Compton scattering (first worked out by Klein & Nishina in 1929 – albeit using a method that makes even the above look relatively straightforward).

Unfortunately, converting the answer to the cross-section is a tedious exercise in the manipulation of γ-matrices, which occupies several pages of algebra; it would not be illuminating to reproduce this here (see e.g. section 5.2.1 of Itzykson & Zuber 1980). The main aim of showing the logical route by which the formula can be calculated has been achieved, and we shall have to be content with quoting the final answers. The angle-dependent cross-section is

$$\frac{d\sigma}{d\Omega} = \frac{3}{32\pi}\,\sigma_{\rm T}\left(\frac{\omega'}{\omega}\right)^2\left(\frac{\omega}{\omega'} + \frac{\omega'}{\omega} - 2 + 4\cos^2\Theta\right), \tag{7.106}$$

where Θ is the angle between the input and output polarization vectors. If the input photon is unpolarized and the detector we use is not polarization-sensitive, then this becomes

$$\frac{d\sigma}{d\Omega} = \frac{3}{8\pi}\,\sigma_{\rm T}\left(\frac{\omega'}{\omega}\right)^2\left(\frac{\omega}{\omega'} + \frac{\omega'}{\omega} - \sin^2\theta\right), \tag{7.107}$$

where θ is the scattering angle, related to the frequency shift through the usual Compton formula $\omega'/\omega = [1 + (\omega/m)(1 - \cos\theta)]^{-1}$.

7.6 Renormalization

So far, we have shown that QED can produce sensible results for first-order processes (Rutherford scattering). With a little more effort, we then showed that second-order processes can yield not only an important interpretation in terms of virtual particles, but also useful quantities such as the Klein–Nishina cross section. These successes might suggest that the main difficulties to be encountered in carrying QED perturbation theory through to higher orders will be algebraic. Any such hopes come unstuck at a very early stage; a recurring problem in not only QED but throughout quantum field theory is the fact that calculations designed to give predictions for concrete quantities such as scattering cross-sections produce results that are infinite. In retrospect, this is perhaps not so surprising, since we have already seen that quantum field theory gets itself into trouble over the energy density of the vacuum. A theory that cannot yield well-defined results when no particles at all are present might well be expected to have difficulty in producing finite answers in more complicated situations.

The vacuum-energy problem and the divergences in field theory are both traceable to the properties of virtual particles, and the fact that these are able to have infinitely large energies. Attitudes to the severity of these problems vary. Some (including such figures as Dirac) have taken them as an indication that the whole programme of quantum field theory is on completely the wrong conceptual footing. Others have taken the view that the theory is basically sound but requires modification at high energies and small distances. This point of view must contain some truth: a theory designed as a marriage of special relativity and quantum mechanics must have limited validity. The dependence on special relativity can be removed in the semiclassical approach to quantum gravity where one studies quantum fields in a non-flat background spacetime. This extension must, however, break down when we reach the Planck energy, since the concept of a fixed background metric becomes invalid (see chapter 8). It is therefore at least plausible that high-energy virtual photons do not make the actual contributions predicted in straightforward applications of quantum field theory. The theory of superstrings is one

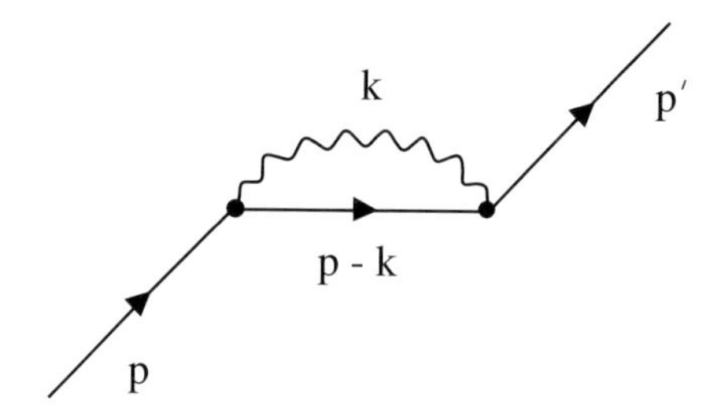

Figure 7.7. The Feynman diagram representing the self-energy of the electron. The electron propagator is modified by the introduction of a virtual photon of 4-momentum k; allowing the energy of these photons to tend to infinity causes a divergent correction to the apparent mass of the electron.

example of an explicit modification of the theory on Planck-energy scales that yields finite results.

It should therefore be possible to extract non-divergent results results from QED. Surprisingly, the results of excluding the high-energy virtual particles turn out not to be sensitive to how this is done. One might have expected that low-energy observations would measure any high-energy cutoff, since results can be strongly sensitive to where this cutoff is. The miracle of **renormalization** is that all these effects turn out to be rendered unobservable. To get an idea of how this happens, consider the simplest example of a divergent process in QED, the self-energy.

THE SELF-ENERGY OF THE ELECTRON Virtual particles can arise not only as mediators of interactions between different particles, but even if there is only a single particle present. Consider the diagram in figure 7.7. This represents a second-order process whereby an initial electron of 4-momentum p is annihilated, creating a virtual photon with 4-momentum k and an electron with 4-momentum $p-k$ (4-momentum still being conserved at vertices), both of which are subsequently annihilated to yield an output electron of 4-momentum p'. The Feynman rules give us the S-matrix for this process:

$$S_{fi} = -i\,(2\pi)^4 \sqrt{\frac{m}{EV}\,\frac{m}{E'V}}\;\delta^{(4)}(p'-p)\,\bar{u}'\Sigma u, \tag{7.108}$$

where the matrix Σ is given by

$$\Sigma = -ie^2 \int \frac{d^4k}{(2\pi)^4}\,\frac{i}{k^2+i\epsilon}\,\gamma^\mu\,\frac{(\not{p}-\not{k})+m}{(p-k)^2-m^2+i\epsilon}\,\gamma_\mu. \tag{7.109}$$

This expression immediately looks likely to cause trouble: for large k, we have an integrand that is $\propto k^{-3}$, and $\int k^{-3}\,d^4k$ seems set to diverge linearly with k.

Before going any further, we should ask what physical interpretation should be given to S_{fi} in this case. This S-matrix element is identical to what would result from the following interaction Hamiltonian:

$$\mathscr{H}_{\rm I} = \delta m\,\bar{\psi}\psi, \tag{7.110}$$

where Σ plays the role of the factor δm. We write it as a perturbation to the mass by analogy with the usual fermionic mass term that we would expect to find in a Lagrangian: $m\bar{\psi}\psi$. This is why this diagram is called the **self-energy** of the electron: it represents a change to the apparent mass of the particle. In other words, the number we measure when we observe an electron gyrating in a magnetic field and infer its mass is not m; the latter is just a parameter in QED. What has happened is that the cloud of virtual particles that surrounds an electron has the effect of making its measured **experimental mass** different from its **bare mass**. We are now able to see the main idea of renormalization: although the correction δm to the total mass is ill defined (depending on some upper cutoff to virtual momenta), δm is not observable. All we can measure is the sum of δm and a bare mass that is also unknown.

In the case of the self-energy, then, it is possible to hide the divergent result by absorbing it in the definition of the experimental mass. This means that calculations can effectively be restricted to **tree level** (where, as the name implies, the diagrams involved have the topologies of trees and branches); the closed self-energy loops need not be considered directly. In fact, the electron self-energy does not diverge quite as badly as our simple counting of powers seemed to imply. If the integral is truncated at $k = \Lambda$, the result can be shown to be (e.g. section 4.9.1 of Renton 1990)

$$\frac{\delta m}{m} = \frac{3\alpha}{2\pi} \ln\left(\frac{\Lambda}{m}\right), \tag{7.111}$$

give or take a finite additive term that depends on exactly how the truncation is done. We thus do not have to face the apparent absurdity of a bare mass that goes to $-\infty$ as the cutoff $\Lambda \to \infty$. In QED, the experimental mass differs from the bare mass by a term of order $1/137$ (a 20% effect even if Λ is set to the quantum-gravity Planck energy of $\sim 10^{19}$ GeV).

This is all very well, but the self-energy arises from just one rather simple diagram. Renormalization can only be a generally useful method if all other divergent diagrams can similarly be incorporated into some observable property of the particle. This may seem rather improbable, given that we have an infinite number of orders of perturbation theory to deal with, but it turns out to be straightforward to see which theories can and cannot be treated in this fashion.

CRITERIA FOR RENORMALIZATION Integrals that diverge are a common feature of quantum field theory, and there exists a simple procedure for determining whether a given diagram is likely to give trouble. The idea is to determine the **superficial degree of divergence** by looking at the behaviour of the diagram at high virtual momenta. Formally, one rescales all internal momenta, $k \to \lambda k$, and looks for the asymptotic index of the power-law behaviour,

$$S \propto \lambda^{\omega} \quad \text{as} \quad \lambda \to \infty. \tag{7.112}$$

The task is just to add up the contributions: each internal integration d^4k gives $\omega = 4$; each internal line gives either $\omega = 2$ (boson propagator) or $\omega = 1$ (fermion propagator). The only complication is that each vertex will also give a contribution to ω, depending on the structure of the interaction Lagrangian (if it includes derivatives). If it does not, then

$$\omega = 4L - 2I_{\rm B} - I_{\rm F}, \tag{7.113}$$

where L and I denote loops and internal lines respectively. Since a little thought should convince one that $L = I - V + 1$ (in terms of the number of vertices) and that the number of internal lines can be given in terms of a sum over the number of lines attached to a given vertex, we obtain

$$\omega - 4 = \sum_v (\omega_v - 4), \tag{7.114}$$

where the term ω_v is the number of boson lines attached to a vertex plus 3/2 times the number of fermion lines (excluding external lines).

Intuition suggests that a diagram will be finite if $\omega < 0$, although ω is not always a perfect guide to the behaviour (cf. the above example of the electron self-energy: the divergence is only logarithmic, although $\omega = 1$ in this case). Certainly, $\omega \gg 1$ is very bad news; the above analysis therefore lets us divide field theories into two important classes.

(1) Non-renormalizable theories with $\omega_v > 4$. For these, higher orders of perturbation theory bring diagrams that are successively more badly behaved. If such theories are to yield finite answers, they will have to be obtained in a non-perturbative way.

(2) Renormalizable theories with $\omega_v \leq 4$. The theories apparently realized in nature, such as QED, sit on the boundary with $\omega_v = 4$: divergences occur at all orders of perturbation theory, but can be tamed by redefinition of parameters as above. Super-renormalizable theories with $\omega_v < 4$ (which give finite results at high orders) are not found. Nature seems to prefer to live as dangerously as possible.

It is now straightforward to see whether a given theory is a candidate for renormalization. A convenient way of summarizing the above results is in terms of the dimension of the coupling constant. Suppose there is a term in the Lagrangian that looks like

$$\mathscr{L} = g\,[\phi]^a\,[A^\mu]^b\,[\psi]^c. \tag{7.115}$$

In natural units, the dimensions of scalar and vector fields are [energy], whereas for spinor fields the dimension is $[\text{energy}]^{3/2}$. Since the dimension of $\mathscr{L}$ is $[\text{energy}]^4$, this lets us deduce the dimension of g, in terms of which the renormalizability requirement $\omega_v < 4$ becomes

$$\boxed{[g] < 0 \Rightarrow \text{non-renormalizable.}} \tag{7.116}$$

As an important application of this rule, any theory that contains a power of a boson field higher than the fourth will be non-renormalizable.

REGULARIZATION We should say something about how Feynman integrals are to be dealt with in practice. First note that the integrals involved ($\int d^4k$) are over the non-Euclidean metric of special-relativity spacetime. To deal with this, we can resort to the trick often used in introductory treatments of relativity: define an imaginary time coordinate to make the integral into one over a 4D Euclidean space. This process often goes by the names of **Wick rotation** or **Euclidean continuation**.

The crudest way to prevent the divergence of Feynman integrals at large k is just to truncate the integration at $k = \Lambda$; more commonly, the cutoff is made 'soft' by multiplying the integrand by a function that is unity for small k but which declines sufficiently fast at high k to cause convergence, e.g.

$$\int d^4k \to \int d^4k \, \frac{\Lambda^2}{\Lambda^2 - k^2}. \tag{7.117}$$

A marginally more appealing trick is that of subtracting a **counterterm** that is negligible at small k, but which cancels the divergence at large k. This is to be thought of as modifying a propagator, for example

$$\frac{1}{k^2 - m^2} \to \frac{1}{k^2 - m^2} - \frac{1}{k^2 - \Lambda^2}. \tag{7.118}$$

This sort of explicit truncation is less popular in modern work because it destroys the gauge invariance of the theory (gauge fields must be massless, whereas the cutoff gives the appearance of a characteristic mass). A more appealing alternative is **dimensional regularization**, where one recognizes that the degree of divergence of a given diagram depends on the dimensionality of spacetime. Replacing d^4k by d^dk will normally cause the Feynman integrals to become ultraviolet convergent for small enough d. Expressing the deviation from $d = 4$ as $d = 4 - \epsilon$, the integrals will typically reduce to a constant term plus one that scales as $1/\epsilon$. The constant is what is of interest, whereas the divergent term can be cancelled by adding a counterterm to the Lagrangian (see e.g. section 11.2 of Weinberg 1996).

All these procedures are arbitrary, and they can only be justified if parameters like Λ eventually disappear from the calculated predictions for observational results. There are two ways in which such a thing might occur, the simplest being that there is a group of divergent diagrams whose sum is actually finite. In this case, a well-defined result is produced as we integrate the total to large k, so it is reasonable that regularizing the separate contributions, evaluating the separate integrals, and then taking the limit $\Lambda \to \infty$ gives a result that is independent of Λ. In fact, on first encountering divergent expressions, one's hope might have been that overall QED results would always end up being finite in this sense. However, this is not the case; we have had to resort to the much more serious procedure of renormalization to render our results independent of the regularization.

THE RENORMALIZATION GROUP Do we have any evidence about whether any finite cutoff scale Λ exists at all? It should be clear that the properties of any theory will be greatly modified at high energies $E \sim \Lambda$, by comparison with a theory with a higher cutoff $\Lambda \gg E$. This means that the renormalized (i.e. observable) properties of a particle will be a function of energy, and so we might hope to detect a cutoff in this way. In practice, this only becomes feasible rather close to the cutoff. At energies $E \ll \Lambda$ there is still some dependence on energy, but we cannot use this as a diagnostic because there are two unknowns: the bare parameter and the cutoff. This problem is discussed further in the next chapter, where it is shown that a coupling constant g measured at an energy E is expected to obey the **renormalization group equation**. This expresses the energy dependence of the coupling constant in a way that is independent of the unknown bare parameter and cutoff:

$$\frac{d}{d \ln E} \, g(E) = \beta\big(g(E)\big), \tag{7.119}$$

where β is some function that depends on the field theory under consideration. This idea, that particle properties may change with energy, is a very important one, and is one of the guiding pieces of motivation for grand unified theories (see chapter 8); we can measure in the laboratory trends of coupling constants with scale which, when extrapolated, indicate an equality of the strengths of all fundamental interactions at some very high energy scale.

7.7 Path integrals

Another innovation bequeathed to quantum field theory by Feynman is the formulation of the subject in terms of **path integrals**. As with the diagrammatic approach, this is rather more than just a formalism that is also useful for doing calculations. The functional methods used here have applications in many areas of physics, wherever systems with infinitely many degrees of freedom are encountered. Unfortunately, this area is one where the tendency of the mathematics to obscure the meaning is particularly marked; moreover, the validity of these techniques is something that has caused some dispute (parallelling the debate in the mathematical community over Dirac's δ-function). We shall therefore take as intuitive an approach as possible.

PATH INTEGRALS IN QUANTUM MECHANICS Feynman's achievement was to show that standard nonrelativistic quantum mechanics could be presented in terms of infinite-dimensional integrals. Consider the probability for finding a particle at position q' at time t', given that it was at q at time t. Feynman's expression for this was

$$\boxed{\langle q', t' \mid q, t \rangle \propto \int \mathcal{D}q \, \mathcal{D}p \, \exp\left[\frac{i}{\hbar} \int_t^{t'} L(q, \dot{q}) \, dt\right].} \tag{7.120}$$

Deriving this expression is a somewhat technical exercise, so we start by explaining the notation and attempting to provide some physical insight into the meaning of this equation.

The most unfamiliar feature of this expression is the integration over the **path** of the particle, $q(t)$; how can we integrate over a path when it contains an infinite number of degrees of freedom? As usual, the answer is to approach infinity by means of a limit, and to divide the path between the initial and final points into n time intervals. We then consider the position of the particle at the end of each of these intervals. All possible paths between $q_i(t_i)$ and $q_f(t_f)$ are then described by the $n-1$ intermediate values of $q(t)$, so the idea of integrating over all possible paths reduces to integrating over these intermediate values. The path integral is then the limit of this process,

$$\int \mathcal{D}q \propto \lim_{n\to\infty} \int \prod_{i=1}^{n-1} dq_i; \tag{7.121}$$

a similar result holds for the momentum integral. It is important to note that the path does *not* approach any sort of continuous trajectory as $n \to \infty$; each of the q_i is varied independently, so that we have something that is rather more fractal-like, with structure on all scales.

Now, Feynman's expression tells us that we take all such possible paths between start and finish, however wildly fluctuating, and add them with a particular weighting to obtain the transition probability. The weights are the exponentials of imaginary numbers and so in general oscillate as the path is changed. We now see why the abstract construction of the path integral is of such physical importance: the paths that contribute most strongly to the transition probability are those for which the weight is stationary with respect to variations in $q(t)$, implying

$$\boxed{\delta \int L \, dt = 0.} \tag{7.122}$$

We have thus derived the classical principle of stationary action as a quantum-mechanical limit. The classical trajectory is singled out by constructive interference of matter waves in the same way as constructive interference of light leads to **Fermat's principle** of stationary phase in optics.

So much for physical motivation; it remains to derive Feynman's path-integral formula. Since this is a technical exercise that can obscure the intuitive appeal of the result, it is relegated to the problem section [problem 7.6].

FUNCTIONAL METHODS The idea of dealing with expressions that depend on functions rather than on a finite number of variables is a rather useful one, which occurs in many branches of physics. It will therefore be appropriate to cover a few general points about **functionals**. Ordinary functions map one number onto another, whereas functionals map an entire function onto a number. A function $f(x)$ produces (in principle) a different number for each value of x fed to it, and none of these numbers depend on each other: specifying $f(2)$ does not tell us anything about $f(1)$. In contrast, a functional such as $F[f] = \int_a^b f(x) \, dx$ produces a single number that 'knows' about the values of $f(x)$ over the entire range of integration. The use of square brackets distinguishes functionals from ordinary functions $g(f(x))$.

This sensitivity to the whole function can be made explicit by defining **functional differentiation**, where we ask how the functional $F[f]$ depends on the value of f at point y. If f is changed only at one point, this of course has no effect on an integral, so the nearest analogue of an infinitesimal change is to perturb the function by a delta function of small weight:

$$\frac{dF[f]}{df(y)} \equiv \lim_{\epsilon \to 0} \frac{F[f(x) + \epsilon\delta(x-y)] - F[f(x)]}{\epsilon}. \tag{7.123}$$

This definition produces sensible answers. Suppose that $F[f]$ involves integration over some given kernel K: $F[f] = \int K(x) f(x) \, dx$. Application of the differentiation rule gives $dF[f]/df(y) = K(y)$.

PATH INTEGRALS IN FIELD THEORY So far, the path-integral approach may seem useful for deep physical insight, but a rather cumbersome means of carrying out calculations that we already know perfectly well how to perform. However, the importance of these methods in field theory is that they provide a method of generating the Green functions needed in perturbative calculations. We illustrate this for the case of scalar fields. Consider the analogue of Feynman's equation for the evolution of a scalar field ϕ (π denotes the conjugate 'momentum' field):

$$\langle \phi'(\mathbf{x}), t' \mid \phi(\mathbf{x}), t \rangle \propto \int \mathscr{D}\phi\, \mathscr{D}\pi \, \exp\left[\frac{i}{\hbar} \int_t^{t'} \int \mathscr{L}(\phi, \pi) \, dt \, d^3x \right]. \tag{7.124}$$

We now perform some minor surgery to produce the **generating functional**

$$W[J] \propto \int \mathscr{D}\phi\, \mathscr{D}\pi \, \exp\left[\frac{i}{\hbar} \int (\mathscr{L} + J\phi) \, d^4x^\mu \right]. \tag{7.125}$$

The reason for the name is readily seen by performing functional differentiation with respect to J:

$$\mathscr{G}^{(n)} \equiv \langle 0 | T[\phi(x_1), \cdots, \phi(x_n)] | 0 \rangle = \left. \frac{(-i\hbar)^n \, \delta^n W[J]}{\delta J(x_1) \cdots \delta J(x_n)} \right|_{J=0}. \tag{7.126}$$

The path integral $W[J]$ thus contains all orders of Green function required in field theory, which should make clear the appeal of functional methods in this area. The modern tendency is to approach field theory immediately through path integrals, and many graduate-level texts barely mention the canonical formalism that has been the basis of the present treatment. However, the canonical approach connects much more easily with undergraduate quantum mechanics, and this is justification enough for having taken a more traditional approach here. For more advanced studies, the path-integral approach is the one that should be followed. Apart from being the default language of the professional in field theory, this method has many practical advantages, particularly in dealing with the difficulties of quantizing gauge fields (see e.g. chapter 7 of Ryder 1985).

Problems

(7.1) Show that the first-order perturbation term for quantum mechanics with an electromagnetic field, $(e/m)\mathbf{A} \cdot \mathbf{p}$, is proportional to the electric dipole moment. What is the interpretation of the A^2 term?

(7.2) Prove the relation $\delta\sqrt{-g} = \frac{1}{2}\sqrt{-g}\, g^{\mu\nu} \delta g_{\mu\nu}$, and hence show that the Hilbert Lagrangian does yield the Einstein field equations.

(7.3) Consider a set of classical Grassmann variables, which are assumed to anticommute: $\{x_i, x_j\} = 0$. Show that derivatives with respect to these variables also anticommute, and that integration is identical to differentiation for Grassmann variables.

(7.4) Derive the asymptotic expansion

$$\begin{aligned} \mathrm{erfc}(x) &= \frac{2}{\sqrt{\pi}} \int_x^\infty e^{-t^2} \, dt \\ &= \frac{e^{-x^2}}{\sqrt{\pi}x} \left[1 - \frac{1}{2x^2} + \frac{1 \times 3}{(2x^2)^2} - \frac{1 \times 3 \times 5}{(2x^2)^3} + \cdots \right]. \end{aligned} \tag{7.127}$$

Show that the series is divergent; comment on the numerical accuracy obtained by retaining the first term only.

(7.5) By using dimensional arguments or otherwise, show that the contribution to the vacuum energy density due to virtual particles of mass m must be

$$\rho_{\rm vac} \sim m^4 c^3/\hbar^3. \tag{7.128}$$

Show that the 'natural' gravitational mass scale of $m = m_{\rm Planck} \simeq 10^{19}$ GeV produces a density more than 10^{120} times as large as observational limits.

(7.6) Derive Feynman's expression for the path integral that gives the propagator in nonrelativistic quantum mechanics.

8 The standard model and beyond

8.1 Elementary particles and fields

The apparatus of quantum field theory is an effective tool for calculating the rates of physical processes. All that is needed is a Lagrangian, and then there is a relatively standard procedure to follow. The trouble with this generality is that there is no way of understanding why certain Lagrangians are found in nature while others are not. The present chapter is concerned with the progress that has been made in solving this problem. The last four decades have seen tremendous developments in understanding: a few simple principles end up specifying much of the way in which elementary particles interact, and point the way to a potential unification of the fundamental interactions. These developments have led to the **standard model** of particle physics, which in principle allows any observable quantity related to the interactions of particles to be calculated in a consistent way.

The standard model asserts that the building blocks of physics are a certain set of fundamental particles from which the composite particles seen in experiments are constructed; their properties are listed in table 8.1 (see the references to the Particle Data Group in the bibliography). This set is much larger than the electron, proton and photon, which were all that was needed in 1920, but it is believed to be complete. The standard model explains some but not all of the features of this list of particles, and it is the task of current research in unified theories to explain exactly why matter should be formed from ingredients that are ordered in this particular way.

There are thus two sets of spin-1/2 fermions (the leptons and quarks), one set of spin-1 bosons, and a single scalar spin-0 boson, the **Higgs boson**. Apart from the τ-neutrino and the gluon, this is the only particle in the above list yet to be observed directly. However, the $W^{\pm}$ and Z bosons and the charmed, top and bottom quarks were all predicted to exist in advance and were eventually found with the expected properties, so the standard model has a very impressive history of experimental verification. The search for the Higgs particle therefore now dominates high-energy experimental physics, and is attracting the investment of colossal sums of money.

It is to be regretted that, in a brief overview like this, there is no space to do justice to the outstanding experiments that have produced our current understanding. This is a particular problem with **quarks** (named from a quote in James Joyce's *Finnigan's Wake*, in which they rhyme with 'Mark'). Quarks were first introduced by Gell-Mann and Zweig as a book-keeping device designed to generate the known hadrons via the variation of internal quantum numbers. Only much later did *ep* **deep inelastic scattering** produce evidence for point-like constituents within hadrons, and it took considerable time before

Table 8.1. The particles of the standard model

particle	$q/\|e\|$	mass
leptons		
e (electron)	-1	0.511 MeV
ν_e (e-neutrino)	0	< 15 eV
μ (muon)	-1	0.106 GeV
ν_μ (μ-neutrino)	0	< 0.17 MeV
τ (tau)	-1	1.777 GeV
ν_τ (τ-neutrino)	0	< 24 MeV
quarks		
u (up)	2/3	6 MeV
d (down)	$-1/3$	10 MeV
s (strange)	$-1/3$	0.25 GeV
c (charm)	2/3	1.2 GeV
b (bottom)	$-1/3$	4.3 GeV
t (top)	2/3	180 GeV
vector bosons		
γ (photon)	0	0
$W^\pm$ (W boson)	± 1	80.3 GeV
Z^0 (Z boson)	0	91.2 GeV
g (gluon)	0	0
Higgs boson		
H (Higgs)	0	< 1 TeV

the present accepted position was reached, in which these **partons** are identified with elementary quarks. The story of the construction of the standard model is one of a dialogue between experiment and theory, where experiment sometimes led in forcing the theorists to postulate new particles in order to explain the fact that certain reactions occur, whereas other apparently similar ones do not. Baroque names, such as the strange quark, clearly betray their origin in experimental puzzles. Nevertheless, theory has returned this experimental input with interest through its prediction of particles needed to complete the pattern that experiment has suggested: the Ω^- in the original quark model, and the $W^\pm$ and Z^0 in the electroweak theory. For some illuminating accounts of this fruitful interaction, see e.g. Cahn & Goldhaber (1989); Llewellyn-Smith (1990).

8.2 Gauge symmetries and conservation laws

The key concept used in understanding the structure of the standard model is that of **gauge symmetry**. The term 'gauge' in general is used to indicate a hidden freedom in a theory: the ability in electromagnetism to make the transformation $\mathbf{A} \rightarrow \mathbf{A} + \boldsymbol{\nabla}\chi$ is the most familiar example of this. In quantum mechanics, the meaning of the term is simpler: the phase of the wave function is unobservable. This should seem obvious in nonrelativistic quantum mechanics, where the probability density is $|\psi|^2$. In general, we will be saying that the differential equations (or Lagrangians) describing a quantum field will be **form invariant** under the transformation

$$\psi \to \psi e^{i\theta}. \tag{8.1}$$

In the case where θ is just some constant angle, this is called a **global gauge transformation**. As usual, any such transformation that leaves the system invariant is called a **symmetry**. The origin of the term 'gauge' is historical: in the 1930s there was considerable interest in theories in which length units (as measured by atomic standards) might vary from place to place, hence the railway-track terminology.

NOETHER'S THEOREM The existence of global symmetries is closely connected with conservation laws in physics. In classical mechanics, conservation of energy and momentum arise by considering Euler's equation

$$\frac{d}{dt}\left(\frac{\partial L}{\partial \dot{q}_i}\right) - \frac{\partial L}{\partial q_i} = 0, \tag{8.2}$$

where $L = \sum_i T_i - V_i$ is a sum over the difference in kinetic and potential energies for the particles in a system. If L is independent of all the position coordinates q_i, then we obtain conservation of momentum (or angular momentum, if q is an angular coordinate): $p_i \equiv \partial L/\partial \dot{q}_i =$ constant for each particle. More realistically, the potential will depend on the q_i, but homogeneity of space says that the Lagrangian as a whole will be unchanged by a translation. Using Euler's equation this gives conservation of total momentum:

$$dL = \sum_i \frac{\partial L}{\partial q_i}\, dq \quad \Rightarrow \quad \frac{d}{dt}\sum_i p_i = 0. \tag{8.3}$$

If L has no explicit dependence on t, then

$$\frac{dL}{dt} = \sum_i \left(\frac{\partial L}{\partial q_i}\dot{q}_i + \frac{\partial L}{\partial \dot{q}_i}\ddot{q}\right) = \sum_i (\dot{p}_i \dot{q}_i + p_i \ddot{q}_i), \tag{8.4}$$

which leads us to define the **Hamiltonian** as a further constant of the motion

$$H \equiv \sum_i p_i \dot{q}_i - L = \text{constant}. \tag{8.5}$$

Something rather similar happens in the case of quantum (or classical) field theory: the existence of a global symmetry leads directly to a conservation law. The difference between discrete dynamics and field dynamics, where the Lagrangian is a *density*, is that the result is expressed as a **conserved current** rather than a simple constant of the motion. In what follows, we symbolize the field by ϕ, but this is not to imply that there is any restriction to scalar fields. If several fields are involved (e.g. ϕ and ϕ^* for a complex scalar field), they should be summed over.

Suppose the Lagrangian has no explicit dependence on spacetime (i.e. it depends on x^μ only implicitly through the fields and their 4-derivatives). The algebra is similar to that above, and results in [problem 8.1]

$$\boxed{\frac{d}{dx_\nu}\left[\frac{\partial \mathcal{L}}{\partial(\partial^\nu \phi)}\frac{\partial \phi}{\partial x^\mu} - \mathcal{L} g_{\mu\nu}\right] = 0.} \tag{8.6}$$

We have produced a conserved tensor: the term in square brackets is therefore plausibly to be identified with the energy–momentum tensor of the field, $T_{\mu\nu}$ (cf. constant H in the above reasoning). Note immediately the important consequence for cosmology: a potential term $-V(\phi)$ in the Lagrangian produces $T^{\mu\nu} = V(\phi)g^{\mu\nu}$. This is the $p = -\rho$ equation of state characteristic of the cosmological constant. This relation between particle physics and cosmology is the basis of the 'inflationary universe', in which a non-zero cosmological constant has its origin in field theory.

This procedure for relating global symmetries to currents works in other cases. Suppose there is a symmetry transformation governed by some parameter α (e.g. θ in the above phase transformation). In this case, $\mathscr{L}$ is independent of α, so that

$$\boxed{\frac{\partial \mathscr{L}}{\partial \alpha} = \frac{\partial}{\partial x^\nu}\left[\frac{\partial \mathscr{L}}{\partial(\partial_\nu \phi)}\,\frac{\partial \phi}{\partial \alpha}\right] = 0.} \tag{8.7}$$

Here, the term in square brackets is a conserved current J^μ. The interpretation of this depends on the nature of α; consider phase transformations as an example. For the Dirac Lagrangian, we get $J^\mu = -\overline{\psi}\gamma^\mu\psi$, which is within a sign of the probability 4-current. For a complex scalar field, we get $J^\mu = i(\phi\partial^\mu\phi^* - \phi^*\partial^\mu\phi)$, which is proportional to the Klein–Gordon current, and this is known not to be positive definite. The only remaining consistent interpretation is that J^μ is the charge 4-current. This is a remarkable outcome: conservation of charge emerges simply from requiring that quantum mechanics be invariant under changes of phase of the wave function. This suggests that the wave function for an uncharged particle must be *real*, so that it cannot undergo phase transformations (cf. Majorana particles, mentioned in chapter 6).

This general association of a conserved current with a global symmetry of $\mathscr{L}$ is called **Noether's theorem**. Other symmetries discussed below will include the $SU(2)$ and $SU(3)$ groups, which are important in the weak and strong interactions respectively; their associated conserved quantum numbers are lepton number and baryon number. This approach also works for approximate symmetries such as isospin; these generate quantum numbers that are not absolutely conserved (unlike charge) but which will be conserved for reactions in the regime where the symmetry concerned holds good.

LOCAL GAUGE INVARIANCE The gauge freedom in quantum theory discussed above differs from that in electromagnetism in one crucial respect. In electromagnetism, the value of $\boldsymbol{\nabla}\cdot\mathbf{A}$ is unmeasurable and can be made to vary at will from place to place by a suitable transformation: this is a **local** gauge invariance. In quantum theory, conversely, we have so far considered only a change of phase that is the same for all observers. This seems an unreasonable restriction: how could an observer on a distant galaxy tell us what origin of phase to use? Surely phase should also be *locally* unobservable? This suggests we consider the transformation

$$\psi \to \psi\, e^{i\theta(x^\mu)}, \tag{8.8}$$

but at first things look unpromising. The problem arises when we start to differentiate ψ: unwanted terms containing derivatives of θ will crop up, and in general they do not cancel. For example, consider the Dirac Lagrangian, $\mathscr{L} = i\overline{\psi}\gamma^\mu\partial_\mu\psi - m^2\overline{\psi}\psi$. Under the above local gauge transformation, this will transform as follows:

$$\mathscr{L} \to \mathscr{L} - \overline{\psi}\gamma^\mu(\partial_\mu\theta)\psi. \tag{8.9}$$

The Lagrangian, and hence the wave equation, depends on the derivatives of θ; the effects of the gauge transformation will therefore have observable consequences. In short, the theory is not gauge invariant. We need not be specific about the Lagrangian – just consider the effect on the derivatives of ψ:

$$\partial^\mu \psi \to e^{i\theta}[\partial^\mu \psi + i\psi(\partial^\mu \theta)]. \tag{8.10}$$

The second term will almost inevitably spoil gauge invariance; what we really want is to be able to commute the $e^{i\theta}$ term with derivatives, in the way we could if θ were constant. The way of achieving this is to cheat: add a new term to the Lagrangian whose purpose is specifically to erase the undesired effect of θ. If we replace ∂^μ everywhere by $\partial^\mu + V^\mu$, then gauge invariance can be obtained if we make the simultaneous transformations

$$\begin{aligned} \psi &\to e^{i\theta}\psi; \\ V^\mu &\to V^\mu - i\partial^\mu \theta. \end{aligned} \tag{8.11}$$

This is saying that we can have local gauge invariance if we are prepared to postulate a new field V^μ. If this is to be anything like other fields we are familiar with, it should have a 'kinetic' contribution to the Lagrangian that depends on derivatives of V^μ. The only guide we have to the form of this is that it should not now spoil the gauge invariance we have tried so hard to protect. Thus, derivatives must occur in the antisymmetric combination $F^{\mu\nu} \equiv \partial^\mu V^\nu - \partial^\nu V^\mu$, so that the simplest kinetic term is

$$\mathcal{L}_V \propto F^{\mu\nu} F_{\mu\nu}. \tag{8.12}$$

What we certainly *cannot* have is a mass term $m^2 V^\mu V_\mu$: this would violate gauge invariance.

Simply by requiring that phase in quantum mechanics should be locally unobservable, we have therefore been forced to introduce a new field, whose interaction and kinetic energy terms can be recognized as exactly those of electromagnetism (give or take constants of proportionality, whose appearance simply corresponds to introducing the empirical value of the electronic charge). Furthermore, this line of reasoning tells us that the photon is forced to have zero mass. The Lagrangian for a massive fermion field then has a simplest possible form, which is the Lagrangian of quantum electrodynamics:

$$\boxed{\mathcal{L} = i\bar{\psi}\gamma^\mu \mathcal{D}_\mu \psi - m^2 \bar{\psi}\psi - \tfrac{1}{4} F^{\mu\nu} F_{\mu\nu},} \tag{8.13}$$

where $\mathcal{D}_\mu$ is the **gauge-covariant derivative**

$$\mathcal{D}_\mu \equiv \partial_\mu + ieA_\mu. \tag{8.14}$$

The similarity of name to the covariant derivative in general relativity is no accident: in both subjects the derivative has had added to it an extra term whose function is to swallow up unsightly terms arising from differentiating the coefficients of a coordinate transformation.

THE GAUGE PRINCIPLE With only a slight degree of extension, this line of argument becomes one of the fundamentals of particle physics. It is generally believed that all the interactions of nature are described by fields that are introduced to allow the operation of various gauge symmetries. In every context, we will be doing essentially the same thing: every time a local parameter θ needs to be hidden, we will introduce a vector field to 'eat' the unwanted 4-vector $\partial^\mu \theta$. This will give terms in the Lagrangian that are linear in the new field and hence correspond to the internal lines characteristic of a particle that mediates an interaction. Because they are vector fields, all these gauge fields will have corresponding quanta that are spin-1: **gauge bosons** are the particles that mediate interactions. The case of gravity is different, because the parameters to be hidden are the coefficients of a Lorentz transformation. Nevertheless, in this sense gravity is also a gauge field, with the spin-2 graviton as its gauge boson.

It is probably best to view arguments based on gauge invariance as analogous to dimensional analysis: certainly, there is the same magical sense of getting 'something for nothing'. We can use gauge arguments to deduce that the Lagrangian of a free particle is incomplete, and can determine most (but not all) of the properties of the missing terms. Although we have tried above to suggest plausible reasons *why* local gauge invariance should be so basic, it is of course true that gauge invariance was discovered historically by the study of electromagnetism, rather than vice versa. Nevertheless, the current standard model of the weak and strong nuclear interactions was generated by straightforward application of the same principles, and has proved a breathtakingly accurate match to reality.

GAUGE DIFFICULTIES WITH QED Although the gauge principle gives an appealingly simple explanation for the structure of the QED Lagrangian, the gauge nature of the electromagnetic field also causes some peculiar difficulties in quantum electrodynamics. Consider the free Maxwell Lagrangian $\mathscr{L} = -F^{\mu\nu}F_{\mu\nu}/4$, and ask what happens when we come to evaluate the 4-momentum conjugate to A_μ:

$$\pi^\mu = \frac{\partial \mathscr{L}}{\partial \dot{A}_\mu} = -\partial^0 A^\mu + \partial^\mu A^0. \tag{8.15}$$

The 4-momentum component π^0 thus vanishes exactly, and we cannot impose the normal commutator between this and A^0. This is a reflection of the fact that a massless field like electromagnetism has fewer degrees of freedom than a massive vector field (longitudinal photons are unphysical). The way to deal with this is to modify the Lagrangian. The massive degree of freedom is absorbed into the gauge invariance, and we have seen that the equations of electromagnetism look simplest when the Lorentz condition $\partial_\mu A^\mu = 0$ is applied, yielding $\Box A^\mu = 0$. These equations also result if we add a **gauge-fixing term** to the Lagrangian, so that

$$\mathscr{L} = -\tfrac{1}{4}F^{\mu\nu}F_{\mu\nu} - \tfrac{1}{2}(\partial_\mu A^\mu)^2. \tag{8.16}$$

It is this difference that explains why the propagator for the photon is not the $m \to 0$ limit of the propagator for a massive vector particle.

8.3 The weak interaction

Quantum electrodynamics had its origin in the classical Maxwell theory, which was first created in the form of a gauge theory. The formulation of the field theory of the weak interaction was more complicated, beginning with a model that works well at low energies, and is still the basis for practical calculations in nuclear physics.

FERMI'S THEORY Fermi's Lagrangian for the weak interaction was based on a hint from QED, where the interaction Lagrangian is of the form $\mathcal{L} \propto J^\mu A_\mu$; here the fermion current is $J^\mu = \bar{\psi}\gamma^\mu\psi$. The essential feature required of the weak Lagrangian is a product of four fermion fields: $\bar{\psi}_p\psi_n\bar{\psi}_e\psi_\nu$. Viewing these as containing creation and annihilation operators, the lowest-order processes in the theory then describe the beta-decay reactions such as $n \to p + e + \bar{\nu}_e$. In order to cast this product into a Lorentz invariant, a current–current product $J^\mu J_\mu$ is suggested. Experimental data eventually forced the form of the **V–A interaction** (vector–axial, referring to the fact that $\bar{\psi}\gamma^\mu\gamma^5\psi$ yields an axial vector):

$$\mathcal{L}_{\text{V-A}} = -\frac{G_\text{F}}{\sqrt{2}} \left[\bar{\psi}_p\gamma^\mu(1 - g\gamma^5)\psi_n\right] \left[\bar{\psi}_e\gamma_\mu(1 - \gamma^5)\psi_\nu\right], \tag{8.17}$$

where the **Fermi constant** takes the numerical value

$$G_\text{F}/(\hbar c)^3 = 1.166 \times 10^{-5}\ \text{GeV}^{-2}. \tag{8.18}$$

The V–A form is not quite exact, having $g \simeq 1.26$ instead of the neater value unity. This reflects the fact that protons and neutrons are composite particles; the weak interactions of quarks obey $g = 1$ in the limit where they are approximated by the Fermi theory.

PARITY VIOLATION The radical feature of the above Lagrangian is the terms involving γ^5, which violate parity. Fermi did not originally include these, as there was at that time (1934) no experimental evidence for parity violation; it was assumed that C, P and T were separately exact symmetries. This situation changed with the suggestion by Lee & Yang (1956), subsequently confirmed experimentally by Wu *et al.* (1957), that the weak interactions violated parity (see the discussion in chapter 6). Moreover, the violation is as large as it could be: the combination $(1 - \gamma^5)\psi_\nu$ is involved, which is (twice) the state of left-handed chirality. It was shown in chapter 6 that chirality and helicity become the same quantity for particles of zero mass, and so we see that the Fermi Lagrangian is only capable of creating neutrinos with left-handed helicity (i.e. spins antiparallel to their momenta). This does not mean that right-handed neutrinos are impossible, only that they are produced by weak interactions in vanishingly small numbers as the neutrino mass tends to zero. The fact that the universe distinguishes so completely between left- and right-handed particles in this way is something that is not understood: it has to be enforced by hand in the standard model, and does not appear to arise in an inevitable fashion from grand unified theories.

THE NEED FOR VECTOR BOSONS Although satisfactory for practical calculations of weak interaction rates, the Fermi Lagrangian cannot be an exact theory of the weak interaction because it is not renormalizable. The coupling constant has dimensions of $[\text{energy}]^{-2}$, whereas the discussion in chapter 7 showed that renormalizable theories must involve dimensionless coupling constants. This is a relief, because the Fermi Lagrangian is in any case peculiar in describing a zero-range interaction: there is no field for a particle to mediate the weak interaction.

The natural way out of this impasse is to guess that the Fermi theory is a low-energy limit of a theory that does involve an **intermediate vector boson**, whose field is symbolized by W^μ. This must be a charged particle, since we need to effectively convert between leptons and neutrinos in the weak interactions, and so a complex field must be

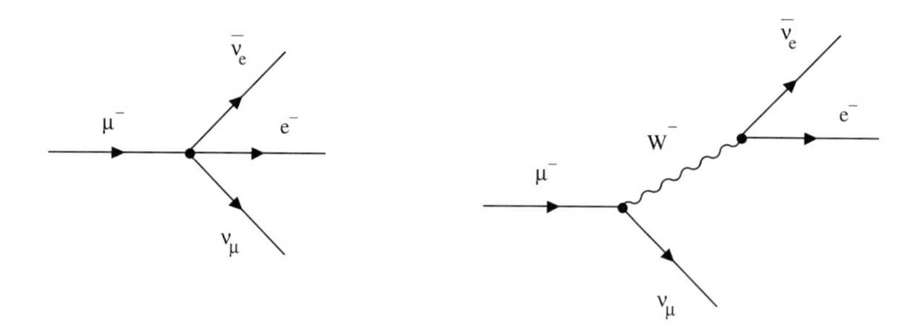

Figure 8.1. The Feynman diagrams for a weak-interaction process (muon decay), in the zero-range Fermi theory and the intermediate vector boson theory, to which the Fermi theory is an approximation at energies below the W mass.

involved. A sensible candidate theory would then be

$$\mathscr{L} = gJ^\mu W_\mu^\dagger + \text{Hermitian conjugate}, \tag{8.19}$$

where the current J^μ is now a sum of terms for each lepton–neutrino pair: $\bar{\psi}_e\gamma_\mu(1-\gamma^5)\psi_{\nu_e}$, etc. The coupling constant g would be dimensionless, and is related to the Fermi constant by

$$\boxed{\frac{G_{\mathrm{F}}}{\sqrt{2}} = \frac{g^2}{M_W^2}.} \tag{8.20}$$

To obtain this relation, and to show that the vector-boson theory reduces to Fermi's theory at low energies, we need to calculate matrix elements in the two theories (whose Feynman diagrams are contrasted in figure 8.1). In the Fermi case, these are first-order elements, which can be read off from the Fermi Lagrangian in the form

$$\mathscr{M} \propto \frac{G_{\mathrm{F}}}{\sqrt{2}}\, a\,\gamma^\mu(1-\gamma^5)\,b\,c\,\gamma_\mu(1-\gamma^5)\,d, \tag{8.21}$$

where a, b, c, d are some set of spinors, ψ or $\bar{\psi}$, for the various particles involved. For the vector theory, the interesting processes are second order, and will involve the propagator for a massive vector field

$$\mathscr{M} \propto g^2\, a\,\gamma_\mu(1-\gamma^5)\,b\left(\frac{g^{\mu\nu} - k^\mu k^\nu/M_W^2}{k^2 - M_W^2}\right)c\,\gamma_\nu(1-\gamma^5)\,d. \tag{8.22}$$

At energies $k \ll M_W$, these two expressions become identical, allowing the coupling-constant relation to be read off.

This is not the end of the story, since it is not yet guaranteed that the theory is renormalizable – all we know is that it definitely would not be so if the coupling constant had the dimensions of a negative power of energy. In fact, the vector-boson theory would only be renormalizable in the case where $M_W = 0$, so that the weak interaction would be a force of infinite range. To cure this problem involves making the W field a gauge field, analogous to the position of the photon in electromagnetism.

8.4 Non-Abelian gauge symmetries

The gauge theories for nuclear interactions are relatively simple extensions of the procedure that applies to electromagnetism. However, a full treatment of this subject really necessitates some understanding of group theory, and this is a convenient point at which to give a very brief summary of the required basics. For more details, a variety of texts exist; Joshi (1982) is one of the more approachable for those preferring a physicist's level of rigour.

DISCRETE GROUPS A group formally consists of a set of elements $\{g_i\}$ and an operation of combination, often written $\circ$. This combination of elements satisfies four rules that form the biblical mnemonic CAIN: *closure*, $g_i \circ g_j = g_k$; *associativity*, $g_i \circ (g_j \circ g_k) = (g_i \circ g_j) \circ g_k$; *inverse element*, for every g_i there exists g_i^{-1}: $g_i \circ g_i^{-1} = e$; *neutral element*, there exists an element e for which $g_i \circ e = g_i$ for all i [problem 8.2]. These abstract rules are introduced because they govern the operation of many **discrete symmetries**. A simple example is a restricted set of operations in two dimensions: reflection in one of two perpendicular mirrors (M_1, M_2), plus rotation through multiples of 180° (R, $R^2 = I$). This group is called C_{2v} and is responsible for a famous paradox [problem 8.3].

Two special classes of groups are of particular importance. These are the **Abelian groups** where all the elements commute: $g_i \circ g_j = g_j \circ g_i$ (named not for Cain's brother, but for the mathematician Abel), and the **cyclic groups**, where all the elements are constructed as combinations of one or more **generators**. A very simple example of the latter is C_2: the rotational part of C_{2v} above (R, $R^2 = I$). This is also an example of a common situation where a restricted set of group elements form a group in their own right: C_2 is a **subgroup** of C_{2v}. It is often the case that a group may be specified entirely by its subgroups, taking the form of a **direct product**:

$$G = G_1 \otimes G_2 \otimes G_3 \cdots \tag{8.23}$$

What this means is that there are several small groups that share the identity element, but which are otherwise distinct sets of elements: $G_1 = (I, a_2, a_3, \ldots)$, $G_2 = (I, b_2, b_3, \ldots)$. The overall group consists of the list of all the distinct terms produced by picking one element from each of the subgroups and multiplying all these together. This concept of direct-product groups is important in the area of symmetry breaking in physics. It is often possible for the Hamiltonian of a given system to be invariant under a group of symmetry transformations, and yet for the energetically favourable ground state to be one that breaks that symmetry. A trivial classical example is a ball placed at the centre of a hemispherical dome; the ball rolls downhill, but in a direction that is not predictable by the symmetry of the situation. It is possible for this to happen in a way that does not completely destroy the validity of the symmetry group, but which removes some of the subgroups, leaving a group of lower symmetry.

REPRESENTATIONS The algebra of possibly non-commuting quantities has a natural counterpart in matrix algebra, e.g. rotations can be associated with antisymmetric matrices. The order-N cyclic group in two dimensions consisting of rotations through some multiple of $\epsilon = 2\pi/N$ is generated by the elements

$$\begin{pmatrix} 1 & 0 \\ 0 & 1 \end{pmatrix}, \quad \begin{pmatrix} \cos\epsilon & -\sin\epsilon \\ \sin\epsilon & \cos\epsilon \end{pmatrix}. \tag{8.24}$$

It is, however, important to be clear that groups are abstract in nature, and that the association of particular group elements with particular matrices is not unique, nor is it even necessary. A set of algebraic quantities a_i will represent the group elements g_i if the products $a_i a_j$ give quantities that respect the same rules as the group: $g_i \circ g_j = g_k$ requires $a_i a_j = a_k$. The complex numbers $1, \exp(i\epsilon)$ would therefore function equally well as a representation of the cyclic rotations.

Similarly, the 2×2 matrices could be extended to much larger matrices by making them of block diagonal form with the 2×2 part at the top left, followed by elements equal to unity down the diagonal. This is a special case of a group representation in the form of a **direct sum**:

$$M = M_1 \oplus M_2 \oplus M_3 \cdots, \tag{8.25}$$

which is shorthand for a matrix in block diagonal form with matrices M_1, M_2 etc. down the diagonal. Clearly each of these matrices needs to be a valid representation of the group, and the example of extending the matrix with a unit diagonal shows that unity is a representation of any group, albeit not a very useful one. For practical purposes, it is preferable to work with matrices that are as small as possible, but where there are as many independent matrices as there are group elements.

CONTINUOUS GROUPS The groups of importance in quantum field theory are those obtained in the continuum limit when the **order** of the group (the number of elements) becomes infinite. Although this may sound more complicated than when the order is finite, the infinitely many elements may in fact be related to each other. A trivial example is the above set of rotations in the limit $\epsilon \to 0$; there are infinitely many rotations, and yet all are contained within the one-parameter expression $\exp i\theta$. In general, the **order of a continuous group** is the minimum number of independent parameters necessary for all the group elements to be expressed as a function of the parameters.

Of greatest interest are the groups where the group elements are *differentiable* functions of the parameters. These **Lie groups** (pronounced 'lee') are named for the Norwegian mathematician Sophus Lie. The group elements here are generated in a very simple way: consider a Taylor expansion of the element g about the identity,

$$g \simeq I + \frac{dg}{dp_i}\, p_i, \tag{8.26}$$

where the p_i are the parameters of the group. The element g is 'near' to the identity, but elements a significant distance away can be produced by taking the product g^N, where N is some large number, since closure says that g^N must also be in the group. What is the parameterization of g^N? Repeated application of g moves us along a locus in the parameter space of the group, but there seems no guarantee that the locus will be a radial line. In other words, we can produce g by making a small change in just one parameter p_i and keeping the others zero, but g^N might in general correspond to non-zero values of several parameters. However, we can easily choose things so that this is not so, because the number of steps along the locus itself defines a parameterization of the locus. If we thus write $p_i = N\epsilon_i$, then the group element at finite p_i can be defined to be

$$g(p_i) \equiv g(\epsilon_i)^N \simeq \left(I + \frac{dg}{dp_i}\frac{p_i}{N}\right)^N. \tag{8.27}$$

It will only be correct to neglect higher-order terms in the Taylor series in the $N \to \infty$ limit. If we then use the identity

$$\lim_{N\to\infty} (1 + x/N)^N = \exp x, \tag{8.28}$$

proved by taking logs and considering the Taylor series of $\ln(1+y)$, then we obtain the important result

$$\boxed{g(p_i) = \exp\left(\frac{dg}{dp_i}\, p_i\right).} \tag{8.29}$$

This says that the group elements of a Lie group can be obtained by taking the exponential (defined as a Taylor-series sum) of a weighted sum of a finite number of **generators**, where the number of generators is the same as the order of the group. This form has already been seen in the above example of 2D rotations, where $g = \exp(i\theta)$. For simplicity, generators G are defined with a factor of i, so that the generator is just unity in the case of 2D rotations:

$$G_i \equiv -i \left(\frac{d\ln g}{dp_i}\right). \tag{8.30}$$

IMPORTANT LIE GROUPS The Lie groups of particular interest for physics are as follows:

$O(n)$ The rotation group in n dimensions. O stands for orthogonal, and denotes the fact that the matrices representing rotations have transposes as their inverses.

$SO(n)$ The special orthogonal group. The term 'special' denotes a restriction to matrices with unit determinant; this is therefore a subgroup of $O(n)$. This restriction forbids the appearance of **improper rotations**, those transformations that correspond to rotation plus coordinate inversion.

$U(n)$ The unitary group in n dimensions. It is common in quantum mechanics to consider a linear transformation of some multicomponent wave function $\psi \to U\psi$, where U is some matrix. Maintaining the normalizability of ψ requires $\psi^\dagger\psi$ to be unchanged, hence $U^\dagger U = 1$: the matrix must be unitary.

$SU(n)$ The special unitary group, which is the subgroup of $U(n)$ with unit determinant.

$L(n)$ The Lorentz or Poincaré group, which is the group of spacetime 'rotations' generated by infinitesimal Lorentz transformations, plus spacetime translations.

It is easy enough to see what the generators will be in some of these cases. For example, in the case of $SU(n)$, the unitarity requirement may be written for small parameter θ as

$$I = (I + i\theta G)(I - i\theta G^\dagger) \simeq I + i\theta(G - G^\dagger), \tag{8.31}$$

which is satisfied if the generators are Hermitian, so that $G = G^\dagger$. Similarly, the requirement for unit determinant of the matrix $I + i\theta G$ requires G to be traceless. A convenient set of matrices satisfying this criterion in the case of $SU(2)$ is thus the Pauli spin matrices, σ_i.

YANG–MILLS THEORIES So far, the only Lie group that has appeared is the trivial case $U(1)$, which can be represented perfectly well without using matrices. All the other groups listed above are, in contrast, non-Abelian, and the question is whether they are of any relevance for physics. The first suggestion that non-Abelian gauge symmetries might be of interest was made by Yang & Mills (1954). To implement non-Abelian symmetries in quantum physics, it is clear that some sort of multi-state system is needed; the archetype for this is probably the wave function for a particle with spin. This provides a good way of visualizing what is meant in some of the later more abstract examples. With spin, we take the different particles 'electron with spin up' and 'electron with spin down' not as fundamentally distinct entities, but as two different states of the same thing. This seems trivially obvious here, but the whole essence of unified field theories is based on conjecturing that particles that are apparently very different may actually be just different manifestations of some common underlying object.

Historically, the first manifestation of this idea was **isospin**. Here one notes that the masses of the neutron and proton are almost identical, suggesting that we construct a two-state wave function $\psi = \binom{n}{p}$, in a way that is analogous to spin (hence the name *isotopic spin*).

$SU(2)$ AND WEAK INTERACTIONS The more fundamental example of a two-state system comes in the weak interaction. Particles can be grouped into families that consist of pairs of particles differing by one unit of charge:

$$\begin{aligned} &\text{leptons} \quad \begin{pmatrix} e \\ \nu_e \end{pmatrix} \begin{pmatrix} \mu \\ \nu_\mu \end{pmatrix} \begin{pmatrix} \tau \\ \nu_\tau \end{pmatrix}; \\ &\text{quarks} \quad \begin{pmatrix} u \\ d \end{pmatrix} \begin{pmatrix} s \\ c \end{pmatrix} \begin{pmatrix} t \\ b \end{pmatrix}. \end{aligned} \tag{8.32}$$

The strongly interacting hadrons are built from quarks, as discussed below, but the quarks also participate in the weak interaction. The multiplet structure is a hint that we might want to consider e.g. the e, ν_e pair as two possible states between which we can flip. In practice, this would clearly be accomplished by the exchange of a charged particle, which would mediate a **charge-exchange interaction**. This particle had in fact been predicted in the early days of weak interactions (before the discovery of quarks), and had received the name of the **intermediate vector boson**, symbolized $W^\pm$ (see figure 8.1).

With such a system, it is natural to ask about the observability not only of changes in phase of the wave function (rotations in the Argand plane), but also of the analogues of rotations in the internal space. This suggests the transformation

$$\psi \to e^{i\theta \mathbf{M}} \psi, \tag{8.33}$$

where $\mathbf{M}$ is some 2×2 matrix. As above, leaving $\psi^\dagger\psi$ unchanged implies that the **generator** $\mathbf{M}$ must be Hermitian. If we consider separately the pure phase transformations (with $\mathbf{M}$ the identity matrix) corresponding to the group $U(1)$, then we are left with generators that are Hermitian and traceless, corresponding to the group $SU(2)$. There are only three possible such matrices, and we may take them as being the Pauli spin matrices. As these do not commute, this is a non-Abelian group.

So, a reasonable guess in the case of the weak interaction might be that quantum mechanics (i.e. $\mathcal{L}_{\text{weak}}$) should be invariant under local $SU(2)$ transformations. The general expression for a group element, $g = \exp(\theta_i G_i)$, shows that each of the three generators G_i comes accompanied by some arbitrary angle θ_i. It is therefore clear that we will need to introduce *three* gauge fields if we wish to hide all trace of these angles. This involves the gauge-covariant derivative

$$\mathscr{D}_\mu \equiv \partial_\mu + ig \sum_j W_\mu^{(j)} \mathbf{M}^{(j)}, \tag{8.34}$$

where g is the **gauge coupling constant** (not to be confused with a group element), which governs the strength of the interaction. Already, the power of the gauge method reveals itself: we might have expected only two fields, corresponding to the positive and negative charged intermediate vector bosons. In fact, we have just shown that there must be an additional particle involved, whose existence was not previously expected.

By analogy with $U(1)$, we might now expect the transformation

$$W_\mu^{(j)} \to W_\mu^{(j)} - \partial_\mu \theta^{(j)}/g \tag{8.35}$$

to ensure gauge invariance, but this does not work because of the non-Abelian nature of the group. The purpose of introducing the covariant derivative is to make the derivative commute with the phase factor:

$$\mathscr{D}_\mu \exp\left[i \sum_j \theta^{(j)} \mathbf{M}^{(j)}\right] = \exp\left[i \sum_j \theta^{(j)} \mathbf{M}^{(j)}\right] \mathscr{D}_\mu. \tag{8.36}$$

The transformation $W_\mu^{(j)} \to W_\mu^{(j)} - \partial_\mu \theta^{(j)}/g$ deals with the effect of ∂_μ on the phase factor, but the commutator of the covariant derivative and the phase factor contains (for small θ) a commutator of the generators. To remove this, the transformation law for the gauge field must be changed as follows:

$$W_\mu^{(j)} \to W_\mu^{(j)} - \frac{1}{g}\partial_\mu \theta^{(j)} + f_{jkl} W_\mu^{(k)} \theta^{(l)}. \tag{8.37}$$

The terms f_{jkl} are called the **structure constants** of the group. These are defined in terms of the commutators of the generators: $[G_i, G_j] = 2if_{ijk}G_k$. For example, in the case of $SU(2)$ the generators are the Pauli matrices σ_i, and $f_{ijk} = \epsilon_{ijk}$.

This more complicated transformation law for non-Abelian gauge fields has an important consequence when we come to consider the kinetic energy of the gauge fields. Because the transformation law mixes the various fields, any gauge-invariant kinetic term will also contain products of the various $W_\mu^{(i)}$:

$$\mathscr{L}_{\text{kinetic}} = -\tfrac{1}{4} F_{\mu\nu}^{(i)} F^{(i)\mu\nu}, \quad F_{\mu\nu}^{(i)} = \partial_\mu W_\nu^{(j)} - \partial_\nu W_\mu^{(j)} - g f_{jkl} W_\mu^{(k)} W_\nu^{(l)}. \tag{8.38}$$

This means that the gauge bosons interact amongst themselves. The Feynman diagram shown in figure 8.2 exemplifies this unusual aspect of non-Abelian theories.

THE STRONG INTERACTION One can clearly imagine extensions of the above algebra to larger numbers of states, with symmetry under $SU(n)$ being applied. There will be $n^2 - 1$ associated gauge bosons, one for each independent generator (for $n \times n$ traceless Hermitian matrices, there are $n-1$ free diagonal real numbers, and $n(n-1)/2$ off-diagonal

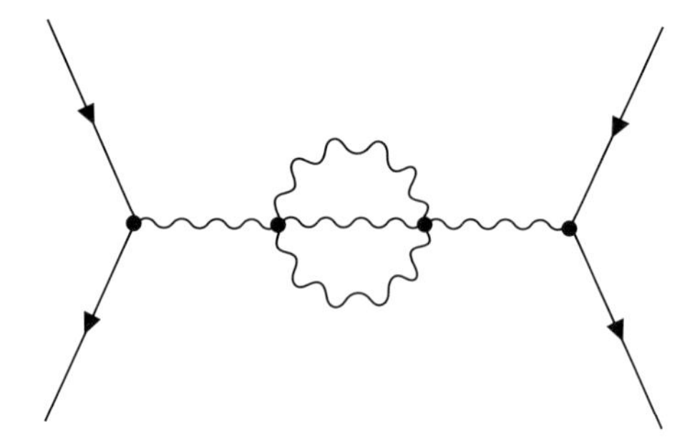

Figure 8.2. A characteristic Feynman diagram for a non-Abelian theory. The fact that linear superposition of fields holds in electromagnetism arises purely because the corresponding gauge group $U(1)$ is Abelian, so that such complicated diagrams are impossible in the electromagnetic case.

complex numbers, making a total of $n^2 - 1$ degrees of freedom). Such a system applies to the strong interaction, with $n = 3$. At the time quarks were first postulated, it was believed that there were only three, and an $SU(3)$-invariant theory based on

$$\psi = \begin{pmatrix} u \\ d \\ s \end{pmatrix} \tag{8.39}$$

was suggested. As extra conserved quantum numbers became apparent, the number of quarks necessary increased.

In fact, the present strong-interaction component of the standard model of particle physics is still based on $SU(3)$, but from the point of view of symmetry with respect to the additional *colour* quantum number. Thus, one considers

$$\psi = \begin{pmatrix} q_{\rm R} \\ q_{\rm G} \\ q_{\rm B} \end{pmatrix}, \tag{8.40}$$

and the theory is referred to as **quantum chromodynamics**, or QCD for short. The colour force has to be mediated by eight gauge bosons, known as **gluons**. QCD is discussed in more detail below.

8.5 Spontaneous symmetry breaking

The picture of the universe according to gauge theories is one of gross social inequality between three different **sectors** of the particle-physics world. The 'bosses' in this society consist of fermions, who interact via exchange of their servants – the *vector* (spin-1) bosons. Scalars form an underclass, and have so far lacked a function in society. They now enter because the above formalism has a critical problem: we have shown that gauge invariance forces all gauge bosons to be massless. This is good news for electromagnetism, but very bad for other interactions, which have a finite range fixed by the mass of the mediating particle. We can illustrate this by looking for a time-independent solution of the Klein–Gordon equation, satisfying

$$(-\nabla^2 + \mu^2)\psi = 0. \tag{8.41}$$

A solution that looks like a modified form of the inverse-square law is

$$\psi \propto \frac{e^{-\mu r}}{r}; \tag{8.42}$$

this corresponds to an interaction that cuts off at a range $\mu^{-1} = \hbar/mc$. This is Yukawa's original argument constraining the mass of the particle that mediates the strong interaction, and is just the uncertainty principle saying that we can 'borrow' the energy mc^2 for a time $t \lesssim \hbar/mc^2$, in which time the particle can travel no further than ct, limiting the range of the interaction.

THE 'MEXICAN HAT' POTENTIAL Because of this problem, Yang–Mills theories seemed for about a decade to be no more than an elegant curiosity, until a way out was found by Higgs (1964; see also Englert & Brout 1964; Kibble 1967). The argument is in fact similar to the one that leads to the introduction of gauge fields: the quantum properties of the system are obviously wrong, so we must have left a term (i.e. a new field) out of the Lagrangian. The field in this case is from the class we have so far not considered: scalar fields. These are the simplest quantum fields; the Lagrangian for a charged scalar field would normally be taken as

$$\mathcal{L} = \partial_\mu \phi \, \partial^\mu \phi^* - V(\phi), \tag{8.43}$$

where the **potential** $V(\phi)$ is just a mass term $V = m^2|\phi|^2$. The system can be made more complicated by adding in a term that looks like a self-interaction: $V(\phi) = m^2|\phi|^2 + \lambda|\phi|^4$. This equation has undergone considerable investigation, because the $\lambda|\phi|^4$ interaction is the highest power compatible with renormalization.

The key to progress is now to suspend physical interpretation temporarily and start playing mathematical games. Apart from renormalizability, there seems nothing to prevent us treating the potential as an arbitrary function. In particular, if we avoid thinking about imaginary masses, why not choose the number m^2 to be negative?:

$$\boxed{V(\phi) = -\mu^2|\phi|^2 + \lambda|\phi|^4.} \tag{8.44}$$

Such a choice yields the potential shown in figure 8.3, which is named for its resemblance to a Mexican hat (or alternatively to the base of a wine bottle). This has many interesting features. The ground state (minimum V) is degenerate, and occupies a circle on the Argand plane, at

$$|\phi|_{\rm GS} = u \equiv \sqrt{\frac{-\mu^2}{2\lambda}}, \tag{8.45}$$

with a 'depth' relative to $|\phi| = 0$ of

$$V_{\rm min} = -\frac{\mu^4}{4\lambda}. \tag{8.46}$$

CLASSICAL AND QUANTUM SYMMETRY BREAKING Now, we know what to expect for a classical system with such a potential function. A ball placed at rest in the centre of a

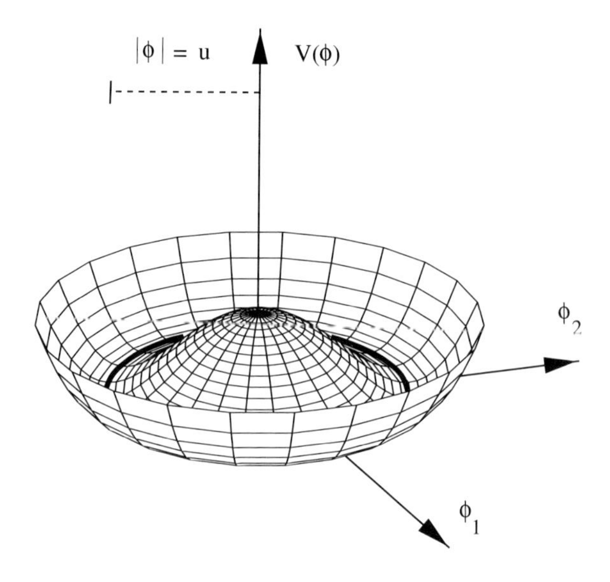

Figure 8.3. The Higgs potential for a complex scalar field, $V(\phi) = -|\mu^2|\,|\phi|^2 + \lambda|\phi|^4$. When the number λ is chosen to be positive, the ground state energy occurs along a circle of $|\phi| \neq 0$. The phase of the minimum is arbitrary, however.

Mexican hat would be unstable and roll in some random direction downhill until, given a little friction, it settled somewhere on the circle of minimum V. The radius is predictable, but the azimuthal angle is not. This should cause no intellectual discomfort, except that it violates one of the basic methods often used in physics: symmetry arguments. For example, if we are given a spherically symmetric charge distribution, we assume almost without question that the field that solves Maxwell's equations for this case will also be spherically symmetric. The Mexican hat potential, however, generates equations of motion that are rotationally symmetric, and yet the solution (the position of the ball) violates this symmetry. What this tells us is that we always need to ask whether the symmetric solution is *stable*.

If we now go back to the quantum system of the scalar field, the equivalent of the classical statement that the field will evolve to lie at the minimum of $V(\phi)$ cannot be true, as the field will experience quantum fluctuations. Nevertheless, a similar statement will apply to the **vacuum expectation value** (vev) of ϕ:

$$\boxed{\langle 0|\phi|0\rangle = ue^{i\theta}.} \tag{8.47}$$

The angle θ is arbitrary; as in the classical case, the ground state will pick out one value of θ, and so the original $U(1)$ symmetry will no longer be obeyed. This is called **spontaneous symmetry breaking** (SSB).

The consequences of the broken symmetry are profound if the symmetry is a local one. In the simple Abelian $U(1)$ case we have been considering, the gauge-invariant Lagrangian is

$$\mathcal{L} = (\partial_\mu + ieA_\mu)\phi(\partial^\mu - ieA^\mu)\phi^* - V(\phi) - \tfrac{1}{4}F_{\mu\nu}F^{\mu\nu}. \tag{8.48}$$

Now suppose we make a change of variables to describe the field ϕ in terms of displacements from the ground state:

$$\phi = (u + R)e^{i\theta}. \tag{8.49}$$

Because of the presence of the gauge field, we know that the angle θ is unobservable and can be taken equal to zero. The physical justification for shifting origin to the minimum in V is that, in practice, in quantum physics we usually solve problems by perturbation theory. It clearly makes more sense to expand about the ground state than $\phi = 0$, which is always unlikely to be close to the actual solution for ϕ. After this substitution, we have a system that looks very different:

$$\mathcal{L} = (\partial_\mu R)(\partial^\mu R) + 2\mu^2 R^2 + e^2u^2 A_\mu A^\mu - V(R) - \tfrac{1}{4}F_{\mu\nu}F^{\mu\nu} + \text{interaction terms}. \tag{8.50}$$

This now looks like the Lagrangian for two fields R and A^μ, both of which are massive (note the quadratic terms). The spontaneous symmetry breaking has given the gauge field A^μ a mass. This has been achieved by introducing the scalar field ϕ, whose corresponding particle also gains a mass. This particle is known as the **Higgs boson**, and the search for it is presently one of the prime goals of experimental particle physics.

It is important to realize that nothing has changed here. The gauge symmetry still applies absolutely, but has been converted to a **hidden symmetry** by our change of variables. Thus, the same system can have several very different physical interpretations, which arise only from simple algebraic rearrangement. In this sense, there need be nothing spontaneous about the symmetry breaking: we do it by hand. The terminology is only good in the classical limit, where the dynamics of the field will cause it to choose a definite ground state in which $\langle 0|\phi|0\rangle$ is non-zero, as discussed above.

GOLDSTONE BOSONS Historically, the study of SSB in field theory first arose in the context of *global* symmetries, without the presence of gauge fields. The above example is now simpler in the sense that $A_\mu = 0$, but more complicated because $\theta = 0$ cannot be assumed. Expanded to second order in R and θ, the Lagrangian is

$$\mathcal{L} = (\partial_\mu R)(\partial^\mu R) + 2\mu^2 R^2 + u^2(\partial_\mu\theta)(\partial^\mu\theta) - V(R). \tag{8.51}$$

There is now a contribution from the θ field (scaled by u). Since this degree of freedom corresponds to rolling around the degenerate minimum, it is not so surprising that there is no associated potential term. The spontaneous breaking of the global $U(1)$ symmetry has resulted in a **Goldstone boson**, which is the quantum of this new massless field. Massless particles of this sort are always produced when a non-gauge symmetry is broken. What happens in the gauge case is sometimes described as follows: the gauge field eats the Goldstone boson and acquires a mass in the process. This is correct from the point of view of counting degrees of freedom, since the Goldstone degree of freedom is replaced by the new longitudinal degree of freedom for the massive photon, but otherwise it is not clear how useful the analogy is.

8.6 The electroweak model

All the pieces are now in place for constructing a model that is not simply an illustrative 'toy', but appears to describe exactly the universe we inhabit. This is a gauge theory that includes the electromagnetic interaction and the weak interaction in a single unified framework. It was first suggested independently by Weinberg (1967) and Salam (1968), although the theory had been proposed earlier by Glashow (1961) without the critical element of mass generation via the Higgs mechanism.

GAUGE GROUP AND NEUTRAL CURRENTS Since the weak interaction involves doublets like (e, ν_e), the natural gauge group to think about is one that will deal with rotations in this 2D space: $SU(2)$. If we want to retain the existing apparatus of QED, this suggests a theory that obeys invariance under $SU(2) \otimes U(1)$. Being more specific, we know that the theory must empirically violate parity; defining left- and right-handed components via

$$\begin{aligned} \psi_{\rm L} &\equiv \tfrac{1}{2}(1-\gamma^5)\psi \\ \psi_{\rm R} &\equiv \tfrac{1}{2}(1+\gamma^5)\psi, \end{aligned} \tag{8.52}$$

we assume that only the left-handed multiplets like

$$L_e \equiv \begin{pmatrix} e^- \\ \nu_e \end{pmatrix}_{\rm L} \tag{8.53}$$

obey the $SU(2)$ transformation. Or, in shorthand, we talk of invariance under $SU(2)_{\rm L} \otimes U(1)$.

This choice of the gauge group immediately has a startling consequence. $SU(2)$ has three generators, and $U(1)$ one, so the theory must contain a total of four gauge fields. The discussion of QED and weak interactions to date has included only three: the photon and the $W^\pm$. There must be a hitherto unsuspected ingredient: a single extra degree of freedom (uncharged therefore), which is the Z^0 boson. This in turn means that there must exist **neutral currents**: weak interactions that are not accompanied by charge exchange, such as lepton–neutrino scattering. It was the detection of such processes at CERN in 1973 that marked the first triumph of the gauge principle in predicting the structure of the fundamental interactions.

MASSES AND UNIFICATION The Higgs mechanism has allowed us to introduce gauge fields and give them mass while maintaining gauge invariance, but we do not wish to be too successful at this. Experimentally, the photon appears to be massless; the explanation of this via unbroken gauge invariance is simple and appealing, so that the Higgs mechanism seems a backwards step in this case. The door can be left open for a massless photon by taking a doublet of Higgs fields of the form

$$\phi = \begin{pmatrix} \phi_1 \\ \phi_2 \end{pmatrix}, \tag{8.54}$$

in terms of which the Mexican hat potential becomes

$$V(\phi) = -\mu^2 \phi^\dagger \phi + \lambda(\phi^\dagger \phi)^2. \tag{8.55}$$

The symmetry breaking requires a vacuum expectation value for ϕ, but the fields can be defined so that this lies entirely along ϕ_2 [problem 8.4]:

$$\langle 0|\phi|0\rangle = \begin{pmatrix} 0 \\ v/\sqrt{2} \end{pmatrix}, \tag{8.56}$$

where $v = (\mu^2/\lambda)^{1/2}$. This allows a $U(1)$ symmetry to be retained in the broken theory via phase rotations in the ϕ_1 part.

As an aside, it may be wondered why the SSB trickery always has to be played with scalar fields. It is possible to show that the electroweak Lagrangian without scalar fields is inconsistent. There are processes, such as $W^+W^- \to W^+W^-$, for which the amplitude diverges. A common solution for such difficulties is to add some extra 'counterterm' to absorb the unwanted divergence. In this case, this corresponds to saying that there must be some additional particle that weakly interacting particles can exchange. For bosons, angular momentum conservation says that this particle must also be a boson, and the case of vector fields has already been covered by the introduction of the W and Z fields themselves, so the next natural case to consider is the scalar-field sector.

The field content of the theory is now complete, and it is helpful to summarize how things will change under gauge transformations involving spatially varying phase angles θ and $\boldsymbol{\theta}$ for the $U(1)$ and $SU(2)$ parts respectively. The transformations of the fields are, for $SU(2)$,

$$\begin{aligned} L &\to \exp(-ig\boldsymbol{\sigma}\cdot\boldsymbol{\theta}/2)\, L \\ R &\to R \\ \phi &\to \exp(-ig\boldsymbol{\sigma}\cdot\boldsymbol{\theta})\, \phi, \end{aligned} \tag{8.57}$$

and, for $U(1)$,

$$\begin{aligned} L &\to \exp(ig'\theta/2)\, L \\ R &\to \exp(ig'\theta)\, R \\ \phi &\to \exp(-ig'\theta/2)\, \phi, \end{aligned} \tag{8.58}$$

where the σ_i are the Pauli matrices. This defines the two coupling constants, g and g', of the two groups involved. These transformations are a little more complicated than might have been expected in the $U(1)$ case, and require comment. Different coefficients of θ are allowed for the different fields, if they have independent contributions to the Lagrangian; if there is an interaction term, the coefficients must be related. The particular coefficient ratios here are again the result of looking ahead and requiring a particle resembling the photon to emerge at the end.

The Lagrangian invariant under these transformations can now be written down, and is a sum of four terms: the contributions of the fermions, the gauge bosons, the Higgs field, and an interaction term:

$$\mathcal{L} = \mathcal{L}_\psi + \mathcal{L}_{A^\mu} + \mathcal{L}_\phi + \mathcal{L}_{\rm interaction}. \tag{8.59}$$

If we call the three $SU(2)$ gauge fields A_i^μ and the $U(1)$ field B^μ, then the fermion part is

$$\mathcal{L}_\psi = \bar{\psi} i\gamma^\mu \mathcal{D}_\mu \psi \tag{8.60}$$

(summed over fermions). The covariant derivative for left-handed fields is

$$\mathcal{D}_\mu = \partial_\mu - i\frac{g}{2}\boldsymbol{\sigma}\cdot\mathbf{A}_\mu + i\frac{g'}{2}B_\mu, \tag{8.61}$$

but just

$$\mathscr{D}_\mu = \partial_\mu + ig' B_\mu. \tag{8.62}$$

for right-handed lepton fields (right-handed neutrinos are not included). The gauge Lagrangian is just the kinetic energy

$$\mathscr{L}_{A^\mu} = -\frac{1}{4} F_{i\mu\nu} F_i^{\mu\nu} - \frac{1}{4} G_{\mu\nu} G^{\mu\nu}, \tag{8.63}$$

where the field tensor $G^{\mu\nu} = \partial_\mu B_\nu - \partial_\nu B_\mu$ as usual, and

$$F_{i\mu\nu} = \partial_\mu A_{i\nu} - \partial_\mu A_{i\nu} - g\epsilon_{ijk} A_{j\mu} A_{k\nu}. \tag{8.64}$$

The Higgs Lagrangian is

$$(\mathscr{D}_\mu \phi)^\dagger (\mathscr{D}^\mu \phi) - V(\phi), \tag{8.65}$$

where the covariant derivative in this case is

$$\mathscr{D}_\mu = \partial_\mu - i\frac{g}{2} \boldsymbol{\sigma} \cdot \mathbf{A}_\mu - i\frac{g'}{2} B_\mu. \tag{8.66}$$

The interaction term will not be of great interest here; it is of the form of field products such as $L\phi R$, which would not have been gauge invariant without the special form of the $U(1)$ transformation above. This is a **Yukawa coupling** – the general term for an interaction term between fermions and scalars of the form $\bar{\psi}\psi\phi$.

Having specified the Lagrangian, all that remains is to break the symmetry by the transformation

$$\phi \rightarrow \begin{pmatrix} 0 \\ (v + \eta)/\sqrt{2} \end{pmatrix}, \tag{8.67}$$

where η is the field corresponding to the Higgs boson. This is easy enough to say, but in practice a large amount of algebra is generated – which can be checked by those readers with sufficient stamina. For the present purpose, the most important feature of the results is the term that comes from the vector boson couplings in $\mathscr{L}_\phi$:

$$\mathscr{L} = \frac{v^2}{8} \left(g^2 \mathbf{A}_\mu \cdot \mathbf{A}^\mu + g'^2 B_\mu B^\mu + 2gg' B_\mu A_3^\mu \right). \tag{8.68}$$

This says that there is a non-diagonal mass matrix between B^μ and A_3^μ, given by

$$M_{AB} = \frac{v^2}{8} \begin{pmatrix} g'^2 & gg' \\ gg' & g^2 \end{pmatrix}, \tag{8.69}$$

which can be diagonalized by making a rotation to new field combinations:

$$\boxed{\begin{aligned} A'^\mu &= (\cos\theta_{\rm W}) B^\mu + (\sin\theta_{\rm W}) A_3^\mu \\ Z^\mu &= (\cos\theta_{\rm W}) A_3^\mu - (\sin\theta_{\rm W}) B^\mu. \end{aligned}} \tag{8.70}$$

These are chosen so that the A'^μ field is massless and corresponds to the photon; the rotation angle is the **Weinberg angle**, given by

$$\tan\theta_{\rm W} = g'/g. \tag{8.71}$$

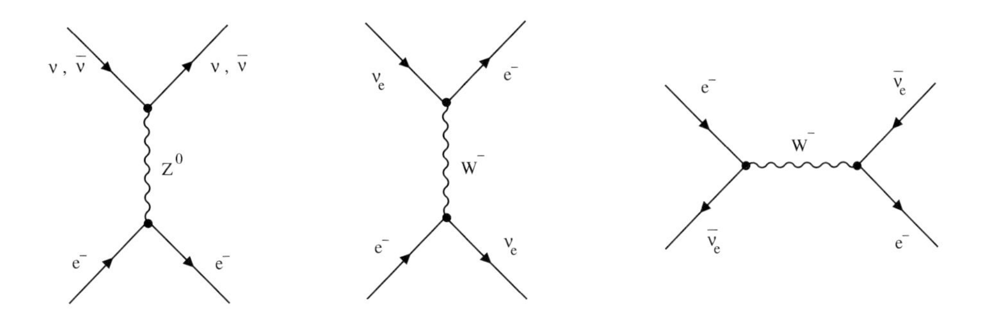

Figure 8.4. Feynman diagrams for electron–neutrino scattering in the electroweak theory. The crucial new diagram is the first one, involving neutral currents. Unlike the diagrams involving the W, interaction via a Z is not an exchange reaction. Neutral currents were unsuspected prior to their prediction via the arguments of gauge invariance.

The vector boson masses are related to the energy scale of symmetry breaking, and are in a ratio that depends only on the Weinberg angle:

$$\begin{aligned} M_W &= gv/2 \\ M_Z &= M_W/\cos\theta_{\rm W}. \end{aligned} \tag{8.72}$$

The Weinberg angle can be determined from the strength of neutral-current interactions, which give a new contribution to the scattering of electrons and neutrinos (figure 8.4). These yield

$$\boxed{\sin^2\theta_{\rm W} \simeq 0.23,} \tag{8.73}$$

or $\theta_{\rm W} \simeq 29^\circ$.

Once the Weinberg angle is known, there is almost no freedom left in the model. The coupling constants can also be related to the electric charge by looking at the interaction terms involving the new photon field or the $W^\pm$ field. What this gives is

$$e = g\sin\theta_{\rm W} \tag{8.74}$$

and

$$\frac{G_{\rm F}}{\sqrt{2}} = \frac{g^2}{8M_W^2} \tag{8.75}$$

(a factor of 8 different from the relation given previously, because the definition of g is different). The relation to e (where $e^2/4\pi \simeq 1/137$ in natural units) gives $g \simeq 0.65$ (and $g' \simeq 0.36$), and also tells us a number of great importance, which is the energy scale of the symmetry breaking:

$$\boxed{v = 246\,\text{GeV}.} \tag{8.76}$$

Lastly, these relations predict the masses of the gauge bosons to be $M_W \simeq 80\,\text{GeV}$ and $M_Z \simeq 90\,\text{GeV}$. In 1983, particles with approximately these masses were discovered at CERN. The verification of the electroweak prediction went so smoothly that it could almost seem an anticlimax; this would be grossly unfair to what must rank as one of the greatest intellectual achievements of the twentieth century.

The only number in the electroweak model that remains to be measured is the mass of the Higgs boson:

$$M_{\rm H} = \sqrt{2}\,\mu = \sqrt{2\lambda}\,v. \tag{8.77}$$

This is not constrained by any relation to other parameters in the theory, and will have to be measured experimentally. It can be argued that the mass should be less than about 1 TeV, basically because the strength of interactions involving the Higgs increase with increasing Higgs mass. If the mass is too high, it violates the **unitarity limit**, meaning that the first-order scattering probabilities exceed unity (see e.g. section 4.12 of Ross 1985). Of course the overall theory must still conserve probability, but this either means that the Higgs mass cannot be this high, or that the nature of interactions involving the other particles in the electroweak model must alter at this energy. Either way, there must be 'new physics' at 1 TeV, which is reassuring knowledge for accelerator builders. Given the success of the electroweak model to date in predicting the existence and mass of the Z^0, it might be thought that the eventual appearance of the Higgs is a fairly safe bet; there is nevertheless a high level of unease among theorists about the idea of the Higgs as a normal elementary particle. Partly this is an argument by analogy: in other areas where symmetry breaking occurs, the potential is not put in by hand, but instead arises via more complicated dynamical effects. The analogues of the Higgs boson in these cases are not elementary particles but composite (pions in the case of chiral symmetry in the strong interaction; see below). Particle-physics models in which the Higgs is a composite entity do exist, under the name **technicolour**, although they are not without problems (see chapter 8 of Collins, Martin & Squires 1989). Efforts to detect the Higgs in accelerators therefore have a good deal more at stake than simply pinning down the mass of the particle. If an elementary Higgs does not exist, particle physics could take a new and unpredictable turning.

That completes this brief survey of the electroweak model. As a closing remark, note that it is not really a unified theory, since it contains two distinct coupling constants with no particular relation between them. Although the electromagnetic interaction emerges as a mixture of the two gauge interactions, this does not mean that the two interactions are really unified. This is only possible in the true unified theories discussed below, in which the parameters of the electroweak model would be related through some higher symmetry.

8.7 Quantum chromodynamics

The strongly interacting quarks are part of the electroweak model, as they also occur in left-handed $SU(2)$ doublets:

$$\begin{pmatrix} u \\ d \end{pmatrix} \begin{pmatrix} s \\ c \end{pmatrix} \begin{pmatrix} t \\ b \end{pmatrix}. \tag{8.78}$$

To extend the electroweak theory of leptons and quarks to encompass the strong interactions of quarks involves introducing additional degrees of freedom, so that these column wave functions gain an additional dimension. In fact, they are already implicitly

2D (with the spin degree of freedom suppressed), so this means that, in the discussion below, ψ really stands for a 3D matrix.

COLOUR AND $SU(3)$ As mentioned earlier, each of the six different **flavours** of quark comes in three possible versions allowing for an extra quantum number called **colour**, labelled by the primary colours: red, green and blue. The quantum field theory of the strong interaction is thus called **quantum chromodynamics**, and the appropriate gauge group is based upon transformations between the different colour states – colour $SU(3)$ (as distinct from flavour $SU(3)$, which was an approximate symmetry suggested when only three quarks were known).

Why is the colour quantum number necessary? Historically, the evidence was provided by the Δ^{++} particle, which consists of three u quarks, all with spin up and in an $L = 0$ state. Since all three parts of the wave function (space, spin, flavour) must be symmetric in this case, the Pauli principle is apparently violated. It can be rescued by introducing the new quantum number of colour, and requiring that the colour part of the wave function be antisymmetric in the case of the Δ^{++}. This may seem rather weak grounds for introducing a fundamental new quantum number; perhaps a good analogy is with Pauli's invention of the neutrino, where again a radical hypothesis was required to save a cherished principle, to be verified subsequently by experiment. Certainly, colour states can be regarded as having been observed, as they change the rate at which decays such as $\pi^0 \rightarrow \gamma\gamma$ occur.

Unlike some of the more fanciful quark names, 'colour' is a good name, since it never appears in isolation. Just as a superposition of the primary colours yields white, so particles in nature must have a neutral colour. This can happen in two ways:

$$\psi = \begin{cases} \frac{1}{\sqrt{6}}\epsilon_{ijk}q_1^i q_2^j q_3^k & \text{(baryon)} \\ \frac{1}{\sqrt{2}}q_1^i \bar{q}_2^i & \text{(meson).} \end{cases} \tag{8.79}$$

Pure coloured states (such as free quarks) do not exist. We shall see the reason for this (**asymptotic freedom**) after looking at some of the details of QCD.

This completes the list of ingredients needed for the standard model. The overall gauge group that includes the electroweak model and QCD is

$$U(1) \otimes SU(2)_{\mathrm{L}} \otimes SU(3)_{\mathrm{C}}, \tag{8.80}$$

where the L and C subscripts remind us that the labelled groups operate on left-handed multiplets and coloured multiplets respectively. The direct-product nature of the group means that (save for the complication of electroweak mixing) the three parts of the fundamental interactions inhabit rather different worlds, so that QCD can to some extent be considered in isolation. This is particularly clear in the electroweak symmetry breaking, which leaves the $SU(3)$ part of the group unchanged:

$$U(1) \otimes SU(2)_{\mathrm{L}} \otimes SU(3)_{\mathrm{C}} \quad \rightarrow \quad U(1) \otimes SU(3)_{\mathrm{C}}. \tag{8.81}$$

THE QCD LAGRANGIAN The gauge structure of QCD is an extension of the same ideas as in the smaller gauge groups. The Lagrangian of QCD is

$$\boxed{\mathscr{L}_{\mathrm{QCD}} = \sum_{\text{flavours}} \bar{\psi}(i\gamma^\mu \mathscr{D}_\mu + m)\psi - \tfrac{1}{2}\mathrm{Tr}\, F_{\mu\nu}F^{\mu\nu},} \tag{8.82}$$

where the colour degree of freedom is hidden in the notation; we have used ψ to denote a column vector of three Dirac spinors:

$$\psi \equiv \begin{pmatrix} \psi_{\rm R} \\ \psi_{\rm G} \\ \psi_{\rm B} \end{pmatrix}. \tag{8.83}$$

To operate on this, the covariant derivative must be a 3×3 matrix

$$\mathscr{D}_\mu = \partial_\mu + igG_\mu, \tag{8.84}$$

where G_μ is a combination of vector gauge fields called **gluons** and the eight $SU(3)$ generators Λ^i:

$$G_\mu \equiv \tfrac{1}{2} G^i_\mu \Lambda^i. \tag{8.85}$$

As before, the gauge-invariant field tensor is given by

$$F_{\mu\nu} = [\mathscr{D}_\mu, \mathscr{D}_\nu]/(ig) = \partial_\mu G_\nu - \partial_\nu G_\mu + ig[G_\mu, G_\nu]. \tag{8.86}$$

The gauge-field part of the Lagrangian may appear to be a factor 2 different from the $-\frac{1}{4}F^{\mu\nu}F_{\mu\nu}$ familiar from QED. However, this is just a consequence of the properties of the generators, which satisfy

$$\begin{aligned} [\Lambda^a, \Lambda^b] &= 2i f_{abc} \Lambda^c \\ \mathrm{Tr}\, \Lambda^a \Lambda^b &= 2\delta^{ab}. \end{aligned} \tag{8.87}$$

CHIRAL SYMMETRY The QCD Lagrangian contains a number of parameters, including the masses of the quarks. However, these masses are small by comparison with typical masses of hadrons; one might therefore expect that some of the main features of QCD might be demonstrated by ignoring the quark masses in the first instance. If we do this, and consider the limit $m_q \to 0$, a new symmetry of QCD becomes apparent, called **chiral symmetry** from the Greek for 'handedness'. Historically, this was of importance because it provided the first evidence that the strong force was indeed vector in nature (mediated by a spin-1 particle).

We start by defining the left- and right-handed chiral fields

$$\begin{aligned} q_{\rm L} &= \tfrac{1}{2}(1 - \gamma^5) q \\ q_{\rm R} &= \tfrac{1}{2}(1 + \gamma^5) q. \end{aligned} \tag{8.88}$$

In terms of these fields, and with zero quark masses, the quark part of the Lagrangian splits into two distinct parts:

$$\mathscr{L} = \bar{q}_{\rm L} i\gamma^\mu \mathscr{D}_\mu q_{\rm L} + \bar{q}_{\rm R} i\gamma^\mu \mathscr{D}_\mu q_{\rm R}. \tag{8.89}$$

Critically, the cross terms between left- and right-handed fields vanish, owing to the commutation properties of the γ matrices (particularly $\{\gamma^\mu, \gamma^5\} = 0$). There are therefore two distinct symmetries, corresponding to rotations in flavour space of the left- and right-handed parts. As in the original quark model, this approach works best with the three lightest quarks (u, d, s) only, so that there is an approximate $U(3)_{\rm L} \otimes U(3)_{\rm R}$ symmetry. Note that this symmetry would not exist in even an approximate form if the strong interaction were mediated by a scalar field: the Lagrangian would then contain a term looking like $\bar{q}_{\rm L} q_{\rm R} \phi$ that mixes left- and right-handed fields. It is only the vector nature of the coupling, bringing in γ matrices, that keeps the two hands separate.

Now, the problem with chiral symmetry is that it is not directly manifested in nature. We would expect there to be many hadrons with mass approximately degenerate with the proton, not just one (the neutron). As usual in particle physics, the solution of broken symmetry may immediately suggest itself. This is indeed the case: chiral symmetry is broken at laboratory energies, but not by the direct Higgs route of introducing a new scalar field. The same effect can be achieved more subtly by creating scalar particles that are bound states of quark–antiquark pairs. The resulting pseudoscalar mesons (π, K, η) are fields that can acquire a vacuum expectation value, producing **dynamical breaking** of chiral symmetry. It is easy enough to see in outline how this happens, since the form of the mass term that has been neglected is

$$\mathcal{L} = m_u \bar{u}u + m_d \bar{d}d + m_s \bar{s}s, \tag{8.90}$$

which classically looks like the sum of three meson fields. The breaking of chiral symmetry can then be thought of approximately as arising because these fields gain a non-zero vacuum expectation value: $\langle 0|\bar{u}u|0\rangle \neq 0$ etc. (see e.g. chapter 5 of Cheng & Li 1984 for details). Since the symmetry being broken is global rather than local, we expect the appearance of Goldstone bosons; indeed, if chiral symmetry were exact, all the above mesons would be massless. The lightest of the set is the pion, so it is helpful to regard this particle as the Goldstone boson of QCD. Something very similar happens in superconductivity, where the bogus scalar particles are Cooper pairs of correlated electrons (see e.g. Feynman 1972).

QUARK MIXING The quarks also participate in weak interactions, and in the standard model are arranged in $SU(2)$ doublets, like the lepton–neutrino pairings. Unfortunately, there is no reason why the states that are recognized as distinct ‘flavours’ of quark by the strong interaction should also be those that are distinguished by the weak interaction, and the former fields are in fact linear combinations of the latter. This was first discovered in the case of the d and s quarks, through the existence of weak-interaction processes that appeared not to conserve strangeness. Consider the low-energy Fermi effective Lagrangian

$$\mathcal{L} = \frac{G_{\rm F}}{\sqrt{2}} J^\mu J^\dagger_\mu, \tag{8.91}$$

where the contribution to the current from the u, d, s, c quarks is

$$J^\mu = \bar{u}[\gamma^\mu(1-\gamma^5)]d' + \bar{c}[\gamma^\mu(1-\gamma^5)]s', \tag{8.92}$$

which is of the expected form except for a rotation

$$\begin{pmatrix} d' \\ s' \end{pmatrix} = \begin{pmatrix} \cos\theta_{\rm C} & \sin\theta_{\rm C} \\ -\sin\theta_{\rm C} & \cos\theta_{\rm C} \end{pmatrix} \begin{pmatrix} d \\ s \end{pmatrix}, \tag{8.93}$$

where $\theta_{\rm C} \simeq 13°$ is the **Cabbibo angle**.

With more quarks, there are more mixing possibilities. The general form of the current has to be a combination of all possible $\bar{\psi}_{\rm L}\gamma^\mu\psi_{\rm L}$ terms:

$$\boxed{J^\mu = (\,\bar{u}\;\bar{c}\;\bar{t}\,)_{\rm L}\;\gamma^\mu\;V \begin{pmatrix} d \\ s \\ b \end{pmatrix}_{\rm L},} \tag{8.94}$$

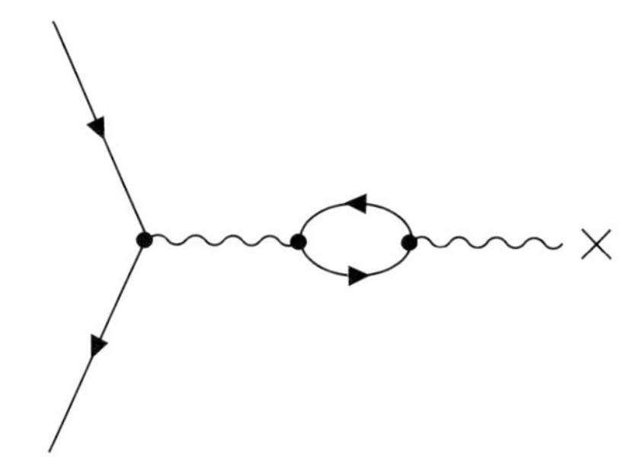

Figure 8.5. A vacuum polarization diagram. This effectively shows a virtual photon (or gluon, in the case of QCD) temporarily splitting into two particles, to form an internal loop in the diagram. For QED, the photon can only split into an electron–positron pair; for non-Abelian theories, the gauge particles can interact with themselves.

where V is the **Kobayashi–Maskawa matrix** (KM matrix). The term from $d-s$ mixing is the largest off-diagonal contribution, the others being smaller than 0.05. It would have been convenient if nature had chosen this matrix to be diagonal, but it is not. This is one respect in which the standard model can hardly be said to be elegant; an eventual final theory would have to account for the particular non-zero mixings observed. However, interesting physics results from these extra terms: they are responsible for CP violation. This phenomenon is in many ways analogous to the case of neutrino oscillations described below: because the CP eigenstates of the theory are not also mass eigenstates, they have different time evolution, leading a given initial state to evolve towards a mixture. This violation of T symmetry, combined with CPT as an exact symmetry, shows why this process also violates CP.

VACUUM POLARIZATION AND ASYMPTOTIC FREEDOM The interactions between particles become progressively more complicated beyond second order, and diagrams such as those shown in figure 8.5 have to be faced. These diagrams and their higher-order counterparts have the effect of modifying the apparent strength of the interaction from its bare value, and moreover in a way that in principle depends on energy. The effect is termed **vacuum polarization** because of the analogy with a charge placed in a dielectric medium, for which the electric field is reduced to $1/\epsilon$ times its value in a vacuum, owing to the effect the electric field has on the polarizable molecules.

A proper calculation of vacuum polarization through Feynman rules is a technical challenge that will have to be omitted here (see e.g. chapter 13 of Itzykson & Zuber 1980), but some important general points can be made. The contributions of internal loops, with virtual 4-momentum q, have caused trouble in other cases, and will be divergent. The standard way of dealing with this is by means of renormalization, and it is particularly clear in this case that this is a natural procedure: the vacuum polarization diagrams affect directly the measured strength of an interaction, as quantified by its dimensionless coupling constant g. The lowest-order correction to the coupling constant will then look like

$$g(k^2) = g_0 \left[1 + g_0^2 \int I(k,q)\, d^4q \right], \tag{8.95}$$

where g_0 is the bare coupling, $g(k^2)$ is the empirical value probed by particles of 4-momentum norm k^2, and I is some integrand obtained from the Feynman rules of the theory. It is inevitable on dimensional grounds that this correction must diverge logarithmically: since g is dimensionless, I must scale as q^{-4} at large loop 4-momenta. If we cannot get a sensible relation between $g(k^2)$ and g_0, it is nevertheless possible to relate $g(k_1^2)$ to $g(k_2^2)$, by writing the equation for g twice, differencing, and replacing g_0 by $g(k_1^2)$ (to which it is equivalent, to lowest order). This gives the following relation between $g(k_2^2)$ and $g(k_1^2)$:

$$g(k_2^2) = g(k_1^2) + [g(k_1^2)]^3 \int [I(k_2,q) - I(k_1,q)]\, d^4q. \tag{8.96}$$

Any cutoff at very large q is now invisible, since the integrand is only significant for q^2 between k_1^2 and k_2^2. This is a very much more satisfactory approach, since it takes the fact that coupling constants are observed to be finite, but adds the idea that quantum corrections may change the result that is observed if we look at a different energy. Arguing by dimensions once again, the change with energy must satisfy the **renormalization group equation**

$$\frac{dg(k^2)}{d\ln k^2} = \beta\big(g(k^2)\big), \tag{8.97}$$

i.e. the quantum effects operating around a given energy know only about the effective strength of the interaction at that energy, there being no other dimensionless numbers in the problem. The perturbation approach above would suggest that $\beta(g) \propto g^3$ in lowest order, in which case the equation integrates to

$$\boxed{g(k^2) = \frac{\pm 1}{\ln(k^2/\lambda^2)}.} \tag{8.98}$$

The apparent interaction strength therefore changes logarithmically with energy. The sign of this vacuum-polarization change depends on the sign of the constant of proportionality in the leading-order dependence of $\beta(g)$. There is no simple general argument for predicting which way this will go, but the results of an exhaustive set of calculations for the ingredients of the standard model are at least simple to remember: β is positive for the Abelian $U(1)$ component of the standard model, but negative for the non-Abelian parts. The former result is in accord with our experience of electrodynamics: we might expect that a vacuum of virtual pairs would act a little like a plasma in shielding a charge, so that the effective g would increase as the energy increased, allowing a smaller de Broglie wavelength to resolve distances closer to an electron. For the non-Abelian case, intuition offers no such assistance; see e.g. chapter 13 of Itzykson & Zuber (1980).

Expressing the coupling constant in terms of a critical energy scale λ is not very sensible in the case of QED, since practical energies are always $\ll \lambda$. For QCD, however,

$$\boxed{\lambda_{\rm QCD} \simeq 200\ \mathrm{MeV}} \tag{8.99}$$

is a scale of critical importance, since it marks the energy above which the strong interaction moves into the regime at which a perturbative approach should yield reasonable answers. This is the idea of **asymptotic freedom**. At the opposite extreme of low energies, hadrons consist of bound quarks that are never seen separately. In the current approach, this may be understood as follows: the effective strength of the interaction binding the quarks increases without limit as the distance between them increases, so that free single quarks are effectively forbidden.

Nevertheless, if a hadron is bombarded with particles having energies $\gg \lambda_{\rm QCD}$, the experiment will appear to probe a set of relatively weakly interacting particles, and perturbative calculations can be used to probe the internal constitution of the hadrons.

THE STRONG CP PROBLEM AND THE AXION The discussion of QCD to date has ignored one additional possible gauge-invariant contribution to the Lagrangian, which takes the form

$$\mathcal{L} = \frac{g^2}{16\pi^2}\,\theta\,\mathrm{Tr}\,F^{\mu\nu}\tilde{F}_{\mu\nu}, \tag{8.100}$$

where the **dual field tensor** is

$$\tilde{F}_{\mu\nu} \equiv \tfrac{1}{2}\epsilon_{\mu\nu\alpha\beta}F^{\alpha\beta}. \tag{8.101}$$

The presence of this term violates CP (because ϵ is a pseudotensor that changes sign under reflection, whereas C is not violated). Now, the observed degree of CP violation in the strong interaction is small; one key diagnostic would be a non-zero magnetic moment for the neutron, which has yet to be detected. From this, the limit

$$|\theta| \lesssim 10^{-9} \tag{8.102}$$

has ben deduced. If this term is so small, then why worry about it? The answer is that θ is to be regarded not just as some number that nature chooses for reasons of its own. This would be bad enough: we would have another parameter in the standard model whose numerical value we could not explain, and which has to be added to the list of those problems awaiting explanation. In fact, QCD processes produce corrections to any 'bare' θ value, so that a small value of θ is **unnatural**, in the sense that fine tuning is required to produce a small residual value.

The most commonly accepted idea for solving this **strong CP problem** was proposed by Peccei & Quinn (1977): θ is driven to zero through the breaking of a new symmetry. If this symmetry is global, its breaking will as usual produce a Goldstone boson, the **axion**, which is one of the most commonly cited candidates for dark matter in cosmology. The **Peccei–Quinn symmetry** is an axial version of a simple phase rotation, and so is denoted $U(1)_{\rm PQ}$. Quark and scalar fields behave as follows:

$$\begin{aligned} q &\to e^{i\beta\gamma^5}q \\ \phi &\to e^{-2i\beta}\phi. \end{aligned} \tag{8.103}$$

For the quarks, this is the same as chiral symmetry, which only applies in the limit of zero quark masses. Because real quarks do have mass, the extra transformation of the newly introduced Higgs field is needed to keep QCD gauge invariant. Although we shall not go into the details, the effect of breaking this symmetry is to add a term to the Lagrangian of the same form as the CP-violating term, but in which θ is now the phase of ϕ. In the normal Higgs mechanism, the phase of ϕ is left free; this would not help here, since we want to force θ to a particular value. A potential that achieves this is

$$V(\phi) = \frac{1}{2}\lambda(|\phi|^2 - v_{\rm PQ}^2)^2 + m_a^2 v_{\rm PQ}^2(1 - \cos\theta). \tag{8.104}$$

Here, θ is the phase of ϕ, choosing the origin so that $\theta = 0$ corresponds to the value desired for solving the strong CP problem. What this says is that, rather than being a degenerate circle of constant energy, the $|\phi| = v_{\rm PQ}$ circle is slightly tilted, so that there is a preferred phase. The first part of the potential forces $|\phi| = v_{\rm PQ}$; for small θ, there is a small transverse component of ϕ, $\phi_\perp \simeq \theta|\phi|$, and the second term in the potential is just a mass term for this degree of freedom, $V \simeq m_a^2 \phi_\perp^2/2$.

This sounds monumentally contrived, but things are better than they seem, because this potential does not have to be put in by hand. Although it would unfortunately get too messy to prove it here, QCD can generate the required potential automatically via **instanton** effects, which correspond to quantum tunnelling between different vacuua in a way that cannot be treated by perturbation theory (see e.g. chapter 5 of Collins, Martin & Squires 1989; Turner, Wilczek & Zee 1983). These effects arise as part of the quark–hadron phase transition, at energies of $\lambda_{\rm QCD} \simeq 200\,{\rm MeV}$. The dimensional part of the θ-dependent potential can depend only on this energy scale, i.e.

$$m_a^2 v_{\rm PQ}^2(\cos\theta - 1) \sim \lambda_{\rm QCD}^4(\cos\theta - 1), \tag{8.105}$$

so that the axion mass can be very small if $v_{\rm PQ}$ is assumed to correspond to new physics at $\gg \lambda_{\rm QCD}$.

$$\boxed{m_a \sim \frac{\lambda_{\rm QCD}^2}{v_{\rm PQ}}.} \tag{8.106}$$

Since the axion has a very small mass, it may seem that it should be completely unimportant in cosmology, in the sense of contributing $\Omega \ll 1$, but this is not so. The reason is that the axion is very weakly coupled; when the axion mass term switches on, the field never comes to thermal equilibrium and so a density cannot be obtained by the usual route of multiplying a neutrino-like number density by the axion mass. In fact, it is more reasonable to think of the axion as a classical field performing coherent oscillations in its potential. The coherence is ensured if inflation has occurred: throughout the universe, the Higgs field then lies at some angle of order unity from the minimum of the potential (without inflation, this would be a different angle in different parts of space – see chapters 10 and 11). The subsequent evolution is reminiscent of what happens in inflation: the field obeys an equation of motion set by the Lagrangian, and 'rolls down' the potential. In a parabolic potential, the classical solution is simple harmonic motion with a period $\sim m_a^{-1}$, although these oscillations are damped away by the expansion of the universe, leaving the field sitting at the $\theta = 0$ minimum of the potential. The initial energy density is $V(\phi) \sim m_a^2 v_{\rm PQ}^2$ (if the initial angle is of order unity, the transverse field is $\sim v_{\rm PQ}$), and so the proper number density of axions is $V(\phi)$ divided by the mass:

$$n_a \sim m_a v_{\rm PQ}^2. \tag{8.107}$$

To see if this corresponds to a classical field, we have to ask whether there are many quanta per cubic wavelength. If inflation did happen, the wavelength is as good as infinite, so this is certainly satisfied. The worst case would be for an axion momentum

$p_a \sim m_a c$, so that the wavelength in natural units is m_a^{-1} and there are $N \sim (v_{\rm PQ}/m_a)^2$ quanta per cubic wavelength; the axion field therefore produces a Bose condensate, which will act as cold dark matter.

Before using the above number density with $m_a \sim \lambda_{\rm QCD}^2/v_{\rm PQ}$ in order to get the density of relic axions, we need to take into account one subtlety. The axion mass term switches on at $T \sim \lambda_{\rm QCD}$, but this does not happen instantaneously; the field can start to move down the potential as soon as there is any mass term at all, and the above oscillatory behaviour can happen before the mass has built up to its full value. This phenomenon is analysed in detail in chapter 10 of Kolb & Turner (1990), but the main effect arises by arguing that the relevant effective axion mass must be such that the period of the oscillations is of order the age of the universe at $T \sim \lambda_{\rm QCD}$ (these are the most rapid oscillations for which the axion field has had time to 'see' the potential and move significantly). In natural units, the radiation-dominated time–temperature relation is

$$t/{\rm GeV}^{-1} = g_*^{-1/2}\,(T/10^{9.283}\,{\rm GeV})^{-2}, \tag{8.108}$$

where g_* is the number of equivalent relativistic boson species – about 60 just prior to $\lambda_{\rm QCD}$ (see chapter 9). Using $t(\lambda_{\rm QCD})$ for the period of oscillation of the axion field gives $m_a^{\rm eff} \simeq 10^{-10.1}$ eV. This is the mass that must be inserted in the expression for the number density; scaling this as $a(t)^{-3}$ to the present and multiplying by the low-temperature axion mass gives the present-day mass density

$$\rho_0 \sim (2.73\,{\rm K}/\Lambda_{\rm QCD})^3\, m_a\, m_a^{\rm eff}\, v_{\rm PQ}^2, \tag{8.109}$$

which scales as m_a^{-1} if we eliminate $v_{\rm PQ}$, because $m_a \sim \lambda_{\rm QCD}^2/v_{\rm PQ}$. Putting numbers into this equation to deduce Ωh^2 gives a critical m_a of $\sim 10^{-3}$ eV for $\Omega h^2 = 1$. Exact calculations give a somewhat smaller number:

$$\boxed{\Omega h^2 \simeq \left(\frac{m_a}{10^{-5}\,{\rm eV}}\right)^{-1}.} \tag{8.110}$$

A cosmologically interesting axion mass therefore requires a rather specific energy scale for symmetry breaking: $v_{\rm PQ} \simeq 4\times 10^{12}$ GeV. It is worth noting that the scale cannot be much less than this, since astrophysical arguments such as the possibility that axions are emitted by stars limit the axion mass to $\lesssim 0.01$ eV (again, see chapter 10 of Kolb & Turner 1990).

Finally, the axion provides one of the few examples of a reasonably well-motivated model that generates **isocurvature** density fluctuations. These are modes orthogonal to the usual adiabatic density fluctuations, for which there are equal fractional perturbations in number density for both radiation and nonrelativistic matter. A universe with adiabatic perturbations is one in which the equation of state is constant, so that perturbations of the total energy density exist, leading to perturbations in the spatial curvature. In contrast, isocurvature initial conditions start with a constant energy density, with equal and opposite fluctuations in matter and radiation

$$\delta\rho_{\rm matter} = -\delta\rho_{\rm rad}. \tag{8.111}$$

Rather than an initial perturbation to the spatial curvature, there is a perturbation to the equation of state, corresponding to a spatially varying entropy per baryon (leading to the alternative name of **entropy perturbations**.

It is relatively straightforward to see that such fluctuations will be generated in the axion model, provided inflation occurred. In chapter 11, it is proved that all quantum fields emerge from inflation bearing a spectrum of irregularities that are a frozen-in form of the quantum fluctuations that exist on small scales during the early parts of the inflationary phase. These are approximately **scale invariant** in the sense that, when the perturbations are decomposed into plane waves, the rms fluctuation from modes in unit range of log wavelength is independent of wavelength:

$$\delta\phi = \frac{H}{2\pi}, \tag{8.112}$$

where H is the Hubble parameter during inflation. This argument should apply to the perpendicular part of the Peccei–Quinn Higgs field, so that the axion field starts moving towards $\theta = 0$ from different values of θ in different locations:

$$\delta\theta = \frac{\delta\phi_\perp}{v_{\rm PQ}} = \frac{H}{2\pi\, v_{\rm PQ}}. \tag{8.113}$$

Since the energy density looks like $V = m_a^2 v_{\rm PQ}^2 \theta^2/2$, we see that $\delta V/V = \delta\theta/\theta$, which is the same as $\delta\theta$, because the starting angle is random and therefore differs from zero by typically ~ 1 radian. These fractional energy fluctuations arise as the mass switches on at late times, and so causality dictates that they must be isocurvature in nature. The fractional fluctuation in axion number density is the same as this energy fluctuation:

$$\boxed{\frac{\delta n_a}{n_a} = \frac{H}{2\pi v_{\rm PQ}}.} \tag{8.114}$$

In contrast to adiabatic fluctuations, which produce a scale-invariant spectrum of *potential* fluctuations, these are scale-invariant density fluctuations. However, the two modes evolve differently, with the large-scale isocurvature modes suffering no gravitational growth, so that the matter power spectra in both cases are rather similar by the time of matter–radiation equality (see chapter 15). This means that $\delta n/n$ should be of the order of 10^{-5} if the model is to be viable. In natural units, $H = \sqrt{V_{\rm I}}/E_{\rm P}$, where $E_{\rm P}$ is the Planck energy, and $V_{\rm I} \sim E_{\rm I}^4$ is the inflationary energy density, in terms of the energy scale of inflation. With the $\Omega h^2 = 1$ figure for the Peccei–Quinn scale, sensible density fluctuations require $E_{\rm I} \sim 5 \times 10^{13}$ GeV. As shown below, this is somewhat on the low side, but it is clearly encouraging that the basic idea comes at all close to giving sensible numbers. The detailed shape of the isocurvature spectrum is confronted with observation in chapter 16 and fares reasonably well. The main difficulty for what is otherwise a very attractive model comes with the microwave background. The need for compensated perturbations yields an extra source of radiation fluctuations, and it appears that these are ruled out (see chapter 18).

8.8 Beyond the standard model

The standard model of particle physics described in the previous chapter is an enormous achievement, but few if any physicists would say that it can be the last word in the understanding of fundamental processes. There are two distinct outstanding questions.

(1) Gravity. The gravitational field is not included in the calculational apparatus of field theory, except as an external purely classical interaction. This must be incorrect in two ways: there will be times when the classical field is strong enough that the usual formalism based on Minkowski spacetime becomes inappropriate. Furthermore, the gravitational field itself should be quantized in some way, by analogy with all other fields. The full theory of quantum gravity still presents formidable difficulties, but some hints are available about the features of any eventual theory, and some of these are explored below. Particularly successful has been the half-way house of treating quantum fields in curved space, where the gravitational field is still unquantized, but no longer weak. This area is of especial importance for cosmology through the phenomenon of Hawking radiation, first made familiar through Hawking's extremely influential discovery that black holes emit particles.

(2) Elegance. Even excluding gravity altogether, many physicists feel that the standard model is incomplete on aesthetic grounds. This is a less straightforward objection; the standard model achieves all that can be demanded from a theory, in that it allows the outcome of any experiment to be calculated (in principle). However, there are many features of the model that seem to cry out for a deeper explanation, in much the same way as did the similarities between electricity and magnetism prior to Maxwell. The guess of many physicists is that the standard model is only the low-energy manifestation of some deeper and simpler theory. Again, such models are of cosmological importance, through their relevance to the matter–antimatter asymmetry in the universe.

It is entirely possible that any satisfactory solution to these objections will be holistic and solve both in an interlinked fashion. In the absence of such a solution, it is usual to keep them separate initially. We therefore start with possible unified generalizations of the standard model, before bringing in gravity.

GRAND UNIFIED THEORIES The task of **GUTs** is to provide some explanation for the form of the standard model, which has many features that it would be nice to explain, rather than assume arbitrarily.

(1) Why are there three distinct coupling constants in the model, and what sets their relative strengths?

(2) Why are there three families or generations of quarks and leptons, when any number of either would seem possible?

(3) Why is the model (and the real world) left/right asymmetric?

(4) Why is charge quantized? There seems no reason why the charges of hadrons (i.e. of quarks) and of leptons should be related, since they lie in different multiplets.

(5) What is the origin of all the other parameters of the model? Apart from the three coupling constants, there are 16 other numbers that must be set by hand (nine fermion masses, four mixing parameters in the KM matrix, two Higgs potential parameters and the QCD θ). There is a widespread feeling that this is too many degrees of freedom for an ultimate theory (although different individuals will differ over whether a satisfactory number would be zero, one, or a few).

The main idea behind grand unification is to work by analogy with the success of the standard model. We postulate that the laws governing particle physics obey some as yet undiscovered symmetry, which is hidden in the present universe through the Higgs mechanism in which a scalar field (or set of fields) acquires a non-zero vacuum expectation value. What do we mean by a symmetry in this context? In the $SU(2)$ part of the standard model, we write wave functions as multiplets such as

$$\psi = \begin{pmatrix} e^- \\ \nu_e \end{pmatrix}, \tag{8.115}$$

which corresponds to the assumption that the electron and its neutrino are merely two different states of the same underlying entity, in the same way that electrons of spin up and spin down are related; similar arguments apply to multiplets of quarks.

However, although the standard model consists of three families each containing one lepton multiplet and one quark multiplet, the leptons and hadrons are distinct, and only the hadron multiplets appear in the part of the Lagrangian devoted to $SU(3)$ and the colour symmetry. In conventional particle physics, this is dignified by the conservation of baryon number: leptons and quarks cannot be transmuted into each other. Nevertheless, this conservation law has never had the same status as conservation of charge or mass, both of which are generated by simple global symmetries of the Lagrangian and relate to the production of long-range forces. There has thus been an increasing willingness to breach what may be an artificial barrier and contemplate the possibility that quarks and leptons may be related. The basic idea of **grand unification** is then to write, symbolically,

$$\psi = \begin{pmatrix} \text{quark} \\ \text{lepton} \end{pmatrix}. \tag{8.116}$$

The idea now is to find some way of accommodating all the various particles of the standard model into a set of multiplets and to write down a Lagrangian that is invariant under the operation of some group G. To complete the recipe, scalar Higgs fields are then added as required in order to break the symmetry, leaving the lower symmetry of the standard model:

$$G \to SU(3) \otimes SU(2) \otimes U(1). \tag{8.117}$$

To avoid a conflict with the existing body of accelerator data, any such symmetry-breaking scale must lie well above the maximum energy studied to date: $E_{\rm GUT} \gg M_W \sim$ 100 GeV.

EMPIRICAL EVIDENCE FOR UNIFICATION Such unification may seem a rather wild notion, so it is worth pointing out that we do have an empirical hint that the phenomenon might exist. This takes the form of the variation with energy of the coupling constants in the standard model, which are logarithmically varying functions described by renormalization group equations of the form

$$\frac{1}{\alpha(E_1)} = \frac{1}{\alpha(E_2)} + \frac{b}{2\pi} \ln\left(\frac{E_2}{E_1}\right), \tag{8.118}$$

where

$$\boxed{\alpha \equiv \frac{g^2}{4\pi}} \tag{8.119}$$

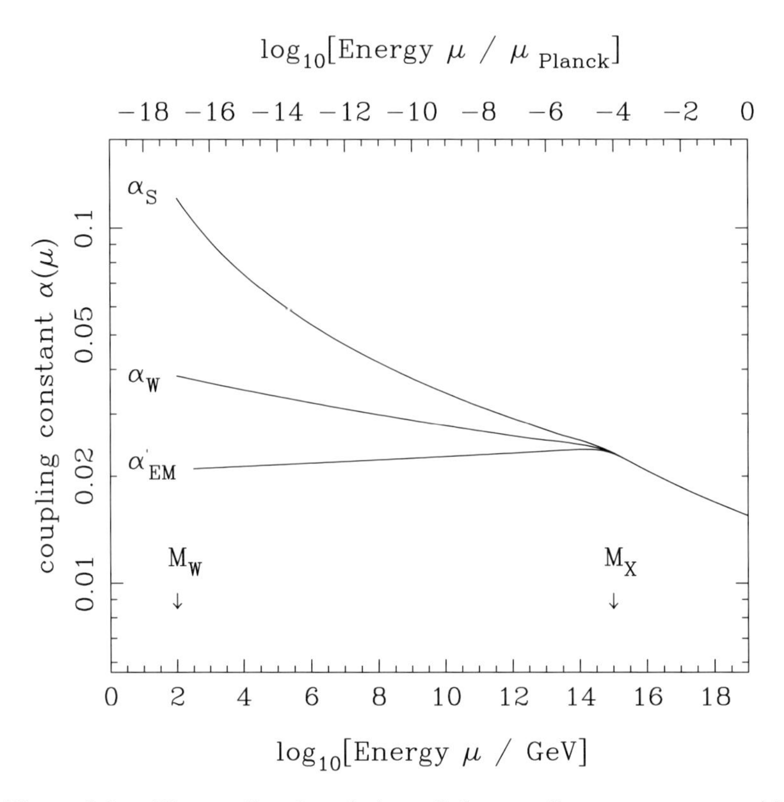

Figure 8.6. The predicted variation of the coupling constants α_i with energy scale (here denoted by μ), according to the $SU(5)$ GUT. All three interactions reach comparable strength at a scale around $M_X \sim 10^{15}$ GeV (adapted from figure 2 of Iliopoulos 1990).

is an alternative way of quoting the strength of a given interaction, in terms of its coupling constant. For laboratory energies near M_W, the measured values of α are rather different: roughly 137, 25, 10. However, their energy dependences are different also, the electromagnetic coupling increasing (i.e. $1/\alpha$ declining) with energy, but the two non-Abelian couplings changing in the opposite sense. The b coefficients are (e.g. chapter 7 of Collins, Martin & Squires 1989)

$$\begin{aligned} b[U(1)] &= -\tfrac{4}{3} N_g \\ b[SU(N)] &= \tfrac{11}{3} N - \tfrac{4}{3} N_g, \end{aligned} \tag{8.120}$$

where N_g is the number of distinct families (three in the standard model). The renormalization group equation thus implies a linear dependence of $1/\alpha$ on ln(energy), whose slope around $E \lesssim M_W$ is known just from the low-energy structure of the theory. Given three coupling constant values $\alpha_i(M_W)$ and extending them linearly to higher energies, we would not in general expect the lines to intersect, but they do (see figure 8.6). This fact strongly suggests that there is something special about the energy of the intersection (M_X): at this point, the strengths of all three fundamental interactions become equal:

$$\boxed{\alpha_i(M_X) \simeq 1/40, \quad M_X \simeq 10^{14} - 10^{15}\ \text{GeV}.} \tag{8.121}$$

This intersection is easily understood only if the present low-energy values of the coupling constants do indeed arise from grand unification, and provides the best indication beyond sheer aesthetics that the concept is valid. In detail, the way the constants join at high energy depends on the GUT model. It has been suggested that the coupling constants extrapolated from laboratory energies do not join at precisely the same energy, indicating the need for extra physics such as supersymmetry (Amaldi *et al.* 1991). However, this does not alter the basic point that the strengths of the interactions become very similar as the M_X scale is approached.

The vast energy scale of unification makes it no surprise that direct evidence of GUT phenomena has yet to turn up in the laboratory. That such high energies are involved is almost inevitable given the disparity in coupling constants, plus the fact that they change only logarithmically with energy. This is nevertheless a radical prediction, which may seem unrealistic in the context of experience at lower energies. Here, almost every decade of energy from the mass of the electron (5×10^{-4} GeV) to the mass of the Z boson (91 GeV) contains new particles. One might then expect that TeV energies would open up new physical phenomena, and so on indefinitely. The prediction of the GUT idea is that, instead, we next encounter a **desert** of at least 12 powers of 10 in energy in which nothing interesting happens. This would certainly be convenient, given that it is hard to see how the maximum energy of accelerators can affordably be increased much further, barring a totally new technology.

The other suspicious aspect of the GUT hypothesis is that the unification scale is so close to the Planck scale. This can be argued both ways: $10^{-4}E_{\text{Planck}}$ might be made to sound like an energy at which quantum gravity can be neglected, but one can also ask why the two scales are so close if they are not connected. The great impact of GUTs on cosmology has been the opening of the possibility that the great mysteries of the initial conditions of the big bang might be solved without recourse to quantum gravity, but there is not much room for manoeuvre in raising the GUT energy, should more detailed modelling require this.

SIMPLE GUT MODELS What are the criteria for a viable GUT model? The main thing that has to be decided is the group to be used. Following that, it is a question of working out how all the the known particles can be accommodated into multiplets that correspond to the group representations, adding a Higgs sector to break the symmetry, and then examining the low-energy structure of the model to see if it reduces to known particle physics.

One immediate problem is that there are infinitely many Lie groups that could be chosen. These have been classified by Cartan into four distinct categories of **simple groups** (those that cannot be expressed as a direct product of other groups). The first two are the familiar **special orthogonal group** and the **special unitary group** $SO(n)$ and $SU(n)$, which leave invariant the scalar product of two n-dimensional vectors that are respectively real and complex. There is in addition the **symplectic group** $Sp(2n)$, which leaves invariant the skew-product of two $2n$-dimensional real vectors:

$$\tilde{\mathbf{y}} \cdot \mathbf{S} \cdot \mathbf{x} = \text{invariant}$$
$$\mathbf{S} = \text{diag}\left(\begin{bmatrix} 0 & 1 \\ -1 & 0 \end{bmatrix} \begin{bmatrix} 0 & 1 \\ -1 & 0 \end{bmatrix} \cdots \right). \tag{8.122}$$

The fourth category is the catch-all for those that do not fit into the first three: the **exceptional groups**, of which there are five: G_2, F_4, E_6, E_7 and E_8.

As well as the **order** of a group (the number of generators), a group can also be quantified by its **rank**, which is the number of generators that can be diagonalized simultaneously. The rank of the standard model is four, so we seek a group of at least this rank, and there are five possibilities: $SU(5)$, $SO(8)$, $SO(9)$, $Sp(8)$ and F_4 (more if we extend the list to include direct products of the same group; it is then possible to argue that there should still only be a single gauge coupling constant, which is what gives the requirement for a simple group). All these five are possible minimal candidates for the GUT group, but it is possible to narrow the list to one by demanding that the group must admit **complex representations**; this turns out to be satisfied for $SU(5)$ alone out of this list. This requirement is introduced by the known existence of parity violation: in the standard model, left- and right-handed fermions are assigned to different representations ($SU(2)$ doublets and singlets, respectively). The operation of charge conjugation (which changes the helicity) involves complex conjugation, and so complex representations are required in order to describe a theory in which the left- and right-handed parts are to be distinct. Thus, $SU(5)$ is the only possible rank-4 GUT, and so it is the natural model to consider if we follow the usual line of making things as simple as they possibly can be.

We start by outlining the main properties of the $SU(5)$ model. There are $5^2-1 = 24$ distinct generators (which can be represented by 5×5 matrices), and so there must be this number of gauge bosons in the theory. The standard model contains 12 bosons: eight gluons, $W^\pm$, Z^0 and the photon. There must therefore be 12 new bosons, which comes down to two: the X and Y bosons, together with their antiparticles, all in three different colour versions. The Lagrangian for the theory can be written down almost immediately along the usual lines, once we have decided where the fermions are to go. This is something of a contortion, given that we are dealing with representations involving either 5D vectors or 5×5 matrices. The particles from one family are contained in a mixture of these:

$$\psi \equiv \begin{pmatrix} d_1^c \\ d_2^c \\ d_3^c \\ e^- \\ -\nu_e \end{pmatrix}_{\rm L}, \quad \chi \equiv \frac{1}{\sqrt{2}} \begin{pmatrix} 0 & u_3^c & -u_2^c & -u_1 & -d_1 \\ -u_3^c & 0 & u_1^c & -u_2 & -d_2 \\ u_2^c & -u_1^c & 0 & -u_3 & -d_3 \\ u_1 & u_2 & u_3 & 0 & -e^+ \\ d_1 & d_2 & d_3 & e^+ & 0 \end{pmatrix}_{\rm L}, \tag{8.123}$$

where the numerical indices denote colour labels, and the superfix c denotes charge conjugation.

Having specified the particle content, the construction of the model proceeds in an analogous way to the electroweak theory, but there is no space here to justify the details (see e.g. chapter 7 of Collins, Martin & Squires 1989). The main feature is the Higgs sector, in which ϕ is a 5×5 matrix that acquires the vacuum expectation value

$$\phi \propto \text{diag}(1, 1, 1, -3/4, -3/2), \tag{8.124}$$

in order to break $SU(5)$ but leave the residual symmetry of the standard model.

SUCCESSES AND FAILURES OF GUT MODELS The $SU(5)$ model makes many predictions, most of which are generic to GUTs. GUTs are theories that explicitly violate conservation of baryon number, allowing the interesting new possibility of **proton decay**:

$$p \to e^+\pi^0 \to e^+\gamma\gamma. \tag{8.125}$$

The rate for such a process can be calculated in a way analogous to the weak decay of the neutron (see chapter 9), where τ_n in natural units is $\sim g_{\text{weak}}^{-4} M_W^4/m_e^5$. In this case, the masses of interest are the GUT mass and the mass of the heavier decay product (which is the π^0, but this mass is of order the proton mass). We therefore expect

$$\boxed{\tau_p \sim \frac{(\hbar/c^2)}{\alpha_{\text{GUT}}^2} \frac{M_{\text{GUT}}^4}{m_p^5} \sim 10^{30}\ \text{yr}.} \tag{8.126}$$

Perhaps surprisingly, it is feasible to see such extremely rare decays with about 1000 tonnes of water, by looking for the Cerenkov radiation from the photons generated by the final pion decay. Such experiments have set limits on the proton lifetime of $\tau_p \gtrsim 10^{32}$ yr. The prediction for the proton lifetime seems quite generic and not tied to a specific GUT group, so failure to see decays is something of a setback for the unification idea. Of course, models can be constructed in which the lifetime is increased by a moderate amount, and supersymmetric GUT models are capable of evading the lifetime limit. The failure of simple models such as $SU(5)$ can therefore be taken as pointing to the need for some extra ingredient such as supersymmetry (see below).

More generally, there are structural reasons why GUTs must still be a long way from being a convincing ultimate theory. The main success of GUTs is to explain why the electric charges of e.g. electrons and protons are the same, and to reduce somewhat the number of free parameters, but they give no answers to some other simple features of the standard model. Perhaps the greatest defect of GUTs is that they provide no natural explanation for the violation of parity. This has to be imposed by hand in weeding out groups that lack complex representations. Left–right asymmetry therefore remains one of the most outstanding mysteries of the universe. Neither do GUTs account for the fact that there are three generations.

In any case, as shown by the lack of proton decay, $SU(5)$ GUTs do not correspond to the real world. Therefore, the 'keep it as simple as possible' principle – which worked so brilliantly in allowing fundamental achievements such as Einstein's field equations or Dirac's wave equation – suffers its first serious failure. This has not prevented interest in the GUT concept, as there are plenty of groups of higher rank; for example, simple rank-5 groups with complex representations are $SU(6)$, $SO(10)$. However, it will now be difficult to argue that the simplest of these that is consistent with observation must be correct; some higher principle will be needed to select the true theory.

8.9 Neutrino masses and mixing

Another intriguing practical consequence of GUTs is that they allow neutrinos to have the non-zero masses that are forbidden in the standard model. For a single neutrino species, the mass term in the Lagrangian is $\mathcal{L} = -m\bar{\nu}\nu$ (writing ν for the neutrino wave function). In terms of the chiral fields

$$\nu_{\rm R,L} = \frac{1 \pm \gamma^5}{2}\,\nu, \tag{8.127}$$

this is $\mathscr{L} = -m(\bar{\nu}_{\rm L}\nu_{\rm R} + \bar{\nu}_{\rm R}\nu_{\rm L})$. What happened to the $\bar{\nu}_{\rm L}\nu_{\rm L}$ and $\bar{\nu}_{\rm R}\nu_{\rm R}$ terms? These vanish identically: $\bar{\nu}_{\rm L} = \bar{\nu}(1+\gamma^5)/2$, so $\bar{\nu}_{\rm L}\nu_{\rm L} \propto (1+\gamma^5)(1-\gamma^5)$, but $\gamma^5\gamma^5 = 1$. The generalized mass term for the case of many neutrino species is written as the **Dirac mass term**

$$\mathscr{L} = -\bar{\nu}_{\rm L}\cdot\mathscr{M}\cdot\nu_{\rm R} + \text{Hermitian conjugate}, \tag{8.128}$$

where now

$$\nu = \begin{pmatrix}\nu_e\\ \nu_\mu\\ \nu_\tau\end{pmatrix} \tag{8.129}$$

and the 3×3 Hermitian matrix $\mathscr{M}$ is the **mass matrix**. For Majorana neutrinos, $\nu_{\rm R} \to C\nu_{\rm L}$ (where C is the charge-conjugation operator) and the Lagrangian is multiplied by a factor 1/2.

NEUTRINO OSCILLATIONS Since there is apparently nothing that prevents the mass matrix being non-diagonal, the mass eigenstates involved may not be the states that we recognize in experiments as the distinct neutrino types. This phenomenon has already been seen with quark mixing, where the weak and strong interactions effectively have different definitions of quarks. As long as the neutrinos are massless, any mixing can be defined away by making new linear combinations of fields; once they have mass, the mixing will have observable consequences. For simplicity, suppose just the e and μ neutrinos mix, so that we can write

$$\begin{pmatrix}\nu_e\\ \nu_\mu\end{pmatrix} = \begin{pmatrix}\cos\theta & \sin\theta\\ -\sin\theta & \cos\theta\end{pmatrix}\begin{pmatrix}\nu_1\\ \nu_2\end{pmatrix}, \tag{8.130}$$

where ν_1 and ν_2 are the mass eigenstates

$$\nu_i(t) = \nu_i(0)\exp(-iE_i t/\hbar). \tag{8.131}$$

In the case of perfect mixing ($\theta = 45^\circ$), it is easy to see that this means that the apparent type of neutrino will oscillate with time. When $|E_1 - E_2|t/\hbar = \pi$, the relative signs of ν_1 and ν_2 will reverse and initial e neutrinos will now be detected as μ neutrinos.

If there is only partial mixing, there are still oscillations, but they do not change the neutrino type completely. Suppose we start with a pure ν_e state, so that the state at time t is

$$\psi(t) = \nu_1(0)\cos\theta\exp(-iE_1 t/\hbar) + \nu_2(0)\sin\theta\exp(-iE_2 t/\hbar). \tag{8.132}$$

From the definition of the initial mixed states, the matrix element for observing ν_μ at some later time is

$$\langle\nu_\mu(0)|\psi(t)\rangle = \sin\theta\,\cos\theta\,[-\exp(-iE_1 t/\hbar) + \exp(-iE_2 t/\hbar)], \tag{8.133}$$

so that the probability of observing an oscillation is

$$\boxed{P(\nu_e \to \nu_\mu) = \tfrac{1}{2}\sin^2(2\theta)\,\{1 - \cos[(E_1 - E_2)t/\hbar]\}\,;} \tag{8.134}$$

sensibly enough, oscillation is improbable in the limit of small mixing.

Since neutrinos energetic enough to be observed are almost certainly ultrarelativistic, we can expand the energy as follows:

$$E = (p^2c^2 + m^2c^4)^{1/2} \simeq pc + \frac{m^2c^3}{2p} \simeq pc + \frac{m^2c^4}{2E}. \qquad (8.135)$$

Since the momentum does not change as the state oscillates, ΔE measures $\Delta(m^2)$ (not to be confused with $(\Delta m)^2$). Rather than seeing oscillations in time, it is more practical to view ct as the travel time from the source that generates the neutrino to the experiment that detects the possible oscillations. We have therefore shown that the spatial wavelength of any oscillations will be

$$\boxed{L = \frac{4\pi E\hbar}{c^3\Delta(m^2)} \simeq 2.5\,\mathrm{m}\ \frac{E/\mathrm{MeV}}{\Delta(m^2)/(\mathrm{eV})^2}.} \qquad (8.136)$$

For moderate mass differences, oscillations can be looked for in reactor-based experiments. The claim by Reines *et al.* (1980) of a positive detection with $\Delta(m^2) \simeq 0.7\,\mathrm{eV}^2$ and near-maximum mixing (subsequently shown to be incorrect) was historically an extremely important stimulus for cosmologists to contemplate universes dominated by nonbaryonic dark matter. Present experiments yield only null results (see e.g. chapter 7 of Kim & Pevsner 1993). For $\sin^2 2\theta = 1$, the limits for ν_e and ν_μ mixing are roughly $\Delta(m^2) \lesssim 0.02\,\mathrm{eV}^2$ and $\Delta(m^2) \lesssim 0.1\,\mathrm{eV}^2$, increasing by approximately a factor of ten at the smallest angles probed ($\sin^2 2\theta \simeq 0.01$).

SOLAR NEUTRINOS Smaller mass-squared differences can be probed by longer baselines, and the detection of neutrinos emitted from the sun is a subject with a 20-year history. Right from the start, these neutrino measurements have failed to match astrophysical predictions, and this subject is presently hardening into a consensus that a non-zero neutrino mass is the best way of explaining the results (see Bahcall 1989; Raghavan 1995; chapter 15 of Kim & Pevsner 1993).

Although we lack direct information on the core of the sun, where neutrinos would be generated, some aspects of the neutrino emission can be predicted with reasonable confidence. The energy output of the sun is known, and so the flux of neutrinos from the main energy-generating process (*pp* fusion) can be predicted directly (barring the silly possibility that nuclear reactions ceased a few decades ago, leaving the sun to shine by outward diffusion of photons). There are several other more obscure nuclear reactions that are important in generating the more energetic neutrinos, which are the easiest to detect, and it is the differing sensitivities of different experiments to these different contributions that has sharpened the Solar neutrino debate. The three principal mechanisms are listed in table 8.2: *pp* fusion; beryllium electron capture; boron decay.

Notice the different characteristic energies of these processes, and that Be neutrinos are monoenergetic (because the final state involves only two bodies), whereas the other two have continuous spectra. Also shown in the table are capture rates in **Solar neutrino units** (1 SNU $= 10^{-36}$ captures per nucleus per second) for two detector nuclei of practical interest. These are predicted from a detailed 'standard' Solar model (Bahcall 1989). Other processes make some contribution, but the above reactions make up respectively 93% and 85% of the predicted chlorine (threshold energy 0.814 MeV) and gallium (threshold energy 0.235 MeV) rates.

Table 8.2. Predicted Solar neutrino signals

Reaction	E/MeV	^{37}Cl/SNU	^{71}Ga/SNU
$p + p \to D + e^+ + \nu_e$	< 0.42	0.0	70
$^7\mathrm{Be} + e^- \to {}^7\mathrm{Li} + \nu_e$	0.86	1.2	36
$^8\mathrm{B} \to 2^4\mathrm{He} + e^+ + \nu_e$	< 14	6.2	14

The first detection of Solar neutrinos was via the chlorine process, based on 615 tons of cleaning fluid (C_2Cl_4) in the Homestake mine experiment of R. Davis. This triumph of experimental physics detects fewer than 200 neutrinos per year by trapping individual argon atoms, and has converged on a rate of 2.3 ± 0.3 SNU (as against a prediction of 8 SNU). This discrepancy of a factor 3 is what would be expected if electron neutrinos oscillate incoherently over the Sun–Earth distance, ending as an equal mixture of the three types. However, because the signal is mainly from boron neutrinos, it is also possible that a small change in the Solar models is indicated: only 0.1% of the Sun's energy is generated in this process.

In contrast, gallium experiments have 70 of a predicted 132 SNU from the less energetic *pp* neutrinos, giving a more direct probe of the main energy-generating reactions. However, again a deficit is seen, with an observed signal of 78 ± 10 SNU, barely allowing any contribution other than the *pp* neutrinos (70 SNU). In particular, the expected 35 SNU from Be neutrinos must be suppressed by at least a factor 5. This is a critical sign that neutrino physics, rather than astrophysics, is the solution to the Solar neutrino problem: changes in the Solar model would tend to change the B and Be rates in tandem.

Finally, Solar neutrinos are also detected by several large experiments that are primarily aimed at detecting proton decay. These consist of several thousand tons of water in underground caverns lined with photomultiplier tubes aiming to detect Cerenkov radiation from decay-produced positrons. For neutrino energies of several MeV, the process $p + \bar{\nu}_e \to n + e^+$ also gives an observable signature. The Kamiokande experiment, based on 3000 tons of water, has measured the neutrino flux at 7.5 MeV (entirely B neutrinos), and finds a signal that is $51 \pm 7\%$ of the model prediction. So, not only is the Be/B ratio incorrect, even the B flux suppression is a function of energy.

MSW EFFECT One way of accounting for these observations is via oscillations in which the oscillation length at 7.5 MeV is approaching one astronomical unit (a.u.), so that these high-energy neutrinos are only moderately affected; this implies $\Delta(m^2) \simeq 10^{-11}\ \mathrm{eV}^2$. Large mixing angles are also needed in order to achieve the large Be suppression: $\sin^2(2\theta) \simeq 0.8$. However, there exists an alternative and more attractive explanation, which is that the usual oscillation analysis may be invalid because it treats the oscillations as occurring in a vacuum. This is fair enough for ν_μ and ν_τ, but the Sun contains a large density of electrons and protons, which can interact with the ν_e. This is the Mikheyev–Smirnov–Wolfenstein (MSW) effect, which in outline goes as follows (see chapter 9 of Bahcall 1989 for more details).

The weak interaction gives a phase change that can be read off immediately from the Fermi Lagrangian:

$$\hbar \frac{d\phi}{dt} = \langle |H_{\rm F}| \rangle = \sqrt{2} G_{\rm F} n_e, \tag{8.137}$$

and this can be incorporated into the equation of motion for the neutrino mixture. Being explicit about the difference between neutrino states and expansion coefficients, write

$$|\nu(t)\rangle = a_e(t)|\nu_e\rangle + a_\mu(t)|\nu_\mu\rangle, \tag{8.138}$$

where the coefficients a_i obey the oscillation equation of motion

$$i\hbar \begin{pmatrix} \dot{a}_e \\ \dot{a}_\mu \end{pmatrix} = \frac{\pi\hbar c}{L} \begin{pmatrix} \cos 2\theta & -\sin 2\theta \\ -\sin 2\theta & -\cos 2\theta \end{pmatrix} \begin{pmatrix} a_e \\ a_\mu \end{pmatrix}, \tag{8.139}$$

L being the oscillation length defined earlier. The effect of the matter-induced MSW phase rotations is to add to the matrix on the rhs an additional term that equals $(\pi\hbar c/L_{\rm M})\,{\rm diag}(1, -1)$, choosing to eliminate an unmeasurable overall phase. The matter oscillation length is, from above,

$$L_{\rm M} = \frac{\sqrt{2}\pi\hbar c}{G_{\rm F} n_e}, \tag{8.140}$$

which is of order 10^5 m in the core of the Sun, so that matter oscillation effects can be important. The equation of motion has to be integrated through the varying density of the Sun, but it should be obvious that complete conversion of ν_e to ν_μ will be possible even in the case of very small mixing. This is so because the sign of L actually matters: it is negative if $m(\nu_e) < m(\nu_\mu)$, so that the diagonal components of the matrix will vanish if $|L| = L_{\rm M} \cos 2\theta$, and ν_e will start to convert directly to ν_μ.

Applying this insight to the Solar neutrino data opens up another branch of solutions, with $\Delta(m^2) \simeq 10^{-5}\,{\rm eV}^2$ and $\sin^2 2\theta \simeq 10^{-2}$. Future experiments with larger detectors will attempt to discriminate between this solution and simple long-wavelength oscillation, which is unaffected by the MSW process because L is so large. Probably the simplest discriminator will be the elliptical nature of the Earth's orbit: because the oscillatory damping of the monoenergetic Be flux is so carefully tuned by matching L to 1 a.u., there should be seasonal variations in the neutrino flux. Failure to observe any such change would firmly favour the MSW solution to the Solar neutrino problem.

If we take the idea of $\Delta(m^2) \simeq 10^{-5}\,{\rm eV}^2$ seriously, what is the interpretation? The line of least resistance is to assume that the mixing is mainly with the ν_μ, so that $m(\nu_\mu) \simeq 0.003\,{\rm eV}$, and the ν_e mass is much smaller. To justify the assumption that the neutrino masses are unlikely to be all nearly equal, we can appeal to an illustrative means of generating neutrino masses, known as the **seesaw mechanism**, which can be realized in GUT models. In this argument, it is supposed that the neutrino mass (for a single generation) is a combination of the Dirac and Majorana forms:

$$\mathcal{L} = \begin{pmatrix} \bar{\nu}_{\rm L} & \bar{\nu}_{\rm L}^c \end{pmatrix} \begin{pmatrix} 0 & m \\ m & M \end{pmatrix} \begin{pmatrix} \nu_{\rm R}^c \\ \nu_{\rm R} \end{pmatrix}. \tag{8.141}$$

Here, the top left entry in the mass matrix is hypothesized to be explicitly absent. The Dirac mass, m, is generated in the same way as the leptons, and is therefore of order the quark mass in that generation. The mass M must be large to explain why right-handed neutrinos are not seen, and could easily be of order the GUT mass. The interesting thing about this mass matrix is the dynamic range of its eigenvalues, which are approximately M and $-m^2/M$. The unexpected minus sign in the second case can easily be amended by inserting a factor i in the definition of the corresponding eigenfield. The small mass can then easily be a small fraction of an eV for quark masses in the GeV range and $M \sim 10^{15}$ GeV. More precisely, the seesaw mechanism predicts the scaling

$$\boxed{m(\nu_e) : m(\nu_\mu) : m(\nu_\tau) = m_u^2 : m_c^2 : m_t^2.} \tag{8.142}$$

Using this with the above mass for the ν_μ, a τ neutrino mass of around 10 eV would be indicated. As discussed at length in chapters 9, 12 and 15, such a mass would be of the greatest importance for cosmology, as neutrinos of this mass would contribute roughly a closure density in the form of hot dark matter.

SUPERNOVA NEUTRINOS A different way of observing neutrino masses was provided by **supernova 1987A**. Because neutrinos in this case are emitted in a sharp pulse, and because they travel over extremely long paths, it is worth considering the small deviation of their velocities from c:

$$\frac{v}{c} = \left(1 - \frac{1}{\gamma^2}\right)^{1/2} \simeq 1 - \frac{1}{2\gamma^2} = 1 - \frac{m^2c^4}{2E^2}. \tag{8.143}$$

Over a path L, the difference in arrival times is therefore

$$\Delta t = \frac{L}{2c}\Delta\left(\frac{m^2c^4}{E^2}\right), \tag{8.144}$$

which is again sensitive to differences in squared mass for a given energy. Since supernovae in the Milky Way occur on the order of once per century, no sensible government would fund an experiment designed specifically to await the neutrinos from a supernova, but again proton-decay experiments come to the rescue. SN1987A occurred in the Large Magellanic Cloud, approximately 51 kpc distant, which is roughly 10^{10} times as far as the Sun. Nevertheless, the intensity of a supernova is such that the neutrino flux briefly raised itself above the background, resulting in the detection of 24 neutrinos in the space of approximately 10 seconds, with energies in the range 7–35 MeV. These results caused some controversy, and have been interpreted in a variety of ways, ranging from an upper limit of $m(\nu_e) < 30\,\mathrm{eV}$ to claimed detections of non-zero masses ranging from 3 to 10 eV (see e.g. Bahcall 1989). Perhaps the most remarkable aspect of all this is how close these figures are to the laboratory limit of $m(\nu_e) < 15\,\mathrm{eV}$. SN1987A was a triumph for neutrino astronomy, but it is frustrating that it was not just a little closer – unless we are very lucky, it is likely that the issue of neutrino mass will have been settled by the time the next close supernova comes along.

8.10 Quantum gravity

We now have to face the much delayed question of how the ideas of quantum field theory might be applied to gravity itself. The subject of quantum gravity has generated a colossal amount of formalism over the years, but the absence of anything resembling a definitive theory means that it would not be productive to dig too deeply into this. We shall be content here to give just an indication of the deep and intractable difficulties that are encountered on the very first few steps along the quantum-gravity road.

First note the characteristic scale at which the phenomenon must become important. The idea of a flat background spacetime inhabited by quantum fields will not be adequate for particles that are so energetic that their de Broglie wavelength is smaller than their Schwarzschild radius. Equating $2\pi\hbar/(mc)$ to $2Gm/c^2$ yields a characteristic mass for quantum gravity, $m_{\rm P}$; this mass, and the corresponding length $\hbar/(m_{\rm P}c)$ and time $\ell_{\rm P}/c$ form the system of **Planck units** that can be constructed from G, $\hbar$ and c:

$$\boxed{\begin{aligned} m_{\rm P} &\equiv \sqrt{\frac{\hbar c}{G}} \simeq 10^{19}\,{\rm GeV} \\ \ell_{\rm P} &\equiv \sqrt{\frac{\hbar G}{c^3}} \simeq 10^{-35}\,{\rm m} \\ t_{\rm P} &\equiv \sqrt{\frac{\hbar G}{c^5}} \simeq 10^{-43}\,{\rm s.} \end{aligned}} \tag{8.145}$$

The remoteness of these scales from laboratory observations is what allows quantum gravity to be ignored in most contexts.

Faced with a new field theory, the standard method for including it in the quantum fold starts with the Lagrangian: what we have to do is construct a Lagrangian that is of the form appropriate for a general action principle, i.e. $\propto \sqrt{-g}\,\times$ some invariant scalar. For general relativity, we already have this in the form of the Hilbert Lagrangian, $\mathscr{L} \propto \sqrt{-g}\,R$, where R is the Ricci scalar. This is a rather nonlinear function of the field $g^{\mu\nu}$ (terms up to fourth order in g or its derivatives).

To this can be added terms for other fields, which can usually be bootstrapped from a knowledge of the Lagrangian in Minkowski space. For example, the Lagrangian for gravity plus a charged scalar field is virtually the same as the Minkowski form, apart from the $\sqrt{-g}$ factor:

$$\mathscr{L} = \sqrt{-g}\left\{\frac{1}{2\kappa}R + \left[\partial^\mu\phi\partial_\mu\phi^* - (m^2 + \xi R^2)|\phi|^2\right]\right\}. \tag{8.146}$$

Here and below, we shall use the gravitational coupling constant

$$\kappa \equiv \frac{8\pi G}{c^4}. \tag{8.147}$$

Since the field ϕ is a Lorentz scalar, coordinate derivatives are the same as covariant ones in this case: $\phi_{,\mu}$ is a general 4-vector, identical to $\phi_{;\mu}$. The only surprise here is the addition of a term proportional to R^2, which represents an explicit interaction between ϕ and gravity in addition to those included implicitly via the $\sqrt{-g}$ term. Clearly such a term cannot be ruled out, since it vanishes in Minkowski space. The best we can do is to limit the unknown to the above form on dimensional grounds: ξ must be just some number, since we do not want to introduce a new dimensional constant into physics. It is neglected in what follows.

SEMICLASSICAL QUANTUM GRAVITY As a first stab at quantizing gravity, consider the case of weak fields: many of the features of any full theory ought to reveal themselves in this limit. As a slight variant on classical weak-field theory, we define

$$g^{\mu\nu} = \eta^{\mu\nu} + \sqrt{\kappa}\, h^{\mu\nu}, \qquad \sqrt{\kappa}\, h^{\mu\nu} \ll 1 \tag{8.148}$$

(this gives $h^{\mu\nu}$ the dimensions of energy, as with other bosonic fields). Using

$$\bar{h}^{\mu\nu} = h^{\mu\nu} - \tfrac{1}{2}\eta^{\mu\nu} h \tag{8.149}$$

($h = h^{\mu}_{\mu} = -\bar{h}^{\mu}_{\mu}$) plus the general relativity equivalent of the Lorentz gauge, $\bar{h}^{\mu}_{\nu,\mu} = 0$, the low-order Lagrangian becomes

$$\mathcal{L} = \bar{h}^{\mu\nu,\lambda}\bar{h}_{\mu\nu,\lambda} + \left(\partial^{\mu}\phi\partial_{\mu}\phi^{*} - m^{2}|\phi|^{2}\right) + \kappa\left(2\bar{h}^{\mu\nu}\partial^{\mu}\phi\partial^{\nu}\phi^{*} - \bar{h}\, m^{2}\,|\phi|^{2}\right) \tag{8.150}$$

(see e.g. Feynman, Morinigo & Wagner 1995). At the level of special relativity, this is the Lagrangian for a consistent field theory, and it tells us several useful things. Because of the tensor nature of the $\bar{h}_{\mu\nu}$ field, we can use the general idea of the representation of angular momentum to deduce that the quantum of the field (the **graviton**) must have spin 2. The fact that the field is massless (no term in $\bar{h}^{2}$) tells us that the field will have gauge degrees of freedom, and hence that there will be only two physical helicity states, ± 2. Since the linearized theory will be indistinguishable from Einstein's full theory in laboratory physics, it is hard to see how these conclusions can fail to have general validity.

There is another conclusion to be drawn from the linearized theory, which is less appealing but is equally unavoidable: the theory is not renormalizable. This can be seen immediately by looking at the last term in the Lagrangian and using the rule of thumb discussed in the section on renormalization: the coupling constant needs to have dimensions (energy)^{d}, where $d \geq 0$. This rule is violated here, since the dimensions of κ are

$$[\kappa] = (\text{energy})^{-2}. \tag{8.151}$$

Gravitation thus presents a severe problem for the traditional programme of field theory at the very first step. Things get even worse when we go to the full theory, since there are terms of higher order that are even more strongly divergent.

The lowest-order gravitational terms in the linearized gravitational Lagrangian are those that couple linearly to the matter field. These correspond to Feynman diagrams that contain no internal graviton lines, and which thus appear as they would if gravity were to be treated as an external classical interaction. A first step along the road to quantum gravity is therefore to ignore the second-order terms: this approximation is known as **one-loop quantum gravity** or **semiclassical quantum gravity**. The latter terminology arises by analogy with quantum electrodynamics: there, ignoring the quantum nature of the electromagnetic field and applying full quantum mechanics only to matter fields gives correct results in some circumstances. The hope is that the same approach, of developing quantum field theory in a classical spacetime that is not Minkowski, will yield a first view of some of the results of full quantum gravity. This method is discussed in detail in the textbook by Birrell & Davies (1982); some of the most important results are outlined below.

From a mathematical point of view, the generalization of field theory from flat space to curved is not so difficult, as we have seen; the difficulties lie in quantizing the Lagrangian once it is written down. In Minkowski space, the standard procedure is to expand ϕ in some complete set and relate the expansion coefficients to creation and annihilation operators. The plane-wave states are a sensible set for all inertial observers to use, since one such mode maps onto another under a Lorentz transformation. In curved space, however, there can never be this degree of agreement, with the profound result that different observers can disagree about simple questions such as whether particles are present at all. For quantum gravity, this initial stumbling block is a fundamental barrier to progress. It is profoundly anti-relativistic to single out any particular metric about which to perform a field expansion; in full quantum gravity, the very notion of a classical background spacetime ceases to exist. This is why it has not so far been possible to go beyond the case of quantum fields in a curved background.

HAWKING RADIATION The most well-known manifestation of the subtleties of curved-space quantum fields is Hawking's (1974) proof that black holes emit thermal radiation. In essence, the same mechanism is responsible both for this effect and for the density fluctuations generated during any inflationary phase in the early universe.

Even expressed as simply as possible, it is hard to avoid technicalities altogether in this topic. We build up to the full result by dealing with a situation first analysed by Davies (1975) and Unruh (1976) – the uniformly accelerated observer. This model contains all the essential physics that is at work in more general situations, but is a little easier to comprehend, because there is a global inertial frame and Newtonian intuition can thus be used. In fact, this simplicity helps to illuminate some of the peculiar features of the situation. It might be thought that nothing interesting would happen to an observer who accelerates in a vacuum. Since this is a quantum calculation, we recognize that the vacuum cannot simply be defined as corresponding to zero energy–momentum tensor; however, the vacuum state $|0\rangle$ can be defined to satisfy

$$\langle 0|T^{\mu\nu}|0\rangle = 0, \tag{8.152}$$

assuming that any Λ-like term with $\rho = -p$ has been subtracted away. This is usually thought of as a definition of the absence of particles, and we would thus expect that an accelerated observer would perceive no particles, since the transformed matrix element will also be zero. However, this conclusion is *false*: an accelerated observer will in fact detect thermal radiation that is a direct result of their motion. This was first shown in outline by Davies (1975) and analysed in detail by Unruh (1976); the radiation is thus often referred to as **Unruh radiation**.

The solution to this paradox lies in recognizing that the concept of a particle has to be defined operationally. The simplest model of a detector is a device with internal energy levels E_i that can be characterized by a 'detector field' D; the device will detect particles through the usual interaction product term in the Lagrangian

$$\mathcal{L}_{\mathrm{I}} = g\, D(\tau)\, \phi\big(x^\mu(\tau)\big), \tag{8.153}$$

where τ is the detector's proper time and we shall assume a scalar field to keep things as simple as possible. The definition of particle detection is that the detector becomes excited, the first-order transition amplitude for which is

$$A(a \to b) = ig \left\langle b \left| \int D\, \phi\, d\tau \right| a \right\rangle. \tag{8.154}$$

We are working in the Heisenberg picture with a time-dependent detector operator

$D(\tau) = [\exp(iH\tau)]D(0)[\exp(-iH\tau)]$, which gives the usual form of expression from time-dependent perturbation theory

$$A(a \to b) = ig \langle b|D(0)|a\rangle \int e^{i(E_b - E_a)\tau} \langle b|\phi(\tau)|a\rangle \, d\tau. \qquad (8.155)$$

The particle detection rate will vanish only if the integral in this expression vanishes – but this is not guaranteed for an arbitrary detector trajectory.

Consider first what will happen if the detector is held stationary, but immersed in a bath of thermal radiation. The field ϕ is expanded as usual in terms of creation and annihilation operators,

$$\phi = \sum_k a_k u_k(x^\mu) + a_k^\dagger u_k^*(x^\mu), \qquad (8.156)$$

where the plane-wave eigenfunctions are normalized over a box of volume V:

$$u_k = \frac{1}{\sqrt{2\omega V}} e^{i\mathbf{k}\cdot\mathbf{x} - i\omega t}. \qquad (8.157)$$

The matrix element $\langle b|\phi|a\rangle$ is then only non-zero when the occupation number of some ϕ state changes by ± 1. Let the ϕ part of the initial $|a\rangle$ state be $|n\rangle$; we then have

$$\langle b|\phi|a\rangle = (2\omega V)^{-1/2} \begin{cases} e^{i\mathbf{k}\cdot\mathbf{x} - i\omega t} \sqrt{n+1} & \text{(emission)} \\ e^{i\omega t - i\mathbf{k}\cdot\mathbf{x}} \sqrt{n} & \text{(absorption).} \end{cases} \qquad (8.158)$$

The transition amplitude can now be evaluated for a detector at rest at the origin:

$$A(a \to b) = ig \langle b|D(0)|a\rangle (2\omega V)^{-1/2} \int e^{i(E_b - E_a \pm \omega)\tau} \, d\tau \times \begin{cases} \sqrt{n+1} & \text{(emission)} \\ \sqrt{n} & \text{(absorption).} \end{cases} \qquad (8.159)$$

The integral gives a δ-function $2\pi\delta(E_b - E_a \pm \omega)$, so that energy conservation is automatically included, and the detector will move to states of higher or lower energy depending on whether it is emitting or absorbing (detecting) particles. If the initial state is the vacuum, $|a\rangle = |0\rangle$, absorption is impossible and the detector will thus not register the presence of particles. If the detector is placed in a thermal bath with $n = (\exp \hbar\omega/kT + 1)^{-1}$, the detection rate becomes

$$\frac{dp(a \to b)}{dt} = g^2 \, |\langle b|D(0)|a\rangle|^2 \, \frac{(\omega V/\pi)^{-1} \delta(E_b - E_a - \omega)}{e^{\hbar\omega/kT} - 1}. \qquad (8.160)$$

This is very similar to the usual treatment of time-dependent perturbation theory: when squaring to get the transition probability, we encounter $|\int \exp(\Delta E - \omega)\, d\tau|^2$, where the limits of integration are formally $\pm\infty$. This is dealt with by saying that one time integral gives a δ-function, in which case the other may be replaced by $\int d\tau$ and just gives T, the time interval under consideration, and thus a well-defined transition rate. To finish the job and obtain the Golden Rule, we need only integrate over a density of states in ω-space.

Now, what happens if the initial state is the vacuum but the detector accelerates? The trajectory of a detector undergoing constant proper acceleration a is [problem 1.2]

$$\begin{aligned} x &= a^{-1}[\cosh(a\tau) - 1] \\ t &= a^{-1} \sinh(a\tau), \end{aligned} \qquad (8.161)$$

and observers comoving with the detector inhabit **Rindler space**. Rather than trying to evaluate the matrix element for this trajectory, consider the transition probability:

$$p(a \to b) = g^2 \, |\langle b|D(0)|a\rangle|^2 \iint d\tau \, d\tau' \; G^+(x[\tau], x[\tau']) \, e^{i(E_a - E_b)(\tau - \tau')}. \tag{8.162}$$

Here, G^+ is the Green function of the ϕ field (see below), and is a function of separation: $G(x, x') = G(x - x')$. We can therefore absorb one of the time integrals into the transition rate, as in the above example:

$$\frac{dp(a \to b)}{dt} = g^2 \, |\langle b|D(0)|a\rangle|^2 \int d\Delta\tau \; G^+(\Delta\tau) \, e^{i(E_a - E_b)\Delta\tau}, \tag{8.163}$$

so that the result depends on the Fourier transform of the Green function. This is related to field matrix elements via

$$G^+(x, x') = \langle 0|\phi(x)\phi(x')|0\rangle \tag{8.164}$$

(similar to the Feynman propagator, which would involve a time-ordered product of fields). How did the Green function get involved? In taking the modulus squared of the transition amplitude, we introduced the combination $\langle 0|\phi|b\rangle\langle b|\phi|0\rangle$, where $|b\rangle$ is the state of the ϕ field after the detector has absorbed a quantum. Since the $|b\rangle$ states are a complete set, we can use the usual completeness relation $\sum_b |b\rangle\langle b| = 1$, which then shows why the transition rate is related to the Fourier transform of the Green function. The calculation is now in principle almost done. Unfortunately, the remaining details are lengthy, and it would not assist our aim of illuminating the main concepts to give all the steps here; section 3.3 of Birrell & Davies (1982) may be consulted. The final result is

$$dp(a \to b) = \frac{g^2}{2\pi} \, |\langle b|D(0)|a\rangle|^2 \, \frac{\Delta E}{e^{2\pi \Delta E/a} - 1}, \tag{8.165}$$

and comparison with the previous result for a detector in a thermal bath shows that particles are absorbed as if the accelerated detector were experiencing radiation with temperature

$$\boxed{T_{\text{accel}} = \left(\frac{\hbar}{ck}\right) \frac{a}{2\pi}.} \tag{8.166}$$

In the limit of a stationary detector, no particles are detected, of course, but the detection rate rises with the acceleration. The effect is not exactly dramatic in everyday physics: for an acceleration of 1g, the temperature is 4×10^{-20} K. However, very much larger accelerations are routinely generated in particle accelerators: for a particle moving near the speed of light around a circle of radius R, the temperature experienced in the rest frame is

$$\frac{T}{\mathrm{K}} \simeq \frac{3.7 \times 10^{-7} \, \gamma^2}{R/\mathrm{km}}. \tag{8.167}$$

Since the maximum Lorentz factors generated can be as high as 10^5 ($e^\pm$ pairs in a 100 GeV head-on collision), the temperatures 'seen' by electrons in accelerators are scarcely negligible. Indeed, experimental effects are seen at high energies that can be interpreted directly in terms of Unruh radiation in the rest frame of the rotating electron (Bell & Leinaas 1987), so there can be little doubt of the reality of the effect.

FIELD THEORY AT NON-ZERO TEMPERATURE Although heavily simplified and truncated, the above 'derivation' of Hawking radiation in an accelerated reference frame is sufficiently complex that it gives little intuitive idea of how and why the effect arises. The appearance of a Planckian spectrum at the end seems little short of a miracle, and it would be nice if there was some more direct way of seeing that the situation must involve thermal radiation, and of obtaining the temperature.

The way to achieve insight into when there will be the appearance of thermal fluctuations in a quantum system is to start with the effect of a true thermal bath. Each eigenstate of the Hamiltonian in such a system will have a statistical weight given by the Boltzmann factor $\exp(-E/kT)$, and so the expectation value for any operator will be

$$\langle O \rangle = \frac{1}{Z} \sum_i \langle i|O|i\rangle \exp\left(\frac{-E_i}{kT}\right), \tag{8.168}$$

where Z is the partition function. This can be written more compactly [problem 6.3] in terms of the **density matrix** ρ:

$$\boxed{\begin{aligned} \langle O \rangle &= \frac{\mathrm{Tr}(\rho\, O)}{\mathrm{Tr}\,\rho}, \\ \rho &= \exp\left(\frac{-H}{kT}\right). \end{aligned}} \tag{8.169}$$

Now, quantum mechanics more commonly involves exponentials of the Hamiltonian in the form of time-evolution factors $\exp(iHt)$ – so the thermal average starts to resemble a time translation by the imaginary interval i/kT. This is another example of the concept of **Euclidean time**, $t \to it'$, which is commonly encountered in field theory. When applied to Green functions, the thermal averaging procedure says that the Green function is periodic in Euclidean time:

$$\begin{aligned} \langle G^+(x,y)\rangle &= \langle \phi(x)\phi(y)\rangle \\ &= \mathrm{Tr}\,[e^{-H/kT}\phi(x)\phi(y)]/\mathrm{Tr}(\rho) \\ &= \mathrm{Tr}\,[e^{-H/kT}\phi(x)e^{H/kT}e^{-H/kT}\phi(y)]/\mathrm{Tr}(\rho) \\ &= \langle \phi(y)\phi(x; t \to t + i/kT)\rangle \\ &= \langle G^-(x; t \to t + i/kT, y)\rangle, \end{aligned} \tag{8.170}$$

where the cyclic property of traces has been used. This relation between G^+ and G^- means that the Green function $G = G^+ + G^-$ is periodic (see section 2.7 of Birrell & Davies 1982).

As it stands, this is a rather abstract result, but what it means is the following: if a metric is periodic in Euclidean time, so will be the Green functions and hence there will be thermal fluctuations whose temperature can be read off from the periodicity. We can apply this reasoning immediately to the accelerated observer: consider Minkowski spacetime under the coordinate transformation $x = r\cosh a\tau$, $t = r\sinh a\tau$:

$$ds^2 = a^2 r^2\, d\tau^2 - dr^2 - dy^2 - dz^2. \tag{8.171}$$

Under the Euclidean transformation $\tau \to i\tau$, this becomes

$$-ds^2 = (dr^2 + a^2 r^2\, d\tau^2) + dy^2 + dz^2. \tag{8.172}$$

The part in brackets is just the metric of 2D space expressed in polar coordinates, with τ acting as the polar angle, so we conclude that Euclidean time is periodic with period $2\pi/a$ and hence that the temperature is $kT = a/2\pi$, as above.

The analytic continuation method also shows that de Sitter space contains thermal fluctuations. Write the metric in the Robertson–Walker form

$$ds^2 = dt^2 - \cosh^2(Ht)[dr^2 + S_k^2(r)\, d\psi^2]; \tag{8.173}$$

under $t \to it$, $\cosh Ht \to \cos Ht$, so the period in Euclidean time is $2\pi/H$, and hence the temperature perceived by an observer in de Sitter space is

$$\boxed{T_{\text{deSitter}} = \left(\frac{\hbar}{k}\right)\frac{H}{2\pi}.} \tag{8.174}$$

BLACK HOLES We can now apply the above results to the original example considered by Hawking: radiation generated in the field-theoretic 'atmosphere' above the event horizon of a black hole. It should by now seem likely that such radiation exists, just from the equivalence principle. A freely falling observer sees no radiation, but a stationary observer accelerates relative to this local inertial frame and should perceive radiation determined by the above analysis. The main quantity that we need is thus the local gravitational acceleration perceived by an observer at rest in the gravitational field of a black hole. For the case of the Schwarzschild metric, this is straightforward: the proper acceleration for an observer at constant (r, θ, ϕ) is [problem 8.5]

$$a = \frac{GM}{r^2}\left(1 - \frac{2GM}{c^2 r}\right)^{-1/2}. \tag{8.175}$$

The factor in parentheses is just the time-dilation factor at radius r. Hence, Unruh radiation with $T(r) = a(r)/2\pi$ emerges to infinity with a temperature $T = GM/2\pi r^2$. If we pretend that the radiation is in fact emitted only near the event horizon, we get the **Hawking temperature**

$$\boxed{T_{\text{H}} = \left(\frac{\hbar c^3}{k}\right)\frac{1}{8\pi GM}.} \tag{8.176}$$

Why does the Unruh argument give us the wrong temperature unless it is applied at the horizon? To deal with the black hole properly, we should expand the quantum field of interest in the eigenfunctions of the wave equation in the Schwarzschild metric. A slightly less intimidating answer is obtained if we notice that the accelerating frame contains an event horizon at a distance $r_a = c^2/a$. For large distances from the black hole, $a \propto 1/r^2$ and so $r_a \gg r$ for large r: the black hole is easily 'seen' in the accelerating frame, and so it is hardly surprising that the Unruh argument (transforming Minkowski spacetime) is inapplicable. However, as the acceleration rises closer to the Schwarzschild horizon and the effective flat-space horizon shrinks, it is perhaps not so surprising that the right answer is obtained. The temperature can also be obtained via the periodicity argument – see section 9.6 of Narlikar & Padmanabhan (1986).

What is the luminosity emitted by the black hole? A naive calculation would multiply the flux density σT^4 by the area of the event horizon, and this must be correct to order of magnitude by dimensions:

$$L_{\rm BH} = \Gamma\, 4\pi \left(\frac{2GM}{c^2}\right)^2 \sigma T_{\rm BH}^4 = \Gamma\, \frac{\hbar c^2}{15\,360\pi} \left(\frac{GM}{c^2}\right)^{-2}, \tag{8.177}$$

where Γ is a 'fudge factor' to be found by reference to an exact calculation (note the factor of $15\,360\pi$ in the denominator of the above expression; dimensional arguments omit factors that are sometimes impressively large). The value of Γ depends on the temperature of the hole: for small holes, the temperature is high enough for pair emission to occur. For those with $T_{\rm BH} \lesssim 10^{10}$ K, the appropriate value of Γ is 14.0; 1.4% of the energy escapes in gravitons, 11.9% in photons, and the remainder in neutrinos (see chapter 8 of Thorne *et al.* 1986). For high energies, the emitted spectrum is purely thermal, with an area corresponding not to that of the event horizon, but $A = 27\pi(GM/c^2)^2$, which comes from the cross-section for particle capture. The low-energy photons, with wavelengths close to the horizon size, are reduced in intensity.

The energy to supply the detected photon flux at infinity has to come from somewhere, and the obvious source is the mass energy of the hole itself; **black hole evaporation** is therefore predicted, in which the mass of the hole is reduced to zero in a finite time. Ignoring the late stages of pair production, we have $\dot{M} = -L \equiv \alpha M^{-2}$, so that $M = (-3\alpha t)^{1/3}$ if the hole disappears at $t = 0$. The lifetime of a hole is then $M^3/3\alpha$, i.e.

$$\boxed{t_{\rm BH} = \frac{5120\pi c^2}{\Gamma \hbar G} \left(\frac{GM}{c^2}\right)^3 = 1.50 \times 10^{66} \left(\frac{M}{M_\odot}\right)^3 \text{ yr.}} \tag{8.178}$$

Any primordial black holes remaining from the early phases of the big bang thus must have a mass of at least $2 \times 10^{-19}\, M_\odot$.

The very late phases of the evaporation are a matter for speculation, since the semiclassical analysis breaks down when the horizon size reaches the Planck length. Possibilities include total evaporation, a stable Planck-mass relic, or a naked singularity. Without a complete theory, no decision between these alternatives can be made.

ALTERNATIVE INSIGHTS There are a variety of ways in which the above results on Hawking radiation can be made more intuitively reasonable. Perhaps the major aspect that has not been clearly exposed is the role of event horizons; these have been present in all the examples we have considered, and this seems likely to be more than a coincidence. From the point of view of the uncertainty principle, it is the presence of the horizon that leads to the thermal radiation: the existence of a horizon prevents the observer from performing a sufficiently global measurement of the state of the field to verify that it inhabits the Minkowski vacuum state. In all cases, the induced temperature is of order the reciprocal of the horizon size (in natural units).

A further argument in this direction concerns virtual particles. Even a Minkowski vacuum acts as a constant source of particle–antiparticle pairs, but the uncertainty principle limits the time for which a pair of energy E can exist to $\Delta t \lesssim \hbar/E$, and hence a maximum separation of $\Delta x \lesssim c\hbar/E$ is possible. The only way these fluctuations can attain a permanent existence is if they can generate an energy income from the gravitational field that is large enough to pay their debts to the uncertainty principle. If there is a tidal gravitational force F between a pair at separation Δx, then the criterion is $F\,\Delta x \sim E$; adding the uncertainty constraint gives $E^2 \lesssim (c\hbar)F$. Now, the tidal force is $F \sim \Delta x\, GM(E/c^2)/r^3 \lesssim (\hbar/c)GM/r^3$, which has a maximum value at the horizon. This argument then says that the gravitational field close to the surface of the black hole can create particles up to an energy

$$E_{\rm max} \sim (\hbar c^3)\,(GM)^{-1}, \tag{8.179}$$

which is $E \sim kT_{\rm BH}$, to an order of magnitude. Classically, it should not be surprising that particles can be created at the black hole event horizon, since the total energy of a particle vanishes at that point; there is no energetic objection to creating a particle from nothing, since the gravitational field is able to supply all the required energy.

8.11 Kaluza–Klein models

With certain exceptions (e.g. dimensional regularization in quantum field theory), most discussion in fundamental physics takes as given the $(3+1)$-dimensional nature of spacetime. However, since there are some aspects of 3D space that are peculiar to that number of dimensions (e.g. the existence of the vector product), it is interesting to ask how physics would look if the number of dimensions in spacetime were to be varied. In the case of theories with larger numbers of dimensions, it is possible that unification of the interactions can result. This goes back to the idea of Kaluza (1921), later elaborated by Klein (1926), that the appearance of electromagnetism may be viewed as an artefact produced by the existence of a 5D spacetime that contains a version of general relativity.

LOWER-DIMENSIONAL GRAVITY First, however, it is interesting to note that $(3+1)$-dimensional spacetime seems to be the smallest in which general relativity is possible at all. In the normal theory, curvature enters through a non-zero Riemann tensor; even though the Ricci tensor vanishes in empty space, this does not imply that the Riemann tensor needs to vanish. However, this is a dimension-dependent statement. In n dimensions, the number of independent components of the Riemann tensor and the Ricci tensor are respectively $n^2(n^2-1)/12$ and $n(n+1)/2$ (for $n > 2$: in the two-dimensional case there is just one independent number that matters: the Gaussian curvature). These numbers of degrees of freedom are equal at six each for $n = 3$, implying that it should be possible to obtain the Riemann tensor from the Ricci tensor. In general, the Riemann tensor can be decomposed into parts respectively dependent on and independent of the Ricci tensor, as follows:

$$\begin{aligned} R_{\alpha\beta\gamma\delta} = &\frac{1}{n-2}(g_{\alpha\gamma}R_{\beta\delta} + g_{\beta\delta}R_{\alpha\gamma} - g_{\alpha\delta}R_{\beta\gamma} - g_{\beta\gamma}R_{\alpha\delta}) \\ &- \frac{R}{(n-1)(n-2)}(g_{\alpha\gamma}g_{\beta\delta} - g_{\alpha\delta}g_{\beta\gamma}) + C_{\alpha\beta\gamma\delta}. \end{aligned} \tag{8.180}$$

The tensor $C_{\alpha\beta\gamma\delta}$ is the **Weyl tensor** and vanishes for $n \leq 3$ (see p. 145 of Weinberg 1972). For $n = 2$ the equation becomes $R_{\alpha\beta\gamma\delta} = (R/2)[g_{\alpha\delta}g_{\beta\gamma} - g_{\alpha\gamma}g_{\beta\delta}]$. We therefore see that, only for $n \geq 4$ is it possible to have long-range gravitational forces (i.e. a non-zero Riemann tensor in regions of empty space governed by the Einstein field equation $R^{\mu\nu} = 0$.

THE KALUZA–KLEIN ARGUMENT Moving in the other direction, Kaluza (1921) considered the case of a $(4+1)$-dimensional spacetime. If we write the corresponding 5-vectors as (x^μ, y), then the (symmetric) metric tensor of the space can be written in the general form

$$g^{(5)} = \phi^{-1/3} \begin{pmatrix} g^{\mu\nu} + \phi A^\mu A^\nu & \phi A^\mu \\ \phi A^\nu & \phi \end{pmatrix}. \tag{8.181}$$

At this stage, the 4×4 matrix $g^{\mu\nu}$, the vector A^μ and the scalar ϕ are arbitrary. Now write down the 5D action of general relativity:

$$S = -\frac{1}{16\pi G^{(5)}} \int R^{(5)} \sqrt{g^{(5)}} d^4x^\mu \, dy. \tag{8.182}$$

If we rewrite this in terms of the above decomposition of the metric tensor, the action can be rearranged into

$$S = -\frac{1}{16\pi G} \int \left(R + \frac{1}{4}\phi F^{\mu\nu} F_{\mu\nu} + \frac{1}{6\phi^2} \partial^\mu \phi \partial_\mu \phi \right) \sqrt{g} \, d^4x^\mu, \tag{8.183}$$

where $F^{\mu\nu} = \partial^\mu A^\nu - \partial^\nu A^\mu$. To obtain this expression, it is necessary to assume that there is no dependence on y (the 'cylinder condition'), and to skip over much algebraic spadework (e.g. chapter 13 of Collins, Martin & Squires 1989). The physical content of the result is astonishing: we have normal general relativity, plus electromagnetism, plus a scalar field, whose quantum is known as the **dilaton**. In some ways this is one of the most radical innovations of twentieth-century physics: a result that is apparently of very deep significance, but whose place in any ultimate theory remains unclear. The basic idea has been incorporated into a variety of GUTs, which extend the total number of dimensions (so far the record is 26). In both these and the original case, the main question is the physical significance of the extra dimensions: do they have the same 'reality' as the usual four, or are they simply a mathematical device that should not be interpreted in this way?

The most common assumption is that, if the extra dimensions are real, they must be **compactified** – i.e. the universe must be very highly curved so that its typical extent in the extra directions is of order the Planck length. This would explain why it is not possible for people to disappear entirely from view by a displacement in one of the new directions, and also why the extra dimensions do not manifest themselves in everyday physics – they are hidden degrees of freedom, analogous to the axial rotations of molecules, which do not contribute to specific heats because their activation temperature is so high. Of course, this approach begs many questions: why are just four of the many dimensions very much larger than the Planck scale? A solution may be sought in anisotropic cosmological models where some of the dimensions expand to large size, perhaps driven by some version of inflation, whereas others turn round and collapse. Although models can be constructed in which this happens, it is fair to say that these require special initial conditions that anticipate the required answer: if small extra dimensions do exist, we suspect there must be some extra physical principle that produces the interesting signature $(+,-,-,-)$ for the non-negligible dimensions. A further problem is that not only must the extra dimensions be small, they must also be kept *static*. In the above discussion of the original Kaluza–Klein argument, we have seen that the dilaton field (which derives from the hidden part of the metric) controls the effective parameters of the $(3+1)$-dimensional laboratory physics; the 4D gravitational theory that results is in fact a scalar-tensor theory of Brans–Dicke form, rather than being precisely Einstein's theory. Expansion or contraction of the hidden dimensions will then lead to changes in the fundamental constants, contrary to observation. Since attempts to construct static solutions in the style of the Einstein universe will be unstable, the choice of the Planck scale for compactification seems inevitable; only when quantum gravity is becoming important will classical arguments about the possibility of static models be inapplicable. Thus, the issue of extra dimensions becomes bound up in the issue of quantum gravity. This is a depressingly common conclusion: all too often, our attempts to explain features of the universe with only low-energy physics find themselves pushed back against this barrier.

8.12 Supersymmetry and beyond

The theme of the forms of unification discussed in the standard model and in GUTs is to present different particles as transformed versions of each other – extensions of the isospin idea. However, this progress has really been confined to the fermions, so that in GUTs quarks and leptons are seen as different states of the same underlying 'something'. This approach fails to embrace the two other areas of particle society; the gauge bosons and the Higgs scalars are left as separate classes. The idea of supersymmetry (often called SUSY by its friends) is to erase this division and take the theme of unification to its logical extreme. This means introducing an operator that can transform between fermions and bosons:

$$\begin{aligned} Q|\text{boson}\rangle &= |\text{fermion}\rangle \\ Q|\text{fermion}\rangle &= |\text{boson}\rangle. \end{aligned} \tag{8.184}$$

The full mathematical implementation of this idea is rather technical, and we shall concentrate here on just a few of the key issues (see Bailin & Love 1994 for further reading).

There is an immediate problem with the phenomenology of supersymmetric theories. In the standard model, there are 28 independent boson degrees of freedom and 90 fermion states (the numbers of particles and antiparticles, times helicity and colour combinations). The implementation of supersymmetry means that new particles must appear; in fact, detailed model-building efforts only make progress if there are no pairings between known particles. The unification aims of supersymmetry therefore score an immediate failure in that they do not associate existing particles with each other, but pair them with particles yet to be observed. Each known particle must have a **superpartner**: boson A implies the fermion A-ino and fermion B implies the boson s-B, populating the universe with novel 'sparticles' such as 'squarks', and 'photinos'. The superpartners are denoted by a tilde: $\tilde{q}$, $\tilde{\gamma}$ etc.

This seems unpromising, but supersymmetry in some form is commonly thought to be inevitable, because of the **gauge hierarchy problem**, which asks how it can be that the real world lives so far below the GUT scale – i.e. why is $M_W/M_{\rm GUT} \sim 10^{-13}$? Now of course nature can choose small numbers sometimes, but in this case such a number is **unnatural**: quantum corrections would be expected to perturb the bare value of the W mass by a significant fraction of the GUT scale. The only way in which the situation can be saved without an unsatisfying fine tuning between the bare value and the corrections is if there is automatically some means for the corrections to be cancelled. This is provided in supersymmetry because many diagrams gain a minus sign if fermions are involved, leading to a reasonable means of understanding how such corrections can add up to zero. In the absence of any other idea for sustaining the gauge hierarchy, supersymmetry will be taken as a very real possibility by particle physicists.

The ideas of supersymmetry are not only relevant at the GUT scale, but should be testable in a number of ways. In the accelerator context, we have seen that the expectation is of new physics at energies around 1TeV, and it is expected by many theorists that the **large hadron collider** (LHC) at CERN will reveal supersymmetric particles (see e.g. Treille 1994). This machine is expected to begin operation around 2005, probing centre-of-mass energies $\simeq$ 16 TeV. In the meantime, supersymmetry provides a possible answer to the cosmological dark-matter problem. Most supersymmetric particles will be unstable, but supersymmetric theories generally forbid processes in which a sparticle decays to normal particles only (although this need not be forbidden on grounds of energy or angular momentum). There will therefore usually be a lightest supersymmetric particle (**LSP**) that is stable – probably the photino $\tilde{\gamma}$. The lack of any experimental detection of sparticles means that they must be massive, $M \gtrsim 10$ GeV, and so the LSP would be a good candidate for cold dark matter. A thorough review of supersymmetric candidates for dark matter is given by Jungman, Kamionkowski & Griest (1996).

SUPERSYMMETRY AND THE Λ PROBLEM As a particular example of the operation of boson–fermion cancellation in supersymmetry, consider the following (flawed) explanation for why the vacuum energy must be zero. We have considered contributions to the vacuum energy arising from the zero-point energies of field wave modes treated as harmonic oscillators:

$$E = \sum (n + \tfrac{1}{2})\hbar\omega \quad \Rightarrow \quad E_{\rm vac} = \sum \tfrac{1}{2}\hbar\omega. \tag{8.185}$$

This result applies for bosonic fields where we quantize via the commutator $[a, a^\dagger] = 1$. If supersymmetry were to be an exact symmetry, there would have to exist a fermionic field for each bosonic field. These would be quantized via *anticommutators*, $\{a, a^\dagger\} = 1$, which implies a field energy

$$E = \sum \left(n - \tfrac{1}{2}\right) \hbar\omega. \tag{8.186}$$

The zero-point energy for a fermionic harmonic oscillator is thus $-\hbar\omega/2$, and there is an automatic cancellation of the total (fermion + boson) zero-point energy for each wave mode. This argument possesses exactly the sort of simplicity and inevitability we would expect for the explanation of such a fundamental property of the universe as $\Lambda = 0$, but sadly it cannot be correct. Supersymmetry, if it exists at all, is clearly a broken symmetry at present-day energies; there appears to be no natural way of achieving this breaking while retaining the attractive consequence of a zero cosmological constant, and so the Λ problem remains as puzzling as ever.

SUPERGRAVITY, SUPERSTRINGS AND THE FUTURE The basic idea of supersymmetry can be extended in a number of ways. Rather than a single operator Q, it is possible to imagine the existence of N independent operators Q_i, each of which changes the spin of the particles to which it is applied by $1/2$. This will produce $N + 1$ different helicity states, so that a maximum value of $N = 8$ is allowed if the resulting theory is to contain no particles beyond the spin-2 graviton.

Another basic idea is the distinction between local and global supersymmetry. By analogy with everything that has gone before, we would expect supersymmetry to be local, which has an interesting consequence. Consider the simplest possible model Lagrangian, in which there is a charged scalar field and a Majorana spinor (both thus having two degrees of freedom, so they can be superpartners):

$$\mathcal{L} = \partial^\mu \phi^* \partial_\mu \phi + \frac{i}{2} \bar{\psi} \gamma^\mu \partial_\mu \psi. \tag{8.187}$$

If we write the scalar field as $\phi = \phi_1 + i\phi_2$, then it can be verified that the following infinitesimal supersymmetry transformation changes the Lagrangian only by a term that is a total derivative, which is as usual ignored by setting a suitable boundary at infinity:

$$\begin{aligned} \delta\phi_1 &= \bar{\epsilon}\psi \\ \delta\phi_2 &= i\bar{\epsilon}\gamma^5\psi \\ \delta\psi &= -i\gamma^\mu [\partial_\mu(\phi_1 + i\gamma^5\psi_2)]\,\epsilon, \end{aligned} \tag{8.188}$$

where ϵ is a spinor that generates the transformation. As usual, if ϵ is spatially varying, the gauge invariance will be spoiled by terms involving $\partial^\mu\epsilon$. The solution is familiar: add a new gauge field χ^μ whose transformation 'eats' the derivative term:

$$\delta\chi^\mu \propto \partial^\mu\epsilon. \tag{8.189}$$

This is a spinor with an index, and must therefore be a Rarita–Schwinger spin-3/2 field. Worse still, because of supersymmetry, we can generate a superpartner for the χ particle whose spin is 1/2 greater. We conclude that the local form of supersymmetry forces the introduction of a spin-2 field, giving the theory the clearly merited name of **supergravity**.

Theories of **extended supergravity** (local supersymmetry with N generators) therefore provide two ways of incorporating gravitation into the framework of unified theories, and a huge amount of effort has gone into exploring such theories. Nevertheless, the basic difficulty of quantum gravity remains: a naive argument from the dimension of the coupling constant suggests that gravity is not a renormalizable theory, and this remains true even when the tensor interaction is introduced in a gauge fashion through local supersymmetry. A cure for this problem probably requires radical surgery, related to the meaning of the particle concept itself. Recall that we do not need quantum field theory in order to get divergences, since even the classical electromagnetic energy of a point-like electron is infinite, and this problem is only barely tamed in quantum theory through renormalization. In the language of quantum fields, the classical particle concept is replaced by interaction at a point, where a field and its conjugate momentum satisfy the equal-time commutator

$$[\phi(\mathbf{x}), \pi(\mathbf{x}')] = i\hbar\delta(\mathbf{x} - \mathbf{x}'). \tag{8.190}$$

To remove divergences, it is necessary to do away with the particle concept, that is to scrap the whole apparatus of quantum fields dealt with so far. As an approach towards this target, much effort has been invested in the theory of **superstrings**, which involves performing (supersymmetric) quantum mechanics on entities that are extended in one dimension rather than being point-like. The hope here is to achieve a theory that will be free of divergences and yet which can be well approximated by a GUT field apparatus at energies far below the Planck scale. A great deal has been written on the progress to date (e.g. Green, Schwartz & Witten 1987; Bailin & Love 1994), but it is far too early to say whether a credible 'theory of everything' will ever emerge by this route.

As we approach the working face of particle physics, things inevitably become murky; the present uncertainty may yet give way to a simple and inevitable new structure, and certainly there is no shortage of ideas. Nevertheless, it is possible to be a little depressed by the prospects. The standard model is a structure that would never have been produced by pure thought – rather, it was enforced by the sometimes quirky results of many carefully crafted experiments. If the next generation of accelerators detects an elementary Higgs boson, but no other new physics, then our understanding of particle physics would be in danger of stagnating. The only hope for understanding the behaviour of fields and forces at GUT scales would then come through the study of the early universe.

Problems

(8.1) Derive the Noether current and energy–momentum tensor for a Lagrangian that is invariant under internal and spacetime translation symmetry.

(8.2) Show that the neutral or identity element in a group is always unique.

(8.3) Why, when you look in a mirror, do you appear with left and right sides transposed, but not upside-down?

(8.4) In the electroweak model, the Higgs potential is

$$V(\phi) = \mu^2 \phi^\dagger \phi + \lambda(\phi^\dagger \phi)^2. \tag{8.191}$$

Show that the potential minimum can always be chosen as $\langle 0|\phi|0\rangle = (0, v/\sqrt{2})$. Is this choice consistent with the numbers of degrees of freedom in the model?

(8.5) Show that the proper acceleration perceived by a stationary observer above a Schwarzschild black hole is

$$a = \frac{GM}{r}\left(1 - \frac{2GM}{c^2 r}\right)^{-1/2}, \tag{8.192}$$

and hence that the Hawking temperature seen at infinity is that of Unruh radiation redshifted from the event horizon.

(8.6) Consider a quantum-mechanical system with a Hamiltonian in the form of a sum over a set of simple harmonic oscillators

$$H = \sum_{\text{modes}} \tfrac{1}{2}\hbar\omega(aa^\dagger + a^\dagger a). \tag{8.193}$$

Quantize this in the usual way with the commutator $[a, a^\dagger] = 1$, and show that $H = \sum(n + \frac{1}{2})\hbar\omega$. Thus show that the total zero-point energy diverges unless the sum is truncated at some energy. Now quantize with the *anti*commutator $\{a, a^\dagger\} = 1$. Prove than $n = 0$ or 1, as expected for fermions, and that $H = \sum(n - \frac{1}{2})\hbar\omega$. Hence show that the vacuum energy is exactly zero in a universe with unbroken supersymmetry.

PART 4 THE EARLY UNIVERSE

9 The hot big bang

Having given an overview of the relevant parts of particle physics, this section of the book now discusses in some detail the application of some of these fundamental processes in the early universe. The next three chapters increase in energy, starting here with 'normal' physics at temperatures up to about 10^{10} K, and moving on to more exotic processes in chapters 10 and 11.

9.1 Thermodynamics in the big bang

ADIABATIC EXPANSION What was the state of matter in the early phases of the big bang? Since the present-day expansion will cause the density to decline in the future, conditions in the past must have corresponded to high density – and thus to high temperature. We can deal with this quantitatively by looking at the thermodynamics of the fluids that make up a uniform cosmological model.

The expansion is clearly **adiathermal**, since the symmetry means that there can be no net heat flow through any surface. If the expansion is also reversible, then we can go one step further, because entropy change is defined in terms of the heat that flows during a reversible change. If no heat flows during a reversible change, then entropy must be conserved, and the expansion will be **adiabatic**. This can only be an approximation, since there will exist irreversible microscopic processes. In practice, however, it will be shown below that the effects of these processes are overwhelmed by the entropy of thermal background radiation in the universe. It will therefore be an excellent approximation to treat the universe as if the matter content were a simple dissipationless fluid undergoing a reversible expansion. This means that, for a ratio of specific heats Γ, we get the usual adiabatic behaviour

$$T \propto R^{-3(\Gamma-1)}. \tag{9.1}$$

For radiation, $\Gamma = 4/3$ and we get just $T \propto 1/R$. A simple model for the energy content of the universe is to distinguish pressureless 'dust-like' matter (in the sense that $p \ll \rho c^2$) from relativistic 'radiation-like' matter (photons plus neutrinos). If these are assumed not to interact, then the energy densities scale as

$$\begin{aligned} \rho_m &\propto R^{-3} \\ \rho_r &\propto R^{-4}. \end{aligned} \tag{9.2}$$

As argued in chapter 3, a direct route to the latter expression is to multiply a conserved photon density scaling as R^{-3} by a photon energy scaling as R^{-1}.

The universe must therefore have been **radiation dominated** at some time in the past, when the densities of matter and radiation crossed over. To anticipate, we know that the current radiation density corresponds to thermal radiation with $T \simeq 2.73\text{K}$. We shall shortly show that one expects to find, in addition to this **cosmic microwave background** (CMB), a background in neutrinos that has an energy density 0.68 times that from the photons (if the neutrinos are massless and therefore relativistic). If there are no other contributions to the energy density from relativistic particles, then the total effective radiation density is $\Omega_r h^2 \simeq 4.2 \times 10^{-5}$ and the redshift of **matter–radiation equality** is

$$\boxed{1 + z_{\text{eq}} = 23\,900\,\Omega h^2\,(T/2.73\text{K})^{-4}.} \tag{9.3}$$

The time of this change in the global equation of state is one of the key epochs in determining the appearance of the present-day universe. Prior to matter–radiation equality, the speed of expansion is so high that fluctuations in matter density cannot collapse under their own gravity; only at later times can cosmological structures start to form. As discussed in chapter 15, the age of the universe at the equality epoch imprints a characteristic scale that can be measured in galaxy clustering today.

Of course, radiation and matter are coupled by Thomson scattering, so the collisionless derivation of $T(R)$ is not really valid. This is why it is better to take the thermodynamic approach and treat everything as a single radiation-dominated fluid with $P = \rho c^2/3$; conservation of energy then says $-P dV = dE = d(\rho c^2 V)$, implying $d\rho/\rho = -(4/3)dV/V$, or $T \propto R^{-1}$, as required. For matter with $\Gamma = 5/3$ alone, the temperature should fall faster: $T \propto R^{-2}$. However, when matter and radiation are coupled, it is the $T_r \propto 1/R$ law that dominates, because the number density of photons is very much greater than that of nucleons ($n_\gamma/n_p \sim 10^9$). The two only part company at $z \lesssim 100$ (see the discussion of radiation drag in chapter 17).

QUANTUM GRAVITY LIMIT In principle, $T \to \infty$ as $R \to 0$, but there comes a point at which this extrapolation of classical physics breaks down. This is where the thermal energy of typical particles is such that their de Broglie wavelength is smaller than their Schwarzschild radius: quantum black holes clearly cause difficulties with the usual concept of background spacetime. Equating $2\pi\hbar/(mc)$ to $2Gm/c^2$ yields a characteristic mass for quantum gravity known as the **Planck mass**. This mass, and the corresponding length $\hbar/(m_{\text{P}}c)$ and time ℓ_{P}/c form the system of **Planck units** discussed in chapter 8:

$$\begin{aligned} m_{\text{P}} &\equiv \sqrt{\frac{\hbar c}{G}} \simeq 10^{19}\text{GeV} \\ \ell_{\text{P}} &\equiv \sqrt{\frac{\hbar G}{c^3}} \simeq 10^{-35}\text{m} \\ t_{\text{P}} &\equiv \sqrt{\frac{\hbar G}{c^5}} \simeq 10^{-43}\text{s}. \end{aligned} \tag{9.4}$$

The Planck time therefore sets the origin of time for the classical phase of the big bang. It was recognized at an early stage that there are some peculiar aspects of the initial conditions of the universe at t_{P}: these are the 'fine-tuning' problems that motivate modern inflationary cosmology and are discussed in chapter 11. The whole aim of inflation is to understand the classical initial conditions for the big bang in terms of physics operating below the Planck scale, but even so energies of $\sim 10^{-4} E_{\rm P}$ are involved.

In any case, it is incorrect to extend the classical solution to $R = 0$ and conclude that the universe began in a singularity of infinite density. A common question about the big bang is 'what happened at $t < 0$?', but in fact it is not even possible to get to zero time without adding new physical laws. The initial singularity does not indicate some fatal flaw with the whole big bang idea; rather, we should be reassured that the model gives sensible results everywhere except the one place where we know in advance that it will be invalid.

COLLISIONLESS EQUILIBRIUM BACKGROUNDS The study of matter under the extremes of pressure and temperature expected in the early phases of the expanding universe might be expected to be a difficult task. In fact, the challenge is greatly simplified in two ways.

(1) The universe expands very fast in its early stages, and the dynamical time $t \sim (G\rho)^{-1/2}$ becomes very short for high density. Is there enough time to establish thermal equilibrium? In almost all cases, yes. Two-body reactions will have rates scaling as $n_1 n_2$, and so the reaction time for a given particle will go as the reciprocal of the number density of the species with which it is trying to react. For a thermal background, this changes as $n \propto R^{-3}$ (see below); by comparison, the dynamical timescales as R^{-2} for radiation domination, and so thermal equilibrium should be an increasingly good approximation at early times.

(2) For a background of weakly interacting particles such as neutrinos, it is reasonable to use the perfect gas approximation. More surprisingly, this also applies for charged particles (and not just in cosmology: the usual analysis of degeneracy in stellar structure treats electrons as a non-interacting perfect gas [problem 9.1]).

So, we need the thermodynamics of a possibly relativistic perfect gas. The basic tools for this analysis are very simple. We consider some box of volume $V = L^3$, and say that we will analyse the quantum mechanics of particles in the box by taking the system to be periodic on scale L. Quantum fields in the box are expanded in plane waves (see chapter 7) with allowed wavenumbers $k_x = n\,2\pi/L$ etc.; these **harmonic boundary conditions** for the allowed eigenstates in the box lead to the density of states in k-space:

$$dN = g\,\frac{V}{(2\pi)^3}\,d^3k \tag{9.5}$$

(where g is a degeneracy factor for spin etc.). This expression is nice because it is **extensive** ($N \propto V$) and hence the number density n is independent of V. The equilibrium **occupation number** for a quantum state of energy ϵ is given generally [problem 9.2] by

$$\langle f \rangle = \left[e^{(\epsilon-\mu)/kT} \pm 1 \right]^{-1} \tag{9.6}$$

($+$ for fermions, $-$ for bosons). Now, for a thermal radiation background, the **chemical potential** μ is always zero. The reason for this is quite simple: μ appears in the first law of thermodynamics as the change in energy associated with a change in particle number, $dE = T dS - P dV + \mu dN$. So, as N adjusts to its equilibrium value, we expect that the system will be stationary with respect to small changes in N. More formally, the

Helmholtz free energy $F = E - TS$ is minimized in equilibrium for a system at constant temperature and volume. Since $dF = -S\,dT - P\,dV + \mu\,dN$, $dF/dN = 0 \Rightarrow \mu = 0$. Thus, in terms of momentum space, the thermal equilibrium **background number density** of particles is

$$\boxed{n = g\,\frac{1}{(2\pi\hbar)^3}\int_0^\infty \frac{4\pi p^2\,dp}{e^{\epsilon(p)/kT} \pm 1},} \tag{9.7}$$

where $\epsilon = \sqrt{m^2c^4 + p^2c^2}$ and g is the degeneracy factor. There are two interesting limits of this expression.

(1) Ultrarelativistic limit. For $kT \gg mc^2$ the particles behave as if they were massless, and we get

$$n = \left(\frac{kT}{c}\right)^3 \frac{4\pi g}{(2\pi\hbar)^3}\int_0^\infty \frac{y^2\,dy}{e^y \pm 1}. \tag{9.8}$$

(2) Nonrelativistic limit. Here we can neglect the ± 1 in the occupation number, in which case

$$n = e^{-mc^2/kT}(2mkT)^{3/2}\frac{4\pi g}{(2\pi\hbar)^3}\int_0^\infty e^{-y^2}\,y^2\,dy. \tag{9.9}$$

This shows us that the background 'switches on' at about $kT \sim mc^2$; at this energy, photons and other species in equilibrium will have sufficient energy to create particle–antiparticle pairs, which is how such an equilibrium background would be created. The point at which $kT \sim mc^2$ for some particle is known as a **threshold**.

Similar reasoning gives the energy density of the background, since it is only necessary to multiply the integrand by a factor $\epsilon(p)$ for the energy in each mode:

$$\boxed{u = \rho\,c^2 = g\,\frac{1}{(2\pi\hbar)^3}\int_0^\infty \frac{4\pi p^2\,dp}{e^{\epsilon(p)/kT} \pm 1}\,\epsilon(p).} \tag{9.10}$$

In the same way, we can get the pressure from kinetic theory: $P = n\langle pv\rangle/3 = n\langle p^2c^2/\epsilon\rangle/3$, where v is the particle velocity, and is related to its momentum and energy by $p = (\epsilon/c^2)v$. The pressure is therefore given by the following integral:

$$\boxed{P = g\,\frac{1}{(2\pi\hbar)^3}\int_0^\infty \frac{4\pi p^2\,dp}{e^{\epsilon/kT} \pm 1}\,\frac{p^2c^2}{3\epsilon}.} \tag{9.11}$$

Clearly, in the ultrarelativistic limit where $\epsilon \simeq pc$, the pressure obeys $P = \rho c^2/3$. In the nonrelativistic limit, the pressure is just $P = nkT$ (see below), whereas the density is dominated by the rest mass: $\rho = mn$, and therefore $P \ll \rho c^2/3$. At a threshold, the equation of state thus departs slightly from $P = \rho c^2/3$, even if the universe is radiation dominated on either side of the critical temperature.

ENTROPY OF THE BACKGROUND One quantity that is of considerable importance is the **entropy** of a thermal background. This may be derived in several ways. The most direct is to note that both energy and entropy are extensive quantities for a thermal background. Thus, writing the first law for $\mu = 0$ and using $\partial S/\partial V = S/V$ etc. for extensive quantities,

$$dE = T\,dS - P\,dV \quad \Rightarrow \quad \left(\frac{E}{V}\,dV + \frac{\partial E}{\partial T}\,dT\right) = \left(T\frac{S}{V}\,dV + T\frac{\partial S}{\partial T}\,dT\right) - P\,dV. \tag{9.12}$$

Equating the dV and dS parts gives the familiar relation $\partial E/\partial T = T\,\partial S/\partial T$ and

$$\boxed{S = \frac{E + PV}{T}.} \tag{9.13}$$

Using the above integral for the pressure, the entropy is

$$S = \frac{4\pi g V}{(2\pi\hbar)^3}\int_0^\infty \frac{p^2\,dp}{e^{\epsilon/kT} \pm 1}\left(\frac{\epsilon}{T} + \frac{p^2c^2}{3\epsilon T}\right), \tag{9.14}$$

which becomes $S = 3.602Nk$ (bosons) or $4.202Nk$ (fermions) in the ultrarelativistic limit and $S = (mc^2/kT)Nk$ in the nonrelativistic limit. So, for radiation, the entropy is just proportional to the number of particles. For this reason, the ratio of photon to baryon number densities $n_\gamma/n_{\rm B}$ is sometimes called the **entropy per baryon**.

The radiation entropy density $s = (u + P)/T = (4/3)u/T$ is impressively large in practice. The energy density is $u = aT^4$, where $a = 7.56 \times 10^{-16}$ J K^{-1}m^{-3}. For $T = 2.728$ K, inserting the cosmological density in terms of Ω and h gives the entropy per unit mass:

$$\frac{S}{M} = 1.09 \times 10^{12}\,(\Omega h^2)^{-1}\ {\rm JK^{-1}kg^{-1}}. \tag{9.15}$$

By comparison, S/M for water at 300 K warmed by 1 K is only 14 J K^{-1}kg^{-1}. Although we do our best on a local scale to obey the second law and cause entropy to increase, these efforts are swamped by the immense quantity of entropy that resides in the microwave background. This is why it is a very good approximation to treat the expansion of the universe as adiabatic and reversible.

The same result for the entropy arises from a more general approach that uses the first law and a Maxwell relation to rewrite the differential for S:

$$\begin{aligned} dS &= \left(\frac{\partial S}{\partial T}\right)_V dT + \left(\frac{\partial S}{\partial V}\right)_T dV \\ &= \frac{1}{T}\left(\frac{\partial E}{\partial T}\right)_V dT + \left(\frac{\partial P}{\partial T}\right)_V dV, \end{aligned} \tag{9.16}$$

which can be integrated once $E(T)$ and $P(T)$ are given as integrals over momentum space.

Alternative methods of deriving the entropy illuminate some interesting points in statistical mechanics. One general way of obtaining thermodynamic functions starts from the free energy using the first law:

$$P = -\left(\frac{\partial F}{\partial V}\right)_{T,N}$$
$$S = -\left(\frac{\partial F}{\partial T}\right)_{V,N} \qquad (9.17)$$
$$\mu = \left(\frac{\partial F}{\partial N}\right)_{T,V}.$$

The free energy F can be expressed [problem 9.2] in terms of the **partition function**, Z:

$$F = -kT \ln Z, \qquad (9.18)$$

where $Z = \sum \exp(-E_i/kT)$ is the sum of Boltzmann factors over the accessible states of the system. The problem with this is that finding Z can be hard: the energy of a given state is the sum over the occupations of all energy levels $E = \sum n_j \epsilon_j$, and in general the n_j are not independent, because of the exclusion principle, in the case of fermions, or because of the constraint of fixed total number of particles, N. This route is only useful in the case of a classical gas, where all the occupation numbers are $\ll 1$. Each particle then has effectively the whole of momentum space available to it, and the partition function factorizes:

$$Z = \frac{1}{N!}\left[\frac{V}{(2\pi\hbar)^3}\int e^{-\epsilon/kT}\, 4\pi p^2\, dp\right]^N \qquad (9.19)$$

(the factor of $N!$ is to avoid Gibbs' paradox, which results from the incorrect multiple counting of states involving identical particles). Thus, the pressure of a perfect classical gas is

$$\boxed{P = \left(\frac{\partial F}{\partial V}\right)_T = \frac{NkT}{V}.} \qquad (9.20)$$

The interesting point here is that the perfect-gas law has arisen without specifying the energy–momentum relation, $\epsilon(p)$. The result thus applies even to a mildly relativistic gas, which is far from obvious if one pursues a derivation using kinetic theory.

In the non-classical case, it is easier to proceed by considering a system with variable particle number. Here, the probability of a given state is given by the **Gibbs' factor**, $p \propto \exp[-(E - \mu N)/kT]$, and the free energy becomes

$$F = -kT \ln Z_{\rm G} + \mu N, \qquad (9.21)$$

analogously to the expression for F in terms of the Boltzmann partition function. As the individual occupation numbers are now unconstrained except by quantum statistics, we can express $Z_{\rm G}$ as a product of terms for each momentum state:

$$Z_{\rm G} = \prod_{\rm states} \sum_{n=0}^{n_{\rm max}} e^{-(\epsilon-\mu)n/kT}$$
$$= \begin{cases} \prod \left[1 + e^{-(\epsilon-\mu)/kT}\right] \equiv \prod f_{\rm F} & \text{(fermions)} \\ \prod \left[1 - e^{-(\epsilon-\mu)/kT}\right]^{-1} \equiv \prod f_{\rm B} & \text{(bosons)} \end{cases} \qquad (9.22)$$

This is still not the end, however, as μ cannot be treated as a constant if we attempt to insert this expression for F into partial derivatives like $P = -(\partial F/\partial V)_{T,N}$. The way out is to define the **grand potential**, $\Phi = F - \mu N = -kT \ln Z_{\rm G}$. Now we have derivatives that *are* easy to evaluate, e.g. $P = -(\partial \Phi/\partial V)_{\mu,T}$:

$$\boxed{P = -\frac{\Phi}{V} = g\frac{4\pi kT}{(2\pi\hbar)^3}\int p^2\, dp\, \ln f_{\rm F,B}.} \tag{9.23}$$

The entropy is $S = -(\partial \Phi/\partial T)_{\mu,V}$:

$$S = g\frac{4\pi kV}{(2\pi\hbar)^3}\left[\int p^2\, dp\, \ln f_{\rm F,B} + \int p^2\, dp\, \frac{(\epsilon - \mu)/kT}{e^{(\epsilon-\mu)/kT} \pm 1}\right], \tag{9.24}$$

and it is an exercise in integration by parts to show that, for the thermal $\mu = 0$ case, this reduces to the earlier expression for the entropy. This only works for energy–momentum relations that obey $d\epsilon/dp = \epsilon/p$), so thermodynamics sets constraints on relativistic dynamics.

Note that, for an ultrarelativistic gas, the above implies $PV \simeq 0.9NkT$ (bosons) or $1.05NkT$ (fermions). Equilibrium backgrounds of photons and neutrinos thus behave slightly differently from classical gases, as expected, given that their occupation numbers are not negligibly small.

FORMULAE FOR ULTRARELATIVISTIC BACKGROUNDS We now summarize the most useful results from this discussion, which are the thermodynamic quantities for massless particles. These formulae are required time and time again in calculations of conditions in the early universe. Consider first bosons, such as the microwave background. Evaluating the dimensionless integrals encountered earlier (only possible numerically in the case of n) gives energy, number and entropy densities:

$$\boxed{\begin{aligned} u &= g\,\frac{\pi^2}{30}\,kT\left(\frac{kT}{\hbar c}\right)^3 = 3P \\ \frac{s}{k} &= g\,\frac{2\pi^2}{45}\left(\frac{kT}{\hbar c}\right)^3 = 3.602n \end{aligned}} \tag{9.25}$$

(remember that $g = 2$ for photons).

It is also expected that there will be a fermionic relic background of neutrinos left over from the big bang. Assume for now that the neutrinos are massless. In this case, the thermodynamic properties can be obtained from those of black body radiation by the following trick:

$$\frac{1}{e^x + 1} = \frac{1}{e^x - 1} - \frac{2}{e^{2x} - 1}. \tag{9.26}$$

Thus, a gas of fermions looks like a mixture of bosons at two different temperatures. Knowing that boson number density and energy density scale as $n \propto T^3$ and $u \propto T^4$, we then get the corresponding fermionic results. The entropy requires just a little more care. Although we have said that entropy density is proportional to number density, in fact the entropy density for an ultrarelativistic gas was shown above to be $s = (4/3)u/T$, and so the fermionic factor is the same as for energy density:

$$\boxed{\begin{aligned} n_{\rm F} &= \frac{3}{4}\frac{g_{\rm F}}{g_{\rm B}} n_{\rm B} \\ u_{\rm F} &= \frac{7}{8}\frac{g_{\rm F}}{g_{\rm B}} u_{\rm B} \\ s_{\rm F} &= \frac{7}{8}\frac{g_{\rm F}}{g_{\rm B}} s_{\rm B}. \end{aligned}} \qquad (9.27)$$

Using these rules, it is usually possible to forget about the precise nature of the relativistic background in the universe and count bosonic degrees of freedom, given the effective degeneracy factor for u or s:

$$g_* \equiv \sum_{\rm bosons} g_i + \frac{7}{8} \sum_{\rm fermions} g_j, \qquad (9.28)$$

although this definition needs to be modified if some species have different temperatures (see below).

NEUTRINO DECOUPLING At the later stages of the big bang, energies are such that only light particles survive in equilibrium: γ, ν and the three leptons e, μ, τ. If neutrinos could be maintained in equilibrium, the lepton–antilepton pairs would annihilate as the temperature fell still further ($T_\tau = 10^{13.3}$ K, $T_\mu = 10^{12.1}$ K, $T_e = 10^{9.7}$ K), and the end result would be that the products of these annihilations would be shared among the only massless particles. However, in practice the weak reactions that maintain the neutrinos in thermal equilibrium 'switch off' at $T \simeq 10^{10}$ K. This **decoupling** is discussed in more detail below, but is a general cosmological phenomenon, which arises whenever the interaction timescales exceed the local Hubble time, leaving behind abundances of particles frozen at the values they had when last in thermal equilibrium. Two-body reaction rates scale in proportion to density, times a cross-section that is often a declining function of energy, so that the interaction time changes at least as fast as R^{-3}. In contrast, the Hubble time changes no faster than R^{-2} (in the radiation era), so that there is inevitably a crossover. For neutrinos, this point occurs at a redshift of $\sim 10^{10}$, whereas the photons of the microwave background typically last interacted with matter at $z \simeq 1000$ (see chapter 18).

The effect of the electron–positron annihilation is therefore to enhance the numbers of photons relative to neutrinos. It is easy to see what quantitative effect this has: although we may talk loosely about the energy of $e^\pm$ annihilation going into photons, what is actually conserved is the *entropy*. The entropy of an $e^\pm + \gamma$ gas is easily found by remembering that it is proportional to the number density, and that all three particle species have $g = 2$ (polarization or spin). The total is then

$$s(\gamma + e^+ + e^-) = \frac{11}{4} s(\gamma). \qquad (9.29)$$

Equating this to photon entropy at a new temperature gives the factor by which the photon temperature is enhanced with respect to that of the neutrinos. Equivalently, given the observed photon temperature today, we infer the existence of a neutrino background with a temperature

$$T_\nu = \left(\frac{4}{11}\right)^{1/3} T_\gamma = 1.95\,\mathrm{K}, \tag{9.30}$$

for $T_\gamma = 2.73$ K. Although it is hard to see how such low-energy neutrinos could ever be detected directly, their gravitation is certainly not negligible: they contribute an energy density that is a factor $(7/8) \times (4/11)^{4/3}$ times that of the photons (the fact that neutrinos have $g = 1$ whereas photons have $g = 2$ is cancelled by the fact that neutrinos and antineutrinos are distinguishable particles). For three neutrino species, this enhances the energy density in relativistic particles by a factor 1.68.

MASSIVE NEUTRINOS Although for many years the conventional wisdom was that neutrinos were massless, this assumption began to be increasingly challenged around the end of the 1970s. Theoretical progress in understanding the origin of masses in particle physics meant that it was no longer natural for the neutrino to be completely devoid of mass. Also, experimental evidence (Reines *et al.* 1980), which in fact turned out to be erroneous, seemed to imply a non-zero mass of $m \sim 10$ eV for the electron neutrino. The consequences of this for cosmology could be quite profound, as relic neutrinos are expected to be very abundant. The above section showed that $n(\nu + \bar{\nu}) = (3/4)n(\gamma;\ T = 1.95\,\mathrm{K})$. That yields a total of 113 relic neutrinos in every cm^3 for each species.

The consequences of giving these particles a mass are easily worked out provided the mass is small enough. If this is the case, then the neutrinos were ultrarelativistic at decoupling and their statistics were those of massless particles. As the universe expands to $kT < m_\nu c^2$, the total number of neutrinos is preserved. Furthermore, their momentum redshifts as $p \propto 1/R$, so that the momentum-space distribution today will just be a redshifted version of the ultrarelativistic form. As discussed more fully below, the momentum-space distribution stays exactly what would have been expected for thermal-equilibrium neutrinos, even though they have long since decoupled. However, this illusion is broken once the temperature falls below $kT < m_\nu c^2$, because the effect of the rest-mass energy on the equilibrium occupation number causes the nonrelativistic momentum distribution to differ from the relativistic one. We therefore obtain the present-day mass density in neutrinos just by multiplying the zero-mass number density by m_ν, and the consequences for the cosmological density are easily worked out to be

$$\Omega h^2 = \frac{\sum m_i}{93.5\,\mathrm{eV}}. \tag{9.31}$$

For a low Hubble parameter $h \simeq 0.5$, an average mass of only 8 eV will suffice to close the universe. In contrast, the current laboratory limits to the neutrino masses are

$$\begin{aligned} \nu_e &\lesssim 15\,\mathrm{eV} \\ \nu_\mu &\lesssim 0.17\,\mathrm{MeV} \\ \nu_\tau &\lesssim 24\,\mathrm{MeV}. \end{aligned} \tag{9.32}$$

(see chapter 8). The more complicated case of neutrinos that decouple when they are already nonrelativistic is studied in chapter 12.

9.2 Relics of the big bang

The massive neutrino is the simplest example of a relic of the big bang: a particle that once existed in equilibrium, but which has decoupled and thus preserves a 'snapshot' of the properties of the universe at the time the particle was last in thermal equilibrium. The aim of this section is to give a little more detail on the processes that determine the final abundance of these relics.

So far, we have used a simple argument, which decrees that **freeze-out** has occurred when expansion and interaction timescales are equal. To do better than this, it is necessary to look at the differential equation that governs the abundance of particle species in the expanding universe. This is the **Boltzmann equation**, which considers the **phase-space density**: the joint probability density for finding a particle in a given volume element and in a given range of momentum, denoted by $f(\mathbf{x}, \mathbf{p})$. The general form of this equation is

$$\boxed{\frac{\partial f}{\partial t} + (\dot{\mathbf{x}} \cdot \boldsymbol{\nabla}_{\mathbf{x}}) f + (\dot{\mathbf{p}} \cdot \boldsymbol{\nabla}_{\mathbf{p}}) f = \dot{f}_c.} \qquad (9.33)$$

The lhs is just the fluid-dynamical convective derivative of the phase-space density, generalized to 6D space. The rhs is the collisional term, and the equation therefore just says that groups of particles maintain their phase-space density as they stream through phase space, unless modified by collisions (**Liouville's theorem**). The truth of this theorem is easily seen informally in one spatial dimension: a small square element $dx\, dv_x$ becomes sheared to a parallelogram of unchanged area, and so the phase-space density is unaltered. In the 3D case, write conservation of particle number as the phase-space continuity equation

$$\frac{\partial f}{\partial t} + \boldsymbol{\nabla}_{\mathbf{x}} \cdot (\dot{\mathbf{x}} f) + \boldsymbol{\nabla}_{\mathbf{p}} \cdot (\dot{\mathbf{p}} f) = 0. \qquad (9.34)$$

Now, since $\mathbf{p}$ and $\mathbf{x}$ are taken as independent variables, spatial derivatives of velocity and momentum derivatives of a spatial quantity such as the force vanish – leaving the Boltzmann equation.

The Boltzmann equation is a completely general formulation, and is of fundamental importance in most fields of physics. For example, fluid mechanics follows by taking the **moment equations**, in which f is integrated over momentum to give functions of position only. Define the mean number density as $n = \int f\, d^3p$ and the mean velocity as $\mathbf{u} = n^{-1} \int \dot{\mathbf{x}} f\, d^3p$. Integrating the Boltzmann equation just gives the usual equation for number conservation: $\dot{n} + \boldsymbol{\nabla} \cdot (n\mathbf{u}) = 0$. There are a few subtleties in obtaining this, however: since the collisional term redistributes momentum but conserves particles, it must integrate to zero; in the second term, because x and $\dot{x}$ are treated as independent, $\boldsymbol{\nabla}_x$ can be taken outside the integral; finally, integration by parts shows that

$$\int X\, (\dot{\mathbf{p}} \cdot \boldsymbol{\nabla}_{\mathbf{p}}) f\, d^3p = -\dot{\mathbf{p}} \cdot \int (\boldsymbol{\nabla}_{\mathbf{p}} X)\, f\, d^3p, \qquad (9.35)$$

where X is any quantity to be averaged. In the simplest case, $X = 1$ and so this term vanishes. Similarly, the moment equations obtained by multiplying the Boltzmann equation by $\dot{x}_i$ and by $(\dot{x})^2/2$ give conservation of momentum and of energy.

The Boltzmann equation has been written with respect to a fixed system of laboratory coordinates, but it is quite easily adapted to the expanding universe. We should now interpret the particle velocities as being relative to a set of uniformly expanding observers. A particle that sets off from $r = 0$ with some velocity will effectively slow down as it tries to overtake distant receding observers. After time t, the particle will have travelled $x = vt$, and so encountered an observer with velocity $dv = Hx$. According to this observer, the particle's momentum is now reduced by $dp = m dv = mHvt = Hpt$. There is therefore the appearance of a **Hubble drag** force:

$$\frac{\dot{p}}{p} = -H. \tag{9.36}$$

In the presence of density fluctuations, this needs to be supplemented by gravitational forces, which as usual manifest themselves through the affine connection [problem 9.4]. In a sense, most of cosmological theory comes down to solving the Boltzmann equation for photons plus neutrinos plus collisionless dark matter, coupled to the matter fluid via gravity in all cases and also by Thomson scattering in the case of photons. Since the interesting processes are operating at early times when the density fluctuations are small, this is an exercise in first-order relativistic perturbation theory. The technical difficulties in detail mean this will have to be omitted here (see Peebles 1980; Efstathiou 1990); when discussing perturbations, the main results can usually be understood in terms of a fluid approximation, and this approach is pursued in chapter 15.

Things are much easier in the case of homogeneous backgrounds, where spatial derivatives can be neglected. The Boltzmann equation then has the simple form

$$\frac{\partial f}{\partial t} - Hp\frac{\partial f}{\partial p} = \dot{f}_c. \tag{9.37}$$

The collision term is usually dominated by particle–antiparticle annihilations (assuming for the moment that the numbers of each are identical, so that there is no significant asymmetry):

$$\dot{f}_c = -\int \langle \sigma v \rangle f \bar{f} \; d^3\bar{p}, \tag{9.38}$$

where $\langle \sigma v \rangle$ is the velocity-averaged product of the cross-section and the velocity. We can take $\langle \sigma v \rangle$ outside the integral even if it is not constant provided it is evaluated at some suitable average energy. Integrating over momentum then gives the moment equation for the number density,

$$\dot{n} + 3Hn = -\langle \sigma v \rangle n^2 + S, \tag{9.39}$$

where S is a source term added to represent the production of particles from thermal processes – effectively pair creation. This term is fixed by a thermodynamic equilibrium argument: for a non-expanding universe, n will be constant at the equilibrium value for that temperature, n_T, showing that

$$S = \langle \sigma v \rangle n_T^2. \tag{9.40}$$

If we define comoving number densities $N \equiv a^3 n$, the rate equation can be rewritten in the simple form

$$\boxed{\frac{d\ln N}{d\ln a} = -\frac{\Gamma}{H}\left[1 - \left(\frac{N_T}{N}\right)^2\right],} \tag{9.41}$$

where $\Gamma = n\langle\sigma v\rangle$ is the interaction rate experienced by the particles.

Unfortunately, this equation must be solved numerically. The main features are easy enough to see, however. Suppose first that the universe is sustaining a population in approximate thermal equilibrium, $N \simeq N_T$. If the population under study is relativistic, N_T does not change with time, because $n_T \propto T^3$ and $T \propto a^{-1}$. This means that it is possible to keep $N = N_T$ exactly, whatever Γ/H. It would however be grossly incorrect to conclude from this that the population stays in thermal equilibrium: if $\Gamma/H \ll 1$, a typical particle suffers no interactions even while the universe doubles in size, halving the temperature. A good example is the microwave background, whose photons last interacted with matter at $z \simeq 1000$. The CMB nevertheless still appears to be equilibrium black body radiation because the number density of photons has fallen by the right amount to compensate for the redshifting of photon energy. It sounds like an incredible coincidence, but is in fact quite inevitable when looked at from the quantum-mechanical point of view. This says that the occupation number of a given mode, $= (\exp\hbar\omega/kT - 1)^{-1}$ for thermal radiation, is an adiabatic invariant that does not change as the universe expands – only the frequency alters, and thus the apparent temperature.

Now consider the opposite case, where the thermal solution would be nonrelativistic, with $N_T \propto T^{-3/2}\exp(-mc^2/kT)$. If the background is to stay at the equilibrium value, the lhs of the rate equation must therefore be $\gg -1$. This is consistent if $\Gamma/H \gg 1$, because then the $(N_T/N)^2$ term on the rhs can still be close to unity. However, if $\Gamma/H \ll 1$, there must be a deviation from equilibrium. When N_T changes sufficiently fast with a, the actual abundance cannot keep up, so that the $(N_T/N)^2$ term on the rhs becomes negligible and $d\ln N/d\ln a \simeq -\Gamma/H$, which is $\ll 1$. There is therefore a critical time at which the reaction rate drops low enough that particles are simply conserved as the universe expands – the population has **frozen out**. This provides a more detailed justification for the intuitive rule-of-thumb used above to define decoupling,

$$N(a \to \infty) = N(\Gamma/H = 1). \tag{9.42}$$

Exact numerical solutions of the rate equation almost always turn out very close to this simple rule (see chapter 5 of Kolb & Turner 1990).

9.3 The physics of recombination

One of the critical epochs in the evolution of the universe is reached when the temperature drops to the point ($T \sim 1000$ K) where it is thermodynamically favourable for the ionized plasma to form neutral atoms. This process is known as **recombination**: a complete misnomer, as the plasma has always been completely ionized up to this time. The historical origin of the name lies in HII regions: interstellar plasma is ionized by ultraviolet radiation from a hot star, and the region emits characteristic **recombination radiation** as the electrons and ions continuously re-form atoms before being photoionized once more. In the cosmological context, the process might seem simpler owing to the absence of a photoionizing source. In practice, things turn out to be more complex, but from these complications a wonderful observational simplicity eventually emerges.

EQUILIBRIUM (SAHA) THEORY The first question to ask is how the ionization of a plasma should change in thermal equilibrium. From the point of view of 'freeze-out', the interaction timescale in the plasma is so short that the expansion of the universe can initially be regarded as quasistatic. This situation is described by the **Saha equation**, which may be derived in two different ways.

(1) Consider a single hydrogen atom, for which we want to know the relative probabilities that the electron is bound or free. In equilibrium, with the probability of a given state being proportional to the Boltzmann factor, this must be given by the ratio of partition functions:

$$\frac{P_{\rm free}}{P_{\rm bound}} = \frac{\sum_{\rm free} g_i e^{-E_i/kT}}{\sum_{\rm bound} g_i e^{-E_i/kT}} = \frac{Z_{\rm free}}{Z_{\rm bound}}. \tag{9.43}$$

In the case of this *single* object (the electron) the degeneracy factor $g = 2$, from spin. Now, a good approximation is that $Z_{\rm bound} = g \exp(\chi/kT)$ where χ is the **ionization potential** of hydrogen. This will be valid at low temperatures $kT \lesssim \chi$, where the contribution of excited states to Z is negligible. Now, in a box of volume V,

$$Z_{\rm free} = g \frac{V}{(2\pi)^3} \int E^{-p^2/2m} \frac{d^3p}{\hbar^3} = (2\pi mkT)^{3/2} \frac{gV}{(2\pi\hbar)^3}. \tag{9.44}$$

To use this expression, we have to decide what V is; do we set $V \to \infty$ in an infinite universe, and so conclude that the atom is always ionized, whatever T may be? Clearly not: the volume 'available' to each electron is effectively $1/n_e$ – the reciprocal of the electron number density. Using this, and defining the fractional ionization x ($n_e = xn$, where n is the total nucleon number density, $n_{\rm H} + n_p$), we get the **Saha equation**

$$\boxed{\frac{x^2}{1-x} = \frac{(2\pi m_e kT)^{3/2}}{n(2\pi\hbar)^3} e^{-\chi/kT}.} \tag{9.45}$$

(2) The above argument illustrates quite well what is going on, but is clearly only heuristic, since it evades having to consider the multiparticle nature of the gas. More rigorously, we need to minimize the total free energy of the system, $F = -kT \ln Z_T$, where the overall partition function is

$$Z_T = \frac{Z_e^{N_e}}{N_e!} \frac{Z_p^{N_p}}{N_p!} \frac{Z_{\rm H}^{N_{\rm H}}}{N_{\rm H}!} \tag{9.46}$$

(for independent indistinguishable particles where the density is low enough to neglect degeneracy effects). If we apply Stirling's approximation to the factorial terms and set $(\partial/\partial N_e) \ln Z_T = 0$, then the relation

$$\frac{N_e^2}{N - N_e} = \frac{Z_e Z_p}{Z_{\rm H}} \tag{9.47}$$

is obtained. The volume V appearing in the single-particle partition functions is now just the total volume of the system under consideration, and so it may be eliminated in a more acceptable fashion via $n = N/V$ etc. This yields exactly the same equation for the fractional ionization as was obtained via the single-particle approach.

The Saha result is quite interesting: x is not simply a function of temperature and does depend on the density also. However, the $\exp(-\chi/kT)$ term ensures that x in practice always diverges from 1 when $kT \sim \chi$.

NON-EQUILIBRIUM REALITY The problem with the Saha approach is that, although it describes the initial phase of the departure from complete ionization, the assumption of equilibrium rapidly ceases to be valid. This may seem paradoxical: processes with small cross-sections such as weak interactions may freeze out, but surely not the electromagnetic interaction? In fact, the problem is just the reverse: electromagnetic interactions are too fast. Consider a single recombination; if this were to occur directly to the ground state, a photon with $\hbar\omega > \chi$ would be produced. Such photons are clearly bad news for recombination: they travel until they encounter a neutral atom, whereupon they ionize it. Recombination in an infinite universe therefore has to proceed via a set of smaller steps.

Even so, there is a problem: highly excited atoms can be produced by a series of small transitions, but to reach the ground state requires the production of photons at least as energetic as the $2P \rightarrow 1S$ spacing (Lyman α, with $\lambda = 1216$ Å). Multiple absorption of these photons will cause reionization once they become abundant, so it would now appear that recombination can never occur at all (unlike in a finite HII region, where the Lyα photons can escape; see e.g. Osterbrock 1974). There is a way out, however, using **two-photon emission**. The $2S \rightarrow 1S$ transition is strictly forbidden at first order and one can only conserve energy and angular momentum in the transition by emitting a *pair* of photons. This gives the mechanism we need for transferring the ionization energy into photons with $\lambda > \lambda_{\mathrm{Ly}\alpha}$. Being a process of second order in perturbation theory, this is slow (lifetime $\simeq 0.1$ s); because recombination has to pass through this bottleneck, it actually proceeds at a rate completely different from the Saha prediction.

A highly stripped-down analysis of events simplifies the hydrogen atom to just two levels ($1S$ and $2S$). Any chain of recombinations that reaches the ground state can be ignored through the above argument: these reactions produce photons that are immediately re-absorbed elsewhere, so they have no effect on the ionization balance. The main chance of reaching the ground state comes through the recombinations that reach the $2S$ state, since some fraction of the atoms that reach that state will suffer two-photon decay before being re-excited. The rate equation for the fractional ionization is thus

$$\frac{d(nx)}{dt} = -R\,(nx)^2\,\frac{\Lambda_{2\gamma}}{\Lambda_{2\gamma} + \Lambda_{\mathrm{U}}(T)}, \tag{9.48}$$

where n is the number density of protons, x is the fractional ionization, R is the recombination coefficient ($R \simeq 3 \times 10^{-17} T^{-1/2}\,\mathrm{m^3 s^{-1}}$), $\Lambda_{2\gamma}$ is the two-photon decay rate, and $\Lambda_{\mathrm{U}}(T)$ is the stimulated transition rate upwards from the $2S$ state. This equation just says that recombinations are a two-body process, which create excited states that cascade down to the $2S$ level, from whence a competition between the upward and downward transition rates determines the fraction that make the downward transition. A fuller discussion (see chapter 6 of Peebles 1993) would include a number of other processes: depopulation of the ground state by inverse two-photon absorption; redshifting of Lyα photons due to the universal expansion, which can prevent their being re-absorbed.

However, at the redshifts of practical interest (1000 to 10), the simplified equation captures the main effect.

An important point about the rate equation is that it is only necessary to solve it once, and the results can then be scaled immediately to some other cosmological model. Consider the rhs: both R and $\Lambda_{\rm U}(T)$ are functions of temperature, and thus of redshift only, so that any parameter dependence is carried just by n^2, which scales $\propto (\Omega_{\rm B} h^2)^2$; here $\Omega_{\rm B}$ is the baryonic density parameter. Similarly, the lhs depends on $\Omega_{\rm B} h^2$ through n; the other parameter dependence comes if we convert time derivatives to derivatives with respect to redshift:

$$\frac{dt}{dz} \simeq -3.09 \times 10^{17} (\Omega h^2)^{-1/2} z^{-5/2} \text{ s}, \tag{9.49}$$

for a matter-dominated model at large redshift (Ω is the total density parameter). Putting these together, the fractional ionization must scale as

$$\boxed{x(z) \propto \frac{(\Omega h^2)^{1/2}}{\Omega_{\rm B} h^2}.} \tag{9.50}$$

This is a very different scaling from the prediction of the Saha equation.

The simplest case for solution of the rate equation is at late times, where the recombination rate is slow enough for the two-photon process to cope with the influx of excited atoms, and where the universe is cool enough for excitations from $2S$ to be neglected. The evolution equation is then

$$\frac{d \ln x}{d \ln z} = 60 x z \frac{\Omega_{\rm B} h^2}{(\Omega h^2)^{1/2}}. \tag{9.51}$$

The criterion neglects the expansion, and so it is invalid if the lhs is smaller than unity. It defines the freeze-out level of ionization at a given redshift; typically, the full non-expanding solution for $x(z)$ falls below this level at redshifts of a few hundred, so that x never falls below $\sim 10^{-4}$.

THE LAST-SCATTERING SHELL Putting in all the relevant processes, Jones & Wyse (1985) found the fractional ionization x near $z = 1000$ to be well approximated by

$$x(z) = 2.4 \times 10^{-3} \frac{(\Omega h^2)^{1/2}}{\Omega_B h^2} \left(\frac{z}{1000} \right)^{12.75}. \tag{9.52}$$

The scaling with Ω and h has a marvellous consequence. If we work out the optical depth to Thomson scattering, $\tau = \int n_e x \sigma_{\rm T} dr_{\rm prop}$, we find just

$$\boxed{\tau(z) = 0.37 \left(\frac{z}{1000} \right)^{14.25},} \tag{9.53}$$

which is independent of cosmological parameters. The rate equation causes $x(z)$ to scale in just the right way that the optical depth is a completely robust quantity. Because τ changes rapidly with redshift, the distribution function for the redshift at which photons were last scattered, $e^{-\tau} d\tau/dz$, is sharply peaked, and is well fitted by a Gaussian of mean redshift 1065 and standard deviation in redshift 80. Thus, when we look at the sky, we can expect to see in all directions photons that originate from a **last-scattering surface** at $z \simeq 1065$. The redshifted radiation from this photosphere is the 2.73 K microwave background found by Penzias & Wilson (1965). This independence of parameters is not quite exact in detail, however, and very accurate work needs to solve the evolution equations exactly (e.g. appendix C of Hu & Sugiyama 1995). Nevertheless, the fact that the properties of the last-scattering surface are almost independent of all the unknowns in cosmology is immensely satisfying, and gives us at least one relatively solid piece of ground to act as a base in exploring the trackless swamp of cosmology.

It is important to emphasize that the surface is rather fuzzy, and that the photons we see were scattered from a range of redshifts. Any fluctuations in temperature at last scattering that are of short wavelength are therefore going to be quite heavily attenuated by averaging through the last-scattering shell. This smearing effect turns out to be of great importance in understanding the microwave sky; without it, anisotropies would be much easier to detect (see chapter 18).

Two last remarks are in order here. First, the cosmic microwave background radiation (CMB) provides the answer to Olbers' paradox about the darkness of the night sky. Also, historically, Gamow and collaborators used the nucleosynthesis argument given below to *predict* a background of this kind (Gamow 1946; Alpher, Bethe & Gamow 1948) – a prediction that unfortunately was not taken very seriously at the time.

9.4 The microwave background

OBSERVATIONS In a famous piece of serendipity, the CMB radiation was discovered in 1965, by Penzias and Wilson, who located an unaccounted-for source of noise in a radio telescope intended for studying our own galaxy. The timing of the discovery was especially ironic, given that experiments were under way at that time to test the theoretical prediction that such a background should exist – the progress of science is rarely a tidy business. Since the initial detection of the microwave background at $\lambda = 7.3$ cm, measurements of the spectrum have been made over an enormous range of wavelengths, from the depths of the Rayleigh–Jeans regime at 74 cm to well into the Wien tail at 0.5 mm. Prior to 1990, observations in the Rayleigh–Jeans portion of the spectrum had established the constancy of the temperature for $\lambda \gtrsim 1$ mm, and yielded the overall average $T = 2.735 \pm 0.017\,\mathrm{K}$ (Gush *et al.* 1990). There were however suggestions of deviations from a Planck spectrum at short wavelengths in the Wien tail, based on short-lived balloon or rocket experiments. Given that the flux is falling rapidly here, there were many worries about systematics, and these worries were seen to be well founded following the 1989 launch of **COBE** – the cosmic background explorer satellite. Early data showed the spectrum to be very close to a pure Planck function (Mather *et al.* 1990), and the final result verifies the lack of any distortion with breathtaking precision (figure 9.1).

The COBE temperature measurement and 95% confidence range of

$$\boxed{T = 2.728 \pm 0.004\,\mathrm{K}} \tag{9.54}$$

improves significantly on the ground-based experiments, and indeed is limited in accuracy mainly by the systematics involved in making a good comparison black body. The lack of distortion in the shape of the spectrum is astonishing, and limits the chemical potential

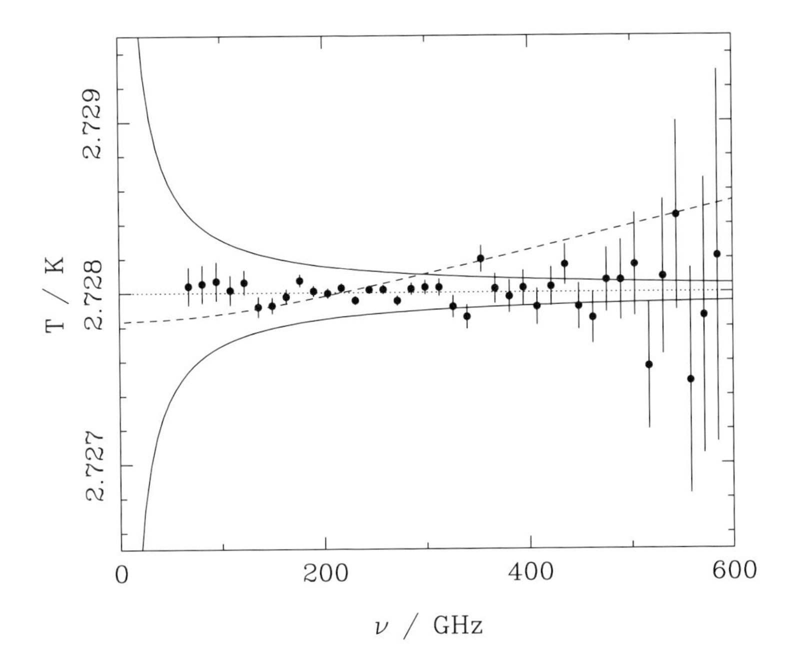

Figure 9.1. The spectrum of the microwave background as observed by the FIRAS experiment on the COBE satellite (Fixsen *et al.* 1996). The overall temperature scale has a systematic uncertainty of up to 0.004 K, but the deviations in shape are very much smaller even than this. The lines show the sort of distortions that might be expected from Compton scattering (solid) or non-zero chemical potential (broken).

to $|\mu| < 9 \times 10^{-5}$ (Fixsen *et al.* 1996). These results also allow the limit $y \lesssim 1.5 \times 10^{-5}$ to be set on the Compton-scattering distortion parameter (as discussed in the section on the X-ray background in chapter 12). These limits are so stringent that many competing cosmological models can be eliminated. Those with a distaste for the idea of a hot big bang suggested that the CMB could be due to starlight reprocessed by dust grains – but this would lead to a mixture of Planck functions because radiation would be received from a variety of redshifts. What the COBE result says is that the radiation must come from a shell within which temperature T scales accurately as $1/R(t)$ (so that the observed temperature of each shell is constant), which is something that only arises in the standard big bang picture. The lack of any distortion from a Planck function moreover limits a number of possibilities for exotic processes in the later phases of the big bang. Although the CMB last scattered at $z \simeq 1000$, it was in one sense already a frozen-out background: Compton scattering conserves photon number, but full thermalization requires emission and absorption processes such as bremsstrahlung (where an electron is caused to radiate by acceleration in the electric field of an ion; see chapter 5 of Rybicki & Lightman 1979). These freeze out at a higher redshift – between 10^6 and 10^7 (see chapter 5 of Partridge 1995). If significant amounts of energy were generated in the factor 1000 expansion between this era and last scattering (for example, by the decay of some exotic particle), it would not be possible to thermalize the spectrum, and a large chemical potential would be observed. The lack of any such distortion means that decays at $z \lesssim 10^6$ must contribute an energy density $\lesssim 10^{-4}$ of that of the CMB.

It should be noted that a significant problem in making these accurate measurements of the CMB is the emission from the Milky Way. This has an obvious quadrupolar signature in the form of the galactic plane, but since we live near to the centre, there is inevitably some isotropic component. To subtract this, we need to look at the frequency dependence of the emission. At low frequencies, we see $\nu^{-0.8}$ synchrotron emission, whereas at high frequency, thermal emission from galactic dust at a few tens of kelvins starts to dominate. The 'cleanest' wavelength, where these effects are minimized, is at $\lambda \simeq 5$ mm; the worst wavelength is around 20 μm, where the dust emission peaks. This foreground is not so important for determining the overall CMB spectrum, but when it comes to measuring *fluctuations* in the background at a fractional precision of 10^{-5} or less, the modelling and subtraction of galactic emission can become a dominant source of uncertainty. The practical problems of extracting cosmological signals from the real sky are discussed at length by Partridge (1995).

THE DIPOLE ANISOTROPY The topic of deviations from isotropy in the microwave background is one of the dominant aspects of modern cosmology, and it is therefore treated separately in chapter 18. However, this is a good place to study one important special case: the effects of the earth's absolute motion. The effect of the observer's motion on the specific intensity of the CMB can be calculated by using

$$\frac{I_\nu}{\nu^3} = \text{invariant}. \tag{9.55}$$

This is a general result that we have seen illustrated in the redshift dependence of surface brightness in cosmology (see chapter 3). It applies because the invariant is proportional to the **phase-space density** of the CMB photons: $\nu^2\, d\nu$ is proportional to the volume element in momentum space, so that the number of photons is given by

$$N_\gamma = \frac{c^2}{(2\pi\hbar)^4} \frac{I_\nu}{\nu^3}\, d^3x\, d^3p, \tag{9.56}$$

which justifies the statement that I_ν/ν^3 is proportional to the phase-space density. Now, it is easy to show through direct application of the Lorentz transformation that the phase-space volume element is a relativistic invariant (dx gets length-contracted, but dp_x gets boosted); the phase-space density must then also be invariant, which completes the result we needed.

Applying this to black body radiation, we find (of course) that the phase-space density is just proportional to the occupation number

$$\frac{I_\nu}{\nu^3} \propto \left(e^{h\nu/kT} - 1\right)^{-1}. \tag{9.57}$$

Now, we observe at some frequency ν, which has been Doppler-boosted from frequency ν_0 by our motion ($\nu = \mathcal{D}\, \nu_0$, where $\mathcal{D}$ denotes the Doppler factor). The invariant occupation number is

$$n = \frac{1}{\exp[h(\nu/\mathcal{D})/kT_0] - 1}, \tag{9.58}$$

where T_0 is the unboosted temperature. This tells us that boosted thermal radiation still appears *exactly thermal*, with a new temperature boosted in exactly the same way as frequency:

$$T = \mathscr{D}\, T_0. \tag{9.59}$$

There is thus no prospect of deducing our absolute space velocity by measuring deviations from a Planckian spectrum.

The Doppler factor is easily evaluated by writing the t- and x- components of the 4-frequency k^μ for photons travelling towards the observer at an angle θ to the x-axis:

$$k^\mu = \frac{\omega}{c}(1, -\cos\theta). \tag{9.60}$$

We now transform *out* of the moving frame: $c\,dt' = \gamma[c\,dt + (v/c)dx]$. This yields

$$\boxed{T_{\rm obs} = \frac{T_0}{\gamma[1 - (v/c)\cos\theta]},} \tag{9.61}$$

which can be expanded as a series of multipoles:

$$T_{\rm obs} = T_0\left[1 + \frac{v}{c}\,\cos\theta + \frac{1}{2}\left(\frac{v}{c}\right)^2 \cos 2\theta + O(v^3)\right]. \tag{9.62}$$

The dipole is the only term that has been detected; the induced quadrupole is rather below current sensitivity. The dipole allows the absolute space velocity of the Earth to be measured, and this was first achieved by Smoot, Gorenstein & Muller (1977). The COBE measurement of this motion is

$$v_\oplus = 371 \pm 1\,{\rm km\,s^{-1}} \quad \text{towards} \quad (\ell, b) = (264°, 48°) \tag{9.63}$$

(Fixsen *et al.* 1996). This velocity measurement has been possible despite the invariant character of thermal radiation because of a leap of faith concerning the CMB: that it is very nearly isotropic. An observed dipole could of course be intrinsic to the universe (we cannot distinguish between this and any effect due to our motion). However, because the quadrupole term is only $\sim 1\%$ of the dipole, it is almost universally agreed that the observed dipole is entirely due to the motion of the Earth.

Of course, it was known long before the discovery of the CMB that the Earth is unlikely to be a sensible rest frame to adopt. There is the rotation of the Earth about the Milky Way; this is usually taken, according to the IAU convention, to be $300\,{\rm km\,s^{-1}}$ towards $(\ell, b) = (90°, 0)$ – i.e. the galactic rotation has a left-hand screw. There are other corrections for the motion of the Milky Way relative to other members of the local group, but this is the dominant term. It was therefore predicted that a microwave dipole due to this motion would be seen. Imagine the surprise when the observed dipole turned out to be in a completely different direction! Correcting for the motion of the Earth with respect to the local group actually *increases* the velocity: the local group has the approximate motion

$$\boxed{v_{\rm LG} \simeq 600\,{\rm km\,s^{-1}} \quad \text{towards} \quad (\ell, b) \simeq (270°, 30°).} \tag{9.64}$$

The obvious interpretation of this motion is that it is caused by the perturbing gravity of large-scale structures beyond the local group. The search for these perturbations has been one of the most interesting stories in recent cosmological research. Understanding the peculiar velocity field in the universe is an important adjunct to the study of anisotropies in the CMB.

9.5 Primordial nucleosynthesis

At sufficiently early times, the temperature of the universe reached that of the centre of the Sun (1.55×10^7 K), where we know that nuclear reactions occur. Nuclear reactions were also important in the early universe, but conditions differed from those in stars in several ways. The cosmological density at this time was dominated by radiation, and the matter density was much lower than in the Sun (roughly 10^{-5} versus 10^{+5} kg m^{-3} respectively). Nuclear reactions in the early universe were thus ineffectively slow at these temperatures, and only reached the critical freeze-out rate at rather higher temperatures ($\sim 10^9$ K).

The abundance of light elements that resulted from these early reactions is fixed by an argument that can be outlined quite simply. In equilibrium, the numbers of neutrons and protons should vary as

$$\frac{N_n}{N_p} = e^{-\Delta mc^2/kT} \simeq e^{-1.5(10^{10}\,\mathrm{K}/T)}. \tag{9.65}$$

The reason that neutrons exist today is that the timescale for the weak interactions needed to keep this equilibrium set-up eventually becomes longer than the expansion timescale. The reactions thus rapidly cease, and the neutron–proton ratio undergoes **freeze-out** at some characteristic value. In practice this occurs at $N_n/N_p \simeq 1/6$. If most of the neutrons ended up in ^{4}He, we would then expect 25% He by mass – which is very nearly what we see. One of the critical calculations in cosmology is therefore to calculate this freeze-out process in detail. As the following outline of the analysis will show, the result is due to a complex interplay of processes, and the fact that there is a significant primordial abundance of anything other than hydrogen is a consequence of a number of coincidences.

NEUTRON FREEZE-OUT The problem of following all the nuclear reactions that could potentially take place is daunting, but the problem is simpler at early times, when the temperatures were so high that all possible nuclei have binding energies less than typical photon energies. The only nuclear reactions that matter initially are the weak interactions that cause conversions between protons and neutrons:

$$\begin{aligned} p + e &\leftrightarrow n + \nu \\ p + \bar{\nu} &\leftrightarrow n + e^+. \end{aligned} \tag{9.66}$$

The main systematics of the relic abundances of the light elements can be understood by looking at the neutron-to-proton ratio and how it evolves.

The number density of neutrons obeys the kinetic equation

$$\frac{dn_n}{dt} = (\Lambda_{pe} + \Lambda_{p\nu})\, n_p - (\Lambda_{ne} + \Lambda_{n\nu})\, n_n, \tag{9.67}$$

where the rate coefficients Λ_i refer to the four possible processes given above. To these two-body processes should also be added the spontaneous decay of the neutron, which has the e-folding lifetime

$$\boxed{\tau_n = 887 \pm 2 \text{ s}} \tag{9.68}$$

(according to the Particle Data Group; see the bibliography). We neglect this for now, as it turns out that neutron freeze-out happens at slightly earlier times.

The quantum-field calculation of the rate coefficients is not difficult, because of the simple form of the Fermi Lagrangian for the weak interaction. Recall that this is proportional to the Fermi constant ($G_{\rm F}$) times the fields of the various particles that participate, and that each field is to be thought of as a sum of creation and annihilation operators. This means that the matrix elements for all processes where various particles change their occupation numbers by one are the same, and the rate coefficients differ only by virtue of integration over states for the particles involved.

For example, the rate for neutron decay is

$$\tau_n^{-1} \propto G_{\rm F}^2 \left[\frac{1}{(2\pi\hbar)^3}\right]^2 2 \int d^3p_e\, d^3p_\nu\, \delta(\epsilon_e + \epsilon_\nu - Q), \tag{9.69}$$

where Q is the neutron–proton energy difference. This expression can be more or less written down at sight. The delta function expresses conservation of energy and as usual arises automatically from integration over space, rather than having to be put in by hand. The factor of 2 before the integral allows for electron helicity, but there is no need to be concerned with the overall constant of proportionality. By performing the integral, this expression can be reduced [problem 9.6] to

$$\tau_n^{-1} \propto \frac{G_{\rm F}^2}{2\pi^4}\, 1.636\, m_e^5 \quad \text{(natural units).} \tag{9.70}$$

The reason why the overall constant of proportionality is not needed is that all other related weak processes have rates of the same form. The only difference is that, unlike the free decay in a vacuum just analysed, we need to include the probabilities that the initial and final states are occupied. For example, in $n + \nu \rightarrow p + e$, we need a factor n_ν for the initial neutrino state and $1 - n_e$ for the final electron state (effectively to allow for the fermionic equivalent of stimulated plus spontaneous emission; if the particles involved were bosons, this would become $1 + n$). The rate for this process is therefore

$$\Lambda_{n\nu} = (1.636 m_e^5 \tau_n)^{-1} \int_0^\infty \frac{p_e\, \epsilon_e\, p_\nu^2\, dp_\nu}{[1 + \exp(p_\nu/kT)]\,[1 + \exp(-\epsilon_e/kT)]}, \tag{9.71}$$

where $\epsilon_e = p_\nu + Q$. Very similar integrals can be written down for the other processes involved.

Since $Q \sim m_e c^2$, it is clear that at high temperatures $kT \gg m_e c^2$, all the rate coefficients will be of the same form; both p_e and ϵ_e can be replaced by p_ν in the integral, leaving a single rate coefficient to be determined by numerical integration:

$$\boxed{\Lambda = 13.893\, \tau_n^{-1} \left(\frac{kT}{m_e c^2}\right)^5 = \left(\frac{10^{10.135}\,\text{K}}{T}\right)^{-5} \text{s}^{-1}.} \tag{9.72}$$

Since the number density of the thermal background of neutrinos and electrons is proportional to T^3, this says that the effective cross-section for these weak interactions scales as (energy)^2. The radiation-dominated era has

$$t = \sqrt{\frac{3}{32\pi G\rho}} = \left(\frac{10^{10.125}\,\mathrm{K}}{T}\right)^2 \mathrm{s}, \tag{9.73}$$

allowing for three massless neutrinos. The T^{-2} dependence of the expansion timescale is much slower than the interaction timescale, which changes as T^{-5}, so there is a quite sudden transition between thermal equilibrium and freeze-out, suggesting that weak interactions switch off, freezing the neutron abundance at a temperature of $T \simeq 10^{10.142}$ K. This implies an equilibrium neutron–proton ratio of

$$\frac{N_n}{N_p} = e^{-Q/kT} = e^{-10^{10.176\,\mathrm{K}}/T} \simeq 0.34. \tag{9.74}$$

This obviously cannot be a precisely correct result, because the freeze-out condition was calculated assuming a temperature well above the electron mass threshold, whereas it appears that freeze-out actually occurs at about this critical temperature. The rate Λ is in fact a little larger at the threshold than the high-temperature extrapolation would suggest, so the neutron abundance is in practice lower.

NEUTRINO FREEZE-OUT There is a yet more serious complication, because another important process that occurs around the same time is neutrino decoupling. The weak reaction that keeps neutrino numbers in equilibrium at this late time is $\nu + \bar{\nu} \leftrightarrow e^+ + e^-$. The rate for this process can be found in exactly the same way as above. At high energies, the result is identical, save only that the squared matrix element involved is smaller by a factor of about 5 because of the axial coupling of nucleons: the neutrino rate scales as G_{F}^2, as opposed to $G_{\mathrm{F}}^2(1 + 2g_{\mathrm{A}}^2)$ for nucleons. This leads to the neutrinos decoupling at a slightly higher temperature:

$$\boxed{T(\nu\ \mathrm{decoupling}) \simeq 10^{10.5}\ \mathrm{K}.} \tag{9.75}$$

This is uncomfortably close to the electron mass threshold, but just sufficiently higher that it is not a bad approximation to say that all e^+e^- annihilations go into photons rather than neutrinos. During the time at which nucleons are decoupling, the neutrino and photon temperatures are therefore becoming different, and a detailed calculation must account for this. The resulting freeze-out temperature is very close to 10^{10} K, at which point the neutron-to-proton ratio is about 1:3.

Unfortunately, we are still not finished, because neutrons are not stable. It does not matter what abundance of them freezes out: unless they can be locked away in nuclei before $t = 887$ s, the relic abundance will decay freely to zero. The freeze-out point occurs at an age of a few seconds, so there are only a few e-foldings of expansion available in which to salvage some neutrons. So far, a remarkable sequence of coincidences has been assembled, in that the freeze-out of neutrinos and nucleons happens at about the same time as e^+e^- annihilation, which is also a time of order τ_n. It may seem implausible that we can add yet more – i.e. that nuclear reactions will become important at about the same time – but this is just what does happen.

CONSTRUCTION OF NUCLEONS This coincidence is not surprising, since the deuteron binding energy of 2.225 MeV is only 4.3 times larger than $m_e c^2$ and only 1.7 times larger than the neutron–proton mass difference. At higher temperatures, the strong interaction $n + p = \mathrm{D} + \gamma$ is fast enough to produce deuterium, but thermal equilibrium favours a small deuterium fraction – i.e. typical photons are energetic enough to disrupt deuterium nuclei very easily. Furthermore, because of the large photon-to-baryon ratio, the photons can keep the deuterium from forming until the temperature has dropped well below the binding energy of the deuteron. The equilibrium abundance of deuterium is set in much the same way as the ionization abundance of hydrogen, and so obeys an equation that is identical (apart from spin-degeneracy factors) to the Saha equation for hydrogen ionization:

$$\frac{n_{\mathrm{D}}}{n_p n_n} = \frac{3}{4} \frac{(2\pi\hbar)^3}{(2\pi k T m_p m_n / m_{\mathrm{D}})^{3/2}} \exp(\chi/kT), \tag{9.76}$$

where χ is the binding energy. As with hydrogen ionization, this defines an abrupt transition between the situations where deuterium is rare and where it dominates the equilibrium. The terms outside the exponential keep the deuterium density low until $kT \ll \chi$: the $n_{\mathrm{D}} = n_n$ crossover occurs at

$$\boxed{T_{\mathrm{deuteron}} \simeq 10^{8.9}\ \mathrm{K},} \tag{9.77}$$

or a time of about 3 minutes. An exact integration of the weak-interaction kinetic equation for the neutron abundance at $n_{\mathrm{D}} = n_n$ (including free neutron decay, which is significant) gives

$$\boxed{\frac{n_n}{n_p} \simeq 0.163\ (\Omega_{\mathrm{B}} h^2)^{0.04}\ (N_\nu/3)^{0.2}} \tag{9.78}$$

(see e.g. chapter 4 of Kolb & Turner 1990; chapter 6 of Peebles 1993). The dependences on the baryon density and on the number of neutrino species are easily understood. A high baryon density means that the Saha equation gives a higher deuterium abundance, increasing the temperature at which nuclei finally form and giving a higher neutron abundance because fewer of them have decayed. The effect of extra neutrino species is to increase the overall rate of expansion, so that neutron freeze-out happens earlier, again raising the abundance.

THE PRIMORDIAL HELIUM ABUNDANCE The argument so far has produced a universe consisting of just hydrogen and deuterium, but this is not realistic, as one would expect ^{4}He to be preferred on thermodynamic grounds, owing to its greater binding energy per nucleon (7 MeV, as opposed to 1.1 MeV for deuterium). To see this in more detail, it is worth giving a different derivation of the Saha abundances of nuclei. In the nonrelativistic limit, the equilibrium number density of a given species will depend on its chemical potential:

$$n = g \frac{(2\pi m k T)^{3/2}}{(2\pi\hbar)^3} \exp\left(\frac{\mu - mc^2}{kT}\right). \tag{9.79}$$

In equilibrium, the chemical potential of a nucleon will equal that of the sum of its constituents: $\mu = Z\mu_p + (A - Z)\mu_n$, for atomic number Z and mass number A, and the proton and neutron potentials can be eliminated in terms of their number densities. This allows the number density of the nucleus to be rewritten as

$$n = \frac{g}{2^A} \left(\frac{m}{m_p^Z m_n^{A-Z}} \right)^{3/2} \left(\frac{2\pi\hbar^2}{kT} \right)^{3(A-1)/2} n_p^Z\, n_n^{A-Z} \exp\left(\frac{\chi}{kT} \right), \tag{9.80}$$

where the binding energy has been written as χ. As expected, this says that nuclei with a large binding energy are favoured; however, if $n_p \hbar^3 (m_p kT)^{-3/2} \ll 1$ (which is always true for a nonrelativistic species; cf. the expression for the thermal background in this limit), then large A for a given χ will be disfavoured, because this factor occurs to the Ath power (assuming $n_n \sim n_p$). The larger binding energy wins over this term in the case of ^{4}He, but not for more massive nuclei. The result is that, in equilibrium, the first nuclei to come into existence in significant numbers should be ^{4}He: the abundance of ^{4}He relative to protons should reach unity at an energy of about 0.3 MeV, at which point the relative abundance of deuterium is only $\sim 10^{-12}$ (see chapter 4 of Kolb & Turner 1990).

Since the simplest way to synthesize helium is by fusing deuterium, it is no surprise that the equilibrium prediction fails miserably in the expanding universe: the production of helium must await the synthesis of significant quantities of deuterium, which we have seen happens at a temperature roughly one-third that at which helium would be expected to dominate. What the thermodynamic argument does show, however, is that it is expected that the deuterium will be rapidly converted to helium once significant nucleosynthesis begins. This argument is what allows us to expect that the helium abundance can be calculated from the final n/p ratio. If all neutrons are incorporated into ^{4}He, then the number density of hydrogen is set by the remaining protons: $n_{\rm H} = n_p - n_n$. The mass fraction of helium, Y, is unity minus the hydrogen fraction, so that

$$\boxed{Y = 1 - \frac{n_p - n_n}{n_p + n_n} = 2\left(1 + \frac{n_p}{n_n}\right)^{-1}.} \tag{9.81}$$

For the earlier n/p ratio of 0.163, this gives $Y = 0.28$.

The 'observed' value of Y is in the region 0.22 to 0.23 (e.g. Pagel 1994), and there exists something of a difference of opinion on whether this is marvellously close agreement, or evidence for something seriously wrong with the standard model. The reason for the quotation marks is that, even leaving aside the physical modelling needed to turn astronomical emission-line strengths into helium abundances, we only see the actual helium that exists; how do we know it is primordial? The route to proving this is to demonstrate that there exists a correlation between Y and the mass fraction in heavier elements, Z, and extrapolate to find the intercept at $Z = 0$. Taking this process seriously would imply that $n/p \simeq 0.125$ is needed, which would in turn imply $\Omega_{\rm B} h^2 \sim 10^{-3}$ for three neutrino species. A baryon-dominated universe with critical density is thus in serious difficulty. It is shown below that the abundances of other light elements favour the slightly higher figure $\Omega_{\rm B} h^2 \simeq 0.015$, which would predict $Y = 0.242$. This remains 8% higher than the best observational figure, and it seems likely that controversy will continue over whether this is a real discrepancy. Certainly, if we are confident of the correctness of the basic big bang framework, Y is a very bad number from which to measure the baryon density, because of its relative insensitivity to density.

THE NUMBER OF PARTICLE GENERATIONS Increasing the number of neutrino species widens the gap between theory and observation by $\Delta Y \simeq 0.01$ for each additional neutrino species. It is therefore clear that strong limits can be set on the number of unobserved species, and thus on the number of possible additional families in particle physics. For many years, these nucleosynthesis limits were stronger than those that existed from particle physics experiments. This changed in 1990, with a critical series of experiments carried out in the **LEP** (large electron–positron) collider at CERN, which was the first experiment to produce Z^0 particles in large numbers. The particles are not seen directly, but their presence is inferred by detecting a peak in the energy-dependent cross-sections for producing pairs of leptons (l) or hadrons (h). The interpretation is that the peak is a 'resonance' due to the production of a Z^0 as an intermediate state, and that the energy of the peak measures the Z^0 mass:

$$e^+ + e^- \to Z^0 \to \begin{cases} l, \bar{l} \\ h, \bar{h}. \end{cases} \tag{9.82}$$

The width of the peak measures the Z^0 lifetime, through the uncertainty principle, and this gives a means of counting the numbers of neutrino species. The Z^0 can decay to pairs of neutrinos so long as their rest mass sums to less than 91.2 GeV; more species increase the decay rate, and increase the Z^0 width, which measures the total decay rate.

Since 1990, these arguments have required N to be very close to 3 (see the Opal consortium, 1990); it is a matter of detailed argument over the helium data as to whether $N = 4$ was ruled out from cosmology prior to this. In any case, it is worth noting that these two routes do not measure exactly the same thing: both are sensitive only to relativistic particles, with upper mass scales of about 1 MeV and 100 GeV in the cosmological and accelerator cases respectively. If LEP had measured $N = 5$, that would have indicated extra species of rather massive neutrinos. The fact that both limits in fact agree is therefore good evidence for the correctness of the standard model, containing only three families.

OTHER LIGHT-ELEMENT ABUNDANCES The same thermodynamic arguments that say that helium should be favoured at temperatures around 0.1 MeV say that more massive nuclei would be preferred in equilibrium at lower temperatures. However, by the time helium synthesis is accomplished, the density and temperature are too low for significant synthesis of heavier nuclei to proceed: the lower density means that reactions tend to freeze out, even for a constant cross-section, and the need for penetration of the nuclear Coulomb barrier means that cross-sections decline rapidly as the temperature decreases [problem 12.4].

Apart from helium, the main nuclear residue of the big bang is therefore those deuterium nuclei that escape being mopped up into helium, plus a trace of ^{3}He, which is produced en route to ^{4}He: $\mathrm{D} + p \to {}^3\mathrm{He}$, followed by ${}^3\mathrm{He} + n \to {}^4\mathrm{He}$ (the alternative route, of first $\mathrm{D} + n$ then p, also happens, but the intermediate tritium is not so strongly bound). There also exist extremely small fractions of other elements: ^{7}Li ($\sim 10^{-9}$ by mass) and ^{7}Be ($\sim 10^{-11}$). Unlike helium, the critical feature of these abundances is that they are rather sensitive to density. One of the major achievements of big bang cosmology is that it can account simultaneously for the abundances of H, ^{2}D, ^{3}He, ^{4}He and ^{7}Li – but *only for a low-density universe*. A proper understanding of the abundances really requires a numerical solution of the coupled rate equations for all the nuclear reactions, putting in the temperature-dependent cross-sections. This careful piece of numerical physics was first carried out impressively soon after the discovery of the microwave background, by Wagoner, Fowler & Hoyle (1967). At least the sense of the answer can be understood intuitively, however. We have seen that helium formation occurs at very nearly a fixed temperature depending only weakly on density or neutrino species. The residual deuterium will therefore freeze out at about this temperature, leaving a number density fixed at whatever sets the reaction rate low enough to survive for a Hubble time. Since this density is a fixed quantity, the *proportion* of the baryonic density that survives as deuterium (or ^{3}He) should thus decline roughly as the reciprocal of density.

This provides a relatively sensitive means of weighing the baryonic content of the universe. A key event in the development of cosmology was thus the determination of the D/H ratio in the interstellar medium, carried out by the COPERNICUS UV satellite in the early 1970s (Rogerson & York 1973). This gave $D/H \simeq 2 \times 10^{-5}$, providing the first evidence for a low baryonic density, as follows. Figure 9.2 shows how the abundances of light elements vary with the cosmological density, according to detailed calculations. The baryonic density in these calculations is traditionally quoted in the field as the reciprocal of the entropy per baryon:

$$\boxed{\eta \equiv (n_p + n_n)/n_\gamma = 2.74 \times 10^{-8} (T/2.73\,\mathrm{K})^{-3} \Omega_B h^2.} \tag{9.83}$$

Figure 9.2 shows that this deuterium abundance favours a low density, $\Omega_B h^2 \simeq 0.02$, and data on other elements give answers close to this. The constraint obtained from a comparison between nucleosynthesis predictions and observational data is rather tight:

$$\boxed{0.010 \lesssim \Omega_B h^2 \lesssim 0.015} \tag{9.84}$$

(e.g. Walker *et al.* 1991; Smith, Kawano & Malaney 1993). This comparison is of course non-trivial, because we can only observe abundances in present-day stars and gas, rather than in primordial material. Nevertheless, making the best allowances possible for the production or destruction of light elements in the course of stellar evolution, the conclusions obtained from different species are in remarkable agreement: baryons cannot close the universe. If $\Omega = 1$, the dark matter must be non-baryonic.

This result is one of the cornerstones of the big bang, and is often assumed elsewhere e.g. in setting the gas density at high redshifts when discussing galaxy formation. It has therefore been subjected to great scrutiny (see e.g. Walker *et al.* 1991; Boesgaard & Steigman 1985; chapter 4 of Kolb & Turner 1990). The abundances of only four species are involved, each of which is subject to some uncertainty over its production or destruction through stellar evolution, and it may be that the concordance range for η is fortuitously narrow. For example, the single most robust datum involved is arguably the abundance of $D + {}^3$He, since one will burn to the other in stars. This abundance in the Solar system is 3.6×10^{-5}, indicating $\Omega_B h^2 \simeq 0.025$ from this result alone. This is in good agreement with the result from the deuterium abundance inferred in high-redshift quasar absorption clouds by Tytler, Fan & Burles (1996): assuming the observed figure to be primordial, they infer $\Omega_B h^2 = 0.024 \pm 0.006$. As we have seen, the ^{4}He results prefer a rather lower figure, helping to dictate the compromise figure of Walker *et al.* (1991).

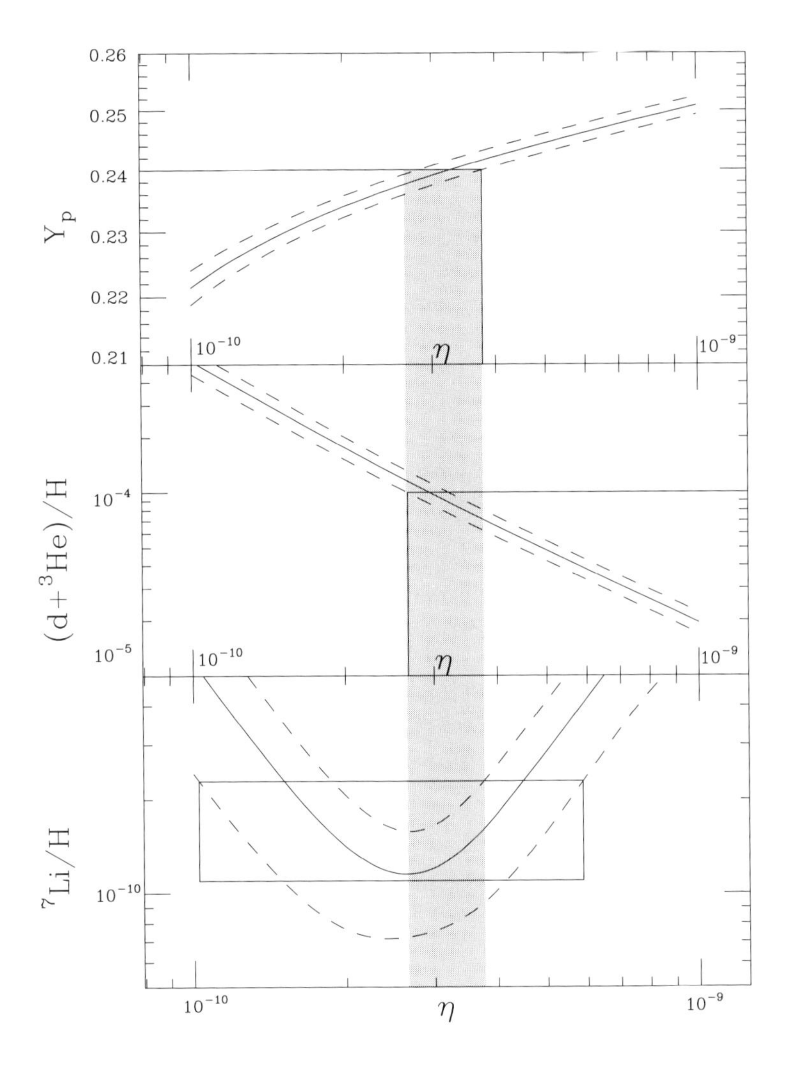

Figure 9.2. The predicted primordial abundances of the light elements, as a function of the baryon-to-photon ratio η (Smith, Kawano & Malaney 1993). For a microwave-background temperature of 2.73 K, this is related to the baryonic density parameter via $\Omega_B h^2 = \eta/(2.74 \times 10^{-8})$. Concordance with the data is found only for $\eta \simeq 3 \times 10^{-10}$, shown by the shaded strip.

The existence of a small systematic error in the ^{4}He results might therefore allow the density to increase slightly, and it would be unwise to rule out $\Omega_B h^2 = 0.03$; but a critical density in baryons is still very far off. In the mid-1980s, strenuous efforts were made to evade this conclusion via **inhomogeneous nucleosynthesis**, the assumption that a possibly first-order quark–hadron phase transition might yield a baryon distribution with gross small-scale nonuniformity in density. However, even this step does not allow a baryonic critical density to be reached (e.g. Mathews, Kajino & Orito 1996).

9.6 Baryogenesis

In discussing nucleosynthesis, we have taken it for granted that photons outnumber baryons in the universe by a large factor. Since baryons and antibaryons will annihilate at low temperatures, this is reasonable; but in that case why are there any baryons at all? A thermal background at high enough temperatures will contain equal numbers of protons and antiprotons, and this symmetry will be maintained as the particles annihilate. As usual, there would be some low frozen-out abundance of both species at late times, but the universe would display a **matter–antimatter symmetry**. A recurring theme in cosmological debate has been to ask what happened to the antimatter. It is perhaps just possible to imagine that other galaxies might be made of antimatter, but this is contrived – and unrealistic in detail given the non-zero density of the intergalactic gas that connects all objects. The inevitable conclusion is that the universe began with a very slight asymmetry between matter and antimatter: at high temperatures there were $1 + O(10^{-9})$ protons for every antiproton. If baryon number is conserved, this imbalance cannot be altered once it is set in the initial conditions; but what generates it? One attractive idea is that the matter-antimatter asymmetry may have been set in the GUT era, before the universe cooled below the critical temperature of $\sim 10^{15}$ GeV.

The main idea to be exploited is that GUTs erase the distinction between baryons and leptons, treating them as different states of the same underlying unity. This raises the conceptual possibility of reactions that can generate a net baryon asymmetry, so long as the temperature is high enough that the GUT symmetry is not broken. The particles that will be involved are the gauge bosons of the GUT, since these are the ones that mediate the baryon $\leftrightarrow$ lepton exchange reactions – e.g. the X and Y bosons of $SU(5)$. In particular, decay processes such as

$$X^{4/3} \rightarrow \begin{cases} e^+ + \bar{d} \\ 2u \end{cases} \tag{9.85}$$

are a promising direct source of baryon-number violation.

These mechanisms provide the first of three general **Sakharov conditions** for baryosynthesis (published in 1967, well before the invention of GUTs):

(1) $\Delta B \neq 0$ reactions;

(2) CP violation;

(3) non-equilibrium conditions.

The second condition requires an asymmetry between particles and antiparticles. Recall what the symmetries C and P mean: a given observed reaction is possible (and will proceed with the same rate) if particles are replaced by antiparticles (C) or viewed in a mirror (P). Although P is violated by the weak interaction to the extent that only left-handed neutrinos are produced, the combined symmetry CP is obeyed in almost all cases. Consider the effect on the above X-boson decay if CP were to hold exactly: if a fraction f of decays produce $e^+\bar{d}$, then decays of $\bar{X}$ will produce the opposite baryon number via e^-d for the same fraction of the time, in which case no net asymmetry can be created. What we need is for the two fractions to be different, and such a process is observed in the laboratory in the form of the neutral kaons: both K_0 and $\overline{K}_0$ decay to either two or three pions, but with branching ratios that differ at the 10^{-3} level.

The third condition is necessary in order to prevent reverse reactions from erasing any baryon asymmetry as soon as it has been created. As in many cases in the expanding universe, the crucial physical results are contained in the ability of reactions to freeze out.

The challenge of baryosynthesis is to predict the observed asymmetry (in the form of a baryon-to-photon ratio $n_{\rm B}/n_\gamma \sim 10^{-9}$. In principle, this can be done once the GUT model is given, and a connection can be made between laboratory measurements of CP violation and the baryon content of the universe. The simplest model for baryosynthesis would consist of a single Majorana particle, whose decays favour the production of baryons over antibaryons:

$$\begin{aligned}\Gamma(X \to \Delta B = +1) &= \tfrac{1}{2}(1+\epsilon)\,\Gamma \\ \Gamma(X \to \Delta B = -1) &= \tfrac{1}{2}(1-\epsilon)\,\Gamma.\end{aligned} \tag{9.86}$$

Here, $\Gamma = 1/\tau$ is the decay rate, and ϵ parameterizes the CP violation. The kinetic equations governing the effect of decays on the X number density and on baryon number B can be written down immediately following our earlier discussion of the Boltzmann equation:

$$\begin{aligned}\dot{n}_X + 3Hn_X &= -\Gamma\,(n_X - n_X^T) \\ \dot{n}_{\rm B} + 3Hn_{\rm B} &= \epsilon\Gamma\,(n_X - n_X^T).\end{aligned} \tag{9.87}$$

The terms involving the thermal-equilibrium X density, n_X^T, allow for inverse processes, and are deduced by asking what source term makes the lhs vanish in equilibrium, as before. What they say is that it is impossible for the X boson to decay when it is relativistic; we have to wait until $kT < mc^2$, so that n_X^T is suppressed. Note that the second equation explicitly makes clear the third Sakharov criterion for a violation of equilibrium. These equations are incomplete, as they neglect two-body processes that would contribute to the changing X abundance, such as $X\bar{X}$ annihilation. The simplified form will apply after the X abundance has frozen out. In terms of comoving densities $N \equiv a^3 n$, the equation for baryon number is just

$$\dot{N}_{\rm B} = -\epsilon\dot{N}_X \quad \Rightarrow \quad N_{\rm B} = \epsilon(N_X^{\rm init} - N_X), \tag{9.88}$$

so that the final baryon comoving density tends to $\epsilon N_X^{\rm init}$ as the X's decay away. The baryon-to-entropy ratio produced by this process is then

$$\frac{N_{\rm B}}{s} \simeq \frac{\epsilon}{g_*}\exp\left(-\frac{m_X c^2}{kT_f}\right), \tag{9.89}$$

where the entropy density is defined here as g_* times the photon density, and the last term allows for the freeze-out suppression of the X density relative to massless backgrounds. Since the required ratio is $\sim 10^{-9}$ and $g_* \sim 100$ at early times, we need $\epsilon \gtrsim 10^{-7}$. Note that the freeze-out point cannot be very late: even for $\epsilon = 1$, $kT \gtrsim m_X c^2/16$ is needed.

To perform this calculation for a proper model is no more complicated than this in principle, but messy in detail. The same point holds: the final asymmetry is a mixture of the intrinsic CP violation in the theory and the non-equilibrium effects that depend on the strengths of the GUT interactions. At present, lacking the 'true' GUT, the best that can be said is that it is not implausible that both the large entropy per baryon and the small CP violation seen in the neutral kaons may be traceable to the same cause. It is certainly interesting to speculate on how the argument might have developed if cosmologists had possessed the confidence to argue for the existence of CP violation prior to its experimental detection (Christenson *et al.* 1964).

OTHER MECHANISMS Baryosynthesis via GUT decay as above is the simplest mechanism, but there are other possibilities. First note that the above picture may well be inconsistent with an inflationary origin for the universe. Inflation generally involves a GUT-scale phase transition that leaves the universe reheated to a temperature somewhat below that of the GUT scale. Any pre-existing baryon asymmetry would be rendered irrelevant by the inflationary expansion, and things would not be hot enough afterwards for GUT processes to operate. One way out of this difficulty is for the baryon asymmetry to be generated during the process of re-heating itself – i.e. the physics of the scalar field that drives inflation would have to violate CP.

As a simple illustration of the differences this could make, consider again the earlier example, but now suppose that the X field is also that which drives inflation. At the end of inflation, the universe contains nothing except X quanta. These reheat the universe via decay (although this is not the only mechanism – see chapter 11), and the entropy is produced entirely by the X decay. In natural units, we get an energy density of $u \sim g_* T^4 = m_X n_X$ from decay, from which the entropy density is $s \sim g_* T^3$. The net baryon density produced is $n_{\rm B} \sim \epsilon n_X$, so the baryon-to-entropy ratio is then

$$\frac{n_{\rm B}}{s} \sim \frac{\epsilon n_X}{g_* T^3} \sim \epsilon \, \frac{T}{m_X}. \tag{9.90}$$

This shows that the reheating temperature must exceed $10^{-9} M_X$.

It is possible that baryosynthesis may proceed at still lower temperatures, since baryon non-conserving processes may even occur as part of the electroweak phase transition at $T \sim 200\,\mathrm{GeV}$. This is a surprise, since the electroweak Lagrangian contains no terms that would violate baryon number: leptons and hadrons are explicitly contained in different multiplets. This constraint may possibly be evaded by quantum tunnelling, but the exact extent to which baryon number violation may be realized in practice within the standard model is still a matter of debate (see e.g. section 6.8 of Kolb & Turner 1990; Grigoriev *et al.* 1992; Moore 1996; Ellis 1997).

LEPTON ASYMMETRY The universe is also asymmetric in the observed leptons: if there had been no primordial e^-e^+ asymmetry, the universe would have a net charge due to its proton content and its global acceleration would grossly exceed observation. We can therefore be certain that the initial conditions contained almost exactly the same asymmetry in electrons as in protons:

$$\frac{n_{e^-} - n_{e^+}}{n_{e^-} + n_{e^+}} = \frac{n_p - n_{\bar{p}}}{n_p + n_{\bar{p}}} \simeq 10^{-9}. \tag{9.91}$$

This state of affairs would arise naturally in GUTs such as $SU(5)$, which conserves $B-L$ even though it violates B.

It need not always be the case that $B - L = 0$, and it is possible that the universe has a large net lepton number. Because of the charge neutrality constraint, this would have to reside in the unobservable neutrinos. There are some attractive games to be played here, given the problems we have seen with nucleosynthesis. The helium abundance certainly disfavours extra neutrino species, and would be more comfortable with $N = 2$ than $N = 3$. Is it possible that there is a gross asymmetry, with antineutrinos absent, thus reducing the relativistic background density? This does not work if there was ever a time at which the neutrinos were in thermal equilibrium. An initial asymmetry would mean that the equilibrium was reached with a non-zero chemical potential. Now, if there is a reaction of the form $\sum a_i \leftrightarrow \sum b_i$, then it must be true that the sums of the chemical potentials are the same on either side of the equation (so that $dF = 0$, as before). Since neutrino–antineutrino pairs can annihilate, this means that

$$\mu(\nu) = -\mu(\bar{\nu}). \tag{9.92}$$

Although a non-zero chemical potential can suppress the energy density of antineutrinos, this is more than compensated by a boost in the neutrino density. The total occupation number is

$$\langle f \rangle = \frac{1}{\exp[(\epsilon - \mu)/kT] + 1} + \frac{1}{\exp[(\epsilon + \mu)/kT] + 1}, \tag{9.93}$$

which has a minimum at $\mu = 0$. Degenerate neutrinos therefore cannot help nucleosynthesis in this way, but the same principle has another effect. Since the weak interactions such as $p + e^- \rightarrow n + \nu_e$ were once in equilibrium, we expect $\mu_n - \mu_p = \mu_e - \mu_\nu \simeq -\mu_\nu$, where the last step follows because the neutrality argument already eliminates a large chemical potential for electrons. This modifies the expression for the freeze-out neutron-to-proton ratio, which assumed $\mu = 0$:

$$\boxed{\frac{n_n}{n_p} = \exp\left(-\frac{Q}{kT} - \frac{\mu_\nu}{kT}\right).} \tag{9.94}$$

Neutrino degeneracy therefore suppresses the relic neutron abundance and greatly weakens the constraints from primordial nucleosynthesis. Apart from arguments of simplicity, this is a possible loophole that is difficult to reject.

Problems

(9.1) Why is it acceptable to treat a system of charged particles as though they were a non-interacting ideal gas?

(9.2) Show that the probability for a given state to be occupied is given by the **Gibbs' factor** $p \propto \exp[-(E - \mu N)/kT]$, and that the free energy is

$$F = -kT \ln Z_G + \mu N. \tag{9.95}$$

(9.3) Suppose the universe contained only a comoving density n of particles with mass m, which have a lifetime τ and which decay to relativistic products. Calculate the entropy density produced when the particles have finished decaying and compare with the naive answer obtained from assuming instantaneous decay at $t = \tau$.

(9.4) Show that the general relativistic form of the Boltzmann equation for particles affected by gravitational forces and collisions is

$$\left(p^\mu \frac{\partial}{\partial x^\mu} - \Gamma^\mu_{\alpha\beta} p^\alpha p^\beta \frac{\partial}{\partial p^\mu}\right) f = C. \tag{9.96}$$

(9.5) Use the following method to derive the relativistic transformation law for specific intensity I_ν. Regard the photons travelling in some range of solid angle $d\Omega$ and frequency range $d\nu$ as a cloud of particles with density $I_\nu \, d\nu \, d\Omega$. Derive the transformation laws for particle flux, frequency and solid angle; use these to obtain the transformation law for I_ν and show that I_ν / ν^3 is an invariant.

(9.6) Show that the momentum-space integrals in the expression for the neutron decay rate reduce to the integral

$$\int_{m_e}^{Q} p_e \, \epsilon_e \, (Q - \epsilon_e)^2 \, d\epsilon_e = 1.636 \, m_e^5. \tag{9.97}$$

10 Topological defects

The temperatures and densities of the nucleosynthesis era are remote from everyday experience, but the picture of the big bang up to $T \sim 10^{10}$ K stands a fair chance of being correct, since it is based on well-established nuclear physics. The next two chapters will be much more speculative. The frontier of cosmology from the 1980s onwards consisted of looking at exotic physics and asking whether the state of the universe at very high redshift could have differed radically from a simple radiation-dominated plasma. Chapters 10 and 11 look at different aspects of such high-energy phase changes.

10.1 Phase transitions in cosmology

There are several phase transitions of potential importance in cosmology that may have left observable signatures in the present. In descending order of energy, these are:

(1) **The GUT transition**, $E \sim 10^{15}$ GeV. Above this temperature, all interactions except gravity had equal strength and the universe had no net baryon number. Below this temperature, the symmetry is broken via the Higgs mechanism so that the gauge group of particle physics degenerates from the grand-unified G to the usual $SU(3) \otimes SU(2) \otimes U(1)$ of the standard model. Baryon non-conserving processes can now operate; this may have generated the present-day excess of matter over antimatter.

(2) **The electroweak transition**, $E \simeq 300$ GeV. At this energy scale, the Higgs mechanism again breaks the $SU(2) \otimes U(1)$ part of the theory to yield the apparently distinct electromagnetic and weak interactions.

(3) **The quark–hadron transition**, $E \simeq 0.2$ GeV. This is a somewhat more complex topic, as several related effects are expected to occur at this temperature. The main event is **quark confinement**. At low temperatures, isolated quarks are thought never to exist: the energy of an isolated coloured particle is infinite, so that things only become simple and perturbative when quarks are close together in nucleons. At high temperatures and densities, this is no longer true, and the natural state is of a plasma of free quarks and gluons. Two other events are associated with this transition: the breaking of QCD chiral symmetry and the acquisition of a mass for the axion, if it exists.

The effect of a phase transition depends on its thermodynamic properties. The most familiar category of transitions are those of **first order**, such as the melting of a solid or evaporation of a liquid. The critical feature of these is the **latent heat**: the specific heat is a well-behaved slowly varying function of temperature on either side of the transition, but possesses a δ-function at the transition temperature so that a finite amount of energy $\int C\,dT$ is required to produce an infinitesimal increase in temperature past the transition. The transitions involving the Higgs mechanism are inevitably first order, since the acquisition of a non-zero vacuum expectation value $\langle|\phi|\rangle$ involves a change in energy density of the scalar field. This is the latent heat of the transition, which appears as radiation – as the universe is **reheated** (see chapter 11 for more detail on this aspect).

Not all transitions are of first order, however. In higher-order transitions, the specific heat will be finite at the critical temperature but will exhibit discontinuities in its higher derivatives. Clearly, very high orders in this scheme will be hard to detect, so attention in practice concentrates on **second-order transitions**. Examples of these include order–disorder transitions in metallic alloys, ferromagnetic transitions, superconducting and superfluid transitions. At first sight, it may seem that second-order transitions should not exist, through the following argument. A first-order phase transition may be thought of in terms of the variation of the free energy F of a given phase with temperature. If the $F(T)$ curves for two phases intersect, there will be a first-order transition if the gradients of the two lines differ. A second-order transition thus apparently requires the curves to meet so that they differ only in curvature at the touching point. This arrangement is unstable in that any slight perturbation of the system (due to impurities, for example) is likely to shift the curves, either converting the transition to first order or removing it altogether. The error here is to imagine that the two phases must always be distinguishable, whereas in second-order transitions the difference vanishes at the transition point: for example, consider order–disorder transitions in alloys such as brass. Here, the positions of atoms are random above the critical temperature but start to show a preference for atoms of one type being at specific lattice sites for lower temperatures; however, perfect order is not achieved unless the temperature is well below critical (see e.g. chapter 9 of Pippard 1957).

In the cosmological context, phase transitions are governed by a scalar field ϕ and its potential $V(\phi)$. We have discussed fields in the language of Lagrangian densities $\mathscr{L} = K(\partial^\mu \phi) - V(\phi)$, so that there are kinetic terms that depend on field derivatives and a potential that depends on the field. The potential determines the classical time dependence of ϕ, in the sense that the field will attempt to minimize $V(\phi)$ (equations of motion for the field will be studied in chapter 11). This is much the same as a ball rolling in a trough of given cross-section, and figure 10.1 shows some examples of potentials that could be of cosmological interest. The first is a Higgs-like potential and corresponds to a first-order transition; the other is second order. The second potential will be more relevant for inflationary models; the first is more interesting for this chapter, since it gives a vacuum expectation value $\langle 0|\phi|0\rangle \neq 0$. Such transitions can leave behind peculiar field configurations called **topological defects**, which are another class of possible big bang relics. This chapter outlines their main properties; Vilenkin & Shellard (1994) is an entire monograph devoted to the subject.

10.2 Classes of topological defect

The type of defect produced in a symmetry-breaking phase transition depends on the symmetry being broken. To illustrate the various possibilities, consider the case of an

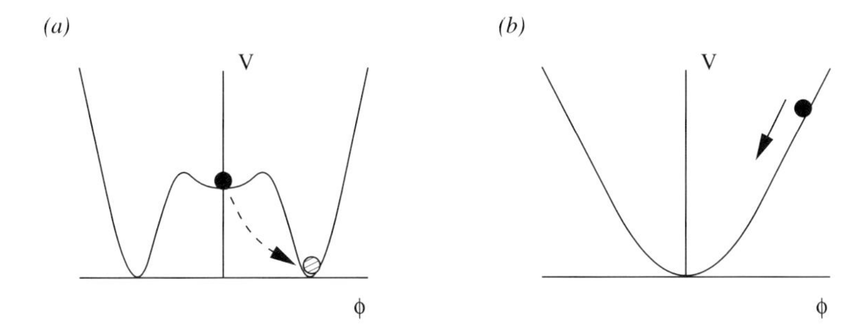

Figure 10.1. Illustrative scalar-field potentials for cosmological phase transitions of (a) first order; (b) second order. The dynamics of ϕ are the same as those of a ball rolling in a trough, trying to reach the ground state of minimum $V(\phi)$. In the first case, the field is trapped near the origin and the transition occurs abruptly by quantum tunnelling. In the second case, the field 'rolls' smoothly down the potential.

$O(n)$ symmetry: the field ϕ that acquires a non-zero vacuum expectation value under the transition is a n-dimensional vector, and the ground state is invariant under rotation of this vector. What happens as the temperature falls below critical is familiar from paramagnetism: below the Curie temperature, magnetization occurs spontaneously, but in different directions in different places, leading to magnetic domains. The same happens in cosmology; the direction of ϕ within its internal n-dimensional space is selected at random independently at each point in space. This is not a minimum-energy state, because the field derivatives are non-zero, and so the fields in different places will attempt to align themselves as the horizon expands following the phase transition. This is the **Kibble mechanism** for defect generation (Kibble 1976); causality and the size of the horizon at the time of the phase transition mean that the field configurations at different places cannot 'know' about each other. As these areas come into causal contact and attempt to align their fields, this process may not be able to go to completion if there are configurations that cannot be removed by any small local changes of the field. The easiest case in which to see this is $n = 2$. Suppose that there exists a loop in real 3D space such that the field $\phi(\mathbf{x})$ makes one or more rotations in in its internal 2D space as the point $\mathbf{x}$ is moved around the loop; this is a topological configuration that cannot be erased by a continuous change in the phase of ϕ. As we shrink the loop to zero size, the field derivatives increase, and so some energy is associated with the field configuration: this is a **topological defect**. Which kind of defect we get in any particular case depends on the internal dimension and the spatial dimension. The possibilities in 2D space are shown in figure 10.2; in 3D, there are four different possibilities:

$n = 1$ **Domain walls**. The symmetry consists of only discrete states (analogous to spin up and spin down). Space divides into connected regions in one or other phase, separated by walls with a certain energy per unit area.

$n = 2$ **Strings**. This is the case we considered above: linear defects, where the 'phase' of ϕ changes by multiples of 2π in making one loop about the string. These defects are characterized by some mass per unit length.

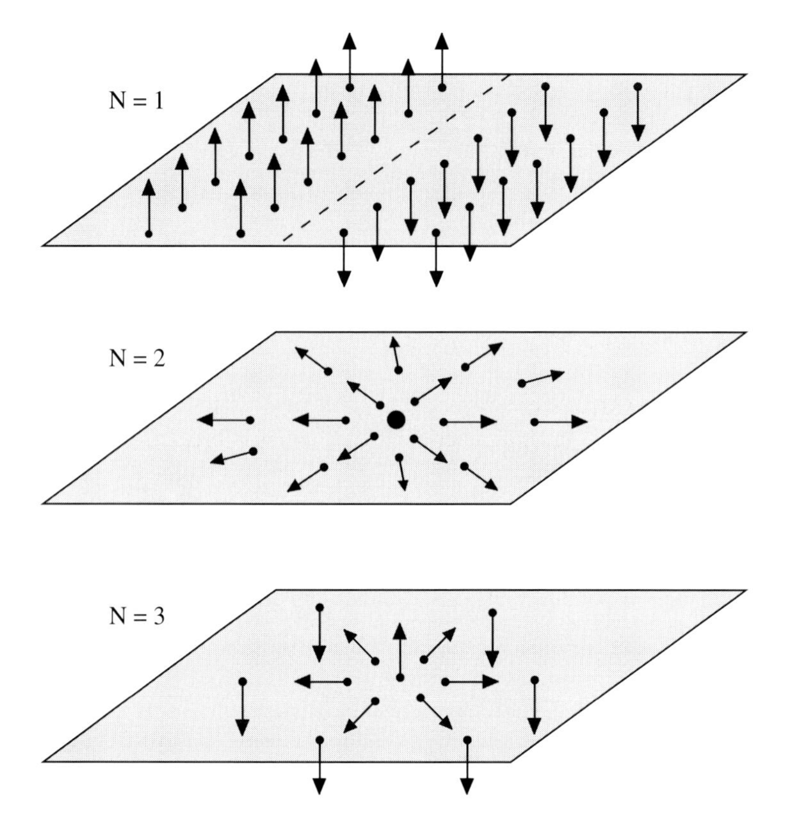

Figure 10.2. This figure (see Turok 1991) illustrates topological defects in two spatial dimensions, as a function of the internal dimensionality of the scalar field ϕ. For a 1D field (spin up or down), domains of different 'spin' are separated by line defects. For a 2D field, point defects exist, for which the field makes one internal rotation as the point of interest travels on a loop about the defect (the 2D analogues of strings). For a 3D field, we have the analogue of textures.

$n = 3$ **Monopoles**. These are point defects, where the field ϕ points radially away from the defect, which has a characteristic mass.

$n = 4$ **Textures**. These are not so easy to visualize, except through a lower-dimensional analogy. Consider figure 10.2: in two spatial dimensions, textures correspond to field configurations where ϕ makes a rotation of π in its 3D internal space in going from $r = 0$ to $r = \infty$. In three spatial dimensions, the effect is similar, except that ϕ is then a 4D vector that moves from its north to its south pole.

We have illustrated topological defects for a specific type of symmetry, but they clearly arise in other cases: for example, strings exist in a model with broken $U(1)$ symmetry. The technical tools needed to make a decision on whether defects will exist are called the homotopy groups of the vacuum manifold, denoted π_n; defects of a given order exist if the nth homotopy group is 'non-trivial'. We shall not need to delve into the exact meaning of this terminology, fortunately.

These arguments are general, and apply to field theory in almost any situation. For example, analogues to the cosmological defects discussed here can be found in a range of solid-state phenomena, allowing the detailed nonlinear dynamics of these objects to be studied in the laboratory (Chuang *et al.* 1991). However, it is important to emphasize that not all models in particle physics contain topological defects: *there are none in the standard model.* The importance of these objects for cosmology therefore rests on the presently speculative foundations of GUTs. A more optimistic way of putting this is that observations proving the existence of (say) cosmic strings would tell us a great deal about particle physics at extremely high energies.

CHARACTERISTIC MASSES AND DENSITIES The scale of the defects is defined by the energy or mass scale m that characterizes the symmetry breaking. The cores of the defects will have sufficiently large derivative energy densities that there will be a residual unbroken high-energy state. Using dimensional arguments in natural units, we see that the characteristic length over which the symmetry remains unbroken must be $\sim m^{-1}$, and within this region the energy density is $\sim m^4$. Thus, the monopole mass is $\sim m$, i.e. it is around the GUT energy of $\simeq 10^{15}$ GeV. Similarly, a string has a mass per unit length $\mu \sim m^2$ and a wall a mass per unit area $\sim m^3$, in natural units. The dimensionless combination $G\mu/c^2$ is usually adopted as the parameter to describe strings; in natural units, this is $\sim (m/m_{\rm P})^2$, where $m_{\rm P}$ is the Planck mass.

If defects do arise, their densities can be estimated straightforwardly. In the Kibble mechanism, the defect density will be of order one per horizon volume at the time of formation. In natural units, the horizon size $ct \sim 1/\sqrt{G\rho}$ is $ct \sim T_{\rm P}/\sqrt{\rho} \sim T_{\rm P}/T_{\rm GUT}^2$ ($T_{\rm P}$ is the Planck temperature, which is equal to $m_{\rm P}$ in natural units). The energy densities for monopoles, strings and walls are then respectively $T_{\rm GUT}/(ct)^3$, $(ct)T_{\rm GUT}^2/(ct)^3$ and $(ct)^2 T_{\rm GUT}^3/(ct)^3$ (mass per unit length times horizon length and mass per unit area times horizon area in the last two cases). Textures are not expected to leave a residual density today; although they are field configurations that are prevented by topology from being simply transformed away, they are unstable. Consider the lowest panel of figure 10.2: the texture consists of one point with 'spin up', with a short transition zone to the rest of the universe, which has 'spin down'. The field at the discrepant point can then tunnel through to the majority direction, so that the entire texture unwinds and radiates away its energy. At any time, we would therefore expect only one texture per horizon volume.

Once the defect population is set up, the universe then expands to the present, and the defect energy density scales in a way that depends on the dimensionality of the defect. The monopole energy density just scales as a^{-3} (conserved number density, and conserved mass); strings may be expected to scale as a^{-2} and walls as a^{-1}, if the expansion simply increases respectively the length and area of the defect networks in these cases. Putting this together predicts the following present energy densities:

$$\begin{aligned} \text{monopoles} \quad & \rho \simeq (T_{\rm GUT}^7/T_{\rm P}^3)(T_{\rm CMB}/T_{\rm GUT})^3 \simeq (T_{\rm GUT}/10^{11.1}\,{\rm GeV})^4\, T_{\rm CMB}^4 \\ \text{strings} \quad & \rho \simeq (T_{\rm GUT}^6/T_{\rm P}^2)(T_{\rm CMB}/T_{\rm GUT})^2 \simeq (T_{\rm GUT}/10^{3.2}\,{\rm GeV})^4\, T_{\rm CMB}^4 \\ \text{walls} \quad & \rho \simeq (T_{\rm GUT}^5/T_{\rm P})(T_{\rm CMB}/T_{\rm GUT}) \simeq (T_{\rm GUT}/10^{1.3}\,{\rm keV})^4\, T_{\rm CMB}^4. \end{aligned} \tag{10.1}$$

In natural units, $T_{\rm CMB}^4$ is the present cosmic microwave background (CMB) energy density ($\Omega_{\rm CMB} \sim 10^{-4}$), and these figures therefore predict a disastrously high relic density unless the GUT energy scale is extremely low, ranging from 10^{12} GeV for monopoles to a mere 100 keV for domain walls. This looks like a strong argument against GUTs that allow these sort of topological defects, but there are some escape routes. In the case of strings, the evolution of the network turns out to be more complicated than the naive argument given here: because strings can cut each other, the network does not increase its length with the scale factor. The string model is thus the most attractive of the defect theories, and the study of its properties takes up most of the remainder of this chapter. For monopoles and walls, the estimates stand: either nature does not permit this sort of defect in practice, or the initial conditions assumed in the Kibble mechanism are not applicable.

Unlike other classes of defect, monopoles are expected to arise in almost any GUT. The predicted number density should therefore be taken very seriously, and it is a great difficulty that the observational limits conflict with the prediction by so many orders of magnitude. This **monopole problem** was one of the early motivations for inflationary cosmology: the monopole density can only be reduced to manageable proportions if the Higgs field can be aligned on super-horizon scales.

GAUGE VERSUS GLOBAL DEFECTS An important distinction has to be made between defects that arise from the breaking of a global symmetry or a gauge one. In either case, there is an energy associated with the field derivatives in the core of the defect. However, in the gauge case both scalar and gauge fields can adjust in order to minimize the energy, as we shall see below in the case of the monopole. The result is that the defect energy is more localized in the gauge case than for global defects; for example, the energy per unit length for a gauge string is perfectly well defined, but for a global string the field energy falls as $\rho \propto 1/r^2$ at large distances from the string, so that the mass per unit length $M = \int \rho \; 2\pi r \; dr$ diverges logarithmically.

As a further result of this self-shielding property of gauge defects, the large-scale interactions of gauge defects are much weaker than for global defects. The Kibble mechanism will seed one defect per horizon volume at the time of the phase transition, but this number may subsequently be reduced in the case of global defects, which can seek each other out over large separations and undergo annihilation. This means that gauge defects are much more likely to have left behind relics observable in the present-day universe.

10.3 Magnetic monopoles

The point-like gauge defects in GUTs have magnetic field configurations at infinity that look like that of a **magnetic monopole**:

$$\mathbf{B} = \frac{\mu_0 M}{4\pi \, r^2} \, \hat{\mathbf{r}}. \tag{10.2}$$

Here, M denotes not the mass but the magnetic charge. These monopoles are the objects required to make electromagnetism symmetric between electric and magnetic fields; for example, monopoles experience an analogue of the Lorentz force: $\mathbf{F} = M(\mathbf{B} + \mathbf{v} \wedge \mathbf{E})$.

Dirac had an interesting argument concerning magnetic monopoles. Although they do not exist in classical electromagnetism, if they did it would solve one of the greatest mysteries in physics: the quantization of charge. Consider a particle of mass m

and charge q in a circular orbit about a monopole. Since the centripetal force will come from the normal Lorentz force $\mathbf{F} = q\mathbf{v}\wedge\mathbf{B}$, the orbital velocity is $v = \mu_0 qM/(4\pi mr)$. Now, angular momentum quantization tells us that $mvr = n\hbar$; using this eliminates both v and r, leaving

$$qM = n\,\frac{4\pi\hbar}{\mu_0} \tag{10.3}$$

One monopole in the universe thus implies quantization of charge; conversely, the observed quantization of charge requires magnetic monopole strength to be quantized in units of $M = 4\pi\hbar/(\mu_0 e)$.

One may be suspicious of this semiclassical argument because it involves a spin-half electron; it is therefore arguable that an angular-momentum quantization of $\hbar/2$ should be used, giving a monopole strength half as large. An independent argument for this smaller value comes from conservation of magnetic flux. A monopole produces a flux of $4\pi r^2 B = \mu_0 M$, and this flux could be supplied within conventional electromagnetism by attaching a long thin solenoid to the monopole – the **Dirac string**. To complete the conjuring trick, the solenoid must be made unobservable. Leaving aside gravity (the string would have a divergent energy per unit length if its width went to zero), the least we can ask is that the field trapped inside the solenoid is not detectable. This is like the Aharonov–Boehm experiment, where the phase change over a loop around a solenoid is

$$\delta\phi = \frac{e}{\hbar}\oint \mathbf{A}\cdot d\boldsymbol{\ell} = \frac{e}{\hbar}\int \mathbf{B}\cdot d\mathbf{S} = \frac{e\mu_0 M}{\hbar}. \tag{10.4}$$

If the phase is a multiple of 2π, then the implied monopole strength is

$$\boxed{M = \frac{2\pi\hbar}{\mu_0 e}\,n,} \tag{10.5}$$

a factor of 2 smaller than the value given by the semiclassical argument.

THE 'T HOOFT–POLYAKOV MONOPOLE There exists a general proof due to 't Hooft and Polyakov that monopoles will exist in most spontaneously broken non-Abelian theories. However, the structure of a gauge defect can be most easily illustrated in a simple specific model. For a monopole, we are interested in a Higgs field with a 3D internal space ϕ^i, $i = 1, 2, 3$. The gauge group to be broken is $SO(3)$, so that there are also three gauge fields A^i_μ, $i = 1, 2, 3$. The part of the Lagrangian involving the Higgs and the gauge bosons only is

$$\mathcal{L} = \tfrac{1}{2}\mathcal{D}^\mu\phi^i\mathcal{D}_\mu\phi^i - \tfrac{1}{4}F^i_{\mu\nu}F^{i\mu\nu} - V(\phi), \tag{10.6}$$

where the field tensors are

$$F^i_{\mu\nu} = \partial_\mu A^i_\nu - \partial_\nu A^i_\mu - e\epsilon^{ijk}A^j_\mu A^k_\nu, \tag{10.7}$$

and the covariant derivatives are

$$\mathcal{D}_\mu\phi^i = \partial_\mu\phi^i - e\epsilon^{ijk}A^j_\mu\phi^k. \tag{10.8}$$

Suppose the potential $V(\phi)$ is arranged to give symmetry breaking with $|\phi| = v$, along some arbitrary direction. At a given point in space, we can choose this as the 3-axis of the Higgs field's internal space. If the field is given time to align itself then, in the absence of defects, the Higgs fields at all point of real space will all be parallel in their internal space. In this case, there exists an internal $U(1)$ symmetry of rotations about the Higgs axis that remains unbroken: the corresponding gauge field, A^3_μ, would therefore be the electromagnetic 4-potential. This model is not the real world, of course, but it gives another example of a model that yields the $U(1)$ of electromagnetism after symmetry breaking.

Consider now the case of the **hedgehog solution**, in which the internal direction of the Higgs vector is coupled with real spatial position so that the Higgs field 'points radially':

$$\phi^i = v\,\frac{r_i}{r}. \tag{10.9}$$

This is clearly a topological defect in the sense that there is a winding of the Higgs field as one performs circles about the origin. It is also referred to as a **monopole solution**, for the following reason.

We want to calculate how the Higgs and gauge fields vary with distance from $r = 0$. Assuming a time-independent solution, the equations of motion obtained from the Lagrangian can in principle be integrated to achieve this. Here, it will suffice to look at the behaviour at large radii. Consider first just the Higgs fields, for which the time-independent Lagrangian is $\mathscr{L} = \nabla_j\phi^i\nabla_j\phi^i/2 - V$. We emphasize that the following equations cover the spatial parts only, and so have Roman subscripts and superscripts. For the radially directed Higgs field, we have

$$\nabla_j\phi^i = v\left(\frac{\delta_{ij}}{r} - \frac{r_i r_j}{r^3}\right), \tag{10.10}$$

so that the kinetic part of the energy density is $\mathscr{L} = v^2/r^2$ and the integrated total energy associated with the defect diverges linearly with radius. This disaster is in danger of being repeated in the gauge case because the covariant derivative $\mathscr{D}\phi$ contains a contribution from the ordinary derivatives of ϕ. The way to prevent this is to ensure that the gauge fields adjust themselves in order to cancel the diverging term, so that $\mathscr{D}\phi^i \to 0$ at large distances; clearly, the equations of motion will enforce such a minimization of the total energy. It may be seen by substitution that the following solution for the gauge fields satisfies the required condition:

$$A^j_i = \epsilon_{ijk}\,\frac{r_k}{er^2} \quad\Rightarrow\quad \nabla_\ell\,\phi^i = e\epsilon_{ijk}A^j_\ell\phi^k \quad \text{if} \quad \phi^i = v\,\frac{r_i}{r}, \tag{10.11}$$

(in proving this, the relation for a triple vector product will be needed: $\epsilon_{ijk}\epsilon_{i\ell m} = \delta_{j\ell}\delta_{km} - \delta_{jm}\delta_{k\ell}$). This equation covers the spatial parts; because of time independence, $A^i_0 = 0$. This shows that $\mathscr{D}_\mu\phi^i$ vanishes, as required to remove the divergent energy at large r.

We now need to identify the magnetic field in the gauge fields. The earlier discussion showed that, if a uniform Higgs field pointed along the 3-axis, then A^3_μ would play the role of the electromagnetic 4-potential. At large radii, the Higgs field does become uniform, so that we can always perform a rotation so that the radius vector becomes the third axis. Care is needed with this process, since just taking the curl of $\mathbf{A}^3$ gives the wrong answer; this will only give the magnetic field when the Higgs field is perfectly aligned. The correct place to find the magnetic field is in the 3-field tensor, which is more complicated than the simple electromagnetic $F^i_{\mu\nu} = \partial_\mu A^i_\nu - \partial_\nu A^i_\mu$ because the symmetry is non-Abelian. Rather than $B_i = \epsilon_{ijk}\nabla_j A^3_k$, we therefore want $B_i = \frac{1}{2}\epsilon_{ijk}F^3_{jk}$, which takes account of all the gauge fields present before making the rotation to align the 3-axis. Inserting the solution for A_i^j into this definition and manipulating ϵ's gives

$$\boxed{B_i = \frac{r_i}{er^3}.} \tag{10.12}$$

This (given the use of natural units where $\hbar = \mu_0 = 1$) is just the semiclassical Dirac monopole charge discussed earlier. The more complex gauge structure in this example has supplied the analogue of the Dirac string in generating a field that is monopole-like at large distances. The deviation from normal monopole-free electromagnetism occurs only in the centre of the defect, where the symmetry of the Higgs fields remains unbroken, so there is no conflict with the successful operation of Maxwell's equations in the low-energy world.

10.4 Cosmic strings and structure formation

STRING SPACETIME The spacetime near a cosmic string has some interesting properties. A string has a very large tension:

$$T^{\mu\nu} = \text{diag}(\rho, -\rho, 0, 0), \tag{10.13}$$

if the string is in the x-direction. The easy way to see this is to make the same argument that gives us $p = -\rho$ for the vacuum: the string vacuum must be invariant under Lorentz transformations along the string (although not perpendicular to it, clearly). Viewing the string as an exotic piano wire, this says that the speed of propagation of transverse waves along the string is the speed of light (for small disturbances). In general, then, we expect strings to be moving at close to the speed of light.

The gravitational properties of such a spacetime are remarkable, since the 'active' source term for the Newtonian potential vanishes:

$$\nabla^2\Phi = 4\pi G\, T^\mu_\mu = 0. \tag{10.14}$$

While this is true locally (an observer close to a cosmic string would feel no acceleration, unlike an observer close to a simple collapsed pressureless object with the same mass per unit length), the global situation is more complicated. The metric of spacetime around a string, i.e. the solution of Einstein's equations for the above energy–momentum tensor, may be verified to be

$$d\tau^2 = dt^2 - dz^2 - dr^2 - r^2(1 - \epsilon/\pi)\, d\theta^2, \tag{10.15}$$

where the string lies along the z-axis, and r and θ are the polar coordinates in the xy-plane. This looks like flat spacetime (cf. the Minkowskian form of the slice at constant θ), but with a **deficit angle** ϵ that is related to the string parameter via

$$\boxed{\epsilon = 8\pi \frac{G\mu}{c^2}.} \tag{10.16}$$

A cross-section through the string thus looks as shown in figure 10.3: a circle with a sector subtending angle ϵ cut from it.

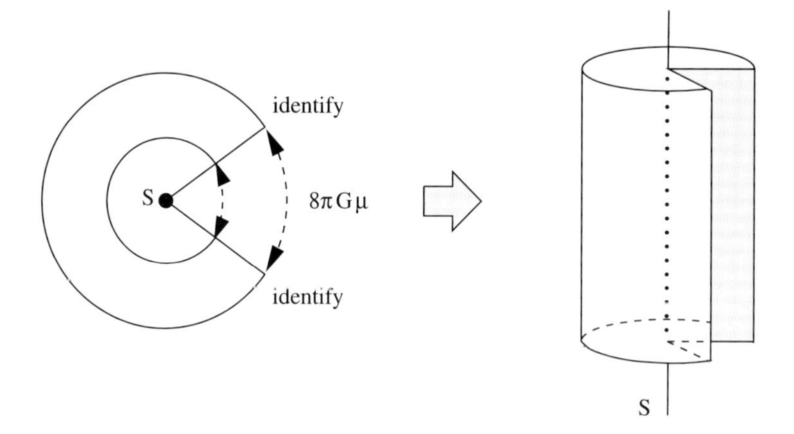

Figure 10.3. Illustrating the deficit angle in spacetime near a string. The two sides of the 'cut' are to be identified, and spacetime is locally Minkowski. However, in order to execute a complete revolution about the string, a polar angle of $2\pi - 8\pi G\mu$ is required. The spacetime around the string thus resembles a cheddar cheese: a cylinder with a missing wedge.

SCALING OF STRING NETWORKS How does a network of cosmic string evolve? At first sight, we might worry that it would come to dominate the universe. The high tension of cosmic strings means that exactly enough work is done in stretching them to compensate for the change in length, and so the energy per unit length is constant:

$$d(\mu L) = T\, dL \quad \Rightarrow \quad d\mu = 0 \quad \text{if} \quad T = \mu. \tag{10.17}$$

This is the same argument that is used to prove that the vacuum energy is unchanged as the universe expands. So, if we had a string network that maintained its shape as the universe expanded, the comoving density in strings would scale as $R(t)$ and the proper density due to strings would decline as R^{-2}, more slowly than matter or radiation. Averaged over a large scale, the universe would appear to contain a fluid with the peculiar equation of state $p = -\rho/3$. The active gravitational mass density vanishes, leading to an undecelerated expansion $R \propto t$ (see e.g. Kolb 1989). However, strings turn out to be more well mannered than this: the string network thins itself out as it evolves, so that it never comes to dominate the expansion. The basic mechanism is the generation of string loops: when (relativistically) moving strings cross, they **intercommute** and join ends to form string loops (figure 10.4).

The loop density then scales as R^{-3} and maintains a constant ratio to the matter density. Again, causality shows that this self-destruction process will be able to operate on scales up to roughly that of the horizon at a given time, so that there will be of order one long string threading the present horizon volume. The exact abundance of long strings needs to be found via numerical simulation, which is a difficult business. High resolution is needed, as the propagation of sharp kinks in the string can alter the answer if these are not followed accurately. The simulations are expected to settle down into a **scaling solution** in which the network always appears the same when viewed on the scale of the horizon, and there is some indication that this has been achieved. The long strings

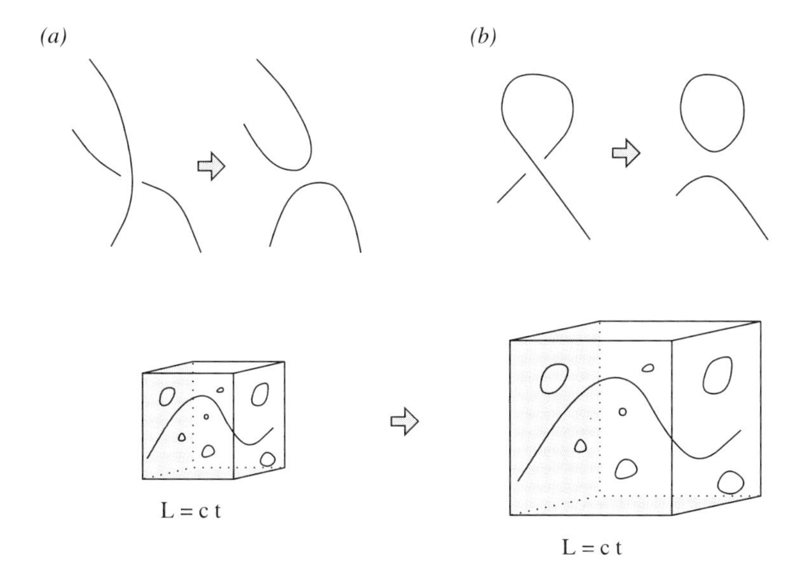

Figure 10.4. This figure illustrates the intercommuting effects that occur when strings try to cross each other. In (*a*), two independent long strings intersect and reconnect into two shorter strings. In (*b*), a string loop is produced. These 'self-mutilating' effects of the string network means that it is able to evolve in a self-similar fashion: the string pattern always has the same character when viewed on the scale of the horizon.

then make a fixed contribution to the density parameter, which is determined to within 20%:

$$\Omega_{\rm long\ string} \simeq \begin{cases} 500G\mu/c^2 & \text{(radiation dominated)} \\ 100G\mu/c^2 & \text{(matter dominated).} \end{cases} \tag{10.18}$$

Alternatively, we can write

$$\rho_{\rm long\ string} = \alpha\mu(ct)^{-2}, \tag{10.19}$$

where $\alpha \simeq 15$ in the radiation era, and is roughly three times smaller in the matter era. What about the string loops? These too settle down into a scaling solution, as follows. As a given time $t_{\rm form}$, the universe will have generated loops of size $L \sim ct_{\rm form}$, whose number density will be of order $(ct_{\rm form})^{-3}$, subsequently scaling as R^{-3}. Each time the universe expands by a factor e, a few new loops will thus be created in a roughly unit range of $\ln L$, centred on the characteristic scale $L \sim ct_{\rm form}$. At some later time, the number density of loops is therefore

$$\frac{dn}{d\ln L} \sim (ct_{\rm form})^{-3}\left(\frac{R}{R_{\rm form}}\right)^{-3} = \nu(ct)^{-3}\left(\frac{ct}{L}\right)^{3/2}, \tag{10.20}$$

where the last equality applies in the radiation era and defines the parameter ν, which is close to unity in practice. The total energy density appears to diverge at the small-loop end, but in fact it is cut off by gravitational radiation. The minimum loop size is [problem 10.2]

$$L_{\rm min} \simeq \Gamma \left(\frac{G\mu}{c^2} \right) ct, \tag{10.21}$$

where numerically $\Gamma \simeq 100$. This minimum loop radius is in the region of 0.1 Mpc today. The net result is a loop density that scales in the same way as for long strings and contains a comparable density.

Cosmic strings are therefore perhaps the most interesting and plausible defects from the point of view of cosmology and structure formation, and they provide one of the principal competitors to inflationary models in generating cosmic structure. Suppose the phase transition occurs in a universe that is completely uniform (either the idea of inflation is incorrect, or it fails to produce relic fluctuations of an interesting magnitude). How large are the density fluctuations produced by the strings? A rough estimate comes from assuming as above that the network of strings evolves to produce just one string on the scale of the horizon at a given time. We then get

$$\frac{\delta\rho}{\rho} \sim \frac{\mu\, ct}{\rho\, (ct)^3} \sim 30\, \frac{G\mu}{c^2}, \tag{10.22}$$

where the last relation follows from the density in the radiation-dominated regime, $\rho = 3/(32\pi G t^2)$. Since we already have the estimate $G\mu/c^2 \sim m^2/m_{\rm P}^2$, this gives the following simple relation for the horizon-scale amplitude:

$$\boxed{\left(\frac{\delta\rho}{\rho} \right)_{\rm H} \sim \left(\frac{E_{\rm GUT}}{E_{\rm P}} \right)^2 .} \tag{10.23}$$

It is clear that numbers close to the characteristic $\delta_{\rm H} \sim 10^{-5}$ can arise quite naturally, and the ability to explain this critical number is an attractive feature of the string picture.

There is a problem of principle to be solved in using a string network to generate density structure. Since the active mass of a string vanishes, and no gravitational forces are generated on test particles near to strings, there is a temptation to say that the effects should be small. However, this ignores the peculiar global nature of the string geometry. There is a mechanism for the generation of substantial density structure: it is the **accretion wakes** generated by moving strings. In the string frame, the two streams of material that pass on either side of the string at a velocity v (generally of order c), receive a relative velocity of

$$\Delta v = \frac{8\pi G\mu}{c^2}\, v, \tag{10.24}$$

i.e. a few $\rm km\,s^{-1}$ for typical parameters. This may not sound spectacular, but the effect is generated at high redshift and produces density perturbations that can grow from $z \simeq 1000$ onwards. Since the horizon size at that time is ~ 100 Mpc and since velocity perturbations grow as $v \propto R^{1/2}$, we see that present-day peculiar velocities of hundreds of $\rm km\,s^{-1}$ on supercluster scales are not implausible in this picture.

The distinguishing feature of structure formation with cosmic strings is that it is a **non-Gaussian model**. Although the density perturbations generated by strings may be Fourier-analysed in the usual way, there will be correlations between the phases of different components that would not be present in a model where the perturbations originated via quantum fluctuations in an inflationary model. However, the differences may not be so marked. By the present, the density field consists of the contributions from many overlapping wakes, so that the central limit theorem implies that the field may be rather close to Gaussian (see chapter 16). Because of the way in which the string network scales, the density field that is generated has a power spectrum that is approximately **scale invariant** (equal power on the scale of the horizon at all times). Thus, if the strings are supplemented by an appropriate kind of dark matter (CDM, massive neutrinos, ...), they may generate a present-day universe that is very similar to that predicted by inflationary models. The way to tell the difference is at high redshift: because the strings are available to seed structure, it is possible to have objects forming at early times in a way that would not be feasible in Gaussian models. Similarly, the most clear-cut signature of strings probably comes in small-scale microwave anisotropies (see below). These are generated almost at last scattering, and so the central limit theorem has had no time to wash out their non-Gaussian nature.

GRAVITATIONAL LENSING BY STRINGS It seems from their peculiar conical spacetime that the lensing properties of cosmic strings should be very different from those of simple linear masses of the same μ but zero tension. However, this turns out not to be so: the effect of the tension is in fact not detectable in this way. To see this, calculate the Newtonian light deflection properties of a line mass. The radial acceleration is (employing Gauss's theorem) $a = 2G\mu/r$; for the bend angle, we use the usual impulsive approximation

$$\alpha = 2 \int_{-\infty}^{\infty} a_\perp \, d\ell/c, \tag{10.25}$$

where we recognize the factor 2 for general-relativistic light deflection. It is easy to show from this that the bend angle is independent of impact parameter:

$$\alpha = 4\pi \frac{G\mu}{c^2} = 2.6 \, \frac{G\mu/c^2}{10^{-6}} \text{ arcsec.} \tag{10.26}$$

The relative deflection in this analysis for two rays passing either side of the string is thus $8\pi G\mu/c^2$, exactly as obtained for the string case (see figure 10.5), although the tension is now zero. If a cosmic string is ever detected, verification that it has the correct tension will thus not be easy.

The above example is something of a special case, since it assumes that the string is perpendicular to the line of sight. If the string is instead orientated at an angle of θ to the plane of the sky, the analysis of the Newtonian line mass would suggest that the bend angle should depend on the projected mass per unit length, and so become larger by a factor $(\cos\theta)^{-1}$. In fact [problem 10.1], the opposite happens and the bend angle becomes

$$\alpha = 4\pi \frac{G\mu}{c^2} \cos\theta, \tag{10.27}$$

(Vilenkin 1984).

Lensing offers in principle a very clean test of the existence of strings. The constancy of the deflection means that there is no effect other than an unobservable translation at moderate distances from the string (at least for straight strings). However, for images that lie behind the string, there is the possibility of multiple imaging. The usual geometrical construction of gravitational lensing shows that pairs of images are

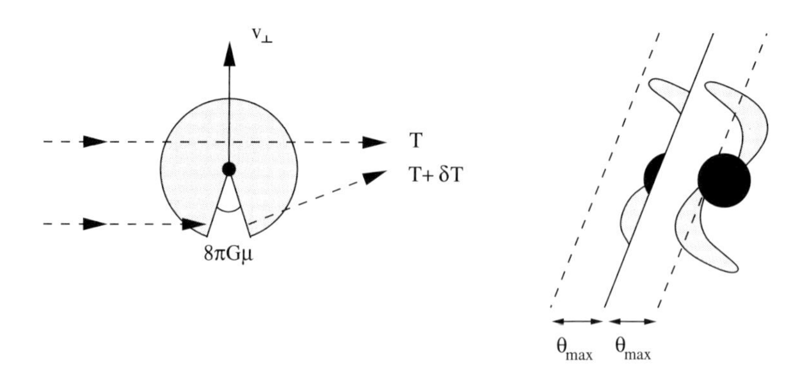

Figure 10.5. Illustrating the gravitational lensing effects of string spacetime. The broken lines show two light rays, which propagate as in Minkowski spacetime owing to the absence of local gravitational forces. However, the two sides of the 'cut' are to be identified, so two light beams passing either side of the string suffer a relative angular deflection of $\epsilon = 8\pi G\mu/c^2$. If the string has a velocity in the plane of the sky $v_\perp$, then one ray receives a Doppler boost relative to the other, causing a step in the CMB temperature. To see this, go to a frame in which the string is at rest and the observer moves sideways at $v_\perp$. On either side of the string, light from a given point on the last-scattering surface reaches the observer via two rays separated by an angle $\Delta\theta$, which is just the deficit angle ϵ if the string is at low z. There is thus a differential Doppler shift of $\delta v/v = \epsilon v_\perp/c$ for small velocities, or a factor γ larger if the velocity is substantial. The relative light deflection causes any image within a critical strip on the sky to be multiply-imaged without distortion or magnification. However, unless the string is exactly perpendicular to the line of sight, its motion causes the line of multiple images to be sheared with respect to their separation.

split by an angle $\Delta\theta$, where

$$\Delta\theta = \frac{8\pi G\mu}{c^2}\,\frac{D_{\rm LO}}{D_{\rm O}}, \tag{10.28}$$

and $D_{\rm O}$ and $D_{\rm LO}$ are respectively the angular-diameter distance of the object and the angular-diameter distance between the lens and object (see chapter 4). Multiple images are produced if the intrinsic position of the object on the sky lies within $\Delta\theta$ of the string. For $G\mu/c^2 = 10^{-6}$, then, the signature of the string would be a lane on the sky 10.4 arcsec across, containing aligned pairs of images separated by up to half this amount. As shown in figure 10.5, the galaxies towards the edges of this lane would be 'cut' in a characteristic way: only part of the galaxy would be multiply-imaged, yielding sharp edges to the partial images. Given a high background density of faint target galaxies, this clear signature gives a good prospect of searching for strings and finding them, or ruling out their existence at an interesting level of $G\mu/c^2$. Apart from the monumental consequences of such a detection, a further bonus of this observation would be to determine the geometry of the universe. The distance ratios that determine the image separations as a function of redshift are sensitive to Ω, whereas the lensing situation has only two parameters: redshift and mass per unit length for the string. Redshifts for two galaxy pairs suffice to determine these and a third galaxy gives Ω; subsequent observations then over-determine the solution and give a consistency check.

One complicating feature, which makes it more difficult in practice to search for the lensing effects of strings, is that the strings are expected to move at close to the speed of light; the lensing properties are then affected by aberration effects (Vilenkin 1986). First of all, the observed angular splitting is altered. This can be analysed by using the invariant scalar product of the photon 4-momenta, $k^\mu(1)k_\mu(2) = \omega_1\omega_2(1-\cos\Delta\theta)$. Since the observed frequencies are related to those in the string frame by $\omega_{\rm O}/\omega_{\rm S} = \gamma(1-\mathbf{v}\cdot\hat{\mathbf{r}})$, we get

$$\Delta\theta \rightarrow \frac{\Delta\theta}{\gamma(1-\mathbf{v}\cdot\hat{\mathbf{r}})} \tag{10.29}$$

(in fact, the two frequencies differ by an amount of order ϵ; see below). This effect merely alters the effective lens strength, and can be beneficial since it allows some strings to produce greater splittings. More serious is the effect of differential light travel time. As the string intersects photon trajectories and deflects them, it produces a splitting that is normally perpendicular to the projected position of the string on the sky. However, if the string is moving and not in the plane of the sky, its projection will appear rotated. The simple signature of a line of images delineating a rectangular strip on the sky will be transformed into a parallelogram whose shear depends on the string velocity; this is illustrated in figure 10.5. This is easily calculated: all that matters is the component of the string velocity in the plane of the sky, $v_\perp$. Consider two points along the string separated by a distance L: the differential light travel time from them to the observer is $\Delta t = L\sin\theta/c$, and the furthermost of the two points will appear displaced on the sky by a distance $v_\perp\Delta t$. Since the projected separation on the sky without time-delay effects is $L\cos\theta$, we deduce that the shear angle, χ, is given by

$$\tan\chi = v_\perp \tan\theta. \tag{10.30}$$

Finally, there is a potentially nasty complication that is less easy to quantify: **string microstructure**. We have treated the strings as though they were line segments that can be treated as straight over the distances involved in gravitational lensing. Image separations of up to 10 arcsec at a redshift of up to ~ 1 are relevant, so the string needs to be straight on scales of $\sim 100h^{-1}$ kpc. It is not so clear that this will be so, which complicates lensing searches. The problem is that kinks are introduced into strings when they intercommute, and these kinks will propagate non-dispersively along the string, much like waves on a piano wire. Numerical string simulations are of limited help here, since the $100h^{-1}$ kpc scale is below their numerical resolution (the simulations must follow the evolution in a horizon volume with a 3D mesh, which is normally limited to something like 128^3). The critical scale is comparable to the size of the smallest string loops (see above), so gravity-wave damping will clearly help smooth the strings to some extent, but the exact behaviour is not known. The main effect of microstructure would be to remove some of the large-scale coherence in position angle that is the main lensing signature, so this is an issue that demands further study (see de Laix, Krauss & Vachaspati 1997).

HUNTING STRINGS Owing to the self-destructive nature of the network evolution, strings are much rarer beasts today than when they seeded the structure we now observe, and gravitational lensing is the only possible route to detection. Hindmarsh (1990) discusses the prospects for direct detection of relic string via gravitational lensing. Using the above infinite long-string density, plus the matter-era loop distribution

$$\frac{dn}{d\ln L} = \nu(ct)^{-3}\left(\frac{ct}{L}\right)^2, \tag{10.31}$$

with $\nu \simeq 1$, it is possible to calculate the total length of string in the sky. This of course depends on the redshift out to which the search is performed; if we just use a Euclidean approximation for the geometry, the result is

$$L_{\rm string} \simeq 10^4\, z^2 \text{ degrees} \tag{10.32}$$

over the whole sky, approximately equally in long strings and in loops, for which the minimum angular size will be of order 1 arcmin. How large an area should we search in order to be sure that some string must be present? Let us consider $z \simeq 0.5$ and adopt angular lengths on the sky of $L = 1000$ degrees of string in each of long string and loops. If we survey a circular patch of angular radius R radians, the field centre must lie within a strip $2R$ wide about the string for success. Ignoring strip overlap, the Poisson probability of this is $p = 1 - \exp(-2RL/4\pi)$. A 50% success probability thus requires a radius of 14°. For a given area of sky, it is better to observe a larger number of smaller fields: $p = 1 - [\exp(-2RLN^{-1/2}/4\pi)]^N$. For the same chance of success, the individual field radius can be a factor $N^{-1/2}$ smaller. For the loops, there will be an angular size distribution similar in form to the linear one: $dn/dL \propto L^{-2}$. If we normalize this to 1000 degrees of string loop over the range 1 sr to 1 arcmin, then the number of loops on the sky larger than $L_{\rm min}$ is

$$N_{\rm loop} \simeq 2\, L_{\rm min}^{-1}. \tag{10.33}$$

To have one loop per Schmidt telescope plate (25 square degrees) thus involves a limiting radius of curvature of about 1 arcminute. Hunting for strings is thus very much a needle-in-haystack job: only a very few special parts of the sky are affected and so thorough searching of very large faint-galaxy databases is required. However, the potential importance for physics of a detection makes the effort more than worthwhile.

CMB ANISOTROPIES WITH STRINGS In string-dominated models, the structure of the microwave background differs quite radically from that in pure gravitational-collapse models, as discussed in chapter 18. Perturbations due to string wakes are set up in a causal fashion, which means that the large-scale distribution of matter and radiation at last scattering can contain no fluctuations on scales larger than the horizon at that time, $183(\Omega h^2)^{-1/2}$ Mpc. Such fluctuations subtend angles of $\sim 1°$, so it may appear that string models will be lacking in large-angle CMB fluctuations. However, there are extra sources of anisotropy in a string-dominated model, the principal effect coming from gravitational lensing. This may seem unpromising, since lensing conserves surface brightness; light deflection operating on the CMB will not generate any new anisotropies. This would be so if the strings were stationary, but their motion means that different redshifts can be imparted to different parts of the sky, generating anisotropic intensity through conservation of I_ν/ν^3. The effect is shown in figure 10.5; the result is that there is a temperature discontinuity at the string of amplitude

$$\frac{\delta T}{T} = \Delta\theta\, \gamma \frac{v_\perp}{c} \tag{10.34}$$

(Kaiser & Stebbins 1984). Since the string velocities are $\sim c$, the temperature anisotropies are of order $8\pi G\mu/c^2$. The sky contains a highly non-Gaussian pattern in which the microwave background is tessellated into patches of different temperature bounded by strings (Bouchet, Bennett & Stebbins 1988).

What about the angular dependence of the resulting fluctuations? The string network evolves in a self-similar fashion, such that there is of the order of one straight string crossing the horizon at a given time. The perturbation field seen on a given angular scale therefore derives from strings at the redshift at which the horizon subtends the angle in question. The horizon length at high redshift is $r_{\rm H} \simeq 6000(\Omega h^2 z)^{-1/2}$ Mpc, so that every time the redshift changes by a factor 4, the string pattern projected onto the sky changes scale by a factor 2 (because the comoving angular-diameter distance to large redshifts is very nearly independent of redshift). The string sky therefore consists of a hierarchy of temperature patterns that are superimposed, but all of which cause the same rms temperature fluctuations. The end result is therefore an angular power spectrum for the temperature fluctuations that is approximately 'scale invariant' (the technical meaning of this term is discussed in detail in chapters 15 and 18). It is amusing that the outcome is so similar to the effect of the scale-invariant primordial fluctuations in gravitational potential that emerge from inflation. In the potential-fluctuation case, the large-scale CMB anisotropies are generated by density perturbations $z \simeq 1000$, and these structures in turn have existed from the very earliest epochs of the big bang – veritable 'rosetta stones' of cosmology. In the string case, however, these fluctuations were generated in the relatively recent past and in local regions of space: the significance of the CMB data is completely altered.

Problems

(10.1) Show that the light deflection angle produced by a cosmic string at an angle θ to the plane of the sky is smaller than the deflection that arises when the string lies in the plane of the sky:

$$\alpha = 4\pi \frac{G\mu}{c^2} \cos\theta. \tag{10.35}$$

(10.2) Show that gravitational radiation by a distorted oscillating loop of cosmic string will produce a emitted power $\sim G\mu^2$, and hence that the smallest surviving loops will have sizes $\sim G\mu t$.

(10.3) In some models, intercommutation may not chop a string network into loops. Show that, in such cases, the density due to strings declines more slowly than that of either matter or radiation, leading to the possibility of a **string-dominated universe**. In such a model, prove that the comoving distance–redshift relation is

$$R_0 r(z) = \frac{c}{H_0} \ln(1+z). \tag{10.36}$$

(10.4) Long strings create perturbations in the density field through which they move by generating wakes. Show that a string with velocity v and Lorentz factor γ causes a velocity perturbation of $\delta v = 4\pi G\mu\gamma v$ perpendicular to its line of travel. If this perturbation is imposed at time t_i, show that the thickness X and surface density Σ of the wake at a later time t are

$$
\begin{aligned}
X &= \frac{12\,\delta v\,t_i}{5}\left(\frac{t}{t_i}\right)^{4/3} \\
\Sigma &= \frac{2\,\delta v}{5\pi G\,t_i}\left(\frac{t}{t_i}\right)^{-2/3} .
\end{aligned}
\tag{10.37}
$$

11 Inflationary cosmology

11.1 General arguments for inflation

The standard isotropic cosmology is a very successful framework for interpreting observations, but prior to the early 1980s there were certain questions that had to be avoided. The initial conditions of the big bang appear to be odd in a number of ways; these puzzles are encapsulated in a set of classical 'problems', as follows.

THE HORIZON PROBLEM Standard cosmology contains a particle horizon of comoving radius

$$r_{\rm H} = \int_0^t \frac{c\,dt}{R(t)}, \tag{11.1}$$

which converges because $R \propto t^{1/2}$ in the early radiation-dominated phase. At late times, the integral is largely determined by the matter-dominated phase, for which

$$D_{\rm H} = R_0 r_{\rm H} \simeq \frac{6000}{\sqrt{\Omega z}}\, h^{-1}\,\text{Mpc}. \tag{11.2}$$

The horizon at last scattering ($z \sim 1000$) was thus only ~ 100 Mpc in size, subtending an angle of about 1 degree. Why then are the large number of causally disconnected regions we see on the microwave sky all at the same temperature? In other words, why do we live in a nearly homogeneous universe, even though there has been no time for different parts to 'compare notes' over quantities like the mean density?

THE FLATNESS PROBLEM The $\Omega = 1$ universe is unstable:

$$[1 - 1/\Omega(z)] = f(z)\,[1 - 1/\Omega], \tag{11.3}$$

where $f(z) = (1+z)^{-1}$ in the matter-dominated era and $f(z) \propto (1+z)^{-2}$ for radiation domination, so that $f(z) \simeq (1+z_{\rm eq})/(1+z)^2$ at early times. To get $\Omega \simeq 1$ today requires a **fine tuning of Ω** in the past, which becomes more and more precisely constrained as we increase the redshift at which the initial conditions are presumed to have been imposed. Ignoring annihilation effects, $1 + z = T_{\rm init}/2.7$ K and $1 + z_{\rm eq} \simeq 10^4$, so that the required fine tuning is

$$|\Omega(t_{\rm init}) - 1| \lesssim 10^{-22}\,(E_{\rm init}/\text{GeV})^{-2}. \tag{11.4}$$

At the Planck epoch, which is the natural initial time, this requires a deviation of only 1 part in 10^{60}. This is satisfied if the spatial curvature vanishes ($k = 0$), since $\Omega = 1$

exactly in that case, but a mechanism is still required to set up such an initial state. This equation is especially puzzling if $\Omega \neq 1$ today: how could the universe 'know' that it should start with a deviation from $\Omega = 1$ just so tuned that the curvature starts to become important only now after so many e-foldings of the expansion?

THE ANTIMATTER PROBLEM At $kT \gtrsim m_p c^2$, there exist in equilibrium roughly equal numbers of photons, protons and antiprotons. Today, $N_p/N_\gamma \sim 10^{-9}$, but $N_{\bar{p}} \simeq 0$. Conservation of baryon number would imply that $N_p/N_{\bar{p}} = 1 + O(10^{-9})$ at early times Where did this initial asymmetry come from?

THE STRUCTURE PROBLEM The universe is not precisely homogeneous. We generally presume that galaxies and clusters grew via gravitational instability from some initial perturbations. What is the origin of these?

This list can be extended to problems that come closer to astrophysics than cosmology *per se*. There is the question of the dark matter and its composition, for example. However, the list as it stands encompasses problems that go back to the very early stages of the big bang. It seems clear that conventional cosmological models need to be set up in an extremely special configuration, and this is certainly a deficiency of the theory. Critics can point out with some force that the big bang model explains nothing about the origin of the universe as we now perceive it, because all the most important features are 'predestined' by virtue of being built into the assumed initial conditions near to $t = 0$.

THE EXPANSION PROBLEM Even the most obvious fact of the cosmological expansion is unexplained. Although general relativity forbids a static universe, this is not enough to understand the expansion. As shown in chapter 3, the gravitational dynamics of the cosmological scale factor $R(t)$ are just those of a cannonball travelling vertically in the Earth's gravity. Suppose we see a cannonball rising at a given time $t = t_0$: it may be true to say that it has $r = r_0$ and $v = v_0$ at this time because at a time Δt earlier it had $r = r - v_0 \Delta t$ and $v = v_0 + g\Delta t$, but this is hardly a satisfying explanation for the motion of a cannonball that was in fact fired by a cannon. Nevertheless, this is the only level of explanation that classical cosmology offers: the universe expands now because it did so in the past. Although it is not usually included in the list, one might thus with justice add an 'expansion problem' as perhaps the most fundamental in the catalogue of classical cosmological problems. Certainly, early generations of cosmologists were convinced that some specific mechanism was required in order to explain how the universe was set in motion.

For many years, it was assumed that any solution to these difficulties would have to await a theory of quantum gravity. The classical singularity can be approached no closer than the Planck time of $\sim 10^{-43}$ s, and so the initial conditions for the classical evolution following this time must have emerged from behind the presently impenetrable barrier of the quantum gravity epoch. There remains a significant possibility that this policy of blaming everything on quantum gravity may be correct, and this is what lies behind modern developments in **quantum cosmology**. This field is not really suitable for treatment at the present level, and it deals with the subtlest possible questions concerning the meaning of the wave function for the entire universe. The eventual aim is to understand whether there could be a way in which the universe could have been spontaneously created as a quantum-mechanical fluctuation, and if so whether it would have the initial properties that observations seem to require. Since this programme has to face up to the challenge of quantizing gravity, it is fair to say that there are as yet no definitive answers, despite much thought-provoking work. See chapter 11 of Kolb & Turner (1990) or Halliwell (1991) for an introduction; Wu (1993) may be consulted for more technical details.

However, the great development of cosmology in the 1980s was the realization that the explanation of the initial-condition puzzles might involve physics at lower energies: 'only' 10^{15} GeV. Although this idea, now known as inflation, cannot be considered to be firmly established, the ability to treat gravity classically puts the discussion on a much less speculative foundation. What has emerged is a general picture of the early universe that has compelling simplicity, which moreover may be subject to observational verification. What follows is an outline of the main features of inflation; for more details see e.g. chapter 8 of Kolb & Turner (1990); Brandenberger (1990); Liddle & Lyth (1993).

11.2 An overview of inflation

EQUATION OF STATE FOR INFLATION The above list of problems with conventional cosmology provides a strong hint that the equation of state of the universe may have been very different at very early times. To solve the horizon problem and allow causal contact over the whole of the region observed at last scattering requires a universe that expands 'faster than light' near $t = 0$: $R \propto t^\alpha$, with $\alpha > 1$. If such a phase had existed, the integral for the comoving horizon would have diverged, and there would be no difficulty in understanding the overall homogeneity of the universe – this could then be established by causal processes. Indeed, it is tempting to assert that the observed homogeneity *proves* that such causal contact must once have occurred. This phase of accelerated expansion is the most general feature of what has become known as the **inflationary universe**.

What condition does this place on the equation of state? In the integral for $r_{\rm H}$, we can replace dt by $dR/\dot{R}$, which the Friedmann equation says is $\propto dR/\sqrt{\rho R^2}$ at early times. Thus, the horizon diverges provided the equation of state is such that ρR^2 vanishes or is finite as $R \to 0$. For a perfect fluid with $p \equiv (\Gamma - 1)\epsilon$ as the relation between pressure and energy density, we have the adiabatic dependence $p \propto R^{-3\Gamma}$, and the same dependence for ρ if the rest-mass density is negligible. A period of inflation therefore needs

$$\boxed{\Gamma < 2/3 \quad \Rightarrow \quad \rho c^2 + 3p < 0.} \tag{11.5}$$

An alternative way of seeing that this criterion is sensible is that the 'active mass density' $\rho + 3p/c^2$ then vanishes. Since this quantity forms the rhs of Poisson's equation generalized to relativistic fluids, it is no surprise that the vanishing of $\rho + 3p/c^2$ allows a coasting solution with $R \propto t$.

Such a criterion can also solve the flatness problem. Consider the Friedmann equation,

$$\dot{R}^2 = \frac{8\pi G \rho R^2}{3} - kc^2. \tag{11.6}$$

As we have seen, the density term on the rhs must exceed the curvature term by a factor of at least 10^{60} at the Planck time, and yet a more natural initial condition might be to have the matter and curvature terms being of comparable order of magnitude. However, an inflationary phase in which ρR^2 increases as the universe expands can clearly make the curvature term relatively as small as required, provided inflation persists for sufficiently long.

DE SITTER SPACE AND INFLATION We have seen that inflation will require an equation of state with negative pressure, and the only familiar example of this is the $p = -\rho c^2$ relation that applies for vacuum energy; in other words, we are led to consider inflation as happening in a universe dominated by a cosmological constant. As usual, any initial expansion will redshift away matter and radiation contributions to the density, leading to increasing dominance by the vacuum term. If the radiation and vacuum densities are initially of comparable magnitude, we quickly reach a state where the vacuum term dominates. The Friedmann equation in the vacuum-dominated case has three solutions:

$$R \propto \begin{cases} \sinh Ht & (k=-1) \\ \cosh Ht & (k=+1) \\ \exp Ht & (k=0), \end{cases} \tag{11.7}$$

where $H = \sqrt{\Lambda c^2/3} = \sqrt{8\pi G \rho_{\rm vac}/3}$; all solutions evolve towards the exponential $k = 0$ solution, known as **de Sitter space**. Note that H is not the Hubble parameter at an arbitrary time (unless $k = 0$), but it becomes so exponentially fast as the hyperbolic trigonometric functions tend to the exponential.

Because de Sitter space clearly has H^2 and ρ in the right ratio for $\Omega = 1$ (this is obvious, since $k = 0$), the density parameter in all models tends to unity as the Hubble parameter tends to H. If we assume that the initial conditions are not fine tuned (i.e. $\Omega = O(1)$ initially), then maintaining the expansion for a factor f produces

$$\Omega = 1 + O(f^{-2}). \tag{11.8}$$

This can solve the flatness problem, provided f is large enough. To obtain Ω of order unity today requires $|\Omega - 1| \lesssim 10^{-52}$ at the GUT epoch, and so

$$\boxed{\ln f \gtrsim 60} \tag{11.9}$$

e-foldings of expansion are needed; it will be proved below that this is also exactly the number needed to solve the horizon problem. It then seems almost inevitable that the process should go to completion and yield $\Omega = 1$ to measurable accuracy today. There is only a rather small range of e-foldings (60 ± 2, say) around the critical value for which Ω today can be of order unity without its being equal to unity to within the tolerance set by density fluctuations ($\pm 10^{-5}$), and it would constitute an unattractive fine tuning to require that the expansion hit this narrow window exactly.

This gives the first of two strong **predictions of inflation**, that the universe must be spatially flat:

$$\boxed{\text{inflation} \quad \Rightarrow \quad k = 0.} \tag{11.10}$$

Note that this need not mean the Einstein–de Sitter model; the alternative possibility is that a vacuum contribution is significant in addition to matter, so that $\Omega_m + \Omega_v = 1$.

Astrophysical difficulties in finding evidence for $\Omega_m = 1$ are thus one of the major motivations, through inflation, for taking the idea of a large cosmological constant seriously. Despite the apparent inevitability of this prediction of spatial flatness, there has been some experimentation with models that use inflationary concepts to generate an open universe. There could be two periods of inflation, one long, the second short; the first would solve the homogeneity problem, so that the second need not produce flatness (e.g. Bucher, Goldhaber & Turok 1995; Górski *et al.* 1995). Many researchers in inflation do not find such models attractive, although they would have to be considered if observational evidence were to prove the universe to be open.

REHEATING FROM INFLATION The discussion so far indicates a possible route to solving the problems of the initial conditions in conventional cosmology, but has a critical missing ingredient. The idea of inflation is to set the universe expanding towards an effective $k = 0$ state by using the repulsive gravitational force of vacuum energy or some other unknown state of matter that satisfies $p < -\rho c^2/3$. There remains the difficulty of returning to a normal equation of state: the universe is required to undergo a **cosmological phase transition**. Such a suggestion would have seemed highly *ad hoc* in the 1960s when the horizon and flatness problems were first clearly articulated by Dicke (1961). The invention of inflation by Guth (1981) had to await developments in quantum field theory that provided a plausible basis for this phase transition. Detailed mechanisms for the phase transition will be discussed below; they have deliberately been put off so far to emphasize the general character of many of the arguments for inflation.

What will emerge is that it is possible for inflation to erase its tracks in a very neat way. If we are dealing with quantum fields at a temperature T, then an energy density $\sim T^4$ in natural units is expected in the form of vacuum energy. The vacuum-driven expansion produces a universe that is essentially devoid of normal matter and radiation; these are all redshifted away by the expansion, and so the temperature of the universe becomes $\ll T$. A phase transition to a state of zero vacuum energy, if instantaneous, would transfer the energy T^4 to normal matter and radiation as a latent heat. The universe would therefore be **reheated**: it returns to the temperature T at which inflation was initiated, but with the correct special initial conditions for the expansion. The transition in practical models is not instantaneous, however, and so the reheating temperature is lower than the temperature prior to inflation.

QUANTUM FLUCTUATIONS Note that de Sitter space contains an **event horizon**, in that the comoving distance that particles can travel between a time t_0 and $t = \infty$ is finite,

$$r_{\rm EH} = \int_{t_0}^{\infty} \frac{c\, dt}{R(t)}; \tag{11.11}$$

this is not to be confused with the particle horizon, where the upper limit for the integral would be t_0. With $R \propto \exp(Ht)$, the proper radius of the horizon is given by $R_0 r_{\rm EH} = c/H$. Figure 11.1 illustrates the situation. The exponential expansion literally makes distant regions of space move faster than light, so that points separated by $> c/H$ can never communicate with each other.

As with black holes, it therefore follows that thermal **Hawking radiation** will be created. These quantum fluctuations in de Sitter spacetime provide the seeds for what will eventually become galaxies and clusters. The second main prediction of inflation is

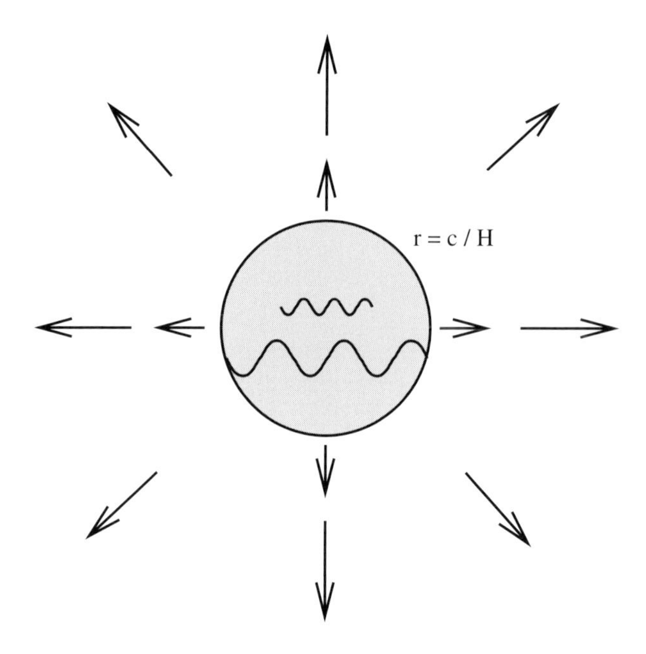

Figure 11.1. The event horizon in de Sitter space. Particles outside the sphere at $r = c/H$ can never receive light signals from the origin, nor can an observer at the origin receive information from outside the sphere. The exponential expansion quickly accelerates any freely falling observers to the point where their recession from the origin is effectively superluminal. The wave trains represent the generation of fluctuations in this spacetime. Waves with $\lambda \ll c/H$ effectively occupy flat space, and so undergo the normal quantum fluctuations for a vacuum state. As these modes (of fixed comoving wavelength) are expanded to sizes $\gg c/H$, causality forces the quantum fluctuation to become frozen as a classical amplitude that can seed large-scale structure.

that such fluctuations should exist in all fields, in particular that there should exist a background of gravitational waves left as a relic of inflation.

The possibility that all the structure in the universe (including ourselves) in fact originated in quantum fluctuations is an idea of tremendous appeal. If it could be shown to be correct, it would rank as one of the greatest possible intellectual advances. We now have to look at some of the practical details to see how this concept might be made to function in practice, and how it may be tested.

11.3 Inflation field dynamics

The general concept of inflation rests on being able to achieve a negative-pressure equation of state. This can be realized in a natural way by quantum fields in the early universe.

QUANTUM FIELDS AT HIGH TEMPERATURES The critical fact we shall need from quantum field theory is that quantum fields can produce an energy density that mimics a cosmological constant. The discussion will be restricted to the case of a scalar field ϕ (complex in general, but often illustrated using the case of a single real field). The restriction to scalar fields is not simply for reasons of simplicity, but because the scalar sector of particle physics is relatively unexplored. While vector fields such as electromagnetism are well understood, it is expected in many theories of unification that additional scalar fields such as the Higgs field will exist. We now need to look at what these can do for cosmology.

The Lagrangian density for a scalar field is as usual of the form of a kinetic minus a potential term:

$$\mathscr{L} = \tfrac{1}{2}\partial_\mu\phi\,\partial^\mu\phi - V(\phi). \tag{11.12}$$

In familiar examples of quantum fields, the potential would be

$$V(\phi) = \tfrac{1}{2}\,m^2\phi^2, \tag{11.13}$$

where m is the mass of the field in natural units. However, it will be better to keep the potential function general at this stage. As usual, Noether's theorem gives the energy–momentum tensor for the field as

$$T^{\mu\nu} = \partial^\mu\phi\partial^\nu\phi - g^{\mu\nu}\mathscr{L}. \tag{11.14}$$

From this, we can read off the energy density and pressure:

$$\begin{aligned} \rho &= \tfrac{1}{2}\dot{\phi}^2 + V(\phi) + \tfrac{1}{2}(\nabla\phi)^2 \\ p &= \tfrac{1}{2}\dot{\phi}^2 - V(\phi) - \tfrac{1}{6}(\nabla\phi)^2. \end{aligned} \tag{11.15}$$

If the field is constant both spatially and temporally, the equation of state is then $p = -\rho$, as required if the scalar field is to act as a cosmological constant; note that derivatives of the field spoil this identification.

If ϕ is a (complex) Higgs field, then the symmetry-breaking Mexican hat potential might be assumed:

$$V(\phi) = -\mu^2|\phi|^2 + \lambda|\phi|^4. \tag{11.16}$$

At the classical level, such potentials determine where $|\phi|$ will be found in equilibrium: at the potential minimum. In quantum terms, this goes over to the **vacuum expectation value** $\langle 0|\phi|0\rangle$. However, these potentials do not include the inevitable fluctuations that will arise in thermal equilibrium. We know how to treat these in classical systems: at non-zero temperature a system of fixed volume will minimize not its potential energy, but the **Helmholtz free energy** $F = V - TS$, S being the entropy. The calculation of the entropy is technically complex, since it involves allowance for quantum interactions with a thermal bath of background particles. However, the main result can be justified, as follows. The effect of the thermal interaction must be to add an interaction term to the Lagrangian $\mathscr{L}_{\rm int}(\phi, \psi)$, where ψ is a thermally fluctuating field that corresponds to the heat bath. In general, we would expect $\mathscr{L}_{\rm int}$ to have a quadratic dependence on $|\phi|$ around the origin, $\mathscr{L}_{\rm int} \propto |\phi|^2$ (otherwise we would need to explain why the second derivative either vanishes or diverges); the coefficient of proportionality will be the square of an effective mass that depends on the thermal fluctuations in ψ. On dimensional grounds, this coefficient must be proportional to T^2, although a more detailed analysis would be required to obtain the constant of proportionality.

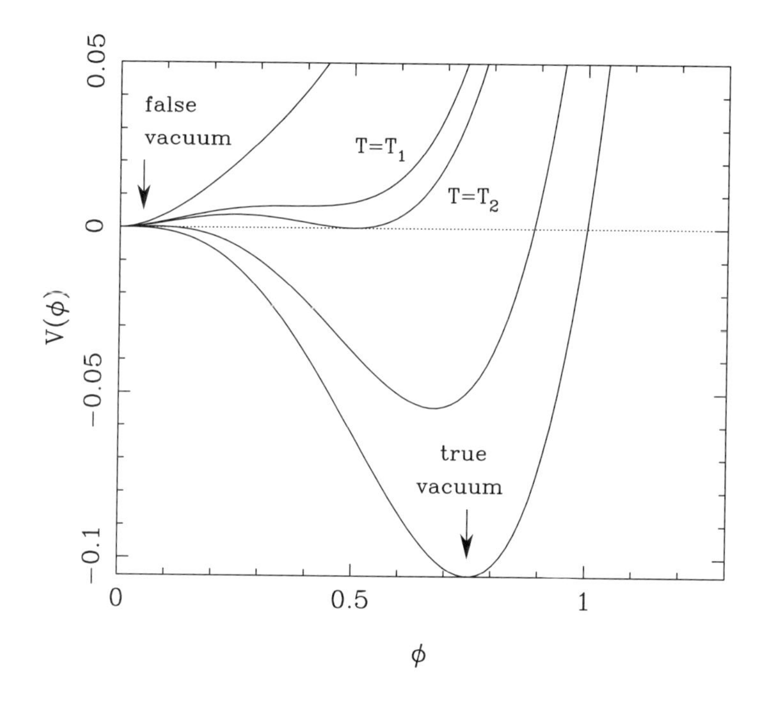

Figure 11.2. The temperature-dependent effective potential $V = T^2|\phi|^2 - |\phi|^3 + |\phi|^4$, illustrated at several temperatures: $T^2 = 0.5$, 9/32, 1/4, 0.1, 0. For $T > T_1 = (9/32)^{1/2} \simeq 0.53$, only the false vacuum is available; for $T < T_2 = 1/2$ the true vacuum is energetically favoured and the potential approaches the zero-temperature form.

There is thus a temperature-dependent **effective potential** that we have to minimize:

$$V_{\text{eff}}(\phi, T) = V(\phi, 0) + aT^2|\phi|^2. \tag{11.17}$$

The effect of this on the symmetry-breaking potential depends on the form of the zero-temperature $V(\phi)$. If the function is taken to be the simple Higgs form $V = -\mu^2 + \lambda\phi^4$, then the temperature-dependent part simply modifies the effective value of μ^2: $\mu^2_{\text{eff}} = \mu^2 - aT^2$. At very high temperatures, the potential will be parabolic, with a minimum at $|\phi| = 0$; below the critical temperature, $T_c = \mu/\sqrt{a}$, the ground state is at $|\phi| = [\mu^2_{\text{eff}}/(2\lambda)]^{1/2}$ and the symmetry is broken. At any given time, there is only a single minimum, and so this is a second-order phase transition.

It is easy enough to envisage more complicated behaviour, as illustrated in figure 11.2. This plots the potential

$$V_{\text{eff}}(\phi, T) = \lambda|\phi|^4 - b|\phi|^3 + aT^2|\phi|^2, \tag{11.18}$$

which displays two critical temperatures [problem 11.2]. At very high temperatures, the potential will have a parabolic minimum at $|\phi| = 0$; at T_1, a second minimum appears in V_{eff} at $|\phi| \neq 0$, and this will be the global minimum for some $T_2 < T_1$. For $T < T_2$, the state at $|\phi| = 0$ is known as the **false vacuum**, whereas the global minimum is known as the **true vacuum**. For this particular form of potential, the second minimum around

$\phi = 0$ always exists, so that there is a potential barrier preventing a transition to the false vacuum. This can be overcome by adding a small $-\mu^2|\phi|^2$ component to the potential, so that there will be a third critical temperature at which the curvature around the origin changes sign, leaving only one minimum in the potential. Alternatively, once the barrier is small enough, quantum tunnelling can take place and free ϕ to move. The universe is no longer trapped in the false vacuum and can make a first-order phase transition to the true vacuum state.

The crucial point to note for cosmology is that there is an energy-density difference between the two vacuum states:

$$\Delta V = \frac{\mu^4}{2\lambda}. \tag{11.19}$$

If we say that the zero of energy is such that $V = 0$ in the true vacuum, this implies that the false-vacuum symmetric state displays an effective cosmological constant. On dimensional grounds, this must be an energy density $\sim m^4$ in natural units, where m is the energy at which the phase transition occurs. For GUTs, $m \simeq 10^{15}$ GeV; in laboratory units, this implies

$$\rho_{\rm vac} = \frac{(10^{15}\mathrm{GeV})^4}{\hbar^3 c^5} \simeq 10^{80}\ \mathrm{kg\,m^{-3}}. \tag{11.20}$$

The inevitability of such a colossal vacuum energy in models with GUT-scale symmetry breaking was the major motivation for the concept of inflation as originally envisaged by Guth (1981). At first sight, the overall package looks highly appealing, since the phase transition from false to true vacuum both terminates inflation and also reheats the universe to the GUT temperature, allowing the possibility that GUT-based reactions that violate baryon-number conservation can generate the observed matter–antimatter asymmetry. Because the transition is first order, the original inflation model is known as **first-order inflation**.

However, while a workable inflationary cosmology will very probably deploy the three basic elements of vacuum-driven expansion, fluctuation generation and reheating, it has become clear that such a model must be more complex than Guth's initial proposal. To explain where the problems arise, we need to look in more detail at the functioning of the inflation mechanism.

DYNAMICS OF THE INFLATION FIELD Treating the field classically (i.e. considering the expectation value $\langle\phi\rangle$, we get from energy–momentum conservation ($T^{\mu\nu}_{\ ;\nu} = 0$) the equation of motion

$$\boxed{\ddot{\phi} + 3H\dot{\phi} - \nabla^2\phi + dV/d\phi = 0.} \tag{11.21}$$

This can also be derived more easily by the direct route of writing down the action $S = \int \mathcal{L}\sqrt{-g}\ d^4x$ and applying the Euler–Lagrange equation that arises from a stationary action ($\sqrt{-g} = R^3(t)$ for an FRW model, which is the origin of the Hubble drag term $3H\dot{\phi}$).

The solution of the equation of motion becomes tractable if we both ignore spatial inhomogeneities in ϕ and make the **slow-rolling approximation** that $|\ddot{\phi}|$ is negligible in comparison with $|3H\dot{\phi}|$ and $|dV/d\phi|$. Both these steps are required in order that inflation can happen; we have shown above that the vacuum equation of state only holds if in some sense ϕ changes slowly both spatially and temporally. Suppose there are characteristic temporal and spatial scales T and X for the scalar field; the conditions for inflation are that the negative-pressure equation of state from $V(\phi)$ must dominate the normal-pressure effects of time and space derivatives:

$$V \gg \phi^2/T^2, \quad V \gg \phi^2/X^2, \tag{11.22}$$

hence $|dV/d\phi| \sim V/\phi$ must be $\gg \phi/T^2 \sim \ddot{\phi}$. The $\ddot{\phi}$ term can therefore be neglected in the equation of motion, which then takes the slow-rolling form for homogeneous fields:

$$\boxed{3H\dot{\phi} = -dV/d\phi.} \tag{11.23}$$

The conditions for inflation can be cast into useful dimensionless forms. The basic condition $V \gg \dot{\phi}^2$ can now be rewritten using the slow-roll relation as

$$\boxed{\epsilon \equiv \frac{m_{\mathrm{P}}^2}{16\pi}\left(\frac{V'}{V}\right)^2 \ll 1.} \tag{11.24}$$

Also, we can differentiate this expression to obtain the criterion $V'' \ll V'/m_{\mathrm{P}}$. Using slow-roll once more gives $3H\dot{\phi}/m_{\mathrm{P}}$ for the rhs, which is in turn $\ll 3H\sqrt{V}/m_{\mathrm{P}}$ because $\dot{\phi}^2 \ll V$, giving finally

$$\boxed{\eta \equiv \frac{m_{\mathrm{P}}^2}{8\pi}\left(\frac{V''}{V}\right) \ll 1} \tag{11.25}$$

(recall that for de Sitter space $H = \sqrt{8\pi G V(\phi)/3} \sim \sqrt{V}/m_{\mathrm{P}}$ in natural units). These two criteria make perfect intuitive sense: the potential must be flat in the sense of having small derivatives if the field is to roll slowly enough for inflation to be possible.

Similar arguments can be made for the spatial parts. However, they are less critical: what matters is the value of $\nabla\phi = \nabla_{\mathrm{comoving}}\,\phi/R$. Since R increases exponentially, these perturbations are damped away: assuming V is large enough for inflation to start in the first place, inhomogeneities rapidly become negligible. This 'stretching' of field gradients as we increase the cosmological horizon beyond the value predicted in classical cosmology also solves a related problem that was historically important in motivating the invention of inflation – the **monopole problem**. As discussed in chapter 10, monopoles are point-like topological defects that would be expected to arise in any phase transition at around the GUT scale ($t \sim 10^{-35}$ s). If they form at approximately one per horizon volume at this time, then it follows that the present universe would contain $\Omega \gg 1$ in monopoles (see chapter 10). This unpleasant conclusion is avoided if the horizon can be made much larger than the classical one at the end of inflation; the GUT fields have then been aligned over a vast scale, so that topological-defect formation becomes extremely rare.

ENDING INFLATION Although spatial derivatives of the scalar field can thus be neglected, the same is not always true for time derivatives. Although they may be negligible initially, the relative importance of time derivatives increases as ϕ rolls down the potential and V approaches zero (leaving aside the subtle question of how we know that the minimum is indeed at zero energy). Even if the potential does not steepen, sooner or later we will have $\epsilon \simeq 1$ or $|\eta| \simeq 1$ and the inflationary phase will cease. Instead of rolling slowly 'downhill', the field will oscillate about the bottom of the potential, with the oscillations becoming damped by the $3H\dot{\phi}$ friction term. Eventually, we will be left with a stationary field that either continues to inflate without end, if $V(\phi = 0) > 0$, or which simply has zero density. This would be a most boring universe to inhabit, but fortunately there is a more realistic way in which inflation can end. We have neglected so far the couplings of the scalar field to matter fields. Such couplings will cause the rapid oscillatory phase to produce particles, leading to **reheating**. Thus, even if the minimum of $V(\phi)$ is at $V = 0$, the universe is left containing roughly the same energy density as it started with, but now in the form of normal matter and radiation – which starts the usual FRW phase, albeit with the desired special 'initial' conditions.

It should help clarify matters to go over a specific solution in some detail. The scalar-field equation of motion is

$$\ddot{\phi} + 3H\dot{\phi} + \Gamma\dot{\phi} + dV/d\phi = 0, \tag{11.26}$$

where the extra term $\Gamma\dot{\phi}$ is often added empirically to represent the effect of particle creation. This form is chosen because it resembles a drag term, so removing energy from the motion of ϕ and dumping it in the form of a radiation background. We shall solve the equation of motion without including this term, since the main features of the solution are still the same: ϕ undergoes oscillations of declining amplitude after the end of inflation, and Γ only changes the rate of damping. For more realistic and detailed reheating models see e.g. Linde (1989) and Kofman, Linde & Starobinsky (1997).

In order to solve the evolution equations for ϕ with a computer, they have to be cast into dimensionless form. Suppose we start inflation at a point specified by the initial values of ϕ, H and V; suitable dimensionless variables for the problem are $\tau \equiv H_i t$, $Y \equiv \phi/\phi_i$, $X \equiv H/H_i$ and $W \equiv V/(\phi_i^2 H_i^2)$. The equation of motion is then

$$Y'' + 3XY' + \frac{dW}{dY} = 0, \qquad X^2 = \frac{W + Y'^2/2}{W_i}, \tag{11.27}$$

where a prime denotes $d/d\tau$. The only parameter that matters is W_i, which sets the amount of inflation. The initial conditions (chosen to be at $\tau = 0$) are $Y = 1$ and the slow-roll condition

$$Y_i' = -\frac{1}{3}\left.\frac{dW}{dy}\right|_i. \tag{11.28}$$

Lastly, the scale factor is determined by $d\ln a/d\ln\tau = X$.

These equations are simple to integrate: set Y and Y' to their starting points, evaluate Y'' from the equation of motion and update the variables for a small change in τ: $Y \to Y + Y'\delta\tau$, $Y' \to Y' + Y''\delta\tau$ etc. Iterating this process solves for $Y(\tau)$ and $a(\tau)$. This solution is illustrated in figure 11.3 for a simple mass-like potential: $W = W_i Y^2$. It is easy to show that, in this case, the inflationary parameters are both equal at $\epsilon = \eta = (2/3)W_i/Y^2$. So, provided W_i is chosen sufficiently small, inflation will proceed quite happily until ϕ has fallen to a small fraction of its initial value.

As well as being of interest for completing the picture of inflation, it is essential to realize that these closing stages of inflation are the *only* ones of observational relevance. Inflation might well continue for a huge number of e-foldings, all but the last few satisfying $\epsilon, \eta \ll 1$. However, the scales that left the de Sitter horizon at these early

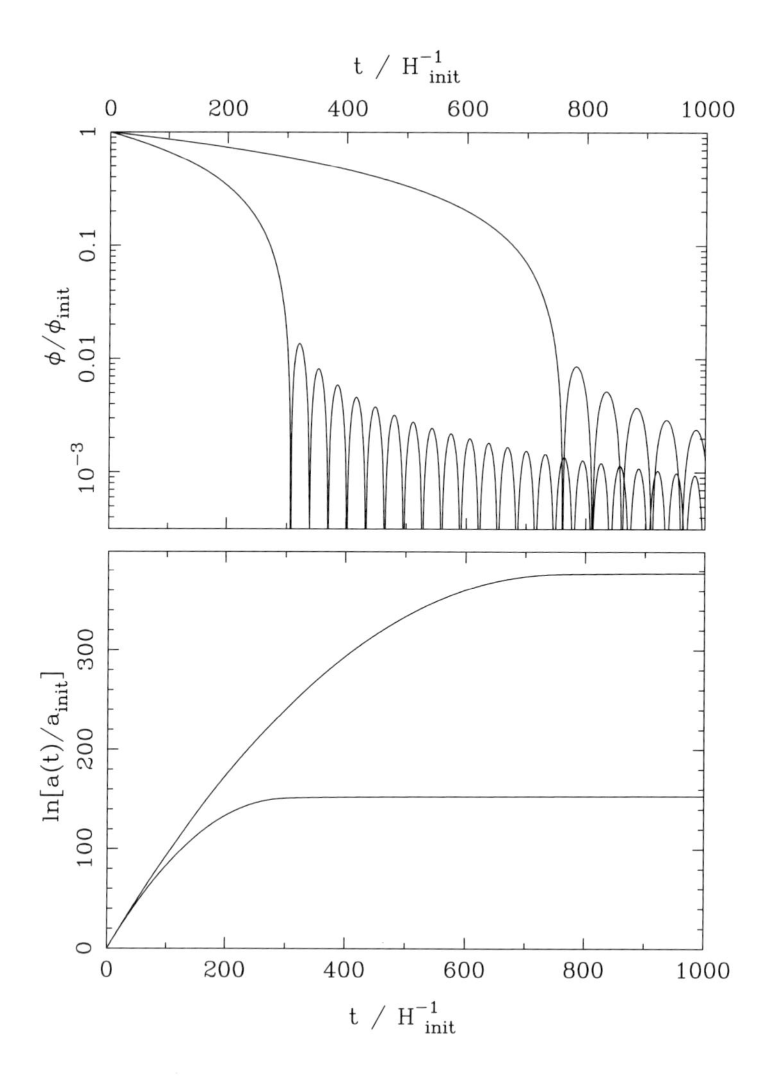

Figure 11.3. A plot of the exact solution for the scalar field in a model with a $V \propto \phi^2$ potential. The top panel shows how the absolute value of ϕ falls smoothly with time during the inflationary phase, and then starts to oscillate when inflation ends. The bottom panel shows the evolution of the scale factor. We see the initial exponential behaviour flattening as the vacuum energy ceases to dominate. The two models shown have starting points of $W_i \equiv V_i/(\phi_i^2 H_i^2) = 0.002$ and 0.005; the former (upper lines in each panel) gives about 380 e-foldings of inflation, the latter (lower lines) only 150. According to the $\epsilon = \eta = 1$ criterion, inflation in these models ends at respectively $t = 730$ and 240. The observationally relevant part of inflation is the last 60 e-foldings, and the behaviour of the scale factor is significantly non-exponential in this regime.

times are now vastly greater than our observable horizon, c/H_0, which exceeds the de Sitter horizon by only a finite factor. If inflation terminated by reheating to the GUT temperature, then the expansion factor required to reach the present epoch is

$$a_{\rm GUT}^{-1} \simeq E_{\rm GUT}/E_\gamma. \tag{11.29}$$

The comoving horizon size at the end of inflation was therefore

$$d_{\rm H}(t_{\rm GUT}) \simeq (c/H_{\rm GUT})\, a_{\rm GUT}^{-1} \simeq (E_{\rm P}/E_\gamma)\, E_{\rm GUT}^{-1}, \tag{11.30}$$

where the last expression in natural units uses $H \simeq \sqrt{V}/E_{\rm P} \simeq E_{\rm GUT}^2/E_{\rm P}$. For a GUT energy of 10^{15}GeV, this is about 10 m. This is a sobering illustration of the magnitude of the horizon problem; if we relied on causal processes at the GUT era to produce homogeneity, then the universe would only be smooth in patches a few comoving metres across. To solve the problem, we need enough e-foldings of inflation to have stretched this GUT-scale horizon to the present horizon size

$$\boxed{N_{\rm obs} = \ln\left[\frac{3000h^{-1}\,{\rm Mpc}}{(E_{\rm P}/E_\gamma)E_{\rm GUT}^{-1}}\right] \simeq 60.} \tag{11.31}$$

By construction, this is enough to solve the horizon problem, and it is also the number of e-foldings needed to solve the flatness problem. This is no coincidence, since we saw earlier that the criterion in this case was

$$N \gtrsim \frac{1}{2}\,\ln\left(\frac{a_{\rm eq}}{a_{\rm GUT}^2}\right). \tag{11.32}$$

Now, $a_{\rm eq} = \rho_\gamma/\rho$, and $\rho = 3H^2\Omega/(8\pi G)$. In natural units, this translates to $\rho \sim E_{\rm P}^2(c/H_0)^{-2}$, or $a_{\rm eq}^{-1} \sim E_{\rm P}^2(c/H_0)^{-2}E_\gamma^{-4}$. The expression for N is then identical to that in the case of the horizon problem: the same number of e-folds will always solve both.

Realizing that the observational regime corresponds only to the terminal phases of inflation is both depressing and stimulating: depressing, because ϕ may well not move very much during the last phases – our observations relate only to a small piece of the potential, and we cannot hope to recover its form without substantial *a priori* knowledge; stimulating, because observations even on very large scales must relate to a period where the simple concepts of exponential inflation and scale-invariant density fluctuations were coming close to breaking down. This opens the possibility of testing inflation theories in a way that would not be possible with data relating to only the simpler early phases. These tests take the form of tilt and gravitational waves in the final perturbation spectrum, to be discussed further below.

11.4 Inflation models

EARLY INFLATION MODELS These general principles contrast sharply with Guth's initial idea, where the potential was trapped at $\phi = 0$, and eventually underwent a first-order phase transition. This model suffers from the problem that it predicts residual inhomogeneities after inflation is over that are far too large. This is easily seen: because the transition is first-order, it proceeds by **bubble nucleation**, where the vacuum tunnels between false and true vacua. However, the region occupied by these bubbles will grow as a causal process, whereas outside the bubbles the exponential expansion of inflation continues. This means that it is very difficult for the bubbles to percolate and eliminate the false vacuum everywhere, as is needed for an end to inflation. Instead, inflation continues indefinitely, with the bubbles of true vacuum having only a small filling factor at any time. This **graceful exit problem** motivated variants in which the potential is flatter near the origin, so that the phase transition is second order and can proceed smoothly everywhere.

However, there is also a more general problem with Guth's model and its variants. If the initial conditions are at a temperature $T_{\rm GUT}$, we expect thermal fluctuations in ϕ; the potential should generally differ from its minimum by an amount $V \sim T_{\rm GUT}^4$, which is of the order of the difference between true and false vacua. How then is the special case needed to trap the potential near $\phi = 0$ to arise? We have returned to the sort of fine-tuned initial conditions from which inflation was designed to save us.

Combined with the difficulties in achieving small inhomogeneities after inflation is over, Guth's original inflation model thus turned out to have insuperable difficulties. However, for many cosmologists the main concepts of inflation have been too attractive to give up. The price one pays for this is to decouple inflation from standard particle physics (taking the liberty of including GUTs in this category): inflation can in principle be driven by the vacuum energy of any scalar field. The ideological inflationist will then take the position that such a field (the **inflaton**) must have existed, and that it is our task to work out its properties from empirical cosmological evidence, rather than from *a priori* particle-physics considerations. Barring a credible alternative way of understanding the peculiarities of the initial conditions of the big bang, there is much to be said for this point of view.

CHAOTIC INFLATION MODELS Most attention is currently paid to the more general models where the field finds itself some way from its potential minimum. This idea is termed **chaotic inflation** (see e.g. Linde 1989). The name originates because this class of models is also quite different in philosophy from other inflation models; it does not require that there is a single Friedmann model containing an inflation-driving scalar field. Rather, there could be some primordial chaos, within which conditions might vary. Some parts may attain the conditions needed for inflation, in which case they will expand hugely, leaving a universe inside a single bubble – which could be the one we inhabit. In principle this bubble has an edge, but if inflation persists for sufficiently long, the distance to this nastiness is so much greater than the current particle horizon that its existence has no testable consequences.

A wide range of inflation models of this kind is possible, illustrating the freedom that arises once the parameters of the theory are constrained only by the requirement that inflation be produced. Things become even less constrained once it is realized that inflation need not correspond to de Sitter space, even though this was taken for granted in early discussions. As discussed earlier, it is only necessary that the universe enter a phase of 'superluminal' expansion in which the equation of state satisfies $p < -\rho c^2/3$. For a pure static field, we will have the usual $p = -\rho c^2$ vacuum equation of state, and so a significant deviation from de Sitter space requires a large contribution from $\dot{\phi}$ terms (although the slow-roll conditions will often still be satisfied). Intuitively, this corresponds to a potential that must be steep in some sense that is determined by the desired time dependence of the scale factor. Three special cases are of particular interest.

(1) **Polynomial inflation.** If the potential is taken to be $V \propto \phi^\alpha$, then the scale-factor behaviour is very close to exponential. This becomes less true as α increases, but investigations are usually limited to ϕ^2 and ϕ^4 potentials on the grounds that higher powers are nonrenormalizable.

(2) **Power-law inflation.** Nevertheless, $a(t) \propto t^p$ would suffice, provided $p > 1$. The potential required to produce this behaviour is [problem 11.1]

$$V(\phi) \propto \exp\left(\sqrt{\frac{16\pi}{p\, m_{\rm P}^2}}\, \phi\right). \tag{11.33}$$

(3) **Intermediate inflation.** Another simple time dependence that suffices for inflation is $a(t) \propto \exp[(t/t_0)^f]$. In the slow-roll approximation, the required potential here is $V(\phi) \propto \phi^{-\beta}$, where $\beta = 4(f^{-1} - 1)$.

There are in addition a plethora of more specific models with various degrees of particle-physics motivation. Since at the time of writing none of these seem at all certain to become permanent fixtures, they will mostly not be described in detail. The above examples are more than enough to illustrate the wide range of choice available.

CRITERIA FOR INFLATION Successful inflation in any of these models requires > 60 e-foldings of the expansion. The implications of this are easily calculated using the slow-roll equation, which gives the number of e-foldings between ϕ_1 and ϕ_2 as

$$N = \int H\, dt = -\frac{8\pi}{m_{\rm P}^2} \int_{\phi_1}^{\phi_2} \frac{V}{V'}\, d\phi. \tag{11.34}$$

For any potential that is relatively smooth, $V' \sim V/\phi$, and so we get $N \sim (\phi_{\rm start}/m_{\rm P})^2$, assuming that inflation terminates at a value of ϕ rather smaller than at the start. The criterion for successful inflation is thus that the initial value of the field exceeds the Planck scale:

$$\boxed{\phi_{\rm start} \gg m_{\rm P}.} \tag{11.35}$$

By the same argument, it is easily seen that this is also the criterion needed to make the slow-roll parameters ϵ and $\eta \ll 1$. To summarize, any model in which the potential is sufficiently flat that slow-roll inflation can commence will probably achieve the critical 60 e-foldings. Counterexamples can of course be constructed, but they have to be somewhat special cases.

It is interesting to review this conclusion for some of the specific inflation models listed above. Consider a mass-like potential $V = m^2\phi^2$. If inflation starts near the Planck scale, the fluctuations in V are $\sim m_{\rm P}^4$ and these will drive $\phi_{\rm start}$ to $\phi_{\rm start} \gg m_{\rm P}$ provided $m \ll m_{\rm P}$; similarly, for $V = \lambda\phi^4$, the condition is weak coupling: $\lambda \ll 1$. Any field with a rather flat potential will thus tend to inflate, just because typical fluctuations leave it a long way from home in the form of the potential minimum. In a sense, inflation is realized by means of 'inertial confinement': there is nothing to prevent the scalar field from reaching the minimum of the potential – but it takes a long time to do so, and the universe has meanwhile inflated by a large factor.

This requirement for weak coupling and/or small mass scales near the Planck epoch is suspicious, since quantum corrections will tend to re-introduce the Planck scale.

In this sense, especially with the appearance of the Planck scale as the minimum required field value, it is not clear that the aim of realizing inflation in a classical way distinct from quantum gravity has been fulfilled.

11.5 Relic fluctuations from inflation

MOTIVATION We have seen that de Sitter space contains a true event horizon, of proper size c/H. This suggests that there will be thermal fluctuations present, as with a black hole, for which the **Hawking temperature** is $kT_{\rm H} = \hbar c/(4\pi r_s)$. This analogy is close, but imperfect, and the characteristic temperature of de Sitter space is a factor 2 higher:

$$kT_{\rm deSitter} = \frac{\hbar H}{2\pi} \tag{11.36}$$

(see chapter 8). This existence of thermal fluctuations is one piece of intuitive motivation for expecting fluctuations in the quantum fields that are present in de Sitter space, but is not so useful in detail. In practice, we need a more basic calculation, of how the zero-point fluctuations in small-scale quantum modes freeze out as classical density fluctuations once the modes have been inflated to super-horizon scales.

The details of this calculation are given below. However, we can immediately note that a natural prediction will be a spectrum of perturbations that are nearly *scale invariant*. This means that the metric fluctuations of spacetime receive equal levels of distortion from each decade of perturbation wavelength, and may be quantified in terms of the rms fluctuations, σ, in Newtonian gravitational potential, Φ ($c = 1$):

$$\boxed{\delta_{\rm H}^2 \equiv \Delta_\Phi^2 \equiv \frac{d\,\sigma^2(\Phi)}{d\,\ln k} = \text{constant.}} \tag{11.37}$$

The notation $\delta_{\rm H}$ arises because the potential perturbation is of the same order as the density fluctuation on the scale of the horizon at any given time (see chapter 15).

It is commonly argued that the prediction of scale invariance arises because de Sitter space is invariant under time translation: there is no natural origin of time under exponential expansion. At a given time, the only length scale in the model is the horizon size c/H, so it is inevitable that the fluctuations that exist on this scale are the same at all times. After inflation ceases, the resulting fluctuations (at constant amplitude on the scale of the horizon) give us the **Zeldovich** or **scale-invariant** spectrum. The problem with this argument is that it ignores the issue of how the perturbations evolve while they are outside the horizon; we have only really calculated the amplitude for the last generation of fluctuations – i.e. those that are on the scale of the horizon at the time inflation ends. Fluctuations generated at earlier times will be inflated outside the de Sitter horizon, and will re-enter the FRW horizon at some time after inflation has ceased.

The evolution during this period is a topic where some care is needed, since the description of these large-scale perturbations is sensitive to the gauge freedom in general relativity. A technical discussion is given in e.g. Mukhanov, Feldman & Brandenberger (1992), but there is no space to do this justice here. Chapter 15 presents a rather more intuitive discussion of the gauge issue; for the present, we shall rely on simply motivating the inflationary result, which is that potential perturbations re-enter the horizon with the same amplitude they had on leaving. This may be made reasonable in two ways.

Perturbations outside the horizon are immune to causal effects, so it is hard to see how any large-scale non-flatness in spacetime could 'know' whether it was supposed to grow or decline. More formally, we shall show in chapter 15 that small potential perturbations preserve their value, provided that they are on scales where pressure effects can be neglected and that this critical scale corresponds to the horizon. We therefore argue that the inflationary process produces a universe that is fractal-like in the sense that scale-invariant fluctuations correspond to a metric that has the same 'wrinkliness' per log length-scale. It then suffices to calculate that amplitude on one scale – i.e. the perturbations that are just leaving the horizon at the end of inflation, so that super-horizon evolution is not an issue. It is possible to alter this prediction of scale invariance only if the expansion is non-exponential; we have seen that such deviations plausibly do exist towards the end of inflation, so it is clear that exact scale invariance is not to be expected.

To anticipate the detailed treatment, the inflationary prediction is of a horizon-scale amplitude

$$\boxed{\delta_{\rm H} = \frac{H^2}{2\pi\,\dot{\phi}}} \tag{11.38}$$

which can be understood as follows. Imagine that the main effect of fluctuations is to make different parts of the universe have fields that are perturbed by an amount $\delta\phi$. In other words, we are dealing with various copies of the same rolling-behaviour $\phi(t)$, but viewed at different times

$$\delta t = \frac{\delta\phi}{\dot{\phi}}. \tag{11.39}$$

These universes will then finish inflation at different times, leading to a spread in energy densities (figure 11.4). The horizon-scale density amplitude is given by the different amounts that the universes have expanded following the end of inflation:

$$\delta_{\rm H} \simeq H\,\delta t = \frac{H^2}{2\pi\,\dot{\phi}}, \tag{11.40}$$

where the last step uses the crucial input of quantum field theory, which says that the rms $\delta\phi$ is given by $H/2\pi$. This result will be derived below, but it is immediately reasonable on dimensional grounds (in natural units, the field has the dimensions of temperature).

THE FLUCTUATION SPECTRUM We now need to go over this vital result in rather more detail (see Liddle & Lyth 1993 for a particularly clear treatment). First, consider the equation of motion obeyed by perturbations in the inflaton field. The basic equation of motion is

$$\ddot{\phi} + 3H\dot{\phi} - \nabla^2\phi + V'(\phi) = 0, \tag{11.41}$$

and we seek the corresponding equation for the perturbation $\delta\phi$ obtained by starting inflation with slightly different values of ϕ in different places. Suppose this perturbation takes the form of a comoving plane-wave perturbation of comoving wavenumber k and amplitude A: $\delta\phi = A\exp(i\mathbf{k}\cdot\mathbf{x} - ikt/a)$. If the slow-roll conditions are also assumed, so that V' may be treated as a constant, then the perturbed field $\delta\phi$ obeys the first-order

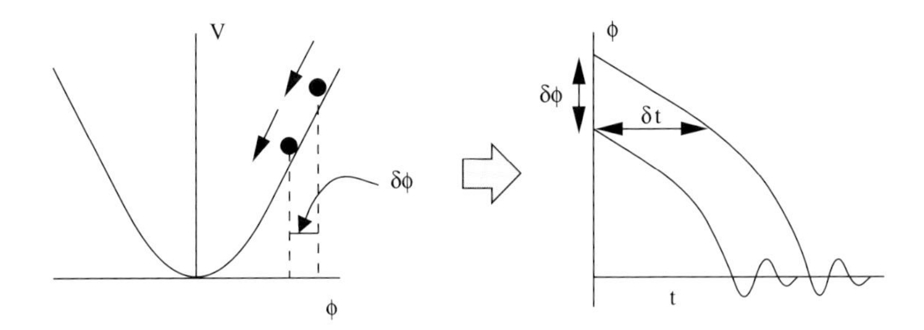

Figure 11.4. This plot shows how fluctuations in the scalar field transform themselves into density fluctuations at the end of inflation. Different points of the universe inflate from points on the potential perturbed by a fluctuation $\delta\phi$, like two balls rolling from different starting points. Inflation finishes at times separated by δt in time for these two points, inducing a density fluctuation $\delta = H\delta t$.

perturbation of the equation of motion for the main field:

$$[\ddot{\delta\phi}] + 3H[\dot{\delta\phi}] + (k/a)^2[\delta\phi] = 0, \tag{11.42}$$

which is a standard wave equation for a massless field evolving in an expanding universe.

Having seen that the inflaton perturbation behaves in this way, it is not much work to obtain the quantum fluctuations that result in the field at late times (i.e. on scales much larger than the de Sitter horizon). First consider the fluctuations in flat space: the field would be expanded as

$$\phi_k = \omega_k a_k + \omega_k^* a_k^\dagger, \tag{11.43}$$

and the field variance would be

$$\langle 0| \, |\phi_k|^2 \, |0\rangle = |\omega_k|^2. \tag{11.44}$$

To solve the general problem, we only need to find how the amplitude ω_k changes as the universe expands. The idea is to start from the situation where we are well inside the horizon ($k/a \gg H$), in which case flat-space quantum theory will apply, and end at the point of interest outside the horizon (where $k/a \ll H$).

Before finishing the calculation, note the critical assumption that the initial state is the vacuum: in the modes that will eventually be relevant for observational cosmology, we start with not even one quantum of the inflaton field present. Is this smuggling fine tuning of the initial conditions in again through the back door? Given our ignorance of the exact conditions in the primordial chaos from which the inflationary phase is supposed to emerge, it is something of a matter of taste whether this is seen as being a problem. Certainly, if the initial state is close to equilibrium at temperature T, this is easily understood, since all initial scales are given in terms of T. In natural units, the energy density in $V(\phi)$ and radiation will be $\sim T^4$, and the proper size of the horizon will be $\sim T^{-2}$. In the initial state, the inflaton occupation number will be $1/2$ for very long wavelengths, and will fall for proper wavelengths $\lesssim T^{-1}$. Now, remember that T is in units of the Planck temperature, so that $T \sim 10^{-4}$ for GUT-scale inflation. That means that perturbations of scale smaller than T times the horizon would start with

$n \simeq 0$ for a thermal state. However, this is only the initial state, and we expect that occupation number will be an adiabatic invariant that is constant for a given comoving wavelength. Thus, after $\ln(1/T)$ e-foldings of inflation (i.e. only a few), every mode that remains inside the horizon will have the required zero occupation number. Of course, the motivation for a thermal initial state is weak, but the main point can be made in terms of energy density. If inflation is to happen at all, $V(\phi)$ must dominate, and it cannot do this if the inflaton fluctuations exist down to zero wavelength because the effective radiation density would then diverge. It thus seems reasonable to treat any initial state that inflates as being one that rapidly enters a vacuum state.

It is also worth noting in passing that these fluctuations in the scalar field can in principle affect the progress of inflation itself. They can be thought of as adding a random-walk element to the classical rolling of the scalar field down the trough defined by $V(\phi)$. In cases where ϕ is too close to the origin for inflation to persist for sufficiently long, it is possible for the quantum fluctuations to push ϕ further out – creating further inflation in a self-sustaining process. This is the concept of **stochastic inflation** (Linde 1986, 1989) [problem 11.3].

Returning now to the calculation, we want to know how the mode amplitude changes as the wavelength passes through the horizon. Initially, we have the flat-space result from chapter 7, which can be rewritten in comoving units as

$$\omega_k = a^{-3/2}\,(2k/a)^{-1/2}\,e^{-ikt/a}. \tag{11.45}$$

The powers of the scale factor, $a(t)$, just allow for expanding the field in comoving wavenumbers k. The field amplitude contains a normalizing factor of $V^{-1/2}$, V being a proper volume; hence the $a^{-3/2}$ factor, if we use comoving $V = 1$. Another way of looking at this is that the proper number density of inflatons goes as a^{-3} as the universe expands. With this boundary condition, it straightforward to check by substitution that the following expression satisfies the evolution equation:

$$\boxed{\omega_k = a^{-3/2}\,(2k/a)^{-1/2}\,e^{-ik/aH}\,(1 + iaH/k);} \tag{11.46}$$

remember that H is a constant, so that $(d/dt)(aH) = H\dot{a} = aH^2$ etc. At early times, when the horizon is much larger than the wavelength, $aH/k \ll 1$, and so ω_k is the flat-space result, except that the time dependence looks a little odd, being $\exp(-ik/aH)$. However, since $(d/dt)(k/aH) = -k/a$, we see that the oscillatory term has a leading dependence on t of the desired kt/a form. In the limit of very early times, the period of oscillation is $\ll H^{-1}$, so a is effectively a constant from the point of view of the epoch where quantum fluctuations dominate.

At the opposite extreme, $aH/k \gg 1$, the fluctuation amplitude becomes frozen out at the value

$$\langle 0|\,|\phi_k|^2\,|0\rangle = \frac{H^2}{2k^3}. \tag{11.47}$$

The initial quantum zero-point fluctuations in the field have been transcribed to a constant classical fluctuation that can eventually manifest itself as large-scale structure. The fluctuations in ϕ depend on k in such a way that the fluctuations per decade are constant:

$$\frac{d\,(\delta\phi)^2}{d\,\ln k} = \frac{4\pi k^3}{(2\pi)^3}\,\langle 0|\,|\phi_k|^2\,|0\rangle = \left(\frac{H}{2\pi}\right)^2 \tag{11.48}$$

(the factor $(2\pi)^{-3}$ comes from the Fourier transform; $4\pi k^2\,dk = 4\pi k^3\,d\ln k$ comes from the k-space volume element). This completes the argument. The rms value of fluctuations in ϕ can be used as above to deduce the power spectrum of mass fluctuations well after inflation is over. In terms of the variance per $\ln k$ in potential perturbations, the answer is

$$\boxed{\begin{aligned} \delta_{\rm H}^2 \equiv \Delta_\Phi^2(k) &= \frac{H^4}{(2\pi\dot{\phi})^2} \\ H^2 &= \frac{8\pi}{3}\frac{V}{m_{\rm P}^2} \\ 3H\dot{\phi} &= -V', \end{aligned}} \qquad (11.49)$$

where we have also written once again the exact relation between H and V and the slow-roll condition, since manipulation of these three equations is often required in derivations.

This result calls for a number of comments. First, if H and $\dot{\phi}$ are both constant then the predicted spectrum is exactly scale invariant, with some characteristic inhomogeneity on the scale of the horizon. As we have seen, exact de Sitter space with constant H will not be strictly correct for most inflationary potentials; nevertheless, in most cases the main points of the analysis still go through. The fluctuations in ϕ start as normal flat-space fluctuations (and so not specific to de Sitter space), which change their character as they are advected beyond the horizon and become frozen-out classical fluctuations. All that matters is that the Hubble parameter is roughly constant for the few e-foldings that are required for this transition to happen. If H does change with time, the number to use is the value at the time that a mode of given k crosses the horizon. Even if H were to be precisely constant, there remains the dependence on $\dot{\phi}$, which again will change as different scales cross the horizon. This means that different inflationary models display different characteristic deviations from a nearly scale-invariant spectrum, and this is discussed in more detail below.

Two other characteristics of the perturbations are more general: they will be Gaussian and adiabatic in nature. A Gaussian density field is one for which the joint probability distribution of the density at any given number of points is a multivariate Gaussian. The easiest way for this to arise in practice is for the density field to be constructed as a superposition of Fourier modes with independent random phases; the Gaussian property then follows from the central limit theorem (see chapter 16). It is easy to see in the case of inflation that this requirement will be satisfied: the quantum commutation relations only apply to modes of the same k, so that modes of different wavelength behave independently and have independent zero-point fluctuations. Finally, the principal result of the inflationary fluctuations in their late-time classical guise is as a perturbation to curvature, and it is not easy to see how to produce the separation in behaviour between photon and matter perturbations that is needed for isocurvature modes. Towards the end of inflation, the universe contains nothing but scalar field and whatever mechanisms that generate the matter–antimatter asymmetry have yet to operate. When they do, the result will be a universal ratio of photons to baryons, but with a total density modulated by the residual inflationary fluctuations – adiabatic initial conditions, in short.

Inflation thus makes a relatively firm prediction about the statistical character of the initial density perturbations, plus a somewhat less firm prediction for their power spectrum. With sufficient ingenuity, the space of predictions can be widened; isocurvature perturbations can be produced at the price of introducing additional inflation fields and carefully adjusting the coupling between them (Kofman & Linde 1987); breaking the Gaussian character of the fluctuations is also possible in such multi-field models (Yi & Vishniac 1993), essentially because all modes in field 2 can respond coherently to a fluctuation in field 1, in much the same way as non-Gaussian perturbations are generated by cosmic strings. However, a nearly scale-free Gaussian adiabatic spectrum is an inevitable result in the simplest models, with a single inflaton; if the theory is to have any predictive power and not to appear contrived, this is the clear prediction of inflation. As we shall see in chapter 16, the true state of affairs seems to be close to this state.

INFLATON COUPLING The calculation of density inhomogeneities sets an important limit on the inflation potential. From the slow-rolling equation, we know that the number of *e*-foldings of inflation is

$$N = \int H\, dt = \int H\, d\phi/\dot{\phi} = \int 3H^2\, d\phi/V'. \tag{11.50}$$

Suppose $V(\phi)$ takes the form $V = \lambda\phi^4$, so that $N = H^2/(2\lambda\phi^2)$. The density perturbations can then be expressed as

$$\delta_{\rm H} \sim \frac{H^2}{\dot{\phi}} = \frac{3H^3}{V'} \sim \lambda^{1/2} N^{3/2}. \tag{11.51}$$

Since $N \gtrsim 60$, the observed $\delta_{\rm H} \sim 10^{-5}$ requires

$$\boxed{\lambda \lesssim 10^{-15}.} \tag{11.52}$$

Alternatively, in the case of $V = m^2\phi^2$, $\delta_{\rm H} = 3H^3/(2m^2\phi)$. Since $H \sim \sqrt{V}/m_{\rm P}$, this gives $\delta_{\rm H} \sim m\phi^2/m_{\rm P}^3 \sim 10^{-5}$. Since we have already seen that $\phi \gtrsim m_{\rm P}$ is needed for inflation, this gives

$$\boxed{m \lesssim 10^{-5} m_{\rm P}.} \tag{11.53}$$

These constraints appear to suggest a defect in inflation, in that we should be able to use the theory to *explain* why $\delta_{\rm H} \sim 10^{-5}$, rather than using this observed fact to constrain the theory. The amplitude of $\delta_{\rm H}$ is one of the most important numbers in cosmology, and it is vital to know whether there is a simple explanation for its magnitude. Such an explanation does seem to exist for theories based on topological defects, where we would have

$$\delta_{\rm H} \sim (E_{\rm GUT}/E_{\rm P})^2 \tag{11.54}$$

(chapter 10). In fact, the situation in inflation is similar, since another way of expressing the horizon-scale amplitude is

$$\delta_{\rm H} \sim \frac{V^{1/2}}{m_{\rm P}^2\, \epsilon^{1/2}}. \tag{11.55}$$

We have argued that inflation will end with ϵ of order unity; if the potential were to have the characteristic value $V \sim E_{\rm GUT}^4$ then this would give the same prediction for $\delta_{\rm H}$ as in defect theories. The appearance of a tunable 'knob' in inflation theories really arises because we need to satisfy $\phi \sim m_{\rm P}$ (for sufficient inflation), while dealing with the characteristic value $V \sim E_{\rm GUT}^4$ (to be fair, this is likely to apply only at the start of inflation, but the potential at 60 e-folds from the end of inflation will not be very different from its starting value unless the total number of e-folds is $\gg 60$). It is therefore reasonable to say that a much smaller horizon-scale amplitude would need $V \ll E_{\rm GUT}^4$, i.e. a smaller $E_{\rm GUT}$ than the conventional value.

This section has demonstrated the cul-de-sac in which inflationary models now find themselves: the field that drives inflation must be very weakly coupled – and effectively undetectable in the laboratory. Instead of Guth's original heroic vision of a theory motivated by particle physics, we have had to introduce a new entity into particle physics that exists only for cosmological purposes. In a sense, then, inflation is a failure. However, the hope of a consistent scheme eventually emerging (plus the lack of any alternative), means that inflationary models continue to be explored with great vigour.

GRAVITY WAVES AND TILT The density perturbations left behind as a residue of the quantum fluctuations in the inflaton field during inflation are an important relic of that epoch, but are not the only one. In principle, a further important test of the inflationary model is that it also predicts a background of gravitational waves, whose properties couple with those of the density fluctuations.

It is easy to see in principle how such waves arise. In linear theory, any quantum field is expanded in a similar way into a sum of oscillators with the usual creation and annihilation operators; the above analysis of quantum fluctuations in a scalar field is thus readily adapted to show that analogous fluctuations will be generated in other fields during inflation. In fact, the linearized contribution of a gravity wave, $h_{\mu\nu}$, to the Lagrangian looks like a scalar field $\phi = (m_{\rm P}/4\sqrt{\pi})\, h_{\mu\nu}$ [problem 11.5], so the expected rms gravity-wave amplitude is

$$h_{\rm rms} \sim H/m_{\rm P}. \tag{11.56}$$

The fluctuations in ϕ are transmuted into density fluctuations, but gravity waves will survive to the present day, albeit redshifted.

This redshifting produces a break in the spectrum of waves. Prior to horizon entry, the gravity waves produce a scale-invariant spectrum of metric distortions, with amplitude $h_{\rm rms}$ per $\ln k$. These distortions are observable via the large-scale CMB anisotropies, where the tensor modes produce a spectrum with the same scale dependence as the Sachs–Wolfe gravitational redshift from scalar metric perturbations. In the scalar case (discussed in chapter 18), we have $\delta T/T \sim \phi/3c^2$, i.e. of the order of the Newtonian metric perturbation; similarly, the tensor effect is

$$\left(\frac{\delta T}{T}\right)_{\rm GW} \sim h_{\rm rms} \lesssim \delta_{\rm H} \sim 10^{-5}, \tag{11.57}$$

where the second step follows because the tensor modes can constitute no more than 100% of the observed CMB anisotropy. The energy density of the waves is $\rho_{\rm GW} \sim m_{\rm P}^2 h^2 k^2$, where $k \sim H(a_{\rm entry})$ is the proper wavenumber of the waves. At horizon entry, we therefore expect

$$\rho_{\rm GW} \sim m_{\rm P}^2\, h_{\rm rms}^2\, H^2(a_{\rm entry}). \tag{11.58}$$

After horizon entry, the waves redshift away like radiation, as a^{-4}, and generate a present-day energy spectrum per $\ln k$ that is constant for modes that entered the horizon while the universe was radiation dominated (because $a \propto t^{1/2} \Rightarrow H^2 a^4 = \text{constant}$). What is the density parameter of these waves? In natural units, $\Omega = \frac{8}{3}\pi\rho/(H^2 m_{\rm P}^2)$, so $\Omega_{\rm GW} \sim h_{\rm rms}^2$ at the time of horizon entry, at which epoch the universe was radiation dominated, with $\Omega_r = 1$ to an excellent approximation. Thereafter, the wave density maintains a constant ratio to the radiation density, since both redshift as a^{-4}, giving the present-day density as

$$\boxed{\Omega_{\rm GW} \sim \Omega_r\,(H/m_{\rm P})^2 \sim 10^{-4}\, V/m_{\rm P}^4.} \tag{11.59}$$

The gravity-wave spectrum therefore displays a break between constant metric fluctuations on super-horizon scales and constant density fluctuations on small scales. As discussed in chapter 15, an analogous break also exists in the spectrum of density perturbations in dark matter. If gravity waves make an important contribution to CMB anisotropies, we must have $h_{\rm rms} \sim 10^{-5}$, and so $\Omega_{\rm GW} \sim 10^{-14}$ is expected.

A gravity-wave background of a similar flat spectrum is also predicted from cosmic strings (see section 10.4 of Vilenkin & Shellard 1994). Here, the prediction is

$$\Omega_{\rm GW} \sim 100\,(G\mu/c^2)\,\Omega_r, \tag{11.60}$$

where μ is the mass per unit length. A viable string cosmology requires $G\mu/c^2 \sim 10^{-5}$, so $\Omega_{\rm GW} \sim 10^{-7}$ is expected – much higher than the inflationary prediction.

The part of the spectrum with periods of order years would perturb the emission from pulsars through fluctuating gravitational redshifts, and the absence of this modulation sets a bound of $\Omega \lesssim 10^{-7}$, so that $V \lesssim 10^{-2} m_{\rm P}^4$. However, this is still a very long way from the interesting inflationary level of $\Omega_{\rm GW} \lesssim 10^{-14}$. The strains implied by this level of relic gravity-wave background are tiny. It is shown in chapter 2 that the energy density of gravity waves of angular frequency ω is

$$\rho_{\rm GW} = \frac{\omega^2}{32\pi G}\, h^2 \quad \Rightarrow \quad \Omega_{\rm GW} = \frac{\omega^2 h^2}{12 H_0^2}, \tag{11.61}$$

so that the typical strain is $h \simeq (\Omega_{\rm GW})^{1/2} \lambda/(c/H_0)$, which is about $10^{-27.5}$ for kHz waves, as opposed to 10^{-20} for single astronomical targets. However, it is not completely inconceivable that space-based versions of the same interferometer technology being used to search for kHz-period gravity waves on Earth might eventually reach the required sensitivity. A direct detection of the gravity-wave background at the expected level would do much the same for the credibility of inflation as was achieved for the big bang itself by Penzias and Wilson in 1965.

An alternative way of presenting the gravity-wave effect on the CMB anisotropies is via the ratio between the tensor effect of gravity waves and the normal scalar Sachs–Wolfe effect, as first analysed in a prescient paper by Starobinsky (1985). Denote the fractional temperature variance per natural logarithm of the angular wavenumber by

Δ^2 (constant for a scale-invariant spectrum). The tensor and scalar contributions are respectively

$$\Delta_{\rm T}^2 \sim h_{\rm rms}^2 \sim \frac{H^2}{m_{\rm P}^2} \sim \frac{V}{m_{\rm P}^4}. \tag{11.62}$$

$$\Delta_{\rm S}^2 \sim \delta_{\rm H}^2 \sim \frac{H^2}{\dot{\phi}} \sim \frac{H^6}{(V')^2} \sim \frac{V^3}{m_{\rm P}^6 \, V'^2}. \tag{11.63}$$

The ratio of the tensor and scalar contributions to the variance of the microwave background anisotropies is therefore proportional to the inflationary parameter ϵ:

$$\frac{\Delta_{\rm T}^2}{\Delta_{\rm S}^2} \simeq 12.4\,\epsilon, \tag{11.64}$$

inserting the exact coefficient from Starobinsky (1985). If it could be measured, the gravity-wave contribution to CMB anisotropies would therefore give a measure of ϵ, one of the dimensionless inflation parameters. The less 'de Sitter-like' the inflationary behaviour, the larger the relative gravitational-wave contribution.

Since deviations from exact exponential expansion also manifest themselves as density fluctuations with spectra that deviate from scale invariance, this suggest a potential test of inflation. Define the **tilt** of the fluctuation spectrum as follows:

$$\boxed{\text{tilt} \equiv 1 - n \equiv -\frac{d \ln \delta_{\rm H}^2}{d \ln k}.} \tag{11.65}$$

We then want to express the tilt in terms of parameters of the inflationary potential, ϵ and η. These are of order unity when inflation terminates; ϵ and η must therefore be evaluated when the observed universe left the horizon, recalling that we only observe the last 60-odd e-foldings of inflation. The way to introduce scale dependence is to write the condition for a mode of given comoving wavenumber to cross the de Sitter horizon,

$$a/k = H^{-1}. \tag{11.66}$$

Since H is nearly constant during the inflationary evolution, we can replace $d/d\ln k$ by $d \ln a$, and use the slow-roll condition to obtain

$$\frac{d}{d \ln k} = a\,\frac{d}{da} = \frac{\dot{\phi}}{H}\,\frac{d}{d\phi} = -\frac{m_{\rm P}^2}{8\pi}\,\frac{V'}{V}\,\frac{d}{d\phi}. \tag{11.67}$$

We can now work out the tilt, since the horizon-scale amplitude is

$$\delta_{\rm H}^2 = \frac{H^4}{(2\pi\dot{\phi})^2} = \frac{128\pi}{3}\left(\frac{V^3}{m_{\rm P}^6\,V'^2}\right), \tag{11.68}$$

and derivatives of V can be expressed in terms of the dimensionless parameters ϵ and η. The tilt of the density perturbation spectrum is thus predicted to be

$$\boxed{1 - n = 6\epsilon - 2\eta.} \tag{11.69}$$

For most models in which the potential is a smooth polynomial-like function, $|\eta| \simeq |\epsilon|$. Since ϵ has the larger coefficient and is positive by definition, a general but not unavoidable prediction of inflation is that the spectrum of scalar perturbations should be slightly tilted, in the sense that n is slightly less than unity. With a similar level of confidence, one can state that there is a coupling between this tilt and the level of the gravity-wave contribution to CMB anisotropies:

$$\frac{\Delta_T^2}{\Delta_S^2} \simeq 6(1-n). \tag{11.70}$$

In principle, this is a distinctive prediction of inflation, but it is a test that loses power the more closely the fluctuations approach scale invariance. Chapter 18 discusses in more detail the prospects for testing inflation through gravity-wave contributions to microwave-background anisotropies.

It is interesting to put flesh on the bones of this general expression and evaluate the tilt for some specific inflationary models. This is easy in the case of power-law inflation with $a \propto t^p$ because the inflation parameters are constant: $\epsilon = \eta/2 = 1/p$, so that the tilt here is always

$$1 - n = 2/p. \tag{11.71}$$

In general, however, the inflation derivatives have to be evaluated explicitly on the largest scales, 60 e-foldings prior to the end of inflation, so that we need to solve

$$60 = \int H\, dt = \frac{8\pi}{m_P^2}\int_{\phi_{\rm end}}^{\phi} \frac{V}{V'}\, d\phi. \tag{11.72}$$

A better-motivated choice than power-law inflation would be a power-law potential $V(\phi) \propto \phi^\alpha$; many chaotic inflation models concentrate on $\alpha = 2$ (mass-like term) or $\alpha = 4$ (highest renormalizable power). Here, $\epsilon = m_P^2\alpha^2/(16\pi\phi^2)$, $\eta = \epsilon \times 2(\alpha - 1)/\alpha$, and

$$60 = \frac{8\pi}{m_P^2}\int_{\phi_{\rm end}}^{\phi} \frac{\phi}{\alpha}\, d\phi = \frac{4\pi}{m_P^2\alpha}(\phi^2 - \phi_{\rm end}^2). \tag{11.73}$$

It is easy to see that $\phi_{\rm end} \ll \phi$ and that $\epsilon = \alpha/240$, leading finally to

$$1 - n = (2 + \alpha)/120. \tag{11.74}$$

The predictions of simple chaotic inflation are thus very close to scale invariance in practice: $n = 0.97$ for $\alpha = 2$ and $n = 0.95$ for $\alpha = 4$. However, such a tilt has a significant effect over the several decades in k from CMB anisotropy measurements to small-scale galaxy clustering. These results are in some sense the default inflationary predictions: exact scale invariance would be surprising, as would large amounts of tilt. Either observation would indicate that the potential must have a more complicated structure, or that the inflationary framework is not correct.

11.6 Conclusions

THE TESTABILITY OF INFLATION This brief summary of inflationary models has presented the 'party line' of many workers in the field for the simplest ways in which inflation could happen. It is a tremendous achievement to have a picture with this level of detail – but is it at all close to the truth?

What are the predictions of inflation? The simplified package is (i) $k = 0$, (ii) scale-invariant, Gaussian fluctuations; these are essentially the only tests discussed in the 1990 NATO symposium on observational tests of inflation (Shanks *et al.* 1991). As the observations have hardened, however, there has been a tendency for the predictions to weaken. The flatness prediction was long taken to favour an $\Omega = 1$ Einstein–de Sitter model, but timescale problems have moved attention to models with $\Omega_{\rm matter}+\Omega_{\rm vacuum} = 1$. More recently, inflationary models have even been proposed that might yield open universes. This is achieved not by fine-tuning the amount of inflation, which would certainly be contrived, but by appealing to the details of the mechanism whereby inflation ends. It has been proposed that quantum tunnelling might create a 'bubble' of open universe in a plausible way (Bucher, Goldhaber & Turok 1995; Górski *et al.* 1995). It seems that inflation can never be disproved by the values of the global cosmological parameters.

A more characteristic inflationary prediction is the gravity-wave background. Although it is not unavoidable, the coupling between tilt and the gravity-wave contribution to CMB anisotropies is a signature expected in many of the simplest realizations of inflation. An observation of such a coupling would certainly constitute very powerful evidence that something like inflation did occur. Sadly, this test loses its power as the degree of tilt becomes smaller, which does seem to be the case observationally (chapter 16). The most convincing verification of inflation would, of course, be direct detection of the predicted flat spectrum of gravity waves in the local universe. However, barring some clever new technique, this will remain technologically unfeasible for the immediate future.

It therefore seems likely that the debate over the truth of inflation will continue without a clean resolution. There are certainly points of internal consistency in the theory that will attract further work: the form of the potential, and whether it can be maintained in the face of quantum corrections. Also, there is one more fundamental difficulty, which has been evaded until now.

THE Λ PROBLEM All our discussion of inflation has implicitly assumed that the zero of energy is set at $\Lambda = 0$ now, but there is no known principle of physics that requires this to be so. By experiment, $\Omega_{\rm vac} \lesssim 1$, which corresponds to a density in the region of 10^{120} times smaller than the GUT value. This fine tuning represents one of the major unsolved problems in physics (Weinberg 1989).

Other ways of looking at the origin of the vacuum energy include thinking of it as arising from a Bose–Einstein condensate, or via contributions from virtual particles. In the latter case, the energy density would be the rest mass times the number density of particles. A guess at this is made by setting the separation of particles at the Compton wavelength, $\hbar/mc$, yielding a density

$$\rho_{\rm vac} \simeq \frac{m^4 c^3}{\hbar^3}. \tag{11.75}$$

This exceeds the cosmological limit unless $m \lesssim 10^{-8} m_e$. One proposal for evading such a nonsensical result was made by Zeldovich: perhaps we cannot observe the rest mass of virtual particles directly, and we should only count the contribution of their gravitational interaction. This gives instead

$$\rho_{\rm vac} \simeq \left(\frac{Gm^2}{c\hbar}\right) \frac{m^4 c^3}{\hbar^3}, \tag{11.76}$$

which is acceptable if $m \lesssim 100 m_e$ – still far short of any plausible GUT-scale or Planck-scale cutoff.

We are left with the strong impression that some vital physical principle is missing. This should make us wary of thinking that inflation, which exploits *changes* in the level of vacuum energy, can be a final answer. It is perhaps just as well that the average taxpayer, who funds research in physics, is unaware of the trouble we have in understanding even nothing at all.

Problems

(11.1) Verify that the potential

$$V(\phi) \propto \exp\left(\sqrt{\frac{16\pi}{p m_{\rm P}^2}}\, \phi\right) \tag{11.77}$$

leads to inflation with a scale-factor dependence $a(t) \propto t^p$. Since this is not de Sitter space, why does the resulting behaviour still lead to zero comoving spatial curvature?

(11.2) For the effective potential $V = aT^2\,|\phi|^2 - b|\phi|^3 + \lambda|\phi|^4$, obtain the critical temperature below which there are two minima, and the critical temperature for which these are of equal V. How high is the energy barrier around the origin as a function of T?

(11.3) Consider the scalar field at a given point in the inflationary universe. Each e-folding of the expansion produces new classical fluctuations, which add incoherently to those previously present. Show that, if the field is sufficiently far from the origin in a polynomial potential, these fluctuations produce a random walk of $\phi(t)$ that overwhelms the classical trajectory in which ϕ tries to roll down the potential. Cast the behaviour as a diffusion process, and thus argue that some parts of the universe will reach infinite ϕ, thus inflating eternally.

(11.4) Consider slow-roll inflation in a polynomial potential, $V \propto \phi^n$. Show that the horizon-scale amplitude varies as a power of the logarithm of comoving wavenumber, $\delta_{\rm H} \propto [\ln(k/k_c)]^\alpha$, and determine the parameters k_c and α in terms of the potential.

(11.5) Show that the linearized contribution of a gravity wave $h_{\mu\nu}$ to the Lagrangian is of the same form as an equivalent scalar field $\phi = (m_{\rm P}/4\sqrt{\pi})\, h_{\mu\nu}$. Argue that inflation should thus leave fluctuations in a gravity-wave background with rms amplitude $h_{\rm rms} \sim H/m_{\rm P}$. Are backgrounds to be expected in other fields?

PART 5 OBSERVATIONAL COSMOLOGY

12 Matter in the universe

One of the fundamental problems in cosmology is to compile a census of the contents of the universe. Material at a range of different densities and temperatures can be detected by emission or absorption somewhere in the electromagnetic spectrum. Gravity, however, detects mass quite independently of its equation of state. In an ideal world, these two routes to the total density would coincide; in practice, the gravitational route is able to detect more mass by a factor of up to ten than can be detected in any other way. This chapter, and the two that follow, summarize some of the methods that have been used to learn about the gas, radiation, dark matter and galaxies that together make up the observed universe.

12.1 Background radiation

We start with the constraints on any large-scale distribution of matter. A near-uniform **intergalactic medium** (IGM) will manifest itself in 'background' radiation that is isotropic on the sky. In many wavebands, the background radiation originates at sufficiently large distances that we are seeing back to a time when there were no discrete objects in existence. However, in other wavebands, the background may consist of the contribution of a large number of discrete sources, which are too faint to be detected individually. Studying backgrounds of this type tells us about the integrated properties of galaxy populations at more recent times, which can also provide crucial cosmological information. Useful reviews of background radiation may be found in e.g. Longair (1978), Ressell & Turner (1990) and Calzetti, Livio & Madau (1995).

Figure 12.1 shows what we shall be dealing with; it is a composite of the cosmological background radiation as a function of frequency. The dominant backgrounds in terms of energy are clearly those in the microwave and infrared regimes. However, we shall see that interesting information may be extracted also from comparatively 'unimportant' regimes such as the X-ray band. This covers electromagnetic radiation only; we shall also be interested in fermionic backgrounds such as the predicted sea of neutrinos. These are more properly termed **relics** of the big bang, but they have much in common with radiation, being a gas of (possibly) massless particles moving at relativistic velocities (see chapter 9).

OLBERS' PARADOX The optical background radiation was the first to be noted – at least by its absence. Olbers is famous for arguing that the night sky should not be dark

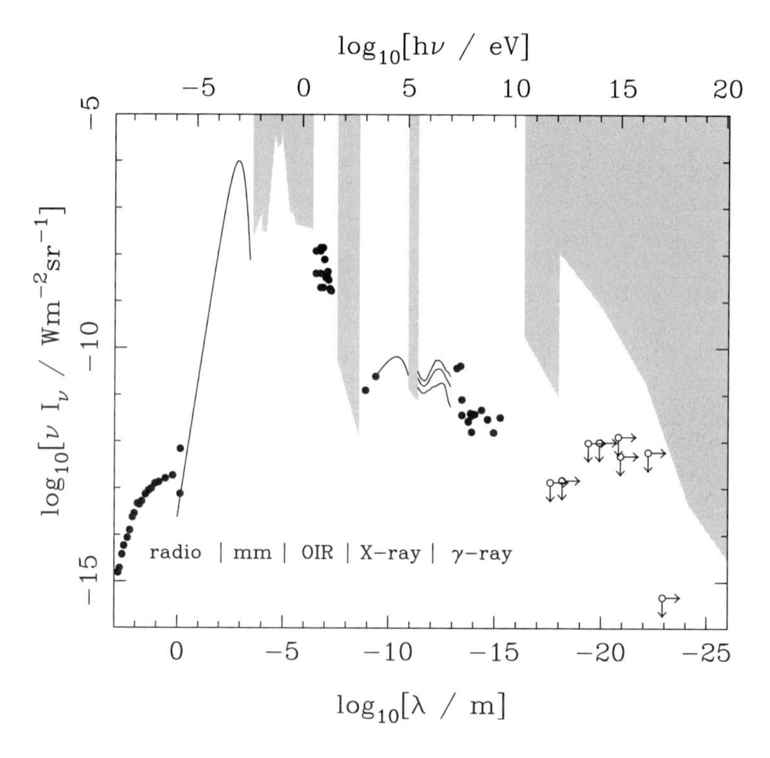

Figure 12.1. The electromagnetic spectrum of cosmological background radiation as a function of frequency, adapted from Ressell & Turner (1990). Plotting νf_ν against $\log \nu$ shows clearly the wavebands at which the density of energy is largest. The shading indicates the regions where only upper limits exist, e.g. because of the dominant foreground emission from the Milky Way.

(in 1826; although he was not the first to do so). The paradox based on this argument (**Olbers' paradox**) has continued to be controversial to this day (e.g. Wesson, Valle & Stabell 1987). The reason for this is that there are actually two versions of the paradox, and the respective resolutions of these are different. We may call these arguments the flux and surface-brightness arguments.

The flux argument considers a universe filled with point sources of radiation. For these, it is easy to show that a uniform distribution of sources in Euclidean space produces the 'three-halves law' for the integral number counts, independent of the distribution of source intrinsic luminosities:

$$N(> S) \propto S^{-1.5}, \tag{12.1}$$

where $N(> S)$ is the number of objects per unit area of sky brighter than S. To show this, consider a given luminosity P: the flux density is $S = P/d^2$, but $N \propto d^3$, where d is distance. The power-law result then follows. If we now construct the sky surface brightness I by integration over the number counts, taking N to be the number of objects per steradian, then

$$I = \int_{S_0}^{\infty} S \frac{dN}{dS} \, dS, \tag{12.2}$$

so that I diverges as the lower flux limit is reduced to zero. A static Euclidean universe thus apparently contains an infinite radiation density. The resolution of the paradox in this form is provided by the non-Euclidean structure of relativistic cosmologies. Neither flux nor volume has the simple power-law dependence on distance given above. The most important of these effects, however, is the volume effect: because a universe that began in a big bang has only a finite age, only a limited volume is accessible to observation (the **horizon**), and hence only a finite number of sources can be seen. Another way of looking at this is that the finite age of the universe does not allow the source population time to produce the infinite energy density demanded by Olbers.

The above is the resolution of Olbers' paradox given by a number of authors. However, the problem is somewhat unrealistic: in practice, one's line of sight must terminate sooner or later on the surface of a finite emitting body. It is furthermore reasonable to suppose that this body will consist of matter hot enough to be ionized and so scatter photons: photospheres at $T \sim 10^3$ K, in short. We can now appeal to a familiar result from Euclidean space: surface brightness is independent of distance. This implies that the sky should everywhere shine as brightly as the surface of the Sun: somewhat less extreme a conclusion than the above infinite brightness, but clearly still hardly a good match to observation! When the paradox is phrased in these terms, the resolution is utterly different. An expanding universe was hot enough in the past to be fully ionized everywhere: all space was one photosphere. The reason we do not see this directly is entirely due to the expansion: the radiation is redshifted to a lower characteristic temperature – just 2.73 K. This gives a Planck function peaking at around 1 mm to produce the **microwave background** (see chapter 9). In this sense, then, Olbers was right: the night sky *does* shine with a uniform surface brightness, and only the expansion of the universe (rather than its finite age) saves us from being cooked.

The microwave background remains the most fundamental of the sources of background radiation, and gives us information on earlier times than data in other wavelength regimes; these form a useful complement in giving us constraints on the composition of the universe at more recent times. Before proceeding to look at each background in turn, we start by giving some useful formulae for dealing with observations of cosmological backgrounds.

FORMULAE FOR ANALYSING BACKGROUNDS Since we will mainly be considering the post-recombination epoch, the following formulae apply for a matter-dominated model only (see chapter 3). The observables are redshift z and angular difference between two points on the sky, $d\psi$. We write the metric in the form

$$c^2 d\tau^2 = c^2 dt^2 - R^2(t) \left[dr^2 + S_k^2(r) \, d\psi^2 \right], \tag{12.3}$$

so that the *comoving* volume element is

$$dV = 4\pi [R_0 S_k(r)]^2 R_0 \, dr. \tag{12.4}$$

The *proper* transverse size is

$$d\ell = d\psi \; R_0 S_k(r)/(1+z) \tag{12.5}$$

and the relation between monochromatic flux density and total luminosity is (assuming isotropic emission)

$$S(\nu) = \frac{L([1+z]\nu)}{4\pi R_0^2 S_k^2(r)(1+z)}. \tag{12.6}$$

Combining these two relations gives the relativistic version of surface-brightness conservation:

$$I_\nu(\nu_0) = \frac{B_\nu([1+z]\nu_0)}{(1+z)^3}. \tag{12.7}$$

The Friedmann equation gives the distance–redshift relation in differential form,

$$R_0\, dr = \frac{c}{H_0} \frac{dz}{(1+z)\sqrt{1+\Omega z}}, \tag{12.8}$$

or in integral form,

$$R_0 S_k(r) = \frac{c}{H_0} \frac{2}{\Omega^2(1+z)} \left[\Omega z + (\Omega - 2)\left(\sqrt{1+\Omega z} - 1\right)\right]. \tag{12.9}$$

This can be related to changes in cosmic time via $c\, dt = R_0\, dr/(1+z)$.

Now suppose we know the emissivity ϵ_ν for some process as a function of frequency and epoch, and want to predict the current background seen at frequency ν_0. The total spectral energy density created during time dt is $\epsilon_\nu([1+z]\nu_0, z)\, dt$; this reaches the present reduced by the volume expansion factor $(1+z)^{-3}$. Note that ϵ_ν is the *total* emissivity, as opposed to the emissivity per solid angle j_ν often encountered in radiative transfer theory. It is hoped that using the total quantity will reduce the chance of errors by a factor 4π. The redshifting of frequency does not matter: ν scales as $1+z$, but so does $d\nu$. The reduction in energy density caused by lowering photon energies is then exactly compensated by a reduced bandwidth, leaving the spectral energy density altered only by the change in proper photon number density. Inserting the redshift–time relation, and multiplying by $c/4\pi$ to obtain the specific intensity, we get

$$\boxed{I_\nu = \frac{1}{4\pi} \frac{c}{H_0} \int \epsilon_\nu([1+z]\nu_0, z) \frac{dz}{(1+z)^5 \sqrt{1+\Omega z}}.} \tag{12.10}$$

This derivation has considered how the energy density evolves at a point in space, even though we know that the photons we see today originated at some large distance. This approach works well enough because of large-scale homogeneity, but it may be clearer to obtain the result for the background directly. Consider a solid angle element $d\Omega$: at redshift z, this area on the sky plus some radial increment dr define a proper volume

$$V = [R_0 S_k(r)]^2\, R_0\, dr\, d\Omega\, (1+z)^{-3}. \tag{12.11}$$

This volume produces a luminosity $V\epsilon_\nu$, from which we can calculate the observed flux density $S_\nu = L_\nu/[4\pi(R_0 S_k)^2(1+z)]$. Since surface brightness is just flux density per unit solid angle, this gives

$$I_\nu = \frac{1}{4\pi} \int \epsilon_\nu \frac{R_0 dr}{(1+z)^4}, \tag{12.12}$$

which is the same result as the one obtained above.

These formulae hold in the limit that the universe is optically thin. If there is significant opacity (as in the case of the UV, for example), then the background will be reduced. We now have to solve the following equation of radiative transfer:

$$\dot{I}_\nu + (3+\alpha)\,\frac{\dot{a}}{a}\,I_\nu = \frac{c j_\nu}{4\pi} - c\kappa I_\nu, \tag{12.13}$$

where α is the effective spectral index of the radiation and κ is the opacity. In the simple case of no expansion and constant κ this would solve to $I_\nu = \int_0^\infty [j_\nu(r)/4\pi]\,\exp(-\kappa r)\,dr$; in other words, the background is contributed mainly by the sources within a distance $1/\kappa$. This is still the content of the radiative transfer equation, but the exact solution depends on how κ and j_ν evolve with time.

THE RADIO BACKGROUND AND NEUTRAL HYDROGEN The extragalactic background at wavelengths longer than about one metre is perhaps the least interesting frequency regime, so we can dispose of it relatively quickly. It is very likely that the background seen here is composed entirely of the integrated contribution of discrete extragalactic synchrotron sources – mainly active galaxies and quasars. However, because the Milky Way itself is a powerful source of synchrotron radiation, subtraction of the foreground emission to reveal the isotropic external background is difficult in this case.

In this waveband, it is common to encounter the flux unit named after Karl Jansky, the inventor of radio astronomy:

$$1\,\mathrm{Jy} \equiv 10^{-26}\,\mathrm{W\,m^{-2}\,Hz^{-1}}. \tag{12.14}$$

The component of the background thought to be extragalactic is approximately

$$I_\nu = 6000(\nu/\mathrm{GHz})^{-0.8}\,\mathrm{Jy\,sr^{-1}}. \tag{12.15}$$

The total observed is roughly three times higher; since the galactic and extragalactic synchrotron has much the same spectral index, and since there may be a component of the galactic synchrotron emission that extends much beyond the Sun's radius, it is clearly hard to be sure exactly how much is genuinely extragalactic.

A further contribution to the radio background comes from neutral intergalactic gas (HI). This emits the **21-cm hyperfine line** with an emissivity of

$$j_{\mathrm{tot}} = \frac{g_2}{g_1+g_2} A_{21}\, n_{\mathrm{HI}}\, h\nu, \tag{12.16}$$

sharply peaked about a frequency of 1420.4 MHz. The spontaneous transition rate is $A_{21} = 2.85 \times 10^{-15}\ \mathrm{s^{-1}}$, and the statistical weights of the upper and lower levels are $g_2 = 3$, $g_1 = 1$; the average power emitted by a single atom is thus $P_0 = 10^{-38.7}$ W.

We now calculate the background this produces. Since we are integrating over frequency, it is clearly allowable to approximate the emission by a δ-function: $\epsilon_\nu = P_0 n_{\mathrm{HI}} \delta(\nu - \nu_{21})$, so that

$$I_\nu = P_0 \int n_{\mathrm{HI}}(z)\,\frac{\delta([1+z]\nu - \nu_{21})}{(1+z)^5\sqrt{1+\Omega z}}\,\frac{d([1+z]\nu)}{\nu} = \frac{P_0 n_{\mathrm{HI}}(z)}{\nu\,(1+z)^5\sqrt{1+\Omega z}}. \tag{12.17}$$

We have not assumed that the proper hydrogen density scales as $(1+z)^3$, because the ionization may be a function of epoch.

This calculation implies a step in intensity at a wavelength of 21 cm, extending to longer wavelengths. Using the **Rayleigh–Jeans law**, we can express the height of the step in terms of brightness temperature:

$$I_\nu = \frac{2\nu^2}{c^2} kT. \tag{12.18}$$

We can convert from $n_{\rm HI}$ to $\Omega_{\rm HI}$ by using the critical density, $\rho_c = 1.88 \times 10^{-26}\,\mathrm{kg\,m^{-3}}$, which gives

$$\Delta T \simeq 0.2\,\Omega_{\rm HI}\ \mathrm{K}. \tag{12.19}$$

This is a frustrating result: it is neither so small as to be completely beyond observation, nor so large that it sets a useful constraint on $\Omega_{\rm HI}$. The fact that no such step has been seen implies

$$\Omega_{\rm HI} \lesssim 0.1; \tag{12.20}$$

however, one can do much better than this ($\Omega_{\rm HI} \lesssim 10^{-5}$) by searching for absorption of ultraviolet radiation from quasars (the **Gunn–Peterson test**).

THE X-RAY BACKGROUND The X-ray background was first detected by satellite in the early 1970s. The spectrum covers the range roughly 1 to 1000 keV in energy. For historical reasons, the keV (1.6022×10^{-16} J) is the preferred energy unit in this waveband; its temperature equivalent is

$$1\ \mathrm{keV} = 1.1605 \times 10^7\ \mathrm{K}. \tag{12.21}$$

The overall background can be fitted with a power law plus a cutoff:

$$\nu I_\nu = 2.7 \times 10^{-11} \left(\frac{E}{3\,\mathrm{keV}}\right)^{0.71} \exp\left(-\frac{E}{kT}\right)\ \mathrm{W\,m^{-2}\,sr^{-1}}, \tag{12.22}$$

with $kT = 40$ keV.

Subtracting point-source contributions down to a source density of $\simeq 400\ \mathrm{deg}^{-2}$ ($2.5 \times 10^{-18}\ \mathrm{W\,m^{-2}}$) accounts for about 60% of the total at 1 to 2 keV. This is an intermediate case between the extremes of microwave and optical backgrounds, in that we cannot be sure whether there is a genuinely smooth component in addition to the contribution of point sources. However, the residual after subtraction of point-source contributions to the deepest level known is a spectrum that is approximately thermal in shape:

$$\nu I_\nu = 1.1 \times 10^{-11} \left(\frac{E}{3\,\mathrm{keV}}\right)^{1-\alpha} \exp\left(-\frac{E}{kT}\right)\ \mathrm{W\,m^{-2}\,sr^{-1}}, \tag{12.23}$$

with $0.2 < \alpha < 0$ and a characteristic temperature in the range 23 to 30 keV, quite unlike the spectral shape of faint quasars. Attempts have therefore been made to fit the background by a uniform, hot, highly ionized gas. We therefore now consider the radiation emitted by such a plasma.

COMPTON LIMITS ON HOT GAS The main cause of radiation from a hot plasma is the acceleration of the electrons by the electric field of the ions, a process known by the German for 'braking radiation': **bremsstrahlung**. The emissivity for a hydrogen plasma is

$$\epsilon_\nu/\mathrm{Wm^{-3}Hz^{-1}} = 6.8 \times 10^{-32}\, T^{-1/2}\, n_e^2\, e^{-h\nu/kT}\, G(\nu, T), \tag{12.24}$$

where G is the Gaunt factor; for the only frequencies of interest ($h\nu \lesssim kT$), it may be approximated to about 20% accuracy by

$$G \simeq 1 + \log_{10}\left(kT/h\nu\right) \tag{12.25}$$

(see e.g. section 5.2 of Rybicki & Lightman 1979). We can now insert this into our fundamental expression for calculating specific intensity from ϵ_ν and obtain a prediction for the X-ray background.

The parameter space for the X-ray background is fairly simple: we have a present-day temperature T_0, plus a gas density, parameterized by $\Omega_g h^2$. According to adiabatic expansion, the epoch dependence of the temperature should be $T = T_0(1+z)^2$ (for $\Gamma = 5/3$). To stop the emission and energy requirements of the model getting out of hand, we therefore also have a **reheating redshift**, at which energy is injected to ionize the neutral gas that had existed since recombination at $z \simeq 1000$. Guilbert & Fabian (1986) find $T_0 \simeq 3.6$ keV and $\Omega_g h^2 \simeq 0.24$ for a reheating redshift of 6. Of course, the density is to be interpreted as an upper limit; because bremsstrahlung is an n^2 process, clumping will increase the emission substantially (see below).

The existence of such a hot diffuse intergalactic medium was for a long time rather hard to test, but now that COBE has shown the microwave background to be thermal to high accuracy, we can eliminate models with too much hot gas. Compton scattering would cause a distortion, depopulating the Rayleigh–Jeans regime in favour of photons in the Wien tail. A dimensionless measure of this distortion is the 'y-parameter', which is roughly the expected fractional frequency change for a photon as it propagates through the hot gas:

$$y \equiv \int \sigma_T \, n_e \frac{kT}{m_e c^2} \, d\ell. \tag{12.26}$$

COBE sets a limit of $y \lesssim 1.5 \times 10^{-5}$ (Fixsen *et al.* 1996).

Let us evaluate y for a uniform cosmology. We have $T = T_0(1+z)^2$, so that T_0 is one parameter. The electron density may be expressed in terms of the density parameter for the gas as follows. For a plasma with 25% He by mass, the total density is 4/3 of the density in hydrogen and the hydrogen number density is 12 times that of helium. It therefore follows that the electron density is

$$n_e = 0.88(\rho_g/m_H) = 9.84\Omega_g h^2(1+z)^3 \ \mathrm{m}^{-3}. \tag{12.27}$$

Putting all this together, we get

$$y = 1.2 \times 10^{-4}\Omega_g h^2 \left(\frac{T_0}{\mathrm{keV}}\right) \int_0^{z_{\max}} \frac{(1+z)^3}{\sqrt{1+\Omega z}} \, dz. \tag{12.28}$$

If we take the Guilbert & Fabian parameters in a flat universe, the COBE limit tells us that $z_{\max} \lesssim 0.1$. However, to account for the X-ray background the hot gas must extend to reheating redshifts > 3, so this model can be strongly excluded.

Things change if the gas is clumped. The X-ray emission scales with $f \equiv \langle n_e^2 \rangle / \langle n_e \rangle^2$, whereas the y-parameter is independent of f. However, large values of the clumping parameter $f \gtrsim 10^4$ are needed to evade the COBE constraint. We are no longer talking about smoothly distributed gas, but clumps with cluster-like overdensity. Our assumptions about how the gas temperature will evolve are now invalid, but we have a worse problem to contend with: the X-ray background is quite smooth. If it is made up of discrete sources, there must be more than 5000 per square degree. Now, the present comoving density of rich clusters is $\sim 10^{-5}h^3$ Mpc^{-3}, so that there are only $\sim 10^7$ clusters in the whole visible universe (even assuming that they continue to exist at high z) – a density of only $\sim$ 200 per square degree. The most likely explanation of the X-ray background is thus that it is generated by a population of numerous, low-luminosity active galaxies.

Whatever the eventual resolution of the puzzle posed by the X-ray background, it is worth emphasizing that we are not able to set very strong constraints on the abundance of ionized gas in the universe. Reducing T_0 to 10^6–10^7 K means that it is possible to have a close to critical density in ionized gas without exceeding either the X-ray or Comptonization constraints. Only by invoking the theory of primordial nucleosynthesis to deduce $\Omega_g h^2 \lesssim 0.02$ can such a possibility be rejected.

12.2 Intervening absorbers

Material that is hard to detect in emission may be more easily seen in absorption. An important sub-discipline of cosmology is therefore the use of luminous active galaxies known as quasars to act as searchlights which illuminate the material that lies between quasar and astronomer. In this way, we can learn about extremely faint galaxies at high redshifts, which would probably not be accessible to observation in any other way. For a useful review of this field, see Blades *et al.* (1988).

VOIGT PROFILES The absorption line profiles expected in intervening absorbers have widths determined by two main causes: intrinsic width due to finite lifetime in an excited state, with subsequent Doppler broadening. Treat the atom as a classical damped oscillator with $x(t) = e^{-i\omega_0 t}e^{-\Gamma t/2}$; the rate of energy loss is then proportional to $\ddot{x}^2$, which is in turn proportional to $x^2(t)$. Hence, by Parseval's theorem, the frequency spectrum of the radiated power is proportional to $|\tilde{x}(\omega)|^2$, where $\tilde{x}(\omega)$ is the Fourier transform of $x(t)$. This gives a probability distribution for emitted frequency [problem 12.1] as a **Lorentzian**:

$$\frac{dp}{d\omega} = \frac{\Gamma/2\pi}{(\omega - \omega_0)^2 + (\Gamma/2)^2}. \tag{12.29}$$

This is the appropriate line profile for spontaneous emission, although we have derived it semiclassically (quantum mechanically, think of a probability decaying, rather than the $|\text{amplitude}|^2$ of a given oscillator). Other factors that limit the lifetime of the excited state, of which the most important is usually collisional broadening, simply have the effect of increasing Γ. If collisions are viewed as producing a set of incoherent wave trains of length τ_i, then the frequency spectrum becomes

$$|\tilde{x}(\omega)|^2 \propto \sum \left| \int_0^{\tau_i} e^{i(\omega-\omega_0)t}\, dt \right|^2 = \sum \left(\frac{\sin \Delta\omega\tau_i/2}{\Delta\omega} \right)^2 \tag{12.30}$$

(because the different wave trains have random phases). Averaging over a Poisson distribution of intercollision times (the probability of a collision in $d\tau$ equals $\exp(-\Gamma_{\rm coll}\tau)$ gives us the Lorentzian profile again, but with an effective decay rate $\Gamma_{\rm eff} = 2\Gamma_{\rm coll}$, which simply adds linearly to the natural decay rate.

This intrinsic line profile can be broadened by convolution with the Gaussian produced via the Doppler effect of thermal motions, producing the **Voigt profile**:

$$\frac{dp}{d\omega} = \frac{\Gamma}{(2\pi)^{3/2}\sigma} \int_{-\infty}^{\infty} \frac{e^{-v^2/2\sigma^2}\, dv}{(\omega - \omega_0 - \omega_0 v/c)^2 + (\Gamma/2)^2}. \tag{12.31}$$

The Doppler width is often much greater than the natural width. We then get a profile that is almost exactly Gaussian in the centre, but with the $1/\omega^2$ wings developing at the point where the Gaussian alone would fall below the Lorentzian distribution at large r.

CURVE OF GROWTH The fact that spectral lines have profiles with non-Doppler wings might be thought unimportant, since the Lorentzian line widths are narrow by comparison with likely Doppler broadening (for Lyα, the natural Γ corresponds to a velocity of 76 m s^{-1}). This is true provided the lines are **optically thin**, so that the observed frequency profile simply mimics the absorption coefficient. Otherwise, the 'depth' of the line is given in terms of the flux density f ($= f_0$ in the absence of absorption) by

$$|f - f_0|/f_0 = 1 - e^{-\tau(\omega)} \tag{12.32}$$

where the frequency-dependent optical depth $\tau(\omega)$ is equal to the cross-section times the **column density**, which is the line integral of the number density of absorbing atoms:

$$\tau(\omega) = \sigma(\omega) \int n\, d\ell. \tag{12.33}$$

The absorption cross-section for a given transition is calculated as follows, using the Einstein A and B coefficients. Consider a transition between an excited state (2) and a lower state (1), which may have corresponding degeneracies g_2 and g_1. The **principle of detailed balance** requires the downwards and upwards transition rates to cancel in equilibrium:

$$(A + BI_\omega)N_2 = (g_2/g_1)N_1 BI_\omega. \tag{12.34}$$

here, I_ω is the **specific intensity** in angular frequency space ($= I_\nu/2\pi$); B may also be defined in terms of I_ν or of energy density, which should cause no problems so long as different definitions are not mixed. The important factor g_2/g_1 on the rhs allows for the fact that stimulated absorption can proceed via g_2 different routes and stimulated emission by g_1 routes, all of which have the same amplitude. This multiplicity is conventionally absorbed into A, which therefore gives the *total* spontaneous decay rate, which is why all the g's appear on the rhs. The usual thermal equilibrium argument then gives $B = \pi^2 c^2 A/(\hbar\omega^3)$. Equating the rate of absorption of energy, $(g_2/g_1)BI_\omega\hbar\omega$, to that expected from a geometrical cross-section, σI_ω, then gives

$$\sigma(\omega) = \hbar\omega \frac{g_2}{g_1} B(\omega) = \frac{\pi^2 c^2}{\omega^2} \frac{g_2}{g_1} A(\omega) = \frac{\lambda^2}{4} \frac{g_2}{g_1} A(\omega). \tag{12.35}$$

An alternative way of writing this relation [problem 12.1] is as

$$\sigma(\omega) = f \left(\frac{\pi \mu_0 e^2 c}{2m_e} \right) \frac{dp}{d\omega}, \tag{12.36}$$

which expresses the cross-section as the product of an **oscillator strength** f and the cross-section for a classical damped oscillator. For Lyα ($2p \to 1s$; $\lambda = 1216$ Å), we have

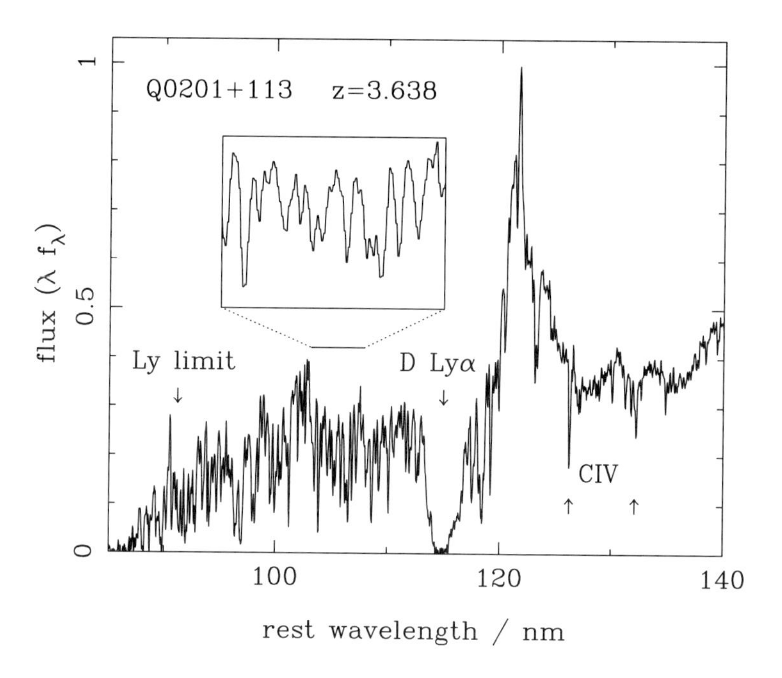

Figure 12.2. A high-resolution spectrum of a quasar in the region between Lyα and the Lyman limit. Note the numerous narrow absorption lines of the 'Lyman alpha forest', the much broader damped system, the cutoff in the spectrum near the Lyman limit (912 Å). Despite all this structure, the continuum between the absorbers appears to rise to the same level as longwards of Lyα, showing no evidence for a homogeneous neutral intergalactic medium.

$A = 6.25 \times 10^8 \mathrm{s}^{-1}$ and $f = 0.416$ so that, via either method, $\sigma = 10^{-5.16}(dp/d\omega)\,\mathrm{m}^2$. For lines where Doppler broadening dominates and the central part of the line looks like a Gaussian with velocity rms σ_v, the central optical depth is

$$\tau_0 = \left(\int n\, d\ell \ / \ 10^{16.27} \mathrm{m}^{-2} \right) \left(\sigma_v / \mathrm{km\,s}^{-1} \right)^{-1}. \qquad (12.37)$$

THE LYMAN ALPHA FOREST The spectra of quasars at wavelengths below that of Lyα (1216 Å) show a very large number of narrow absorption lines (see figure 12.2). The fact that these lines are found uniformly at all redshifts shows that they represent intervening material, rather than material ejected from the quasar. The observed velocity widths of the Lyα lines are a few tens of $\mathrm{km\,s}^{-1}$, so the weakest known lines are sensitive to absorption by column densities of around $10^{16.5}\mathrm{m}^{-2}$. There is a wide distribution of column densities N_c, going roughly as a power law: $P(> N_c) \propto N_c^{-0.75}$. The most spectacular systems go up to $\sim 10^{25}\mathrm{m}^{-2}$, and form a distinct class known as **damped**

Lyman alpha clouds. The reason for this terminology is that, for $N_c \gtrsim 10^{22}\mathrm{m}^{-2}$, the optical depth becomes unity at a point where the profile is dominated by the Lorentzian wings of the 'damped' profile rather than the Gaussian core. The lines are heavily **saturated** with effectively zero central intensity and widths far in excess of the narrow Doppler widths found in low column-density lines, making them easily recognizable (see figure 12.2).

The strength of these absorption lines is commonly expressed in terms of their **equivalent width**; this is calculated by taking the total energy that the line removes from the continuum and asking what wavelength interval of the continuum spectrum at that point would yield the same energy. A similar measure can also be found for emission lines; in both cases, it is important to be clear that the equivalent width is a measure of intensity alone, and should not be confused with velocity width. Consider a Lyα absorption cloud of fixed temperature whose column density is steadily increased. At first, the line is optically thin, so that its velocity width stays constant but its equivalent width increases linearly with column density. When the Gaussian core of the line saturates, however, both velocity width and equivalent width become rather insensitive to column density. Eventually, the damping wings saturate and both measures of width then increase as the square root of the column density. In the linear regime, the equivalent width for a very weak absorber with a column density of $10^{16.5}\,\mathrm{m}^{-2}$ would be about 0.02 Å; for the strongest systems, this can rise to more than 10 Å, which is why damped absorbers stand out so strongly.

The distribution of absorption lines changes with epoch. If the absorbing clouds had constant cross-sections, then the expected number of systems per redshift increment would just depend on the usual element of proper radial distance,

$$\frac{dN}{dz} = \sigma n_0 \frac{c}{H_0} \frac{1+z}{\sqrt{1+\Omega z}}, \tag{12.38}$$

because the proper number density is $n_0(1+z)^3$. In fact, an evolving number density is found, with

$$\frac{dN}{dz} \propto (1+z)^{2.3\pm0.4}, \tag{12.39}$$

i.e. the comoving density is evolving in the sense $n_{\mathrm{comoving}} \propto (1+z)^{1.8\pm0.4}$ (for $\Omega = 1$). The clouds were more numerous and/or had a higher cross-section in the past (e.g. Murdoch *et al.* 1986).

THE GUNN–PETERSON EFFECT The absorption in the Lyα clouds clearly eats away the quasar continuum to the blue of the Lyα emission. This provides us with a very powerful way to estimate the general abundance of neutral hydrogen in the universe. Suppose we observe radiation at some frequency ν_0 that lies bluewards of Lyα in the quasar; this was emitted at $\nu = \nu_0(1+z_{\mathrm{Q}})$, and at redshift z was liable to be absorbed by neutral hydrogen, the appropriate cross-section being $\sigma([1+z]\nu_0)$. The total optical depth is then just the proper line integral of this cross-section times the neutral hydrogen proper density (n_{HI}):

$$\tau = \frac{c}{H_0} \int_0^{z_{\mathrm{Q}}} \frac{\sigma([1+z]\nu_0)\, n_{\mathrm{HI}}(z)\, dz}{(1+z)^2\sqrt{1+\Omega z}}. \tag{12.40}$$

Inserting the cross-section $\sigma = 10^{-5.16}(dp/d\omega)\,\mathrm{m}^2$, and noting that $dp/d\omega$ is sharply peaked about the frequency of Lyα, we find [problem 12.2]

$$\tau = \left[\frac{n_{\rm HI}(z)}{(1+z)\sqrt{1+\Omega z}} \right] \Big/ \; 10^{-4.62}\, h\,{\rm m}^{-3}. \tag{12.41}$$

This applies to all parts of the spectrum to the blue of Lyα, and so a discontinuity should be expected in the level of the continuum about this line; this is known as the **Gunn–Peterson effect**. No such continuum step has been seen, even in quasars up to $z \simeq 4.9$. If we thus take $\tau < 0.1$, for $\Omega = 1$ this implies a *comoving* neutral hydrogen density of $\lesssim 10^{-6.8}\, h\,{\rm m}^{-3}$, i.e. $\Omega_{\rm HI} \lesssim 10^{-7.8} h^{-1}$. Of course, the Gunn–Peterson effect *is* detected if we think of the integrated effect of the many narrow absorption lines: these produce a mean $\tau = 1$ in the Lyα forest at about $z = 3$, and so we have $\Omega_{\rm HI} \simeq 10^{-5.5} h^{-1}$.

There are two things to note: the very low abundance of neutral hydrogen in the general IGM by comparison with that in clumps, and the fact that the total is still very far short of the nucleosynthetic prediction, $\Omega_{\rm B} h^2 \simeq 0.015$. It is hardly reasonable to suppose that star formation was successful in locking up all but $\sim 0.01\%$ of the gas, so this must be telling us that the gas is highly ionized, even at very high redshifts.

PROPERTIES OF THE ABSORBERS The mechanism of ionization is not hard to find, once we recall the high degree of quasar activity at high redshift: the quasar continuum extends well through the ultraviolet into X-rays, so there is a ready supply of photons to produce **photoionization**. Here, the fractional ionization x is set by a balance between the rates of photoabsorption and recombination:

$$x\,\beta n_e = (1-x) \int_{\nu_{\rm min}}^{\infty} \sigma(\nu) \frac{4\pi I_\nu}{h\nu}\, d\nu. \tag{12.42}$$

In this equation, x gives the fraction of hydrogen nuclei that are ions, and each of these can recombine with electrons at a rate βn_e, where the recombination coefficient β is a function of the temperature of the gas. This creation rate of hydrogen atoms must balance the fraction of hydrogen nuclei that are atoms, times the cross section for ionization, times the flux density of ionizing photons.

Clearly, the ionization balance is mainly set by the ratio of the number densities of ionizing photons and electrons, and we therefore define an **ionization parameter**

$$U \equiv \frac{n_\gamma}{n_e} = n_e^{-1} \int \frac{4\pi I_\nu}{ch\nu}\, d\nu \tag{12.43}$$

(sometimes the factor c is omitted to give U the dimensions of velocity). In fact, the temperature dependence of β is not very important, since the temperature is almost always in the region of 10^4 K (because the excess kinetic energy per photoelectron is of order the ionization potential of hydrogen, so $kT \sim 13.6\,{\rm eV}$). For hydrogen, this then gives an invaluable rule of thumb:

$$\frac{x}{1-x} = \frac{U}{10^{-5.2}}. \tag{12.44}$$

For the line-emitting regions in quasar nuclei, $U \sim 0.01$ and we expect a very high degree of ionization (see chapter 14); what about in intergalactic space?

The ionizing background may be estimated from the **proximity effect** whereby clouds near the quasar are ionized by the quasar itself: cloud numbers change significantly when the quasar contributes ionizing radiation of the same order as the background. The known quasar luminosity therefore allows the general background to be estimated. A common notation that absorbs the typical value for the background found at $z = 2.5$ is

$$I_\nu = J_{-21}\,(\nu/\nu_{\rm L})^{-\alpha} \times 10^{-21} {\rm erg\,s^{-1}\,Hz^{-1}\,cm^{-2}\,sr^{-1}}, \tag{12.45}$$

$\nu_{\rm L}$ being the threshold frequency for ionization – the Lyman limit. For a spectral index $\alpha = 1$, the fiducial level of $J_{-21} = 1$ corresponds to a proper density of 63 photons m^{-3}. From the known luminosity function of quasars, one can predict the integrated quasar contribution to this background as a function of redshift. The calculation is complicated by the fact that the Lyα clouds themselves cut out some of the radiation: the answer appears to be lower than the above figure by a factor $\lesssim 10$ (see Meiksin & Madau 1993); there have been suggestions that the extra flux is supplied by star-forming galaxies, or even decaying exotic dark-matter particles. However, it is suggestive that the quasar contribution comes so close – it may well be that quasars do make all the ionizing background, and that there are a few inaccuracies in the calculation.

If we take this figure for n_γ, then the electron density is $n_e \simeq 10^{7.0}(1-x)$, where x is the fractional ionization (assumed close to unity). Using $n_e = 9.8\Omega_{\rm H}h^2(1+z)^3x$ (for 25% He by mass), and the Gunn–Peterson constraint of $\Omega_{\rm H}h(1-x) \lesssim 10^{-7.8}$, we deduce $\Omega_{\rm H}h^{3/2} \lesssim 0.044$ for the general IGM, which is consistent with the nucleosynthetic $\Omega_{\rm B}h^2 \simeq 0.015$. In other words, we would not expect to see a Gunn–Peterson step from a homogeneous IGM. The integrated effect of the Lyα clouds is larger because their higher densities allow them to have a higher neutral fraction.

LYMAN ALPHA CLOUDS AS PROTOGALAXIES How big are the absorbing clouds? The answer to this question depends on having some idea about their density. This is possible for the clouds of high column density, since they contain enough neutral hydrogen to be **self-shielding**: their optical depth to ionizing radiation is $\gg 1$. The cross-section for ionization of hydrogen is $\sigma = 7.88 \times 10^{-22}$ m^2 at the Lyman limit, scaling as ν^{-3} (Rybicki & Lightman 1979). The average optical depth to ionizing photons is

$$\tau = N_{\rm HI} \frac{\int (\sigma J_\nu/h\nu)\, d\nu}{\int (J_\nu/h\nu)\, d\nu}, \tag{12.46}$$

which gives

$$\boxed{\tau = N_{\rm HI}\ /\ 10^{21.71}\ {\rm m}^{-2},} \tag{12.47}$$

assuming a spectral index of unity for the ionizing background. Observationally, this is the threshold column density for a **Lyman-limit system**, which will truncate the spectrum of the quasar at $\lambda_0 < 912$ Å. The damped systems with $N_{\rm HI} \simeq 10^{25}$ m^{-2} should then be largely neutral. Without the complications of ionization, the path length through the absorber must be just $\ell = N_{\rm HI}/n_{\rm H}$. Using a hydrogen density of f_c times the nucleosynthetic $\Omega_{\rm B}h^2 = 0.015$, this gives

$$\ell/{\rm kpc} = (N_{\rm HI}/10^{23.2}\,{\rm m}^{-2})\,(f_c/1000)^{-1} \tag{12.48}$$

at $z = 2.5$. The most spectacular absorption systems therefore have properties consistent with their being of the dimensions of galactic disks.

Conversely, the systems of lowest column density will be optically thin to ionizing radiation, and we need to resort once again to the above calculations of the neutral fraction. Where the fractional ionization is $x \simeq 1$, the typical size is now

$$\ell/\mathrm{kpc} = (N_{\mathrm{HI}}/10^{20.1}\,\mathrm{m}^{-2})\,(f_c/1000)^{-2} \tag{12.49}$$

(at $z = 2.5$, assuming $J_{-21} = 1$). The sensitivity to f_c is now higher, and it is quite easy for sizes to reach scales much larger than a galaxy if the overdensity is only moderate, even for the weakest systems with $N_{\mathrm{HI}} \simeq 10^{16.5}\,\mathrm{m}^{-2}$. There is mounting evidence that this is the case: common absorption lines have been seen in pairs of quasars that lie close together on the sky, implying a transverse absorber size of at least ~ 10 kpc. On a single line of sight, the Lyα lines at $z \simeq 2.5$ appear to be clustered with $\xi = 1$ on separations of $300\,\mathrm{km\,s^{-1}}$ (i.e. $3h^{-1}/\sqrt{1+\Omega z}$ comoving Mpc), which is quite reasonable for low-mass galactic haloes at these redshifts (Fernández-Soto *et al.* 1996). It may thus be an oversimplification to think of the weaker clouds as being isolated objects. Numerical simulations including photoionized gas predict that the gas should display a filamentary structure in which the low-N absorbers trace out the Mpc-scale caustic features, rather than being isolated quasi-spherical objects (Katz *et al.* 1996).

It is not even necessary for the gas to be strongly gravitationally bound, and a good fraction of the gas in such simulations is barely turning around from the Hubble flow. Adiabatic losses would thus keep the clouds from being heated by photoionization, even though there is time for ionization equilibrium to be maintained (Duncan *et al.* 1991). This model has the advantage that it explains reasonably naturally why the Lyα forest evolves: the clouds become more strongly ionized as they expand to lower density. Nevertheless, the gas is far from being primordial, and it appears that both strong and weak systems have metallicities $Z \simeq 10^{-2} Z_\odot$ (Pettini *et al.* 1994; Cowie *et al.* 1995b), so there appears to have been universal enrichment to this level by gas expelled from galaxies at $z > 2.5$.

Finally, it is interesting to compare the total amount of HI gas seen at $z \simeq 2.5$ with the baryonic content of present-day galaxies. The total HI density can be deduced from the observabie $n(N,z)$, which is the number of systems along a given line of sight per dz in the column-density range dN. Now imagine a cylinder of comoving cross-sectional area A and length Δz in the redshift direction. The mean column through the cylinder is observable:

$$\overline{N} = \Delta z \int N n(N,z)\, dN. \tag{12.50}$$

The total comoving number density of HI is therefore $m_{\mathrm{H}} A \overline{N}$ divided by the comoving volume of the cylinder:

$$\rho_{\mathrm{HI}} = \frac{m_{\mathrm{H}}}{dr/dz} \int N n(N,z)\, dN, \tag{12.51}$$

which scales $\propto h$. It is usual to express this comoving density as a density parameter, Ω_{HI}, which is ρ_{HI} divided by the present-day critical density, not the critical density at the redshift of observation. The result is a function of redshift and peaks at about $z = 3$:

$$\boxed{\Omega_{\mathrm{HI}} \simeq 2 \times 10^{-3} h^{-1}} \tag{12.52}$$

(see figure 12.3); the value of Ω_{HI} at low redshifts is roughly ten times smaller than this. The high-redshift value is an interesting number because it is comparable to the contribution that stars in spiral disks presently make to Ω. This is further supporting

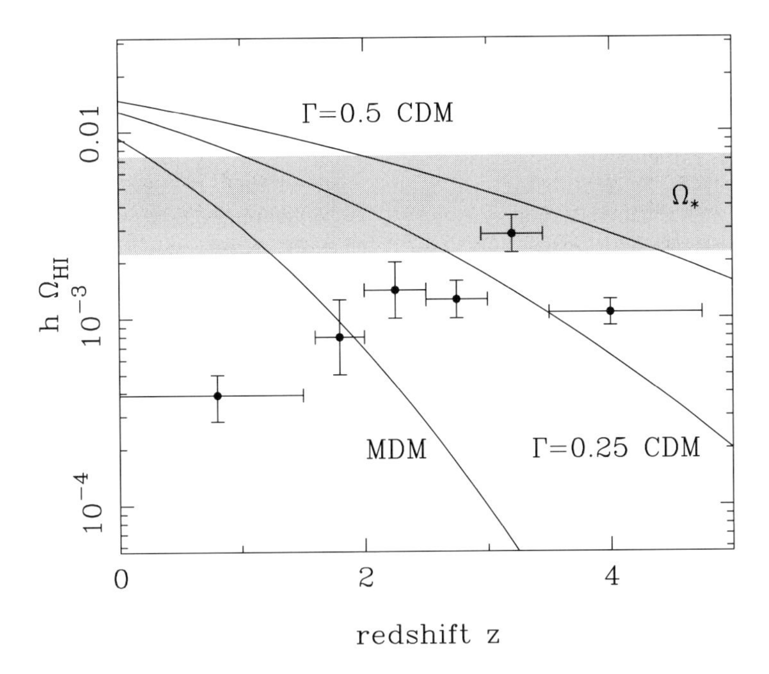

Figure 12.3. The collapsed neutral baryon density parameter as a function of redshift, as deduced from the Lyα forest. The data points are adopted from Lanzetta *et al.* (1993), with the exception of the highest-redshift point, from Storrie-Lombardi *et al.* (1996). The shaded area shows the present-day contribution to Ω from stars, assuming blue M/L ratios in the range 30 to 10. The predictions of two CDM models are shown, together with $h = 0.5$, $\Omega_\nu = 0.25$ mixed dark matter. These take the nucleosynthetic baryon density, $\Omega_B h^2 = 0.0125$. The data points and stellar locus scale as $\Omega \propto h^{-1}$, whereas the lines scale as $\Omega \propto h^{-2}$; they are plotted assuming $h = 0.65$. This diagram demonstrates that a substantial fraction of all present-day stars were created after $z = 3$.

evidence for the arguments given in chapters 13 and 17 that spiral disks were formed at relatively low redshifts: the depletion of HI since $z = 3$ would then be attributed to its destruction by star formation in the disks.

12.3 Evidence for dark matter

So much for the census of visible matter. Despite the ingenuity that has been invested in studying the IGM, it appears that there is a dominant component of the mass distribution that is not detected in this way, but only manifests itself via gravity. This section reviews some of the evidence for its existence, together with the methods whereby the nature of the material may be constrained. Although perhaps not quite such a fundamental problem as the classical cosmological problems (the horizon, flatness and baryon problems), the issue of dark matter arguably constitutes the most *astrophysically* interesting aspect of modern cosmology.

MASS-TO-LIGHT RATIOS AND Ω The importance of being able to determine the total amount of mass in astronomical objects is that it allows us to use a classical technique for weighing the whole universe: '$L \times M/L$'. The overall light density of the universe is reasonably well determined from redshift surveys of galaxies, so that a good determination of mass M and luminosity L for a single object suffices to determine Ω *if* the mass-to-light ratio is universal. Of course, different bodies will have different M/L values: if the Sun has a value of unity (these measurements are always quoted in Solar units), then low-mass stars have values of several tens while a comet of radius 10 km and temperature 100 K would have $M/L \sim 10^{12}$ (assuming M/L scales $\propto R/T^4$). The stellar populations of galaxies typically produce M/L values between 1 and 10.

Galaxy redshift surveys allow us to deduce the galaxy luminosity function, ϕ, which is the comoving number density of galaxies; this may be described by the **Schechter function**, which is a power law with an exponential cutoff:

$$\phi = \phi^* \left(\frac{L}{L^*}\right)^{-\alpha} e^{-L/L^*} \frac{dL}{L^*}. \tag{12.53}$$

The total luminosity density produced by integrating over the function is

$$\rho_L = \phi^* L^* \Gamma(2-\alpha), \tag{12.54}$$

and this tells us the average mass-to-light ratio needed to close the universe. Answers vary (principally owing to uncertainties in ϕ^*; see chapter 13). In blue light, a canonical value for the total luminosity density is $\rho_{\rm L} = 2 \pm 0.7 \times 10^8\ hL_\odot \mathrm{Mpc}^{-3}$ (Efstathiou *et al.* 1988a). Since the critical density is $2.78 \times 10^{11} \Omega h^2 M_\odot \mathrm{Mpc}^{-3}$, the critical M/L for closure is

$$\boxed{\left(\frac{M}{L}\right)_{\rm crit,\,B} = 1390h \pm 35\%.} \tag{12.55}$$

For reference, primordial nucleosynthesis implies a minimum Ωh^2 of 0.01, so we should certainly expect to find bodies with $M/L > 14/h$. We therefore know in advance that there must be more to the universe than normal stellar populations – dark matter of some form is inevitable.

In fact, as detailed below, dynamical determinations of mass on the largest accessible scales consistently yield blue M/L values of at least $300h$, but normally fall short of the closure value. If we want a flat universe, then we need to produce an order-unity contribution to Ω from some uniform background such as vacuum energy, or to assume that there is dark matter that lies outside virialized systems such as clusters. This idea that light may not follow the distribution of mass even on very large scales has become known as **biased galaxy formation**, and is one of the central themes of modern cosmology.

DARK MATTER IN THE MILKY WAY To look for dark matter in our neighbourhood, we study the **Oort limit**; this term implies a comparison of the mass density in known stars and gas with a dynamical determination of the total, to see if there is any discrepancy.

The former exercise is quite straightforward: parallax and proper-motion surveys yield the local **stellar luminosity function**, $\phi(L)$, which is the space density of stars in the range dL (often quoted as the number of stars in unit range of absolute magnitude). Given the well-known mass–luminosity relation, the stellar density, $\rho = \int m(L)\phi(L)dL$, follows. In practice, this integral must be truncated at low masses where stars no longer burn hydrogen (about 0.08 $M_\odot$; see below). Including contributions from gas and dust clouds yields a total density *in luminous material* of

$$\rho_{\rm lum} = 0.1 M_\odot\ {\rm pc}^{-3}; \tag{12.56}$$

see e.g. Freeman (1987).

We now have to compare this with the total density. The approach here is to say that the local density plus some measure of the thickness of the Milky Way determine the acceleration towards the galactic plane (related to the perpendicular 'escape velocity'), which can be measured from the velocity dispersion in stellar orbits. The stars are treated as though they were a fluid in **hydrostatic equilibrium**:

$$g\rho_* = -\frac{\partial}{\partial z}(\rho_* \sigma_v^2), \tag{12.57}$$

relating acceleration perpendicular to the plane, g, to height above the plane, z, and the effective pressure expressed in terms of the rms in stellar radial velocity, σ_v. This ignores the fact that the disk of the galaxy rotates, but this has a relatively minor effect (see Binney & Tremaine 1987). Gauss's theorem now says $\partial g/\partial z = 4\pi G\rho$, where ρ is the total matter density, and so

$$\boxed{4\pi G\rho = -\frac{\partial}{\partial z}\left[\frac{1}{\rho_*}\frac{\partial}{\partial z}\left(\sigma_v^2 \rho_*\right)\right].} \tag{12.58}$$

Note that this expression depends only on the scaling of the stellar density with z, not on its absolute value. Now, suppose we knew the **disk scale height**, h, for the mass, assuming an exponential dependence $\rho_* \propto \exp(-z/h)$; it is then easy to show that $\sigma_v^2(z=0) = 2\pi G\rho_0 h^2$. Typical values are $h \simeq 300$–350 pc (Freeman 1987); comparing with the distance of the sun from the centre of the Galaxy (about 8 kpc) shows why the disk component of the Galaxy is often referred to as the **thin disk**.

So, the study of a 'well-mixed' population of stars that trace the mass distribution can yield both the scale height and velocity dispersion, and reveal the total local density. Another way this is sometimes phrased is to say that one needs to measure the derivative of the gravitational acceleration normal to the plane (often denoted by K_z, rather than g as above). In this case, Poisson's equation is rewritten as

$$\rho_0 = \frac{1}{4\pi G}\left(\frac{\partial K_z}{\partial z}\right)_{z=0}. \tag{12.59}$$

Clearly, ρ_0 is rather sensitive to assumptions about the details of the density distribution above the disk. A more robust quantity is the **surface density** of the disk: $\Sigma = \int_{-\infty}^{\infty} \rho\, dz$. Unlike the density itself, this involves only a single differentiation of the observable $\rho_*\sigma_v^2$, and so is a more stable quantity from a numerical point of view. However one formulates the problem, the difficulty in application is not so much the simplicity of the analysis, but in finding a set of stars that can be confidently assumed to trace the mass. Different sets give different answers (Bahcall 1984; Kuijken & Gilmore 1991) and the situation is uncertain. The dynamical determinations of ρ_0 tend to give more mass than is seen directly, but the discrepancy is a factor two at most, and may not exist. In a sense, it would be quite a triumph if all the dynamically inferred mass could be accounted for: one would then have a test of Newtonian gravity on scales ~ 1 kpc, much larger than the existing tests on Solar-system scales ($\sim 10^{-4}$ pc).

ISOTHERMAL DISTRIBUTIONS A particularly interesting special case of stellar hydrostatic equilibrium occurs for **isothermal mass distributions**, where the 'temperature' σ_v^2 is constant (see chapter 17 for reasons why this could arise; the thermodynamic terminology gives a hint that such distributions might be expected in equilibrium). We can imagine this situation occurring in systems with different dimensionalities of symmetry: plane symmetric ($d = 1$); cylindrically symmetric ($d = 2$); or spherically symmetric ($d = 3$). The equation $\mathbf{g}\rho = -\boldsymbol{\nabla}(\rho\sigma_v^2)$ may be written in d dimensions by taking the divergence and using Gauss's theorem to relate the lhs to the total density. Denoting distance from the origin by r and derivatives by primes, we have the **isothermal equation**,

$$\boxed{\frac{4\pi G}{\sigma_v^2} r^{d-1} \rho = -\left(r^{d-1} \frac{\rho'}{\rho} \right)'.} \tag{12.60}$$

This can be made dimensionless by defining

$$y \equiv \frac{\rho}{\rho_0}, \qquad x \equiv \frac{r}{(4\pi G\rho_0/\sigma_v^2)^{-1/2}}, \tag{12.61}$$

where ρ_0 is the central density. With boundary conditions $y(0) = 1$, $y'(0) = 0$, this has solutions

$$y = \begin{cases} \left[\cosh^2(x/\sqrt{2})\right]^{-1} & (d = 1) \\ \left(1 + x^2/8\right)^{-2} & (d = 2). \end{cases} \tag{12.62}$$

Unfortunately, although we can thus obtain **isothermal sheets** and **isothermal cylinders**, no analytic solution of the isothermal equation is known in three dimensions, apart from the **singular isothermal sphere**:

$$\boxed{\rho = \frac{\sigma_v^2}{2\pi G r^2}.} \tag{12.63}$$

The solution to the isothermal equation with finite central density tends to this singular form at large radii, but departs from it at $x \sim 1$.

DARK MATTER IN GALAXY HALOES If the existence of non-luminous material inside galaxies is still somewhat uncertain, there is no disputing the situation regarding the outer regions of galaxies. The most dramatic evidence that large amounts of dark matter exist is that many galaxies show **flat rotation curves**: rather than the Keplerian fall-off $V \propto 1/r$ that would be expected if the total mass inside r had converged, a roughly constant V implies a linearly divergent $M(< r) \propto r$. It might seem that it would be very difficult to establish the behaviour of the mass at large radii, simply because one

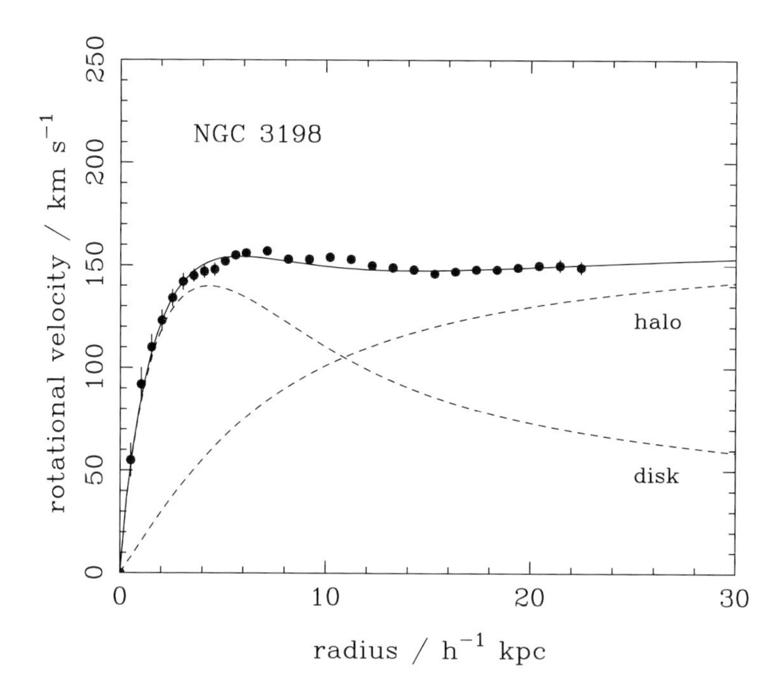

Figure 12.4. The brightness profile and rotation curve of the spiral galaxy NGC3198 (adapted from van Albada *et al.* 1985). Note the flat form of the curve at large radii by comparison with the contribution expected from the luminous disk.

will run out of light. However, it is in fact possible to measure the rotation velocities of galaxies well beyond the point at which the emission from stars becomes undetectable, using 21-cm emission from neutral hydrogen (e.g. figure 12.4). This means that the main evidence for dark haloes exists for spiral galaxies; however, in some cases ellipticals contain sufficient neutral hydrogen at large distances that a similar case for a dark halo can be made (e.g. Schweizer *et al.* 1989).

This discrepancy exists inside the Milky Way. Consider the Sun: it rotates at about 220 km s^{-1} about the centre of the Galaxy, at a radius of about 8 kpc. The mass per unit area from the disk is $\Sigma = 2\rho_0 h \simeq 60 M_\odot \, \mathrm{pc}^{-2}$; is this enough to supply the centripetal force for the galactic rotation? We could solve for the orbital speed in the plane of a thin sheet of given $\Sigma(r)$, but this is a surprisingly nasty potential problem (see section 2.6 of Binney & Tremaine 1987). Instead, we anticipate the answer by imagining that the mass distribution is in fact spherically symmetric. If the rotation curve is flat, then the mass distribution must be that of the singular isothermal sphere, for which it is easy to show that the circular orbital velocity is

$$V_c = \sqrt{2}\sigma_v. \tag{12.64}$$

Now work out the surface density of the isothermal sphere:

$$\Sigma(R) = 2\int_0^\infty \rho\left(r = \sqrt{R^2 + x^2}\right)\, dx = \frac{\sigma_v^2}{\pi G}\int_0^\infty \frac{dx}{x^2 + R^2} = \frac{\sigma_v^2}{2GR}. \tag{12.65}$$

In terms of the circular velocity, the surface density for the isothermal sphere is thus $\Sigma = V_c^2/(4Gr)$. For a thin disk with constant rotation speed, the corresponding expression is very similar: $\Sigma = V_c^2/(2\pi Gr)$. At the Sun's radius, for $V_c = 220\,\mathrm{km\,s^{-1}}$, this works out at about $360 M_\odot\,\mathrm{pc}^{-2}$, far in excess of the disk mass. The Milky Way's rotation therefore needs substantial dark matter, which must exist at distances several kpc above the plane. It is this approximately spherical distribution that is probed by the microlensing experiments described in chapter 4.

It has been suggested that there is something of a conspiracy at work here, in order for the rotation curve to stay flat while the dominant gravitating species changes from stars and gas to something unknown. There seems no reason why the baryonic dominance might not have been greater, leading to a regime in which the expected Keplerian fall-off was seen, before the influence of dark matter made the rotation curve flatten. That such a gap seems not to exist suggests a connection between the dark and visible matter, and has been proposed as evidence that the dark matter is baryonic (Bahcall & Casertano 1985). However, the true explanation is probably the one discussed in chapter 17: that the dissipative settling of baryonic material in the potential well of the dark-matter halo causes the dark-matter density to adjust in such a way as to match on to the visible mass distribution (Flores *et al.* 1993).

The need for massive dark haloes around galaxies was in fact originally suggested before most of the data on galaxy rotation curves were available (Ostriker, Peebles & Yahil 1974). The theoretical argument comes from the stability of self-gravitating disks. Numerical experiments indicate that disks can in many circumstances be prone to the **bar instability**, where perturbations in density that vary with angle as $\cos^2\theta$ grow in amplitude until the centre of the galaxy generates a nearly linear feature. Most spirals are not barred, and one way of understanding this is if the visible matter in the disk merely orbits passively in the potential well defined by an approximately spherical halo of dark matter. There were therefore good reasons to expect to find dark matter, and the results on rotation curves landed in fertile soil. Of course, it is healthy to be suspicious of results that verify preconceptions too neatly. There was therefore a period when it was argued that disks with M/L ratios as the upper end of normal might satisfy the dynamical results (**maximum disk** models). However, such a point of view becomes harder and harder to sustain as the HI data push beyond the edge of the visible disk. Such rotation-curve studies have now demonstrated that the material in the outer parts of galaxies must have $M/L \gtrsim 1000$; within $30h^{-1}$ kpc, the typical values for the whole galaxy are $M/L \simeq 30h$ (e.g. White 1990). Since enclosed mass increases linearly with radius for an isothermal sphere, this requires that the haloes extend to roughly 1 Mpc to achieve closure.

DARK MATTER IN GROUPS AND CLUSTERS OF GALAXIES Historically, the first detection of dark matter was made in 1933 by Zwicky. He looked at the dispersion of velocities in the Coma cluster and deduced $M/L \simeq 300h$ from application of the virial theorem. Modern techniques are able to probe the mass distribution in rather more detail, but end up with very much the same answer.

It is usually assumed that the cluster to be studied is spherically symmetric and in a relaxed equilibrium state. This means that the galaxies act similarly to a fluid in **hydrostatic equilibrium**, their orbits in random directions playing the role of a pressure. The main difference is that the pressure need not be isotropic: the radial velocity dispersion σ_r may differ from the transverse dispersion σ_t. Consider the equilibrium of a fluid element defined by a radial increment dr of a cone of semi-angle ψ; in addition to gravity, the radial acceleration is determined by the net radial component of the force acting on the surface of the element (think of $\rho\sigma^2$ as a pressure):

$$F_r = -\frac{\partial}{\partial r}\left[\pi(\psi r)^2\rho\sigma_r^2\right]\, dr + 2\pi(\psi r)\, dr\ \psi\rho\sigma_t^2. \qquad (12.66)$$

This gives the equilibrium equation

$$-\frac{\partial\Phi}{\partial r} = -\frac{GM(<r)}{r^2} = \frac{1}{\rho}\frac{\partial}{\partial r}\left(\rho\sigma_r^2\right) + \frac{2}{r}\left(\sigma_r^2 - \sigma_t^2\right), \qquad (12.67)$$

where Φ is the gravitational potential and $M(<r)$ is the total mass integrated out to radius r. Remember that ρ is the mass density in the galaxies, which can differ from the total density. Indeed, it is clear from this equation that only the shape of $\rho(r)$ matters, rather than its absolute value.

The velocity dispersions are only constrained by the integrated line-of-sight dispersion

$$\begin{aligned}\sigma_v^2(R) &= \frac{2}{\Sigma(R)}\int_R^\infty \left[\frac{R^2}{r^2}\sigma_t^2 + \left(1-\frac{R^2}{r^2}\right)\sigma_r^2\right]\rho(r)\ \frac{r\,dr}{\sqrt{r^2-R^2}}\\ \Sigma(R) &= 2\int_R^\infty \rho(r)\ \frac{r\,dr}{\sqrt{r^2-R^2}}.\end{aligned} \qquad (12.68)$$

We need to solve these two equations plus the hydrostatic equation for $M(<r)$, given the observables of $\sigma_v^2(R)$ plus the galaxy surface density $\Sigma(R)$. Several approaches of varying complexity are usually considered: (i) light traces mass and orbits are isotropic; (ii) light traces mass, but anisotropy is allowed an arbitrary radial variation; (iii) light does not trace mass, and orbits are isotropic; (iv) light does not trace mass, and isotropy is arbitrary.

If the light-traces-mass approach is taken, then the only freedom is in M/L. The observed $\Sigma_g(R)$ can be deprojected (see chapter 4) to yield $\rho(r)$ up to an unknown factor. With no anisotropy, the shape of the $\sigma_v^2(R)$ profile is predicted, and scaling to the data gives an estimate of M/L. Of course, the shape of the profile is not guaranteed to match observation, and variable anisotropy may be invoked to avoid abandoning the constant M/L assumption. Suppose, for example, the anisotropy was such that there was a change from mainly transverse orbits at small r, to mainly radial ones at large r. This would have the effect of producing a larger gradient in $\sigma_v^2(R)$ for a given $\rho(r)$. This effect of anisotropy is also a problem in interpreting velocity-dispersion profiles in galaxies: a sharp increase in velocity dispersion towards the centre could indicate a central black hole, but it might instead be due to changing anisotropy (Binney & Mamon 1982).

If the constant M/L assumption is abandoned, the equations can be solved exactly for isotropic orbits. The equations for $\rho(r)$ and $\sigma^2(r)$ can both be deprojected. The only subtlety is that $\rho(r)$ in the hydrostatic equation is the *mass* density of the tracers, whereas only the light density is observable. Nevertheless, even if the system as a whole shows large variations in M/L, the tracer population itself usually does not. The M/L ratios measured in the central regions of galaxies tend to have uniform values for galaxies of a given type, ranging from a few Solar units in spirals and low-luminosity galaxies to of order 10 Solar units in massive ellipticals (see e.g. chapter 10 of Binney & Tremaine 1987). Even these modest variations are clearly due to straightforward systematic differences in the stellar-population content of different galaxies, and they correlate well with variations in colour.

Finally, the deduced $M(< r)$ depends on the assumed isotropy, and there is the freedom to insert any σ_t/σ_r ratio that we please. The only way to constrain some of the more extreme possibilities is to go beyond the simple rms statistic $\sigma_v(R)$ and look at the full velocity distribution as a function of R. Reassuringly, these different methods all give rather similar answers for the mass in the central parts of the Coma cluster: about $6 \times 10^{14} h^{-1} M_\odot$ within a radius of $1h^{-1}$ Mpc (e.g. Merritt 1987). Results extending to larger radii can be obtained by superimposing the data from many clusters into a single fictitious mean cluster (Carlberg, Yee & Ellingson 1997); this gives a sufficiently high signal-to-noise ratio that the light and mass can be mapped over a reasonable range of radii, and the result is that the two profiles follow each other quite closely. The implication is a nearly constant mass-to-light ratio, $M/L_B \simeq 300h$–$400h$, implying $\Omega \simeq 0.2$–0.3 if clusters are large enough systems for their M/L ratios to be representative.

MASSES FROM X-RAY GAS A completely different tracer of mass is also available in clusters of galaxies in the form of hot gas that is visible in X-rays. For this material, it is reasonable to suppose that hydrodynamic equilibrium applies (with guaranteed isotropic pressure). The radial run of mass can then be deduced in terms of radial temperature and density gradients,

$$\frac{GM(< r)}{r} = -\frac{kT(r)}{\mu m_p} \left(\frac{d \ln T}{d \ln r} + \frac{d \ln \rho}{d \ln r} \right). \tag{12.69}$$

Unfortunately, the temperature run is often not very well constrained and the cluster is usually assumed to be isothermal. A common model used to fit cluster X-ray data is to assume that the total and gas density profiles satisfy the two-parameter forms

$$\begin{aligned} \rho_m &= \rho_m(0) \left[1 + (r/r_c)^2\right]^{-3/2} \\ \rho_g &= \rho_g(0) \left[1 + (r/r_c)^2\right]^{-3\beta/2}, \end{aligned} \tag{12.70}$$

the former being an analytic approximation to the central part of the isothermal sphere. This is a solution of the joint hydrostatic equations if the parameter β is a constant, satisfying

$$\beta = \frac{\mu m_p \sigma_r^2}{kT}. \tag{12.71}$$

This parameter is just the 'temperature' of the dark-matter distribution divided by the temperature of the gas. The mean gas temperature can be measured from the overall shape of the X-ray spectrum, and the absolute value of the gas density follows from the X-ray bremsstrahlung luminosity. The emissivity scales as ρ^2, so the total X-ray luminosity is $L \propto \rho_g(0)^2 r_c^3$. Since the 'observed' luminosity scales as $L \propto h^{-2}$ and $r_c \propto h^{-1}$, the total inferred gas mass scales as $h^{-5/2}$. Note that all these parameters, including the scale r_c and β, can be determined from the X-ray data alone.

A long-standing problem with this approach was the **beta discrepancy**, in that the fitted value of β consistently came out at $\beta \simeq 0.7$, implying that the gas was hotter than the dark matter. This in isolation would be an interesting observation, but it conflicts with the fact that β can be measured directly, and generally gives values consistent with unity. According to Bahcall & Lubin (1994), the resolution of this puzzle is simply that the model for the mass profile is inadequate in detail. They claim that all existing clusters are consistent with $\beta = 1$, even those where more recent data favour a non-zero temperature gradient. Incidentally, it is worth noting that the temperature gradients are generally modest: $d \ln T / d \ln r \simeq -0.75$. This is a much slower variation than that expected if the gas had simply fallen into the cluster and suffered adiabatic compression, $T \propto \rho^{\Gamma-1}$, which would imply $d \ln T / d \ln r \simeq -2$ if the density fell as r^{-3} in the outer parts.

Again, Coma is a good example of the application of these methods: the gas mass is well constrained in the central $1h^{-1}$ Mpc:

$$M_{\rm g} = 3 \times 10^{13} h^{-5/2}\, M_\odot, \tag{12.72}$$

see e.g. Hughes (1989). If we add this to the mass of stars in Coma ($\simeq 3 \times 10^{13} h^{-1}\, M_\odot$), then we get a measure of the fraction of the mass in Coma that is in the form of baryons:

$$\boxed{\frac{M_{\rm B}}{M_{\rm tot}} \simeq 0.01 + 0.05 h^{-3/2}} \tag{12.73}$$

(White *et al.* 1993). If the overall density is critical, this ratio is far above that predicted by the nucleosynthetic estimate, $\Omega_{\rm B} h^2 \simeq 0.0125$. If we adopt $h = 0.65$, then a total density parameter $\Omega \simeq 0.28$ is implied. Thus, high Ω requires an abnormal concentration of baryons towards clusters, not just a greater star-formation efficiency there. This is a more robust argument in favour of low Ω, since it does not depend on assuming that the stellar populations in clusters are representative (which they are not). In other environments, some fraction of the intracluster gas might have formed stars, but so long as we count the totality of baryons this does not matter. As discussed below, the outer parts of clusters have only recently separated from the Hubble flow, so it is hard to see how the baryon-to-total mass ratio can have been affected.

THE SUNYAEV–ZELDOVICH EFFECT The hot gas that X-ray observations reveal in clusters of galaxies may be detected in ways other than its thermal bremsstrahlung emission. Photons passing through the cluster will undergo **Compton scattering** by collisions with electrons in the ionised IGM, causing distortions in the spectrum of the emergent radiation. The place where one can hope to detect these distortions most readily is in the cosmic microwave background (CMB). To analyse these distortions quantitatively, we need to derive the **transport equation** governing the effect of Compton scattering on the photon energy distribution.

The most important point to realize about this process is that it conserves photons. In the absence of interactions between particles (a good approximation in a sparse astrophysical plasma), photons cannot be absorbed or created – scattering can only redistribute their energies. At first sight, we might expect the differential equation describing such a process to be first order. Call the number density of photons in unit energy interval $N(E)$, and let a given scattering event change the photon energy by an amount Δ, $\langle \Delta \rangle \equiv A$ being the average change of photon energy per photon–electron collision. Regarding AN/t_c as a one-dimensional energy current (t_c being the time between photon-electron collisions) might suggest the conservation law

$$t_c \dot{N} = -(AN)', \tag{12.74}$$

where dot and prime denote time and energy differentiation. In many classes of problem, however, this is not the correct equation to use: we have neglected the fact that scattering also causes a *broadening* of the energy spectrum.

This is analysed as follows. Suppose we know the probability distribution function $p(\Delta|E)$ for the change in energy at a single scattering; considering scattering of photons both into and out of the element dE gives the following expression for the change in the number of electrons in dE:

$$\delta N = \int \left[N(E-\Delta)p(\Delta|E-\Delta) - N(E)p(\Delta|E) \right] d\Delta. \tag{12.75}$$

If we now expand $N(E-\Delta)$ and $p(\Delta|E-\Delta)$, collect terms to second order in Δ and define $B \equiv \langle \Delta^2 \rangle$, then we obtain the general form of the **Fokker–Planck equation**

$$\boxed{t_c \dot{N} = -(AN)' + \tfrac{1}{2}(BN)''.} \tag{12.76}$$

When A vanishes, this is effectively just the diffusion equation. The expansion could in principle be extended to higher orders, but this is rarely necessary. For the 'diffusive' term in the Fokker–Planck equation to become significant, we need $B \sim AE$, i.e. the fractional variance in E per collision needs to be of the same order as the mean fractional change in E. If these quantities are both small, then higher moments will be negligible except in pathological cases.

To apply this formalism to the problem of Compton scattering, we need the appropriate A and B coefficients, assuming that the scattering electrons have a Maxwellian momentum distribution at temperature T [problem 12.3]. For $\hbar\omega \ll m_e c^2$, the result is

$$\begin{aligned} \frac{A}{\hbar\omega} &= \left(4 - \frac{\hbar\omega}{kT}\right)\left(\frac{kT}{m_e c^2}\right) \\ \frac{B}{(\hbar\omega)^2} &= 2\left(\frac{kT}{m_e c^2}\right), \end{aligned} \tag{12.77}$$

and the scattering time is given by $t_c^{-1} = n_e \sigma_T c$. It is better to express the final result in terms not of $N(E)$ but of the photon phase-space occupation number $n(\omega)$, where $N \propto \omega^2 n(\omega)$. At this point, we should consider quantum corrections to the classical Fokker–Planck equation. If the photon occupation numbers are not small, then we must remember that the rate of scattering from E to $E + \Delta$ should actually be proportional to $n(E)[1 + n(E + \Delta)]$. The factor $[1 + n]$ takes care of the fact that bosons like to clump: stimulated transitions ensure that a final state is more probable if it is already occupied. In this case, the Fokker–Planck equation is modified to

$$t_c \dot{N} = -[AN(1+n)]' + \tfrac{1}{2}(BN)''(1+n) - \tfrac{1}{2}BN(1+n)'', \tag{12.78}$$

and so finally we obtain the **Kompaneets equation**:

$$\boxed{\dot{n} = \left(\frac{\sigma_T n_e \hbar}{m_e c}\right) \frac{1}{\omega^2}\frac{\partial}{\partial\omega}\left\{\omega^4 \left[n(1+n) + \frac{kT}{\hbar}\frac{\partial n}{\partial \omega}\right]\right\}.} \tag{12.79}$$

Solving the time-dependent Kompaneets equation in order to study the **Comptonization** of a photon population is hard in general. In the limit of many scatterings, we of course obtain an equilibrium Bose–Einstein distribution with non-zero chemical potential. For X-ray gas in clusters, conversely, the optical depth is low enough that each photon can be considered to scatter only once. In this case, we can integrate through the cluster to find the linear change in n,

$$\delta n = y\nu^{-2}\frac{\partial}{\partial\nu}\left(\nu^4\frac{\partial n}{\partial\nu}\right). \tag{12.80}$$

Here we have defined the dimensionless **Compton y-parameter**,

$$\boxed{y \equiv \int \sigma_{\mathrm{T}} n_e \frac{kT}{m_e c^2}\, d\ell,} \tag{12.81}$$

and used the fact that, provided $T_{\mathrm{gas}} \gg T_{\mathrm{rad}}$ (which is certainly true here), the last term inside the brackets in the Kompaneets equation will dominate. Thus, inserting the black body radiation law and defining $x \equiv \hbar\omega/(kT_{\mathrm{rad}})$, the observed effect is

$$\boxed{\frac{\delta I_\nu}{I_\nu} = \frac{\delta n}{n} = -y\frac{xe^x}{e^x - 1}\left[4 - x\coth\left(\frac{x}{2}\right)\right].} \tag{12.82}$$

This is the frequency-dependent **Sunyaev–Zeldovich effect** (SZ effect).

At low frequencies, in the Rayleigh–Jeans tail, the effect is a fractional cooling of the background by $\delta I/I = \delta T/T = -2y$. It should be clear from the foregoing derivation that this effect is due not to the mean shift in photon energy alone, but to a combination of the shift and the convolving effect of the scattering. At high frequencies, beyond the peak of the CMB spectrum, the SZ effect changes sign, offering a clear observational signature. The low-frequency effect has been detected in a number of clusters (e.g. Jones *et al.* 1993), and is a valuable diagnostic for two reasons. First, being proportional to n_e, it constrains the possible effect of clumping on X-ray emission, which scales as n_e^2. Second, the method has a different dependence on h ($n_e \propto h$ as opposed to $n_e \propto h^{1/2}$ from X-rays); see chapter 5.

CLUSTER MASSES FROM GRAVITATIONAL LENSING In principle, the simplest method of measuring cluster masses is from gravitational deflection of light, using either luminous arcs generated from a single galaxy, or from weak-lensing distortions averaged over a number of galaxies (see chapter 4). How well do these estimates agree with the above methods? A number of attempts have been made to carry out this comparison, with slightly inconsistent results. There have been claims that lensing results give masses that are larger than those from X-rays, by about a factor of 2 (e.g. Miralda-Escudé & Babul 1995). If this turned out to be a general phenomenon, it would require some means of supporting the gas in addition to its thermal pressure, and magnetic fields would be a possibility. A more plausible explanation, however, is that the modelling of the X-ray gas may need to be more sophisticated: in a cluster with cooling flows, there will be a **multiphase IGM**: a mix of temperatures and densities at a given radius. With good spectral data, this can be allowed for, yielding consistent lensing and X-ray masses in cluster cores (Allen, Fabian & Kneib 1996).

On scales beyond 100 kpc, where masses can be probed by weak lensing distortions rather than the strongly lensed luminous arcs, the gas should be better behaved and there is no evidence for any discrepancy between the lensing and X-ray masses (e.g. Squires *et al.* 1996). The lensing results thus so far appear to fall in with the canonical blue $M/L \simeq 300h$ obtained from Coma. The real test will come if the weak lensing measurements can be used to push beyond radii of 1 Mpc, at which point the total cluster light begins to converge. Will the mass density drop also, or will we see the analogue of spiral galaxy rotation curves – i.e. a dark isothermal halo to the entire cluster? The experiment would require careful control of systematics but is possible in principle. If $\Omega = 1$ and biased galaxy formation are to be a reality, this is the outcome that might be predicted.

CLUSTER INFALL Tempting though it is to draw an analogy between dark haloes around galaxies and around clusters, the underlying assumption of relaxed isothermal self-gravitating equilibrium must be invalid in the outer parts of cluster haloes. For example, in the Coma cluster, we saw above that there is a well-determined mass of about $6 \times 10^{14} h^{-1}\, M_\odot$ within $r = 1h^{-1}$ Mpc. This corresponds to

$$\frac{\rho_{\rm Coma}}{\rho_{\rm background}}(r < 1h^{-1}\,{\rm Mpc}) \simeq \frac{530}{\Omega}, \tag{12.83}$$

which is already close to the limit for virialization if $\Omega = 1$. If $\Omega = 1$, the density contrast at which mass shells are just starting to fall into the cluster is 5.55 (see chapter 15). If we assume that the halo profile is close to $\rho \propto r^{-2}$, the predicted turnround radius is

$$r_{\rm turn}^{\rm Coma} \simeq 9.8\; h^{-1}\,{\rm Mpc} \simeq 8\ {\rm degrees}. \tag{12.84}$$

What this says is that, at radii beyond a few Mpc, structure in the outer parts of clusters of galaxies is only just starting to form today; we do not have available the analogue of dissipative processes that allow baryonic material to sink slowly into a stable dark-matter potential well. It is therefore difficult to see how clusters could develop baryon-free dark haloes, justifying the use of the baryon fraction in the centres of clusters as giving the universal value of this ratio.

This transition from coherent infall in the outer parts of clusters to virialized random orbits in the centres should give rise to characteristic features in the distribution of galaxy velocities with radius, allowing the turnround point to be identified and thus Ω to be measured. Such an analysis was performed by Regős & Geller (1989), who claimed that the result was in disagreement with $\Omega = 1$ but in accord with $\Omega \simeq 0.2$. The difference between models is in fact quite subtle, but there is undoubtedly an effect here of potential power. The general idea of using the large-scale peculiar velocity field as a means of measuring Ω has become an important sub-field in cosmology (see chapter 16).

12.4 Baryonic dark matter

Whether or not one takes seriously a factor 2 discrepancy in the Oort limit, dark matter does exist in the outer parts of galaxies. The simplest hypothesis about this is that it is baryonic, in which case the question arises why this matter is not incorporated into stars.

THE IMF Our estimate of the density in stars does not count all objects; it is cut off at some lower luminosity. Whether this matters depends on the mass distribution function, usually referred to as the **initial mass function** or IMF ('initial' since the mass distribution is set during the process of star formation). This is sometimes crudely taken to be a power law with a number density

$$\frac{dn}{d\ln m} \propto m^{-x}. \tag{12.85}$$

An early determination (Salpeter 1955) yielded the **Salpeter IMF**: $x = 1.35$. Modern studies yield a more complex shape (Miller & Scalo 1979; Scalo 1986), but a simple power law will do to illustrate the point. If we integrate to get the total density, $\rho = \int_{m_c}^{\infty} m\, dn$, then to achieve a factor 2 change in density, we need a factor 7.2 change in m_c for $x = 1.35$ or a factor 2 change for $x = 2$. So, reasonable amounts of dark matter can be produced if there are low-mass bodies that we do not count as stars.

THE NUCLEAR BURNING LIMIT How do stars come to shine? If they start life as collapsing gas clouds, they will radiate energy ('cool') on the **Kelvin timescale**, which comes from assuming that the body must radiate away its binding energy:

$$L \sim \frac{\partial}{\partial t}\left(\frac{GM^2}{R}\right) \quad \Rightarrow \quad t_{\rm K}\left(\equiv \frac{R}{|\dot{R}|}\right) \sim \frac{GM^2}{RL}. \tag{12.86}$$

For the Sun, this gives about 3×10^7 years, showing the need for a non-gravitational source of energy. The collapse will cause the star to heat up, as may be seen from the virial theorem: (potential energy) + 2× (kinetic energy) = 0 [problem 17.2]. In terms of the number of particles N and the mean temperature T this gives

$$T = \frac{\alpha GM^2}{3NkR} \propto \frac{M}{R} \tag{12.87}$$

(where α is a constant of order unity; $\alpha = 3/5$ for a uniform-density sphere). So, normally collapse will cause the temperature to rise until nuclear burning can begin. The only way this can be prevented from happening is if electron degeneracy sets in first and halts the collapse. The condition for the onset of degeneracy is just that the interparticle spacing becomes small enough for the uncertainty principle to matter:

$$n^{-1/3} p \lesssim \hbar \quad \Rightarrow \quad \text{degeneracy}, \tag{12.88}$$

where n is the electron number density and p the momentum. We can roughly estimate the condition for this to occur by assuming the body to be of uniform density and temperature. For a given mass, this yields an estimate of the maximum temperature that can be attained before degeneracy becomes important:

$$T_{\rm max} \simeq 6 \times 10^8 \left(\frac{M}{M_\odot}\right)^{4/3} \ {\rm K}. \tag{12.89}$$

Since fusion reaction rates drop exponentially for $T \lesssim 10^6$ K [problem 12.4], a minimum mass for normal stars on the range 0.1–0.01 $M_\odot$ might be expected. The accepted figure for this limit when realistic calculations are performed is in fact $0.08 \pm 0.01 M_\odot$. Objects much less massive than this will generate energy only gravitationally, and will therefore cool to a state of virtual invisibility on the Kelvin timescale. Such objects constitute one of the holy grails of stellar astrophysics, and are known as **brown dwarfs**. The search for these objects has been somewhat controversial, since it is very hard to measure the masses of faint cool stars directly. Systems with surface temperatures between 1000 and 2000 K have spectra that are heavily affected by broad molecular absorption bands, which are very hard to model (see Burrows & Liebert 1993). Despite these uncertainties, some objects are known where the consensus is that the mass is below the nuclear burning limit (e.g. Marley *et al.* 1996).

However, the mass–luminosity relation is sufficiently uncertain in this regime that the low-mass part of the mass function remains weakly constrained even near the Sun, let alone at the sort of distances where large amounts of dark matter are required by the dynamics (Bessell & Stringfellow 1993). The best limit on the possible contribution of low-mass objects comes from gravitational microlensing results (see chapter 4). At present, the results are somewhat confusing. The microlensing rate is consistent with the assumption that roughly half the dynamical mass of the Milky Way is in the form of discrete objects, with a preferred mass around $0.5M_\odot$ (Alcock *et al.* 1996). It appears that objects below the burning limit must give a contribution at the $\lesssim 20\%$ level, rather than constituting a dominant fraction of the dark matter.

THE FRAGMENTATION LIMIT In any case, is it possible to form a brown dwarf? A cloud will collapse under its own gravity if it is above the **Jeans mass**

$$M_J \equiv \rho\lambda_J^3 = \left(\frac{\gamma\pi kT}{G\mu m_H}\right)^{3/2} \rho^{-1/2} \tag{12.90}$$

where $\lambda_S = c_S\sqrt{\pi/(G\rho)}$ gives the length at which the sound-crossing time is equal to the **collapse time**

$$t_{\rm coll} \equiv (G\rho)^{-1/2} \tag{12.91}$$

(see chapter 15 for more details on the Jeans criterion). In practical units, the Jeans mass is

$$M_J = 10^{5.3}(T/\mathrm{K})^{3/2}(n/\mathrm{m}^{-3})^{-1/2}\ M_\odot. \tag{12.92}$$

So, to achieve a Jeans mass of 0.01 $M_\odot$ needs a number density $n \gtrsim 10^{16}\ \mathrm{m}^{-3}$, since the temperature cannot be < 2.7 K. This compares rather poorly with the densest parts of molecular clouds ($n \sim 10^{11}$). The only way to form low-mass objects is via **fragmentation**: allow a massive object to collapse until the density is high enough that internal clumps can grow.

However, in order to achieve much in the way of collapse, there is an energy barrier to be surmounted. Collapse will heat the gas cloud; if no energy can escape, the usual adiabatic $T(\rho)$ law gives $M_J \propto \rho^{3\gamma/2-2}$. So, after the radius has fallen by a factor of order unity, pressure effects will halt the collapse unless the cloud can radiate away the excess heat (a process known as **cooling**, which is also of crucial importance in galaxy formation – see chapter 17). Moreover, the cooling must be *fast*: it is not enough for the entire cloud to cool quasistatically, with pressure always being important, since substructure will never grow in this case. Fragmentation will only take place if the cloud can cool on the **collapse timescale**: $t_{\rm coll} = (G\rho)^{-1/2}$. This allows one to deduce a bound on the mass of fragments that can form, as follows.

The luminosity must be sufficient to account for the rate of increase of binding energy, i.e.

$$L \sim \left(\frac{GM^2}{r}\right) \Big/ (G\rho)^{-1/2}. \tag{12.93}$$

However, the cloud cannot radiate faster than a black body. If the energy density of thermal radiation is $\epsilon = aT^4$, where a is the Stefan–Boltzmann constant, then

$$L \lesssim 4\pi r^2 \frac{ac}{4} T^4. \tag{12.94}$$

Inserting the above expression for L into this inequality, and using the virial theorem $kT = GM^2/(3Nr)$ (valid for deducing the limiting mass, since cooling will only just allow collapse there), we can obtain the **fragmentation limit**:

$$M \gtrsim 6\mu^{-2} \left(\frac{kT}{\mu m_p c^2}\right)^{1/4} \left(\frac{\hbar c}{G m_p^2}\right)^{3/2} m_p \tag{12.95}$$

(where $\mu m_p = M/N$ is the mean mass per particle). The last factor is the same combination as that arising in the case of white dwarfs, which are stars supported by electron degeneracy pressure. For these, there is a maximum possible mass at the **Chandrasekhar limit** (see chapter 2): $(\hbar c/Gm_p^2)^{3/2} m_p = 1.863 M_\odot$; we therefore see that the fragmentation limit is clearly going to be of order the Solar mass. There is some temperature dependence, but only very slow: a 'typical' $T = 100$ K has to be correct to within a factor ~ 3 in $T^{1/4}$. This finally gives us $M \gtrsim 0.01 M_\odot$, suggesting there may be something like a decade of mass below the hydrogen-burning limit in which brown dwarfs may form. It has to be said that making this number more precise is difficult and there is no consensus about how easily fragments smaller than $0.08 M_\odot$ can form. Some clearly do, since planets exist, but these are of little interest from the point of view of baryonic dark matter; we are looking for a way to make 1000 or more Jupiter-sized bodies for every Solar-mass star, and it remains uncertain whether this is possible.

12.5 Nonbaryonic dark matter

Although baryonic dark matter has its attractions, the balance of cosmological evidence favours dark matter that consists of weakly interacting relic particles. There are a number of arguments in favour of this conclusion, but the strongest comes from primordial nucleosynthesis, which estimates a baryonic contribution to the density parameter of $\Omega_B \simeq 0.0125 h^{-2}$. Explicitly, this is the contribution of the protons and neutrons that interacted to fix the light-element abundances at $t \sim 1$ minute. The uncertainties in this number were discussed in chapter 9, and one might contemplate a figure at most twice this, giving $\Omega_B \simeq 0.06$ for $h = 0.65$. Since there is ample evidence that the total Ω is at least 0.2, one is apparently forced towards a form of matter other than that weighed by the nuclear reactions.

In conventional cosmology, the matter content of the universe at nucleosynthesis consisted of baryons (plus electrons; these are included under the heading of 'baryonic material'), photons and three species of neutrinos. There are therefore two distinct ways forward: either the neutrinos have mass, or there must exist some additional particle species that is a frozen-out relic from an earlier stage of the big bang. A small mass for neutrinos would not affect nucleosynthesis, as the neutrinos would be ultrarelativistic at this time (prior to matter–radiation equality). Other relic particles would have to be either very rare or extremely weakly coupled (i.e. more weakly than neutrinos) in order not to affect nucleosynthesis. Either alternative would yield a form of dark matter that is collisionless, and this is the other main argument for nonbaryonic dark matter: the clustering power spectrum appears to be free of the oscillatory features expected from the gravitational growth of perturbations in matter that is able to support sound waves (see chapter 16).

Although perhaps the most natural species of nonbaryonic dark matter to consider, the massive neutrino is far from being the only candidate for a relic particle that could close the universe, and chapter 8 has considered other plausible alternatives that lie outside the standard model: the axion and various supersymmetric particles. A common collective term for all these particles is **WIMP**, standing for weakly interacting massive particle. The inevitable counter to this from the supporters of baryonic dark matter was **MACHO**, massive astrophysical compact halo object. Although this joke is is some danger of outstaying its welcome, these terms seem likely to persist.

The most obvious distinction in WIMP astrophysics is the one made above between neutrinos and the rest, but there are really three generic types to consider, as follows.

HOT DARK MATTER (HDM) These are particles that decouple when relativistic, and which have a number density roughly equal to that of photons; eV-mass neutrinos are clearly the archetype here. Why 'hot'? Consider the average energy for massless thermal quanta:

$$\langle E \rangle = \begin{cases} 2.701kT & \text{(bosons)} \\ 3.151kT & \text{(fermions).} \end{cases} \tag{12.96}$$

For a massless particle, $\langle E \rangle$ is related to the mean momentum via $\langle E \rangle = \langle |p| \rangle c$. As the universe expands and the particles cool and go nonrelativistic, the momentum is redshifted, eventually becoming just $m|v|$. At late times, the neutrino momentum distribution still has the zero-mass form, so we should use the above relations with $E \to m|v|c$. This gives a velocity

$$\langle |v| \rangle = 158(m/\mathrm{eV})^{-1}\,\mathrm{km\,s^{-1}} \tag{12.97}$$

for neutrinos with $T = 1.95$ K, so low-mass relics are indeed hot in the sense of possessing large quasi-thermal velocities. These velocities were larger at high redshifts, leading to major effects on the development of self-gravitating structures.

WARM DARK MATTER (WDM) To reduce the present-day velocity while retaining particles that decouple when relativistic, the present density (and hence temperature) relative to photons must be reduced. This is possible if the particle decouples sufficiently early, since the relative abundance of photons can then be boosted by annihilations other than just $e^{\pm}$. In grand unified theories there are of order 100 distinct particle species, so the critical mass can be boosted to around 1–10 keV. The most commonly encountered dark-matter candidates are not of this form, but it is easy enough to postulate the existence of a particle that would have these properties, even if making it an inevitable consequence of a particle-physics model is a little harder (e.g. Rajagopal, Turner & Wilczek 1991).

COLD DARK MATTER (CDM) If the relic particles decouple while they are nonrelativistic, the number density can be exponentially suppressed and so the mass can apparently be as large as is desired – and the thermal velocities effectively zero, justifying the name 'cold'. If decoupling occurs at very high redshift, the horizon scale at that time is very small and so negligible damping occurs through free streaming. Structure formation in a CDM universe is then a **hierarchical** process in which nonlinear structures grow via the merger of very small initial units. Now, if it is assumed that the CDM particle interacts via the weak interaction in the same way as a neutrino, then the abundance as a function of mass may be calculated following the route described in detail in chapter 9.

The essential features of this calculation are that for small masses we have just the standard massive-neutrino result: the relic density is the product of the conventional number density for massless neutrinos and the particle mass, so $\Omega \propto m$. A critical density thus requires a neutrino mass of order 100 eV, and much higher masses are ruled out:

$$\Omega_\nu h^2 = \frac{\sum m_i}{93.5\,\mathrm{eV}}, \tag{12.98}$$

where the sum is over different neutrino species. This state of affairs continues until masses of a few MeV ($\Omega \sim 10^4$), at which point the neutrinos cease to be relativistic at decoupling. In this mass regime, the relic comoving number density of neutrinos is reduced by annihilation. The essential features of the calculation may be seen in the nonrelativistic limit, where the number density of thermal particles is, using natural units,

$$n = g \left(\frac{mT}{2\pi}\right)^{3/2} e^{-m/T}. \tag{12.99}$$

It appears that the relic abundance should fall exponentially with mass m, but we now show that this is not so: in fact the relic density falls as m^{-2}.

To work out the proper effect of decoupling, we have to start with the criterion $\Gamma/H \sim 1$, where the reaction rate is $\Gamma = n\langle\sigma v\rangle$. The nonrelativistic cross-section can be obtained through a dimensional argument: $\sigma \sim G_{\mathrm{F}}^2 m^2$, where G_{F} is the Fermi constant (a weak cross-section must be proportional to G_{F}^2 for a second-order process). In chapter 9, we saw that the relativistic cross-section was $\sigma \sim G_{\mathrm{F}}^2 E^2$; replacing E by m is the only alternative with sensible dimensions, and can be justified more rigorously by using the arguments of chapter 9. If we ignore the slight deviation of v from c, and other factors of order unity, then the freeze-out criterion is

$$[(mT)^{3/2} e^{-m/T}]\,(G_{\mathrm{F}}^2 m^2)\,H^{-1} \sim 1. \tag{12.100}$$

In natural units, the Hubble parameter in the radiation era is $H = 1.66 g_*^{1/2} T^2 m_{\mathrm{P}}^{-1}$, where g_* is the effective number of bosonic degrees of freedom (see chapter 9), m_{P} is the Planck mass and $G_{\mathrm{F}} = (292.8\,\mathrm{GeV})^{-2}$. Taking $g_* = 100$, the equation to be solved is then

$$\left(\frac{m}{2.2\,\mathrm{MeV}}\right)^3 \left(\frac{m}{T}\right)^{1/2} e^{-m/T} \sim 1, \tag{12.101}$$

which shows that annihilations will start to reduce the neutrino abundance for masses above about 1 MeV. As m is increased above 1 Mev, it is very easy to satisfy this equation through small changes in T, because of the sensitivity of the exponential factor. Consequently, the freeze-out temperature is always close to 1 MeV, varying only logarithmically with mass.

Now consider the freeze-out density, $n \sim (mT)^{3/2} \exp(-m/T)$; the freeze-out equation says that $n \propto T^2/m^2$, and so the number density at freeze-out is very nearly independent of m. The resulting density is thus $\propto m$, which must be scaled by a factor $(T/2.73\,\mathrm{K})^3$ to obtain the present density. The final result is a scaling $\Omega_\nu \propto m^{-2}$, as

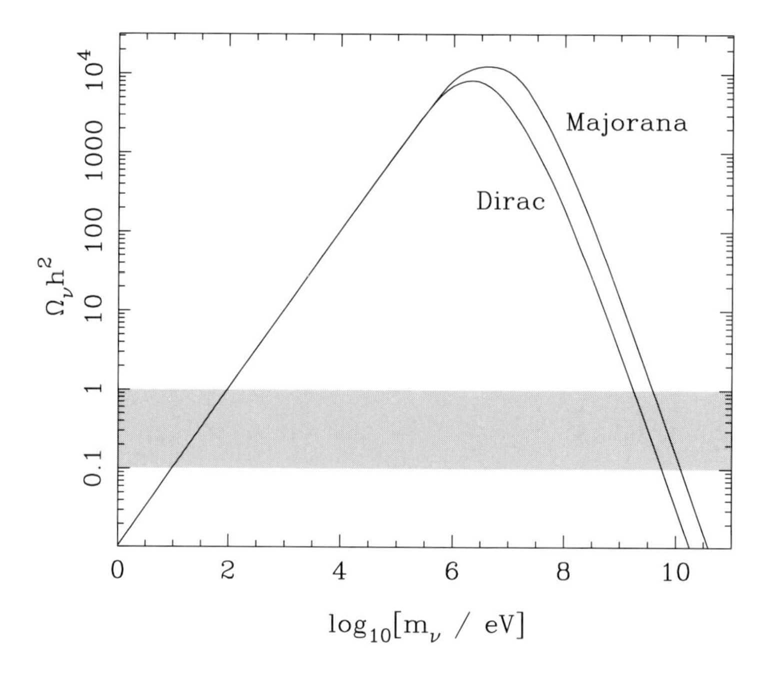

Figure 12.5. The contribution to the density parameter produced by relic neutrinos (or neutrino-like particles) as a function of their rest mass. At low masses, the neutrinos are highly relativistic when they decouple: their abundance takes the zero-mass value, and the density is just proportional to the mass. Above about 1 Mev, the neutrinos are nonrelativistic at decoupling, and their relic density is reduced by annihilation below the linear $\Omega h^2 \propto m$ relation. The total density in fact falls in absolute terms beyond this point, owing to the mass dependence of the interaction cross-section, and the universe would also be closed by particles of mass a few GeV.

promised. The exact curve depends on the order unity factors we have neglected, and in particular on whether the neutrinos are Majorana or Dirac, since the different number of spin states has a small effect on the annihilation cross-section (see section 5.2 of Kolb & Turner 1990 for details). The resulting $\Omega(m)$ plot is shown in figure 12.5. This shows that the universe may also be closed by massive neutrino-like particles with masses around 3 Gev. Lower masses are disallowed (the **Lee–Weinberg limit**; Lee & Weinberg 1977), until we reach the light-neutrino point around 30 eV. It is hard to change this calculation to allow MeV particles: the critical freeze-out temperature scales as $G_F^{-2/3}$, and a reduction of the temperature from 1 MeV to 30 eV (so that neutrinos never yield $\Omega > 1$) requires an effective G_F that is *larger* by a factor 100. Particle physicists are normally happy to invent new particles with very weak interactions, but not the opposite.

A problem for this calculation is that none of the known neutrinos can be as massive as 3 GeV; there would have to exist a new particle, perhaps associated with a fourth 'family' in particle physics. Such a possibility was ruled out in 1990 by LEP at

CERN through Z^0 decay experiments (see chapter 9). The decay rate of this particle (via $Z^0 \to \nu\bar{\nu}$) measures the total number of neutrino species, which was found to be three. Any extra neutrinos must exceed half the Z^0 mass, and so are not relevant here. The ways in which CDM can still be realized in practice are now twofold. The particle may have weaker interactions than a neutrino, which is possible with some of the supersymmetric candidates (in which case GeV-scale masses are still possible), and with the axion (although the mass is very small in this case, as explained in chapter 9). It is also possible that the $\Omega(m)$ curve may reverse once more at higher energies, since cross-sections alter beyond the electroweak scale at $\sim$ 100 Gev. This could conceivably allow neutrino-like WIMPs with masses $\sim$ 1 TeV.

A huge parameter space is therefore available for experimentalists to search in order to detect dark-matter particles. There is a very substantial effort in this area worldwide (see e.g. Smith & Lewin 1990; Spooner 1997), but the chances of success are very hard to estimate. In the end, however, the debate over the nature of dark matter is likely to continue until a laboratory detection is obtained.

PHASE-SPACE CONSTRAINTS In the absence of laboratory constraints, there is one important limit that can be set to the mass of any dark-matter particle, which comes from the fact that the evolution of the dark-matter distribution is collisionless. Unlike the case for baryonic material, there are no dissipative processes that can cause dark matter to radiate away energy and hence become more strongly bound. There is therefore a limit to the density of dark matter that can be accumulated in the centres of astronomical objects. By Liouville's theorem, the evolution of a collisionless system is such that the phase-space density is preserved, so this is the invariant we should use to calculate the degree of dark-matter collapse that is possible.

Suppose initially that the dark matter is HDM. This has a momentum-space distribution of thermal form:

$$dN = g\,\frac{V}{(2\pi\hbar)^3}\,[\exp(E/kT)+1]^{-1}\,d^3p, \tag{12.102}$$

so that the initial phase-space density has a maximum value at low energies, corresponding to an occupation number of $\bar{n} = 1/2$:

$$f_{\rm init} \leq (2\pi\hbar)^{-3} g/2. \tag{12.103}$$

Suppose that the velocity dispersion in an astronomical object has relaxed to a Maxwellian:

$$dp = (2\pi\sigma^2)^{-3/2} \exp(-v^2/2\sigma^2)\, d^3v, \tag{12.104}$$

so that the observed phase-space density now satisfies

$$f_{\rm final} \leq (\rho/m)\,(2\pi\sigma^2)^{-3/2}\, m^{-3}. \tag{12.105}$$

If there are N neutrino species, all of mass m, then the condition that the maximum phase-space density has not increased is the **Tremaine–Gunn limit** (1979):

$$\boxed{m^4 > \frac{\rho}{Ng}\left(\frac{\sqrt{2\pi}\,\hbar}{\sigma}\right)^3.} \tag{12.106}$$

In words, a potential well of given density and velocity dispersion can only trap massive particles.

To make a quantitative estimate of this limit, assume that the density structure is an isothermal sphere, so that $\rho = \sigma^2/(2\pi G r^2)$, and take the simple case $Ng = 1$. Neutrinos then have no difficulty fitting into a cluster of galaxies, since $\sigma = 1000\,\mathrm{km\,s^{-1}}$ and $r = 1$ Mpc implies $m > 1.5$ eV. However, changing the parameters to a typical massive galaxy halo ($\sigma = 150\,\mathrm{km\,s^{-1}}$ and $r = 5$ kpc) gives the much larger limit of $m > 33$ eV. This is close to the result required for a critical-density universe, and neutrinos clearly cannot form the dark matter in much smaller objects. However, there is ample evidence for dark matter even in dwarf galaxies (see Carr 1994), and so it is clear that HDM cannot provide a single solution to the dark-matter problem. Either the dark matter in small galaxies must be baryonic, or the dark-matter particle must be much more massive. It is easy to see how increasing the particle mass solves the problem: if we change the number density of the background particles to a factor F of the thermal density (i.e. boost the mass by a factor $1/F$, for fixed Ω), this corresponds to an effective degeneracy factor $g \to Fg$ in the above. The Tremaine–Gunn mass limit scales as $g^{-1/4}$, so the mass inferred from galaxies changes only slightly, but it can now be easily made compatible with $\Omega \lesssim 1$. The simplest solution is therefore to say that all dark matter is CDM. However, Moore (1994) claims that this would predict density profiles that are more concentrated than those inferred in dwarf galaxies. Since these objects are almost completely dominated by dark matter, this is a serious objection. Moore's preferred solution is keV-scale WDM as the dark matter, and this is clearly an issue worthy of further investigation.

Problems

(12.1) Derive the frequency-dependent electromagnetic absorption cross-section for a classical damped oscillator of mass m and charge e: $\sigma(\omega) = (\pi \mu_0 e^2 c/2m) dp/d\omega$, where $dp/d\omega$ is a Lorentzian probability distribution.

(12.2) Explain how to perform the redshift integration needed to obtain the Gunn–Peterson optical depth. If all absorption is provided by clouds with comoving number density $\propto (1+z)^\gamma$, how does the mean Gunn–Peterson optical depth scale with redshift, i.e. what is the suppression of the continuum by the smoothed-out effect of the Lyα forest as a function of redshift?

(12.3) Derive the mean and variance of the energy of monoenergetic photons that undergo a single Compton scattering by thermal electrons at temperature T, and hence obtain the Fokker–Planck equation for Comptonization. Why are there two terms of opposing sign in the equation for the mean energy?

(12.4) Fusion processes in stars will only proceed if charged particles can approach closely enough ($\sim 10^{-15}$ m) for nuclear forces to be important. At what temperature would typical thermal protons have sufficient energy to overcome their mutual Coulomb repulsion and attain this separation? Show that quantum tunnelling is needed in practice, and estimate the fusion rate as a function of temperature.

13 Galaxies and their evolution

13.1 The galaxy population

GALAXY TYPES For the optical astronomer, the most striking feature of the universe is the fact that stars appear in the discrete groups known as galaxies. There is no danger that the identification of galaxies is a subjective process akin to the grouping of stars in the Milky Way into constellations: the typical distances between galaxies are of order Mpc, and yet their characteristic sizes are a few kpc. Galaxies are thus concentrations of $\gtrsim 10^8$ times the mean stellar density. Why matter in the universe should be organised around such clear characteristic units is one of the most outstanding cosmological questions.

Galaxies come in several clear types, and it is a challenge to account for their distinct properties. At the crudest level, galaxies form a two-parameter family in which the characteristics that vary are the total amount of light and how this is divided between two components, the bulge and the disk.

(1) The **bulge**. This dominates the central portions of galaxies and is distinguished by its stellar populations and dynamics. The component is close to spherically symmetric and, although it rotates in general, it is supported against gravity primarily through stellar 'pressure': the bulge stars have a large radial component to their orbital velocities. The stars are **population II**: a set of stars that have aged sufficiently that short-lifetime stars more massive than the Sun have departed from the main sequence, leaving light that is dominated by the contribution of the giant branch. The other distinguishing feature of this old population is that the stars are of low metallicity by comparison with the Sun.

(2) The **disk**. This is a flattened structure in which the stars move in very nearly circular orbits. The disk is associated with active star formation (young massive stars) and its associated debris (interstellar dust and molecular clouds): **population I** stars. In the Milky Way, we inhabit the disk, and the Sun is a population I star.

The family of galaxies formed by adding bulges and disks in different proportions is called the **Hubble sequence** (see figure 13.1). At one extreme, it consists of **elliptical galaxies**: those with no discernible disks. Elliptical galaxies are usually given a numerical classification that describes the axial ratio of their **isophotes** (contours of constant surface brightness, which are indeed very nearly elliptical on the sky): E1 to E6, with the latter denoting a major-to-minor axis ratio of 6. Massive galaxies are divided between ellipticals and the other major class, **spiral galaxies**. These are named for the beautiful

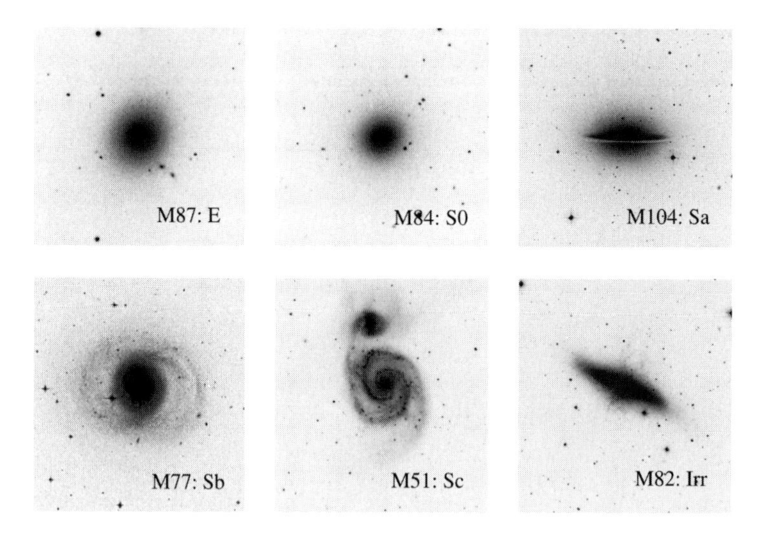

Figure 13.1. Examples of the different types of galaxy, arranged according to the Hubble sequence. As we move from elliptical galaxies to the later spirals and irregulars, this is a progression towards increasing dominance of an old spheroidal component by flattened disks with active star formation.

spiral patterns often evident in their disks, and are divided into the three classes Sa, Sb, Sc – in order of increasing prominence of both the disk and the spiral arms. Between the two lies the transitional class of S0 or **lenticular galaxies**, which possess a disk but no spiral arms. Distinguishing these from true ellipticals can be subjective: many catalogued ellipticals show evidence for a disk if you look hard enough.

This simple sequence of varying bulge-to-disk ratio has spawned a variety of historical terminology. Because Hubble originally thought of his sequence as being an evolutionary path, ellipticals and S0's can be called **early-type galaxies**, and spirals **late-type galaxies**. One can even speak of early-type spirals (i.e. Sa galaxies). The bulge component is sometimes called the **spheroid** (especially in the Milky Way), and so ellipticals may be referred to as 'spheroidals'. Unfortunately, population I and II stars are also called early and late spectral types, so that early-type galaxies are dominated by late-type stars and vice versa.

Aside from their flattening, the bulge and disk components can be distinguished by the radial dependence of their surface brightness (or the dependence of surface brightness on major axis, in the case of elliptical isophotes). The disks are usually well described by a simple exponential law

$$I(r) = I_0 \exp(-r/r_0), \tag{13.1}$$

whereas ellipticals have a brightness distribution that is more extended and does not cut off so abruptly as an exponential disk. A simple analytical law to consider as the next step from an exponential profile is to take $I \propto \exp(-r^\alpha)$ instead of assuming $\alpha = 1$. This in fact works very well, and the version with $\alpha = 0.25$ (although this is not always the best value) has become known as the $\mathbf{r^{1/4}}$ **law**, or **de Vaucouleurs' law**. It is more commonly written in terms of the radius r_e that encloses half the light, and the surface brightness at that point:

$$I(r) = I_e \exp\left\{-7.67\left[(r/r_e)^{1/4} - 1\right]\right\}. \tag{13.2}$$

The total luminosity of such a model converges when integrated to large r:

$$L_{\rm tot} \simeq 22.7\, I_e r_e^2. \tag{13.3}$$

The numbers r_0 and r_e thus give a reasonable definition of the size of a galaxy, and typically come out at a few kpc to tens of kpc. In principle, the sizes of galaxies become much larger than this if defined at a faint enough isophote, but this is rarely done. The main practical restriction is that the typical characteristic surface brightnesses I_0 and I_e do not greatly exceed the surface brightness of the night sky. Mapping galaxies to levels well below this foreground is an observationally challenging task. Furthermore, there is a serious worry that whole classes of galaxies may be missed altogether if their surface brightnesses are sufficiently low. The extent to which this is a problem is a controversy that has rumbled on over the years; see Disney (1976). Certainly some examples of quite massive galaxies with very low surface brightnesses exist; they are probably objects that have not yet processed much of a disk of neutral hydrogen into stars (e.g. Impey & Bothun 1997).

The galaxies that do not fit easily into the above idealized scheme are the **irregulars**, denoted by the Hubble-type shorthand Irr. These are generally low-mass systems such as the Magellanic Clouds, and are often distinguished by active star formation, being closely related to **blue compact galaxies**. It is this aspect in addition to a lack of a well-defined morphology that distinguishes irregulars from other **dwarf galaxies**; dwarf spirals or ellipticals exist that overlap irregulars in luminosity. The irregulars are also distinct in surface brightness: this property decreases with luminosity for ellipticals, whereas for disks it increases. The galaxies of lowest luminosity are thus markedly bimodal in surface brightness.

COLOURS AND METALLICITIES Apart from morphological appearance, the different Hubble types are readily distinguished by means of their spectral energy distributions. Elliptical galaxies generally show spectra that resemble those of a relatively cool star, i.e. an approximately 5000 K black body heavily modified by metallic absorption features. This indicates an age of several Gyr since the last major episode of star formation, so that the main-sequence turnoff point has moved to this temperature. The optical spectrum alone does not tell us whether the light is dominated by those dwarf stars that remain on the main sequence, or by the giant stars that have recently moved off it, since both have similar temperatures. The issue was only settled in favour of the dominance of giants by observations of the characteristic 2.3-μm CO absorption (Frogel *et al.* 1978).

Spirals, conversely, have experienced significant star formation within the past Gyr, and so contain young stars. These hotter stars contain fewer significant absorption features; in the limit of the very actively star-forming Irr galaxies, the overall spectrum is close to flat ($f_\nu \sim$ constant). The presence of hot stars also produces significant amounts of ionizing radiation, generating **HII regions** around the hottest stars. The spectra of late-type galaxies therefore also contain the strong nebular narrow emission lines: most notably Hα at 5653 Å and the OIII lines at 4959 Å and 5007 Å.

Disks and spheroids also differ in their metal content. Disks are actively forming stars today, from gas that is already enriched in metals. The supernovae that they add

Table 13.1. Solar abundance ratios

Element	$\log_{10}(n/n_{\rm H})$
He	−1.01
C	−3.40
N	−4.00
O	−3.07
Ne	−3.91
Mg	−4.42
Si	−4.45
Fe	−4.33

(type II, associated with young massive stars) further enrich the intergalactic medium. Population I stars such as the Sun therefore have high metallicities by comparison with the older stellar populations in spheroids and globular clusters. Various notations exist to quantify metallicity. The symbols X, Y, Z are used to denote the fractions by mass of respectively H, He and all other elements ('metals'). It is also common to quote abundances relative to those in the Solar neighbourhood, for which we have

$$\boxed{\begin{aligned} X_\odot &= 0.70 \\ Y_\odot &= 0.28 \\ Z_\odot &= 0.02 \end{aligned}} \tag{13.4}$$

(e.g. Pagel 1994). It is convenient to have the detailed abundance ratios of other elements, since these do not always follow the Solar pattern, which is given in table 13.1 for some of the most abundant elements (see Grevesse & Anders 1991; Wilson & Rood 1994).

Abundances relative to Solar are given in a bracket notation:

$$[{\rm A/H}] \equiv \log_{10}\left(\frac{n_{\rm A}/n_{\rm H}}{n_{\rm A}^\odot/n_{\rm H}^\odot}\right) \tag{13.5}$$

etc.; because this abundance measure is defined as a ratio to the Solar value, it does not matter whether one deals with ratios of number densities or mass densities.

Abundances in galaxies are measured in practice via a number of absorption indices, which have to be calibrated against synthetic spectra in order to yield abundances (Worthey, Faber & González 1992). This exercise yields important insight into the star-formation history of ellipticals, since it is found that the abundances of the lighter metals relative to iron in ellipticals and bulges are much higher than Solar, typically $[{\rm Mg/Fe}] \simeq 0.2 - 0.5$. This selective enrichment of lighter metals is characteristic of type II supernovae, showing that massive stars were important in the early phases of elliptical evolution. Overall, the metallicity of galaxies tends to increase with galactic mass and with decreasing radius (Dressler 1984; Sandage & Visvanathan 1978), causing galaxies to be redder in the centre. The reasons why deeper potential wells should contain more metals are discussed in chapter 17.

DYNAMICAL PROPERTIES As discussed in chapter 5, further clear systematic properties of galaxies reveal themselves when velocity information is available in the form of rotational velocities in disks or velocity dispersions in bulge and elliptical components. The latter show a correlation between total luminosity, surface brightness and velocity dispersion in the form of the **fundamental plane**:

$$L \propto I_0^x \, \sigma_v^y, \tag{13.6}$$

with $(x, y) \simeq (-0.7, 3)$ (Dressler *et al.* 1987; Djorgovski & Davis 1987; Dressler 1987). It is shown in chapter 5 that such a relation arises inevitably for galaxies that are self-gravitating systems with roughly constant mass-to-light ratios.

Of probably deeper significance is the fact that only a relatively narrow locus on this plane is occupied: in the absence of surface-brightness information, there is still a relation between luminosity and surface brightness:

$$L \propto \sigma_v^\alpha, \tag{13.7}$$

with $\alpha \simeq 3$–4, as shown by Faber & Jackson (1976). There is an rms scatter of about 1 magnitude about this mean relation, as opposed to at most 0.4 magnitudes for the fundamental plane. A similar relation holds between luminosity and rotational velocity for spiral disks, in the form of the **Tully–Fisher relation**. This dynamical similarity of spheroids and disks may not seem so surprising, but what is noteworthy is that the Tully–Fisher relation is much the more well defined, with a scatter similar to that about the fundamental plane. The origin of such a tight relation for spiral disks is still not very well understood, although the similarity to the Faber–Jackson relation argues that both are controlled by the formation of the dark-matter haloes that define galactic potential wells (e.g. Persic & Salucci 1995).

MORPHOLOGICAL SEGREGATION Astronomers are forever examining the observational properties of galaxies in search of clues about how they may have formed. All too often, any such evidence is subtle; but there are a few outstanding trends such as **morphological segregation**. In essence, this principle states that early-type galaxies are far more common than spirals in the centres of rich clusters, whereas the opposite is true in the remainder of the universe (Dressler 1980; see figure 13.2). The effect seems to operate at a point where the density of galaxies is a few hundred times the mean, although it is not as simple as a universal function of density: there are clusters of similar richness that are (relatively) spiral-rich or spiral-poor. In terms of the two-component description of galaxies, what this seems to say is that disks are unable to survive in the environment of rich clusters of galaxies. More weight is given to this suspicion when we add the fact that rich clusters are found to possess a mass exceeding that of stars in the form of **intracluster gas**, which is detected through the X-rays generated by the bremsstrahlung process (see chapter 10). Some of this is very likely to be gas that might in other environments have formed itself into stars in spiral disks.

GALAXY GROUPS AND CLUSTERS By comparison with the existence of galaxies themselves, the distribution of matter on larger scales betrays far fewer obvious features. Once it became generally agreed in the 1920s that 'spiral nebulae' were external stellar systems rather than gaseous objects in the Milky Way (the crucial evidence being Hubble's discovery of Cepheid variable stars in M31), it was immediately clear that galaxies were not distributed at random. Like the planets, many galaxies have small satellite neighbours (the Large and Small Magellanic Clouds fulfil this role for the Milky Way). On a larger scale, it was possible to recognize the general existence of a tendency

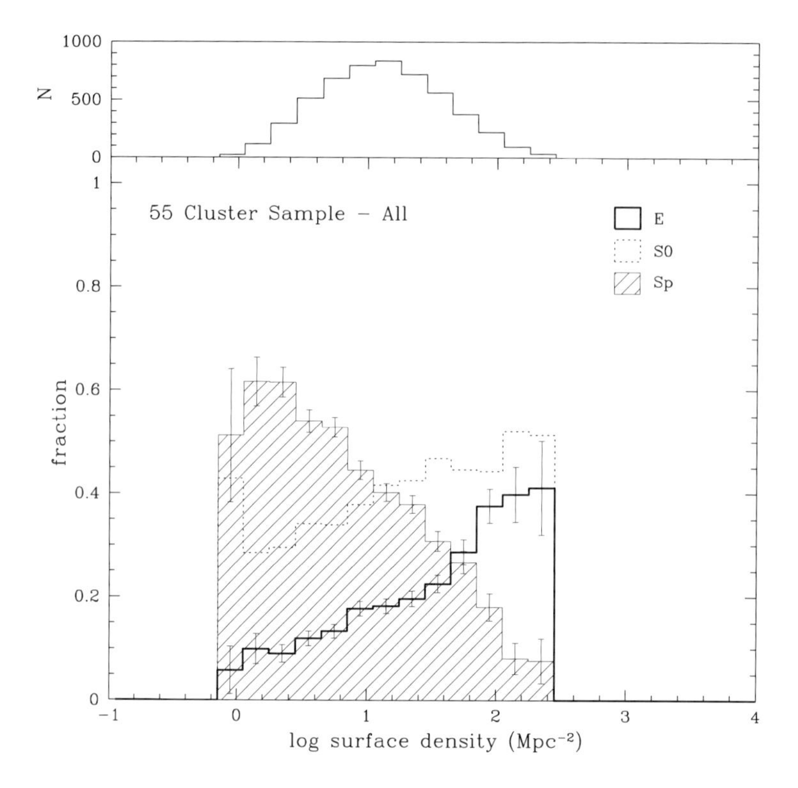

Figure 13.2. The morphology–density relation: a plot of the relative numbers of galaxies of different Hubble type against (projected) density enhancement (an updated version of the relation of Dressler 1980, from Dressler *et al.* 1997). The tendency for spirals to be depleted in clusters relative to ellipticals is one of the clearest systematic trends of the galaxy population.

for galaxies to group together. This discovery too goes back to Hubble, who showed that the 2D **cell counts** (the numbers of galaxies counted per box when a grid is imposed on the sky) obeyed a lognormal distribution with a width greater than would be expected from the simple Poisson fluctuations of galaxies placed at random on the sky.

As measurements of redshifts and imaging surveys progressed, the existence of groups of galaxies of different size became established. The Milky Way is part of the **Local Group**, which is dominated by the Milky Way plus the Andromeda Nebula M31, and includes roughly a dozen smaller galaxies. At the other extreme lie the **rich clusters** such as those in Virgo or Coma, where hundreds of galaxies are gathered within a few Mpc of each other. Morphological segregation gives an extra visual clue about the location of such a concentration of galaxies over and above a simple large excess number of galaxies at a given point on the sky. Even a few elliptical galaxies can draw attention to a particular location very effectively. This means that the first astronomers to have access to the first extensive deep photographic surveys of the sky were able to identify a large population of rich clusters, even though the task is complicated by projection effects: the desire is to find regions of high density in three dimensions, not those where several weaker clumps line up in the same direction on the sky. Two major catalogues were made, based on the survey of the Northern sky using the Palomar Schmidt telescope; this was carried out in the mid-1950s. Fritz Zwicky produced the larger catalogue, but the work of George Abell has been much more influential; Abell clusters remain important objects of cosmological study today, and Abell's approach still influences the new generation of clusters selected by digital means (Dalton *et al.* 1992; Lumsden *et al.* 1992).

The key to Abell's success was that he was able to measure the richness of clusters in a manner that was independent of distance. In principle this is easy once the redshift of some cluster candidate is known. Pick a standard physical radius (the **Abell radius** was set at $1.5h^{-1}$ Mpc) and count the number of galaxies that lie within this radius in projection, down to some fixed limit in luminosity. The difficulty lies in achieving this without time-consuming spectroscopy to determine the distance. Abell's approach was to assume that the luminosity function of cluster galaxies was universal, with a relatively sharp upper cutoff (as could be seen in nearby clusters). The brightest few galaxies in a cluster will then always be of approximately this maximum luminosity, and their apparent magnitudes can be used to estimate a distance. In practice, Abell used the magnitude of the tenth-brightest galaxy, m_{10}, so as to minimize confusion with any foreground galaxies. From this a redshift could be estimated, and hence the angle corresponding to the Abell radius. The number of galaxies in this zone between m_3 (the third-brightest galaxy) and $m_3 + 2$ were counted; once the mean number expected in a random area of this size was subtracted, this count gave the quantitative measure of the richness of the cluster. In practice, Abell divided the results into **richness classes**: $R = 0$ ($N = 30$–49); $R = 1$ ($N = 50$–79); $R = 2$ ($N \geq 80$). The $R = 0$ clusters were never intended to be a complete list, but the richer clusters were. Many of these turned out to have redshifts beyond 0.2, and the richer list was substantially complete to $z = 0.1$. Together with a later Southern counterpart, Abell's catalogues contain over 4000 clusters, and give a remarkably accurate picture of the local high peaks in the density field (Abell 1958; Abell, Corwin & Olowin 1989).

SUPERCLUSTERING The fact that rich clusters of many hundreds of luminous galaxies seemed to be the largest virialized structures in existence encouraged a picture in which there was no place for larger objects. Galaxy populations were divided into those in clusters and the remainder, known as **field galaxies**. This term goes back to Hubble, who presumed that the general population of predominantly spiral galaxies filled the space between clusters more or less uniformly, give or take some clumping into small groups. Some astronomers (de Vaucouleurs being a notable example) argued for the existence of larger-scale patterns in the galaxy distribution at a low level, suggesting in particular that both the local group and the Virgo cluster were embedded in a flattened **local supercluster**, which could be seen as a strip of enhanced galaxy density around an approximate great circle in the sky. This entity was also dubbed the **supergalaxy** by de Vaucouleurs; a name that still lingers on today in the terms **supergalactic plane** and **supergalactic coordinates**, which define a supergalactic plane normal to a Z-axis that points towards $(\ell, b) = (47.37°, 6.32°)$, and an X-axis in the plane towards $(\ell, b) = (137.37°, 0°)$. Here, ℓ and b are galactic longitude and latitude; see the appendix on useful formulae for conversions between galactic and normal **celestial coordinates**. The latter are a system of polar coordinates defined by the rotation axis of the Earth: if the usual polar angles are θ and ϕ, then declination δ is $\pi - \theta$ and right ascension α is ϕ.

The modern view, as expounded in chapters 15 and 16, is that all galaxies are in fact embedded in a overdense filamentary web of large-scale structure. Individual superclusters are difficult to define clearly: regions of around 30 times the mean density tend towards **percolation**, so that there is a single connected structure of galaxies. Much of the volume of the universe is almost entirely empty of galaxies, so that Hubble's idealized uniform field population does not really exist.

13.2 Optical and infrared observations

Before proceeding with quantitative studies of the galaxy population, it will be helpful to summarize some of the methods astronomers use to quantify the energy output from astronomical sources of radiation. Particularly in the area of optical/infrared astronomy, the accumulation of tradition and arcane convention can make the subject seem almost completely impenetrable to the novice.

DEFINITIONS AND UNITS The fundamental object on which all discussions of astronomical radiation are based is the **specific intensity**, I_ν. This measures the energy received per unit time, per unit area of detector, per unit frequency range and from unit solid angle of sky; its units are thus $\mathrm{W\,m^{-2}\,Hz^{-1}\,sr^{-1}}$. It is common enough to measure the total output integrated over the angular extent of the source (unavoidable for unresolved objects), in which case we have the **flux density** S_ν, whose units are $\mathrm{W\,m^{-2}\,Hz^{-1}}$. Because of the small numbers often generated in practice, a very common unit is that named after the radioastronomy pioneer Karl Jansky:

$$\boxed{1\,\mathrm{Jy} \equiv 10^{-26}\,\mathrm{W\,m^{-2}\,Hz^{-1}}.} \tag{13.8}$$

In optical astronomy, it is more common to use the symbol f_ν and the term **spectral energy distribution** instead of S_ν.

It is of course possible to use wavelength instead of frequency as a measure of bandwidth, and there may well be practical reasons for doing so in terms of optical spectrographs, which produce a nearly linear wavelength scale. We then use S_λ, where

$$S_\lambda = \frac{d|\nu|}{d|\lambda|}\,S_\nu = \frac{c}{\lambda^2}S_\nu. \tag{13.9}$$

This is a quantity whose simple SI unit ($\mathrm{W\,m^{-3}}$) ought to impress users of cgs units, who must grapple with monstrosities such as $\mathrm{ergs\,s^{-1}\,cm^{-2}\,\AA^{-1}}$. There is a trap here for the unwary, in that the wavelength of light depends on the refractive index of the medium in which it is propagating. Since frequency does not have this dependence, and because wavelength here is being used just as an alternative frequency scale, the sensible course would be to use vacuum wavelengths. However, for unfortunate historical reasons, most wavelengths referred to in the astronomical literature are as measured under standard laboratory conditions, where wavelengths are typically 0.03% shorter than in vacuum. This should not matter, because the lists of lines used in calibrating arc lamps and the tables of standard astronomical emission lines are usually in the same units, so redshifts are unaffected. However, the convention in the ultraviolet is different: vacuum wavelengths are used for $\lambda < 2000$ Å. Take two lines that are commonly encountered in cosmological studies, Lyα at 1216 Å and OIII at 3727 Å; in the first case, the corresponding frequency is c/λ but in the second this formula gives the wrong answer.

MAGNITUDES In principle, the frequency-dependent flux density is the only tool that is required, but real instruments commonly deal with bandwidths that are large by comparison with features in the spectrum of the source under study. The total energy measured is then the integral under the source flux times some frequency-dependent **effective filter response**. This last quantity includes all the factors that modify the energy arriving at the top of the Earth's atmosphere. The main factor is the instrumental filter, but there are several other important effects: atmospheric absorption; absorption in the telescope; frequency-dependent sensitivity of the detector. A notion of relative brightness can still be maintained by using the instrumental output, O, which is what leads to the notion of magnitude. The definition of magnitude is one of astronomy's unfortunate pieces of historical baggage, since it originates as a quantification of the ancient system where the brightest stars to the eye were 'first magnitude', and fainter stars were of larger magnitude. The eye is a logarithmic detector and the empirical constant of proportionality is 2.5, leading to

$$m = -2.5 \log_{10}\left(\frac{O_{\text{object}}}{O_0}\right). \tag{13.10}$$

A **magnitude system** is then defined by the total effective filter and by the **zero point** (the object of zero magnitude that produces output O_0). Different systems and their zero points are discussed below.

What is frustrating here is that the magnitude definition would have been so much simpler if only natural logarithms had been used instead. All too often it is necessary to remember corrections of a few per cent when converting between increments of magnitude and $\ln S$:

$$\begin{aligned} \Delta \ln S &= 0.92103\, \Delta m \\ \frac{d}{d \ln S} &= 1.08574\, \frac{d}{dm}. \end{aligned} \tag{13.11}$$

The necessity for the minus sign in the definition of magnitude is perhaps more understandable, but it continues to be a source of calculational errors and computer bugs.

ABSOLUTE MAGNITUDE AND K-CORRECTION Absolute magnitude is defined as the apparent magnitude that would be observed if the source lay at a distance of 10 pc; it is just a measure of luminosity. Absolute magnitudes in cosmology are affected by a shift of the spectrum in frequency; The K-correction accounts for this effect, giving the difference between the observed dimming with redshift for a fixed observing waveband, and that expected on bolometric grounds:

$$m = M + 5\log_{10}\left(\frac{D_{\rm L}}{10\ \text{pc}}\right) + K(z), \tag{13.12}$$

where $D_{\rm L}$ is luminosity distance. There is a significant danger of ending up with the wrong sign here: remember that $K(z)$ should be large and positive for a very red galaxy.

For a $\nu^{-\alpha}$ spectrum, recalling the formulae in chapter 3,

$$K(z) = 2.5(\alpha - 1)\log_{10}(1 + z). \tag{13.13}$$

It makes sense that $K(z)$ vanishes for $\alpha = 1$, because νf_ν is then a constant; the quantity $\lambda f_\lambda = \nu f_\nu$ is the flux per $\ln \nu$ and the effect of redshift is merely to translate this in log space. This argument shows how the K-correction should be evaluated for a general spectrum: if we define the logarithmic flux density as $f_{\log} = \lambda f_\lambda$, then we just need to compare the results of multiplying by the effective filter transmission function $T(\lambda)$ and then integrating $d \ln \lambda$, with and without redshift:

$$10^{0.4K(z)} = \frac{\int T(\lambda)\, f_{\log}(\lambda)\, d\ln\lambda}{\int T(\lambda)\, f_{\log}(\lambda/[1+z])\, d\ln\lambda}. \tag{13.14}$$

In this expression, $f_{\log}(\lambda)$ is the observed variation of $f_{\log}$ with wavelength; $f_{\log}(\lambda/[1+z])$ is the observed function shifted to the emitted wavelength scale.

AIRMASS AND COLOUR TERMS To illustrate the use of different magnitude systems, it may be helpful to summarize the procedure used in practice for **photometry**, the measuring of magnitudes. The observer will be equipped with a piece of equipment that returns a signal that can be converted via the $-2.5\log_{10}$ operator to an **instrumental magnitude** $m_{\rm I}$. Apart from targets of unknown magnitude, the observer will also possess a list of **standard stars**: objects of accurately measured magnitude, in some standard system whose response is close to that of the instrument to hand. These will be observed through varying amounts of air, depending on the angle from the zenith, θ. If the atmosphere is plane parallel, the total amount of material will depend on the **airmass**:

$$\text{airmass} = \frac{1}{\cos\theta}. \tag{13.15}$$

The standard procedure is to fit a linear function to the difference between standard and instrumental magnitudes:

$$m_{\rm S} - m_{\rm I} = \alpha + \beta \times \text{airmass} + \gamma \times \text{colour}. \tag{13.16}$$

The **instrumental zero point** α is a measure of the sensitivity of the equipment; β allows for absorption by the atmosphere. The third term on the rhs expresses a pious hope that the response function of the equipment is close to that used to define the standards. Since there will never be a perfect match, the exact flux measured will depend on the steepness of the continuum, which can usually be quantified by a colour measure in the relevant part of the spectrum (e.g. $B - V$ or $V - R$ for V photometry). Provided the targets have a smooth star-like spectrum (not always true), this procedure will often work well – although it requires great care in practice to obtain residuals smaller than 0.01 magnitude.

Having fitted the standards, this equation can be used in two distinct ways to provide magnitudes for non-standard objects. Either measured colours can be used to predict magnitudes of the objects in the system used to define the standards. Alternatively (and this is the only option if colours are lacking) magnitudes can be measured in the **natural photometric system** of the detector by using a colour of zero. This is effectively using the standards to predict the response of the system to an A0 star. The offsets between these two choices can often be substantial, and it is important to be clear about which kind of magnitude is being used.

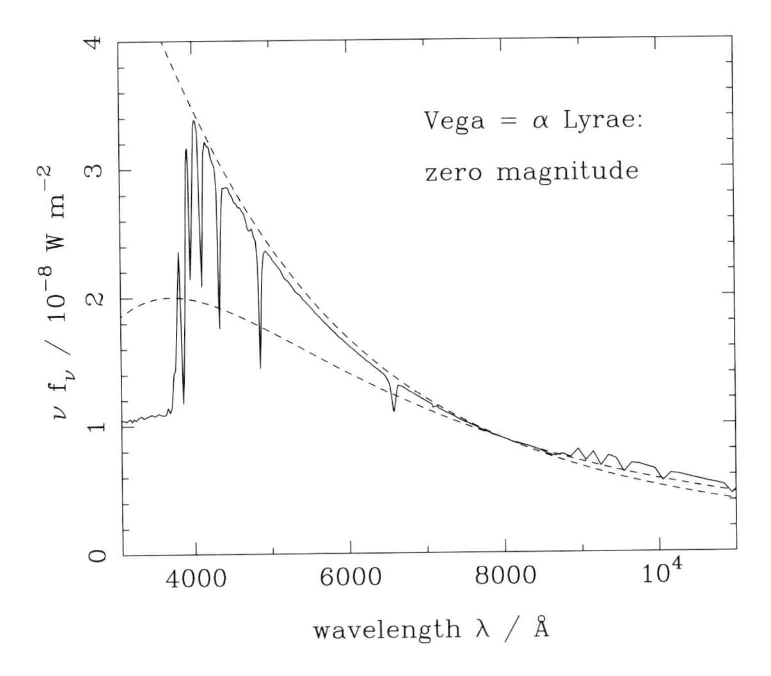

Figure 13.3. The spectrum of Vega, which is the star defined to have zero magnitude in most systems. Note the structure in the spectrum, which illustrates why it can be difficult to compare broad-band photometry taken with different filter systems. The lower broken line shows a black body with $T = 9900$ K, which is the effective temperature that determines the total energy output. However, the higher broken line of 16 000 K is clearly more appropriate, although it is absorbed below 4000 Å by the star's atmosphere.

COMMON MAGNITUDE SYSTEMS The following table gives the effective wavelengths and zero points for a number of common filter systems. The principal ones that may be encountered are the Johnson and Kron-Cousins *UBVRI* magnitudes. These are identical for *UBV*, but significantly different in *R* and *I*. A variety of more modern systems now exists, based on CCD detectors and interference filters, as opposed to the older glass filters plus photocathode. The main sets of these are the 'Mould' or 'KPNO' filters, which are very similar to the Kron-Cousins system (and are often calibrated in practice using the extensive set of equatorial standard stars from Landolt 1983). In the more recent literature, it is these filters that are probably meant by the unqualified use of the term *UBVRI*. All these systems use Vega (α Lyrae; figure 13.3) as their star of zero magnitude, to within occasional few-per-cent deviations. In contrast, the other common CCD-based magnitudes, the Thuan–Gunn *ugri* system, have their own unique standard stars. Lastly, in the near-infrared, there are the *JHKL* filters.

Table 13.2 gives the main parameters of these systems. The filter profiles are conventionally described by their **effective wavelength**, which is just the mean wavelength

Table 13.2 Parameters for common filter systems

Filter	λ_{eff}/nm	$\Delta\lambda$/nm	m(1 Jy)
Johnson			
U	360	65	8.03
B	440	100	9.02
V	540	80	8.91
R	680	95	8.65
I	900	230	8.41
Cousins			
R	640	150	8.74
I	790	150	8.49
Mould			
B	440	110	9.00
V	550	95	8.87
R	650	130	8.71
I	830	180	8.45
Gunn-Thuan			
u	350	40	7.75
g	490	70	8.91
r	650	90	9.13
i	820	130	9.20
photographic			
U	370	60	8.19
J	450	150	8.97
F	650	90	8.69
N	790	160	8.44
infrared			
J	1220	150	8.01
H	1630	170	7.57
K	2190	190	7.07
L	3450	280	6.12

integrated over the (normalized) filter transmission curve:

$$\lambda_{\text{eff}} = \frac{\int \lambda\, T(\lambda)\, d\lambda}{\int T(\lambda)\, d\lambda}. \tag{13.17}$$

Another useful parameter is the effective width $\Delta\lambda$, which is the full range between the points at which the filter falls to half its maximum transmission.

An ideal filter would be a simple top-hat function, and some come close to this shape; in detail, real filter profiles are more complicated and these must be used if accurate synthetic photometry for an object with a rapidly varying spectrum is required (not forgetting that wavelength-dependent atmospheric extinction can alter bandpass shapes significantly in the near-UV). However, for most purposes, results accurate to 0.05 to 0.1 magnitude can be obtained by assuming the approximation that the filter is an ideal top hat. The numbers in the table have been obtained using this approximation. The zero points are quoted in terms of the magnitudes produced by a source with $S_\nu = 1$ Jy ($= 10^{-26}$ W m^{-2} Hz^{-1}) at the effective wavelength, and can be used to convert an observed magnitude to an effective number of janskys. Because of the width of the filters, this is a somewhat imprecise process, and the detailed answer depends on the spectrum. These numbers assume a constant f_λ, and so will be most reliable for objects with a spectral shape similar to this. Somewhat different zero points can be found in the literature, since estimates of the absolute flux of Vega have changed over the years; the results quoted here assume 3570 Jy at 5556 Å (Hayes & Latham 1975). For further reading on the messy details of these conversions, see e.g. Bessell & Brett (1988) and Frei & Gunn (1994).

It would be more convenient if a standard star of constant S_ν existed; the zero points would then all be the same and life would be much simpler. No such star exists, but it is still convenient to define an **AB magnitude** in this way, as was done by Oke (1970):

$$m_{\rm AB} = 8.90 - 2.5 \log_{10} \left(\frac{S_\nu}{\rm Jy} \right). \tag{13.18}$$

Lastly, in some circumstances, the passband used may be sufficiently broad that it includes all the energy from an object (e.g. when a single narrow emission line dominates the output). In this case, what is measured is the **bolometric magnitude**. The zero point for this quantity is again the output of Vega, and the zero point is an energy flux density of 2.52×10^{-8} W m^{-2}. The **bolometric correction** is the difference between the bolometric magnitude and that measured in some finite passband. To complete the catalogue of confusing astronomical notation, absolute bolometric luminosities are often quoted in Solar units ($1 L_\odot = 3.90 \times 10^{26}$ W).

13.3 Luminosity functions

The main tool for a statistical description of the galaxy population is the luminosity function $\phi(M)$, which is the comoving number density of objects in some range of luminosity. Several decades of work have gone into establishing this function, how it varies for different types of galaxy, how it depends on environment and how it changes with cosmic epoch. Short of living long enough to watch a galaxy age, it is this statistical evolution of the galaxy population that must be relied upon for hints about the processes that produced the present galaxies and their properties.

LUMINOSITY FUNCTION ESTIMATORS Given a set of galaxies drawn from a magnitude-limited survey for which redshifts are known, the simplest estimator of the luminosity function is given by binning up the data as a function of absolute magnitude. To investigate evolution with redshift, this binning can be done separately for shells covering different distance ranges. The correct estimator for the density in a given bin is the common-sense sum of the density defined by each object,

$$\hat{\phi} = \sum_i \hat{\phi}_i = \sum_i [V(z_{\rm max}) - V(z_{\rm min})]_i^{-1} \tag{13.19}$$

(Felten 1976). In this **maximum-volume estimator**, $z_{\rm max}$ is the smaller of the maximum redshift within which a given object could have been seen, and the upper limit of the redshift band under consideration; $z_{\rm min}$ is the lower limit of the band. $V(z)$ is the comoving volume out to a given redshift over the relevant area of sky.

Felten's estimator is a **maximum-likelihood** estimator. This is a common statistical technique in astronomy, where datasets are often smaller than one would like, and it is important to extract all the available information. The idea of likelihood is based on the theorem of total probability, which allows us to write two different expressions for the joint probability that the true luminosity function takes a certain value and the data we obtain take a certain value:

$$P(\phi|\text{data})P(\text{data}) = P(\text{data}|\phi)P(\phi). \tag{13.20}$$

Here, the conditional probability $P(\text{data}|\phi)$ is the probability of getting the data, given a certain true luminosity function; this is also known as the **likelihood**. A trivial rearrangement gives

$$\boxed{P(\phi|\text{data}) = \frac{P(\phi)\,P(\text{data}|\phi)}{P(\text{data})};} \tag{13.21}$$

this innocent-looking equation is known as **Bayes' theorem**, and is a perennial source of controversy in statistics. What it says is that the thing we want to know, $P(\phi|\text{data})$ (the **posterior probability** of any true luminosity function given the data we have), is a product of the likelihood and the **prior probability** $P(\phi)$. The term in the denominator of the rhs is just a normalization factor, $P(\text{data}) = \int P(\phi)\,P(\text{data}|\phi)\,d\phi$, but what does $P(\phi)$ actually mean? To assign a probability to a property of the universe is only really possible if we are dealing with an ensemble of universes, so that relative frequencies exist for different outcomes. Since there is only one universe known, arguments about the correct prior (probability) are inevitable. If we neglect this factor entirely for the moment, then the posterior probability is just proportional to the likelihood. This is the justification for the maximum-likelihood method: if all universes are equally likely *a priori*, then the the most likely universe given the data is the one that maximizes the probability of observing the data that were actually obtained.

A good example of the difference that the prior issue can make is provided by estimating the mean and variance from a set of data. Suppose we have a set of N measurements x_i, which we assume are random drawings from a Gaussian distribution with mean μ and variance σ^2. The likelihood $\mathscr{L} \equiv P(\text{data}|\phi)$ is just

$$\ln \mathscr{L} = -\sum_i (x_i - \mu)^2/2\sigma^2 - (N/2)\ln(2\pi\sigma^2). \tag{13.22}$$

Differentiating to maximize the likelihood gives

$$\begin{aligned} \mu_{\rm ML} &= \frac{1}{N}\sum_i x_i \\ \sigma^2_{\rm ML} &= \frac{1}{N}\sum_i (x_i - \mu_{\rm ML})^2. \end{aligned} \tag{13.23}$$

The trouble with this result is that it is known to be biased: $\langle \sigma^2_{\rm ML}\rangle = [(N-1)/N]\sigma^2_{\rm true}$. Here, angle brackets denote the expectation value over an ensemble of different datasets;

for any datum x_i, this is equivalent to integration over the probability distribution, so that $\langle x_i \rangle = \mu$ etc. The way to cure the bias in this case is to realize that ignoring the prior is equivalent to assuming a particular prior, in which all intervals $d\mu \, d\sigma$ are equally likely. Since we know that σ must be positive, it can be argued that this should be replaced by $d\mu \, d\ln\sigma$, allowing an unrestricted range of $\pm\infty$ in each variable. The posterior probability is then $P = \mathcal{L}/\sigma$, and it is easy to show that maximizing this yields the unbiased estimator for σ. The lesson here is clear, since the difference between the two estimators is negligible for large N: if the dataset is large enough, then maximum likelihood will give an answer that is effectively independent of the choice of prior. If N is very small, then care is needed. See Kendall & Stuart (1958) or Box & Tiao (1992) for details of the subtleties in this form of statistical estimation.

Returning to the galaxy luminosity function, it is often convenient to describe the results analytically e.g. via a **Schechter function** fit at each redshift

$$\boxed{\begin{aligned} d\phi &= \phi^*(L/L^*)^\alpha \exp(-L/L^*) \, dL/L^* \\ &= 0.921 \, \phi^*(L/L^*)^{\alpha+1} \exp(-L/L^*) \, dM. \end{aligned}} \tag{13.24}$$

Note the factor 0.921, which is needed if we want numbers per magnitude rather than per $\ln L$. Unfortunately, it is quite common in the literature for this factor to be omitted, even though the Schechter function was originally defined in terms of flux. Worse still, it is often unclear which definition is being used. In the absence of clustering, one would define the likelihood $\mathcal{L}$ by using the joint probability for the observables:

$$\mathcal{L} = \prod_i \frac{d^2p}{dM \, dz}(M_i, z_i). \tag{13.25}$$

It is not necessary to restrict the method to simple analytical forms such as the Schechter function. The value of ϕ in a number of bins can be used as a set of parameters, each of which is solved for directly; this is known as the stepwise maximum-likelihood method (see e.g. Loveday *et al.* 1992).

These methods assume spatial homogeneity, which does not apply to a very local redshift survey, where the presence of clustering renders the vertical normalization of the luminosity function uncertain. It may also affect the shape of the function, but such luminosity segregation has never been demonstrated convincingly. The universality of the luminosity function may well depend on waveband: the known phenomenon of morphological segregation plus the tendency for ellipticals to be redder than spirals should produce some brightwards shift in characteristic luminosity in redder filters – so that a positive density perturbation boosts the number density of bright galaxies in two ways. However, the following method at least avoids the direct density boost, and so should be closer to the average luminosity function. The likelihood method can be applied directly for an infinitesimal redshift band, since only the probability distribution for M at given z is involved and amplitude scalings normalize away:

$$\mathcal{L} = \prod_i \frac{dp}{dM}(M_i \mid z_i). \tag{13.26}$$

This expression can be immediately generalized to a finite redshift range by continuing to use the conditional probability of M at given z – but this must now be normalized

Table 13.3. Type-dependent luminosity function according to Glazebrook *et al.* (1994)

Type	Relative ϕ^*	$M^*_{B_J}(h=1)$
E/S0	0.35	−19.59
Sa	0.07	−19.39
Sb	0.18	−19.39
Sc	0.17	−19.39
Sd	0.15	−18.94
Irr	0.08	−18.94

individually for each z_i of interest. This likelihood estimator is now immune to the effects of inhomogeneities (Sandage, Tammann & Yahil 1979).

THE LOCAL LUMINOSITY FUNCTION Over the years, redshift surveys have consistently found that the Schechter description is a good one, and that the faint end of the luminosity function is rather flat in terms of number per unit magnitude, i.e. that α is close to -1. For example, Loveday *et al.* (1992) give $\alpha = -1.0 \pm 0.15$, and $M_* = -19.50 \pm 0.13$ (assuming $h = 1$) in the photographic B_J waveband. This band is also known as photographic J – see Table 13.2. There is also general agreement that the luminosity function depends on Hubble type, in the sense that characteristic luminosities are higher for the earlier types. The type-dependent luminosity function has been discussed by e.g. King & Ellis (1985), Shanks *et al.* (1984) and Glazebrook *et al.* (1994), whose summary of the results is given in Table 13.3.

Note that these figures minimize the difference between the different types, because the later types are bluer. In red wavebands, the differences in M^* would be more pronounced; this would also be the case if the luminosity function were constructed for the spheroidal components of galaxies only. Finally, it may not be the case that the appropriate Schechter α for all these types is close to -1: there is some suggestion that the overall appearance of $\alpha = -1$ arises because of the range of M^* values, and in fact the more luminous types have luminosity functions that fall below a Schechter function at low luminosities (see Binggeli, Sandage & Tammann 1988).

NORMALIZATION The true local space density of galaxies is set by the parameter ϕ^*, which is usually derived from local redshift surveys along with the rest of the luminosity function; Efstathiou *et al.* (1988a) considered five different surveys and gave a mean ϕ^* of $(0.0156 \pm 0.0034)h^3$ although the dispersion among estimates from the differing surveys amounts to a factor 2. Loveday *et al.* (1992) used the larger APM survey to obtain $\phi^* = (0.0140 \pm 0.0017)h^3$. An alternative approach is to choose a value of ϕ^* that normalizes the predicted number counts to the bright end of the data; this favours values up to 50% larger (see below). The correct local value is likely to remain controversial: bright nearby galaxies suffer from problems with large-scale structure and possible nonlinear magnitude scales. In many ways, it is easier to study the intermediate universe, at redshifts of a few tenths.

Apart from changing the interpretation of the galaxy number counts, the normalization is important because of its impact on the density of the universe. The total luminosity density obtained by integrating the Schechter function down to zero luminosity is

$$\rho_L = \Gamma(\alpha + 2)\, \phi^*\, L^*, \tag{13.27}$$

and so an estimate of the mass density follows by obtaining an M/L ratio for some large object, and assuming that mass follows light. It is of course improbable that this assumption holds (see chapters 12 and 17), but at least a comparison of this estimate of Ω with one obtained in some other way shows what degree of bias is then required in the galaxy distribution. The reluctance of the luminosity density to be pinned down is therefore a fundamental headache.

CLUSTER LUMINOSITY FUNCTIONS For many years, clusters of galaxies were used as a convenient short-cut to the study of galaxy evolution. Redshifts are not needed in order to obtain luminosity functions, because all the galaxies are at the same redshift. Provided a statistical correction can be made for foreground and background galaxies, very large datasets can be amassed at little expense. By looking at the derived properties as a function of cluster richness, the influence of environment on galaxies and their evolution can be assessed. The correction for field galaxies seen in projection is not so simple, however, since they are clustered; the number of background galaxies seen some distance from the cluster will differ from the number actually superimposed on the cluster by a larger factor than would be expected from simple Poisson statistics. Despite these problems, many interesting results have been forthcoming. The Schechter function fits the optical cluster luminosity function data equally as well as it does the data for field galaxies. However, deviations exist at very bright and very faint magnitudes. At the bright end, many clusters contain **cD galaxies** (this name derives from the notation for supergiant stars, and does not stand for 'central dominant' – although this is a fair description). The statistics of cDs are not consistent with their being the most luminous member of the Schechter function – or indeed with a random drawing from *any* luminosity function (Bhavasar & Barrow 1985). This implies that special processes were at work in cD formation (see the section on mergers in chapter 17).

The faint end of the cluster luminosity function shows substantial departures from a single Schechter function, steepening to almost the point at which the total brightness would diverge logarithmically, $\alpha \simeq -1.8$ (Driver *et al.* 1994; see figure 13.4). It has taken some time for this strong dwarf contribution to be accepted, since these are galaxies of low surface brightness and they can easily be missed on images of indifferent quality. It seems increasingly likely that a similar feature exists in the luminosity function of field galaxies, but there is some controversy over this point (Marzke, Huchra & Geller 1994; Ellis *et al.* 1996; Loveday 1997).

The luminosity functions of clusters show no evidence for any variation with richness (e.g. Colless 1989), and this is a major puzzle and constraint on theories of galaxy formation. In order to obtain a biased galaxy distribution (to reconcile the prejudice that $\Omega = 1$ with the observed cluster M/L ratios), many mechanisms predict **luminosity segregation**: an increase of clustering strength with luminosity, or a trend to brighter luminosities in richer clusters (see chapter 17). That this is not observed tells us either that the degree of bias is weak or that the mechanism is subtle.

Early studies of clusters at different redshifts revealed the first evidence for the **Butcher–Oemler (BO) effect** (Butcher & Oemler 1978), which took the form of an increased fraction of blue galaxies in high-redshift clusters. This was the first

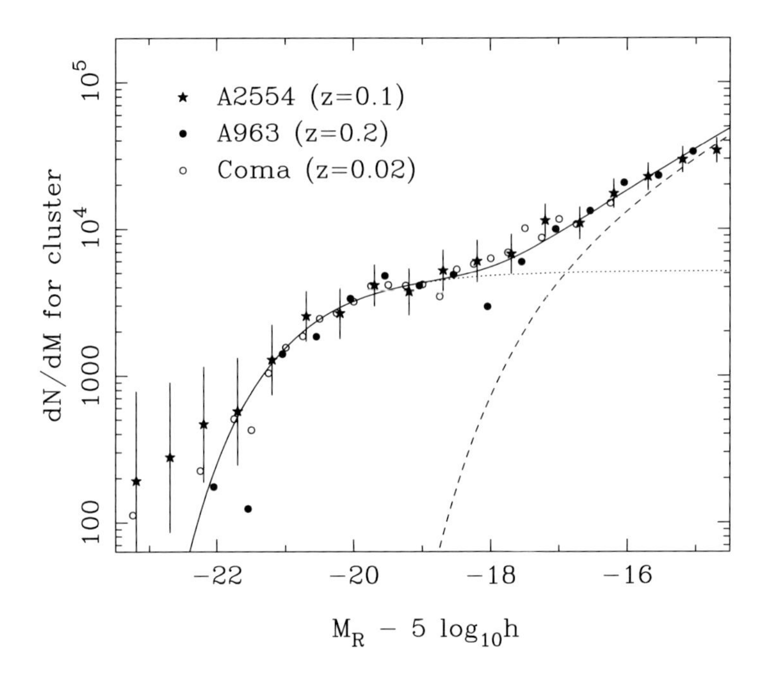

Figure 13.4. The luminosity functions for the clusters A963 and A2554, from Driver *et al.* (1994). A single Schechter function fits the bright end well, but the function steepens when the population of faint, low surface brightness dwarf galaxies becomes important.

indication that galaxies had changed their behaviour in a qualitative way in the past: the galaxy populations in clusters at $z \simeq 0.5$ are not consistent with a simple backwards extrapolation of the present star-formation rates of cluster galaxies. The result took some time to be accepted, particularly because of the difficulty in performing the subtraction of foreground contamination. However, redshift surveys of the clusters confirmed the phenomenon in due course – although it appears to be confined to the richest clusters, with little evolution in blue fraction for weaker groups (Allington-Smith *et al.* 1993). What causes the BO effect? Either the galaxies in the distant clusters are suffering enhanced star formation, or they are a population that has become depleted by the present, perhaps due to mergers. Imaging studies have tended to favour both views: many of the blue galaxies are morphologically normal spirals, but there is also an enhanced population of irregulars (Dressler *et al.* 1994).

13.4 Evolution of galaxy stellar populations

Any interpretation of the spectral data on galaxies requires some means of predicting how galaxies should appear as a function of their star-formation histories. Over a number of years, a semiempirical method has arisen in order to meet this need: **evolutionary spectral synthesis**. What this means is that the theory of stellar evolution is used to predict the

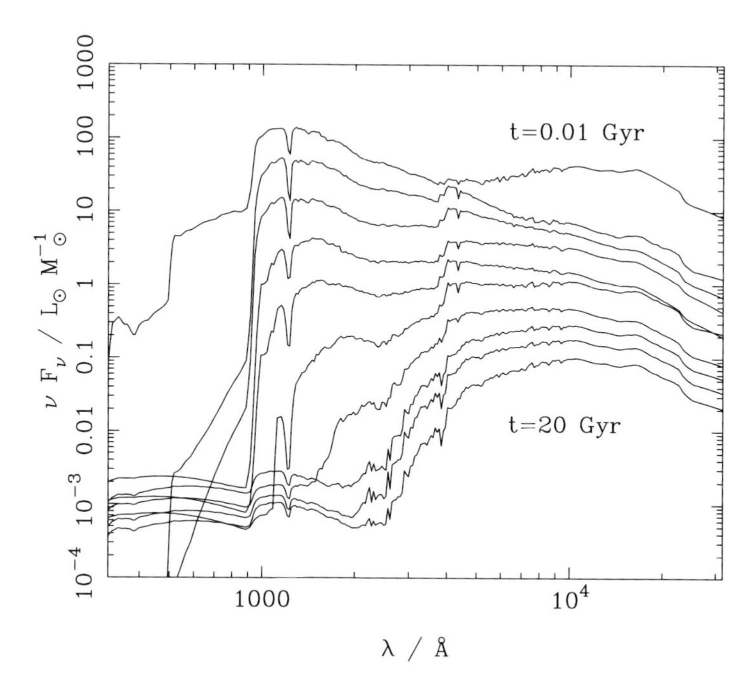

Figure 13.5. A plot of the expected evolution with time of the spectral energy distribution of a burst of star formation. This is as calculated by Bruzual & Charlot (1993), assuming Solar metallicity. The time sampling is logarithmic, from an age of 0.1 Gyr at the top to 20 Gyr at the bottom. Note the ageing away of the blue O and B stars, leading to the establishment of the characteristic red spectrum shortwards of the '4000 Å break' at an age of about 1 Gyr.

population of the HR diagram (the luminosity–temperature plane; see chapter 5) as a function of time, starting from a main sequence with a known probability distribution of stellar mass, the **initial mass function**, or IMF. As stars evolve off the main sequence, the predicted mix of stellar types present becomes more complex than the simple initial range from blue supergiant to red dwarf. Now, given a sufficiently extensive library of observed spectra for all the different kinds of star, the stellar spectra can be added in the theoretically predicted proportions in order to predict the overall spectrum of a galaxy at a given time. The results of one attempt to implement this scheme are shown in figure 13.5.

This evolution of a **starburst** is the Green function for galaxy stellar evolution, since any star-formation history can be accounted for by summing over a series of infinitesimal bursts. Figure 13.5 shows the null hypothesis for galaxy evolution: **passive evolution** following a single burst, in which the galaxy fades in all wavebands once star formation has ceased. Redwards of the 4000 Å break, the rate of evolution is similar at all wavelengths, and this can be understood readily if the light in this region is dominated by giants. Their short lifetimes mean that the stellar luminosity is determined by the rate at which stars evolve off the main sequence. Using this insight, it is straightforward to show [problem 13.1] that the luminosity of a galaxy is expected to decline as a power of time:

$$L_G \propto t^{-\alpha}, \quad \alpha \simeq 0.5. \tag{13.28}$$

Passively evolving galaxies with high formation redshifts are therefore expected to be between 0.4 magnitudes ($\Omega = 0$) and 0.6 magnitudes brighter ($\Omega = 1$) at $z = 1$ than at present.

Passive evolution is a reasonable hypothesis only for the most inactive elliptical galaxies; in general, a simple model of the Hubble sequence can be set up in which there is a declining rate of star formation:

$$\frac{\partial M_{\rm stellar}}{\partial t} \propto \exp\left(\frac{-t}{\tau}\right), \tag{13.29}$$

and the different spectral signatures merely tell us how rapidly the star formation has been truncated. For example, taking an age $t = 15$ Gyr and a timescale of $\tau = 1$ Gyr produces a fair impression of a present-day elliptical galaxy, whereas spirals of different types have $\tau \simeq 3$–10 Gyr, and an irregular can be reproduced by a constant star-formation rate of indefinite age (Rocca-Volmerange & Guiderdoni 1988).

These models can be extended to cover metallicities other than Solar (e.g. Worthey 1994). At the lowest order there is an **age–metallicity degeneracy**, so that the age inferred from a given spectrum scales very roughly as the reciprocal of the metallicity. In other words, for a given age, stars of high metallicity have more pronounced absorption features and so they more readily reproduce the spectral features that arise at late times in these composite stellar models, requiring a lower age to fit a given set of data. In more detail, different absorption features have a different balance between their sensitivities to age and to metals, so that both parameters can be obtained independently, given sufficiently good data.

STARBURSTS AND AFTER A good example of the utility of these synthetic models comes in interpreting the excess of blue galaxies seen in the BO effect. In the course of spectroscopic investigations of BO clusters, a class of galaxies was uncovered which became known as **E+A galaxies** or post-starburst galaxies. The former name indicates that the overall spectral shape is that expected for an elliptical, but that the Balmer absorption lines (particularly Hδ) are much stronger than normal – so that the galaxy appears to contain an excess population of A stars, in which such lines are particularly strong. It was pointed out by Couch & Sharples (1987) that such spectra can be accounted for quantitatively by taking an existing galaxy and either truncating its star formation or adding an additional burst that is then truncated. In either case, the effect after about 1 Gyr is to produce an A-star-dominated spectrum, since at this time all the more massive stars have moved from the main sequence (see figure 13.6).

13.5 Galaxy counts and evolution

The first studies of galaxy evolution concentrated on clusters, since these are convenient 'laboratories' where many galaxies are found together at the same redshift. The study of the general population of galaxies had to await recent technical developments, which have allowed imaging and spectroscopy to be carried out to very faint levels (see e.g.

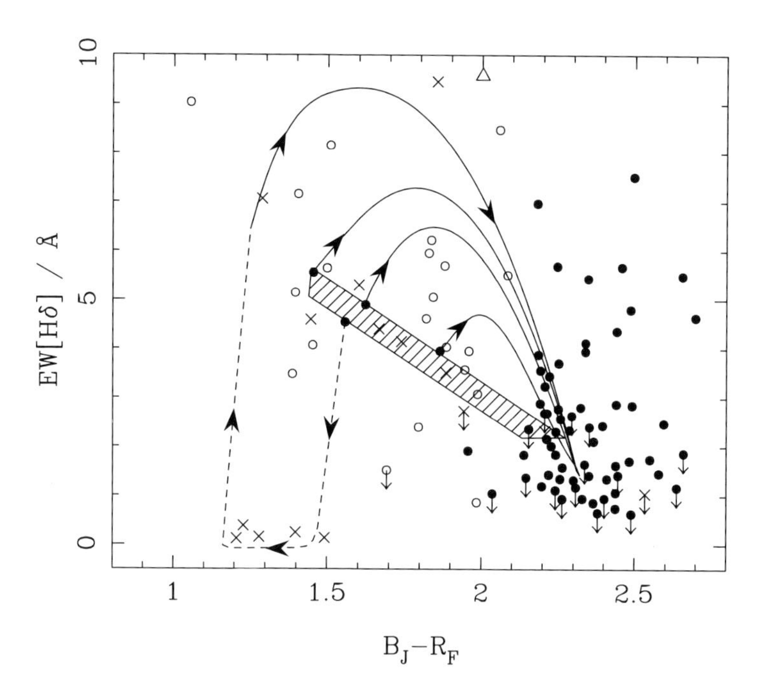

Figure 13.6. The evolution of a starburst, taken from Couch & Sharples (1987). The hatched area denotes a typical relation between $H\delta$ equivalent width and continuum colour for systems undergoing continuous star formation, such as spiral galaxies. The tracks show the behaviour expected following either cessation of star formation (solid lines) or an abrupt starburst that is then terminated (broken line). In either case, the evolution eventually moves to the enhanced A-star contribution to the spectra, characteristic of a post-starburst galaxy. However, all curves eventually terminate in the giant-dominated spectrum characteristic of elliptical galaxies, for which observed data values are shown as filled points.

Lilly, Cowie & Gardner 1991; Cowie, Hu & Songaila 1995a). For many years it had been known from imaging surveys that the numbers of faint galaxies seen in blue light exceeded that expected if there had been no evolution in the population. This was conventionally ascribed to stellar evolution: galaxies were expected to have been more luminous in the past. However, recent spectral and infrared data have shown that what is going on is much more interesting.

NO-EVOLUTION MODELS The first step in trying to make sense of the numbers of galaxies seen in deep surveys is to predict the counts that would be expected if the galaxy population at all redshifts was identical to what we see today. This exhibits clearly the effects of cosmological geometry, and any discrepancy between model prediction and observation indicates the need for evolution, without measuring any redshifts. These no-evolution models require a knowledge of the local luminosity function, plus the spectral shape of galaxies.

In order to convert between absolute magnitude M and apparent magnitude m, we require a knowledge of the luminosity distance D_L, the K-correction $K(z)$, and the aperture correction $A(z)$:

$$M(z) = m - 5\log_{10}(D_L/10\ \mathrm{pc}) - K(z) + A(z). \tag{13.30}$$

For simplicity, we shall throughout quote absolute magnitudes assuming $h = 1$ for the Hubble parameter. The aperture correction converts the observed aperture magnitudes to some proper diameter. This correction really requires a detailed knowledge of the galaxy's light profile, but it is often acceptable to use a simple approximation due to Gunn & Oke (1975), in which the growth in metric luminosity is parameterized as:

$$L(<r) \propto r^{\alpha}. \tag{13.31}$$

For brightest cluster galaxies, $\alpha \simeq 0.7$ (Schneider *et al.* 1983), but lower values ($\alpha \simeq 0.4$) are in general more appropriate for field galaxies. A common choice of standard aperture is

$$D_0 = 20h^{-1}\ \mathrm{kpc}, \tag{13.32}$$

so that the aperture correction is given in terms of the angular-diameter distance $D_A(z)$ as

$$A(z) = 2.5\alpha\log_{10}[\theta D_A(z)/20h^{-1}\ \mathrm{kpc}], \tag{13.33}$$

where θ is the angular diameter actually used for measurement. For work on faint galaxies, this will be only a few arcseconds: for larger apertures, the galaxy profile falls below the sky brightness and the measurement becomes noisy. It would be convenient if this complication could be ignored, and total magnitudes measured for galaxies, but this can only be done by extrapolating the light profile beyond the largest radius where it can be measured. In practice, magnitudes within $20h^{-1}$ kpc are usually within a few tenths of total, but this difficulty in precise magnitude definition is a problem when comparing the results of different authors.

As discussed earlier, the K-correction $K(z)$ is the difference between the dimming with redshift for a fixed observing waveband and that expected on bolometric grounds; it is large and positive for a very red galaxy. For a $f_\nu \propto \nu^{-\alpha}$ spectrum,

$$K(z) = 2.5(\alpha - 1)\log_{10}(1 + z). \tag{13.34}$$

The effective spectral index of galaxies is a function of wavelength. Thinking of galaxies very crudely as black bodies, we expect the Rayleigh-Jeans value $\alpha = -2$ in the far infrared, with α becoming positive at rest wavelengths $\gtrsim 1\ \mu\mathrm{m}$, and returning to zero at very short wavelengths as we encounter the contribution of young blue O and B stars due to current star formation. In the B and K wavebands, α values of about 4 and -1.5 respectively apply, but these are very rough figures. Accurate modelling requires the detailed spectral energy distribution of the galaxies.

THE FAINT BLUE GALAXIES Figure 13.7 gives the results of the no-evolution modelling, and shows that the K counts are a much better match to the non-evolving prediction than are the counts in B. This follows the trend long seen in optical counts in other bands (e.g. Tyson 1988) where the galaxy excess is greatest in B and progressively less in R and I filters. This excess of faint blue galaxies is clearly due to a population of **starburst galaxies** that are undergoing active star formation, so that they are more

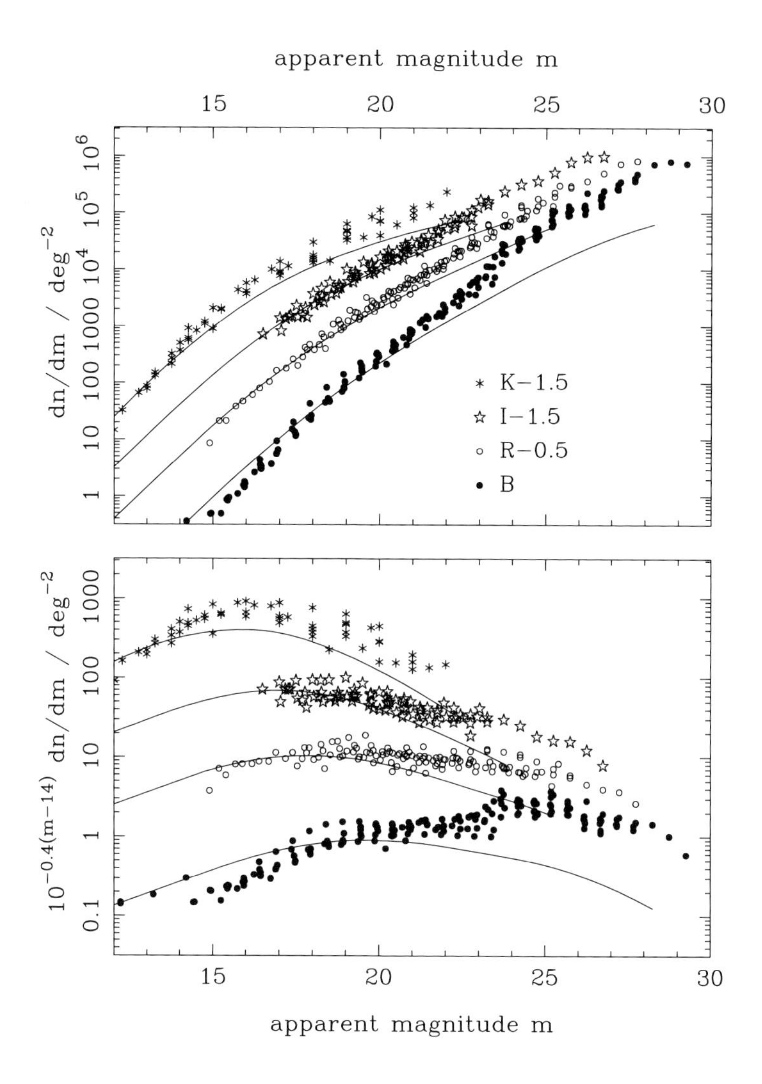

Figure 13.7. Differential counts of galaxies in the B, R, I and K bands. The upper panel shows the raw counts per unit magnitude interval, reaching up to a million galaxies per square degree at the faintest level. The lower panel normalizes the counts by dividing by a count that scales as 1/(flux); this is the count for which the sky brightness diverges logarithmically. In this form, the significance of the deepest points (from the Hubble Deep Field) is revealed as indicating for the first time a convergence in the total amount of star-forming activity in the universe. The lines show no-evolution models, whereby the current galaxy population is transported unchanged to high redshifts and the resulting counts are predicted, assuming $\Omega = 1$. Although both sets of counts have very similar shapes, the different optical and infrared K-corrections mean that the blue counts were expected to cut off more rapidly, but this is not seen. The K counts are nearly consistent with the expectation for a non-evolving galaxy population, whereas there is a great excess of faint blue galaxies. The larger volume elements in low-density models raise the predicted faint counts somewhat, but do not change the qualitative point.

luminous in blue wavebands; the key question is what kinds of galaxies they are, and what triggers the star formation.

A possible but incorrect interpretation of this phenomenon would be that we are seeing back to the era of general galaxy formation, where all high-redshift galaxies are undergoing vigorous starbursts. The predicted number counts will always start to fall below the observed approximate power law when we start to see L^* galaxies to $z \simeq 1$, since it is then that the effects of K-correction and non-Euclidean volume elements start to makes themselves felt. A general increase in activity in all galaxies could therefore only make up the deficiency in faint galaxies by adding many galaxies at $z > 1$, but these were not seen by the first deep spectroscopic surveys to $B \simeq 22$ (Broadhurst *et al.* 1988). This tells us that the bursting galaxies seen at this level must be of sub-L^* luminosity, and this is backed up by a general argument to do with the numbers of galaxies. If the luminosity function is assumed to be a Schechter function with $\alpha \simeq -1$, then the density of galaxies brighter than L changes only logarithmically at low L, and effectively converges to a finite total density. Any galaxy that we see in the B band must lie at $z \lesssim 4$ in order to escape Lyman-limit absorption, so there is only a finite comoving volume to be explored and hence a limit to the number of faint galaxies that can ever be seen. If the total counts exceed this limit (and they do, especially for high Ω where the volume is lower), then there must have been more galaxies in the past.

Possible ways of achieving this increased number density range from 'bursting dwarfs' – a new population of dwarf galaxies undergoing an initial starburst at $z \sim 0.4$ and fading to invisibility by the present day (or at least to very low surface brightness; see Babul & Rees 1992, McGaugh 1994) – to the idea that a large amount of galaxy-galaxy merging has occurred in the field population in the recent past (Broadhurst, Ellis & Glazebrook 1992). To distinguish these alternatives, we need to look at the detailed results of galaxy redshift surveys. First, however, note that the size of the problem depends on ϕ^*: according to Maddox *et al.* (1990a), the slope of the B counts between $B = 15$ and $B = 20$ is very steep, indicating either very strong evolution or a large local hole in the galaxy distribution. However, many workers are uncomfortable with the idea of strong evolution between $z \simeq 0.1$ and the present, and it remains possible that the bright blue counts are incomplete. In this case, $\phi^* = 0.02h^3$–$0.025h^3$ would give a good match to the blue counts around $B = 19$, and the blue excess would only become dominant for magnitudes fainter than $B = 23$. With the lower normalization, there is an apparent excess even at $B = 20$.

EVOLUTION OF LUMINOSITY FUNCTIONS As larger telescopes with more sensitive spectrographs have become available, so the problem of obtaining redshifts of many galaxies to $B = 24$ or fainter has yielded to technology and several thousand faint-galaxy redshifts are now known. At the time of writing, the most extensive surveys are as follows: the Canada–France Redshift Survey (Lilly *et al.* 1995); the Autofib Redshift Survey (Ellis *et al.* 1996); the Hawaii Deep Fields (Cowie *et al.* 1996). These cover a variety of wavebands from the blue to the infrared, and it is important to have this range. Redshift surveys in the infrared are a very clean probe of galaxy evolution: the near-infrared light in galaxies is produced by giants drawn from the population of old evolved stars that dominate the stellar mass in a galaxy. Moreover the K-correction is much better defined for a K-band sample than in the optical. This is again because the spectral slope in the optical is dominated by the star-formation rate; thus spiral and elliptical galaxies have blue K-corrections that differ by 1 magnitude at $z = 0.5$. In

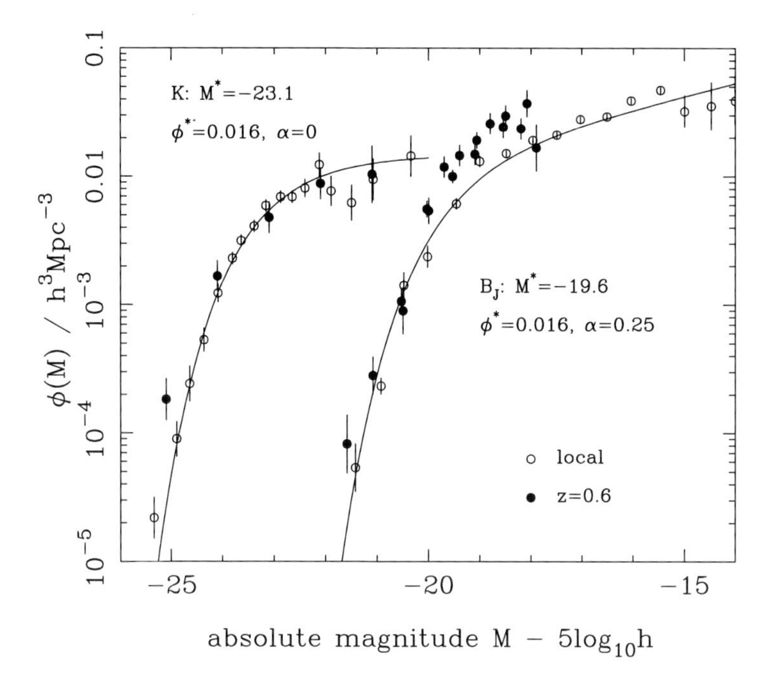

Figure 13.8. The redshift-dependent B_J-band and K-band galaxy luminosity functions, taken from Lilly *et al.* (1995), Ellis *et al.* (1996), Cowie *et al.* (1996), and Gardner *et al.* (1997), assuming $\Omega = 1$. The infrared results show that there is little change in the mass function, but that star-formation activity is enhanced in galaxies of progressively higher mass at high redshift.

contrast, galaxy colours in the near-infrared are dominated by old stars and are uniform across Hubble types, thus yielding a morphological mix that does not change with redshift. Nevertheless, it is important also to have surveys that select in the rest-frame ultraviolet, otherwise we would fail to sample any population of low-mass galaxies with high levels of star-forming activity.

Figure 13.8 shows a summary of the results from these surveys, contrasting the local luminosity functions with their high-redshift counterparts. The infrared results are striking in revealing very little change in the luminosity function between $\bar{z} = 0.6$ and the present. As required by the number counts, the blue luminosity functions show a very different picture, the space density of galaxies of all luminosities being higher in the past. This blue evolution does not take the form of either a simple vertical or horizontal shift. We do not see a translation to bright magnitudes, as would be expected if all galaxies were fading away rapidly from a high-redshift starburst. Instead, the high-luminosity end of the function shows little change, and additional galaxies are present at lower luminosities, resulting in a steeper luminosity function. The bright end of the luminosity function is therefore consistent with a picture in which massive galaxies were already in place well before $z = 1$ and have subsequently undergone little star formation. This would not be expected in a picture of merging: if we simply redistribute starlight, then mergers would be expected to satisfy roughly $\phi^* L^* \simeq$ constant. In order to attain a higher number density in the past, L^* should be fainter at early times, contrary to observation.

What seems to be happening at lower luminosities is that additional star formation at intermediate redshifts is enhancing the blue luminosities of lower-mass galaxies. In a picture where spheroids are old and passively evolving, it would be natural to associate this feature with the epoch of disk formation (e.g. Gunn 1982). This behaviour can be seen directly in the redshift surveys by examining star-formation rates (as deduced from narrow-line luminosities; Kennicutt 1983) as a function of redshift and mass. From these numbers, a characteristic e-folding time for the stellar content can be deduced

$$\tau_* = M_* / \dot{M}_*. \tag{13.35}$$

Today, the only galaxies for which this time is less than the Hubble time are the low-mass irregulars. At $z = 1$, it is less than the local age of the universe, $t(z)$, for all galaxies below L^* (Cowie *et al.* 1996).

Does this mean that massive galaxies form early, whereas low-mass galaxies form late? Not necessarily. High star-formation rates could correspond to the formation of a disk around a pre-existing spheroid of much greater age. The lack of star formation in the most massive galaxies could then be related to morphological segregation: ellipticals are unable to grow disks owing to their environments, and they are also the most massive galaxies. In any case, the cause of the observed star formation remains an open question. In a hierarchical universe, true galaxy formation would probably correspond to a merger, but the non-evolving K data appear to rule out a large degree of merging. Tidal interactions without mergers might, however, still generate starbursts (and this is a possible way of biasing galaxy formation in order to generate more galaxies in clusters: Lacey & Silk 1991).

13.6 Galaxies at high redshift

The systematic evolution of the luminosity function raises as many questions as it answers about the origin of galaxies, and researchers have naturally sought clarification of the issues by detailed study of individual objects at high redshift. The hope is to say whether we have seen the main period of star formation in the evolution at $z \lesssim 1$, or whether there was an earlier period of activity that has not so far been observed.

AGES AND FORMATION REDSHIFTS Because of the ease with which a relatively small burst of star formation can hide an old and red population of stars, attempts to limit the era of galaxy formation have concentrated on the reddest galaxies – the ellipticals. One approach has been to seek out high-redshift ellipticals and estimate their age by fitting synthetic models (Hamilton 1985; Lilly 1988; Dunlop *et al.* 1996). Such studies are most powerful if there exists detailed spectroscopy, since an overall red colour can be due to dust, whereas the absorption-line breaks that develop with increasing age are more robust. Allowance must be made for possible non-Solar metallicity, but even this can be separated from age given sufficient spectral features (Worthey 1994). The best case from this approach has been the galaxy studied by Dunlop *et al.*, which is well described by a single Solar-metal burst of age 3.5 Gyr at $z = 1.55$. At this redshift, the age of the universe ranges from $1.6h^{-1}$ Gyr ($\Omega = 1$) to $2.7h^{-1}$ Gyr ($\Omega = 0.2$, open) to $3.5h^{-1}$ Gyr ($\Omega = 0.2$, flat). What formation redshift this corresponds to depends on the chosen cosmology, and some models (high Ω and/or high h) do not allow a universe that is old enough to contain this galaxy. One is pushed towards formation redshifts of at least 6–8, so that the emission from the first generation of forming galaxies may therefore plausibly be confined to the infrared or longer wavebands.

An alternative argument that points in the same direction is a differential measurement based on the fact that elliptical galaxies have a very tight correlation between colour and luminosity, reflecting the higher metallicities of more massive objects (Bower *et al.* 1992; Ellis *et al.* 1996). The rms scatter about this relation is only a few hundredths of a magnitude, and many star-formation histories would be inconsistent with this observational constraint. Since galaxies redden with age, the thickness of the **colour–magnitude relation** limits the spread in age of ellipticals. It is implausible that the exact formation of individual galaxies could be precisely synchronized, and so there must be an initial dispersion in age, of order the age of the universe at the typical formation epoch. This can be reduced to a small fractional dispersion in age only if the galaxies are observed long after their formation epoch. This argument for the formation redshift is independent of h, since it is couched in terms of the fractional dispersion in age only. The dispersion of the colour–magnitude relation at the present is about 0.03 magnitudes, increasing to 0.07 magnitude at $z = 0.5$. Both figures require a minimum formation redshift of around 3.

Lastly, note that the small scatter in the colour–magnitude relation also sets limits on the importance of mergers in elliptical formation. The slope is about 0.07, in the sense that a galaxy 1 magnitude more luminous will be 0.07 magnitude redder. This slope is of the same order or larger than the scatter, which means that a galaxy could not boost its luminosity at constant colour by more than a factor 2 without moving off the relation. This is a critical constraint: it removes the possibility of making stars in globular-cluster-sized units and later throwing them together to assemble an elliptical galaxy. It seems as if the stars in ellipticals were formed at a time when the depth of potential well that they would eventually inhabit was already determined.

THE RADIO-GALAXY HUBBLE DIAGRAM A good deal of work on high-redshift galaxies has concentrated on those that are radio-loud active galaxies. One advantage of studying radio galaxies is that they form an optically homogeneous set of objects: they are massive giant ellipticals with an upper limit dictated by the cutoff in the Schechter function and a lower limit set by poorly understood processes that favour activity in massive galaxies. Thus, empirically, the **Hubble diagram** (apparent magnitude versus redshift plot) shows a scatter of only about 0.4 magnitude. Radio galaxies have been detected up to $z \simeq 4$ (e.g. McCarthy 1993), so the expected q_0-dependent curvature of the diagram should in principle allow us to measure the geometry of the universe.

Interpreting the data straightforwardly yields a high value of $q_0 \simeq 4$: objects at $z \simeq 1$ are brighter than would be expected in a low-density universe. However, galaxies seen at high redshift are younger than those nearby, which means we need to take into account evolution of the stellar populations in the galaxies. Putting in such corrections with realistic uncertainties, Lilly & Longair (1984) estimated

$$0 \lesssim q_0 \lesssim 1. \tag{13.36}$$

However, the assumption that the mass of these galaxies has remained totally unchanged with cosmic epoch may not be correct: merging is not a process that can be totally neglected, as evidenced by the fact that some low-redshift radio galaxies are cD ellipticals. Finally, the highest redshift galaxies are almost inevitably more radio luminous (since they derive from a flux-limited sample), and we need to be sure that this does not correlate with stellar luminosity. Putting all these uncertainties together, it is clear that testing $q_0 = \Omega/2$ to any degree of precision is not feasible. The best that can be said is that the lack of curvature in the Hubble diagram sets important limits to the total density; in particular, the vacuum contribution must satisfy

$$\Omega_{\rm vac} \lesssim 1. \tag{13.37}$$

This is similar to the constraint that can be set by the supernova Hubble diagram (see chapter 5); these objects have a smaller dispersion in luminosity, but have not yet been seen at such extreme redshifts. A final important fact is that the scatter in the Hubble diagram stays small out to $z \simeq 3$: all the galaxies we see at that epoch must be of similar ages to avoid a large spread in stellar luminosities. Unless one contrives a picture in which galaxy formation occurs in a very narrow redshift range, this argues that the stars in these objects formed at redshifts a good deal higher than $z = 3$.

NUCLEOSYNTHESIS AND THE OPTICAL/IR BACKGROUND Despite the usual way in which Olbers' paradox is stated, the night sky is not entirely dark. Even without a telescope, a general glow can be perceived by a dark-adapted observer at a location well removed from city lights. It is this radiation that makes observational cosmology so hard: faint galaxies are swamped by the random noise of the arrival of sky photons, in the same way as the bright stars become invisible when the sky brightness increases during the day. Almost all this sky background is generated either by atmospheric emission or zodiacal light (sunlight scattered from Solar-system dust). The former varies with the sunspot cycle; some typical numbers for a good ground-based observing site are $B = 22.3\,\mathrm{mag\,arcsec^{-2}}$ and $B = 19.3\,\mathrm{mag\,arcsec^{-2}}$ (these are the flux densities emitted from one square arcsec of sky). The night sky is rather red, owing to the strong OH emission lines from atmospheric water vapour; in space, the blue number is not greatly different, whereas the I-band number drops by about two magnitudes. It is this difference, as much as the extra image sharpness, which explains why the **Hubble Space Telescope** (HST) has had such an impact on studies of galaxy evolution, through projects such as the Hubble Deep Field (Williams *et al.* 1996).

To measure any true extragalactic component of this light, one must attempt to model the atmospheric and zodiacal components. This is not so hard as it sounds: for example, the atmospheric contribution should vary as $1/\cos\theta$, where θ is the angle from the zenith. An alternative is to observe on and off a dark galactic dust cloud, although this is not quite straightforward, since the cloud can also scatter radiation into the line of sight. Both methods give similar limits of about 1% of the total sky brightness:

$$\nu I_\nu \lesssim 3 \times 10^{-8}\ \mathrm{W\,m^{-2}\,sr^{-1}}. \tag{13.38}$$

We now turn to the question of how close this limit lies to the total contribution of faint galaxies.

One important question is whether we have seen the majority of star formation in galaxies. We know how much activity must have occurred through simple energetic arguments, based on the known abundance of 'metals' (elements more massive than helium). The metal density can be estimated by knowing the galaxy luminosity density, plus something about the stellar populations of these galaxies, and hence their metallicities. The answer is, to within a factor $\lesssim 10$

$$\rho Z \simeq 3.6 \times 10^{-31} \, h^2 \, \mathrm{kg\,m^{-3}}. \tag{13.39}$$

The total contributed by enriched gas in clusters is roughly a factor 2 larger than this (Songaila, Cowie & Lilly 1990). Clearly, this density is only a lower limit corresponding to material that has been seen; if the dark matter were baryonic, the figure would be increased by an order of magnitude.

Now, most of these metals are manufactured in stars rather more massive than the sun: O and B stars, with short lifetimes ($\sim 10^7$ years). The mass-to-energy conversion efficiency is fixed by nuclear physics: $\epsilon \simeq 0.01$ (see e.g. Pagel 1997). The known metal density plus this nucleosynthetic efficiency therefore tells us how much energy must have been radiated away by all the stars that have ever existed. Even better, by a piece of good luck in the way the cosmological effects come in, we can make a robust prediction of the optical background that should be observed. Suppose that we know exactly the galaxy luminosity function as a function of redshift, so that $\rho(L, z)$ gives us the epoch dependence of the comoving density of galaxies in the range dL/L. The metal density is the result of integrating the energy output over time,

$$\rho Z = \frac{1}{\epsilon c^2} \iint \rho(L,t)\, L \, \frac{dL}{L} \, dt. \tag{13.40}$$

Now, we add the critical assumption that the star-forming galaxies will have 'flat spectra', f_ν constant, so that the relation between observed flux density and bolometric luminosity is $L = 4\pi BSD^2 (1+z)$, where D is comoving distance and B is a 'bolometric correction', corresponding to the assumption that the spectrum is flat up to the Lyman limit at 912 Å, at which point stellar atmospheres become optically thick to their own radiation. It is easy to show from this [problem 13.3] that the metal density from these galaxies is just directly proportional to the observed surface brightness of the population, integrating over all fluxes and redshifts:

$$\rho Z = \frac{4\pi B}{\epsilon c^3} \iint N(S,z)\, S \, \frac{dS}{S} \, dz = \frac{4\pi B}{\epsilon c^3} I_\nu. \tag{13.41}$$

With the canonical efficiency of $\epsilon \simeq 0.01$ for the operation of nucleosynthesis in stars, this translates to a predicted background of

$$I_\nu = 6.5 \times 10^{-25} \left(\frac{\rho Z}{10^{-31}\,\mathrm{kg\,m^{-3}}} \right) \, \mathrm{W\,Hz^{-1}\,m^{-2}\,sr^{-1}}. \tag{13.42}$$

Songaila, Cowie & Lilly (1990) show that this is very close to the background obtained by integrating over the observed population of faint blue galaxies.

THE GLOBAL HISTORY OF STAR FORMATION This global measure of metal production in the universe can be made even more precise once we have redshifts for a faint galaxy sample. This allows a measurement of the contributions to the total background from different redshift shells – which gives a direct measure of the rate of metal production as a function of time, $\dot{\rho}_Z$. Even better, since the stars that make the metals are very short-lived, there is a direct proportionality between this rate and the total star-formation rate, $\dot{\rho}_*$ (once an IMF shape has been assumed). Such a dissection was first carried out

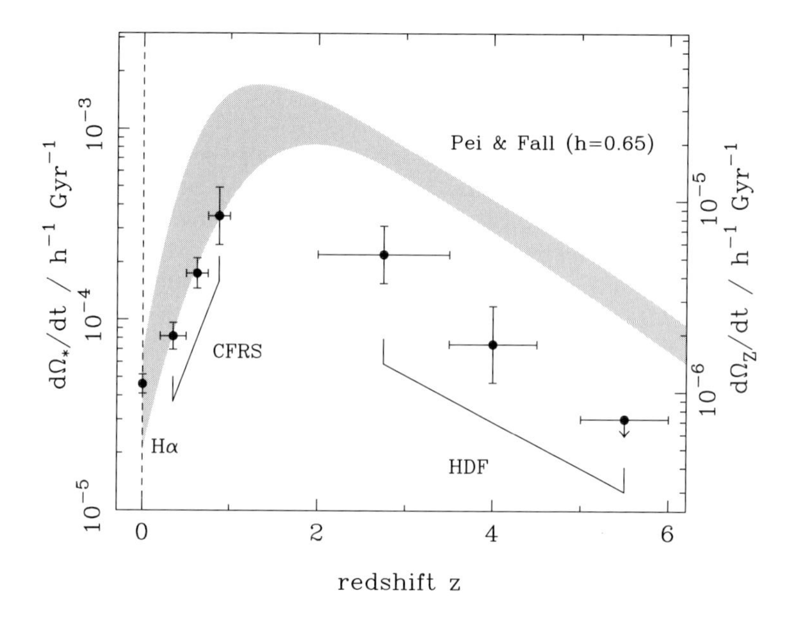

Figure 13.9. The global star-formation history of the universe. The total UV luminosity density of galaxies at a given epoch can be related directly to the formation rate of metals, since both originate in massive stars. Assuming a Salpeter IMF, this can be converted to a total star formation rate. At $z < 1$, this was measured by the Canada–France Redshift Survey of Lilly *et al.* (1996), and at higher redshifts from the Hubble Deep Field (Madau *et al.* 1996). The zero-redshift point was inferred from line-emitting galaxies by Gallego *et al.* (1995). The shaded region shows the prediction due to Pei & Fall (1995). They argued from the observed column densities in QSO absorbers, plus the low metallicities in these systems, that the star-formation rate must have peaked between $z = 1$ and $z = 2$. Although the data lie somewhat below the prediction, a plausible (small) amount of dust in the high-z galaxies would easily close the gap.

by Lilly *et al.* (1996), and it reveals a star-formation rate per unit comoving volume roughly ten times higher at $z = 1$ than at present (figure 13.9).

Beyond $z = 1$, very few galaxies are found in complete magnitude-limited samples, and this redshift regime long belonged only to luminous active galaxies. However, the impasse was broken when it was realized that broad-band colours could be used to select galaxies at extremely high redshifts. An actively star-forming galaxy will have a flat spectrum down to the Lyman limit at 912 Å, beyond which the spectrum will be truncated by intervening absorption. At $z \simeq 3$, this spectral break has moved to the centre of the U band, and will produce galaxies that are red in a colour like $U - B$. For $z > 3.4$, the redshifted Lyman limit risks confusion with the 4000 Å break at low redshift, so that **Lyman-limit galaxies** can be picked out in a narrow redshift window by looking for objects that are red in $U - B$ but flat-spectrum at longer wavelengths, giving a very efficient window on star-forming activity at high redshift (Steidel *et al.* 1996). The critical result of these studies is that no very luminous starbursts are found: the star-formation rates at $z \simeq 3.4$ are typically $10\ M_\odot \mathrm{yr}^{-1}$, only perhaps one-tenth of the value expected when a massive galaxy forms in a single burst. With the ultraviolet capability of the HST, it became possible to perform these studies in other redshift ranges, and so measure the history of the star-formation rate. Figure 13.9 compares results from the Hubble Deep Field with the ground-based data for $z < 1$; it appears that the star-formation rate must have peaked between $z = 1$ and $z = 2$ (Madau *et al.* 1996). The big question now is whether there could have been a second peak at even higher redshift. To settle this, we have to move to longer wavelengths.

DUSTY PRIMAEVAL GALAXIES The global energetics of star formation can be cast in other forms. One version is to say that there is a direct relation between the rate of star formation in a galaxy and its luminosity (assuming the latter to be dominated by the young stars):

$$\boxed{1\ M_\odot\ \mathrm{yr}^{-1} \quad \Rightarrow \quad L = 2.2 \times 10^9 L_\odot = 8.4 \times 10^{35}\ \mathrm{W}} \tag{13.43}$$

(Baron & White 1987). This corresponds to an effective conversion efficiency of mass into energy of 1.5×10^{-4} for material processed by stars. This is much lower than the canonical 1% nucleosynthesis efficiency because massive stars only burn a small fraction of their fuel before leaving the main sequence to ascend the giant branch. This argument allows us to set some limits on how bright a galaxy at high redshift must be. The best one can do is to spread the star formation out over the entire age of the universe at that time; more concentrated star formation will lead to even higher luminosities. At high redshift, the local density parameter will be close to unity, whatever Ω is today, and so the age of the universe will be very nearly

$$t = \frac{2}{3H(z)} = \frac{2}{3H_0(1+z)\sqrt{1+\Omega z}}. \tag{13.44}$$

So, for example, a galaxy with a stellar mass of $10^{11} M_\odot$ at $z = 3$ in a flat universe must have a luminosity of at least $1.8 \times 10^{11} h^{-1} L_\odot = 11hL^*$, roughly ten times the current luminosity of such a galaxy.

Moreover, stellar population models imply that about 10% of this energy will be radiated in the Lyα emission line alone, implying a rest equivalent width of around 100 Å (Kennicutt 1983). A good survey strategy for finding primaeval galaxies at $z \lesssim 4$ is thus to search blank areas of sky in narrow passbands looking for objects whose emission is dominated by only one strong line – but no such objects have ever been found. The reasons for this may be that canonical protogalaxies do not exist, and that star formation mainly takes place in low-mass units and is spread out over the history of the universe. Another possibility is that the galaxies cannot be seen because the initial burst of star formation rapidly became shrouded in dust. The example of low-redshift 'starburst galaxies' such as M82 certainly makes it plausible that this may have happened. In this case, we can still use the nucleosynthesis energy argument, but now the radiation will appear at wavelengths much longer than the optical: we expect an infrared/submillimetre background.

The argument is similar to the above. If all the energy is produced at a redshift z, this produces an energy density of $U = \epsilon \rho Z c^2 (1+z)^3$, and hence a total radiation specific intensity $I_{\mathrm{tot}} = Uc/4\pi$. This will evolve $\propto a^{-4}$ (three powers of scale factor for the effect of the expansion in diluting proper density of photons, one for the redshifting of photon energy), so that the current background is

$$I_{\rm tot} = \frac{\epsilon \rho Z c^3}{4\pi(1+z)} = \frac{2.1 \times 10^{-9}}{1+z} \left(\frac{\rho Z}{10^{-31}\,{\rm kg\,m^{-3}}} \right) {\rm W\,m^{-2}\,sr^{-1}}. \tag{13.45}$$

Currently, the conservative COBE limits on the extragalactic background in the IR/submillimetre region are approximately

$$\nu I_\nu \lesssim 10^{-7}\ {\rm W\,m^{-2}\,sr^{-1}}, \tag{13.46}$$

excluding the region around 12 μm, where the emission from interplanetary dust has a peak. Clearly, the removal of the various galactic sources of emission will have to be done to better than 0.1% in order to reach the predicted levels. However, the COBE database at many frequencies is sufficiently extensive that this may well prove possible; indeed, Puget *et al.* (1996) have claimed a detection of $\nu I_\nu \sim 5 \times 10^{-9}\ {\rm W\,m^{-2}\,sr^{-1}}$ at long wavelengths ($\lambda \sim 300\,\mu$m). If upheld, this will be an important constraint on high-redshift galaxy formation.

Problems

(13.1) Show that if the light from a stellar population is dominated by the emission from short-lived giants then its luminosity is expected to decline as a power of time, approximately as $L \propto t^{-0.5}$.

(13.2) Broadhurst *et al.* (1992) have explored empirical merger models that conserve mass and evolve the luminosity function homologously, so that

$$\phi_* L_* = \text{constant}, \quad \phi_* \propto \exp(Q\tau). \tag{13.47}$$

Here, τ is the normalized look-back time: $\tau \equiv Ht \simeq z$ for $z \ll 1$. Show that very large values of Q are permitted without altering the galaxy number counts significantly from the no-evolution form.

(13.3) Show that, for galaxies that emit with a constant f_ν up to the Lyman limit, the local metal density is directly proportional to the surface brightness of the integrated light seen on the sky from past generations of these galaxies.

(13.4) One way in which actively star-forming galaxies at very high redshift might be detected is through redshifted molecular-line emission. Discuss the feasibility of detecting objects at $z = 10-20$ through observations of the rotational transitions of CO, where the transition from level $J+1$ to J corresponds to a frequency of approximately $\omega = \hbar(J+1)/I \simeq 115(J+1)\,$GHz. Such low-level transitions are observed to have a luminosity in excess of $10^9\,L_\odot$ in nearby starburst galaxies.

14 Active galaxies

Advances in observational astronomy during the 1990s have finally allowed direct study of the population of normal galaxies at high redshifts, as discussed in chapter 13. For more than two decades prior to this, the only objects that could be studied out to cosmologically important distances were 'active' galaxies such as quasars, where the dominant energy output is not due to stars. This chapter attempts to separate out those aspects of active galaxies that are of especial cosmological interest, although any such division is inevitably blurred. Many of the interesting details of the subject will be omitted: good references for digging deeper into this area are Weedman (1986), Blandford, Netzer & Woltjer (1990), Hughes (1991) and Robinson & Terlevich (1994).

14.1 The population of active galaxies

CLASSIFICATION Active galaxies come in a variety of species, many of which overlap. The definition of an active galaxy is one where a significant fraction of the energy output in at least some waveband is not contributed by normal stellar populations or interstellar gas; however, all galaxies emit **non-thermal radiation** at some level, and so classification as active is only a matter of degree. Things are a little more clear cut with **AGN** (active galactic nuclei), where the non-thermal emission comes mainly from the central few pc of the galaxy.

Historically, the most spectacular advance in AGN research was the discovery of the first quasi-stellar objects (QSOs), or **quasars**, which turned up as a result of optical identifications for the early radio surveys (Schmidt 1963). These are the most powerful astronomical sources of radiation known, and their extreme properties caused controversy about their nature – which still persists to some extent even today. Sensitive high-resolution optical imaging is now capable of proving that a quasar is in fact an entire galaxy being out-shone by point-like optical emission from the active nucleus. This would have seemed less surprising in 1963 had research into **Seyfert galaxies** been more advanced. These galaxies are named after their (1943) discoverer, Carl Seyfert (pronounced 'seefurt'), although one nearby example (NGC1068) was noted even in the early days of galaxy spectroscopy, by Slipher in 1918. Seyferts are galaxies with unresolved spikes of nuclear light and strong nuclear emission lines having velocity widths $\gtrsim 1000\,\mathrm{km\,s^{-1}}$. These are exactly the properties that characterize quasars, and indeed, the two classes of object are part of a continuous population. Some active galaxies have strong nuclear emission lines that are much narrower, with velocity widths of a few hundred $\mathrm{km\,s^{-1}}$; these are known as Seyfert 2 galaxies. Sometimes, a continuous classification is attempted in terms of the line width, and workers will speak of a 'Seyfert 1.4'. However, some galaxies can change their line widths during outbursts, and there is evidence that the appearance of Seyferts (and indeed all AGN) may depend on orientation (see below).

Seyferts were effectively rediscovered by Sandage (1965), who demonstrated that selection by **ultraviolet excess** ($U - B \lesssim -0.3$) detected a population of quasars that exceeded more than tenfold those found in radio survey identifications. Ever since, the relation between **radio-quiet quasars** and **radio-loud quasars** has been a source of debate. All quasars emit radio radiation to some extent, and so the distinction is partly a matter of definition. Miller *et al.* (1990) argued that quasars are bimodal in radio output, with a clear desert around $L_\nu(5\,\mathrm{GHz}) \sim 10^{24}h^{-2}\,\mathrm{W\,Hz^{-1}sr^{-1}}$, but much work continues to use Schmidt's (1970) definition, where the division is made at a radio to optical flux-density ratio of $R \sim 10$. Not all AGN are Seyfert-like, however; there exist comparable numbers of radio-loud quasars and **radio galaxies**. These often have the optical appearance of a normal elliptical galaxy, but radio mapping reveals similar structures to those around radio-loud quasars: roughly linear structures of extent often > 100 kpc. The active nucleus reveals itself by the jets that power these structures (see figure 14.1), but is less apparent in direct emission at higher frequencies.

This cast list was extended in the 1980s by the addition of **starburst galaxies**. As the name implies, these are galaxies where a significant fraction of the total stellar mass of the galaxy was generated near to the time of observation, so that the properties associated with active star formation (blue high-mass stars and dust) dominate the observational signature of the galaxy. The appreciation of the importance of this class of galaxies was greatly enhanced after the 1983 mission of the Infrared Astronomy Satellite (**IRAS**). By observing in the 10–100 μm waveband, this was able to pierce dust-shrouded star-forming regions and frequently to reveal activity in galaxies hitherto thought to be normal. The defining characteristic of starburst galaxies is that the far-infrared emission from dust heated by the young stars gives the dominant energy output. From this point of view, many classical quasars and Seyferts are also starbursts. Indeed, one of the key questions being debated in the field is which is the chicken and which the egg, in this case. As will be outlined below, the standard view is that active galaxies are powered by gravitational energy deriving from a massive black hole. However, for low-luminosity AGN, the nuclear energy released in a starburst may be sufficient to generate the observed AGN phenomena without the existence of a black hole (see Terlevich *et al.* 1992).

OBSERVATIONAL PROPERTIES There is far from sufficient room here to do justice to the many striking observational properties of active galaxies, which have been pursued in great detail over the years. Many of these details can be found in the books by Weedman (1986) and Robson (1996). An illustration of some of the facts we need to explain is given in figures 14.1 and 14.2. They may be summarized under the following headings.

(1) Point-like variable nuclear continuum. The defining feature of a quasar is the featureless continuum radiation that originates in the nucleus. This radiation may extend over a colossal range in frequency, from γ-rays down to radio. The overall spectrum shows features, but νS_ν is often remarkably close to a constant: the same output to within a factor 10 over perhaps 15 decades of frequency. The

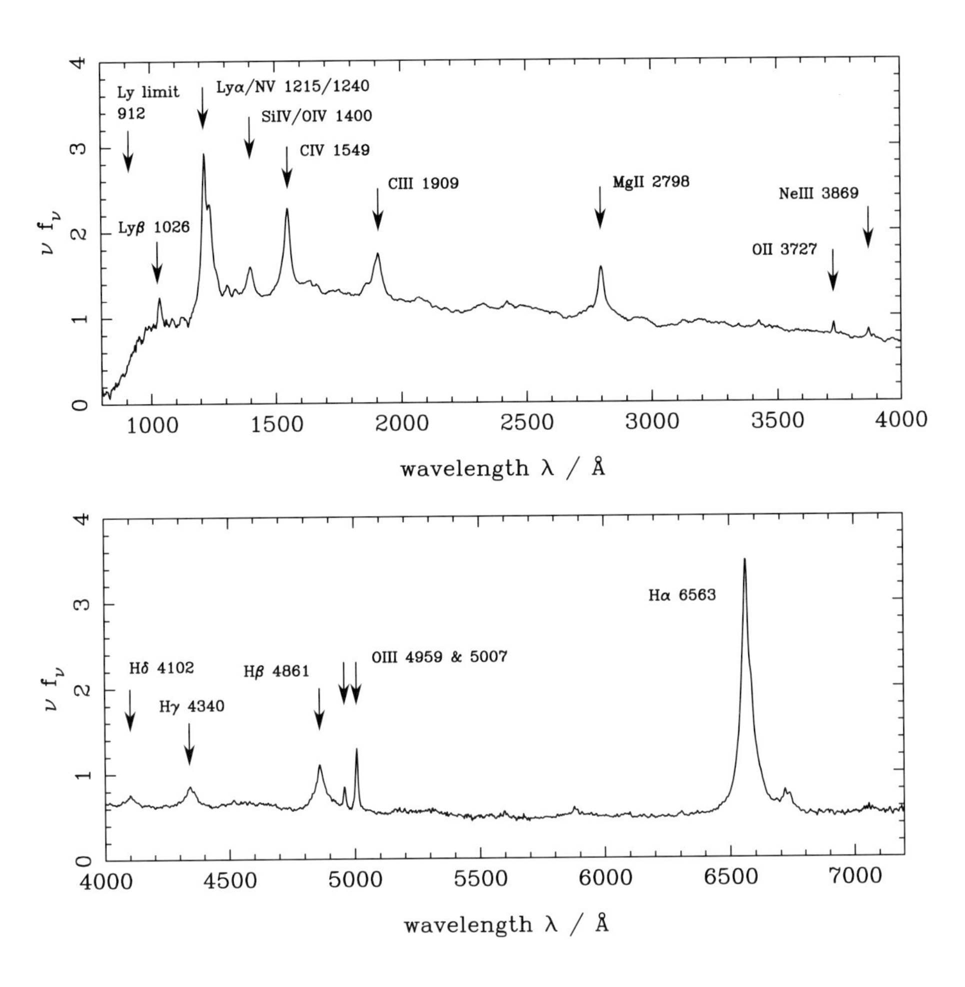

Figure 14.1. The two outstanding characteristics of some of the most energetic active galaxies are shown in this figure and the one following. The optical spectra of quasars display a mixture of broad and narrow emission lines; the broad lines show that there exists a potential well sufficiently deep that some material orbits at a substantial fraction of *c*. This composite quasar spectrum is an extension (courtesy of C.B. Foltz and P.C. Hewett) of that published by Francis *et al.* (1991).

continuum shows variations on all timescales, down to hours in the case of X-ray emission, and is often significantly polarized.

(2) Strong narrow and broad emission lines. The spectra of almost all AGN contain strong narrow emission lines. These are the **nebular lines** (principally from OII and OIII) familiar from HII regions, indicating gas that is photoionized (collisional ionization would not yield such a strong admixture of the more highly ionized species). In addition, some but not all quasars exhibit broad lines with Doppler widths of order several thousand $\mathrm{km\,s^{-1}}$ (principally in lines of hydrogen, carbon and magnesium). The **broad-line region** is clearly compact and close to the nucleus,

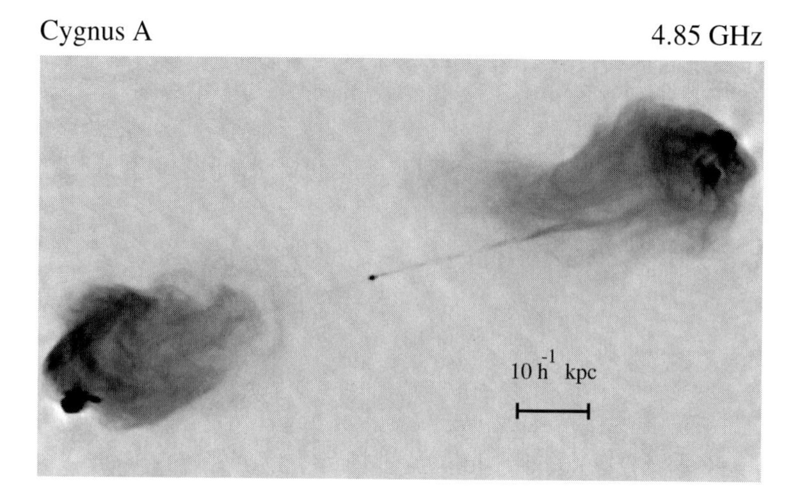

Figure 14.2. An interferometric map of the classic radio galaxy Cygnus A, made with the NRAO Very Large Array (courtesy of C. Carilli). Note the narrow jets, which transport material at possibly relativistic velocities to distances of order 100 kpc from the nucleus. This observation argues for a collapsed massive object, in order both to provide gravitational energy to power the jets and to produce a well-defined stable axis, which could be the rotation axis of a black hole.

if the velocities are due to gravity. The **narrow-line region** is conversely more distant and consists of lower density gas, which is often extended on kpc scales. Two classes of AGN lack measurable emission lines: (i) low-luminosity radio galaxies, where the optical properties are those of an inactive giant elliptical; (ii) **BL Lac objects** or **blazars**, which are radio-loud AGN where the polarized continuum is particularly strong.

(3) Jets. Most radio-loud AGN exhibit structures that are approximately linear and symmetrical, extending well beyond the optical body of the galaxy to distances of tens of kpc up to 1 Mpc. As will be outlined below, it had become clear by around 1970 that emission at these distances required a continuous supply of energy from the nucleus by a 'beam' or 'jet'. The latter term has tended to be more frequently used as interferometer arrays have made ever better maps of the radio emission and revealed emission from narrow features that are almost certainly supplying the required energy via bulk kinetic energy (figure 14.2).

HOST GALAXIES What sort of galaxies host these active nuclei? Astronomers have sought to answer this question in the hope of learning some clue about the causes of the activity, and some interesting patterns have emerged. One of the most striking systematic property of active galaxies is the dependence of galaxy activity on mass. Above a certain level of activity, AGN are almost inevitably sited in the most massive galaxies. The other clear trend is that the *radio-loud* AGN above a certain level are universally found in elliptical (and S0) galaxies. Surveys of nearby E/S0 galaxies allow us to deduce the probability distribution for radio emission as a function of optical luminosity (e.g. Sadler *et al.* 1989). For the most luminous galaxies ($M_B = -21 + 5\log_{10} h$, equivalent to several times L^*), there is a break in the radio distribution around $P = 10^{23.5}$ W Hz^{-1} sr^{-1} at 1.4 GHz. About 30 per cent of galaxies lie above this break, beyond which $F(> P)$ scales roughly as $1/P$, with a shallower dependence at lower powers. For only slightly fainter ellipticals ($M_B = -19$), the situation is completely different, with perhaps only ~ 1 per cent of galaxies exceeding even 10^{21} W Hz^{-1}sr^{-1}. This is the level of emission produced by interstellar synchrotron emission in spiral galaxies, so it seems that the probability of producing a genuine active nucleus is a very strong function of galaxy mass.

Similar biases to high mass obtain for quasars, although this is harder to show since one needs to image the underlying galaxy by subtracting the point-like central continuum source. What is found by doing this (e.g. Smith *et al.* 1986; Taylor *et al.* 1996) is that the underlying galaxies are luminous in all cases, although those underlying radio-loud quasars are 0.5–1 magnitude brighter. This is consistent with the idea that radio-loud quasars are hosted by massive ellipticals also, and that radio-quiet quasars inhabit mainly spirals. It has to be admitted that the causes of this difference are not well understood. The nuclear processes occur on scales very much smaller than the bulk of the galaxy; it is a little surprising that global differences in properties such as gas content and merger history should change the properties of the central engine so clearly.

Worse still, these differences can be traced to even larger scales. Since the early suggestions that radio sources such as Cygnus A might be 'colliding galaxies', astronomers have been interested in learning whether these objects occupy regions of enhanced density where interactions or mergers might be more common. The main quantitative tool used in these studies has been the AGN-galaxy cross-correlation function $\xi(r)$ (see chapter 16). This function is defined to be the excess probability over random of finding a galaxy at a distance r away from the AGN – i.e. it describes the density profile of any cluster around the AGN; it can be estimated in practice by looking for an excess in the numbers of galaxies in the vicinity of the AGN. Using this method, it has been possible to show that radio-loud AGN inhabit regions of space with densities approaching that of Abell richness class 0; this is consistent with their location in luminous elliptical galaxies, which are preferentially found in clusters (e.g. Longair & Seldner 1979; Yates *et al.* 1989). Radio-quiet AGN, conversely, are not located in regions of significant enhancement; this is consistent with the hypothesis that the host galaxies are spirals (Boyle & Couch 1993).

The range of observational phenomena in this summary is very wide, and it is a particular challenge to understand whether a single type of 'central engine' can yield these diverse properties. The first step along the road is to consider the physical means by which AGN are able to shine.

14.2 Emission mechanisms

The emission from AGN is characterized by the presence of relativistic electrons, as may be seen by thermodynamic arguments. The surface brightness of a body in equilibrium at a given temperature cannot exceed the Planck function at that temperature. If we

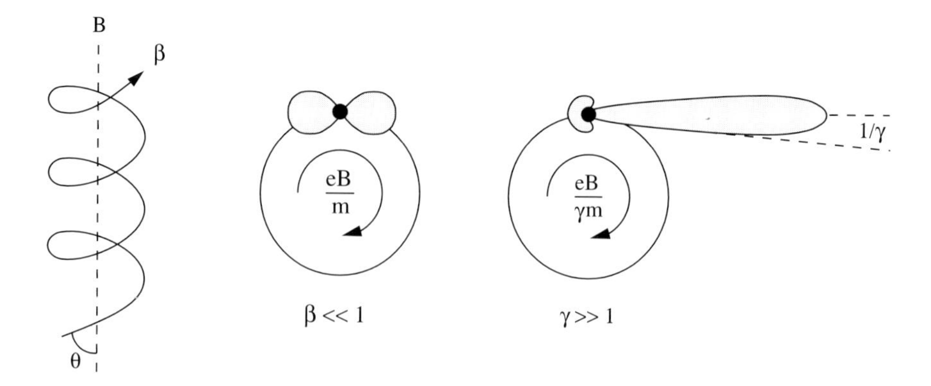

Figure 14.3. Synchrotron radiation is generated by relativistic electrons spiralling around field lines, with pitch angle θ. At low velocities, the gyrofrequency is $\omega = eB/m$ and the power pattern is symmetric about the (radial) acceleration vector. For ultrarelativistic electrons, the angular gyrofrequency is reduced by a factor γ, and the radiation is beamed forwards by relativistic aberration into a cone of semi-angle $\sim 1/\gamma$.

define the **brightness temperature** by analogy with the Rayleigh–Jeans law,

$$\boxed{T_{\rm B} \equiv \frac{c^2 I_\nu}{2k\nu^2},} \tag{14.1}$$

then an observed $T_{\rm B}$ in excess of $10^{9.8}$ K ($kT = m_e c^2$) implies relativistic electrons. As we shall see, brightness temperatures in AGN do often exceed this level, and so we are led to consider emission processes involving relativistic electrons.

SYNCHROTRON RADIATION This is the emission mechanism that characterizes the most spectacularly active AGN, and we shall therefore treat it a little more fully than some of the other possibilities. Synchrotron emission arises via the gyration of relativistic electrons in magnetic fields, as illustrated in figure 14.3; the detailed properties of this radiation can become rather messy, but the basics can be obtained in a simplified but still useful form.

For a relativistic electron, we need the relativistic equation of motion $F^\mu = m\,dU^\mu/d\tau$ (F^μ is 4-force and U^μ 4-velocity). A covariant expression for this in terms of the field tensor $F^{\mu\nu} = \partial^\mu A^\nu - \partial^\nu A^\mu$ is $m\,dU^\mu/d\tau = eF^{\mu\nu}U_\nu$. Since the spatial part of the 4-force is just $\gamma\mathbf{F}$, taking spatial components of the equation of motion proves that the Lorentz force is covariant and gives us

$$\mathbf{F} = m\gamma\ddot{\mathbf{x}} + (\mathbf{F}\cdot\dot{\mathbf{x}})\dot{\mathbf{x}}/c^2 = e(\mathbf{E} + \dot{\mathbf{x}}\wedge\mathbf{B}). \tag{14.2}$$

For force that is purely magnetic, force and velocity are orthogonal, so this simplifies to give helical motion about the field line at the **gyrofrequency**, which is $1/\gamma$ times the nonrelativistic **Larmor frequency** $\omega_{\rm L}$:

$$\omega_g = \frac{eB}{\gamma m} \tag{14.3}$$

(m denotes rest mass throughout). We now need the rate of radiation of energy caused by this motion. In the electron rest frame, this is given by the standard result for the **Hertzian dipole**,

$$-\dot{E} = \frac{|e\ddot{x}|^2}{6\pi\epsilon_0 c^3}. \tag{14.4}$$

To convert this to the observer's frame, we need to work out how $\dot{E}$ and $\ddot{x}$ transform. First consider the 4-vector $dP^\mu/d\tau$: if the rate of change of momentum, $\dot{p}$, is zero (as it will be for a symmetrically radiating electron, ignoring radiation reaction), this tells us that $\dot{E}$ is the same in all frames. The acceleration is transformed using the norm of the 4-acceleration: $\mathscr{A}^\mu \mathscr{A}_\mu = -a_0^2$, where a_0 is the rest-frame acceleration, and the 4-acceleration $\mathscr{A}^\mu$ is not to be confused with the 4-potential A^μ, although the same notation is generally used for both elsewhere in the book. In general, $\mathscr{A}^\mu = \gamma(d/dt)(\gamma, \gamma\dot{\mathbf{x}})$, but γ is a constant for magnetic force, so this simplifies to $\mathscr{A}^\mu = (0, \gamma^2 \mathbf{a})$, implying $a_0^2 = (\gamma^2 a)^2$. Putting all this together gives the observed rate of radiation,

$$\boxed{-\dot{E}_{\rm synch} = 2\sigma_{\rm T}\, \gamma^2 \beta^2 \sin^2\theta\, U_{\rm mag} c,} \tag{14.5}$$

where $\sigma_{\rm T}$ is the Thomson cross-section, $\beta = v/c$, θ is the pitch angle ($\cos\theta = \hat{\mathbf{v}} \cdot \hat{\mathbf{B}}$), and the magnetic energy density $U_{\rm mag} = B^2/2\mu_0$.

The detailed spectrum of the radiation is found by Fourier-transforming the time-dependent emission from the oscillating electron. A simpler alternative is to note that relativistic aberration means that the radiation from the electron will be largely confined to a cone of semi-angle $\sim 1/\gamma$, which sweeps over the observer in a time $\sim (\gamma^3\omega_g)^{-1}$ [problem 14.1]. For a broad spectrum of electron energies, it should therefore be reasonable to treat the emission as a delta function at a frequency $\nu = \gamma^3 \nu_g$. In this case, the **synchrotron emissivity** $\epsilon_\nu = 4\pi j_\nu = -\dot{E}\, N(E)\, dE/d\nu$, where $N(E)$ is the number of emitting electrons in dE. Now consider a power-law spectrum $N(>\gamma) = N_0\gamma^{1-x}$, where N_0 is the total electron number density. This gives $N(E) = N_0(x-1)(mc^2)^{x-1}E^{-x}$, and the approximate synchrotron emissivity becomes

$$\epsilon_\nu = N_0\,(x-1)\,\frac{\mu_0 c e^3}{18\pi m}\left(\frac{e}{m}\right)^{(x-1)/2} B^{(x+1)/2}\,\nu^{(1-x)/2} \tag{14.6}$$

(assuming isotropy, so that $\langle \sin^2\theta \rangle = 2/3$). This gives an important result, the **spectral index** of the emission:

$$\boxed{\epsilon_\nu \propto \nu^{-\alpha}, \qquad \alpha = \frac{x-1}{2}.} \tag{14.7}$$

For a (much longer) derivation of the exact result, consult Rybicki & Lightman (1979). A convenient form of the answer is

$$\epsilon_\nu = 2.14 \times 10^{-30}(x-1)\,N_0\,a(\alpha)\left(\frac{B}{100\ {\rm nT}}\right)^{1+\alpha}\left(\frac{\nu}{4\ {\rm kHz}}\right)^{-\alpha}\ {\rm W\,Hz^{-1}}, \tag{14.8}$$

where $a(\alpha)$ is a function tabulated by the above authors, which is close to 0.08 for spectral indices in the practical range 0.5 to 1. It is clear that this radiation must be strongly polarized, with an **E** vector parallel to the direction of the acceleration (i.e. perpendicular to the magnetic field direction). The polarization would be 100% if all electrons orbited in circles with pitch angle zero, and the magnetic field were perpendicular to the line of sight. For more general configurations, something closer to 50% is expected, unless the field is tangled, causing the signal to average out.

This above expression for the synchrotron emissivity applies so long as the source is **optically thin**: there must be insufficient radiating material to re-absorb any of its own radiation. To see when this is important, consider the **radiative transfer equation**:

$$\boxed{\frac{dI_\nu}{dx} = j_\nu - \kappa I_\nu,} \tag{14.9}$$

which has the solution $I_\nu = (j_\nu/\kappa)[1 - \exp(-\kappa x)]$, for a uniform slab of thickness x. We can obtain an approximate expression for the **absorption coefficient** κ by a thermodynamic argument. An electron 'temperature' is given as a function of energy by $3kT = \gamma mc^2$, and so we expect a limit to the surface brightness at low frequencies to follow the Rayleigh–Jeans law, $I_\nu = 2kT\nu^2/c^2$. The limiting surface brightness is then

$$I_\nu < \frac{2\nu^2}{3c^2}\gamma mc^2 = \frac{2}{3} m\nu_{\mathrm{L}}^{-1/2}\nu^{5/2}, \tag{14.10}$$

where ν_{L} is the Larmor frequency. To interpolate between this expression and the optically thin $\nu^{-\alpha}$ behaviour that holds for high frequencies, the radiative transfer solution must look like

$$I_\nu \propto \left(\frac{\nu}{\nu_0}\right)^{5/2} \left\{1 - \exp\left[-\left(\frac{\nu}{\nu_0}\right)^{-(\alpha+5/2)}\right]\right\}. \tag{14.11}$$

The spectrum thus has a peak at a frequency that depends on the magnetic field and the depth of the source.

EQUIPARTITION FIELD How can we determine the magnetic field strength, on which much of the properties of the synchrotron source depend? The best we can usually do is to make an energy argument, as follows. For a source of fixed volume, V, the total magnetic energy rises as B^2; what about the energy in relativistic electrons? From our previous expression for j_ν, plus $E = mc^2(\nu/\nu_{\mathrm{L}})^{1/2}$, and using the luminosity $P_\nu = V j_\nu$ for low optical depth, we get

$$E_e = \frac{3\pi mc\sqrt{\nu_{\mathrm{L}}}}{\sigma_{\mathrm{T}} U_{\mathrm{mag}}} \int P_\nu \nu^{-1/2}\, d\nu, \tag{14.12}$$

so that particle density scales as $B^{-3/2}$. There is therefore a point of minimum energy where $E_e = (4/3)E_{\mathrm{mag}}$. Invoking this condition allows an estimate of the magnetic field to be made. If $E_e \equiv AB^{-3/2}$, then

$$\begin{aligned} E_{\mathrm{min}} &= A^{4/7}(V/\mu_0)^{3/7} \\ B_{\mathrm{min}} &= (3A\mu_0/2V)^{2/7}. \end{aligned} \tag{14.13}$$

In practical units, this takes the form

$$(B_{\rm min}/{\rm nT})^{7/2} = (V/{\rm kpc}^3)^{-1} \int (P_\nu/10^{21}\ {\rm W\,Hz^{-1}\,sr^{-1}})\, dy/\sqrt{y}, \tag{14.14}$$

where $y \equiv \nu/{\rm GHz}$. Note that this value (sometimes known, slightly inaccurately, as the **equipartition field**) is *not* a minimum value for the magnetic field – merely that which corresponds to the lowest total energy. Whether it is reasonable to expect equipartition to be satisfied depends on how the fields and particles in the radiating object are generated. In some models, where energy is exchanged between the two, approximate equipartition is expected, but this is not always the case.

The above simple formulae differ from their exact equivalents by typically $\lesssim 30\%$, depending on spectral index. This is quite acceptable, given the observational uncertainties and assumptions involved in their applications. We now list these explicitly, since they make some of the parameters inferred for synchrotron sources worryingly uncertain: (i) we have ignored any contribution of nucleons to the particle energy; (ii) we have assumed isotropy of the electron pitch angle distribution; (iii) we have assumed that the radiating volume is homogeneous. These uncertainties all add to the basic worry about whether energy equipartition holds even approximately. By the nature of a minimum, the total energy is not very sensitive to departures from equipartition, but the field strength certainly is. Nevertheless, apart from the limits set by inverse Compton emission (see below) there is no other way of getting even an approximate idea of the likely magnitude of the magnetic field.

INVERSE COMPTON RADIATION A further signature of ultrarelativistic electrons is the radiation produced by the scattering of a population of low-energy photons. To analyse this, we can use the above result for synchrotron radiation: $\dot{E}$ will be an invariant in this case also. So, consider the set of photons moving in the lab frame at an angle θ to the electron's velocity, with a density n. The 4-current is $J^\mu = nc(1, \cos\theta)$ and so the density in the electron's frame is $n' = n\gamma(1 - \beta\cos\theta)$, and the photons have frequency $\omega'/\omega = n'/n$ (transforming the 4-wavevector similarly). The rate of radiation, $-\dot{E}$, is thus $nc\sigma_{\rm T}\hbar\omega\gamma^2(1 - \beta\cos\theta)^2$. Integrating over an isotropic photon distribution gives a total photon number density $n_{\rm tot}$ and a total energy loss rate $-\dot{E} = c\sigma_{\rm T}\gamma^2(1+\beta^2/3)U_{\rm rad}$, where the energy density in radiation is $U_{\rm rad} = n_{\rm tot}\hbar\omega$. We are not quite done, as the electron is also removing energy from the radiation field. In fact, for an isotropic distribution of photons, the electron is perceived to scatter photons at the same rate in the lab and rest frames [problem 14.2], so that the rate of energy removal is just $c\sigma_{\rm T}U_{\rm rad}$. Finally, therefore, we get an equation that is identical in the ultrarelativistic limit to that for synchrotron radiation:

$$-\dot{E}_{\rm ic} = \tfrac{4}{3}\sigma_{\rm T}\,\gamma^2\beta^2\,U_{\rm rad}c. \tag{14.15}$$

What about the spectrum of the radiation? Again, this is complicated in detail; however, the fact that the photon scattering rate is an invariant may be used with the expression for $\dot{E}$ to deduce the average increase in energy:

$$\langle \nu'/\nu \rangle = \gamma^2(1 + \beta^2/3). \tag{14.16}$$

For $\gamma \gg 1$, this is just the same $\nu \propto \gamma^2$ relation as for synchrotron radiation; the spectrum therefore takes the same form, with the initial soft photon frequency playing the role of the gyrofrequency. One way of understanding these close similarities is to say that synchrotron radiation can be thought of as the Compton scattering of virtual photons from the magnetic field. This similarity of the synchrotron and inverse Compton (IC) processes gives the only route by which the magnetic field can be measured; if the source of 'soft photons' for the IC process is known, the ratio of IC to synchrotron emission gives the field strength. The practical application of this idea is the generation of IC X-rays from the microwave background; in a few cases, these have been seen at a level consistent with equipartition fields (Kaneda *et al.* 1995).

The fact that the relative power output of the two mechanisms just depends on the ratio of the energy densities in the magnetic field and in photons can clearly lead to trouble, if we attempt to pack too many electrons into a given space: the density of synchrotron photons can then reach a level where they induce a higher rate of loss through Compton scattering than the synchrotron rate itself. This **inverse Compton catastrophe** sets some interesting limits on compact AGN, as follows. The power ratio may be written as

$$\frac{P_{\rm ic}}{P_{\rm synch}} = \frac{U_\gamma}{U_{\rm mag}} = \frac{(4\pi/c)\int I_\nu\, d\nu}{B^2/(2\mu_0)}. \tag{14.17}$$

To simplify the appearance of this, set $\int I_\nu\, d\nu \equiv I_{\bar\nu}\bar\nu$; $\bar\nu$ is roughly the frequency where νI_ν peaks. For some spectra, there may be no frequency $\bar\nu$ that satisfies this relation exactly, but we shall shortly see that only an order of magnitude equality matters. In terms of brightness temperature, we therefore get

$$\frac{P_{\rm ic}}{P_{\rm synch}} = \frac{16\pi\mu_0 k}{c^3 B^2}\, T_{\bar\nu}\bar\nu^3 \tag{14.18}$$

(and, because we are near the peak in νI_ν, it makes very little difference if we use the values observed at some frequency $\nu \neq \bar\nu$). Now, using the above thermodynamic limit to brightness temperature, we get a limit on the magnetic field,

$$B^{-2} > \left(\frac{3kT_\nu}{mc^2}\right)^4 \nu^{-2} \left(\frac{e}{2\pi m}\right)^2, \tag{14.19}$$

and so

$$\boxed{\frac{P_{\rm ic}}{P_{\rm synch}} > \left(\frac{T_\nu}{10^{12.0}\,{\rm K}}\right)^5 \left(\frac{\nu}{\rm GHz}\right).} \tag{14.20}$$

In practice, the synchrotron brightness temperature therefore cannot exceed 10^{12} K without leading to catastrophic inverse Compton losses. The rapid dependence on T means that any factors of order unity treated loosely in the above argument do not affect the result. It is not always easy to check observationally whether this limit is obeyed, since the implied critical angular size for a 1-Jy source is $\sim 10^{-3}$ arcsec, just within the reach of very long baseline interferometry (**VLBI**). A simple alternative is to estimate source sizes from variability arguments, $\ell \sim c\Delta t$, where Δt is the timescale for flux doubling; this can yield inferred brightness temperatures above the limit. An obvious way in which this process can go wrong is if there is bulk relativistic motion in the source, which compresses the observed timescales. These arguments are one of the lines of evidence that first pointed to relativistic jets in AGN.

LINE RADIATION One of the characteristic features of emission from active galaxies is a rich spectrum of emission lines. The interpretation of these spectra to infer physical conditions in the emitting object can be a complicated task, about which whole books have be written. Nevertheless, a few general points can be made that give a useful feeling for what is going on; see e.g. Davidson & Netzer (1979), Weedman (1986) and Osterbrock (1989) for more details. First, divide the atomic transitions into **permitted** and **forbidden** categories. This is a division according to lifetime for spontaneous emission, and also a division according to density. The forbidden lines cannot be produced in high-density environments, because the ion will be removed from its excited state before it has had a chance to decay. Second, the emission lines are often clearly of two classes in terms of velocity width: **broad lines** ($\sigma_v \gtrsim 1000\,\mathrm{km\,s^{-1}}$) and **narrow lines** ($\sigma_v \sim 100\,\mathrm{km\,s^{-1}}$). It will come as no surprise that the forbidden lines are generally narrow, whereas the permitted transitions are broad: in a picture where there is a central massive object, the velocity width is presumably related to orbital speed, and so the central dense gas is moving more rapidly. This sort of argument gives us one way of weighing the central object (see below).

In a little more detail, the way the density diagnostics work is as follows. Consider for simplicity a two-level ion and write down the condition for equilibrium between the rates of population of the upper level via collisions and depopulation via spontaneous radiation plus collisions:

$$N_1 N_e \Gamma_{12} = N_2 A_{21} + N_2 N_e \Gamma_{21}. \tag{14.21}$$

Here, Γ denotes $\langle v\sigma(v)\rangle$, the mean product of velocity and collisional (de)excitation cross-section; A_{21} is the Einstein emission coefficient; the N_i are the number densities of the species involved. Solve this for N_1/N_2 and note that the line flux equals $N_2 A_{21}\hbar\omega$; for a fixed degree of ionization, the number density of ions in the lower state, N_1, will scale with the electron density, giving

$$\boxed{j_\nu \propto \frac{N_e}{1 + A_{21}/(N_e\Gamma_{21})}.} \tag{14.22}$$

At low densities, the flux therefore scales as N_e^2 – in the same way as for **recombination radiation**, which yields permitted lines. However, once we pass the **critical electron density** at which $N_e = A_{21}/\Gamma_{21}$, forbidden-line emission becomes negligible relative to that from permitted lines. In practice, the most important diagnostic lines are the forbidden [OIII] lines at 4959 Å and 5007 Å, plus the semiforbidden CIII] 1909 Å line. Their critical densities are approximately $10^{14}\mathrm{m}^{-3}$ and $10^{16}\mathrm{m}^{-3}$. Since broad lines are seen in the latter case but not the former, we deduce an electron density in this range.

Now, how can we infer the other physical conditions of the emitting gas (temperature, ionization)? The answer depends on how one thinks the gas is ionized, and there are two possibilities: **photoionization** and **collisional ionization**. If the gas is collisionally ionized, then matters are relatively simple; if everything is in thermal equilibrium, then all line ratios are determined only by the temperature (at low density). This does not mean that the level occupation numbers are given by Boltzmann factors (except at high density as above), but that all the cross-sections for recombination and collisions plus the level lifetimes are known in principle, and so the equilibrium populations can be found. Reading the temperature from the line ratios allows the ionization to be deduced, given the above way of estimating the density. This is clearly quite a formidable calculation in detail, when one considers all the possible transitions that may be of interest. Things may often be simplified somewhat by the use of the so-called **case B approximation**. This refers to a situation where the density is sufficiently low that all lines are optically thin except the hydrogen **Lyman series** (on account of the large H abundance and the large absorption cross-section). For these, the photons will be absorbed by another atom immediately after emission, a process that continues until the excited atom radiates in some other series, eventually reaching the $n = 2$ level. Only Lyα photons can escape degradation in this way, and will eventually diffuse out of the emitting region, leading to the case B rule: every hydrogen recombination yields one Lyα photon plus one Balmer photon. Even with this simplification, the calculation of line ratios is complex (see e.g. Aller 1984 for the details). In practice, it is found that the strengths of the main lines emitted by AGN give a reasonable match to low-density gas at $T \sim 10^4$ K. This is encouraging at first sight, since we expect from a knowledge of the temperature-dependent cooling function that plasma will tend to get trapped at temperatures of either 10^4 or 10^8 K (see the discussion in chapter 17). However, in reality this 'success' is nothing of the sort: quasar spectra contain in addition many highly ionized species that have ionization potentials too high for them to be produced at such a relatively low temperature. The obvious route for producing these is via the photoionization that is to be expected so close to the intense nuclear continuum source.

At least for optically thin gas, the analysis of photoionization equilibria is relatively straightforward. We need to balance the rate of recombination ($\propto N_e N_{\rm ion}$) against the rate of ionization ($\propto N_{\rm atom} J_\gamma$, where J_γ is the flux density of ionizing photons). The result is clearly determined by the value of a dimensionless **ionization parameter**, which is the ratio of ionizing photon density to electron density:

$$\frac{N_{\rm ion}}{N_{\rm atom}} = \frac{U}{U_c}, \qquad U \equiv \frac{J_\gamma / c}{N_e} \tag{14.23}$$

(U is sometimes defined multiplied by c, so that it has dimensions of velocity). Because the recombination coefficient depends on temperature, so does U_c, the critical value of U. However, the temperature is rarely far from 10^4 K: the value of the ionization potential for hydrogen (13.6 eV) means that the excess kinetic energy carried by ionized electrons will correspond to a temperature of this order, which is also a stable value from the point of view of cooling. We then have the invaluable rule of thumb that for hydrogen $U_c \simeq 10^{-5.2}$. For other ions, the critical values are higher ($10^{-3.3}$ for ionization of He^+ to He^{++}, for example). Note that, because the density of ionizing photons depends on the species under consideration, U is defined always in terms of hydrogen-ionizing photons. The fact that highly ionized species like C^{5+} have been seen is an indication that U must be high, and indeed values of $U \sim 0.01$ are inferred from detailed attempts to match quasar spectra.

One surprising aspect of all of this is the metal abundance of the line-emitting gas. This is not very easy to measure directly, mainly because the gas is optically thick to many lines apart from Ly α and so the situation presents a difficult problem in radiative transfer. Nevertheless, all the indications are that the abundances are close to Solar, even in the highest redshift quasars known ($z = 4.9$ at the time of writing). This has major implications for the history of star formation, at least in these objects.

THERMAL RADIATION The final contributors to the continuum emission from AGN are more prosaic. Subtraction of power laws fitted to the optical/IR and radio data often reveals 'bumps' in AGN spectra peaking in the ultraviolet and submillimetre wavebands (say, $\lambda = 500$ Åand 500 μm). Viewed as black bodies, these would correspond to temperatures of respectively $10^{5.5}$ K and $10^{1.5}$ K. The former is reasonable for the emission from a central accretion disk (see below), while the latter could easily correspond to dust.

14.3 Extended radio sources

JETS, HOTSPOTS AND LOBES Although they are a small fraction of all AGN numerically, radio-loud AGN present the greatest physical challenges. All the main features of radio-quiet AGN (nuclear continuum and emission lines) are present, but in addition there is the phenomenon illustrated in figure 14.2: the transport of energy over colossal scales. These are perhaps the largest distinct entities found in nature, and certainly one of the most striking. The ability to obtain spatially resolved data has contributed immensely to their understanding, although even now many questions remain open.

Early radio interferometer observations established the extended nature of many radio sources, in the form of double structure about a central galaxy or quasar. The key question was whether these **radio lobes** were the result of some form of ejection, or were continuously supplied with energy from the nucleus. It was realized that the extended sources came in two basic types or **Fanaroff–Riley classes** (1974). In **FRI** sources, the regions of highest surface brightness lie closest to the nucleus, whereas in **FRII** sources the **hotspots** lie at the outer edges. The synchrotron lifetimes in FRII sources are so short that a continuous supply of energy from the centre was inferred to be necessary in the form of a **beam** (Longair, Ryle & Scheuer 1973). As mentioned above, modern terminology now favours the term **jet**, because better maps have actually revealed these hypothetical supply lines and shown them to contain fluid in motion.

The FR morphological division is also a division in radio power: FRII sources have $P \gtrsim 10^{24} h^{-2}\, \mathrm{W\,Hz^{-1}\,sr^{-1}}$. What this suggests is that jets come with a variety of outputs, and that some are too puny to penetrate the intergalactic medium (IGM) in their vicinity. The jets in FRI sources resemble a turbulent plume, whereas the jets in FRII sources are very narrow, with opening angles of only a few degrees. It is plausible that FRI jets dissipate their kinetic energy by entraining IGM material, whereas FRII jets maintain their integrity until they impact on the IGM in a shock. The hotspots are therefore to be interpreted as a rather complex working surface, whose qualitative features are illustrated in figure 14.4.

SOURCE LIFETIMES The calculation of synchrotron losses given earlier allows important insights into the overall lifetime of a radio-source outburst. An electron with Lorentz factor γ has a characteristic lifetime

$$\tau \equiv E/|\dot{E}| = 10^{11.39} \gamma^{-1} (B/\mathrm{nT})^{-2} \quad \mathrm{yr}. \tag{14.24}$$

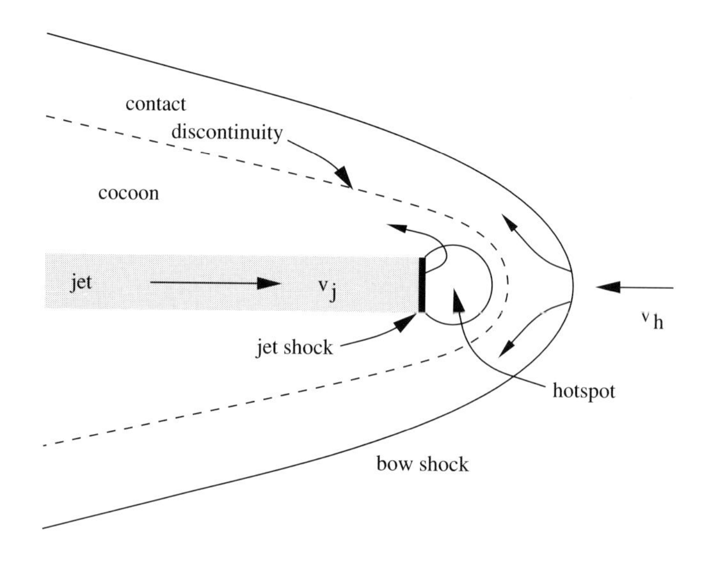

Figure 14.4. Illustrating the main features expected when an extragalactic jet impacts on the IGM. There are two shocks: a bow shock in the IGM and a shock where the jet terminates; the intermediate region is called the hotspot. This view is in the rest frame of the hotspot, so that the IGM appears to flow in from the right at the hotspot velocity v_h. The jet flows with velocity v_j from the left. Shocked jet material flows backwards to form a cocoon around the jet. There is a contact discontinuity between the shocked jet material and the shocked IGM. Particle acceleration is assumed to occur at the jet shock and the hotspot emission occurs in the region downstream of the jet shock.

In practice, we often have $\gamma \gtrsim 10^3$ and $B \sim 1$ nT, so that significant ageing of the electron population can take place in a time $\ll H^{-1}$. The detailed effects on the electron spectrum may be derived by solving the conservation equation for $N(E)$, which is obtained by regarding $\dot{E}$ as a 'velocity' in energy space:

$$\frac{\partial N(E)}{\partial t} = -\frac{\partial}{\partial E}[\dot{E}N(E)] + Q. \tag{14.25}$$

The last term represents injection; if $Q \propto E^{-x}$, then the steady-state solution clearly has $N \propto E^{-(x+1)}$, i.e. ageing steepens the spectrum. We can see this more directly by solving the equation with $Q = 0$. Define $-\dot{E} \equiv bE^2$; the solution for $N \propto E^{-x}$ at $t = 0$ is then [problem 14.4]

$$N(t)/N(t=0) = (1 - Ebt)^{x-2} \qquad \text{for} \quad E < 1/bt. \tag{14.26}$$

There are thus no electrons with E above a cutoff energy; for $x > 2$ (spectral index $\alpha > 0.5$), the spectrum steepens continuously towards high energies. This signature can be used, once we know the magnetic field, to date the emitting electrons. In the case of double radio sources, this means that we can estimate the age of the active phase. The source consists of diffuse material between the active hotspots and the nucleus, and it is plausible that the electrons we see were accelerated in the hotspot and 'dumped' as the source expanded. One often sees a trend of increasing spectral curvature along a path from hotspot to nucleus; if this is interpreted as **synchrotron ageing** and the magnetic field is assumed to take the equipartition value, then a linear age–radius relation tells us both the age of the object and the average rate of expansion. For Cygnus A, the nearest powerful radio galaxy, this yields an age of about $4.5 \times 10^6 h^{-3/7}$ yr and an expansion speed of about $0.03h^{-4/7}c$ (Winter *et al.* 1980). More typical values for the most luminous sources are about $0.1h^{-4/7}c$ (Liu, Pooley & Riley 1992).

We can use the close similarity between the synchrotron and inverse Compton emission mechanisms to improve on the rather uncertain age estimates from simple equipartition. Compton scattering must always be important at some level because of the microwave background. This is equivalent, in energy-loss terms, to synchrotron radiation via isotropic electrons in a magnetic field of

$$B_{\rm cmb} = 0.324(1+z)^2 \ {\rm nT}. \tag{14.27}$$

We now have synchrotron + Compton ages being proportional to $[\gamma(B^2 + B_{\rm cmb}^2)]^{-1}$; $\gamma^2 B$ is a constant that fixes the observed synchrotron frequency. This gives a *maximum* age at $B = B_{\rm cmb}/\sqrt{3}$. This is more like the minimum-energy argument: even if the magnetic field is not well determined, the age is, to first order, insensitive to these uncertainties. In any case, because the result is now an upper limit, we can be very confident in asserting that radio-loud active galaxies shine for a small fraction of the age of the universe.

SPEEDS AND OUTPUT OF JETS A knowledge of the approximate advance speed of hotspots gives some idea about the properties of the jets themselves. A jet can be characterized by its output of mass, momentum and energy:

$$\begin{aligned} \dot M &= A\gamma\rho_r v \\ T &= A\gamma^2(\rho + p/c^2)v^2 + p \\ K &= A\gamma^2(\rho + p/c^2)c^2 v - \dot M c^2. \end{aligned} \tag{14.28}$$

Here, A is the cross-sectional area of the jet, ρ is the total rest-frame density and ρ_r is the rest-frame density of rest mass (i.e. the internal energy density of the fluid is $u = (\rho - \rho_r)c^2$). This is the reason for the second term in K: it says that K is the total available internal energy, which is the total energy delivered minus mc^2 times the number of particles delivered. It is usual to simplify these relations by assuming that jets are nonrelativistic and internally cold. Otherwise, it is hard to see how they can be confined to the observed opening angles of ~ 1 deg; a free jet will expand as a cone of semi-angle $\mathcal{M}^{-1}$, where $\mathcal{M}$ is the Mach number, so a cold highly supersonic jet seems the best bet. In this case, the **jet thrust** is just $T = A\rho v^2$ – pure ram pressure – and the energy flux is $K = Tv/2 = \dot M v^2/2$.

In the frame of the hotspot, the internal pressure of the emitting plasma must balance the ram pressure of the incoming IGM on one side, and the jet on the other:

$$\boxed{2\left(\frac{B^2}{2\mu_0}\right) = \rho_i v_h^2 = \rho_j(v_j - v_h)^2 \quad \Rightarrow \quad \frac{v_j}{v_h} = 1 + \left(\frac{\rho_i}{\rho_j}\right)^{1/2}.} \tag{14.29}$$

The last relation says that the jet velocity will be roughly equal to the hotspot velocity unless the jet is very much lighter than the IGM ($\rho_j \ll \rho_i$). For Cygnus A, a highly relativistic jet, $v_j \sim c$, apparently requires a jet density of $\rho_j < \rho_i/1000$. It is easy enough to check that these pressure-balance sums give sensible results at the outer edge. For

Cygnus, the equipartition hotspot field is $50h^{2/7}$ nT, implying an external proton density of $n \sim 10^{4.2}h^{12/7}\,\mathrm{m}^{-3}$. This is within an order of magnitude what would be expected in the IGM of a cluster or around a massive elliptical (e.g. Sarazin 1988). Obtaining the jet density and hence the speed is harder. The main route is to attempt to estimate the entire energetics of the source, and set $K\tau$ equal to the sum of three terms: (i) the total energy radiated; (ii) the energy presently in relativistic particles and fields; (iii) the $P\,dV$ work in pushing back the IGM. It turns out that the last of these dominates, and is easy to calculate. Suppose the source is idealized as a cylinder of length L and radius R ($L \simeq 88\,h^{-1}$ kpc and $R \simeq 0.05L$ for Cygnus A). If the cylinder expands uniformly over the source lifetime, the work done against the ram pressure is

$$E \simeq \frac{\pi R^2 L \rho_i}{12}\, v_h^2; \tag{14.30}$$

this expression is dominated by the ends of the cylinder, which move fastest. This gives an estimated $K = 10^{38.3}h^{-2}$ W. Since the jet radius is about $1\,h^{-1}$ kpc, this implies a jet speed of $v_j/c \simeq 0.22h^{-4/7}$, and $\dot{M} \simeq 1.5h^{-6/7}M_\odot\mathrm{yr}^{-1}$, using the known thrust. It therefore appears that jets in powerful radio sources are light and marginally relativistic. Although obtained in a simplistic way, this conclusion is consistent with more sophisticated arguments that use synchrotron lifetimes in the emitting regions to limit the jet speed (Meisenheimer *et al.* 1989). These high speeds mean that the efficiency of radio AGN in turning directed kinetic energy into relativistic particles and fields is only $\sim 1\%$ (Scheuer 1974).

Only quite recently has it been possible to check these estimates directly. Proper motions in the jet of M87 (a luminous FRI) have been detected, implying velocities of about $2h^{-1}c$ at 1 kpc, declining to $0.35h^{-1}c$ at greater radii (Biretta, Zhou & Owen 1995). The figure at small radii sounds ridiculous, but is a phenomenon familiar from parsec-scale jets, where projection effects can yield apparent transverse velocities above c if the flow is relativistic (see below). The flow may decelerate through shocks in the case of M87, but remains marginally relativistic. As yet, there is no determination of the variation of velocity along an FRII jet.

GENERATION OF JETS The arguments for the generation of bulk relativistic motion are compelling, so the ultimate challenge is to understand how this colossal kinetic energy flux can be set up within a region of size $\lesssim 1$ pc. The short answer has to be that we do not yet have a certain idea of how the 'central engine' operates, although some interesting pieces of possibly relevant physics do exist.

One very simple way of obtaining bulk relativistic motion was outlined in a classic paper by Blandford and Rees (1974): exploit the large internal energy available from relativistically hot plasma. It was shown in chapter 2 that the relativistic version of **Bernouilli's equation** requires a bulk Lorentz factor of

$$\gamma = (p_0/p)^{1/4} \tag{14.31}$$

for steady flow of an internally relativistic fluid that starts from rest at pressure p_0. In other words, bulk relativistic motion is expected for any such fluid that experiences a significant change in pressure.

The ingredients needed to set up a jet are thus: (i) some ultra-hot plasma; (ii) some cold dense material that can pressure-confine the jet; (iii) a slight asymmetry in pressure gradient that gives the fluid a preferred flow direction. Blandford and Rees suggested that the latter could be provided by a rotationally flattened gas cloud where the plasma would tend to escape along the rotation axis. Suppose that the plasma flows in a one-dimensional manner, the confining cloud defining a 'hosepipe' of radius r. In steady flow, the rate of transport of energy down the jet must be a constant along its length (unlike the rate of transport of momentum, which will change if the walls of the pipe are not parallel; pressure forces then give momentum to the jet). It was shown in chapter 2 that the energy flux density is $\gamma^2 wv$, where w is the specific enthalpy ($w = \rho c^2 + p = 4p$ here). A constant jet power Q then implies a channel radius that varies with pressure:

$$\boxed{Q = 4\gamma^2 p v \pi r^2 \quad \Rightarrow \quad \pi r^2 = \frac{Q/(4p_0 c)}{x(1-x)^{1/2}},} \tag{14.32}$$

where $x \equiv \sqrt{p/p_0}$, and the Bernouilli relation has been used to give $v/c = \sqrt{1-x}$. The channel of the jet therefore forms a **de Laval nozzle**: the radius declines, reaches a minimum at $x = 2/3$, and then flares out. At the minimum, the velocity is $v = c/\sqrt{3}$, which is the sound speed for an ultrarelativistic gas. The minimum of the nozzle defines a transition between subsonic and supersonic flow.

Although we have allowed this to happen by letting the confining cloud adjust its shape, a nozzle must always exist somewhere in order to produce a supersonic jet, as all rocket designers know. The problem is that the particle flux density has a maximum at the sonic point, and so the channel must have a minimum if there is to be a transition to supersonic flow. In nonrelativistic fluid mechanics, this is shown as follows. Define the flux density as $J = \rho v$, and consider Euler's equation for steady flow along a streamline

$$(\mathbf{v} \cdot \boldsymbol{\nabla})\,\mathbf{v} = -\frac{\boldsymbol{\nabla} p}{\rho} \quad \Rightarrow \quad v\,dv = -\frac{dp}{\rho} = -c_s^2\,\frac{d\rho}{\rho}. \tag{14.33}$$

The trick here is to define a vector $d\mathbf{r}$ that is parallel to $\mathbf{v}$, and to note that $(d\mathbf{r} \cdot \boldsymbol{\nabla})X = dX$, for any quantity X. This relation now gives one for the flux density:

$$\frac{dJ}{dv} = \rho\left(\frac{1 - v^2}{c_s^2}\right), \tag{14.34}$$

confirming that the flux density is a maximum at the sonic point.

The problem with the Blandford–Rees mechanism is that VLBI observations show that jets originate on sub-pc scales, whereas the above equation gives an estimate of the size of the collimation region by working out r at $p = p_0$ (although the flow will not be 1D until the pressure has dropped somewhat, of course). For a given jet power, the required pressure scales as $1/r_0^2$, and so very compact jets require a high confining pressure. In practice, this would require other evidence of high-pressure gas such as extremely strong X-rays, which are not seen. Jets are probably pressure-confined at larger radii, and some of their variation in brightness can even be attributed to small departures from confinement, such as occur when they encounter the rapid drop in pressure associated with leaving the body of the galaxy and entering the IGM (Sanders 1983). However, it does seem that the initial direction of the flow has to be set on very small scales by some other mechanism.

The two main possibilities for jet generation that have been discussed here are radiation pressure and magnetic fields; these may both be important. The environment of the central engine is likely to be more complex and chaotic than the simple models that have been worked on so far, but these probably contain a grain of truth in that the central regions will inevitably be made isotropic through rotation. Accreting material will settle into some form of differentially rotating accretion disk (see below). This disk will probably emit high-energy radiation preferentially along its rotation axis (just through geometry), and will freeze the magnetic field lines and wind them up so that there is a strong magnetic field along the rotation axis. Flux freezing will tend to make plasma stream along the field lines, and radiation pressure may provide the initial acceleration needed to produce an outflow.

LIMITING JET VELOCITY There is an interesting limit to the velocity that radiation pressure can produce. Suppose the radiation source has a luminosity such that it produces a radiation pressure force equal to Γ times the gravitational force, so that the net radial acceleration is

$$a = (\Gamma - 1)\, a_g = (\Gamma - 1)\, \frac{GM}{r^2} \tag{14.35}$$

($\Gamma = 1$ defines the **Eddington limit**, the point at which radiation pressure and gravity just balance; see below). There is already a problem here in that a super-Eddington source will tend to blow itself apart; however, if the radiation is anisotropic, it may appear to have $\Gamma > 1$ along the jet, which is all we require. At the Newtonian level, the jet therefore accelerates without limit, and will pick up a velocity of $\Gamma - 1$ times the escape velocity from the nucleus. However, relativistic dynamics change this conclusion, as the radiation-driven and gravitational accelerations have a different dependence on Lorentz factor.

Consider first the effects of radiation pressure, where the momentum transfer is due to Compton scattering. If a blob of jet plasma contains N electrons, and the radiation field has a number density of photons n, then the force is the product of the geometrical cross-section and the momentum flux density: $F = (N\sigma_T) \times (ncp_\gamma)$. This is fine for a stationary plasma; for plasma in relativistic motion, the safe way of calculating the force would be to transform the radiation field to the rest frame, and calculate the force there. However, we can avoid this effort: the probability that a photon will pass through the plasma unscattered cannot be changed by the motion of the plasma, so all that changes is the rate at which photons pass across the boundary of a given blob of plasma, which is clearly $n(c - v)$; this is one case where velocities in relativity add simply. Since the force acts parallel to the velocity, relativistic dynamics says $a = F/(\gamma^3 m)$:

$$a_\gamma = \frac{1-\beta}{\gamma^3}\, \Gamma a_g \quad \to \quad \frac{\Gamma a_g}{2\gamma^5}, \tag{14.36}$$

where the last expression is the ultrarelativistic limit. Relativistic effects thus switch off the radiation pressure pretty effectively. What about the gravitational acceleration? (a_g is still the Newtonian acceleration). Consider the electromagnetic analogy: the radial acceleration of a charge again picks up the γ^{-3} factor: $a = qE/(\gamma^3 m)$. For gravity, the effect is in fact slightly less severe: $a = a_g/\gamma^2$. The one power of γ by which these differ may be thought of as arising because the gravitational force couples to the total mass-energy of the plasma (i.e. $1/\gamma^2 = \gamma/\gamma^3$).

We now see that there is a limiting Lorentz factor at which the radiation-driven and gravitational accelerations balance:

$$\gamma_{\max} = \left(\frac{\Gamma}{2}\right)^{1/3}. \tag{14.37}$$

This looks very bad, since grossly super-Eddington luminosities will be required in order to obtain large Lorentz factors, whereas the indications are (see below) that quasars typically have $\Gamma \sim 0.01$–0.1. However, there is an appealing way out, which is to make a jet out of pair plasma. The large electric fields generated by the magnetosphere of an accretion disk could generate pairs, so there is a natural source of plasma. Radiation pressure is much more effective on a pair plasma, since the force doubles whereas the inertia changes by a factor $2m_e/m_p$. In this case, the speed limit is

$$\boxed{\gamma_{\max} = \left(\frac{\Gamma}{2}\right)^{1/3} \left(\frac{m_p}{m_e}\right)^{1/3},} \tag{14.38}$$

or $\gamma = 9.7$ for a source at the Eddington limit. This is a very natural explanation of the fact that the most powerful AGN jets seem to have Lorentz factors of this order (Abramowicz, Ellis & Lanza 1990). Jets of e^+e^- **pair plasma** are therefore an attractive possibility, and there exist a range of other arguments for their existence (see e.g. Wardle 1977; Ghisellini *et al.* 1992).

14.4 Beaming and unified schemes

One interesting aspect of AGN statistics concerns models where the observed emission is anisotropic: some viewing angles may be affected by obscuration, and others enhanced by 'beaming' from bulk relativistic flows.

SUPERLUMINAL MOTION The evidence for relativistic motion takes the form of apparent motions at speeds in excess of that of light. It is worth noting that this was a theoretical prediction (Rees 1967), based on the very high brightness temperatures inferred for variable AGN. Here, the typical size of the emitting regions was deduced by light travel-time arguments, leading to temperatures above the maximum 10^{15} K. One way of evading this problem was relativistic motion: this increases the observed flux density and reduces the timescale of any intrinsic variability, leading to a large increase in the inferred surface brightness. These issues are analysed in detail below, but we begin with the signature of apparent superluminal expansion.

Imagine an emitting blob ejected from an AGN at an angle θ to the line of sight, with velocity v, and call the distance away from the line of sight y. Clearly $dy/dt = v \sin\theta$, but this is not the projected transverse velocity we observe. As the blob moves, it approaches us and the interval between the reception of two photons is less than the interval between their emission: $\Delta t_{\rm obs} = \Delta t - \Delta x/c = \Delta t(1 - \beta\cos\theta)$, where $\beta \equiv v/c$. We therefore get

$$\left(\frac{dy}{d\,ct}\right)_{\rm obs} = \frac{\beta}{1-\beta\cos\theta}. \tag{14.39}$$

This has a maximum value of $\gamma\beta$ at $\sin\theta = 1/\gamma$. This **superluminal motion** has been seen in many compact radio-loud quasars (e.g. 3C273; see Unwin *et al.* 1985). Observed values of β are typically in the range $3h^{-1}$ to $10h^{-1}$.

RELATIVISTIC BEAMING What is the effect of the relativistic motion on the flux density of the emitting blob? We begin with the relativistic transformation of specific intensity (or surface brightness),

$$I_\nu(\mathscr{D}\nu_0) = \mathscr{D}^3 I_\nu(\nu_0), \tag{14.40}$$

where $\mathscr{D}$ is the Doppler factor caused by relative motion of source and observer. This equation may be derived in an elementary way by transforming photon number densities and energies plus the relativistic aberration of solid angle elements [problem 14.1]. A more direct approach is to note that I_ν/ν^3 is proportional to the photon phase-space density, which is a relativistic invariant. This arises because the phase-space volume element $d^3x\, d^3p$ is unchanged under Lorentz transformations, as can easily be proved for a set of particles of non-zero rest mass. In the rest frame, the dispersion in momentum causes no dispersion in energy (to first order): $dp^\mu = (0, d\mathbf{p})$. Transforming to the lab frame therefore length-contracts distances ($dx' = dx/\gamma$) but boosts momenta ($dp'_x = \gamma\, dp_x$). Transverse components are unaffected, so the phase-space volume element is indeed an invariant. One can now deal with photons by taking the limit of zero rest mass. Finally, the phase-space density is not only invariant, it is also conserved along light rays (by Liouville's theorem). See the material on gravitational lensing in chapter 4 for further discussion of surface brightness transformations.

If we now consider an optically thin (spherical) blob moving with velocity β ($c = 1$) at an angle $\cos^{-1}\mu$ to our line of sight then, if the emission is isotropic in the blob's rest frame, our received flux density also transforms in the same way as surface brightness. This is due to a general theorem that states that moving bodies in special relativity appear rotated but *not* length-contracted (see e.g. section 18 of Rindler 1982). An alternative way of seeing the truth of this statement is as follows. Suppose that, instead of allowing the source to move, we give a boost to the observer; in the best traditions of relativity, it may seem that only the relative velocity matters (this is wrong, but it is still useful to analyse a boosted observer). The element of solid angle seen by the observer is aberrated [problem 14.1], and transforms as

$$d\Omega' = d\Omega \,/\, \mathscr{D}^2. \tag{14.41}$$

Multiplying this by the $\mathscr{D}^{3+\alpha}$ dependence of surface brightness and integrating, we appear to get the weaker dependence $S \propto \mathscr{D}^{1+\alpha}$. The reason why this is the wrong answer is that boosting the source does not change the source–observer distance, but boosting the observer does. The distance can be defined via the light travel time: $D = c\Delta t$, which transforms to $D' - c\Delta t' = \gamma(c\Delta t - \beta\Lambda x)$, where $\Delta x = -D\mu$. Now, the Doppler factor, obtained by transforming the 4-wavevector $k^\mu = (\omega, \mathbf{k})$, is

$$\mathscr{D} = \gamma(1 + \beta\mu) = [\gamma(1 - \beta\mu')]^{-1}, \tag{14.42}$$

and so the source distance perceived by the observer after the boost is *greater*: $D' = \mathscr{D}D$. When we come to calculate the apparent change in source luminosity as a result of the boost, the inverse-square law then brings in a factor of $\mathscr{D}^2$, which exactly cancels the aberration of solid angle: the output of the source is apparently increased by $\mathscr{D}^{3+\alpha}$, the same factor as the surface brightness.

To summarize, for a relativistic blob that emits an isotropic power-law spectrum with flux density $S_\nu \propto \nu^{-\alpha}$, we have $S = S_0(1 - \beta\mu)^{-(3+\alpha)}$, where S_0 is the value at $\mu = 0$ (in the plane of the sky), and μ is now the cosine of the angle between the blob velocity and the line of sight as seen by the observer. For a quasi-continuous jet formed out of finite-lifetime blobs, the number of blobs observed at a given instant scales as $\mathscr{D}^{-1}$ and hence the appropriate index above is reduced to $2+\alpha$:

$$\boxed{S = S_0(1-\beta\mu)^{-(2+\alpha)}.} \tag{14.43}$$

The 'standard' model consists of a pair of such jets oppositely directed; however, life is simpler if we follow Lind & Blandford (1985) and neglect the receding component, whose flux is generally very small. We can then obtain a useful expression for the probability distribution of the beamed flux density. In general there will be an isotropic component in addition to the jet: we define R to be the ratio at $\mu = 0$ of one side of the jet to the isotropic component. The amplification relative to $\mu = 0$ is therefore $A = 1 + R[(1-\beta\mu)^{-(2+\alpha)} - 1]$ and, since μ is uniformly distributed,

$$P(>A) = \frac{(1-\beta)}{\beta}\left[\left(\frac{A_m + R - 1}{A + R - 1}\right)^{1/(2+\alpha)} - 1\right], \tag{14.44}$$

where the maximum amplification $A_m = 1+R[(1-\beta)^{-(2+\alpha)}-1]$. Provided that $A_m \gtrsim 1/R$, there is therefore a 'tail' of probability with a characteristic $A^{-1/(2+\alpha)}$ form at large amplifications. If the numbers of faint sources rise rapidly enough, this can lead to strong selection effects: a flux-limited sample may be dominated by 'beamed' sources, even though the solid angle of the $1/\gamma$ beaming cone is very small.

The above equation is a considerable idealization and some complications are discussed by Lind & Blandford (1985). The most natural of these is to recognize that if the jet is in any way non-steady, the observed emission will arise behind shocks. The effective beaming (post-shock) speed is then lower than the shock speed – which is the pattern speed observed with VLBI. Further complications considered by Lind and Blandford include non-planar shocks and optical depth effects. The latter are especially important: for synchrotron emission, the optical depth scales as $\nu^{-(\alpha+5/2)}$ and the jet is likely to become optically thick when seen end-on. These effects thus tend to broaden the beaming cone beyond the canonical $\theta \sim 1/\gamma$.

OBSCURATION The above analysis assumes that the central source remains visible even when the jet axis is rotated perpendicular to the line of sight ($\mu = 0$). If the ejection axis is surrounded by a obscuring **torus** of material, however, then the central object might only be visible within some critical cone ($\mu > \mu_{\rm crit}$). This can lead to more complex selection effects, where the classification of the object changes completely inside the critical cone. This mechanism may well provide a link between Seyfert 1 and Seyfert 2 galaxies. The latter may possess broad lines that are hidden from view in some orientations. This is certainly so in at least one case: NGC1068. Antonucci & Miller (1985) showed that this object displayed broad lines that were visible in polarized light only. The implication is that broad-line radiation escapes roughly in the plane of the sky, and some part of this radiation is scattered towards the observer once it has escaped the central regions, thus producing the polarization. Given that the radio AGN population also has objects (radio galaxies) without broad lines, the question arises of whether this mechanism operates there also.

UNIFIED MODELS Unified models of the radio AGN population combine the elements of obscuration and relativistic beaming. Many workers have considered the question of what a compact radio source would look like when any putative beamed component

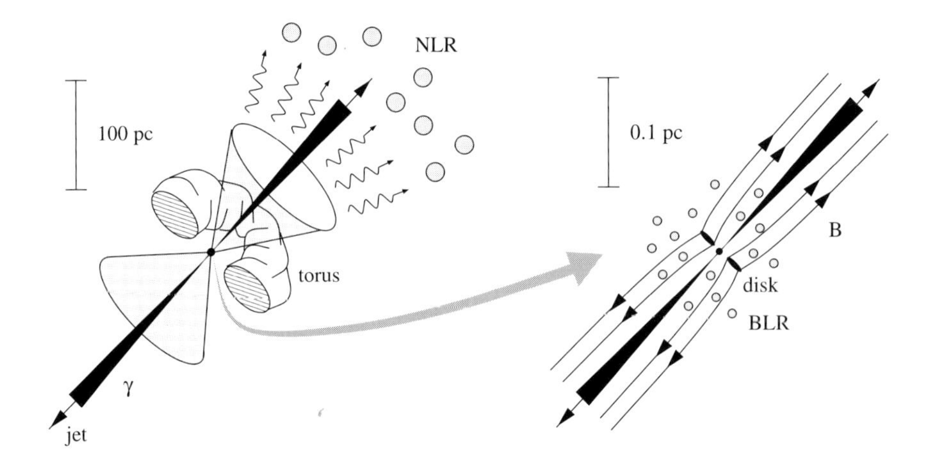

Figure 14.5. A cartoon of the unified model for radio-loud AGN. An obscuring torus and relativistic beaming respectively define cones of semi-angle roughly 45° and 1°. Within the latter, an observer sees a blazar; within the former, a broad-line quasar is seen; at larger angles to the line of sight, the AGN appears as a narrow-line radio galaxy. Evidence of the hidden engine exists in the form of (i) ionized gas in the quasar-like cone; (ii) scattered light from the broad-line region, which is polarized.

is turned away from our line of sight: what is the unbeamed parent population? Orr & Browne (1982) suggested that compact core-dominated quasars were simply double radio quasars seen end-on. A related unified scheme was suggested by Blandford & Rees (1978), and applies to **blazars**: objects with highly polarized ($>$ 3 percent) rapidly variable ($\sim$ 1 day) optical nuclear emission (see Angel & Stockman 1980). Blandford and Rees concluded that these properties made a strong case for bulk relativistic flow, and suggested that blazars might be radio galaxies seen end-on.

The Orr–Browne and Blandford–Rees unified schemes are related because one can regard essentially all flat-spectrum quasars as blazars: in the radio ($\nu > 10$ GHz), they display the characteristic variable polarized emission that defines this class. In the optical/infrared, a variable polarized continuum is seen in $>$ 50% of objects; the exceptions are generally those of high optical luminosity where the synchrotron component is washed out by the 'blue bump' (the presumed accretion-disk feature seen at $\sim$ 2000 Å in the rest frame). The above ideas can be combined in what are known as **unified models** (no relation to the particle physics term). Here, there is postulated a continuous trend from radio galaxy through extended quasar to compact quasar and finally blazar, as a function of orientation angle (Scheuer 1987; Peacock 1987; Barthel 1989); see figure 14.5.

These ideas arise because the extended structures of both radio galaxies and quasars are tantalizingly similar: both possess central cores and one-sided jets, but the latter are a factor $\sim$ 10–100 weaker in radio galaxies, consistent with beaming. The approximate equality of numbers of radio galaxies and extended quasars would imply an obscuration cone of $\mu \simeq 0.5$ for quasars. For the standard model, this implies beaming by a factor $\gtrsim 4$ with respect to $\mu = 0$ (or a factor 2 lower if we include the counterjet), rising rapidly with μ. This makes quite an attractive picture, with galaxies for $\mu < 0.5$, extended quasars for $0.5 < \mu < 0.9$, core-dominated quasars for $\mu > 0.9$, with increasing blazar characteristics as we move towards $\mu = 1$.

These ideas have some impressive observational support. Superluminal motion may be found often in cores of quasars selected via their extended lobes (Barthel 1989); it thus seems that no quasar lies in the plane of the sky, in agreement with the obscuration argument. Wills & Browne (1986) found a strong correlation between broad emission-line velocity widths and radio core dominance, in the sense expected if the line-emitting clouds are confined to a disk: compact quasars have narrower lines. Baker (1997) found evidence that quasars with weak cores were affected by dust extinction, implying that we are seeing the effects of the torus starting to become significant in these objects. Most convincing of all has been the observation that some radio galaxies have extended polarization structure (e.g. Cimatti *et al.* 1993). There is a reflection nebula, caused by scattering of blue continuum light which is not seen directly; the evidence for a hidden blazar in at least some radio galaxies is very strong (although polarization is not seen in all cases; Shaw *et al.* 1995).

This idea of unifying some of the disparate components of the active-galaxy world is clearly very exciting. However, the evidence in favour is still somewhat circumstantial; reality has no obligation to be as simple as our models. Tests of this scheme are likely to remain a highly active area of AGN research, and are developing into an entire field of their own (see Urry & Padovani 1993 for a review)

14.5 Evolution of active galaxies

As with normal galaxies, a dominant aspect of AGN research has been the task of surveying and mapping the territory: understanding the distribution in flux and redshift, and working out what it tells us about the universe.

COUNTS AND LUMINOSITY FUNCTIONS Unlike ordinary galaxies, active galaxies have no characteristic luminosity: there is a very broad spread in activity. It is the existence of this spectacular high-luminosity tail in the distribution that allows some of even the apparently brightest AGN to lie at vast distances. We quantify this by introducing the **luminosity function**, which gives the *comoving* density of objects in some interval of luminosity. Unfortunately, different conventions exist: optical astronomers quote numbers per magnitude and radio astronomers numbers per $\log_{10}$ luminosity. We shall adopt here the compromise convention of using natural log luminosity. The monochromatic luminosity for AGN is denoted by P, and is related to flux density as follows for power-law spectra ($S \propto \nu^{-\alpha}$):

$$P = SD^2(1+z)^{3+\alpha}, \tag{14.45}$$

where D is angular-diameter distance ($\simeq cz/H_0$ for $z \ll 1$). For practical purposes (S in Jy and P in $\mathrm{W\,Hz^{-1}\,sr^{-1}}$) the relevant value of $(c/H_0)^2$ to insert in D^2 is $10^{25.933}h^{-2}$ (defining $h \equiv H_0/100\,\mathrm{km\,s^{-1}Mpc^{-1}}$). Note the units of P; this equation is really a surface brightness transformation, so there is no assumption of isotropic emission and thus no factor 4π.

So, denoting the luminosity function by $\rho(P,z)$, the expected number of objects seen in a range of flux and redshift is

$$\boxed{dN = \rho(P,z)\, d\ln S\; dV(z),} \tag{14.46}$$

because the Jacobian $\partial(\ln S, z)/\partial(\ln P, z)$ is unity. For an area of sky of solid angle A, the differential volume element is $dV = AD^2(1+z)^2 dr$, where dr is the element of comoving distance $dr/dz = (c/H_0)(1+z)^{-1}(1+\Omega z)^{-1/2}$ in Friedmann models. Note that results for different cosmologies scale simply: S and z are observables, so P scales according to D^2 and number density scales inversely with volume element, i.e. $\rho \propto D^{-2}(1+\Omega z)^{1/2}$. At a given redshift the translation to a different Ω simply involves moving the luminosity function horizontally and vertically by the appropriate amounts. Indeed, the translation to more general cosmologies could also be accomplished in this way, which is a justification for considering only Friedmann models: evolutionary studies cannot determine the geometry of the universe.

EUCLIDEAN COUNTS The number–flux relation assumes an important form if space is Euclidean. In this case, the volume element is $V = (4\pi/3)(P/S)^{3/2}$, and integration over V at constant S is simply transformed to integration over P at constant S. Thus, if ρ is a function of P only, the flux distribution becomes

$$\boxed{N(>S) \propto S^{-3/2}.} \tag{14.47}$$

This **Euclidean source count** is the baseline for all realistic surveys, and shows us that faint sources are likely to heavily outnumber bright ones. The relation is one form of **Olbers' paradox**: integration over S implies a divergent sky brightness as we include fainter and fainter sources. Since the universe does not contain an infinite energy density from AGN, it is clear that relativistic effects in the distance–redshift and volume–redshift relations must cause the true counts to lie below the Euclidean prediction. Consider the behaviour of the horizon distance in expanding models: $r_{\rm H} = \int c\, dt/R(t)$. For zero vacuum density, this always converges; there is a finite amount of comoving volume to be explored, however deep a survey we perform. As an example, consider the simplest case where the luminosity function is a δ-function, so that there is a one-to-one relation between flux and redshift. The count slope, β is just

$$\beta \equiv -\frac{d\ln N}{d\ln S} = -\left.\frac{d\ln V}{d\ln z} \right/ \frac{d\ln S}{d\ln z}. \tag{14.48}$$

This is always < 1.5 for any Ω: for $\Omega = 1$ and $z \gg 1$, $d\ln V/d\ln z \simeq 1.5/\sqrt{z}$ and $d\ln S/d\ln z \simeq -(1+\alpha)$. This illustrates the general case that it is the closing down of the volume elements, rather than the non-Euclidean dimming of high-z objects, which produces the low count slope. The only way in which $\beta > 1.5$ can be observed is if the population is evolving, so that ρ was larger in the past.

The question of the slope of the source counts was of great importance in the early 1960s. Prior to the discovery of the microwave background, there was much debate about the possible applicability of the **steady-state universe**, in which all properties such as the luminosity functions were held to be independent of time. It is easy to show that, again, $\beta < 1.5$ for such a model (see chapter 3), so the fact that radio surveys of that time gave $\beta > 1.5$ seemed to argue convincingly for a big bang. The debate was acrimonious, as the steady-state advocates questioned the reliability of the observations. In retrospect, some of these doubts turned out to be correct: β lies close to 1.5 at high flux densities and declines as one goes fainter (though not fast enough to save the steady-state model); see Wall (1994).

DENSITY AND LUMINOSITY EVOLUTION The number counts alone thus showed that the AGN population was more active in the past. A more direct indication that drastic change had occurred in the population of active galaxies came as redshifts became available for some of the first quasars and it turned out that many were as large as $z = 2$. Is this surprising? Suppose we have a power-law luminosity function, $\rho \propto P^{-\gamma}$. At a given flux density, we then expect a redshift distribution

$$\frac{dN}{dz} \propto D^{2-2\gamma}(1+z)^{1-\gamma(3+\alpha)}(1+\Omega z)^{-1/2}. \tag{14.49}$$

At high z, this falls as $(1+z)^{-3/2-\gamma(1+\alpha)}$; since $\gamma > 1$ is required for the total luminosity density to converge, it is clear that high-z objects should be rather rare, unless they have highly inverted spectra with $\alpha < -1$ instead of the more typical $\alpha \simeq 0.8$. The fact that they are not rare implies that the luminosity function must have increased at high z.

Two simple extremes are usually discussed for describing the evolution of a luminosity function. These came into being in the early days where data limitations meant that some assumptions were needed to make progress. The number of objects may be changed by scaling the whole luminosity function either vertically (**density evolution**) or horizontally (**luminosity evolution**):

$$\rho(P, z) = \begin{cases} f(z)\,\rho_0(P) & \text{(density evolution)} \\ \rho_0\big(P/g(z)\big) & \text{(luminosity evolution).} \end{cases} \tag{14.50}$$

Parametric forms for the **evolution functions** $f(z)$ and $g(z)$, such as powers of $1+z$, were tried in first attempts at understanding quasar redshift data, and can still provide a reasonable fit today (see below). Both these alternatives are clearly equivalent for a pure power-law luminosity function, but they can be distinguished if the luminosity function contains any curvature. The physical motivation for either of these descriptions is weak. Density evolution suggests a set of objects of constant output but with a distribution of lifetimes; luminosity evolution suggests an output that declines at the same rate for objects of all powers. However, such simplistic interpretations are to be avoided.

THE $V/V_{\rm max}$ TEST It is obviously desirable to have some form of test for evolution that does not depend on preconceived notions about the form of change. This is provided by the $V/V_{\rm max}$ test (Rowan-Robinson 1968; Schmidt 1968), which is admirably simple both in conception and application. Consider a set of objects all of the same luminosity: for a given flux limit, there will be some maximum distance to which they can be seen. This maximum distance, together with the area of sky surveyed, defines the volume that has been surveyed, $V_{\rm max}$; objects are assumed to be uniformly distributed within this space. So, if we consider the volume V defined by the survey area and the actual redshift at which a given object is found, the quantity $V/V_{\rm max}$ is clearly expected to have a distribution that is uniform between 0 and 1. The same is clearly true for a set of objects

Table 14.1. Luminosity-function parameters

Waveband	α	β	$\rho_0/h^3\text{Gpc}^{-3}$	Reference
Optical	0.3	2.6	$10^{4.1}$	Boyle *et al.* (1987; 1988)
X-ray	0.5	2.3	$10^{3.6}$	Boyle *et al.* (1993)
Radio	0.8	2.1	$10^{3.4}$	Dunlop & Peacock (1990)

with a spread of properties (i.e. different powers and spectra): we just work out V_{max} for each object based on the sample limit and its properties. A variety of statistical tests can be used to see if the observed distribution is indeed flat, but the simplest is to work out the mean and standard error for a set of n observations:

$$\left\langle \frac{V}{V_{\text{max}}} \right\rangle = 0.5 \pm (12n)^{-1/2}. \tag{14.51}$$

Values found in practice with this test vary from sample to sample, often with figures in the region of 0.7. How is this to be interpreted in terms of changes in density? The answer is model dependent. Suppose we assume density evolution, with $\rho \propto V(z)^\beta$:

$$\left\langle \frac{V}{V_{\text{max}}} \right\rangle = \frac{\int V^{\beta+1} dV}{\int V^\beta dV} = \frac{(\beta+1)}{(\beta+2)}, \tag{14.52}$$

so a mean value of 0.7 would imply $\beta \simeq 1.3$. For $\Omega = 1$, this would imply a change in density of a factor ~ 500 between redshifts of 0.2 and 2. For this toy model, the current density is zero and so the answer depends on what redshift is taken as 'local'. Nevertheless, it is clear that the universe was very different at high redshift. Note again that the test uses comoving volumes: quasars at a given power were more numerous in the past in *proper* density by a further factor $(1+z)^3$. At redshift $z = 3$, the nearest quasar would have been roughly fourth magnitude, easily visible to the naked eye (scaling the distance of 3C273, which is twelfth magnitude).

The V/V_{max} test can be immediately generalized in a number of neat ways. Several different surveys can be combined to consider a region of space with different distance limits as a function of direction. Upper and lower redshift limits can be imposed by hand, so that one considers evolution only over some redshift shell. Finally, some assumed density evolution law can be tested by defining the **generalized volume** $V' = \int f(z)\, dV(z)$, and the law tested by requiring $\langle V'/V'_{\text{max}} \rangle$ to be consistent with 0.5 (working in comoving volume is an example of this, of course). See Avni & Bahcall (1980) for details.

THE EVOLVING LUMINOSITY FUNCTION Following the establishment and quantification of AGN evolution by the above methods, the next step is to map out in detail the epoch dependence of the luminosity function. This requires the assembly of large surveys with complete redshift data, and has been a major undertaking, involving a decade of effort by many groups. The final result is rather simple, however, at least for redshifts $\lesssim 2$, as AGN in all wavebands appear to behave similarly. The local luminosity function can be described by a two-power-law form with a characteristic break:

$$\rho(L) = \rho_0[(L/L^*)^\alpha + (L/L^*)^\beta]^{-1}. \tag{14.53}$$

Table 14.1 gives the parameters from this fit that are appropriate to different wavebands.

Evolution at $z < 2$ is parameterized by a simple shift in luminosity: **pure luminosity evolution** (PLE). The critical luminosities at $z = 0$ are approximately $M_B^* = -20.3 - 5\log_{10} h$ ($= 10^{10.3} h^{-2} L_\odot$) in the optical; $L^* = 10^{35.8}$ W in X-rays (0.5–2 keV) and $P^*_{1.4\,\mathrm{GHz}} = 10^{24.6} h^{-2}$ W Hz^{-1} sr^{-1} in the radio. At $z = 2$ they were in all cases a factor of about 30 larger, with $L \propto (1+z)^3$ providing a reasonable description of the evolution.

The above has simplified things a little: in the radio, two populations with grossly different spectral indices exist, and compact flat-spectrum sources with $\alpha < 0.5$ are often treated separately from the steep-spectrum population. Similar behaviour applies for steep- and flat-spectrum AGN, so slight quantitative differences in their properties have been ignored. See Dunlop & Peacock (1990) for details. Also, it now appears that the PLE form for the evolution of optical quasars does not apply exactly. The work of Boyle *et al.* (1987; 1988) combined faint quasar surveys with the Palomar Green (PG) survey, which was a UVX survey for all bright ($B \lesssim 16$) quasars over the northern hemisphere (Schmidt & Green 1983), and it now appears that the PG survey may have been substantially incomplete (Goldschmidt *et al.* 1992). This raises the local space density of luminous quasars and thus reduces their inferred rate of evolution. The resulting picture is one in which the most rapid evolution is at intermediate luminosities $M_B \simeq -26$, with quasars at both lower and higher luminosities changing their space density more slowly (Hewett, Foltz & Chaffee 1993).

THE REDSHIFT CUTOFF A question of particular interest is what happens to the AGN population at very high redshifts $z \gtrsim 2$. This is a different regime in optical quasar surveys, because the **ultraviolet excess** (UVX) method of selection no longer works (the Lyα emission line is redshifted out of the U photometric band). Probing evolution at high redshift is always going to be hard: for $z \gg \Omega^{-1}$, $S \propto (1+z)^{-(1+\alpha)}$, while $dV/dz \propto (1+z)^{-3/2}$. The flux density declines in a quasi-Euclidean manner, but the volume elements at high redshift quickly become too small to offer much encouragement to observers. In addition, high-redshift quasars may be obscured in a variety of ways. Shortward of Lyα, the continuum becomes blanketed by absorption due to intergalactic clouds of neutral hydrogen. Empirically, the optical depth scales roughly as $\tau \propto (1+z)^3$, and reaches unity (i.e. 1 magnitude suppression of the continuum) at $z \simeq 3$; searches in the blue are clearly likely to be inefficient.

Other forms of obscuration are possible, but the strength of the effect is less certain. Ostriker & Heisler (1984) argued that dust in the disks of intervening galaxies may prevent us seeing high-z quasars. For opacity $\propto \lambda^{-1}$, the optical depth to dust is

$$\tau = \tau^* \int \frac{(1+z)^2 dz}{\sqrt{1+\Omega z}}, \tag{14.54}$$

where τ^* may possibly be as high as 0.1–1 in the observed B band (although Fall & Pei 1993 argue for a smaller value). This effect can be constrained to some extent, since it does not affect radio AGN. To the extent that radio surveys can be completely identified, the obscuration of the high-z universe cannot be total – although the steepness of the quasar luminosity function is such that even a few tenths of a magnitude would have a large effect on space densities.

Despite these caveats, evidence has mounted over recent years that quasar evolution switches off and declines at high redshifts, although as we have seen the expected abundance of high-z objects is so low that only a few missed objects at $z \gtrsim 4$

can make a huge difference to the inferred density. In both the radio and optical wavebands, there seems to be a decline in ρ by a factor ~ 3 by $z = 4$ from a peak at $z \simeq 2$. However, there is some evidence that the behaviour is luminosity dependent, the most luminous objects holding their comoving densities more nearly constant (Osmer 1982; Peacock 1985; Dunlop & Peacock 1990; Warren, Hewett & Osmer 1994). At present, both quasars and radio galaxies are known up to $z = 4.5$–5, and it seems likely that all-sky searches for rare bright objects are the technique most likely to extend this record.

WHAT CAUSES AGN EVOLUTION? If AGN had a peak in their activity at redshifts of around 2, why was this? The success of pure luminosity evolution (PLE) models may suggest that we are always looking at the same set of objects, which simply fade slowly over a Hubble time. However, we know this is not the case for extended radio AGN, and it therefore seems a dangerous assumption for radio-quiet quasars. Instead, it is more likely that we see at any given redshift the result of some triggering event in the recent past of the galaxy. In this case, the number density of objects we see at a given epoch measures just how many events were triggered in the lifetime of one outburst: $n \propto B(z)\tau(z)$. It is therefore conceivable that AGN evolution was caused by the lifetime $\tau(z)$ being larger in the past. However (see below), most attempts at an explanation focus on some epoch dependence of the 'birthrate' $B(z)$. One way of distinguishing long- and short-lived models is that they predict that quasars are either confined to a few special galaxies or arise in all galaxies at some time. Searches for black holes in presently inactive galaxies may then tell us whether they might have been active in the past.

The most popular contender for a triggering mechanism has been galaxy mergers or interactions. The exact means by which such an event may introduce fuel into the central engine may not be very well understood, but there is mounting evidence that such events may have been important in many observed AGN (e.g. tidal tails in QSO galaxies), and the whole process is something that is naturally expected to have been more important in the past. The analysis of the merging rate expected between clumps of dark matter in a hierarchical density field is given in chapter 17 (Press–Schechter theory). In outline, the Press–Schechter prediction depends on the rms fractional density contrast, $\sigma(R)$, when the cosmic density field is averaged in a sphere of Lagrangian radius R. The radius R encloses the mass M of interest. At any epoch, there is a cutoff scale M^* at which $\sigma \sim 1$; for much smaller masses, the predicted number density of objects scales as $n \propto 1/\sigma$. This density declines with time, as a result of mergers, at a rate shown in chapter 17 to be at least as fast as $dn/dt \propto (1+z)^{5/2}$. Allowing for the probability that two stellar cores may fail to merge, even though their dark haloes do, can double this redshift dependence. If we assume that AGN are short-lived objects with a constant lifetime, then the comoving density of events should track this rapidly varying rate, and something like density evolution could be understood (see Carlberg 1990).

This sort of picture can also account in a reasonably natural way for the AGN redshift cutoff: at high z, M^* will eventually fall below the size of a massive galaxy, leading to an exponential decline in the number of available AGN sites. We could try to argue that we don't understand the reason that local AGN prefer massive galaxies, and that perhaps this changed at high z; however, the Eddington limit for quasars sets a lower limit to how far we can escape along this route. How long it takes M^* to evolve from the current Abell cluster value to that of a massive galaxy depends on the power spectrum. For cold dark matter, the critical redshift is probably in the region of 3–5

(Efstathiou & Rees 1988), and so we should not be surprised at a lack of very high-z quasars.

These arguments are interesting, but we are far from being sure they are correct; other alternative explanations for AGN evolution exist, some of them radically different. One interesting mechanism at the opposite extreme from mergers is to suppose that AGN could be 'self-financing': fuelled only by the gas lost from their evolving stellar content. Bearing in mind that an $M_B = -25$ quasar needs only $\sim 1 M_\odot c^2$ of fuel per year, this is an attractive possibility. In fact, a rule of thumb for the mass-loss rate of a stellar population of age t is

$$\frac{\dot{M}}{M} \simeq 0.03 \left(\frac{t}{\text{Gyr}} \right)^{-1} \text{Gyr}^{-1}, \tag{14.55}$$

which can certainly provide adequate fuel when the galaxy is massive and relatively young. The fuel supply declines with time as the population ages and the most massive stars disappear from the main sequence. As with merging, the evolution is accelerated further because the gas can be driven out of the galaxy if it is heated by supernovae. The condition for the formation of such a **supernova-heated wind** is

$$\frac{\dot{E}_{\text{SN}}}{\dot{M}_{\text{stars}}} \gtrsim \frac{GM}{r} \tag{14.56}$$

(Bailey & Macdonald 1981); this process is more important at low z, as $\dot{M}$ declines, and this can lead to a much more rapid epoch dependence of the rate at which gas can reach the AGN.

The history of the gas in galaxies is of course bound up with the overall history of star formation, and it is plausible that the periods when a galaxy is gaining gas will be times when both young stars shine and AGN are fuelled. It therefore may be more than a coincidence that both the overall level of AGN activity and the global star-formation rate peak around $z \simeq 2$. However, there clearly remains much work to be done in order to turn this insight into a quantitative understanding of how and why AGN change their numbers in such dramatic fashion.

14.6 Black holes as central engines

The systematics of AGN argue for a source of great energy within the central fraction of a parsec of the host galaxy. The general belief is that this must be a supermassive black hole, for four main reasons: (i) the efficiency with which gravitational energy can be liberated by accretion onto black holes greatly exceeds what can be achieved via nuclear processes; (ii) the existence of 100-kpc jets argues for a stable ejection axis, most naturally supplied by a rotating massive object; (iii) relativistic ejection argues for a potential well that is relativistically deep, in that orbital velocities approach c; (iv) rapid variability limits the size of the engine to close to a Schwarzschild radius. Some of these points are elaborated in the following sections. Rees (1984) gives a detailed review of models for powering active galaxies by black holes.

ACCRETION DISKS It is to be expected that material falling into a black hole will settle into a disk, which will become flattened by rotation. The rotation axis will be set by the mean angular momentum of the material (although, for a rotating hole, **Lense–Thirring precession** will cause the disk to align with the hole's angular momentum). Dissipative processes in the disk will then cause material to spiral inwards and eventually be swallowed by the hole. The properties of such **accretion disks** have been studied in some detail; we concentrate here on a few important essentials that are particularly important to AGN.

We begin with Newtonian disks: material with negligible self-gravity in orbit about a central object of mass M (Pringle 1981). The problem is quite simple if the disk rotates supersonically: the equation of hydrostatic equilibrium perpendicular to the disk is

$$\frac{dp}{dz} = -\rho \frac{dg_\perp}{dz} \simeq -\rho \frac{GMz}{R^3}, \tag{14.57}$$

where R is the radius of the orbit. This equation can be integrated by putting $dp = c_S^2 d\rho$, to give $\rho \propto \exp[-\frac{1}{2}(z/h)^2]$; here, the perpendicular scale height, h, is $h = R/\mathcal{M}$, $\mathcal{M}$ being the orbital Mach number, $\mathcal{M} = v_{\rm orbit}/c_S$. We thus have a **thin accretion disk** if the material can cool efficiently enough that the orbit is supersonic.

Let us now study the properties of a disk that is assumed to be in an equilibrium state of steady accretion. The radial structure depends on the mechanism of dissipative heating of the disk, which we assume can be written in terms of a viscous stress. At a radius R, there is a couple, G, acting between the inner and outer parts of the disk: $G = (2\pi Rh)\,\eta\, R(d\Omega/dR)\, R$ – i.e. couple = area × viscous stress × radius; Ω here denotes angular velocity, not density parameter. Defining the kinematic viscosity $\nu = \eta/\rho$ and Σ as surface density gives $G = 2\pi R^3 \nu \Sigma \Omega'$, where primes denote d/dR. In equilibrium, this must be related to the inward flux of angular momentum, $F = 2\pi R\,\Sigma R^2 \Omega\, v_{\rm R}$ (where $v_{\rm R}$ is the radial velocity), by $F' = G'$. Similarly, conservation of mass says $(2\pi R\Sigma v_{\rm R})' = 0$. Integration gives the two fundamental equations for the accretion disk:

$$\begin{aligned} 2\pi R\Sigma v_{\rm R} &= -\dot{M} \\ R^3 \Sigma \Omega v_{\rm R} &= R^3 \nu \Sigma \Omega' + A, \end{aligned} \tag{14.58}$$

where $\dot{M}$ (the accretion rate) and A are constants. Eliminating $v_{\rm R}$, we find that the surface density vanishes at some critical **accretion radius**

$$\nu\Sigma = \frac{\dot{M}}{3\pi}\left(1 - \sqrt{\frac{R_*}{R}}\right). \tag{14.59}$$

Clearly, R_* corresponds to the radius at which the central object disrupts the slow viscous inward spiral of the disk; for a black hole, we shall see that R_* is of order the Schwarzschild radius.

So much for density; what about the temperature structure of the disk? This is set by the rate of doing work per unit area, which is clearly $W = (G\Omega)'/(2\pi R)$, evaluated in the rest frame of the disk, so that $(G\Omega)' \to G\Omega'$. We therefore get $W = \nu\Sigma(R\Omega')^2$. Using our previous result for $\Sigma(R)$ plus the Keplerian $\Omega = \sqrt{GM/R^3}$, we get a result for the rate of dissipation that is magically independent of the viscosity. Assuming that the dissipated energy is all radiated away at the radius at which is it generated gives us the radial temperature structure, provided the disk is optically thick:

$$\boxed{W = \frac{3GM\dot{M}}{4\pi R^3}\left(1 - \sqrt{\frac{R_*}{R}}\right) = 2\sigma T^4(R)} \tag{14.60}$$

(the opposite extreme, **advection-dominated flow**, where little of the energy is radiated away locally, is discussed by Narayan 1996). Integrating the thin-disk emissivity W over radius gives a total luminosity $L = GM\dot{M}/(2R_*)$, as required by the virial theorem. The emission is clearly dominated by the central parts of the disk, which are also the hottest (the temperature has a maximum at $R/R_* = (7/6)^2$). The spectrum of an accretion disk is therefore close to being that of a black body, with a characteristic temperature given by

$$T_*^4 = \frac{3GM\dot{M}}{8\pi\sigma R_*^3}. \tag{14.61}$$

However, the spectrum is very different from that for a black body at low frequencies, where the superposition of different temperatures changes the power-law index. In the Rayleigh–Jeans limit, we get a flux density $S_\nu \propto \nu^2 \int T R\, dR$. Now use $T/T_* = (R/R_*)^{-3/4}$ as a good approximation to the temperature profile, convert the integral to one over $x \equiv h\nu/kT$ and truncate at $x \sim 1$ where the Rayleigh–Jeans form fails in the cool outer parts; the result is the characteristic nearly flat spectrum for an accretion disk,

$$\boxed{S_\nu \propto \nu^{1/3}.} \tag{14.62}$$

Nevertheless, the peak of the spectrum is at a frequency close to that expected for a black body at temperature T_*. For an accretion rate of $1\, M_\odot \mathrm{yr}^{-1}$ (reasonable for many quasars, as we shall see below), this yields a frequency $\sim 10^{15}(M/10^9 M_\odot)^{-1/2}$ Hz *if* the central object is a black hole, so that $R_* \simeq 2GM/c^2$. The fact that a **blue bump** component (apparently unpolarized) is often seen in quasar continua at about this frequency is taken by some workers as supporting the existence both of accretion disks and black holes. Robinson & Terlevich (1994) contains detailed discussion on the problems in matching disk spectra to observation.

RELATIVISTIC DISKS The above results apply to Newtonian disks only, and will alter in detail if the depth of the potential well becomes relativistic. The major role of exact general relativity calculations is to establish the value of R_*, the inner edge of the disk, and to determine the accretion efficiency. The inner edge is set by the point at which the assumed circular orbits become unstable to small radial perturbations; at this point material will rapidly spiral into the hole without time for further dissipation. This happens at $R_* = 6GM/c^2$ for a Schwarzschild hole, reducing to $R_* = GM/c^2$ for an extreme Kerr hole with the largest allowed angular momentum, as discussed in the section on orbital stability in chapter 2. This distinction matters, because it affects the maximum gravitational redshift that can be obtained. As discussed in chapter 2, the redshift factor is the inverse square root of the lapse function, which may be written in term of Kerr coordinates with $G = c = 1$ as

$$1 + z = \frac{\Sigma}{\Delta^{1/2}\rho} = \sqrt{\frac{1 + a^2/r^2 + 2Ma^2/r^3}{1 + a^2/r^2 - 2M/r}}. \tag{14.63}$$

For a Schwarzschild hole, $a = 0$ and $r = 6M$ gives $1 + z = \sqrt{3/2}$, so that the maximum redshift is rather small, even allowing for additional Doppler effects from the orbital velocities. Conversely, as we approach the extreme Kerr case of $a = M = r$, the redshift diverges. Remarkably, it is possible that this signature has been seen, since some Seyfert galaxies have been observed to have iron X-ray lines that are broad and significantly

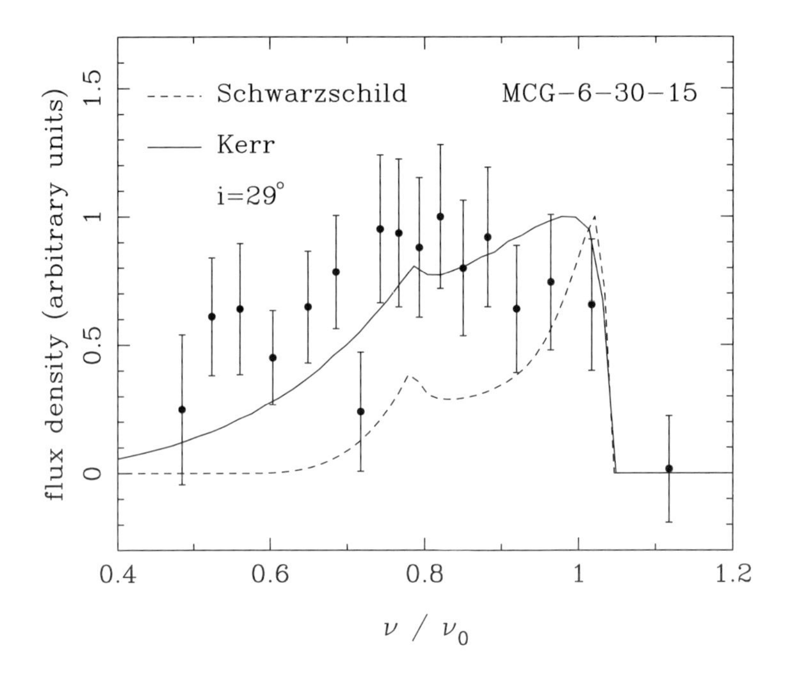

Figure 14.6. The profile of the 6.5-keV iron K line, from the Seyfert galaxy MCG–6–30–15 (Iwasawa *et al.* 1996). The expected profiles from accretion disks around Schwarzschild and extreme Kerr black holes are also shown (Dabrowski *et al.* 1997). The latter matches very well, suggesting that we are seeing emission very close to the event horizon. This was the first test of general relativity in the strong-field regime.

redshifted with respect to the body of the galaxy. In one case (figure 14.6), the redshifted tail of the line extends far beyond what would be expected for a Schwarzschild hole. However, the tail to very high redshifts expected for an extreme Kerr hole provides a good match to the data. This is an extremely important result, since it may be the first genuine strong-field test of general relativity – which the theory appears to pass very well.

The distinction between different kinds of hole also affects the efficiency with which holes can turn fuel into radiation. Chapter 2 also shows that the binding energy at the marginally stable radius is 5.7% of the rest-mass energy for a Schwarzschild hole, increasing to 42% for an extreme Kerr hole. Thus, a black hole can do about 10^4 times better than a star, for which the mass–energy conversion efficiency in nucleosynthesis is $\sim 1\%$, but where only $\sim 1\%$ of the mass is actually burned (for Solar metallicities). Even better, black holes give you two chances to withdraw the mc^2 you invest through the accreting material. The 90% that slips through becomes part of the black hole and is apparently lost, since mass cannot be extracted from a hole other than via Hawking radiation. However, the accreting material increases both the mass and the angular momentum of the black hole; there is a component of energy that corresponds to the rotational energy of the hole, and this can in principle be tapped (Penrose 1969). The simplest way to see this is to realize that a black hole has an entropy proportional to the area of its event horizon. This is easily seen for a Schwarzschild hole, where $E \propto M$ and $T \propto 1/M$: integrating $dE = T\,dS$ implies $S \propto M^2$. Historically, this reasoning was important in leading Hawking to the conclusion that holes emit thermal radiation, and he was able to prove that the horizon area in general is an entropy-like quantity, which never decreases when black holes merge (Hawking 1976). So, for a Kerr hole, this area is

$$A_{\rm H} = 4\pi \left(r_{\rm H}^2 + a^2 \right), \tag{14.64}$$

where $r_{\rm H} = M + \sqrt{M^2 - a^2}$ ($G = c = 1$; see chapter 2). Conserving entropy implies an **irreducible mass** for the $a = 0$ hole of the same area:

$$\boxed{M_{\rm irr} = \tfrac{1}{2} \left[\left(M + \sqrt{M^2 - a^2} \right)^2 + a^2 \right]^{1/2}.} \tag{14.65}$$

The rotational energy that is accessible is then $(M - M_{\rm irr})c^2$. For an extreme Kerr hole ($a = M$), the energy that can be liberated is $(1 - 1/\sqrt{2})Mc^2$, or 29% of the rest-mass energy. A practical means of achieving this, first suggested by Blandford & Znajek (1977), is via **black hole electrodynamics**. High field strengths produced by compression of magnetic fields near the horizon can allow the extraction of e^+e^- pairs, and allow the hole's rotational energy to be converted to an energetic stream of plasma (see Thorne, Price & Macdonald 1986). In practice, the complicated configuration of matter and fields around a black hole, and the associated continuous reprocessing of radiation, probably means that seeing either the accretion or the electrodynamic mechanisms in their pure form is unlikely. Nevertheless, the point about the uniquely high efficiency of energy conversion via relativistically deep potential wells is likely to stand.

Even for these high efficiencies, the implied masses of black holes are impressive. For example, the total power and lifetime of Cygnus A were calculated above, giving a total energy budget of some $10^{52.5} h^{-17/7}$ J. At an efficiency of (say) 10%, this requires a mass of $\sim 10^6\, M_\odot$. The most powerful AGN have energetics that exceed Cygnus A by several powers of 10, so hole masses up to 10^8–$10^9\, M_\odot$ would seem to be possible. We now turn to the means by which such objects can be found experimentally.

14.7 Black hole masses and demographics

EDDINGTON LUMINOSITY For isotropic emission and static emitting material, the **Eddington limit** gives the maximum luminosity that can be radiated by a gravitating body of mass M. This limit arises because radiation pressure from a central source should not exceed gravity: $\sigma \dot{E}/(4\pi r^2 c) < GMm/r^2$ (in the absence of other forces; humans radiate at well above their Eddington limit). If the opacity is formed mainly by electron scattering, then the appropriate cross-section is $\sigma_{\rm T}$ and we get the maximum rate of radiation

$$\boxed{\dot{E}_{\rm Edd} = 4\pi G M \frac{m_e c}{\sigma_{\rm T}} = 10^{39.1} \left(\frac{M}{10^8 M_\odot} \right) \ {\rm W}.} \tag{14.66}$$

As an aside, it is interesting to ask how long it takes an object radiating at the Eddington limit to radiate away all of its mass. The lifetime is then $\sim Mc^2/L_{\rm Edd}$, which is known as the **Salpeter time**:

$$t_{\rm Salpeter} = \frac{\mu_0^2 e^4 c}{24\pi^2 G m_e^3} \simeq 10^{11.9}\ \text{yr}, \tag{14.67}$$

which is *independent* of the mass.

To apply this, we need $\dot{E}$ for quasars, which is usually obtained by taking νS_ν in the optical, applying a rather uncertain **bolometric correction**, g, and assuming isotropy. This gives the total energy output as

$$\dot{E} = 10^{38.5} g \times 10^{-0.4(M_B+25)}\ \text{W}, \tag{14.68}$$

where $g \simeq 10$ is a reasonable value. The central mass can be found once we know what fraction of the Eddington limit this luminosity represents. In pure luminosity evolution models, a fixed set of objects radiate over the entire history of the universe. The total energy emitted, and hence the mass accreted can therefore be found directly. Taking an efficiency ϵ, the implied black hole mass from the luminosity function discussed above (considering $z < 2$ only) is

$$M/M_\odot = 10^{9.3} \epsilon^{-1} h^{-1} g \times 10^{-0.4(M_B+25)} \quad (\Omega = 1), \tag{14.69}$$

the inferred mass being about a factor 2 higher for $\Omega = 0$. For canonical values $(h, g, \epsilon) = (0.5, 10, 0.1)$ this implies that local quasars have $L \sim 10^{-3.2} L_{\rm Edd}$: these objects could have been much more luminous in the past without exceeding the limit.

DIRECT MASS ESTIMATES Direct estimates of the black hole masses depend on gaining an idea of the size of the emitting regions, and this may be done in three ways. The first is the simplest: measure the shortest timescale for variability that is observed in X-rays. We have already seen that these are expected to arise in the innermost parts of an accretion disk, and so can vary on a timescale of around $r_{\rm S}/c = 2GM/c^3$, where $r_{\rm S}$ is the Schwarzschild radius. The other methods utilize the broad-line emitting gas. If we take its velocity width to measure orbital and/or inflow velocity, then $v^2 \simeq GM/r$ and we can infer M once we know how far away from the centre we are; this can be done in two ways. One is to use the photosynthesis models to fix the ionization parameter: $U = N_\gamma/N_e$. Since N_e is given by arguments from forbidden lines, and $N_\gamma \propto L_{\rm quasar}/r^2$, we can observe the quasar luminosity and hence infer r. Since U empirically does not vary greatly with quasar luminosity, we see that the broad-line regions are much larger in the more active objects, varying between a few tenths of a pc for Seyferts and several pc in luminous quasars. A more direct route is to rely on **reverberation mapping**: we observe directly in the AGN continuum fluctuations of the ionizing source, which will be reflected in variations in the intensity of the broad lines, but with some time delay – which tells us the separation of nucleus and broad-line gas.

Somewhat remarkably, all these methods yield results in reasonably good agreement. Wandel & Mushotzky (1986) contrasted the mass estimates from X-ray variation and photoionization models, concluding that quasars over a very wide range of powers emit at about 1% of the Eddington limit, thus requiring holes less than one-tenth the mass implied in the case of pure luminosity evolution (PLE). Time-delay results are available for fewer objects (mainly of low luminosity, as the timescales involved are shorter). The best case is probably NGC4151, with a mass $\sim 10^{8.8} M_\odot$ (Ulrich *et al.*

1984). With $M_B \simeq -19$, this result is not greatly in conflict with PLE, which would imply $M \sim 10^{9.2} M_\odot$. The more recent reverberation results have complicated the picture a little, tending to favour broad-line regions somewhat smaller than those inferred from the pre-existing photoionization modelling. Nevertheless, the broad picture stands, in which AGN radiate at a significant fraction of the Eddington limit of their central object (e.g. Peterson 1993).

THE TOTAL BLACK HOLE DENSITY A more general argument was given by Soltan (1982). He showed that the total emitted energy density of quasars is independent of h and of Ω [problem 14.6]. Taking the $\Omega = 1$ optical luminosity function and integrating over all M_B and $z < 2$, we get a mass density in black holes

$$\boxed{\rho_{\rm BH} = 10^{11.7} g\epsilon^{-1}\ M_\odot {\rm Gpc}^{-3}.} \tag{14.70}$$

To relate this to what we expect to see in galaxies, note that Smith *et al.* (1986) have shown that underlying galaxies in quasars are luminous, with a median $M_B \simeq -22$, and that this conclusion is roughly independent of nuclear luminosity. From Efstathiou *et al.* (1988a), we then deduce a mass per galaxy of $\sim 10^6 g\epsilon^{-1} h^{-3} M_\odot$. This assumes that all luminous galaxies underwent a quasar phase at some stage, and yields a canonical mass for 'dead' quasars of $\sim 10^9 M_\odot$. From above, the current density of quasars with $M_B < -18$ (the faintest limit studied by Smith *et al.*) is ~ 0.5 times that of galaxies with $M_B < -22$. Thus, about one quarter of galaxies with $M_B < -22$ produce quasars, and the figure of $10^9 M_\odot$ is largely independent of whether quasars are long lived or occur briefly in all galaxies.

BLACK HOLES IN GALAXIES If we take the above figures seriously, the implication is that the large hole masses required under PLE are not seen: quasar activity was not confined to a few special objects that faded over a Hubble time. The discrepancy in mass of about a factor 10 fits reasonably well with the fact that about 10% of massive galaxies currently host an active nucleus, implying that the others were active in the past. Given that the hole mass estimates are not very easy to perform, it is prudent to test this picture by looking for the nuclear holes that should exist in the centres of inactive galaxies, if they are really dead quasars. In principle, this may seem easy: if a stellar rotation curve very near the centre can be measured, then $v^2 = GM/r$ and and a central massive object will produce a cusp in the velocity. However, both elliptical galaxies and the bulges of spiral galaxies are not supported by rotation alone, but largely by the 'pressure' of randomly directed stellar orbits (see chapter 13). If these were isotropic with a velocity dispersion $\sigma_v(r)$, the equation of hydrostatic equilibrium would apply:

$$\frac{\partial}{\partial r}\left(\rho\sigma_v^2\right) = -\rho\,\frac{\partial\Phi}{\partial r}. \tag{14.71}$$

If the potential is supplied entirely by the hole, we again expect a cusp $\sigma_v^2 \propto 1/r$, for any power-law density variation. However, as discussed in chapter 12, things can become more complex if the stellar orbits are not isotropic.

Resolving these ambiguities requires high-quality data that go as close to the centre as possible. See Kormendy & Richstone (1995) for a review of all the attempts to detect black holes in galactic nuclei. The best constraints thus come from the nearest galaxies: the Andromeda nebula M31 and its companion M32 appear to possess central masses of respectively $3 \times 10^7\ M_\odot$ and $2 \times 10^6\ M_\odot$; the most massive suggested hole is in the nearby radio galaxy M87, at $3 \times 10^9\ M_\odot$. Most recently, VLBI data on OH masers in NGC4258 has made the best case of all for a central black hole, of mass $4 \times 10^7\ M_\odot$ (Miyoshi *et al.* 1995). M31 makes an interesting comparison with our own galaxy: here the situation has been controversial for some time, and there is no agreement on whether there is a black hole. If there is, however, its mass is no greater than $10^6\ M_\odot$.

Problems

(14.1) Consider the radiation from an accelerated electron, which has an angular dependence $\propto \sin^2\theta$, where θ is the polar angle about the acceleration vector. Obtain the power pattern if the electron now moves with Lorentz factor γ perpendicular to the acceleration vector, and show that most of the energy is confined to a cone of semi-angle $\sim 1/\gamma$ about the forward direction. Derive the relation between the intrinsic and observed times taken for this cone to sweep over the observer.

(14.2) An electron moves relativistically through a radiation field. By transforming the number density of photons moving at an angle θ to the electron velocity, show that the electron is perceived to scatter photons at the same rate in both laboratory and rest frames, provided the laboratory radiation field is isotropic.

(14.3) In the extreme relativistic limit, give a classical derivation for the rate of Compton scattering, $-\dot{E}_{\rm ic} = (4/3)\sigma_{\rm T}\gamma^2\beta^2 U_{\rm rad}c$.

(14.4) Solve the synchrotron loss equation $\partial N(E)/\partial t = -\partial(bE^2N)/\partial E$, subject to the boundary condition $N(E) = N_0 E^{-x}$ at $t = 0$. Show that the spectrum steepens at high energy provided $x > 2$.

(14.5) If galaxies possess a luminosity function of the Schechter form,

$$dn/d\ln L = \phi^*(L/L^*)^{-\alpha}\exp(-L/L^*), \qquad (14.72)$$

and the probability of obtaining an active nucleus rises as $p_{\rm AGN} \propto L^x$, what value of x will produce a scatter of 0.4 magnitude in the luminosities of the AGN hosts?

(14.6) Calculate the density of relic black holes in 'dead' AGN, assuming a knowledge of the luminosity function. Show that the result is model independent – i.e. that it requires only the observable distribution of flux and redshift $N(S, z)$.

PART 6 GALAXY FORMATION AND CLUSTERING

15 Dynamics of structure formation

15.1 Overview

The overall properties of the universe are very close to being homogeneous; and yet telescopes reveal a wealth of detail on scales varying from single galaxies to **large-scale structures** of size exceeding 100 Mpc (see figure 15.1). The existence of these cosmological structures tells us something important about the initial conditions of the big bang, and about the physical processes that have operated subsequently. This chapter deals with the gravitational and hydrodynamical processes that are relevant to structure formation; the following chapters apply these ideas to large-scale structure, galaxy formation and the microwave background. We will now outline the main issues to be covered.

ORIGIN AND GROWTH OF INHOMOGENEITIES The aim of studying cosmological inhomogeneities is to understand the processes that caused the universe to depart from uniform density. Chapters 10 and 11 have discussed at some length the two most promising existing ideas for how this could have happened: either through the amplification of quantum zero-point fluctuations during an inflationary era, or through the effect of topological defects formed in a cosmological phase transition. Neither of these ideas can yet be regarded as established, but it is astonishing that we are able to contemplate the observational consequences of physical processes that occurred at such remote energies.

A problem with testing these models is that many features of structure formation are generic. In studying how matter in the expanding universe responds to its own self-gravity, a linear solution for how the matter behaves can be found by expanding the equations of motion in terms of a dimensionless density perturbation δ:

$$\boxed{1 + \delta(\mathbf{x}) \equiv \rho(\mathbf{x})/\langle\rho\rangle.} \tag{15.1}$$

The density content can be divided into nonrelativistic matter and radiation, and there are two distinct perturbation modes, with different relations between the two density components. Imagine starting with uniform distributions of matter and radiation; the simplest way to perturb the density would be to compress or expand some set of volume elements adiabatically. This would change the matter density and the photon number density by the same factor, and so **adiabatic perturbations** would affect the energy densities of matter and radiation in different ways: $\delta_r = 4\delta_m/3$ (because energy density for radiation is $\propto T^4$, whereas number density is $\propto T^3$). The orthogonal mode

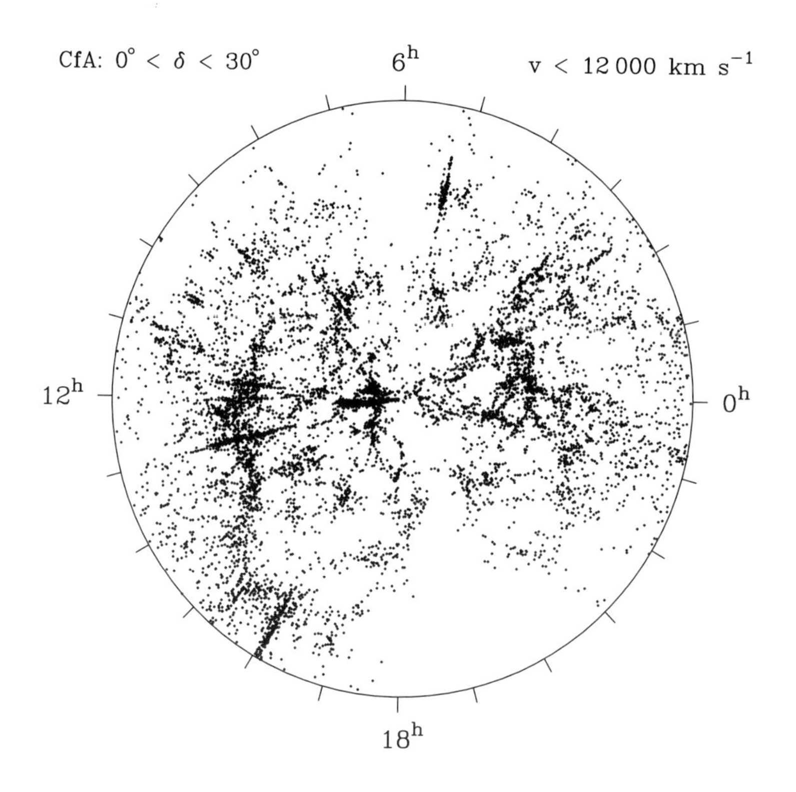

Figure 15.1. One of the most dramatic pictures of the large-scale structure in the galaxy distribution is this slice made from John Huchra's ZCAT compilation of galaxy redshifts. The dataset shown here consists mainly of redshifts from the Harvard-Smithsonian Center for Astrophysics redshift survey to $B \simeq 15.5$. The survey coverage is not quite complete; as well as the holes due to the galactic plane around right ascensions $6^{\rm h}$ and $19^{\rm h}$, the rich clusters are somewhat over-represented with respect to a true random sampling of the galaxy population. Nevertheless, this plot emphasizes nicely both the large-scale features such as the 'great wall' on the left, the totally empty void regions, and the radial 'fingers of God' caused by virialized motions in the clusters. One of the principal challenges in cosmology is to explain this pattern.

of fluctuations would take the opposite approach, perturbing the entropy density but not the energy density. The resulting fluctuations are termed **isocurvature perturbations**; the total density remains homogeneous, and so there is no perturbation to the spatial curvature: $\rho_r\delta_r = -\rho_m\delta_m$. At very early times, when the universe is strongly dominated by radiation, isocurvature initial conditions effectively correspond to a vanishingly small fractional perturbation in the radiation density, with only the matter density varying significantly.

If we wanted to create a non-uniform density field at some given time, isocurvature perturbations would be more natural, since causality would make it impossible to change the mean density on scales larger than the horizon at that time. This is true in models involving a late-time cosmological phase transition (either the topological defect models studied in chapter 10, or the axion model from chapter 8). However, in inflationary models, the nature of the particle horizon is changed at early times so that fluctuations in total density can be produced on scales that vastly exceed c/H_0. If these curvature fluctuations are imprinted prior to the processes that generate the baryon asymmetry of the universe, then adiabatic modes will be the norm: GUT-based processes that violate baryon number conservation will tend to yield a constant entropy per baryon. If we could succeed in proving the existence of adiabatic fluctuations on super-horizon scales, this would be good evidence in favour of inflationary processes having occurred.

MATTER CONTENT AND TRANSFER FUNCTIONS The matter content of the universe can influence structure formation by a variety of processes that modify any primordial density perturbations: growth under self-gravitation; effects of pressure and dissipative processes. In general, modes of short wavelength have their amplitudes reduced relative to those of long wavelength in this way. Linear adiabatic perturbations scale with time as follows:

$$\delta \propto \begin{cases} a(t)^2 & \text{(radiation domination)} \\ a(t) & \text{(matter domination; } \Omega = 1\text{),} \end{cases} \tag{15.2}$$

whereas isocurvature perturbations to the matter are initially constant, and then decline:

$$\delta_m \propto \begin{cases} \text{constant} & \text{(radiation domination)} \\ a(t)^{-1} & \text{(matter domination; } \Omega = 1\text{).} \end{cases} \tag{15.3}$$

In the first case, gravity causes the mode amplitude to increase; in the second, the evolution acts to preserve the initial uniform density. In both cases, the shape of the primordial perturbation spectrum is preserved, and only its amplitude changes with time. However, on small scales, a variety of processes affect the growth law:

(1) Pressure opposes gravity effectively for wavelengths below the **Jeans length**:

$$\lambda_J = c_S \sqrt{\frac{\pi}{G\rho}}. \tag{15.4}$$

While the universe is radiation dominated, $c_S = c/\sqrt{3}$ and so the Jeans length is always close to the size of the **horizon**. The Jeans length then reaches a maximal value at matter–radiation equality, when the sound speed starts to drop. The *comoving* horizon size at z_{eq} is therefore an important scale:

$$\boxed{R_0 r_H(z_{eq}) = 2(\sqrt{2} - 1)(\Omega z_{eq})^{-1/2} \frac{c}{H_0} = \frac{16.0}{\Omega h^2}\,\text{Mpc}.} \tag{15.5}$$

Beyond this scale, perturbations should be affected by gravity only, and we would expect to see a bend in the spectrum where pressure starts to become important: since this scale depends on Ω, the density of the universe should be written on the sky with the galaxy distribution.

(2) A further important scale arises where photon diffusion can erase perturbations in the matter–radiation fluid; this process is named **Silk damping** after the cosmologist Joseph

Silk. The distance travelled by the photon random walk by the time of last scattering is

$$\lambda_S = 2.7 \left(\Omega \Omega_B h^6\right)^{-1/4} \text{ Mpc.} \tag{15.6}$$

In models with dark matter, this effect is of limited importance, since the baryons can fall into the dark matter potential wells after last scattering.

(3) At early times, dark matter particles will undergo **free streaming** at the speed of light, and so erase all scales up to the horizon – a process that only ceases when the particles go nonrelativistic. For light massive neutrinos, this happens at $z_{\rm eq}$ (because the number densities of neutrinos and photons are comparable); all structure up to the horizon-scale power-spectrum break is in fact erased. Hot(cold)-dark-matter models are thus sometimes dubbed large(small)-scale-damping models.

NONLINEAR PROCESSES The task of modern cosmology is to confront the predictions of these models with observations. The problem is that the objects that can be seen by astronomers are the results of nonlinear gravitational evolution: clusters of galaxies are thousands of times denser than the mean, and galaxies themselves have densities a million times the average. It is therefore critical to understand the processes that modify the simple linear evolution growth laws in these cases. This can be achieved by brute force: numerical integration of the equations of motion for a very large number of particles that are given small initial perturbations. It is nevertheless essential to develop some more intuitive tools by means of which the results of the **N-body simulations** can be understood. This chapter derives the most important of these methods, which are applied in chapters 16 and 17.

15.2 Dynamics of linear perturbations

The study of cosmological perturbations can be presented as a complicated exercise in linearized general relativity. Fortunately, much of the essential physics can be extracted from a Newtonian approach (not so surprising when one remembers that small perturbations imply weak gravitational fields). We start by writing down the fundamental equations governing fluid motion (nonrelativistic for now):

$$\begin{aligned} &\text{Euler} && \frac{D\mathbf{v}}{Dt} = -\frac{\boldsymbol{\nabla} p}{\rho} - \boldsymbol{\nabla}\Phi \\ &\text{energy} && \frac{D\rho}{Dt} = -\rho \boldsymbol{\nabla}\cdot\mathbf{v} \\ &\text{Poisson} && \nabla^2\Phi = 4\pi G\rho, \end{aligned} \tag{15.7}$$

where $D/Dt = \partial/\partial t + \mathbf{v}\cdot\boldsymbol{\nabla}$ is the usual convective derivative. We now produce the **linearized equations of motion** by collecting terms of first order in perturbations about a homogeneous background: $\rho = \rho_0 + \delta\rho$ etc. As an example, consider the energy equation:

$$\left[\partial/\partial t + (\mathbf{v}_0 + \delta\mathbf{v})\cdot\boldsymbol{\nabla}\right](\rho_0 + \delta\rho) = -(\rho_0 + \delta\rho)\boldsymbol{\nabla}\cdot(\mathbf{v}_0 + \delta\mathbf{v}). \tag{15.8}$$

For no perturbation, the zero-order equation is $(\partial/\partial t + \mathbf{v}_0\cdot\boldsymbol{\nabla})\rho_0 = -\rho_0\boldsymbol{\nabla}\cdot\mathbf{v}_0$; since ρ_0 is homogeneous and $\mathbf{v}_0 = H\mathbf{x}$ is the Hubble expansion, this just says $\dot\rho_0 = -3H\rho_0$.

Expanding the full equation and subtracting the zeroth-order equation gives the equation for the perturbation:

$$\left(\partial/\partial t + \mathbf{v}_0 \cdot \boldsymbol{\nabla}\right) \delta\rho + \delta\mathbf{v} \cdot \boldsymbol{\nabla}(\rho_0 + \delta_\rho) = -(\rho_0 + \delta\rho)\boldsymbol{\nabla} \cdot \delta\mathbf{v} - \delta\rho\, \boldsymbol{\nabla} \cdot \mathbf{v}_0. \tag{15.9}$$

Now, for sufficiently small perturbations, terms containing a product of perturbations such as $\delta\mathbf{v} \cdot \boldsymbol{\nabla}\delta_\rho$ must be negligible in comparison with the first-order terms. Remembering that ρ_0 is homogeneous leaves the linearized equation

$$\left(\partial/\partial t + \mathbf{v}_0 \cdot \boldsymbol{\nabla}\right) \delta\rho = -\rho_0\, \boldsymbol{\nabla} \cdot \delta\mathbf{v} - \delta\rho\, \boldsymbol{\nabla} \cdot \mathbf{v}_0. \tag{15.10}$$

A key question for this linearized theory is to quantify how small the perturbations need to be in order for the theory to be applicable. Clearly $\delta\rho/\rho \ll 1$ is a minimum criterion, but the linearized equations will still be invalid if the derivatives of $\delta\rho$ are large enough, however small we make $\delta\rho$ itself. This is a difficult issue, which is best approached by comparing linear theory with exact nonlinear models, as is done at the end of this chapter.

It is straightforward to perform the same steps with the other equations; the results look simpler if we define the fractional density perturbation

$$\boxed{\delta \equiv \frac{\delta\rho}{\rho_0}.} \tag{15.11}$$

As above, when dealing with time derivatives of perturbed quantities, the full convective time derivative D/Dt can always be replaced by $d/dt \equiv \partial/\partial t + \mathbf{v}_0 \cdot \boldsymbol{\nabla}$, which is the time derivative for an observer comoving with the unperturbed expansion of the universe. We then can write

$$\begin{aligned} \frac{d}{dt}\delta\mathbf{v} &= -\frac{\boldsymbol{\nabla}\,\delta p}{\rho_0} - \boldsymbol{\nabla}\,\delta\Phi - (\delta\mathbf{v} \cdot \boldsymbol{\nabla})\mathbf{v}_0 \\ \frac{d}{dt}\delta &= -\boldsymbol{\nabla} \cdot \delta\mathbf{v} \\ \nabla^2\delta\Phi &= 4\pi G\rho_0\delta. \end{aligned} \tag{15.12}$$

There is now only one complicated term to be dealt with, $(\delta\mathbf{v} \cdot \boldsymbol{\nabla})\mathbf{v}_0$, on the rhs of the perturbed Euler equation. This is best attacked by writing it in components:

$$[(\delta\mathbf{v} \cdot \boldsymbol{\nabla})\mathbf{v}_0]_j = [\delta v]_i \nabla_i [v_0]_j = H\,[\delta v]_j, \tag{15.13}$$

where the last step follows because $\mathbf{v}_0 = H\,\mathbf{x}_0 \Rightarrow \nabla_i [v_0]_j = H\delta_{ij}$. This leaves a set of equations of motion that have no explicit dependence on the global expansion speed v_0; this is only present implicitly through the use of convective time derivatives d/dt.

These equations of motion are written in **Eulerian coordinates**: proper length units are used, and the Hubble expansion is explicitly present through the velocity $\mathbf{v}_0$. The alternative approach to fluid dynamics is to use **Lagrangian coordinates**, which have the property that the Lagrangian position of a particle or fluid element does not change with time; the Eulerian coordinates at some fixed time t_0 would be a set of Lagrangian coordinates. Lagrangian coordinates in cosmology have already been encountered in the form of **comoving coordinates**; these label observers who follow the Hubble expansion in an unperturbed universe. Once we consider a perturbed universe, however, the comoving coordinates formed by dividing the Eulerian coordinates by the scale factor $a(t)$ are no longer Lagrangian, and this distinction is illustrated in figure 15.2. Gravity will

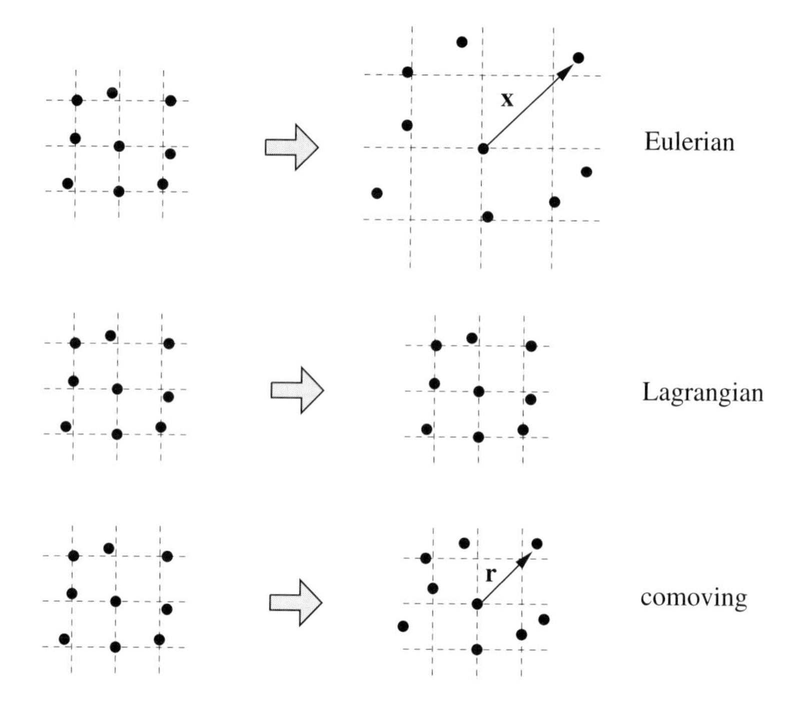

Figure 15.2. Various descriptions of the dynamics of particles in an expanding universe. At some time t_0, let there exist a set of particles displaced slightly from a uniform grid, as shown on the left. At some later time, the expansion will have carried the particles further apart, but gravitational attraction will have increased the degree of irregularity. This is the Eulerian point of view, in which proper length units are used, and the position vector of a particle is **x**. Alternatively, the Lagrangian description uses coordinates that do not change with time; all information about density and velocity fields is contained in the mapping between Eulerian and Lagrangian coordinates. Comoving coordinates are a compromise between these alternatives, in which the large-scale expansion is divided out, but the comoving position vector **r** still changes with time as irregularities grow.

cause a non-uniform distribution of particles to grow increasingly irregular, so that matter undergoes a displacement with respect to the uniformly expanding background. These displacements are the most direct signature of inhomogeneity, and so it makes sense to recast the perturbed equations of motion in comoving units. To achieve this transformation, we need to define comoving spatial coordinates, and to make a similar transformation for the **peculiar velocity**, $\delta\mathbf{v}$:

$$\boxed{\begin{aligned} \mathbf{x}(t) &= a(t)\mathbf{r}(t) \\ \delta\mathbf{v}(t) &= a(t)\mathbf{u}(t). \end{aligned}} \qquad (15.14)$$

The next step is to translate spatial derivatives into comoving coordinates:

$$\boldsymbol{\nabla}_x = \frac{1}{a}\boldsymbol{\nabla}_r. \tag{15.15}$$

To keep the notation simple, subscripts on $\boldsymbol{\nabla}$ will normally be omitted hereafter, and spatial derivatives will be with respect to comoving coordinates. The linearized equations for conservation of momentum and matter as experienced by fundamental observers moving with the Hubble flow then take the following simple forms in comoving units:

$$\boxed{\begin{aligned} \dot{\mathbf{u}} + 2\frac{\dot{a}}{a}\mathbf{u} &= \frac{\mathbf{g}}{a} - \frac{\boldsymbol{\nabla}\,\delta p}{\rho_0} \\ \dot{\delta} &= -\boldsymbol{\nabla}\cdot\mathbf{u}, \end{aligned}} \tag{15.16}$$

where dots stand for d/dt. The peculiar gravitational acceleration $\boldsymbol{\nabla}\delta\Phi/a$ is denoted by $\mathbf{g}$.

Before going on, it is useful to give an alternative derivation of these equations, this time working in comoving length units right from the start. First note that the comoving peculiar velocity $\mathbf{u}$ is just the time derivative of the comoving coordinate $\mathbf{r}$:

$$\dot{\mathbf{x}} = \dot{a}\mathbf{r} + a\dot{\mathbf{r}} = H\mathbf{x} + a\dot{\mathbf{r}}, \tag{15.17}$$

where the rhs must be equal to the Hubble flow $H\mathbf{x}$, plus the peculiar velocity $\delta\mathbf{v} = a\mathbf{u}$. In this equation, dots stand for exact convective time derivatives – i.e. time derivatives measured by an observer who follows a particle's trajectory – rather than partial time derivatives $\partial/\partial t$. This allows us to apply the continuity equation immediately in comoving coordinates, since this equation is simply a statement that particles are conserved, independently of the coordinates used. The exact equation is

$$\frac{D}{Dt}\rho_0(1+\delta) = -\rho_0(1+\delta)\boldsymbol{\nabla}\cdot\mathbf{u}, \tag{15.18}$$

and this is easy to linearize because the background density ρ_0 is independent of time when comoving length units are used. This gives the first-order equation $\dot{\delta} = -\boldsymbol{\nabla}\cdot\mathbf{u}$ immediately. The equation of motion follows from writing the Eulerian equation of motion as $\ddot{\mathbf{x}} = \mathbf{g}_0 + \mathbf{g}$, where $\mathbf{g} = \boldsymbol{\nabla}\delta\Phi/a$ is the peculiar acceleration defined earlier, and $\mathbf{g}_0$ is the acceleration that acts on a particle in a homogeneous universe (neglecting pressure forces, for simplicity). Differentiating $\mathbf{x} = a\mathbf{r}$ twice gives

$$\ddot{\mathbf{x}} = a\dot{\mathbf{u}} + 2\dot{a}\mathbf{u} + \frac{\ddot{a}}{a}\mathbf{x} = \mathbf{g}_0 + \mathbf{g}. \tag{15.19}$$

The unperturbed equation corresponds to zero peculiar velocity and zero peculiar acceleration, $(\ddot{a}/a)\mathbf{x} = \mathbf{g}_0$; subtracting this gives the perturbed equation of motion $\mathbf{u} + 2(\dot{a}/a)\mathbf{u} = \mathbf{g}$, as before. This derivation is rather more direct than the previous route of working in Eulerian space. Also, it emphasizes that the equation of motion is exact, even though it happens to be linear in the perturbed quantities. The only point that needs a little more thought is the nature of the unperturbed equation of motion. This cannot be derived from Newtonian gravity alone, and we saw in chapter 2 that general relativity is needed for a proper derivation of the homogeneous equation of motion. However, as long as we are happy to accept that $\mathbf{g}_0$ is given, then it is a well-defined procedure to add a peculiar acceleration that is the gradient of the potential derived from the density perturbations.

After doing all this, we still have three equations in the four variables δ, $\mathbf{u}$, $\delta\Phi$ and δp. The system needs an equation of state in order to be closed; this may be specified in terms of the sound speed

$$c_s^2 \equiv \frac{\partial p}{\partial \rho}. \tag{15.20}$$

Now think of a plane-wave disturbance $\delta \propto e^{-i\mathbf{k}\cdot\mathbf{r}}$, where $\mathbf{k}$ is a comoving wavevector; in other words, suppose that the wavelength of a single Fourier mode stretches with the universe. All time dependence is carried by the amplitude of the wave, and so the spatial dependence can be factored out of time derivatives in the above equations (which would not be true with a constant comoving wavenumber k/a). An equation for the amplitude of δ can then be obtained by eliminating $\mathbf{u}$ (take the divergence of the perturbed Euler equation and the time derivative of the continuity equation and eliminate $\boldsymbol{\nabla}\cdot\dot{\mathbf{u}}$):

$$\boxed{\ddot{\delta} + 2\frac{\dot{a}}{a}\dot{\delta} = \delta\left(4\pi G\rho_0 - \frac{c_s^2 k^2}{a^2}\right).} \tag{15.21}$$

This equation is the one that governs the gravitational amplification of density perturbations. If we had linearized about a stationary background and taken $\mathbf{k}$ to be a *proper* wavevector, then the equation would have been much easier to derive, and the result would be just $\ddot{\delta} = \delta(4\pi G\rho_0 - c_s^2 k^2)$, which has solutions

$$\delta(t) = e^{\pm t/\tau}, \quad \tau = 1/\sqrt{4\pi G\rho_0 - c_s^2 k^2}. \tag{15.22}$$

As mentioned earlier, there is a critical proper wavelength, known as the **Jeans length**, at which we switch from the possibility of exponential growth for long-wavelength modes to standing sound waves at short wavelengths. This critical length is

$$\boxed{\lambda_{\rm J} = c_s\sqrt{\frac{\pi}{G\rho}},} \tag{15.23}$$

and clearly delineates the scale at which sound waves can cross an object in about the time needed for gravitational free-fall collapse. When considering perturbations in an expanding background, things are more complex. Qualitatively, we expect to have no growth when the 'driving term' on the rhs is negative. However, owing to the expansion, $\lambda_{\rm J}$ will change with time, and so a given perturbation may switch between periods of growth and stasis. These effects help to govern the form of the perturbation spectrum that has propagated to the present universe from early times; they will be considered in detail shortly.

RADIATION-DOMINATED UNIVERSES At early enough times, the universe was radiation dominated ($c_s = c/\sqrt{3}$) and the analysis so far does not apply. It is common to resort to general relativity perturbation theory at this point. However, the fields are still weak, and so it is possible to generate the results we need by using special relativity fluid mechanics and Newtonian gravity with a relativistic source term, as studied in chapter 2. For simplicity, assume that accelerations due to pressure gradients are negligible in comparison with gravitational accelerations (i.e. restrict the analysis to $\lambda \gg \lambda_{\rm J}$ from the start). The basic equations are then a simplified Euler equation and the full energy and gravitational equations:

$$\begin{aligned} &\text{Euler} && \frac{D\mathbf{v}}{Dt} = -\boldsymbol{\nabla}\Phi \\ &\text{energy} && \frac{D}{Dt}\left(\rho + p/c^2\right) = \frac{\partial}{\partial t}\left(p/c^2\right) - \left(\rho + p/c^2\right)\boldsymbol{\nabla}\cdot\mathbf{v} \\ &\text{Poisson} && \nabla^2\Phi = 4\pi G(\rho + 3p/c^2). \end{aligned} \tag{15.24}$$

For total radiation domination, $p = \rho c^2/3$, and it is easy to linearize these equations as before. The main differences come from factors of 2 and 4/3 due to the non-negligible contribution of the pressure. The result is a continuity equation $\boldsymbol{\nabla}\cdot\mathbf{u} = -(3/4)\dot{\delta}$, and the evolution equation for δ:

$$\boxed{\ddot{\delta} + 2\frac{\dot{a}}{a}\dot{\delta} = \frac{32\pi}{3}G\rho_0\delta,} \tag{15.25}$$

so the net result of all the relativistic corrections is a driving term on the rhs that is a factor 8/3 higher than in the matter-dominated case.

In this simplified exercise, we have implicitly used $(\nabla\Phi)_k \propto \rho_k/k$ to argue that all other acceleration terms on the rhs of the Euler equation must be negligible with respect to the gravitational acceleration when k is sufficiently small. The omitted term on the rhs of Euler's equation is $(\boldsymbol{\nabla}p+\mathbf{v}\dot{p})/(\rho+p/c^2)$, and this can be kept in at the cost of expanding the algebra. If this is done, the eventual effect is just to add a term $-k^2c^2\delta/(3a^2)$ to the rhs of the equation for δ. This is identical to the pressure term in the nonrelativistic case, remembering that the sound speed is $c_s = c/\sqrt{3}$ for a radiation-dominated fluid.

SOLUTIONS FOR $\delta(t)$ In both matter- and radiation-dominated universes with $\Omega = 1$, we have $\rho_0 \propto 1/t^2$:

$$\begin{aligned} &\text{matter domination } (a \propto t^{2/3}) && 4\pi G\rho_0 = 2/(3t^2) \\ &\text{radiation domination } (a \propto t^{1/2}) && 32\pi G\rho_0/3 = 1/t^2. \end{aligned} \tag{15.26}$$

Every term in the equation for δ is thus the product of derivatives of δ and powers of t, and a power-law solution is obviously possible. If we try $\delta \propto t^n$, then the result is $n = 2/3$ or -1 for matter domination; for radiation domination, this becomes $n = \pm 1$. For the growing mode, these can be combined rather conveniently using the **conformal time** $\eta \equiv \int dt/a$:

$$\boxed{\delta \propto \eta^2.} \tag{15.27}$$

Recall that η is proportional to the comoving size of the horizon.

One further way of stating this result is that gravitational potential perturbations are independent of time (at least while $\Omega = 1$). Poisson's equation tells us that $-k^2\Phi/a^2 \propto \rho\,\delta$; since $\rho \propto a^{-3}$ for matter domination or a^{-4} for radiation, that gives $\Phi \propto \delta/a$ or δ/a^2 respectively, so that Φ is independent of a in either case. In other words, the metric fluctuations resulting from potential perturbations are frozen, at least for perturbations with wavelengths greater that the horizon size. This conclusion demonstrates the self-consistency of the basic set of fluid equations that were linearized. The equations for momentum and energy conservation are always valid, but the correct relativistic description of linear gravity should be a wave equation, including time derivatives of Φ as well as spatial derivatives. Since we have used only Newtonian gravity, this implies that any solutions of the perturbation equations in which Φ varies will not be valid on super-horizon scales. This criticism does not apply to the growing mode, where Φ is constant, but it does apply to decaying modes (Press & Vishniac 1980). This difficulty concerning perturbations on scales greater than the horizon is related to gauge freedom in general relativity and the fact that the value of density perturbations δ can be altered by a coordinate transformation. The exchange of light signals can be used to establish a 'sensible' coordinate system, but only on sub-horizon scales. Otherwise, the results obtained are gauge dependent. We are implicitly using here what might be called the Newtonian gauge, where the metric is expressed as the Robertson–Walker form with perturbation factors $(1 \pm 2\Phi/c^2)$ in the time and spatial parts respectively:

$$c^2 d\tau^2 = \left(1 + 2\Phi/c^2\right) c^2 dt^2 - \left(1 - 2\Phi/c^2\right) R^2 \left[dr^2 + S_k^2(r) d\psi^2\right]. \tag{15.28}$$

Even for simple perturbations to the density and pressure of a perfect fluid, this is far from being the only possible gauge choice. Different choices of gauge will lead to apparently quite different equations of motion – although the physical content is the same, as indicated by the fact that the growing mode is obtained correctly in the present treatment. The gauge issue is discussed in more detail in chapter 2 and below.

PERTURBATION GROWTH WITH $\Omega \neq 1$ Rather than attempting to solve the differential equation for $\delta(t)$ directly, we can use a powerful trick. When we discussed the solution to the equations of motion for a spherically symmetric universe in chapter 3, we noted that these equations applied equally well to a spherical Newtonian mass. The general growth of a spherical density perturbation can therefore be found in this way. Later, we will use this as a model for the *nonlinear* collapse of structure in the universe. For now, if we follow the motion of two spheres of slightly different rates of expansion, we can obtain the *linear* behaviour in the general case.

Write the parametric evolution of the radius of a spherical mass as

$$\begin{aligned} R &= k\frac{GM}{c^2}[1 - C_k(\eta)] \\ ct &= k\frac{GM}{c^2}[\eta - S_k(\eta)] \end{aligned} \tag{15.29}$$

and consider the difference in density of two models with different mass and radius:

$$\delta \equiv \frac{\delta\rho}{\rho} = \frac{\delta M}{M} - 3\frac{\delta R}{R}. \tag{15.30}$$

The variation δR depends on δM and $\delta\eta$, but $\delta\eta$ is determined in terms of δM by the condition that we should use the same time coordinate for both spheres. Hence, we can obtain $\delta\rho/\rho$ in terms of the time-independent constant $\delta M/M$. This gives us the general solution for matter domination in terms of the conformal time, η:

$$\boxed{\delta \propto \frac{2k[1 - C_k(\eta)]^2 - 3S_k(\eta)[\eta - S_k(\eta)]}{[1 - C_k(\eta)]^2}.} \tag{15.31}$$

This can be cast in terms of observables by inverting the $R(\eta)$ relation:

$$\eta = C_k^{-1}\left[1 - \frac{2(\Omega-1)}{\Omega(1+z)}\right]. \tag{15.32}$$

This is the growing mode; the decaying mode can be obtained by noting that the origin of time is arbitrary: $t = kGM[\eta - S_k(\eta)] + C$ would do just as well. If $R(t, C)$ is a solution to the growth equation, then so is $\partial R/\partial C$ at constant t, since there are no other terms in the differential equation governing R that depend on C. The associated density perturbation is $\delta \propto d\ln R/dC$, which can be evaluated by working out $d\eta/dC$ at constant t. This yields the decaying mode

$$\delta \propto \left(\frac{d\ln R}{dC}\right)_t = \left(\frac{d\ln R}{d\eta}\right)\left(\frac{d\eta}{dC}\right)_t \propto \frac{S_k(\eta)}{[1-C_k(\eta)]^2}. \tag{15.33}$$

Note that this is just the Hubble parameter: $H(t) = \dot{R}/R$ satisfies the perturbation equation for the density.

For an open universe, the growing mode has the limits

$$\begin{aligned} \eta \to 0 \qquad & \delta \to \eta^2/10 \\ \eta \to \infty \qquad & \delta \to 1. \end{aligned} \tag{15.34}$$

Thus, $\delta \propto (1+z)^{-1}$ for large z, as expected, but as $\Omega_0 \to 0$ we get **gravitational freeze-out** and growth ceases. The exact solution can be cast into a useful approximate form by comparing $\delta(z)$ with what would be expected in a flat universe:

$$\frac{\delta(z=0,\Omega)}{\delta(z=0,\Omega=1)} \simeq \Omega^{0.65}. \tag{15.35}$$

Thus, a universe with $\Omega = 0.2$ implies fluctuations at recombination (when $\Omega(z) = 1$) that are roughly a factor 3 higher than in a flat universe. The same formula can be applied at high redshift by using $\Omega(z)$ instead; however, since this quickly goes to unity at high redshifts, any departures from Einstein–de Sitter growth only become important near the current epoch.

With only a little more work, this same method can be extended to cope with a universe containing a mixture of matter and radiation. Take the parametric solutions for $R(t)$ in such a universe, as given in chapter 3. The matter and radiation perturbations are given by

$$\begin{aligned} \delta_m &= -3\frac{\delta R}{R} + \frac{\delta M_m}{M_m} \\ \delta_r &= -4\frac{\delta R}{R} + \frac{\delta M_m R_0}{M_m R_0}. \end{aligned} \tag{15.36}$$

Using $\delta_r = 4\delta_m/3$ for adiabatic perturbations (see below) gives the relation between δM_m and δM_r. Now proceed to eliminate $\delta\eta$ as before; the result for the growing mode comes out to be

$$\begin{aligned} \delta \propto & \Big\{2y_r[kS_k^2 - 2C_k(1-C_k)] + y_m^2[2k(1-C_k)^2 - 3S - k(\eta - S_k)] \\ & + \sqrt{2y_r}y_m[S_k(1-C_k) - 3C_k(\eta - S_k)]\Big\} \\ & \times \left[2y_rS_k^2 + y_m^2(1-C_k)^2 + 2\sqrt{2y_r}y_mS_k(1-C_k)\right]^{-1}, \end{aligned} \tag{15.37}$$

where $y_{m,r} = k\Omega_{m,r}/[2(\Omega - 1)]$, and the notation has been simplified by letting S_k denote $S_k(r)$ etc. For $\Omega = 1$ this goes over to

$$\delta \propto \tau^2 \, \frac{1 + 6\tau/5 + 2\tau^2/5}{(1+\tau)^2}, \qquad \tau = \frac{\Omega_m}{\sqrt{\Omega_r}} \, \eta/4, \tag{15.38}$$

so that growth is not precisely as $\delta \propto \eta^2$ in the transition between radiation domination and matter domination.

THE GENERAL CASE It is also interesting to think about the growth of matter perturbations in universes with non-zero vacuum energy, or even possibly some other exotic background with a peculiar equation of state. The differential equation for δ is as before, but $a(t)$ is altered. The trick of using the spherical model will not generate an analytic solution for δ, because the parametric form for $a(t)$ no longer applies. Nevertheless, the basic concept of treating spherical perturbations as small universes is still valid. Consider the Friedmann equation in the form

$$(\dot{a})^2 = \Omega_0^{\rm tot} H_0^2 a^2 + K, \tag{15.39}$$

where $K = -kc^2/R_0^2$; this emphasizes that K is a constant of integration. A second constant of integration arises in the expression for time:

$$t = \int_0^a \dot{a}^{-1} \, da + C. \tag{15.40}$$

This lets us argue as before in the case of decaying modes: if a solution to the Friedmann equation is $a(t, K, C)$, then valid density perturbations are

$$\delta \propto \left(\frac{\partial \ln a}{\partial K} \right)_t \quad \text{or} \quad \left(\frac{\partial \ln a}{\partial C} \right)_t . \tag{15.41}$$

Since $\partial(\dot{a}^2)/\partial K = 1$, this gives the growing and decaying modes as

$$\delta \propto \begin{cases} (\dot{a}/a) \int_0^a (\dot{a})^{-3} \, da & \text{(growing mode)} \\ \dot{a}/a & \text{(decaying mode).} \end{cases} \tag{15.42}$$

(Heath 1977; see also section 10 of Peebles 1980).

The equation for the growing mode requires numerical integration in general, with $\dot{a}(a)$ given by the Friedmann equation. A very good approximation to the answer is given by Carroll *et al.* (1992):

$$\frac{\delta(z=0,\Omega)}{\delta(z=0,\Omega=1)} \simeq \tfrac{5}{2}\Omega_m \left[\Omega_m^{4/7} - \Omega_v + (1 + \tfrac{1}{2}\Omega_m)(1 + \tfrac{1}{70}\Omega_v) \right]^{-1} . \tag{15.43}$$

This fitting formula for the growth suppression in low-density universes is an invaluable practical tool. For flat models with $\Omega_m + \Omega_v = 1$, it says that the growth suppression is less marked than for an open universe – approximately $\Omega^{0.23}$ as against the $\Omega^{0.65}$ quoted above. This reflects the more rapid variation of Ω_v with redshift; if the cosmological constant is important dynamically, this only became so very recently, and the universe spent more of its history in a nearly Einstein–de Sitter state by comparison with an open universe of the same Ω_m.

MÉSZÁROS EFFECT What about the case of collisionless matter in a radiation background? The fluid treatment is not appropriate here, since the two species of particles can interpenetrate. A particularly interesting limit is for perturbations well inside the horizon: the radiation can then be treated as a smooth, unclustered background that affects only the overall expansion rate. This is analogous to the effect of Λ, but an analytical solution does exist in this case. The perturbation equation is as before

$$\ddot{\delta} + 2\frac{\dot{a}}{a}\dot{\delta} = 4\pi G \rho_m \delta, \tag{15.44}$$

but now $H^2 = 8\pi G(\rho_m + \rho_r)/3$. If we change variable to $y \equiv \rho_m/\rho_r = a/a_{\rm eq}$, and use the Friedmann equation, then the growth equation becomes

$$\delta'' + \frac{2+3y}{2y(1+y)}\,\delta' - \frac{3}{2y(1+y)}\,\delta = 0 \tag{15.45}$$

(for $k = 0$, as appropriate for early times). It may be seen by inspection that a growing solution exists with $\delta'' = 0$:

$$\boxed{\delta \propto y + 2/3.} \tag{15.46}$$

It is also possible to derive the decaying mode. This is simple in the radiation-dominated case ($y \ll 1$): $\delta \propto -\ln y$ is easily seen to be an approximate solution in this limit.

What this says is that, at early times, the dominant energy of radiation drives the universe to expand so fast that the matter has no time to respond, and δ is frozen at a constant value. At late times, the radiation becomes negligible, and the growth increases smoothly to the Einstein–de Sitter $\delta \propto a$ behaviour (Mészáros 1974). The overall behaviour is therefore similar to the effects of pressure on a coupled fluid: for scales greater than the horizon, perturbations in matter and radiation can grow together, but this growth ceases once the perturbations enter the horizon. However, the explanations of these two phenomena are completely different. In the fluid case, the radiation pressure prevents the perturbations from collapsing further; in the collisionless case, the photons have free-streamed away, and the matter perturbation fails to collapse only because radiation domination ensures that the universe expands too quickly for the matter to have time to self-gravitate. Because matter perturbations enter the horizon (at $y = y_{\rm entry}$) with $\dot{\delta} > 0$, δ is not frozen quite at the horizon-entry value, and continues to grow until this initial 'velocity' is redshifted away, giving a total boost factor of roughly $\ln y_{\rm entry}$ [problem 15.5]. This log factor may be seen below in the fitting formulae for the CDM power spectrum.

15.3 The peculiar velocity field

The equations for velocity-field perturbations were developed in section 5.2 as part of the machinery of analysing self-gravitating density fluctuations. There, the velocity field was eliminated, in order to concentrate on the behaviour of density perturbations. However, the peculiar velocity field is of great importance in cosmology, so it is convenient to give a summary that highlights the properties of velocity perturbations.

Consider first a galaxy that moves with some peculiar velocity in an otherwise uniform universe. Even though there is no peculiar gravitational acceleration acting, its velocity will decrease with time as the galaxy attempts to catch up with successively more distant (and therefore more rapidly receding) neighbours. If the proper peculiar velocity is v, then after time dt the galaxy will have moved a proper distance $x = v\,dt$ from its original location. Its near neighbours will now be galaxies with recessional velocities $Hx = Hv\,dt$, relative to which the peculiar velocity will have fallen to $v - Hx$. The equation of motion is therefore just

$$\dot{v} = -Hv = -\frac{\dot{a}}{a}\,v, \tag{15.47}$$

with the solution $v \propto a^{-1}$: peculiar velocities of nonrelativistic objects suffer redshifting by exactly the same factor as photon momenta. It is often convenient to express the peculiar velocity in terms of its comoving equivalent, $\mathbf{v} \equiv a\mathbf{u}$, for which the equation of motion becomes $\dot{u} = -2Hu$. Thus, in the absence of peculiar accelerations and pressure forces, comoving peculiar velocities redshift away through the **Hubble drag** term $2H\mathbf{u}$.

If we now add the effects of peculiar acceleration, this gives the equation of motion

$$\dot{\mathbf{u}} + \frac{2\dot{a}}{a}\mathbf{u} = -\frac{\mathbf{g}}{a}, \tag{15.48}$$

where $\mathbf{g} = \boldsymbol{\nabla}\delta\Phi/a$ is the peculiar gravitational acceleration. Pressure terms have been neglected, so $\lambda \gg \lambda_{\rm J}$. Remember that throughout we are using comoving length units, so that $\nabla_{\rm proper} = \nabla/a$. This equation is the exact equation of motion for a single galaxy, so that the time derivative is $d/dt = \partial/\partial t + \mathbf{u}\cdot\boldsymbol{\nabla}$. In linear theory, the second part of the time derivative can be neglected, and the equation then turns into one that describes the evolution of the linear peculiar velocity field at a fixed point in comoving coordinates.

The solutions for the peculiar velocity field can be decomposed into modes either parallel to $\mathbf{g}$ or independent of $\mathbf{g}$ (these are the homogeneous and inhomogeneous solutions to the equation of motion). The interpretation of these solutions is aided by knowing that the velocity field satisfies the **continuity equation** $\dot{\rho} = -\boldsymbol{\nabla}\cdot(\rho\mathbf{v})$ in proper units, which obviously takes the same form $\dot{\rho} = -\boldsymbol{\nabla}\cdot(\rho\mathbf{u})$ if lengths and densities are in comoving units. If we express the density as $\rho = \rho_0(1+\delta)$ (where in comoving units ρ_0 is just a number independent of time), the continuity equation takes the form

$$\dot{\delta} = -\boldsymbol{\nabla}\cdot[(1+\delta)\mathbf{u}], \tag{15.49}$$

which becomes just

$$\boxed{\boldsymbol{\nabla}\cdot\mathbf{u} = -\dot{\delta}} \tag{15.50}$$

in linear theory when both δ and $\mathbf{u}$ are small. This says that it is possible to have **vorticity modes** with $\boldsymbol{\nabla}\cdot\mathbf{u} = 0$, for which $\dot{\delta}$ vanishes. We have already seen that δ either grows or decays as a power of time, so these modes require zero density perturbation, in which case the associated peculiar gravity also vanishes. These vorticity modes are thus the required homogeneous solutions, and they decay as $v = au \propto a^{-1}$, as with the kinematic analysis for a single particle. For any gravitational-instability theory, in which structure forms via the collapse of small perturbations laid down at very early times, it should therefore be a very good approximation to say that the linear velocity field must be curl-free. This will turn out to be rather important (see chapter 16).

For the growing modes, we want to try looking for a solution $\mathbf{u} = F(t)\mathbf{g}$. Then using continuity plus Gauss's theorem, $\nabla \cdot \mathbf{g} = 4\pi G a \rho \delta$, gives us

$$\delta \mathbf{v} = \frac{2f(\Omega)}{3H\Omega} \mathbf{g}, \tag{15.51}$$

where the function $f(\Omega) \equiv (a/\delta) d\delta/da$. A very good approximation to this (Peebles 1980) is $g \simeq \Omega^{0.6}$ (a result that is almost independent of Λ; Lahav *et al.* 1991). Alternatively, we can work in Fourier terms. This is easy, as $\mathbf{g}$ and $\mathbf{k}$ are parallel, so that $\nabla \cdot \mathbf{u} = -i\mathbf{k} \cdot \mathbf{u} = -iku$. Thus, directly from the continuity equation,

$$\boxed{\delta \mathbf{v}_{\mathbf{k}} = -\frac{iHf(\Omega)a}{k} \delta_k \hat{\mathbf{k}}.} \tag{15.52}$$

The $1/k$ factor tells us that cosmological velocities come predominantly from larger-scale perturbations than those that dominate the density field. Deviations from the Hubble flow are therefore in principle a better probe of the inhomogeneity of the universe than large-scale clustering (see chapter 16).

15.4 Coupled perturbations

We will often be concerned with the evolution of perturbations in a universe that contains several distinct components (radiation, baryons, dark matter). It is easy to treat such a mixture if only gravity is important (i.e. for large wavelengths). Look at the perturbation equation in the form

$$L\delta = \text{driving term}, \qquad L \equiv \frac{\partial^2}{\partial t^2} + \frac{2\dot{a}}{a}\frac{\partial}{\partial t}. \tag{15.53}$$

The rhs represents the effects of gravity, and particles will respond to gravity whatever its source. The coupled equations for several species are thus given by summing the driving terms for all species.

MATTER PLUS RADIATION The only subtlety is that we must take into account the peculiarity that radiation and pressureless matter respond to gravity in different ways, as seen in the equations of fluid mechanics. The coupled equations for perturbation growth are thus

$$L \begin{pmatrix} \delta_m \\ \delta_r \end{pmatrix} = 4\pi G \begin{pmatrix} \rho_m & 2\rho_r \\ 4\rho_m/3 & 8\rho_r/3 \end{pmatrix} \begin{pmatrix} \delta_m \\ \delta_r \end{pmatrix}. \tag{15.54}$$

Solutions to this will be simple if the matrix has time-independent eigenvectors. Only one of these is in fact time independent: $(1, 4/3)$. This is the **adiabatic mode** in which $\delta_r = 4\delta_m/3$ at all times. This corresponds to some initial disturbance in which matter particles and photons are compressed together. The entropy per baryon is unchanged, $\delta(T^3)/T^3 = \delta_m$, hence the name 'adiabatic'. In this case, the perturbation amplitude for both species obeys $L\delta = 4\pi G(\rho_m + 8\rho_r/3)\delta$. We have already seen how to solve this equation by using the spherical model. We also expect the baryons and photons to obey this adiabatic relation very closely even on small scales: the **tight-coupling approximation** says that Thomson scattering is very effective at suppressing the motion of the photon and baryon fluids relative to each other.

ISOCURVATURE MODES The other perturbation mode is harder to see until we realize that, whatever initial conditions we choose for δ_r and δ_m, any subsequent changes to matter and radiation on large scales must be adiabatic (only gravity is acting). Suppose that the radiation field is initially chosen to be uniform; we then have

$$\boxed{\delta_r = \frac{4}{3}(\delta_m - \delta_i),} \tag{15.55}$$

where δ_i is some initial value of δ_m. The equation for δ_m becomes

$$L\delta_m = \frac{32\pi G}{3}\left[\left(\rho_r + \frac{3}{8}\rho_m\right)\delta_m - \rho_r\delta_i\right], \tag{15.56}$$

which is as before if $\delta_i = 0$. The other solution is therefore a particular integral with $\delta \propto \delta_i$. For $\Omega = 1$, the answer can be expressed most neatly in terms of $y \equiv \rho_m/\rho_r$ (Peebles 1987):

$$\begin{aligned} \frac{\delta_m}{\delta_i} &= \frac{4}{y} - \frac{8}{y^2}(\sqrt{1+y} - 1) \simeq 1 - \frac{y}{2} + \cdots \\ \frac{\delta_r}{\delta_i} &= \frac{4}{3}\left(\frac{\delta_m}{\delta_i} - 1\right) \simeq \frac{-2y}{3} + \cdots \end{aligned} \tag{15.57}$$

At late times, $\delta_m \to 0$, while $\delta_r \to -4\delta_i/3$. This mode is called the **isocurvature mode**, since it corresponds to $\delta\rho/\rho \to 0$ as $t \to 0$. Subsequent evolution attempts to preserve constant density by making the matter perturbations decrease while the amplitude of δ_r increases. An alternative name for this mode is an **entropy perturbation**. This reflects the fact that one only perturbs the initial ratio of photon and matter number densities. The late-time evolution is then easily understood: causality requires that, on large scales, the initial entropy perturbation is not altered. Hence, as the universe becomes strongly matter dominated, the entropy perturbation becomes carried entirely by the photons. This leads to an increased amplitude of microwave-background anisotropies in isocurvature models (see chapter 18).

A more old-fashioned piece of terminology concerns **isothermal perturbations**; these have $\delta_r = 0$ at all times (hence the name). It should be clear from the above discussion, however, that a perturbation cannot *remain* isothermal: it is a mixture of adiabatic and isocurvature modes, which evolve differently. An isocurvature perturbation is initially isothermal, but subsequently δ_r becomes negative as δ_m falls. This behaviour continues until the perturbation comes within the horizon scale, after which time the random motions of the photons move them away from the matter perturbation, erasing the radiation perturbation. The subsequent evolution is then purely isothermal, and is described by the Mészáros solution. Once the universe is matter dominated, isocurvature modes thus eventually match onto adiabatic growing modes after entering the horizon. The net result for the matter distribution today is similar to that produced from adiabatic fluctuations, but with different relative amplitudes of modes as a function of wavelength.

BARYONS AND DARK MATTER This case is simpler, because both components have the same equation of state:

$$L\begin{pmatrix}\delta_b \\ \delta_d\end{pmatrix} = \frac{4\pi G\rho}{\Omega}\begin{pmatrix}\Omega_b & \Omega_d \\ \Omega_b & \Omega_d\end{pmatrix}\begin{pmatrix}\delta_b \\ \delta_d\end{pmatrix}. \tag{15.58}$$

Both eigenvectors are time independent: $(1, 1)$ and $(\Omega_d, -\Omega_b)$. The time dependence of these modes is easy to see for an $\Omega = 1$ matter-dominated universe: if we try $\delta \propto t^n$, then we obtain respectively $n = 2/3$ or -1 and $n = 0$ or $-1/3$ for the two modes. Hence, if we set up a perturbation with $\delta_b = 0$, this mixture of the eigenstates will quickly evolve to be dominated by the fastest-growing mode with $\delta_b = \delta_d$: the baryonic matter falls into the dark potential wells. This is one process that allows universes containing dark matter to produce smaller anisotropies in the microwave background: radiation drag allows the dark matter to undergo growth between matter–radiation equality and recombination, while the baryons are held back.

This is the solution on large scales, where pressure effects are negligible. On small scales, the effect of pressure will prevent the baryons from continuing to follow the dark matter. We can analyse this by writing down the coupled equation for the baryons, but now adding in the pressure term (sticking to the matter-dominated era, to keep things simple):

$$L\,\delta_b = L\,\delta_d - k^2 c_S^2 \delta_b / a^2. \tag{15.59}$$

In the limit that dark matter dominates the gravity, the first term on the rhs can be taken as an imposed driving term, of order δ_d/t^2. In the absence of pressure, we saw that δ_b and δ_d grow together, in which case the second term on the rms is smaller than the first if $kc_S t/a \ll 1$. Conversely, for large wavenumbers ($kc_S t/a \gg 1$), baryon pressure causes the growth rates in the baryons and dark matter to differ; the main behaviour of the baryons will then be slowly declining sound waves:

$$\delta_b \propto (ac_S)^{-1/2} \exp\left[\int kc_S \, d\eta\right], \tag{15.60}$$

where η is conformal time [problem 15.4].

This oscillatory behaviour holds so long as pressure forces continue to be important. However, the sound speed drops by a large factor at recombination, and we would then expect the oscillatory mode to match on to a mixture of the pressure-free growing and decaying modes. This behaviour can be illustrated in a simple model where the sound speed is constant until recombination at the conformal time η_r and then instantly drops to zero. The behaviour of the density field before and after t_r may be written as

$$\delta = \begin{cases} \delta_0 \sin(\omega\eta)/(\omega\eta) & (\eta < \eta_r) \\ A\eta^2 + B\eta^{-3} & (\eta > \eta_r), \end{cases} \tag{15.61}$$

where $\omega \equiv kc_S$. Matching δ and its time derivative on either side of the transition allows the decaying component to be eliminated, giving the following relation between the growing-mode amplitude after the transition and the amplitude of the initial oscillation:

$$A\eta_r^2 = \frac{\delta_0}{3} \cos\omega\eta_r. \tag{15.62}$$

The amplitude of the growing mode after recombination depends on the phase of the oscillation at the time of recombination. The output is maximized when the input density perturbation is zero and the wave consists of a pure velocity perturbation; this effect is known as **velocity overshoot**. The post-recombination transfer function will thus display oscillatory features, peaking for wavenumbers that had particularly small amplitudes prior to recombination. Such effects determine the relative positions of small-scale features in the power spectra of matter fluctuations and microwave-background fluctuations (see chapter 18).

15.5 The full treatment

The above treatment of the growth of linear fluctuations is simplified in a number of critical respects, and the aim of this section is to give an overview of the extensions that are required in order to make the theory exact. For reasons of space, and to keep the discussion from becoming too technical, the algebraic details will generally be omitted. The treatment given so far is adequate for achieving a physical understanding of most of the features of the exact theory, and the most important thing is to be aware of where and why the simplified treatment breaks down.

The subject of cosmological fluctuations is really a study of the perturbations of Einstein's field equations, which must satisfy

$$\delta G^{\mu\nu} = -\frac{8\pi G}{c^4}\,\delta T^{\mu\nu}. \tag{15.63}$$

Two key simplifications have been used so far, which affect the left and right sides of this equation separately. On the lhs, the perturbations to the metric are described in the Newtonian gauge, so that they depend only on a single function of space and time, the gravitational potential Φ. On the rhs, the matter content is treated as a mixture of perfect fluids, whereas some constituents of the density field will in fact be a set of collisionless particles. We now discuss these problems in turn, starting with the gauge issue.

GAUGE PROBLEMS IN COSMOLOGY Gauge transformations pose a significant problem in cosmology, where the apparently straightforward task of defining perturbations from the mean density turns out to be highly non-trivial if approached fully relativistically. To see how the problems arise, ask how tensors of different order change under a gauge transformation $x^\mu \to x'^\mu = x^\mu + \epsilon^\mu$. Consider first a scalar quantity S (which might be density, temperature etc.). A scalar quantity in relativity is normally taken to be independent of coordinate frame, but this is only for the case of Lorentz transformations, which do not involve a change of the spacetime origin. A gauge transformation therefore not only induces the usual transformation coefficients dx'^μ/dx^ν, but also involves a translation that relabels spacetime points. The scalar function S therefore appears to alter under the gauge transformation, but only because of this translation: the post-transformation function $S'(y^\mu)$ differs from $S(y^\mu)$ because the actual value of the argument of the function S is being held constant. Since $S'(x^\mu + \epsilon^\mu) = S(x^\mu)$, the rule for the gauge transformation of scalars is

$$S'(x^\mu) = S(x^\mu) - \epsilon^\alpha S_{,\alpha}. \tag{15.64}$$

Similar reasoning but a little more algebra is required to obtain the transformation laws for higher tensors, because we need to account not only for the translation of the origin, but also for the usual effect of the coordinate transformation on the tensor. For a 4-vector, the result is

$$A'_\alpha = A_\alpha - \epsilon^\mu A_{\alpha;\mu} - A_\mu \epsilon^\mu_{;\alpha}. \tag{15.65}$$

Note that covariant derivatives are used, as the result must still be a tensor. However, the terms involving Γ cancel, so the result is equivalent to using coordinate derivatives.

For higher-rank tensors, a number of terms of the form $-T_{\mu\nu}\epsilon^{\mu}_{;\alpha}$ are added, so that $\epsilon^{\mu}_{;\alpha}$ is contracted in turn with each index of the tensor.

Returning to the case of density perturbations, the gauge transformation has a very simple meaning in the case of a uniform universe; here ρ only depends on time, so that

$$\rho' = \rho - \epsilon^0 \dot{\rho}. \tag{15.66}$$

An effective density perturbation is thus produced by a local alteration in the time coordinate. Conversely, any given density perturbation can be cancelled exactly by an appropriate choice of the **hypersurface of simultaneity**. By now, it may seem as though general covariance has led us into a swamp in which all commonsense notions of space and time have been swept away by gauge ambiguities. Surely things can't really be this bad? An ingot of lead is denser than one of iron, even if we choose a set of coordinates in which this appears not to be true. However, this is not so much to say that some gauges are incorrect descriptions of physical reality as that some yield simpler descriptions than others. The way the puzzle is resolved in the case of the laboratory frame is to say that the simplest gauge is the Minkowski spacetime that is our (approximately) inertial local frame. The reason why this metric can be used is that observers are able to construct the coordinate system directly – in principle by using light signals. This shows that the cosmological horizon plays an important role in this topic: perturbations with wavelength $\lambda \lesssim ct$ inhabit a regime in which gauge ambiguities can be resolved directly via common sense. The real difficulties lie in the super-horizon modes with $\lambda \gtrsim ct$. However, at least within inflationary models, these difficulties can be overcome. According to inflation, perturbations on scales greater than the horizon were originally generated via quantum fluctuations on small scales within the horizon of a nearly de Sitter exponential expansion at the earliest times. There is thus no problem in understanding how the initial density field is described, since the simplest coordinate system can once again be constructed directly.

In recent years, there has been an increasing trend towards solving these difficulties in the most direct possible way, by constructing perturbation variables that are explicitly independent of gauge. Comprehensive technical discussions of this method are given by Bardeen (1980), Kodama & Sasaki (1984) and Mukhanov, Feldman & Brandenberger (1992), and we summarize some of the key results here. The starting point for a discussion of metric perturbations is to devise a notation that will classify the possible perturbations. Since the metric is symmetric, there are 10 independent degrees of freedom in $g^{\mu\nu}$; a convenient scheme that captures these possibilities is to write the cosmological metric as

$$d\tau^2 = a^2(\eta)\left\{(1+2\phi)d\eta^2 + 2w_i d\eta\, dx^i - \left[(1-2\psi)\gamma_{ij} + 2h_{ij}\right] dx^i\, dx^j\right\}. \tag{15.67}$$

In this equation, η is conformal time, and γ_{ij} is the comoving spatial part of the Robertson–Walker metric.

The total number of degrees of freedom here is apparently 2 (scalar fields ϕ and ψ) + 3 (3-vector field $\mathbf{w}$) + 6 (symmetric 3-tensor h_{ij}) = 11. To get the right number, the tensor h_{ij} is required to be traceless, $\gamma^{ij}h_{ij} = 0$. The perturbations can thus be split into three classes: **scalar perturbations**, which are described by scalar functions of spacetime coordinate and which correspond to the growing density perturbations studied above; **vector perturbations**, which correspond to vorticity perturbations; and **tensor perturbations**, which correspond to gravitational waves. A thorough discussion of this decomposition is given by Bertschinger (1997). It might be expected that only scalar perturbations would be important for cosmology, since we have seen that vorticity perturbations decay, and gravitational wave energy would be expected to redshift away in the same way as radiation. Nevertheless, tensor perturbations do play a critical role, because large gravity-wave amplitudes are excited during the inflationary era, as described in chapter 11.

Here, we shall concentrate mainly on scalar perturbations. Since vectors and tensors can be generated from derivatives of a scalar function, the most general scalar perturbation actually makes contributions to all the $g_{\mu\nu}$ components in the above expansion:

$$\delta g_{\mu\nu} = a^2 \begin{pmatrix} 2\phi & -B_{,i} \\ -B_{,i} & 2(\psi\delta_{ij} - E_{,ij}) \end{pmatrix}, \tag{15.68}$$

where four scalar functions ϕ, ψ, E and B are involved. For simplicity, we have concentrated on the $k = 0$ case, where $\gamma_{ij} = \delta_{ij}$. As usual, this expansion suffers from the problem that the values of these four functions will alter under a gauge transformation, and the earlier discussion of the gauge transformation properties of tensors can be used to derive explicit expressions for the changes. Although the result will not be proved here, the solution to this problem is to define the following **gauge-invariant variables**, which are linearly independent of the gauge transformation parameters:

$$\boxed{\begin{aligned} \Phi &\equiv \phi + \frac{1}{a}\,[(B - E')a]' \\ \Psi &\equiv \psi - \frac{a'}{a}\,(B - E'), \end{aligned}} \tag{15.69}$$

where primes denote derivatives with respect to conformal time. Non-zero values of these variables is the general indicator of inhomogeneity that we have been seeking.

These gauge-invariant 'potentials' have a fairly direct physical interpretation, since they are closely related to the Newtonian potential. The easiest way to evaluate the gauge-invariant fields is to make a specific gauge choice and work with the **longitudinal gauge** in which E and B vanish, so that $\Phi = \phi$ and $\Psi = \psi$. A second key result is that inserting the longitudinal metric into the Einstein equations shows that ϕ and ψ are identical in the case of fluid-like perturbations where off-diagonal elements of the energy–momentum tensor vanish. The proof of this result is similar to the proof in chapter 2 that the Newtonian gauge applies in the case of a static classical source. In this case, the longitudinal gauge becomes identical to the Newtonian gauge, and perturbations are described by a single scalar field, which is the gravitational potential. The **anisotropic stress** required for ϕ and ψ to differ does exist on small scales, as a result of matter–radiation interactions, but on the largest scales the two functions are effectively equal. The conclusion that the gravitational potential does after all give an effectively gauge-invariant measure of the amplitude of cosmological perturbations on scales of the horizon or above is reassuring, and provides some retrospective justification for the intuitive approach that has been taken on this issue.

THE BOLTZMANN EQUATION We now turn to the question of how to treat the matter source without assuming that it is a fluid. The general approach that should be taken is to consider the phase-space distribution function $f(\mathbf{x}, \mathbf{p})$ – i.e. the product of the particle number density and the probability distribution for momentum. The equation that describes the evolution of f is the Boltzmann equation. This was discussed in chapter 9, where it was shown that the general relativistic form of the equation is

$$\left(p^\mu \frac{\partial}{\partial x^\mu} - \Gamma^\mu_{\alpha\beta} p^\alpha p^\beta \frac{\partial}{\partial p^\mu}\right) f = C. \tag{15.70}$$

This equation is exact for particles affected by gravitational forces and by collisions. The collision term on the rhs, C, has to contain all the appropriate scattering physics (Thomson scattering, in the case of a coupled system of electrons and photons); the gravitational forces are contained implicitly in the connection coefficients. What has to be done is to perturb this equation, using the perturbed metric coefficients, together with their equation of motion derived from the Einstein equations. Although this is easily stated, the detailed algebra of the calculation consumes many pages, and it will not be reproduced here (see Peebles 1980; Efstathiou 1990; Bond 1997). The result is a system of coupled differential equations for the distribution functions of the non-fluid components (photons, neutrinos, plus possibly collisionless dark matter) together with the density and pressure of the collisional baryon fluid. Remembering to include gravitational waves, we then have to integrate the whole system numerically, starting with a single Fourier mode of wavelength much greater than the horizon scale, and evolving to the present. Finally, the present-day perturbations to observational quantities such as density and radiation specific intensity are constructed by adding together modes of all wavelengths (which evolve independently in the linear approximation).

This, then, is a brief summary of the professional approach to cosmological perturbations. A modern cosmological Boltzmann code, such as that described by Seljak & Zaldarriaga (1996), is a large and sophisticated piece of machinery, which is the final outcome of decades of intellectual effort. Although heroic analytical efforts have been made in an attempt to find alternative methods of calculation (e.g. Hu & Sugiyama 1995), results of high precision demand the full approach. For a non-specialist, the best that can be done is to attempt to use simple approximate physical arguments to understand the main features of the results; on large scales, this approach is usually quantitatively successful.

15.6 Transfer functions

Real power spectra result from modifications of any primordial power by a variety of processes: growth under self-gravitation; the effects of pressure; dissipative processes. In general, modes of short wavelength have their amplitudes reduced relative to those of long wavelength in this way. The overall effect is encapsulated in the **transfer function**, which gives the ratio of the late-time amplitude of a mode to its initial value:

$$\boxed{T_k \equiv \frac{\delta_k(z=0)}{\delta_k(z)\, D(z)},} \tag{15.71}$$

where $D(z)$ is the linear growth factor between redshift z and the present. The normalization redshift is arbitrary, so long as it refers to a time before any scale of interest has entered the horizon. Once we possess the transfer function, it is a most valuable tool. The evolution of linear perturbations back to last scattering obeys the simple growth laws summarized above, and it is easy to see how structure in the universe will have changed during the matter-dominated epoch.

As discussed above, to calculate accurate results for transfer functions is a technical challenge, mainly because we have a mixture of matter (both collisionless dark particles and baryonic plasma) and relativistic particles (collisionless neutrinos and collisional photons), which does not behave as a simple fluid. Particular problems are caused by the change in the photon component from being a fluid tightly coupled to the baryons by Thomson scattering, to being collisionless after recombination. Accurate results require a solution of the Boltzmann equation to follow the evolution in detail, as discussed above. The transfer function is thus a byproduct of elaborate numerical calculations of microwave background fluctuations. Here, we shall concentrate on presenting the results, and on trying to understand their scaling properties.

There are in essence two ways in which the power spectrum that exists at early times may differ from that which emerges at the present, both of which correspond to a reduction of small-scale fluctuations:

(1) Jeans mass effects. Prior to matter–radiation equality, we have already seen that perturbations inside the horizon are prevented from growing by radiation pressure. Once $z_{\rm eq}$ is reached, one of two things can happen. If collisionless dark matter dominates, perturbations on all scales can grow. If baryonic gas dominates, the Jeans length remains approximately constant, as follows: The sound speed, $c_{\rm S}^2 = \partial p/\partial \rho$, may be found by thinking about the response of matter and radiation to small adiabatic compressions:

$$\delta p = \frac{4}{9}\rho_r c^2 \frac{\delta V}{V}, \qquad \delta\rho = \left(\rho_m + \frac{4}{3}\rho_r\right)\frac{\delta V}{V}, \tag{15.72}$$

implying

$$c_{\rm S}^2 = c^2\left(3 + \frac{9}{4}\frac{\rho_m}{\rho_r}\right)^{-1} = c^2\left[3 + \frac{9}{4}\left(\frac{1+z_{\rm rad}}{1+z}\right)\right]^{-1}. \tag{15.73}$$

Here, $z_{\rm rad}$ is the redshift of equality between matter and photons; $1+z_{\rm rad} = 1.68(1+z_{\rm eq})$ because of the neutrino contribution. At $z \ll z_{\rm rad}$, we therefore have $c_{\rm S} \propto \sqrt{1+z}$. Since $\rho = (1+z)^3 3\Omega_{\rm B} H_0^2/(8\pi G)$, the *comoving* Jeans length is constant at

$$\lambda_{\rm J} = \frac{c}{H_0}\left[\frac{32\pi^2}{27\Omega_{\rm B}(1+z_{\rm rad})}\right]^{1/2} = 50\,(\Omega_{\rm B}h^2)^{-1}\ \text{Mpc}. \tag{15.74}$$

Thus, in either case, one of the critical length scales for the power spectrum will be the horizon distance at $z_{\rm eq}$ ($= 23\,900\Omega h^2$ for $T = 2.73$ K, counting neutrinos as radiation). In the matter-dominated approximation, we get

$$d_{\rm H} = \frac{2c}{H_0}(\Omega z)^{-1/2} \quad\Rightarrow\quad d_{\rm eq} = 39\,(\Omega h^2)^{-1}\text{Mpc}. \tag{15.75}$$

The exact distance–redshift relation is

$$R_0 dr = \frac{c}{H_0}\frac{dz}{(1+z)\sqrt{1+\Omega_m z + (1+z)^2\Omega_r}}, \tag{15.76}$$

from which it follows that the correct answer for the horizon size including radiation is a factor $\sqrt{2}-1$ smaller: $d_{\rm eq} = 16.0\,(\Omega h^2)^{-1}$ Mpc.

It is easy from the above to see the approximate scaling that must be obeyed by transfer functions. Consider the adiabatic case first. Perturbations with $kd_{\rm eq} \ll 1$ always undergo growth as $\delta \propto d_{\rm H}^2$. Perturbations with larger k enter the horizon when $d_{\rm H} \simeq 1/k$; they are then frozen until $z_{\rm eq}$, at which point they can grow again. The missing growth factor is just the square of the change in $d_{\rm H}$ during this period, which is $\propto k^2$. The approximate limits of an adiabatic transfer function would therefore be

$$T_k \simeq \begin{cases} 1 & (kd_{\rm eq} \ll 1) \\ (kd_{\rm eq})^{-2} & (kd_{\rm eq} \gg 1). \end{cases} \tag{15.77}$$

For isocurvature perturbations, the situation is the opposite. Consider a perturbation of short wavelength: once it comes well inside the horizon, the photons disperse, and so all the perturbation to the entropy density (which must be conserved) is carried by the matter perturbation. The perturbation thus enters the horizon with the original amplitude δ_i. Thereafter, it grows in the same way as an isothermal perturbation. This means there are two regimes, for perturbations that enter the horizon before and after matter–radiation equality. The former match onto the Mészáros solution, and keep their amplitudes constant until they start to grow after $a_{\rm eq}$. The present-day amplitude for these is $\delta/\delta_i = (3/2)(1/a_{\rm eq})$. Perturbations that enter after matter–radiation equality start to grow immediately, so that their present amplitude is $\delta/\delta_i \simeq 1/a_{\rm entry}$. Entry occurs when $kd_{\rm H} \simeq 1$, and the horizon evolves as $d_{\rm H} = (2c/H_0)a^{1/2}$ (assuming $\Omega = 1$). Putting these arguments together, the isocurvature transfer function relative to δ_i is

$$T_k \simeq \begin{cases} (2/15)\,(kc/H_0)^2 & (kd_{\rm eq} \ll 1) \\ (3/2)\,a_{\rm eq}^{-1} & (kd_{\rm eq} \gg 1) \end{cases} \tag{15.78}$$

(a more sophisticated argument is required to obtain the exact factor 2/15 in the long-wavelength limit; see Efstathiou 1990). Since this goes to a constant at high k, it is also common to quote the transfer function relative to this value. This means that $T_k < 1$ at $kd_{\rm eq} \lesssim 1$, and so the isocurvature transfer function is the mirror image of the adiabatic case: one falls where the other rises (see figure 15.3).

(2) Damping. In addition to having their growth retarded, very-small-scale perturbations will be erased entirely; this can happen in one of two ways. For collisionless dark matter, perturbations are erased simply by **free streaming**: random particle velocities cause blobs to disperse. At early times ($kT > mc^2$), the particles will travel at c, and so any perturbation that has entered the horizon will be damped. This process switches off when the particles become nonrelativistic; for massive particles, this happens long before $z_{\rm eq}$ (resulting in **cold dark matter**, CDM). For massive neutrinos, however, it happens *at* $z_{\rm eq}$: only perturbations on very large scales survive in the case of **hot dark matter** (HDM). In a purely baryonic universe, the corresponding process is called **Silk damping**: the mean free path of photons due to scattering by the plasma is non-zero, and so radiation can diffuse out of a perturbation, convecting the plasma with it. The typical distance of a random walk in terms of the diffusion coefficient D is $x \simeq \sqrt{Dt}$, which gives a damping length of

$$\lambda_{\rm S} \simeq \sqrt{\lambda d_{\rm H}}, \tag{15.79}$$

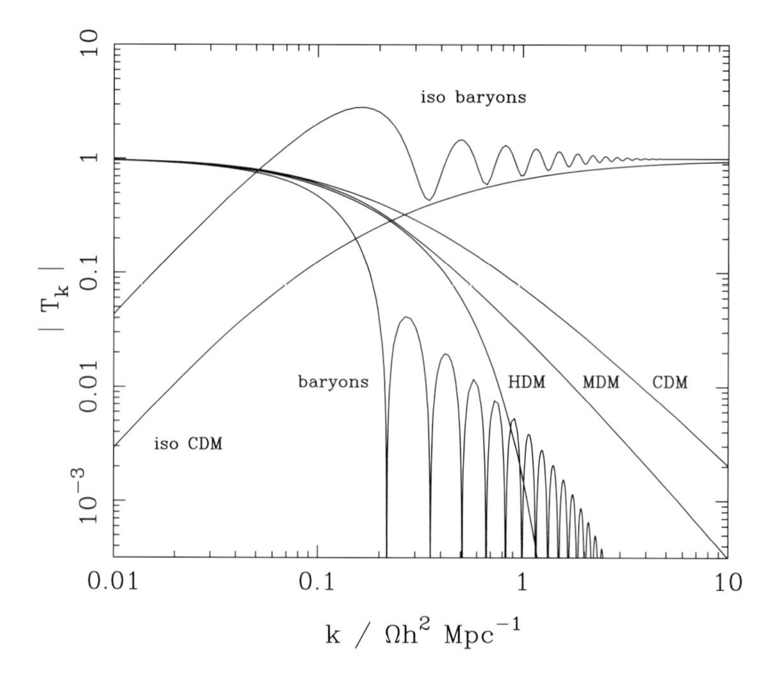

Figure 15.3. A plot of transfer functions for various models. For adiabatic models, $T_k \to 1$ at small k, whereas the opposite is true for isocurvature models. A number of possible matter contents are illustrated: pure baryons; pure CDM; pure HDM; MDM (30% HDM, 70% CDM). For dark-matter models, the characteristic wavenumber scales as Ωh^2. The scaling for baryonic models does not obey this exactly; the plotted cases correspond to $\Omega = 1$, $h = 0.5$.

the geometric mean of the horizon size and the mean free path. Since $\lambda = 1/(n\sigma_{\rm T}) = 44.3(1+z)^{-3}(\Omega_{\rm B}h^2)^{-1}$ proper Gpc, we obtain a comoving damping length of

$$\lambda_{\rm S} = 16.3\,(1+z)^{-5/4}(\Omega_{\rm B}^2\Omega h^6)^{-1/4}\ \text{Gpc}. \tag{15.80}$$

This becomes close to the Jeans length by the time of last scattering, $1+z \simeq 1000$.

FITTING FORMULAE It is invaluable in practice to have some accurate analytic formulae that fit the numerical results for transfer functions. We give below results for some common models of particular interest (illustrated in figure 15.3, along with other cases where a fitting formula is impractical). For the models with collisionless dark matter, $\Omega_{\rm B} \ll \Omega$ is assumed, so that all lengths scale with the horizon size at matter–radiation equality, leading to the definition

$$q \equiv \frac{k}{\Omega h^2\ \text{Mpc}^{-1}}. \tag{15.81}$$

We consider the following cases: (1) adiabatic CDM; (2) adiabatic massive neutrinos (one massive, two massless); (3) isocurvature CDM. These expressions come from Bardeen *et al.* (1986; BBKS). Since the characteristic length scale in the transfer function depends on the horizon size at matter–radiation equality, the temperature of the CMB enters. In the above formulae, it is assumed to be exactly 2.7 K; for other values, the characteristic wavenumbers scale $\propto T^{-2}$. For these purposes massless neutrinos count as radiation, and three species of these contribute a total density that is 0.68 that of the photons.

$$\begin{aligned} &(1) \quad T_k = \frac{\ln(1+2.34q)}{2.34q}\left[1+3.89q+(16.1q)^2+(5.46q)^3+(6.71q)^4\right]^{-1/4} \\ &(2) \quad T_k = \exp(-3.9q-2.1q^2) \\ &(3) \quad T_k = (5.6q)^2\left\{1+\left[15.0q+(0.9q)^{3/2}+(5.6q)^2\right]^{1.24}\right\}^{-1/1.24} \end{aligned} \tag{15.82}$$

The case of **mixed dark matter** (MDM, a mixture of massive neutrinos and CDM) is more complex. Ma (1996) gives the following expression:

$$\frac{T_{\rm MDM}}{T_{\rm CDM}} = \left[\frac{1+(0.081x)^{1.630}+(0.040x)^{3.259}}{1+(2.080x_0)^{3.259}}\right]^{\Omega_\nu^{1.05}/2}, \tag{15.83}$$

where $x \equiv k/\Gamma_\nu$, $\Gamma_\nu \equiv a^{1/2}\Omega_\nu h^2$ and x_0 is the value of x at $a=1$. The scale-factor dependence is such that the damping from neutrino free-streaming is less severe at high redshift, but the spectrum is very nearly of constant shape for $z \lesssim 10$. See Pogosyan & Starobinksy (1995) for a more complicated fit of higher accuracy.

The above expressions assume pure dark matter, which is unrealistic. At least for CDM models, a non-zero baryonic density lowers the apparent dark-matter density parameter. We can define an apparent shape parameter Γ for the transfer function:

$$\boxed{q \equiv \frac{k/h\,{\rm Mpc}^{-1}}{\Gamma},} \tag{15.84}$$

and $\Gamma = \Omega h$ in a model with zero baryon content. This parameter was originally defined by Efstathiou, Bond & White (1992), in terms of a CDM model with $\Omega_{\rm B} = 0.03$. Peacock & Dodds (1994) showed that the effect of increasing $\Omega_{\rm B}$ was to preserve the CDM-style spectrum shape, but to shift to lower values of Γ. This shift was generalized to models with $\Omega \neq 1$ by Sugiyama (1995):

$$\Gamma = \Omega h \exp[-\Omega_{\rm B}(1+\sqrt{2h}/\Omega)]. \tag{15.85}$$

This formula fails if the baryon content is too large, and the transfer function develops oscillations (see Eisenstein & Hu 1997 for a more accurate approximation in this case).

MODELS WITH DECAYING PARTICLES The critical length scale in the transfer function can be increased by delaying the onset of matter–radiation equality. The conventional argument infers the total relativistic density by multiplying the photon density by 1.68, but there is no direct observational proof that this is the correct result for the total density in relativistic particles – what would happen if this number changed? A simple increase of the number of relativistic degrees of freedom would do undesirable violence to primordial nucleosynthesis: any such boost would thus have to be provided by a particle that decays after nucleosynthesis. Moreover, the products of this decay cannot include photons, since we already know the density in these particles. It is most natural to suppose that the decaying particle is a heavy neutrino, so the decays would then have to be to the lighter neutrinos or to other hypothetical particles. Particle-physics motivation for this is weak, but such a possibility cannot be rejected on cosmological grounds.

If the decaying particle is a heavy neutrino, the initial number density of the particle is equal to that for a massless neutrino. The energy density in decaying particles then becomes dominant over the radiation when mc^2 is of the order of the radiation energy kT. Until this point $\rho \propto T^4$ and the age of the universe is $t \sim (G\rho)^{-1/2} \propto T^{-2} \propto m^{-2}$. At this point, the universe becomes matter dominated; however, the equation of state returns to radiation domination after the particle decays, and so in this model there will be *two* eras of matter–radiation equality.

Once they are nonrelativistic, the density of heavy neutrinos scales as $\rho \propto a^{-3}$, whereas the density of radiation in the standard massless neutrino model scales as $\rho \propto a^{-4}$. By the time of decay, the massive neutrinos therefore dominate the conventional relativistic density by a factor

$$\frac{\rho_{\rm decay}}{\rho_{\gamma+3\nu}} \simeq \frac{a_{\rm decay}}{a_{\rm eq1}} \simeq \left(\frac{\tau}{t_{\rm eq1}}\right)^{2/3} \propto m^{4/3}\tau^{2/3}. \tag{15.86}$$

Here $a_{\rm eq1}$ and $a_{\rm decay}$ are the scale factors at the first matter–radiation equality and at decay, $\rho_{\rm decay}$ is the energy density in the relativistic decay products from the massive neutrinos and $\rho_{\gamma+3\nu}$ is the standard energy density for radiation and three light neutrinos. This assumes an effectively instantaneous decay for the particles at time τ and a negligible mass for the two other neutrino species. We can now write the ratio of relativistic energy density in a decaying model to the conventional value as a dimensionless number θ:

$$\theta = \frac{\rho_{\gamma+2\nu} + \rho_{\rm decay}}{\rho_{\gamma+3\nu}} \simeq \frac{1.45}{1.68}\left\{1 + x\left[\left(\frac{m}{\rm keV}\right)^2 \frac{\tau}{\rm yr}\right]^{2/3}\right\}, \tag{15.87}$$

with x a dimensionless constant. By a numerical solution of the full equations describing the problem, Bond & Efstathiou (1991) obtained $x \simeq 0.15$. The apparent value of Ωh is now given by

$$(\Omega h)_{\rm apparent} = \Omega h\, \theta^{-1/2}. \tag{15.88}$$

We therefore see that masses of order 1 keV and lifetimes of order 1 yr for a decaying neutrino could significantly affect the development of cosmological structure by increasing the relativistic energy density by a factor of order unity. Such parameters are not ruled out for the τ neutrino.

15.7 N-body models

The equations of motion are nonlinear, and we have only solved them in the limit of linear perturbations. The exact evolution of the density field is usually performed by means of an **N-body simulation**, in which the density field is represented by the sum of a set of fictitious discrete particles. The equations of motion for each particle depend on solving for the gravitational field due to all the other particles, finding the change in particle positions and velocities over some small time step, moving and accelerating the particles and finally re-calculating the gravitational field to start a new iteration. Using comoving units for length and velocity ($\mathbf{v} = a\mathbf{u}$), we have previously seen the equation of motion

$$\frac{d}{dt}\mathbf{u} = -2\frac{\dot{a}}{a}\mathbf{u} - \frac{1}{a^2}\boldsymbol{\nabla}\Phi, \tag{15.89}$$

where Φ is the Newtonian gravitational potential due to density perturbations. The time derivative is already in the required form of the convective time derivative observed by a particle, rather than the partial $\partial/\partial t$. If we change time variable from t to a, this becomes

$$\frac{d}{d\ln a}(a^2\mathbf{u}) = \frac{a}{H}\mathbf{g} = \frac{G}{aH}\sum_i m_i \frac{\mathbf{x}_i - \mathbf{x}}{|\mathbf{x}_i - \mathbf{x}|^3}. \tag{15.90}$$

Here, the gravitational acceleration has been written exactly by summing over all particles, but this becomes prohibitive for very large numbers of particles. Since the problem is to solve Poisson's equation, a faster approach is to use Fourier methods, since this allows the use of the fast Fourier transform (**FFT**) algorithm (see chapter 13 of Press *et al.* 1992). If the density perturbation field (not assumed small) is expressed as $\delta = \sum \delta_k \exp(-i\mathbf{k}\cdot\mathbf{x})$, then Poisson's equation becomes $-k^2\Phi_k = 4\pi G a^2 \bar{\rho}\,\delta_k$, and the required k-space components of $\boldsymbol{\nabla}\Phi$ are just

$$(\boldsymbol{\nabla}\Phi)_k = -i\Phi_k\mathbf{k} = \frac{-i4\pi G a^2\bar{\rho}}{k^2}\,\delta_k\,\mathbf{k}. \tag{15.91}$$

If we finally eliminate matter density in terms of Ω_m, the equation of motion for a given particle is

$$\frac{d}{d\ln a}(a^2\mathbf{u}) = \sum \mathbf{F}_k \exp(-i\mathbf{k}\cdot\mathbf{x}), \qquad \mathbf{F}_k = -i\mathbf{k}\,\frac{3\Omega_m H a^2}{2k^2}\,\delta_k. \tag{15.92}$$

The use of Fourier methods in the description of cosmological density fields is discussed in detail in chapter 16.

UNITS From a practical point of view, it is convenient to change to a new set of units: these incorporate the size of the computational box, allowing the simulation to be rescaled to different physical situations. Let the side of the box be L; it is clearly convenient to measure length in terms of L and velocities in terms of the expansion velocity across the box:

$$\begin{aligned} \mathbf{X} &= \mathbf{x}/L \\ \mathbf{U} &= \delta\mathbf{v}/(HLa) = \mathbf{u}/HL. \end{aligned} \tag{15.93}$$

For N particles, the density is $\rho = Nm/(aL)^3$, so the mass of the particles and the gravitational constant can be eliminated and the equation of motion can be cast in an attractively dimensionless form:

$$\boxed{\frac{d}{d\ln a}[f(a)\mathbf{U}] = \frac{3}{8\pi}\,\Omega_m(a)f(a)\,\frac{1}{N}\sum_i \frac{\mathbf{X}_i - \mathbf{X}}{|\mathbf{X}_i - \mathbf{X}|^3}.} \tag{15.94}$$

The function $f(a)$ is proportional to $a^2H(a)$, and has an arbitrary normalization – e.g. unity at the initial epoch. If the forces are instead evaluated on a grid, dimensionless wavenumbers $\mathbf{K} = \mathbf{k}\,L$ are used, and the corresponding equation becomes

$$\frac{d}{d\ln a}[f(a)\mathbf{U}] = \sum \mathbf{F}_K \exp(-i\mathbf{K}\cdot\mathbf{X}), \qquad \mathbf{F}_K = -i\mathbf{K}\,\frac{3}{2}\,\frac{\delta_K}{K^2}\,\Omega_m(a)f(a). \tag{15.95}$$

Particles are now moved according to $d\mathbf{x} = \mathbf{u}\,dt$, which becomes

$$d\mathbf{X} = \mathbf{U}\,d\ln a \tag{15.96}$$

in our new units. It only remains to set up the initial conditions; this is easy to do if the initial epoch is at high enough redshift that $\Omega_m = 1$, since then $\mathbf{U} \propto a$ and the initial displacements and velocities are related by

$$\Delta\mathbf{X} = \mathbf{U}. \tag{15.97}$$

In the case where the initial conditions are specified late enough that Ω_m is significantly different from unity, this can be modified by using the linear relation between density and velocity perturbations: for a given $\Delta\mathbf{X}$, the corresponding velocity scales as $\Omega_m^{0.6}$ (see above). It is not in fact critical that the density fluctuations be very small at this time: this is related to a remarkable approximation for nonlinear dynamics due to Zeldovich, which is discussed below.

BOXES AND GRIDS The efficient way of performing the required Fourier transforms is by averaging the density field onto a grid and using the FFT algorithm both to perform the transformation of density and to perform the (three) inverse transforms to obtain the real-space force components from their k-space counterparts. This leads to the simplest N-body algorithm: the **particle–mesh (PM) code**. The only complicated part of the algorithm is the procedure for assigning mass to gridpoints and interpolating the force as evaluated on the grid back onto the particles (for consistency, the same procedure must be used for both these steps). The most naive method is simply to bin the data: i.e. associate a given particle with whatever gridpoint happens to be nearest. There are a variety of more subtle approaches (see Hockney & Eastwood 1988; Efstathiou *et al.* 1985), but whichever strategy is used, the resolution of a PM code is clearly limited to about the size of the mesh. To do better, one can use a **particle–particle–particle–mesh (P^3M) code**, also discussed by the above authors. Here, the direct forces are evaluated between particles in the neighbouring cells, the grid estimate being used only for particles in more distant cells. A similar effect, although without the use of the FFT, is achieved by **tree codes** (e.g. Hernquist, Bouchet & Suto 1991). There is a publicly-available version of the P^3M code, due to Couchman (1991; see also the electronic addresses in the bibliography), which has received widespread use.

In practice, however, the increase in resolution gained from these methods is limited to a factor $\lesssim 10$. This is because each particle in a cosmological N-body simulation in fact stands for a large number of less massive particles. Close encounters of these spuriously large particles can lead to wide-angle scattering, whereas the true physical systems are completely collisionless (see chapter 17). To prevent collisions, the forces must be **softened**, i.e. set to a constant below some critical separation, rather than rising as $1/r^2$. If there are already few particles per PM cell, the softening must be some significant fraction of the cell size, so there is a limit to the gain over pure PM. For example, consider a box of side $50h^{-1}$ Mpc, which is the smallest that can be used to simulate the present-day observed universe without serious loss of power from the omitted long-wavelength modes. A typical calculation might use 128^3 particles on a Fourier mesh of the same size, so that the mean density is one particle per cell, and the cell size is of order that of the core of a rich cluster. To use such a simulation to study cluster cores means we are interested in overdensities of 10^3–10^4, or typical interparticle separations of 0.05–0.1 of a cell. To avoid collisional effects, the pairwise interaction must be softened on this scale. A much larger improvement in resolution is only justified in regions of huge overdensity ($\sim 10^6$ for a 100-fold increase in resolution over PM). The overall message is that N-body simulations in cosmology are severely limited by mass resolution, and that this limits the spatial resolution that can be achieved while modelling the evolution of the true collisionless fluid.

The small-scale results of N-body simulations are also unreliable in detail because they treat the collisionless component only. A minimum of around 10% of the material in the real universe is baryonic and therefore dissipational: in high-density regions, this gas will be heated by shocks and then cool radiatively if it is dense enough. If cooling is efficient, it will form stars, which subsequently make an energy input to the surrounding gas via supernovae and stellar winds. These processes are discussed in some detail in chapter 17, and they are the route by which the dark-matter distribution lights up into the galaxy distribution that we observe. Attempts have been made to treat these effects numerically, by combining numerical hydrodynamics with N-body gravity for the collisionless component, but so far these studies have had little to say about galaxy formation. The reason is that the extra computational demands of treating fluids means that the resolution achieved is inferior to a pure N-body code. Since large-scale structure imposes a minimum box size, the only studies that are possible consider small sub-regions, and it is hard to know how representative they are. Things are better at high redshift, where the nonlinear scale is smaller, allowing smaller boxes to be used. For detailed accounts of cosmological hydrodynamics, see Kang *et al.* (1994); Couchman *et al.* (1995); Katz *et al.* (1996b); Shapiro *et al.* (1996).

Despite these caveats, the results of N-body dynamics paint an impressive picture of the large-scale mass distribution. Consider figure 15.4, which shows slices through the computational density field for two particular sets of initial conditions, with different relative amplitudes of long and short wavelengths, but with the same amplitude for the modes with wavelengths equal to the side of the computational box. Although the small-scale 'lumpiness' is different, both display a similar large-scale network of filaments and voids – bearing a striking resemblance to the features seen in reality (figure 15.1). This large-scale pattern is mainly controlled by the dynamical evolution of the largest-scale modes, as discussed in the following section. Although the probability of forming galaxies within this network may depend on the local density, these pictures nevertheless generate the hope that the main features of the large-scale galaxy distribution may be attributed to nonlinear gravitational instability.

15.8 Nonlinear models

Although the full development of the gravitational instability cannot be solved exactly without resort to N-body techniques, there are some very useful special cases and approximations that help us to understand the general case.

THE ZELDOVICH APPROXIMATION Zeldovich (1970) invented a *kinematical* approach to the formation of structure. In this method, we work out the initial displacement of particles and assume that they continue to move in this initial direction. Thus, we write

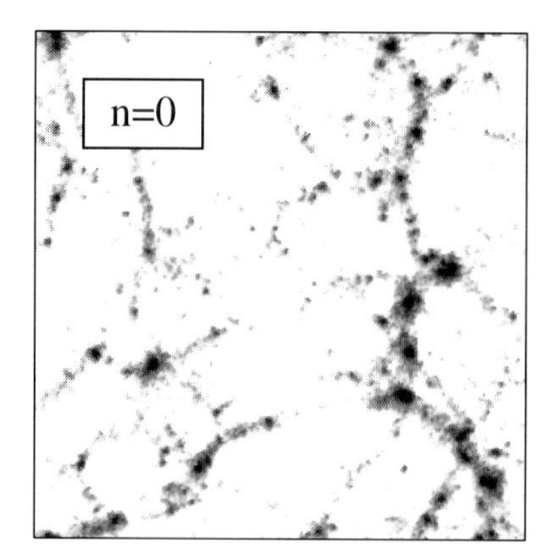

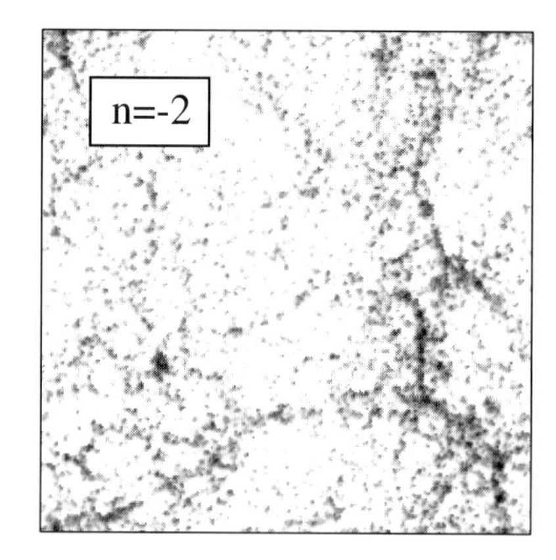

Figure 15.4. Slices through N-body simulations of different power spectra, using the same set of random phases for the modes in both cases. The slices are $1/15$ of the box in thickness, and density from 0.5 to 50 is log encoded. The box-scale power is the same in both cases, and produces much the same large-scale filamentary structure. However, the $n = 0$ spectrum has much more small-scale power, and this manifests itself as far stronger clumping within the overall skeleton of the structure.

for the proper coordinate of a given particle

$$\boxed{\mathbf{x}(t) = a(t)\mathbf{q} + b(t)\mathbf{f}(\mathbf{q}).} \tag{15.98}$$

This looks like Hubble expansion with some perturbation, which will become negligible as $t \to 0$. The coordinates $\mathbf{q}$ are therefore equal to the usual comoving coordinates at $t = 0$, and $b(t)$ is a function that scales the time-independent displacement field $\mathbf{f}(\mathbf{q})$. In fluid-mechanical terminology, $\mathbf{x}$ is said to be the **Eulerian position**, and $\mathbf{q}$ the **Lagrangian position**.

To get the Eulerian density, use the Jacobian of the transformation between $\mathbf{x}$ and $\mathbf{q}$, in which frame ρ is constant:

$$\frac{\rho}{\rho_0} = \left[\left(1 - \frac{b}{a}\alpha\right)\left(1 - \frac{b}{a}\beta\right)\left(1 - \frac{b}{a}\gamma\right)\right]^{-1}, \tag{15.99}$$

where $(-\alpha, -\beta, -\gamma)$ are the eigenvalues of the **strain tensor** or **deformation tensor** $\partial f_i/\partial q_j$. Collapse therefore takes place first along the axis defined by the largest negative eigenvalue. In the case of a triaxial ellipsoid perturbation, this would correspond to collapse along the shortest axis. Gravity thus accentuates asphericity, leading to flattened structures known as **pancakes**. In order to reach this conclusion, it has been assumed that the strain tensor is symmetric, so that we can find a coordinate system in which the tensor is diagonal, making it easy to evaluate the Jacobian. This assumption of symmetry holds provided we assume that the density perturbation originated from a growing mode. The displacement field is then irrotational, so that we can write it in terms of a potential:

$$\mathbf{f}(\mathbf{q}) = \nabla\psi(\mathbf{q}) \quad \Rightarrow \quad \frac{\partial f_i}{\partial q_j} = \frac{\partial^2 \psi}{\partial q_i \partial q_j}, \tag{15.100}$$

making it obvious that the strain tensor is symmetric.

If we linearize the expression for the density, then the relation to the density and velocity perturbations is

$$\begin{aligned} \delta &= -\frac{b}{a}(\alpha + \beta + \gamma) = -\frac{b}{a}\nabla \cdot \mathbf{f} \\ \mathbf{u} &= \frac{1}{a}\left(\dot{\mathbf{x}} - \frac{\dot{a}}{a}\mathbf{x}\right) = \left(\frac{\dot{b}}{a} - \frac{\dot{a}b}{a^2}\right)\mathbf{f}. \end{aligned} \tag{15.101}$$

This satisfies the conservation equation $\nabla \cdot \mathbf{u} = -\dot{\delta}$. Using the Friedmann equation plus the growth equation for δ gives a differential equation for b:

$$\boxed{\frac{\ddot{b}}{b} = -\frac{2\ddot{a}}{a} = \frac{8\pi G\rho_0}{3},} \tag{15.102}$$

which yields the growing mode $b \propto t^{4/3}$ for $\Omega = 1$. Of course, we had no need to solve a differential equation to obtain this result. The Zeldovich approximation is first-order **Lagrangian perturbation theory**, in contrast to the earlier approach, which carried out perturbation theory in Eulerian space (higher-order Lagrangian theory is discussed by Bouchet *et al.* 1995). When the density fluctuations are small, a first-order treatment from either point of view should give the same result. Since the linearized density relation is $\delta = -(b/a)\nabla \cdot \mathbf{f}$, we can tell immediately that $[b(t)/a(t)] = D(t)$, where $D(t)$ is the linear density growth law. Without doing any more work, we therefore know that the first-order form of Lagrangian perturbations must be

$$\mathbf{x}(t) = a(t)[\mathbf{q} + D(t)\mathbf{f}(\mathbf{q})], \tag{15.103}$$

so that $b(t) = a(t)D(t)$. The only advantage of the Zeldovich approximation is that it normally breaks down later than Eulerian linear theory – i.e. first-order Lagrangian perturbation theory can give results comparable in accuracy to Eulerian theory with higher-order terms included. This method is therefore commonly used to set up quasi-linear initial conditions for N-body simulations, as discussed above. The same arguments that we used earlier in discussing peculiar velocities show that the growing-mode comoving displacement field $\mathbf{f}$ is parallel to $\mathbf{k}$ for a given Fourier mode, so that

$$\mathbf{f}_k = -i\,\frac{\delta_k}{k^2}\,\mathbf{k}. \tag{15.104}$$

Given the desired linear density mode amplitudes, the corresponding displacement field can then be constructed.

Why should the Zeldovich approximation work so well? Mainly because the scheme for extrapolating peculiar velocities is exact in one dimension, as is easily seen. Gravitational dynamics in 1D really corresponds to following the motions of parallel sheets of matter. Since the gravitational acceleration towards a sheet is independent of distance, the acceleration experienced by a given sheet depends only on the numbers of sheets to either side. As long as sheets do not cross, the acceleration stays constant, and the full equation of motion can be extrapolated from the initial displacement. In 3D, the collapse of aspherical structures tends to accentuate any asphericity, leading to initial collapse into a **pancake**, a flattened structure that forms where the largest eigenvalue of the deformation tensor reaches the critical value of unity. The fact that this is a sheet-like structure means that the final stages of the collapse are nearly one dimensional, and it is thus not so surprising that they are well described by the Zeldovich approach. Conversely, the Zeldovich approximation fares least well in situations of exact spherical symmetry. It implies that singularities are produced when the linearly evolved density contrast reaches a critical value of 3 that is independent of the value of Ω (all eigenvalues of curvature are equal in this case), whereas we now show that the correct result is a linear contrast of about 1.7.

THE SPHERICAL MODEL An overdense sphere is a very useful nonlinear model, as it behaves in exactly the same way as a closed sub-universe. The density perturbation need not be a uniform sphere: any spherically symmetric perturbation will clearly evolve at a given radius in the same way as a uniform sphere containing the same amount of mass. In what follows, therefore, density refers to the *mean* density inside a given sphere. The equations of motion are the same as for the scale factor, and we can therefore write down the cycloid solution immediately. For a matter-dominated universe, the relation between the proper radius of the sphere and time is

$$\begin{aligned} r &= A(1 - \cos\theta) \\ t &= B(\theta - \sin\theta), \end{aligned} \tag{15.105}$$

and $A^3 = GMB^2$, just from $\ddot{r} = -GM/r^2$. Expanding these relations up to order θ^5 gives $r(t)$ for small t:

$$r \simeq \frac{A}{2}\left(\frac{6t}{B}\right)^{2/3}\left[1 - \frac{1}{20}\left(\frac{6t}{B}\right)^{2/3}\right], \tag{15.106}$$

and we can identify the density perturbation within the sphere:

$$\delta \simeq \frac{3}{20}\left(\frac{6t}{B}\right)^{2/3}. \tag{15.107}$$

This all agrees with what we knew already: at early times the sphere expands with the $a \propto t^{2/3}$ Hubble flow and density perturbations grow proportional to a.

We can now see how linear theory breaks down as the perturbation evolves. There are three interesting epochs in the final stages of its development, which we can read directly from the above solutions. Here, to keep things simple, we compare only with linear theory for an $\Omega = 1$ background.

(1) **Turnround**. The sphere breaks away from the general expansion and reaches a maximum radius at $\theta = \pi$, $t = \pi B$. At this point, the true density enhancement with respect to the background is just $[(A/2)(6t/B)^{2/3}]^3 r^{-3} = 9\pi^2/16 \simeq 5.55$. By comparison, extrapolation of linear $\delta \propto t^{2/3}$ theory predicts $\delta_{\rm lin} = (3/20)(6\pi)^{2/3} \simeq 1.06$.

(2) **Collapse**. If only gravity operates, then the sphere will collapse to a singularity at $\theta = 2\pi$. This occurs when $\delta_{\rm lin} = (3/20)(12\pi)^{2/3} \simeq 1.69$.

(3) **Virialization**. Clearly, collapse will never occur in practice; dissipative physics will eventually intervene and convert the kinetic energy of collapse into random motions. How dense will the resulting body be? Consider the time at which the sphere has collapsed by a factor 2 from maximum expansion. At this point, its kinetic energy K is related to its potential energy V by $V = -2K$. This is the condition for equilibrium, according to the **virial theorem**. For this reason, many workers take this epoch as indicating the sort of density contrast to be expected as the endpoint of gravitational collapse. This occurs at $\theta = 3\pi/2$, and the corresponding density enhancement is $(9\pi + 6)^2/8 \simeq 147$, with $\delta_{\rm lin} \simeq 1.58$. Some authors prefer to assume that this virialized size is eventually achieved only at collapse, in which case the contrast becomes $(6\pi)^2/2 \simeq 178$.

These calculations are the basis for a common 'rule of thumb', whereby one assumes that linear theory applies until $\delta_{\rm lin}$ is equal to some δ_c a little greater than unity, at which point virialization is deemed to have occurred. Although the above only applies for $\Omega = 1$, analogous results can be worked out from the full $\delta_{\rm lin}(z, \Omega)$ and $t(z, \Omega)$ relations; $\delta_{\rm lin} \simeq 1$ is a good criterion for collapse for any value of Ω likely to be of practical relevance. The full density contrast at virialization may be approximated by [problem 15.6]

$$1 + \delta_{\rm vir} \simeq 178\,\Omega^{-0.7} \tag{15.108}$$

(although flat Λ-dominated models show less dependence on Ω; Eke *et al.* 1996), The faster expansion of open universes means that, by the time a perturbation has turned round and collapsed to its final radius, a larger density contrast has been produced.

This procedure should therefore be quite accurate as far as determining the time of collapse is concerned. What is much less certain is the density contrast achieved. Even with exact spherical symmetry, one can clearly not do much better than argue for a contrast of order a few hundred. Of course, real objects are not exactly symmetric; figure 15.5 shows the complex and anisotropic sequence of events leading to the formation of the N-body version of a cluster of galaxies. Nevertheless, the basic idea that an overdensity of a few hundred distinguishes collapsed systems from those still in the process of infall does work well in picking out cluster-like structures from simulations.

COSMOLOGICAL INFALL Once a nonlinear body has formed in the universe, this is not the end of the story, as it will continue to attract matter in its neighbourhood and its mass will grow by accretion of new material. A good deal of insight into this process can be gained by considering the spherically symmetric case. This was first studied in a seminal paper by Gunn & Gott (1972), with further important extensions by Bertschinger (1985b). The basic picture is thus one in which we have a spherically symmetric run of density perturbation $\delta(r)$ at some given initial time; the problem is to calculate the density profile at some later time. An interesting case is where the initial disturbance is confined to small radii, so that the initial condition at large r is $\delta = 0$ and the mean density perturbation interior to r declines as $\bar{\delta} \propto r^{-3}$. In this situation, each shell of matter responds only to the mass interior to it. Provided $\bar{\delta}$ declines monotonically, shells will not cross and each evolves independently until the density first becomes infinite in the centre. What happens next depends on how exact the symmetry is: in principle, a black hole forms in the centre, and shells that fall in subsequently accrete onto the hole. In practice, small deviations from symmetry mean that each infalling shell will tend to

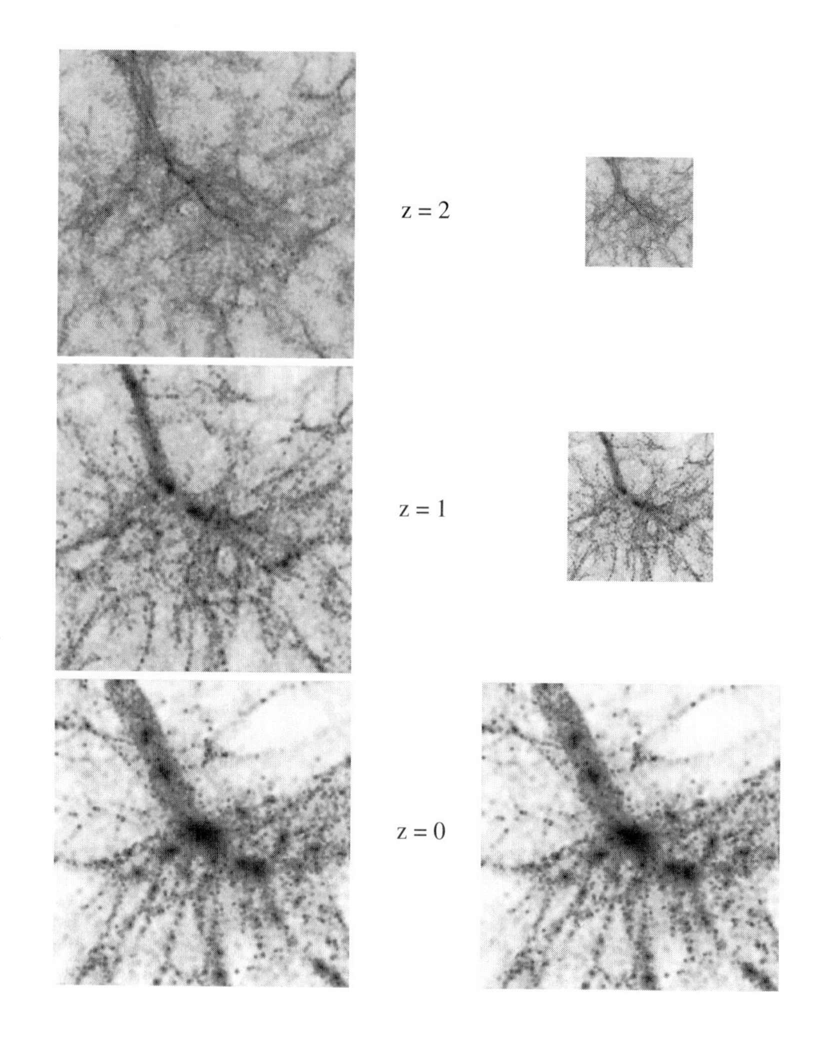

Figure 15.5. The evolution of a rich cluster of galaxies, according to an N-body integration of a $\Gamma = 0.2$ CDM universe, adapted from Colberg *et al.* (1997). The left-hand column shows the evolution in comoving coordinates, for a slice $21\,h^{-1}$ Mpc across and $8\,h^{-1}$ Mpc thick. The right-hand column shows the evolution in proper coordinates, emphasizing that the protocluster initially expands, prior to turning round from the expansion and virializing.

virialize (i.e. randomize its inward radial motion), building up a gravitating system with a self-similar profile.

Consider once again the spherical model, where the radius–time relation for shells of matter is $r = A(1 - \cos\theta)$, $t = B(\theta - \sin\theta)$. Collapse proceeds with shells reaching their maximum or **turnround radius** of $r = 2A$ at parametric time $\theta = \pi$ and then contracting to a singularity of infinite density at $\theta = 2\pi$. The first thing we can work out is the density profile at the instant that the object first forms a central singularity. For shells close to the centre, $\theta = 2\pi - \epsilon$, so that

$$\begin{aligned} r &\simeq A\epsilon^2/2 \\ t &\simeq B(2\pi - \epsilon^3/6). \end{aligned} \tag{15.109}$$

Now, if the initial perturbation lacked a singularity, the B parameter (which is proportional to the reciprocal square root of the density at turnround) will in general have a quadratic dependence on turnround radius, $B \propto 1 + \alpha A^2$; this just says that the initial density perturbation had some parabolic maximum. At the given time at which the first shell becomes singular, we therefore get $\epsilon^3/6 \propto A^2$, and $r \propto A^{7/3}$. Since the enclosed mass will go proportional to A^3 near the centre, this implies a cusped density profile

$$\rho(r) \propto r^{-12/7}, \tag{15.110}$$

a result first derived by Penston (1969).

SIMILARITY SOLUTION It is perhaps more interesting to ask what happens at later times, when the collapse has gone completely nonlinear. On general grounds, we may expect the infall structure to approach a **self-similar** form that is independent of the initial conditions. The reason for expecting this is that there is no preferred time in the $\Omega = 1$ universe: density perturbations grow $\propto t^{2/3}$ in the linear regime, until a given shell turns round, which always occurs at a mean density that is a fixed multiple of the background density (a factor 5.55). If the initial perturbations, defined as a function of r at time t_i, display no characteristic scale (so that the profile has the power-law form $\bar{\delta}_i \propto r_i^{-\alpha}$), then it should be clear that the density law will obey some universal function, subject only to power-law scalings with t:

$$\bar{\rho}(r \mid t) = 5.55\rho_b(t)\, f\big(r/r_{\rm turn}(t)\big). \tag{15.111}$$

The length $r_{\rm turn}$ is the radius of the shell that is just breaking away from the Hubble expansion. As this increases with time, so the scale of the collapsed object grows. The assumption of a power-law initial density profile may seem contrived, but with $\alpha = 3$ this is exactly what results if $\delta = 0$ beyond some initial zone of disturbance. The function $f(r)$, which obeys $f(1) = 1$ and tends at large r to describe the still-linear relics of the initial perturbations, may be deduced by following the collapse of a single shell, as follows.

Start by relating the parameters in the exact spherical collapse to the initial conditions at t_i. By expanding to order θ^5 and looking at how $r(t)$ departs from $t^{2/3}$ growth, we get successively the initial radius

$$r_i = \frac{A}{2}\left(\frac{6t_i}{B}\right)^{2/3} \tag{15.112}$$

and the initial density perturbation

$$\bar{\delta}_i = \frac{3}{20}\left(\frac{6t_i}{B}\right)^{2/3}. \tag{15.113}$$

These are therefore related via $r_i = (10/3)\bar{\delta}_i A$. Consider now the point at which a given shell turns round and virializes. Turnround occurs at $\theta = \pi$ and virialization (taken to correspond to collapse by a factor 2 from maximum expansion) at $\theta = 3\pi/2$. The radius and time at virialization are thus

$$\boxed{\begin{aligned} r_{\rm vir} &= \frac{3}{10}\left(\frac{r_i}{\bar{\delta}_i}\right) \\ t_{\rm vir} &= (9\pi + 6)\left(\frac{20\bar{\delta}_i}{3}\right)^{-3/2} t_i. \end{aligned}} \tag{15.114}$$

Now, if we insert the power-law dependence $\bar{\delta}_i \propto r_i^{-\alpha}$, this gives

$$r_{\rm vir} \propto t_{\rm vir}^{2(\alpha+1)/3\alpha}. \tag{15.115}$$

In the most natural case, $\alpha = 3$ and the time dependence is $r \propto t^{8/9}$. This is the proper radius; the comoving radius grows as $t^{2/9}$, and so the accreted mass grows as $t^{2/3}$.

The density profile depends on what happens to the mass shells after they turn round. In reality, there must be some asphericity in the initial conditions, which is magnified under collapse: spherical symmetry will be destroyed and we should expect each collapsing sphere to virialize and attain a final size of about $r_{\rm vir}$. In this case, the density within a given $r_{\rm vir}$ is some constant multiple (about 200) times the background density at virialization. Hence, the density profile expected is

$$\bar{\rho} \propto t_{\rm vir}^{-2} \propto r_{\rm vir}^{-3\alpha/(\alpha+1)}. \tag{15.116}$$

For accretion onto some initial seed mass, we therefore expect that a power-law density profile will develop:

$$\boxed{\rho(r) \propto r^{-9/4}.} \tag{15.117}$$

Notice that this is tantalizingly close to the 'thermodynamic' equilibrium density profile of the isothermal sphere, which has $\rho \propto r^{-2}$. Galaxy rotation curves that are very nearly flat thus arise inevitably through secondary infall in a universe with critical density.

INFALL ONTO DENSITY PEAKS The above discussion can be made more realistic by considering the initial conditions for infall that are likely to arise in a random density field (see e.g. Hoffman 1988 for more details). Rather than postulate some localized density perturbation, we can imagine infall proceeding following the collapse of some region of high overdensity – i.e. some peak in the initial density perturbation field. These perturbations will possess long-range correlations, which will govern the density profile around the peak. For fields with Gaussian statistics, it is possible to calculate the peak profile and hence the initial conditions for infall. The statistics of density peaks are discussed in chapter 16, and the full treatment can be rather messy (consult Bardeen *et al.* 1986 for the details). However, the essence can be recovered by considering a randomly selected point of high overdensity, since this is almost certain to be a peak if a high threshold is chosen. It is then easily shown [problem 16.4] that the mean density profile has the shape of the correlation function, $\xi(r) \equiv \langle \delta(x)\delta(x+r) \rangle$:

$$\langle\, \delta(r) \,\rangle = \delta(0)\,\frac{\xi(r)}{\xi(0)} \propto r^{-(n+3)}, \tag{15.118}$$

where we have assumed a power-law power spectrum $|\delta_k|^2 \propto k^n$ (see chapter 16). The density profile from infall is now

$$\rho(r) \propto r^{-(3n+9)/(n+4)}, \tag{15.119}$$

Notice that this is slightly steeper than the nonlinear slope of the evolved correlation function, which is shown in chapter 16 to scale as $r^{-(3n+9)/(n+5)}$. The difference arises because (not surprisingly), the density gradient is steeper around a high peak than it is around an average point. Something odd happens to the above expression at $n = -1$. The initial density profile has $\bar{\delta} \propto r^{-2}$, which is also the slope of the final nonlinear profile. This means that the excess mass in the perturbation grows $\propto r$, so that the binding energy is independent of r. For $n < -1$, the binding energy increases outwards so that the shells that fall in later are more strongly bound. It is clearly not a stable configuration to have these outside the more weakly bound inner shells. This then leads to the conjecture that the density profiles of virialized haloes will never be flatter than r^{-2}, i.e. we get $\rho \propto r^{-2}$ from infall if $n < -1$, a result that appears to agree with numerical simulation (Hoffman 1988).

For completeness, it is worth mentioning what happens if we maintain the assumption of exact spherical symmetry. When the first singularity forms at $r = 0$, a black hole results, into which subsequent mass shells are accreted. This alters the density profile somewhat in the nonlinear regime. The problem is now similar to the one we studied above: we need to consider shells close to total collapse, where $\theta = 2\pi - \epsilon$. In the same way as we obtained $t_{\rm vir}$ above, the time is

$$t = (12\pi - \epsilon^3)\left(\frac{20\bar{\delta}_i}{3}\right)^{-3/2} t_i, \tag{15.120}$$

and we have

$$1 - \frac{\epsilon^3}{12\pi} \propto \bar{\delta}_i^{3/2} \propto r_i^{-3\alpha/2}. \tag{15.121}$$

The radius at the present time is $\propto \epsilon^2$. Using the above relation to form the density, we obtain

$$\rho(r) \propto \frac{r_i^2}{r^2}\frac{dr_i}{dr} \propto r^{-1.5}, \tag{15.122}$$

which is independent of α. Infall into a central black hole thus leads to a less extreme density cusp in the somewhat idealized totally symmetrical case.

Problems

(15.1) Use the spherical model to describe the exact nonlinear evolution of a cosmological structure (cluster or void). Show that the following formula is an excellent approximation for the relation between true density contrast δ and linearly extrapolated density contrast $\delta_{\rm lin}$:

$$1 + \delta = (1 - 2\delta_{\rm lin}/3)^{-3/2} \tag{15.123}$$

(Bernardeau 1994a). How do the predictions of the Zeldovich approximation compare in this case?

(15.2) Consider a critical-density universe in which massive neutrinos contribute Ω_ν to the density parameter. Show that, on scales smaller than the neutrino Jeans length, perturbations in the remaining cold component grow as $\delta \propto t^\alpha$, where $\alpha = (\sqrt{25 - 24\Omega_\nu} - 1)/6$.

(15.3) Show that the equation of motion for particles in the expanding universe can be obtained from an action principle, $\delta \int L \, dt = 0$, and give an expression for the Lagrangian, L.

(15.4) Show how the WKB approximation can be used to solve the equation for the growth of a coupled system of dark matter and baryons. If at early times the baryons and dark matter have the same overdensity, show how the effect of pressure causes δ_b/δ_d to decrease with time. Do the two perturbations ever have opposite signs?

(15.5) Consider the growth of a perturbation in cold collisionless material in the radiation-dominated era. By matching a growing mode outside the horizon onto the sub-horizon growing and decaying modes, show that the small-scale behaviour of the transfer function is approximately $T_k = (kr_{\rm H})^{-2} \ln(kr_{\rm H})$, where $r_{\rm H}$ is the comoving horizon size at matter–radiation equality.

(15.6) A spherically symmetric perturbation collapses in a matter-dominated universe of arbitrary Ω. Derive the exact density contrast at virialization, and compare with the rule of thumb $1 + \delta = 178\,\Omega^{-0.7}$.

16 Cosmological density fields

16.1 Preamble

Chapter 15 has shown how gravitational instability is expected to yield patterns of inhomogeneity in the universe, with a characteristic dependence on scale depending on the precise matter content. The next step is to see how these ideas can be confronted with statistical measures of the observed matter distribution, and this chapter attempts to summarize what is known about the dimensionless **density perturbation field**

$$\boxed{\delta(\mathbf{x}) \equiv \frac{\rho(\mathbf{x}) - \langle\rho\rangle}{\langle\rho\rangle}.} \tag{16.1}$$

This is a quantity that has featured in the sections on cosmological perturbations in chapter 15, but it need not be assumed to be small. Indeed, some of the most interesting issues arise in understanding the evolution of the density field to large values of δ.

A critical feature of the δ field is that it inhabits a universe that is isotropic and homogeneous in its large-scale properties. This suggests that the statistical properties of δ should also be homogeneous, even though it is a field that describes inhomogeneities. This statement sounds contradictory, and yet it makes perfect sense if there exists an **ensemble of universes**. The concept of an ensemble is used every time we apply probability theory to an event such as tossing a coin: we imagine an infinite sequence of repeated trials, half of which result in heads, half in tails. To say that the probability of heads is 1/2 means that the coin lands heads up in half the members of this ensemble of universes. The analogy of coin tossing in cosmology is that the density at a given point in space will have different values in each member of the ensemble, with some overall variance $\langle\delta^2\rangle$ between members of the ensemble. Statistical homogeneity of the δ field then means that this variance must be independent of position. The actual field found in a given member of the ensemble is a **realization** of the statistical process.

There are two problems with this line of argument: (i) we have no evidence that the ensemble exists; (ii) in any case, we are only able to observe one realization, so how is the variance $\langle\delta^2\rangle$ to be measured? The first objection applies to coin tossing, and may be evaded if we understand the physics that generates the statistical process – we only need to *imagine* tossing the coin many times and do not actually need to perform the exercise. The best that can be done in answering the second objection is to look at widely separated parts of space, since the δ fields there should be causally unconnected; this is therefore as good as taking measurements from two different members of the ensemble.

In other words, if we measure the variance $\langle\delta^2\rangle$ by averaging over a sufficiently large volume, the results would be expected to approach the true ensemble variance, and the averaging operator $\langle\cdots\rangle$ is often used without being specific about which kind of average is intended. Fields that satisfy this property, whereby

$$\text{volume average} \quad \leftrightarrow \quad \text{ensemble average} \tag{16.2}$$

are termed **ergodic**. Giving a formal proof of ergodicity for a random process is not always easy (Adler 1981); in cosmology it is perhaps best regarded as a commonsense axiom.

16.2 Fourier analysis of density fluctuations

It is often convenient to consider building up a general field by the superposition of many modes. For a flat comoving geometry, the natural tool for achieving this is via Fourier analysis. For other models, plane waves are not a complete set and one should use instead the eigenfunctions of the wave equation in a curved space. Normally this complication is neglected: even in an open universe, the difference only matters on scales of order the present-day horizon.

How do we make a Fourier expansion of the density field in an infinite universe? If the field were periodic within some box of side L, then we would just have a sum over wave modes:

$$F(\mathbf{x}) = \sum F_{\mathbf{k}} e^{-i\mathbf{k}\cdot\mathbf{x}}. \tag{16.3}$$

The requirement of periodicity restricts the allowed wavenumbers to **harmonic boundary conditions**

$$k_x = n\,\frac{2\pi}{L}, \qquad n = 1, 2 \ldots, \tag{16.4}$$

with similar expressions for k_y and k_z. Now, if we let the box become arbitrarily large, then the sum will go over to an integral that incorporates the density of states in k-space, exactly as in statistical mechanics; this is how the general idea of the Fourier transform is derived. The Fourier relations in n dimensions are thus

$$\boxed{\begin{aligned} F(x) &= \left(\frac{L}{2\pi}\right)^n \int F_k(k) \exp(-i\mathbf{k}\cdot\mathbf{x})\; d^n k \\ F_k(k) &= \left(\frac{1}{L}\right)^n \int F(x) \exp(i\mathbf{k}\cdot\mathbf{x})\; d^n x. \end{aligned}} \tag{16.5}$$

One advantage of this particular Fourier convention is that the definition of convolution involves just a simple volume average, with no gratuitous factors of $(2\pi)^{-1/2}$:

$$f * g \equiv \frac{1}{L^n} \int f(\mathbf{x}-\mathbf{y}) g(\mathbf{y}) d^n y. \tag{16.6}$$

Although one can make all the manipulations on density fields that follow using either the integral or sum formulations, it is usually easier to use the sum. This saves having to introduce δ-functions in k-space. For example, if we have $f = \sum f_k \exp(-ikx)$, the obvious way to extract f_k is via $f_k = (1/L) \int f \exp(ikx)\, dx$: because of the harmonic boundary conditions, all oscillatory terms in the sum integrate to zero, leaving only f_k to be integrated from 0 to L. There is less chance of committing errors of factors of 2π in this way than considering $f = (L/2\pi) \int f_k \exp(-ikx)\, dk$ and then using $\int \exp[i(k-K)x]\, dx = 2\pi\delta_{\rm D}(k-K)$.

CORRELATION FUNCTIONS AND POWER SPECTRA As an immediate example of the Fourier machinery in action, consider the important quantity

$$\boxed{\xi(\mathbf{r}) \equiv \langle \delta(\mathbf{x})\delta(\mathbf{x}+\mathbf{r}) \rangle\,,} \tag{16.7}$$

which is the autocorrelation function of the density field – usually referred to simply as the **correlation function**. The angle brackets indicate an averaging over the normalization volume V. Now express δ as a sum and note that it is real, so that we can replace one of the two δ's by its complex conjugate, obtaining

$$\xi = \left\langle \sum_{\mathbf{k}} \sum_{\mathbf{k}'} \delta_{\mathbf{k}} \delta^*_{\mathbf{k}'} e^{i(\mathbf{k}'-\mathbf{k})\cdot\mathbf{x}} e^{-i\mathbf{k}\cdot\mathbf{r}} \right\rangle . \tag{16.8}$$

Alternatively, this sum can be obtained, without replacing $\langle\delta\delta\rangle$ by $\langle\delta\delta^*\rangle$, from the relation between modes with opposite wavevectors that holds for any real field: $\delta_{\mathbf{k}}(-\mathbf{k}) = \delta^*_{\mathbf{k}}(\mathbf{k})$. Now, by the periodic boundary conditions, all the cross terms with $\mathbf{k}' \neq \mathbf{k}$ average to zero. Expressing the remaining sum as an integral, we have

$$\boxed{\xi(\mathbf{r}) = \frac{V}{(2\pi)^3} \int |\delta_{\mathbf{k}}|^2 e^{-i\mathbf{k}\cdot\mathbf{r}} d^3k.} \tag{16.9}$$

In short, the correlation function is the Fourier transform of the **power spectrum**. This relation has been obtained by volume averaging, so it applies to the specific mode amplitudes and correlation function measured in any given realization of the density field. Taking ensemble averages of each side, the relation clearly also holds for the ensemble average power and correlations – which are really the quantities that cosmological studies aim to measure. We shall hereafter often use the alternative notation

$$\boxed{P(k) \equiv \langle |\delta_k|^2 \rangle} \tag{16.10}$$

for the ensemble-average power (although this only applies for a Fourier series with discrete modes; see problem 16.1). The distinction between the ensemble average and the actual power measured in a realization is clarified below in the section on Gaussian fields.

In an isotropic universe, the density perturbation spectrum cannot contain a preferred direction, and so we must have an **isotropic power spectrum**: $\langle |\delta_{\mathbf{k}}|^2(\mathbf{k}) \rangle = |\delta_k|^2(k)$. The angular part of the k-space integral can therefore be performed immediately: introduce spherical polars with the polar axis along $\mathbf{k}$ and use the reality of ξ, so that $e^{-i\mathbf{k}\cdot\mathbf{x}} \to \cos(kr\cos\theta)$. In three dimensions, this yields

$$\xi(r) = \frac{V}{(2\pi)^3} \int P(k)\, \frac{\sin kr}{kr}\, 4\pi k^2\, dk. \tag{16.11}$$

The 2D analogue of this formula is

$$\xi(r) = \frac{A}{(2\pi)^2} \int P(k)\, J_0(kr)\, 2\pi k\, dk. \tag{16.12}$$

We shall usually express the power spectrum in dimensionless form, as the variance per $\ln k$ ($\Delta^2(k) = d\langle\delta^2\rangle/d\ln k \propto k^3 P[k]$):

$$\boxed{\Delta^2(k) \equiv \frac{V}{(2\pi)^3}\, 4\pi k^3\, P(k) = \frac{2}{\pi} k^3 \int_0^\infty \xi(r)\, \frac{\sin kr}{kr}\, r^2\, dr.} \tag{16.13}$$

This gives a more easily visualizable meaning to the power spectrum than does the quantity $VP(k)$, which has dimensions of volume: $\Delta^2(k) = 1$ means that there are order-unity density fluctuations from modes in the logarithmic bin around wavenumber k. $\Delta^2(k)$ is therefore the natural choice for a Fourier-space counterpart to the dimensionless quantity $\xi(r)$.

POWER-LAW SPECTRA The above shows that the power spectrum is a central quantity in cosmology, but how can we predict its functional form? For decades, this was thought to be impossible, and so a minimal set of assumptions was investigated. In the absence of a physical theory, we should not assume that the spectrum contains any preferred length scale, otherwise we should then be compelled to explain this feature. Consequently, the spectrum must be a featureless power law:

$$\boxed{\left\langle |\delta_k|^2 \right\rangle \propto k^n.} \tag{16.14}$$

The index n governs the balance between large- and small-scale power. The meaning of different values of n can be seen by imagining the results of filtering the density field by passing over it a box of some characteristic comoving size x and averaging the density over the box. This will filter out waves with $k \gtrsim 1/x$, leaving a variance $\langle\delta^2\rangle \propto \int_0^{1/x} k^n 4\pi k^2 dk \propto x^{-(n+3)}$. Hence, in terms of a mass $M \propto x^3$, we have

$$\delta_{\rm rms} \propto M^{-(n+3)/6}. \tag{16.15}$$

Similarly, a power-law spectrum implies a power-law correlation function. If $\xi(r) = (r/r_0)^{-\gamma}$, with $\gamma = n + 3$, the corresponding 3D power spectrum is

$$\Delta^2(k) = \frac{2}{\pi}\,(kr_0)^\gamma\, \Gamma(2-\gamma)\, \sin\frac{(2-\gamma)\pi}{2} \equiv \beta(kr_0)^\gamma \tag{16.16}$$

($= 0.903(kr_0)^{1.8}$ if $\gamma = 1.8$). This expression is only valid for $n < 0$ ($\gamma < 3$); for larger values of n, ξ must become negative at large r, because $P(0)$ must vanish, implying $\int_0^\infty \xi(r)\, r^2\, dr = 0$. A cutoff in the spectrum at large k is needed to obtain physically sensible results.

What general constraints can we set on the value of n? Asymptotic homogeneity clearly requires $n > -3$. An upper limit of $n < 4$ comes from an argument due to Zeldovich. Suppose we begin with a totally uniform matter distribution and then group it into discrete chunks as uniformly as possible. It can be shown [problem 16.3] that conservation of momentum in this process means that we cannot create a power spectrum that goes to zero at small wavelengths more rapidly than $\delta_k \propto k^2$. Thus, discreteness of matter produces the **minimal spectrum**, $n = 4$. More plausible alternatives lie between these extremes. The value $n = 0$ corresponds to **white noise**, the same power at all wavelengths. This is also known as the **Poisson** power spectrum, because it corresponds to fluctuations between different cells that scale as $1/\sqrt{M_{\rm cell}}$ (see below). A density field created by throwing down a large number of point masses at random would therefore consist of white noise. Particles placed at random within cells, one per cell, create an $n = 2$ spectrum on large scales [problem 16.3].

Practical spectra in cosmology, conversely, have negative effective values of n over a large range of wavenumber. For many years, the data on the galaxy correlation function were consistent with a single power law:

$$\boxed{\xi_g(r) \simeq \left(\frac{r}{5\,h^{-1}\,\text{Mpc}}\right)^{-1.8} \qquad \left(1 \lesssim \xi \lesssim 10^4\right);} \tag{16.17}$$

see Peebles (1980), Davis & Peebles (1983). This corresponds to $n \simeq -1.2$. By contrast with the above examples, large-scale structure is 'real', rather than reflecting the low-k Fourier coefficients of some small-scale process.

THE ZELDOVICH SPECTRUM Most important of all is the **scale-invariant spectrum**, which corresponds to the value $n = 1$, i.e. $\Delta^2 \propto k^4$. To see how the name arises, consider a perturbation $\delta\Phi$ in the gravitational potential:

$$\nabla^2 \delta\Phi = 4\pi G\rho_0\delta \quad \Rightarrow \quad \delta\Phi_k = -4\pi G\rho_0\delta_k/k^2. \tag{16.18}$$

The two powers of k pulled down by ∇^2 mean that, if $\Delta^2 \propto k^4$ for the power spectrum of density fluctuations, then Δ^2_Φ is a constant. Since potential perturbations govern the flatness of spacetime, this says that the scale-invariant spectrum corresponds to a metric that is a **fractal**: spacetime has the same degree of 'wrinkliness' on each resolution scale. The total curvature fluctuations diverge, but only logarithmically at either extreme of wavelength.

Another way of looking at this spectrum is in terms of the perturbation growth balancing the scale dependence of δ: $\delta \propto x^{-(n+3)/2}$. We know that δ viewed on a given comoving scale will increase with the size of the horizon: $\delta \propto r_{\rm H}^2$. At an arbitrary time, though, the only natural length provided by the universe (in the absence of non-gravitational effects) is the horizon itself:

$$\delta(r_{\rm H}) \propto r_{\rm H}^2 r_{\rm H}^{-(n+3)/2} = r_{\rm H}^{-(n-1)/2}. \tag{16.19}$$

Thus, if $n = 1$, the growths of $r_{\rm H}$ and δ with time have cancelling effects, so that the universe always looks the same when viewed on the scale of the horizon; such a universe is self-similar in the sense of always appearing the same under the magnification of cosmological expansion. This spectrum is often known as the **Zeldovich spectrum** (sometimes hyphenated with Harrison and Peebles, who invented it independently).

FILTERING AND MOMENTS A common concept in the manipulation of cosmological density fields is that of **filtering**, where the density field is convolved with some **window**

function: $\delta \to \delta * f$. Many observable results can be expressed in this form. Some common 3D filter functions are as follows:

$$\begin{aligned} &\text{Gaussian} \quad f = \frac{V}{(2\pi)^{3/2} R_{\rm G}^3} e^{-r^2/2R_{\rm G}^2} \Rightarrow f_k = e^{-k^2 R_{\rm G}^2/2} \\ &\text{top-hat} \quad f = \frac{3V}{4\pi R_{\rm T}^3} \quad (r < R_{\rm T}) \;\Rightarrow\; f_k = \frac{3}{y^3}(\sin y - y\cos y) \quad (y \equiv kR_{\rm T}). \end{aligned} \tag{16.20}$$

Note the factor of V in the definition of f; this is needed to cancel the $1/V$ in the definition of convolution. For some power spectra, the difference in these filter functions at large k is unimportant, and we can relate them by equating the expansions near $k = 0$, where $1 - |f_k|^2 \propto k^2$. This equality requires

$$R_{\rm T} = \sqrt{5}\, R_{\rm G}. \tag{16.21}$$

We are often interested not in the convolved field itself, but in its variance, for use as a statistic (e.g. to measure the rms fluctuations in the number of objects in a cell). By the convolution theorem, this means we are interested in a **moment** of the power spectrum times the squared filter transform. We shall generally use the following notation:

$$\boxed{\sigma_n^2 \equiv \frac{V}{(2\pi)^3} \int P(k)\, |f_k|^2\, k^{2n}\, d^3k;} \tag{16.22}$$

the filtered variance is thus σ_0^2 (often denoted by just σ^2). Clustering results are often published in the form of **cell variances** of δ as a function of scale, σ^2, using either cubical cells of side ℓ (Efstathiou *et al.* 1990) or Gaussian spheres of radius $R_{\rm G}$ (Saunders *et al.* 1991). For a power-law spectrum ($\Delta^2 \propto k^{n+3}$), we have for the Gaussian sphere

$$\sigma^2 = \Delta^2 \left(k = \left[\frac{1}{2} \left(\frac{n+1}{2} \right)! \right]^{1/(n+3)} R_{\rm G}^{-1} \right). \tag{16.23}$$

For $n \lesssim 0$, this formula also gives a good approximation to the case of cubical cells, with $R_{\rm G} \to \ell/\sqrt{12}$. The result is rather insensitive to assumptions about the power spectrum, and just says that the variance in a cell is mainly probing waves with $\lambda \simeq 2\ell$.

Moments may also be expressed in terms of the correlation function over the sample volume:

$$\sigma^2 = \int\!\!\int \xi(|\mathbf{x} - \mathbf{x}'|) f(\mathbf{x}) f(\mathbf{x}')\, d^3x\, d^3x'. \tag{16.24}$$

To prove this, it is easiest to start from the definition of σ^2 as an integral over the power spectrum times $|f_k|^2$, write out the Fourier representations of P and f_k and use $\int \exp[i\mathbf{k} \cdot (\mathbf{x} - \mathbf{x}' + \mathbf{r})]\, d^3k = (2\pi)^3 \delta_{\rm D}^{(3)}(\mathbf{x} - \mathbf{x}' + \mathbf{r})$. Finally, it is also sometimes convenient to express things in terms of derivatives of the correlation function at zero lag. Odd derivatives vanish, but even derivatives give

$$\xi^{(2n)}(0) = (-1)^n \, \frac{\sigma_n^2}{2n+1}. \tag{16.25}$$

NORMALIZATION For scale-invariant spectra, a natural amplitude measure is the variance in gravitational potential per unit $\ln k$, which is a constant, independent of scale:

$$\epsilon^2 \equiv \frac{\Delta_\Phi^2}{c^4} = \frac{9}{4}\left(\frac{ck}{H_0}\right)^{-4} \Delta^2(k). \tag{16.26}$$

Another two commonly encountered measures relate to the clustering field around 10 Mpc. One is σ_8, the rms density variation when smoothed with a **top-hat filter** (sphere of uniform weight) of radius $8h^{-1}$ Mpc; this is observed to be very close to unity. The other is an integral over the correlation function:

$$J_3 \equiv \int_0^r \xi(y)\, y^2 dy = \int \Delta^2(k) W(k)\, \frac{dk}{k}, \tag{16.27}$$

where $W(k) = (\sin kr - kr\cos kr)/k^3$. The canonical value of this is $J_3(10\,h^{-1}\,\text{Mpc}) = 277h^{-3}\,\text{Mpc}^3$ (from the CfA survey; see Davis & Peebles 1983). It is sometimes more usual to use instead the dimensionless **volume-averaged correlation function** $\bar{\xi}$:

$$\bar{\xi}(r) = \frac{3}{4\pi r^3}\int_0^r \xi(x)\, 4\pi x^2\, dx = \frac{3}{r^3} J_3(r). \tag{16.28}$$

The canonical value then becomes $\bar{\xi}(10\,h^{-1}\,\text{Mpc}) = 0.83$; this measure is clearly very close in content to $\sigma_8 = 1$.

A point to beware of is that the normalization of a theory is often quoted in terms of a value of these parameters extrapolated according to *linear* time evolution. Since the observed values are clearly nonlinear, there is no reason why theory and observation should match exactly. Even more confusingly, it is quite common in the literature to find the linear value of σ_8 called $1/b$, where b is a **bias parameter**. The implication is that $b \neq 1$ means that light does not follow mass; this may well be true in reality, but with this definition, nonlinearities will produce $b \neq 1$ even in models where mass traces light exactly. Use of this convention is not recommended.

N-POINT CORRELATIONS An alternative definition of the autocorrelation function is as the **two-point correlation function**, which gives the excess probability for finding a neighbour a distance r from a given galaxy (see figure 16.1). By regarding this as the probability of finding a pair with one object in each of the volume elements dV_1 and dV_2,

$$\boxed{dP = \rho_0^2\,[1 + \xi(r)]\, dV_1\, dV_2,} \tag{16.29}$$

this is easily seen to be equivalent to the autocorrelation definition of ξ: $\xi = \langle\delta(x_1)\delta(x_2)\rangle$. A related quantity is the **cross-correlation function**. Here, one considers two different classes of object (a and b, say), and the cross-correlation function ξ_{ab} is defined as the (symmetric) probability of finding a pair in which dV_1 is occupied by an object from the first catalogue and dV_2 by one from the second:

$$dP = \rho_a \rho_b\,[1 + \xi_{ab}(r)]\, dV_1\, dV_2. \tag{16.30}$$

In terms of density fields, clearly $\xi_{ab} = \langle\delta_a(x_1)\delta_b(x_2)\rangle$. Cross-correlations give information about the density profile around objects; for example, ξ_{gc} between galaxies and clusters

Figure 16.1. The N-point correlation functions of a density field consisting of a set of particles are calculated by looking at a set of cells of volume dV (so small that they effectively only ever contain 0 or 1 particles). The Poisson probability that two cells at separation r_{12} are both occupied is $\rho_0^2\, dV_1\, dV_2$; with clustering, this is modified by a factor $1 + \xi(r_{12})$, where ξ is the two-point correlation function. Similarly, the probability of finding a triplet of occupied cells is a factor $1 + \xi(r_{12}, r_{13}, r_{23})$ times the random probability; this defines the three-point correlation function.

measures the average galaxy density profile around clusters (at least out to radii where clusters overlap).

There is a straightforward extension of two-point statistics to larger numbers of points: the probability of finding an n-tuplet of galaxies in n specified volumes is

$$dP = \rho_0^n [1 + \xi^{(n)}]\, dV_1 \cdots dV_n. \tag{16.31}$$

As with the two-point function, the probability is proportional to the product of the density field at the n points, and so

$$1 + \xi^{(n)} = \left\langle \prod_i (1 + \delta_i) \right\rangle. \tag{16.32}$$

Expanding the product gives a sequence of terms. For $n = 3$, for example,

$$\xi^{(n)} = \xi(r_{12}) + \xi(r_{23}) + \xi(r_{31}) + \zeta(\mathbf{r}_1, \mathbf{r}_2, \mathbf{r}_3). \tag{16.33}$$

The term $\zeta = \langle \delta_1 \delta_2 \delta_3 \rangle$, which represents any excess correlation over that described by the two-point contributions we already know about, is called the **reduced three-point correlation function**. Similar expressions exist for larger numbers of points; the reduced n-point correlation function is $\langle \prod \delta_i \rangle$.

Observationally, the three-point function is non-zero. It has been suggested (Groth & Peebles 1977) that the results fit the so-called hierarchical form

$$\zeta = Q(\xi_{12}\xi_{23} + \xi_{23}\xi_{31} + \xi_{31}\xi_{12}), \tag{16.34}$$

with $Q \simeq 1$. This has been generalized to the **hierarchical ansatz**, which assumes that the reduced correlations can all be expressed in terms of two-point functions in this way (e.g. Balian & Schaeffer 1989). However, no convincing derivation of such a situation starting from linear initial conditions has ever been given.

The reduced functions are sometimes called the **connected correlation functions** by analogy with Green functions in particle physics. The analogy allows diagrammatic techniques borrowed from that subject to be used to aid calculation (Grinstein & Wise 1986; Matsubara 1995). To describe all the statistical properties of the density field requires the whole hierarchy of correlation functions (White 1979), which is a problem since only the first few have been measured. However, this is not a problem for Gaussian fields (see below), where all reduced correlations above the two-point level either vanish, if odd, or are expressible in terms of two-point functions, if even. It is because of this relation to the two-point function, and hence to the power spectrum, that the latter is often described as the 'holy grail' of cosmology: it tells us all that there is to know about the statistical properties of the density field.

16.3 Gaussian density fields

Apart from statistical isotropy of the fluctuation field, there is another reasonable assumption we might make: that the phases of the different Fourier modes δ_k are uncorrelated and random. This corresponds to treating the initial disturbances as some form of random noise, analogous to Johnson noise in electrical circuits; indeed, many mathematical tools that have become invaluable in cosmology were first established with applications to communication circuits in mind (e.g. Rice 1954). The random-phase approximation has a powerful consequence, which derives from the **central limit theorem**: loosely, the sum of a large number of independent random variables will tend to be normally distributed. This will be true not just for the field δ; all quantities that are derived from linear sums over waves (such as field derivatives) will have a joint normal distribution. The result is a **Gaussian random field**, whose properties are characterized entirely by its power spectrum.

The central limit theorem is a rather general result, whose full proof is involved (e.g. Kendall & Stuart 1958). However, the proof is simple for a restricted class of probability distributions. Suppose we make n measurements of a variable x, whose probability distribution is $f(x)$; the critical simplification will be to assume that all the moments $\int f(x)\, x^m\, dx$ are finite. Now introduce the **complementary function**, which is the Fourier transform of f: $C(k) = \int f(x)\, \exp(-ikx)\, dx$. This function obeys the key result that the complementary function for the distribution of the sum of two random variables x and y, with distribution functions f and g and complementary functions C_1 and C_2, is the product of the individual complementary functions:

$$C_+(k) = \int f(x)g(y)\, \exp[-ik(x+y)]\, dx\, dy = C_1(k)C_2(k). \tag{16.35}$$

For the sum of n drawings from the same distribution, $C_+(k) = C(k)^n$. Now consider a Taylor expansion of C, which involves the mean $\mu = \langle x \rangle$ and variance $\sigma^2 = \langle x^2 \rangle - \mu^2$:

$$C(k) = 1 - i\mu k - (\sigma^2 + \mu^2)k^2/2 + O(k^3) = \exp\left[-i\mu k - \sigma^2 k^2/2 + O(k^3)\right]. \tag{16.36}$$

Raising this to the nth power and defining $k' = k/n$, so that we deal with the distribution for the mean of the n observations, rather than the sum, gives

$$C_{\rm mean}(k') = \exp\left[-i\mu k' - (\sigma^2/n)k'^2/2 + O(k'^3/n^2)\right]. \tag{16.37}$$

The higher powers of $1/n$ in the terms beyond k'^2 mean that these terms can be neglected for large enough n, so long as the higher-order moments are finite. What remains is the complementary function of a Gaussian, as required. A fuller proof shows that the theorem in fact holds even when these moments diverge; all that is needed is a finite second moment. The simplest counterexample to the theorem is thus the **Cauchy distribution**, $f \propto 1/(1+x^2)$.

REALIZATIONS An immediate example of a Gaussian variable related to a Gaussian field arises when we consider one realization of such a field: a sample within a finite volume. The Fourier coefficients for the field within the box are linear sums over the field, and hence will themselves be (two-dimensional, complex) Gaussian variables. Call these a_k, where $\delta = \sum a_k \exp(-i\mathbf{k}\cdot\mathbf{x})$. If we call the realization volume $V_{\rm R}$ to distinguish it from the arbitrary normalization volume, the expectation value of the realization power follows from fixing the power in a given region of k-space in the continuum limit:

$$\langle |a_k|^2 \rangle = \frac{V}{V_{\rm R}} P(k). \tag{16.38}$$

The power in a given mode will not take this expectation value, but will be exponentially distributed about it (a **Rayleigh distribution** for the mode amplitude; the real and imaginary parts of a_k will be Gaussian):

$$P\left(|a_k|^2 > X\right) = \exp\left(-X^2/\langle |a_k|^2 \rangle\right). \tag{16.39}$$

We now have the prescription needed to set up the conditions in k-space for creating a Gaussian realization with a desired power spectrum: assign Fourier amplitudes randomly with the above distribution and assign phases randomly between 0 and 2π. The real-space counterpart may then be found efficiently via the fast Fourier transform (FFT) algorithm. This is how the set of mode amplitudes and phases may be chosen for N-body initial conditions (see chapter 15). In fact, the central limit theorem means that, even if the dispersion in amplitudes is neglected, the field will still be closely Gaussian if many modes are used. However, its statistical properties may be suspect on scales approaching the box size, where there are few modes.

An alternative, simpler, method is to create a spatial array of white noise: Gaussian density variations in which each pixel is independent and has unit variance. This field can then be given the desired statistical properties by convolving it with a function f that is the 'square root' of the desired correlation function, i.e. $f * f = \xi(r)$. The squared Fourier transform of f is thus just the desired power spectrum; by the convolution theorem, this is also the power spectrum of the convolved white-noise field. From the computational point of view, this is not competitive with the FFT approach in terms of speed, although it may be simpler to code. However, it is often a useful concept to imagine random fields being generated in this two-step manner.

DENSITY MAXIMA To see the power of the Gaussian-field approach, consider the question of **density peaks**: regions of local maxima in $\delta(\mathbf{x})$. These are the points that will be the first to reach nonlinear density contrasts as the perturbation field develops. If we adopt the 'collapse at $\delta = \delta_c$' prescription from chapter 15, then the distribution of collapse redshifts can be deduced once we know the probability distribution for the value of $\delta(\mathbf{x})$ at the location of density maxima. This can be found as follows; we shall give only the 1D case in detail, because larger numbers of dimensions become messier to handle. See Peacock & Heavens (1985) and Bardeen *et al.* (1986; BBKS) for details of the 3D case. The field and its derivatives are jointly Gaussian

$$p(\delta, \delta', \delta'', \ldots) = \frac{|\mathbf{M}|^{1/2}}{(2\pi)^{m/2}} \exp(-\tfrac{1}{2}\mathbf{V}^{\rm T}\mathbf{M}\mathbf{V}), \tag{16.40}$$

for m variables in a vector $\mathbf{V}$; $\mathbf{M} \equiv \mathbf{C}^{-1}$ is the inverse of the **covariance matrix**. The elements of the covariance matrix depend only on moments over the power spectrum; in the particular case where our vector is $\mathbf{V}^{\mathrm{T}} = (\delta\, \delta''\, \delta')$, the covariance matrix is

$$\mathbf{C} = \begin{pmatrix} \sigma_0^2 & -\sigma_1^2 & \\ -\sigma_1^2 & \sigma_2^2 & \\ & & \sigma_1^2 \end{pmatrix}, \tag{16.41}$$

where

$$\sigma_m^2 \equiv \left(\frac{L}{2\pi}\right)^n \int |\delta_k|^2 k^{2m}\, d^n k. \tag{16.42}$$

This says that the field derivatives are independent of the values of the field and its second derivatives, but that the latter two quantities are correlated. We will be interested in points with $\delta' = 0$. At first sight, these may appear not to exist, since the probability that δ' is exactly zero vanishes. The way out of this paradox is to take the 'volume element' $d\delta\, d\delta''\, d\delta'$ and use the Jacobian to replace δ' by position, giving $d\delta\, d\delta''\, |\delta''|\, d^n x$; in n dimensions, $|\delta''|$ denotes the Jacobian determinant of the **Hessian matrix**, $\partial^2\delta/\partial x_i \partial x_j$. We then have only to integrate over the region of δ'' space corresponding to peaks, and we are done. To express the answers neatly, Bardeen *et al.* (1986) defined some (mainly dimensionless) parameters as follows:

$$\boxed{\nu \equiv \frac{\delta}{\sigma_0}, \quad \gamma \equiv \frac{\sigma_1^2}{\sigma_0 \sigma_2}, \quad R_* \equiv \sqrt{n}\,\frac{\sigma_1}{\sigma_2}.} \tag{16.43}$$

The meaning of these variables is as follows: ν is the 'height' of the field, in units of the rms; R_* is a measure of the coherence scale in the field; γ is a measure of the width of the power spectrum, with $\gamma = 1$ corresponding to a shell in k-space.

If we also put $x \equiv -\delta''/\sigma_2$, then the number density of stationary points in one dimension can be deduced immediately from the Gaussian distribution,

$$dN = \frac{e^{-Q/2}}{(2\pi)^{3/2}(1-\gamma^2)^{1/2}\, R_*}\, |x|\, dx\, d\nu, \qquad Q = \frac{(\nu - \gamma x)^2}{1-\gamma^2} + x^2, \tag{16.44}$$

with peaks corresponding to $x > 0$. Doing the ν integral first, followed by that over x, gives the total peak density as

$$N_{\rm pk} = \frac{1}{2\pi R_*}. \tag{16.45}$$

Doing the x-integral first is a tedious operation; nevertheless, the result can in fact be integrated again to yield the integral peak density (Cartwright & Longuet-Higgins 1956):

$$P(>\nu) = \frac{1}{2}\left(\operatorname{erfc}\left[\frac{\nu}{\sqrt{2(1-\gamma^2)}}\right] + \gamma e^{-\nu^2/2}\left\{1 + \operatorname{erf}\left[\frac{\gamma\nu}{\sqrt{2(1-\gamma^2)}}\right]\right\}\right). \tag{16.46}$$

The corresponding probability density for higher numbers of dimensions is quite easily deduced: the correlation between δ and δ'' means that, for very high peaks, all principal second derivatives tend to become equal and scale $\propto \nu$. It is then improbable that a stationary point is anything other than a peak, so the integration over second derivatives is unrestricted, leading to

$$N_{\rm pk}(>\nu) = \frac{1}{\sqrt{2\pi}} \left(\frac{\gamma}{\sqrt{2\pi} R_*} \right)^n \nu^{n-1} e^{-\nu^2/2}. \qquad (16.47)$$

In general, allowing for the constraint that all second derivatives must be negative is quite involved; details are given by Bardeen *et al.* (1986) and Bond & Efstathiou (1987). The final result can be expressed as

$$\frac{dN_{\rm pk}}{d\nu} = \frac{1}{(2\pi)^{(n+1)/2}} R_*^n e^{-\nu^2/2} G(\gamma, \gamma\nu), \qquad (16.48)$$

where

$$G(\gamma, \gamma\nu) = \int_0^\infty F(x) \, \frac{\exp\left[-\frac{1}{2}(x - \gamma\nu)^2/(1-\gamma^2)\right]}{[2\pi(1-\gamma^2)]^{1/2}} \, dx. \qquad (16.49)$$

In one dimension, $F(x) = x$ (Rice 1954), whereas $F(x) = x^2 + \exp(-x^2) - 1$ in two dimensions (Bond & Efstathiou 1987). $F(x)$ is given for $n = 3$ by equations (A1.9) and (4.5) of Bardeen *et al.* (1986). The integral densities in higher dimensions come out to be

$$\begin{aligned} N_{\rm pk} &= \frac{1}{4\pi\sqrt{3} R_*^2} \quad \text{(2D)} \\ N_{\rm pk} &= \frac{29 - 6\sqrt{6}}{8\pi^2 5^{3/2} R_*^3} \simeq 0.016 R_*^{-3} \quad \text{(3D)}. \end{aligned} \qquad (16.50)$$

The above results are illustrated in figure 16.2. We see that the height distribution for peaks is much narrower than the unconstrained positive tail of a Gaussian. Since we may use the spherical model to justify taking $\delta \simeq 1$ to mark the creation of a bound object, and since δ grows as $(1+z)^{-1}$, we can convert a distribution of ν to a distribution of collapse redshifts. The relatively small fractional dispersion in ν for peaks means that it is possible to speak of a reasonably well-defined epoch of galaxy formation.

CLUSTERING OF PEAKS Another important calculation that can be performed with density peaks is to estimate the clustering of cosmological objects. Peaks have some inbuilt clustering as a result of the statistics of the linear density field: they are 'born clustered'. For galaxies, this clustering amplitude is greatly altered by the subsequent dynamical evolution of the density field, but this is not true for clusters of galaxies, which are the largest nonlinear systems at the current epoch. We recognize clusters simply because they are the most spectacularly large galaxy systems to have undergone gravitational collapse; this has an important consequence, as first realized by Kaiser (1984). The requirement that these systems have become nonlinear by the present means that they must have been associated with particularly high peaks in the initial conditions. If we thus confine ourselves to peaks above some **density threshold** in ν, the statistical correlations can be very strong – especially for the richer clusters corresponding to high peaks.

The main effect is easy to work out, using the **peak–background split**. Here, one conceptually decomposes the density field into short-wavelength terms, which generate the peaks, plus terms of much longer wavelength, which modulate the peak number density. Consider the large-wavelength field as if it were some extra perturbation δ_+; if

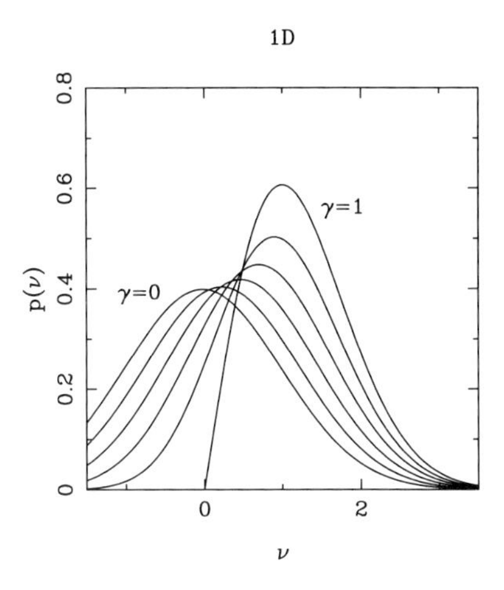

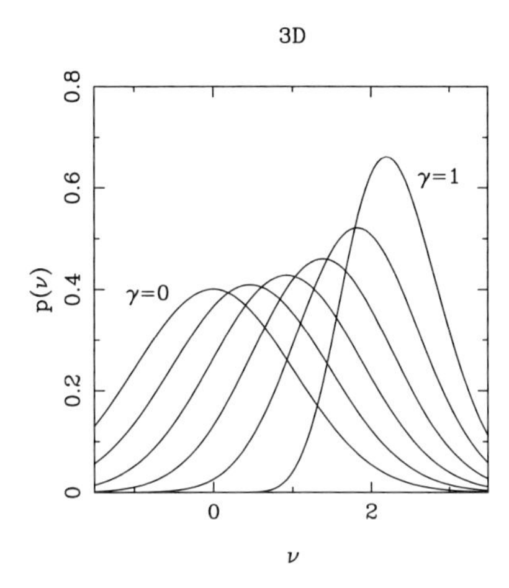

Figure 16.2. The distribution of the 'height' (relative to the rms) of maxima in Gaussian density fields, in one and three dimensions, as a function of the spectral parameter γ, for $\gamma = 0$, 0.2, 0.4, 0.6, 0.8, 1. As $\gamma \to 0$, we recover the unconstrained Gaussian case. For γ close to unity, most peaks are quite overdense ($\nu \simeq 2$) and the fractional dispersion in ν, which controls the dispersion in collapse times, is much smaller than in the unconstrained case.

we select all peaks above a threshold ν in the final field, this corresponds to taking all peaks above $\delta = \nu\sigma_0 - \delta_+$ in the initial field. This varying effective threshold will now produce more peaks in the regions of high δ_+, leading to amplification of the clustering pattern. For high peaks, $P(>\nu) \propto \nu^2 e^{-\nu^2/2}$; the exponential is the most important term, leading to a perturbation $\delta P/P \simeq \nu(\delta_+/\sigma_0)$. Hence, we obtain the high-peak amplification factor for the correlation function:

$$\boxed{\xi_{\rm pk}(r) \simeq \frac{\nu^2}{\sigma_0^2}\, \xi_{\rm mass}(r).} \tag{16.51}$$

It is important to realize that the process as described need have nothing to do with biased galaxy formation; it works perfectly well if galaxy light traces mass exactly in the universe. Clusters occur at special places in the mass distribution, so there is no reason to expect their correlations to be the same as those of the mass field.

In more detail, the exact clustering of peaks is just an extension of the calculation of the number density of peaks. We want to find the density of peaks of height ν_2 at a distance r from a peak of height ν_1. This involves a 6×6 covariance matrix for the fields and first and second derivatives even in 1D (20×20 in 3D). Moreover, most of the elements in this matrix are non-zero, so that the analytical calculation of ξ is sadly not feasible (see Lumsden, Heavens & Peacock 1989). However, a closely related calculation is easier to solve: the correlations of **thresholded regions**. Assume that objects form with unit probability in all regions whose density exceeds some threshold value, so that we need to deal with the correlation function of a modified density field that is constant above the threshold and zero elsewhere. This is

$$1 + \xi_{>\nu}(r) = \frac{1}{[P(>\nu)]^2} \int_\nu^\infty \int_\nu^\infty \frac{dx\, dy}{2\pi[1-\psi^2(r)]^{1/2}} \times \exp\left\{ -\frac{x^2 + y^2 - 2xy\psi(r)}{2[1-\psi^2(r)]} \right\} \qquad (16.52)$$

(Kaiser 1984). For high thresholds, this should be very close to the correlation function of peaks. The complete solution of this equation is given by Jensen & Szalay (1986) (see also Kashlinsky 1991 for the extension to the cross-correlation of fields above different thresholds). A good approximation, which extends Kaiser's original result, is

$$1 + \xi_{>\nu} \simeq \exp\left(1 + \frac{\nu^2}{\sigma_0^2} \xi_{\rm mass}\right). \qquad (16.53)$$

There remains the question of the inclusion of dynamics into the above treatment. As the density field evolves, density peaks will move from their initial locations, and the clustering will alter. The general problem is rather nasty (see Bardeen *et al.* 1986), but things are relatively straightforward in the linear regime where the mass fluctuations are small. If the statistical enhancement of correlations produces a fractional perturbation in the numbers of thresholded objects of $\delta_{\rm statistical} = f\,\delta_{\rm mass}$, then the effect of allowing weak dynamical evolution is just

$$\delta_{\rm obs} = \delta_{\rm statistical} + \delta_{\rm mass}. \qquad (16.54)$$

To see this, think of density perturbations arising as in the Zeldovich approximation, via objects moving closer together. Density peaks will be convected with the flow and compressed in number density in the same way as for any other particle. Thus, the effective enhancement ends up as $f \to f + 1$. We can deduce the value of f for Kaiser's model by looking at the expression for the correlation function in the limit of small correlations: $f \simeq \nu/\sigma_0$. So, for large-scale correlations of high peaks, we expect

$$\xi_{\rm pk} \simeq \left(1 + \frac{\nu}{\sigma_0}\right)^2 \xi_{\rm mass}. \qquad (16.55)$$

This idea of obtaining enhanced correlations by means of a threshold in density has been highly influential in cosmology. As well as the original application to clusters, attempts have also been made to use this mechanism to explain why galaxies might have clustering properties that differ from those of the mass. These issues are discussed in chapter 17.

APPLICATION TO GALAXY CLUSTERS This is the class of object that forms the main application of the peak clustering method. The rich **Abell clusters** were discussed in chapter 13; for those of richness class $R \geq 1$, the comoving density is $6 \times 10^{-6} h^3$ Mpc^{-3}. In order to model these systems as density peaks, it is necessary to specify a filter radius and a threshold; once we choose a filter radius to select cluster-sized fluctuations, the threshold is then fixed mainly by the number density (although altering the power-spectrum model also has a slight influence through γ). For Gaussian filtering, the conventional choice of R_f for clusters is $5h^{-1}$ Mpc. For $h = 1/2$ cold dark matter with $\gamma = 0.74$ on this scale, the required threshold is $\nu = 2.81$. These figures seem quite reasonable: Abell clusters are the rare high peaks of the mass distribution, and collapsed only recently. The reason for setting any threshold at all is the requirement of gravitational collapse by the present, so it is inevitable that $\nu \sim 1$.

The observations of the spatial correlations of clusters are somewhat controversial. The correlation function found by most workers is consistent with a scaled version of the galaxy function, $\xi = (r/r_0)^{-1.8}$, but values of r_0 vary. The original value found by Bahcall & Soneira (1983) was $25\,h^{-1}$ Mpc, but later work favoured values in the range 15–20 h^{-1} Mpc (Sutherland 1988; Dalton *et al.* 1992; Peacock & West 1992). The enhancement with respect to ξ for galaxies is thus a factor $\simeq 10$. Since σ_0 is close to unity for this smoothing, the simple asymptotic scaling would imply a threshold $\nu \simeq 3$, which is reasonable for moderately rare peaks.

16.4 Nonlinear clustering evolution

Observations of galaxy clustering extend into the highly nonlinear regime, $\xi \lesssim 10^4$, so it is essential to understand how this nonlinear clustering relates to the linear-theory initial conditions. A useful trick for dealing with this problem is to think of the density field under full nonlinear evolution as consisting of a set of collapsed, virialized clusters. What is the density profile of one of these objects? At least at separations smaller than the clump separation, the density profile of the clusters is directly related to the correlation function, since this just measures the number density of neighbours of a given galaxy. For a very steep cluster profile, $\rho \propto r^{-\epsilon}$, most galaxies will lie near the centres of clusters, and the correlation function will be a power law, $\xi(r) \propto r^{-\gamma}$, with $\gamma = \epsilon$. In general, because the correlation function is the convolution of the density field with itself, the two slopes differ. In the limit that clusters do not overlap, the relation is $\gamma = 2\epsilon - 3$ (for $3/2 < \epsilon < 3$; see Peebles 1974 or McClelland & Silk 1977). In any case, the critical point is that the correlation function may be be thought of as arising directly from the density profiles of clumps in the density field.

In this picture, it is easy to see how ξ will evolve with redshift, since clusters are virialized objects that do not expand. The hypothesis of **stable clustering** states that although the separation of clusters will alter as the universe expands, their internal density structure will stay constant with time. This hypothesis clearly breaks down in the outer regions of clusters, where the density contrast is small and linear theory applies, but it should be applicable to small-scale clustering. Regarding ξ as a density profile, its small-scale shape should therefore be fixed in *proper* coordinates, and its amplitude should scale as $(1+z)^{-3}$ owing to the changing mean density of unclustered galaxies, which dilute the clustering at high redshift. Thus, with $\xi \propto r^{-\gamma}$, we obtain the comoving evolution

$$\boxed{\xi(r,z) \propto (1+z)^{\gamma-3} \quad \text{(nonlinear).}} \tag{16.56}$$

Since the observed $\gamma \simeq 1.8$, this implies slower evolution than is expected in the linear regime:

$$\boxed{\xi(r,z) \propto (1+z)^{-2} g(\Omega) \quad \text{(linear).}} \tag{16.57}$$

This argument does not so far give a relation between the nonlinear slope γ and the index n of the linear spectrum. However, the linear and nonlinear regimes match at the scale of quasilinearity, i.e. $\xi(r_0) = 1$; each regime must make the same prediction for how this break point evolves. The linear and nonlinear predictions for the evolution of r_0 are respectively $r_0 \propto (1+z)^{-2/(n+3)}$ and $r_0 \propto (1+z)^{-(3-\gamma)/\gamma}$, so that $\gamma = (3n+9)/(n+5)$. In terms of an effective index $\gamma = 3 + n_{\rm NL}$, this becomes

$$\boxed{n_{\rm NL} = -\frac{6}{5+n}.} \tag{16.58}$$

The power spectrum resulting from power-law initial conditions will evolve self-similarly with this index. Note the narrow range predicted: $-2 < n_{\rm NL} < -1$ for $-2 < n < +1$, the $n = -2$ spectrum having the same shape in both linear and nonlinear regimes.

Indications from the angular clustering of faint galaxies (Efstathiou *et al.* 1991) and directly from redshift surveys (Le Fèvre *et al.* 1996) are that the observed clustering of galaxies evolves at about the linear-theory rate, rather more rapidly than the scaling solution would indicate. However, any interpretation of such data needs to assume that galaxies are unbiased tracers of the mass, whereas the observed high amplitude of clustering of quasars at $z \simeq 1$ ($r_0 \simeq 7\,h^{-1}$ Mpc; see Shanks *et al.* 1987, Shanks & Boyle 1994) warns that at least some high-redshift objects have clustering that is apparently not due to gravity alone. See chapter 17 for further discussion.

For many years it was thought that only these limiting cases of extreme linearity or nonlinearity could be dealt with analytically, but in a marvellous piece of ingenuity, Hamilton *et al.* (1991; HKLM) suggested a general way of understanding the linear $\leftrightarrow$ nonlinear mapping. The conceptual basis of their method can be understood with reference to the spherical collapse model. For $\Omega = 1$, a spherical clump virializes at a density contrast of order 100 when the linear contrast is of order unity. The trick now is to think about the density contrast in two distinct ways. To make a connection with the statistics of the density field, the correlation function $\xi(r)$ may be taken as giving a typical clump profile. What matters for collapse is that the integrated overdensity within a given radius reaches a critical value, so one should work with the volume-averaged correlation function $\bar{\xi}(r)$:

$$\bar{\xi}(R) \equiv \frac{3}{4\pi R^3} \int_0^R \xi(r)\, 4\pi r^2\, dr. \tag{16.59}$$

A density contrast of $1+\delta$ can also be thought of as arising through collapse by a factor $(1+\delta)^{1/3}$ in radius, which suggests that a given nonlinear correlation $\bar{\xi}_{\rm NL}(r_{\rm NL})$ should be thought of as resulting from linear correlations on a linear scale:

$$r_{\rm L} = [1 + \bar{\xi}_{\rm NL}(r_{\rm NL})]^{1/3} r_{\rm NL}. \tag{16.60}$$

This is the first part of the **HKLM procedure**. Once this translation of scales has been performed, the second step is to conjecture that the nonlinear correlations are a universal function of the linear ones:

$$\bar{\xi}_{\rm NL}(r_{\rm NL}) = f_{\rm NL}\big(\bar{\xi}_{\rm L}(r_{\rm L})\big). \tag{16.61}$$

The asymptotics of the function can be deduced readily. For small arguments $x \ll 1$, $f_{\rm NL}(x) \simeq x$; the spherical collapse argument suggests $f_{\rm NL}(1) \simeq 10^2$. Following collapse, $\bar{\xi}_{\rm NL}$ depends on scale factor as a^3 (stable clustering), whereas $\bar{\xi}_{\rm L} \propto a^2$; the large-$x$ limit is therefore $f_{\rm NL}(x) \propto x^{3/2}$. HKLM deduced from numerical experiments a numerical fit that interpolated between these two regimes, in a manner that empirically showed negligible dependence on the power spectrum.

However, these equations are often difficult to use stably for numerical evaluation; it is better to work directly in terms of power spectra. The key idea here is that $\bar{\xi}(r)$ can often be thought of as measuring the power at some effective wavenumber: it is obtained as an integral of the product of $\Delta^2(k)$, which is often a rapidly rising function, and a window function that cuts off rapidly at $k \gtrsim 1/r$:

$$\bar{\xi}(r) = \Delta^2(k_{\rm eff}), \qquad k_{\rm eff} \simeq 2/r, \tag{16.62}$$

where n is the effective power-law index of the power spectrum. This approximation for the effective wavenumber is within 20 per cent of the exact answer over the range $-2 < n < 0$. In most circumstances, it is therefore an excellent approximation to use the HKLM formulae directly to scale wavenumbers and powers:

$$\boxed{\Delta^2_{\rm NL}(k_{\rm NL}) = f_{\rm NL}\left(\Delta^2_{\rm L}(k_{\rm L})\right), \quad k_{\rm L} = [1 + \Delta^2_{\rm NL}(k_{\rm NL})]^{-1/3} k_{\rm NL}.} \tag{16.63}$$

What about models with $\Omega \neq 1$? The argument that leads to the $f_{\rm NL}(x) \propto x^{3/2}$ asymptote in the nonlinear transformation is just that linear and nonlinear correlations behave as a^2 and a^3 respectively following collapse. If collapse occurs at high redshift, then $\Omega = 1$ may be assumed at that time, and the nonlinear correlations still obey the a^3 scaling to low redshift. All that has changed is that the linear growth is suppressed by some Ω-dependent factor $g(\Omega)$. According to Carroll, Press & Turner (1992), the required factor may be approximated almost exactly by

$$g(\Omega) = \tfrac{5}{2}\Omega_m \left[\Omega_m^{4/7} - \Omega_v + (1 + \tfrac{1}{2}\Omega_m)(1 + \tfrac{1}{70}\Omega_v)\right]^{-1}. \tag{16.64}$$

where we have distinguished matter (m) and vacuum (v) contributions to the density parameter explicitly. It then follows that the large-x asymptote of the nonlinear function is

$$f_{\rm NL}(x) \propto [g(\Omega)]^{-3}\, x^{3/2}. \tag{16.65}$$

This says that the amplitude of highly nonlinear clustering is greater for low-density universes, which is as expected from chapter 15, where it was shown that collapsed systems virialize at a greater density contrast in low-density universes.

The suggestion of HKLM was that $f_{\rm NL}$ might be independent of the form of the linear spectrum, based on N-body data from Efstathiou *et al.* (1988b), but Jain, Mo & White (1995) showed that this is not true, especially when the linear spectrum is rather flat ($n \lesssim -1.5$). Peacock & Dodds (1996) suggested that the HKLM method should be generalized by using the following fitting formula for the n-dependent nonlinear function (strictly, the one that applies to the power spectrum, rather than to $\bar{\xi}$):

$$f_{\rm NL}(x) = x \left\{ \frac{1 + B\beta x + (Ax)^{\alpha\beta}}{1 + [(Ax)^{\alpha} g^3(\Omega)/(V x^{1/2})]^{\beta}} \right\}^{1/\beta}. \tag{16.66}$$

B describes a second-order deviation from linear growth; A and α parameterize the power law that dominates the function in the quasilinear regime; V is the virialization parameter that gives the amplitude of the $f_{\rm NL}(x) \propto x^{3/2}$ asymptote; β softens the transition between these regimes. An excellent fit to the N-body data (illustrated in figure 16.3) is given by

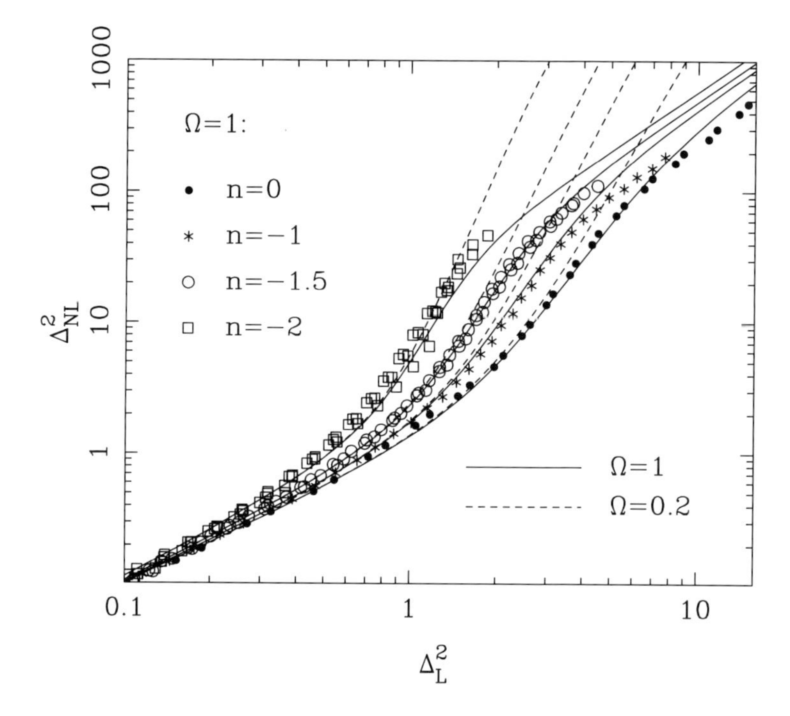

Figure 16.3. The generalization of the HKLM function relating nonlinear power to linear power, for the cases $n = 0, -1, -1.5$ and -2. Data points are shown for the case $\Omega = 1$ only, with the corresponding fitting formulae shown as solid lines. This diagram clearly displays three regimes: (i) linear ($\Delta^2_{NL} \lesssim 1$); (ii) quasilinear ($1 \lesssim \Delta^2_{NL} \lesssim 100$); (iii) stable-clustering ($\Delta^2_{NL} \gtrsim 100$). For a given linear power, the nonlinear power increases for more negative n. There is also a greater nonlinear response in the case of an open universe with $\Omega = 0.2$, indicated by the broken lines. The fitting formula is shown for models with zero vacuum energy only, but what matters in general is the Ω-dependent linear growth suppression factor $g(\Omega)$.

the following spectrum dependence of the expansion coefficients:

$$\begin{aligned} A &= 0.542\,(1+n/3)^{-0.685} \\ B &= 0.097\,(1+n/3)^{-0.224} \\ \alpha &= 3.235\,(1+n/3)^{-0.236} \\ \beta &= 0.659\,(1+n/3)^{-0.356} \\ V &= 11.54\,(1+n/3)^{-0.371}. \end{aligned} \tag{16.67}$$

The more general case of curved spectra can be dealt with very well by using the tangent spectral index at each linear wavenumber:

$$n_{\rm eff} \equiv \frac{d\ln P}{d\ln k}, \tag{16.68}$$

although evaluating $n_{\rm eff}$ at a wavenumber of $k/2$ gives even better results.

Note that the cosmological model does not enter anywhere in these parameters. It is present in the fitting formula only through the growth factor g, which governs the amplitude of the virialized portion of the spectrum. This says that all the quasilinear features of the power spectrum are independent of the cosmological model, and only know about the overall level of power. This is not surprising, given that quasilinear evolution is well described by the Zeldovich approximation, in which the final positions of particles are obtained by extrapolating their initial displacements. In this case, the final density field depends only on the total size of the displacement, not on its growth history. Models in which the size of the linear displacement field are normalized to be equal have the same nonlinear effects, which are completely independent of Ω.

EVOLUTION OF CLUSTERING The discussion of nonlinear evolution has revealed that in practice the regime $1 \lesssim \Delta^2 \lesssim 100$ is dominated by the steep **quasilinear transition** where $f_{\rm NL}(x) \propto x^\alpha$, $\alpha \simeq 3.5$–4.5. This turns out to predict a rate of evolution that is very different from the extremes of linear evolution or stable clustering. For $\Delta^2 \gg 1$, the transitional spectrum scales as

$$\begin{aligned} k_{\rm NL} &\simeq [\Delta^2_{\rm NL}]^{1/3} k_{\rm L} \\ \Delta^2_{\rm NL} &\propto [D^2(a)\,\Delta^2_{\rm L}]^{1+\alpha}, \end{aligned} \tag{16.69}$$

where $D(a)$ is the linear growth law for density perturbations. For a power-law linear spectrum, $\Delta^2 \propto k^{3+n}$, this predicts a quasilinear power law

$$\Delta^2_{\rm NL} \propto D^{(6-2\gamma)(1+\alpha)/3}\, k^\gamma_{\rm NL}, \tag{16.70}$$

where the nonlinear power-law index depends as follows on the slope of the linear spectrum:

$$\gamma = \frac{3(3+n)(1+\alpha)}{3+(3+n)(1+\alpha)}. \tag{16.71}$$

For the observed index of $\gamma \simeq 1.8$, this would require $n \simeq -2.2$, very different from the $n = 0$ that would give $\gamma = 1.8$ in the virialized regime.

We can now summarize the rate of evolution of clustering in the three different regimes:

$$\begin{array}{ll} \text{linear} & \xi(r,z) \propto [D(z)]^2 \\ \text{quasilinear} & \xi(r,z) \propto [D(z)]^{(6-2\gamma)(1+\alpha)/3} \\ \text{nonlinear} & \xi(r,z) \propto (1+z)^{-(3-\gamma)}, \end{array} \tag{16.72}$$

where γ is the power-law slope in the relevant regime. For $\alpha = 4$, $\gamma = 1.7$, this gives $\xi \propto D^{4.3}$ for the quasilinear evolution; this is more than twice as fast as the linear evolution, and over three times the rate of stable-clustering evolution if $\Omega = 1$, so that $D(z) = 1/(1+z)$. The conclusion is that clustering in the regime where most data exist is expected to evolve very rapidly with redshift, unless Ω is low. We discuss below whether this effect has been seen.

16.5 Redshift-space effects

Although a huge amount of astronomical effort has been invested in galaxy redshift surveys, with the aim of mapping the three-dimensional galaxy distribution, the results are not a true 3D picture. The radial axis of redshift is modified by the Doppler effects of peculiar velocity: $1 + z \to (1 + z)(1 + v/c)$. Since the peculiar velocities arise from the clustering itself, the apparent clustering pattern in **redshift space**, where redshift is assumed to arise from ideal Hubble expansion only, differs systematically from that in **real space**. In the following discussion, it will often be necessary to distinguish quantities measured in redshift space from their real-space counterparts; this will be accomplished by using an s subscript for redshift space and an r subscript for real space.

The distortions caused by working in redshift space are relatively simple to analyse if we assume we are dealing with a distant region of space that subtends a small angle, so that radial distortions can be considered as happening along one Cartesian axis. In this case, the apparent amplitude of any linear density disturbance is readily obtained from the usual linear relation between displacement and velocity (Kaiser 1987). Chapter 15 showed that the relation between the linear displacement field $\mathbf{x}$ and the velocity field $\mathbf{u}$ is $\mathbf{u} = Hf(\Omega)\,\mathbf{x}$, where the function $f(\Omega) \simeq \Omega^{0.6}$ is the well-known velocity-suppression factor due to Peebles, which is in practice a function of Ω_m only, with negligible dependence on the vacuum density (Lahav *et al.* 1991). If we work in redshift space, then

$$\mathbf{r}_{\text{apparent}} = \mathbf{r} + (\hat{\mathbf{r}} \cdot \mathbf{u}/H)\,\hat{\mathbf{r}} = \mathbf{r} + (\mu u/H)\,\hat{\mathbf{r}}, \tag{16.73}$$

where μ is the cosine of the angle between the velocity vector and the line of sight ($\mu = \hat{\mathbf{r}} \cdot \hat{\mathbf{k}}$). Now consider a plane-wave disturbance running at some angle to the line of sight, producing a displacement field $\mathbf{x}$ parallel to $\mathbf{k}$. The apparent displacement produced is $\mathbf{x} + f(\Omega)\mu x\,\hat{\mathbf{r}}$, but what counts in determining the apparent amplitude of the density disturbance is the component of this along the wavevector, which is $x + f(\Omega)\mu^2 x$. Provided that gradients in the mean space density caused by observational selection can be ignored, we saw in discussing the Zeldovich approximation in chapter 15 that the apparent density perturbation due to this mode is directly proportional to the amplitude of the apparent displacement. The redshift-space and real-space density fields are therefore related via $\delta_s^{\text{mass}} = \delta_r^{\text{mass}}[1 + f(\Omega)\mu^2]$.

As usual, what is observable is not the perturbation to the mass, but the perturbation to the light. The relation between light and mass is determined by complex physical processes discussed in chapter 17 and below. However, it is reasonable to assume that these effects will lead to some linear response of the galaxy-formation process to small density perturbations, so that in real space we should be able to write

$$\delta^{\text{light}} = b\,\delta^{\text{mass}} = \delta^{\text{mass}} + (b-1)\,\delta^{\text{mass}}, \tag{16.74}$$

whenever the fluctuations are small. The trivial rearrangement emphasizes that the observed density fluctuation must be a mixture of the dynamically generated density fluctuation, plus the additional term due to bias, which populates different regions of space in different ways. The first term is associated with peculiar velocities, but the second is not – the enhancements in galaxy densities are just some additional pattern. In redshift space, we therefore add the anisotropic perturbations due to the dynamical component to the isotropic biased component, obtaining

$$\boxed{\delta_s^{\rm light} = \delta_r^{\rm mass}\left[1 + f(\Omega)\mu^2\right] + (b-1)\,\delta_r^{\rm mass} = \delta_r^{\rm light}\left[1 + \frac{f(\Omega)\mu^2}{b}\right].} \tag{16.75}$$

Redshift-space effects thus give a characteristic anisotropy of clustering, which can be used to measure the parameter

$$\boxed{\beta \equiv \Omega^{0.6}/b.} \tag{16.76}$$

The way to achieve this is to write the redshift-space power spectrum as

$$\frac{P_s}{P_r} = (1+\beta\mu^2)^2 = \left(1 + \frac{2}{3}\beta + \frac{1}{5}\beta^2\right) + \left(\frac{4}{3}\beta + \frac{4}{7}\beta^2\right) P_2(\mu) + \frac{8}{35}\beta^2 P_4(\mu), \tag{16.77}$$

where P_2 and P_4 are two of the orthogonal zero-mean Legendre polynomials. The main effects are (i) a boost in average power; (ii) a quadrupole term. The ratio of the quadrupole to the monopole is a monotonic function of β that rises from zero at $\beta = 1$ to just over unity at $\beta = 1$ (see Hamilton 1993a for the real-space equivalent of this analysis). Both these effects provide a possible way to estimate β, and recent results have tended towards a consensus figure of around $\beta = 0.5$ for optically-selected galaxies (Cole, Fisher & Weinberg 1994, 1995; Loveday *et al.* 1996; Ratcliffe *et al.* 1996). A practical complication is that these results do not apply to surveys of large solid angle, since μ then varies between galaxies; a fuller analysis is given by Heavens & Taylor (1995).

On small scales, this linear analysis is not valid. The main effect here is to reduce power through the radial smearing due to virialized motions and the associated 'finger of God' effect. This is hard to treat exactly because of the small-scale velocity correlations. A simplified model treats the small-scale velocity field as an incoherent Gaussian scatter with 1D rms dispersion σ. This turns out to be quite a reasonable approximation, because the observed pairwise velocity dispersion is a very slow function of separation, and is all the more accurate if the redshift data are afflicted by significant measurement errors (which should be included in σ). This model is just a radial convolution, and so the k-space effect is

$$\delta_k \to \delta_k \, \exp\left(-k^2\mu^2\sigma^2/2\right). \tag{16.78}$$

A slightly better alternative is to use the observation that the pairwise distribution of velocity differences is well fitted by an exponential (probably a superposition of Gaussians of different widths from different clumps). For this, the k-space damping function is a Lorentzian:

$$\delta_k \to \delta_k \, \left(1 + k^2\mu^2\sigma^2/2\right)^{-1}. \tag{16.79}$$

The overall power is then

$$\frac{P_s}{P_r} = \frac{(1+\beta\mu^2)^2}{(1+k^2\mu^2\sigma^2/2)^2}. \tag{16.80}$$

Note that both the damping factor and the $(1+\beta\mu^2)^2$ Kaiser factor are anisotropic in k-space and they interfere when averaging to get the mean power. What happens in this case is that the linear Kaiser boost to the power is lost at large k, where the result tends to $P_s/P_r = 2^{-3/2}\pi(k\sigma)^{-1}$ (because the main contribution at large k comes from small μ). The overall power is thus only damped mildly, by one power of k.

In practice, the relevant value of σ to choose is approximately $1/\sqrt{2}$ times the **pairwise velocity dispersion** σ_p seen in galaxy redshift surveys (to this should be added in quadrature any errors in measured velocities). The dispersion is measured in practice by comparing the correlation function measured in the radial direction with that measured transversely (see below); at $1h^{-1}$ Mpc separation, the pairwise dispersion is approximately

$$\sigma_p \simeq 300 - 400\,\mathrm{km\,s^{-1}} \tag{16.81}$$

(Davis & Peebles 1983; Mo, Jing & Börner 1993; Fisher *et al.* 1994). We therefore expect wavenumbers $k \gtrsim 0.3\,h\,\mathrm{Mpc^{-1}}$ to be seriously affected by redshift-space smearing.

REAL-SPACE CLUSTERING It is possible to avoid the complications of redshift space. One can deal with pure two-dimensional projected clustering, as discussed in the next section. Alternatively, peculiar velocities may be dealt with by using the correlation function evaluated explicitly as a 2D function of transverse ($r_\perp$) and radial ($r_\parallel$) separation. Integrating along the redshift axis then gives the **projected correlation function**, which is independent of the velocities:

$$\boxed{w_p(r_\perp) \equiv \int_{-\infty}^{\infty} \xi(r_\perp, r_\parallel)\, dr_\parallel = 2\int_{r_\perp}^{\infty} \xi(r)\, \frac{r\,dr}{(r^2 - r_\perp^2)^{1/2}}.} \tag{16.82}$$

In principle, this statistic can be used to recover the real-space correlation function by using the inverse relation for the **Abel integral equation**:

$$\xi(r) = -\frac{1}{\pi}\int_r^{\infty} w_p'(y)\, \frac{dy}{(y^2 - r^2)^{1/2}}. \tag{16.83}$$

An alternative notation for the projected correlation function is $\Xi(r_\perp)$ (Saunders, Rowan-Robinson & Lawrence 1992). Note that the projected correlation function is not dimensionless, but has dimensions of length. The quantity $\Xi(r_\perp)/r_\perp$ is more convenient to use in practice as the projected analogue of $\xi(r)$.

The reason that $w_p(r_\perp)$ is independent of redshift-space distortions is that peculiar velocities simply move pairs of points in $r_\parallel$, but not in $r_\perp$, and the expected pair count is just proportional to $2\pi r_\perp\, dr_\perp\, dr_\parallel$. Suppose we ignore the linear-theory velocities (which are more easily treated in Fourier space as above), and just consider the effect of a small-scale velocity dispersion. The correlation function is then convolved in the radial direction:

$$\begin{aligned}\xi(r_\perp, r_\parallel) &= \int_{-\infty}^{\infty} \xi_{\rm true}(r_\perp, r)\, f(r_\parallel - r)\, dr \\ &= \frac{r_0^\gamma}{\sqrt{2\pi}\,\sigma_v}\int_{-\infty}^{\infty} [r_\perp^2 + (r_\parallel - x)^2]^{-\gamma/2}\, e^{-x^2/2\sigma_v^2}\, dx,\end{aligned} \tag{16.84}$$

where the latter expression applies for power-law clustering and a Gaussian velocity dispersion. Looking at the function in the redshift direction thus allows the pairwise velocity dispersion to be estimated; this is the origin of the above estimate of σ_p. See Fisher (1995) for more discussion of this method.

Sometimes these complications are neglected, and correlations are calculated in redshift space assuming isotropy. The result is a small increase in scale length, as power on small scales is transferred to separations of order the velocity smearing. The result is a scale length around $7h^{-1}$ Mpc for the redshift-space $\xi(s)$ as opposed to the $5h^{-1}$ Mpc that applies for $\xi(r)$.

16.6 Low-dimensional density fields

It is common to encounter datasets that are projections of a 3D density field, and there are various tools that are used to extract the desired 3D information.

SKEWERS AND SLICES The simplest case is a skewer, where we have data along a line in 3D space. The 1D power spectrum becomes a projection of the 3D power: in terms of the variance per logarithmic interval of wave number $\Delta^2_{1D} \equiv d\sigma^2_{1D}/d\ln k = kLP_{1D}/\pi$ (note the factor 2 from $\pm k$ in one dimension), the result is [problem 16.5]

$$\Delta^2_{1D}(k) = k \int_k^\infty \Delta^2_{3D}(y) y^{-2} dy. \tag{16.85}$$

As expected, 1D modes of wavevector k receive contributions from all 3D modes with wavevectors $\leq k$; this just says that a long-wavelength disturbance along a skewer can be produced by a short-wavelength mode directed nearly perpendicular to the skewer. An interesting point is that P_{3D} is just proportional to the derivative of P_{1D}; the latter must therefore be a monotonically decreasing function, since P_{3D} is positive definite.

Provided the 3D power spectrum has an effective index $n > -2$ as $k \to 0$, and assuming the integral converges at large k, the large-wavelength terms in the 1D power spectrum are dominated by power projected from short wavelengths in 3D, and are highly insensitive to the true 3D power on large scales for most spectra. The 1D power goes to a constant as $k \to 0$, so the power spectrum in the skewer is Poisson in form, indicating that the power is dominated by the quasi-independent positions of the dense small-scale clumps. This projection of spurious large-scale power is an inevitable effect in lower-dimensional redshift surveys and will always weaken any conclusions that can be drawn concerning large-scale density perturbations. The effect for a two-dimensional slice is similar to that in one dimension:

$$\Delta^2_{2D} = k^2 \int_k^\infty \Delta^2_{3D}(y) \frac{y^{-2} dy}{\sqrt{y^2 - k^2}}. \tag{16.86}$$

Again, an effective $n = 0$ portion is produced at low k.

To treat properly the realistic case where the skewer or slice has a non-zero thickness, the above formulae are modified as follows:

$$\Delta^2_{3D}(y) \to \Delta^2_{3D}(y) |\tilde{W}|^2 \left(\sqrt{y^2 - k^2}\right), \tag{16.87}$$

where $\tilde{W}$ is the 1D Fourier transform of the convolving function used to thicken the skewer or slice. For example, if a skewer is thickened to a cylinder of radius R, $\tilde{W}(x) = 2J_1(xR)/(xR)$; if a slice is thickened to a slab of depth $2L$, $\tilde{W}(x) = \sin(xL)/(xL)$.

As an illustration of these points, consider the redshift surveys compiled by

Broadhurst *et al.* (1990). These had a geometry that was roughly a tube of diameter $6h^{-1}$ Mpc, over a very large baseline. The redshift histogram appeared strikingly periodic, with a high spike in the 1D power spectrum at a wavelength of $128h^{-1}$ Mpc. This observation caused great excitement, apparently challenging fundamental concepts about galaxy clustering and homogeneity. However, from the formula for Δ^2_{1D}, we see that the power at large wavelengths will be dominated by the 3D power around the tube size (which cuts off the integral). This is of order unity, so the appearance of enormous large-scale power is readily understood. The lack of small-scale 1D power, which makes the spike all the more impressive, is to be attributed to a combination of redshift errors and random velocities (Kaiser & Peacock 1991). The conclusion is that quantitative investigations of clustering on some scale λ require a fully 3D sample, which is at least λ in its shortest dimension. Nevertheless, pencil beams are a valuable probe of large-scale structure because of what they tell us about the *texture* of the clustering pattern. They emphasize the predominance of voids and the concentration of galaxies towards high-density regions.

PROJECTION ON THE SKY A more common situation is where we lack any distance data; we then deal with a projection on the sky of a magnitude-limited set of galaxies at different depths. The statistic that is observable is the angular correlation function, $w(\theta)$, or its angular power spectrum Δ^2_θ. If the sky were flat, the relation between these would be the usual **Hankel transform** pair:

$$\boxed{\begin{aligned} w(\theta) &= \int_0^\infty \Delta^2_\theta \, J_0(K\theta)\, dK/K \\ \Delta^2_\theta &= K^2 \int_0^\infty w(\theta)\, J_0(K\theta)\, \theta\, d\theta. \end{aligned}} \tag{16.88}$$

For power-law clustering, $w(\theta) = (\theta/\theta_0)^{-\epsilon}$, this gives

$$\Delta^2_\theta(K) = (K\theta_0)^\epsilon \, 2^{1-\epsilon} \, \frac{\Gamma(1-\epsilon/2)}{\Gamma(\epsilon/2)}, \tag{16.89}$$

which is equal to $0.77(K\theta_0)^\epsilon$ for $\epsilon = 0.8$. At large angles, these relations are not quite correct. We should really expand the sky distribution in **spherical harmonics**:

$$\delta(\hat{\mathbf{q}}) = \sum a_\ell^m \, Y_{\ell m}(\hat{\mathbf{q}}), \tag{16.90}$$

where $\hat{\mathbf{q}}$ is a unit vector that specifies direction on the sky. The functions $Y_{\ell m}$ are the eigenfunctions of the angular part of the ∇^2 operator: $Y_{\ell m}(\theta, \phi) \propto \exp(im\phi) P_\ell^m(\cos\theta)$, where P_ℓ^m are the **associated Legendre polynomials** (see e.g. section 6.8 of Press *et al.* 1992). Since the spherical harmonics satisfy the orthonormality relation $\int Y_{\ell m} Y^*_{\ell' m'} \, d^2q = \delta_{\ell\ell'}\delta_{mm'}$, the inverse relation is

$$a_\ell^m = \int \delta(\hat{\mathbf{q}}) \, Y^*_{\ell m} \, d^2q. \tag{16.91}$$

The analogues of the Fourier relations for the correlation function and power spectrum are

$$\boxed{\begin{aligned} w(\theta) &= \frac{1}{4\pi}\sum_\ell \sum_{m=-\ell}^{m=+\ell} |a_\ell^m|^2\, P_\ell(\cos\theta) \\ |a_\ell^m|^2 &= 2\pi \int_{-1}^{1} w(\theta)\, P_\ell(\cos\theta)\, d\cos\theta. \end{aligned}} \tag{16.92}$$

For small θ and large ℓ, these go over to a form that looks like a flat sky, as follows. Consider the asymptotic forms for the Legendre polynomials and the J_0 Bessel function:

$$\begin{aligned} P_\ell(\cos\theta) &\simeq \sqrt{\frac{2}{\pi\ell \sin\theta}}\, \cos\left[\left(\ell + \tfrac{1}{2}\right)\theta - \tfrac{1}{4}\pi\right] \\ J_0(z) &\simeq \sqrt{\frac{2}{\pi z}}\, \cos\left[z - \tfrac{1}{4}\pi\right], \end{aligned} \tag{16.93}$$

for respectively $\ell \to \infty$, $z \to \infty$; see chapters 8 and 9 of Abramowitz & Stegun 1965. This shows that, for $\ell \gg 1$, we can approximate the small-angle correlation function in the usual way in terms of an angular power spectrum Δ_θ^2 and angular wavenumber K:

$$w(\theta) = \int_0^\infty \Delta_\theta^2(K)\, J_0(K\theta)\, \frac{dK}{K}, \qquad \Delta_\theta^2\left(K = \ell + \tfrac{1}{2}\right) = \frac{2\ell+1}{8\pi} \sum_m |a_\ell^m|^2. \tag{16.94}$$

An important relation is that between the angular and spatial power spectra. In outline, this is derived as follows. The perturbation seen on the sky is

$$\delta(\hat{\mathbf{q}}) = \int_0^\infty \delta(\mathbf{y})\, y^2 \phi(y)\, dy, \tag{16.95}$$

where $\phi(y)$ is the **selection function**, normalized such that $\int y^2\phi(y)\, dy = 1$, and y is comoving distance. The function ϕ is the comoving density of objects in the survey, which is given by the integrated luminosity function down to the luminosity limit corresponding to the limiting flux of the survey seen at different redshifts; a flat universe ($\Omega = 1$) is assumed for now. Now write down the Fourier expansion of δ. The plane waves may be related to spherical harmonics via the expansion of a plane wave in **spherical Bessel functions** $j_\ell(x) = (\pi/2x)^{1/2} J_{n+1/2}(x)$ (see chapter 10 of Abramowitz & Stegun 1965 or section 6.7 of Press *et al.* 1992):

$$e^{ikr\cos\theta} = \sum_0^\infty (2\ell+1)\, i^\ell\, P_\ell(\cos\theta)\, j_\ell(kr), \tag{16.96}$$

plus the spherical harmonic addition theorem

$$P_\ell(\cos\theta) = \frac{4\pi}{2\ell+1} \sum_{m=-\ell}^{m=+\ell} Y_{\ell m}^*(\hat{\mathbf{q}}) Y_{\ell m}(\hat{\mathbf{q}}'), \tag{16.97}$$

where $\hat{\mathbf{q}} \cdot \hat{\mathbf{q}}' = \cos\theta$. These relations allow us to take the angular correlation function $w(\theta) = \langle \delta(\hat{\mathbf{q}})\delta(\hat{\mathbf{q}}')\rangle$ and transform it to give the angular power spectrum coefficients. The actual manipulations involved are not as intimidating as they may appear, but they are left as an exercise and we simply quote the final result (Peebles 1973):

$$\langle |a_\ell^m|^2 \rangle = 4\pi \int \Delta^2(k)\, \frac{dk}{k} \left[\int y^2 \phi(y)\, j_\ell(ky)\, dy \right]^2 . \tag{16.98}$$

What is the analogue of this formula for small angles? Rather than manipulating large-ℓ Bessel functions, it is easier to start again from the correlation function. By writing as above the overdensity observed at a particular direction on the sky as a radial integral over the spatial overdensity, with a weighting of $y^2\phi(y)$, we see that the angular correlation function is

$$\langle \delta(\hat{\mathbf{q}}_1)\delta(\hat{\mathbf{q}}_2) \rangle = \iint \langle \delta(\mathbf{y}_1)\delta(\mathbf{y}_2) \rangle\; y_1^2 y_2^2 \phi(y_1)\phi(y_2)\, dy_1\, dy_2. \tag{16.99}$$

We now change variables to the mean and difference of the radii, $y \equiv (y_1 + y_2)/2$; $x \equiv (y_1 - y_2)$. If the depth of the survey is larger than any correlation length, we only get a signal when $y_1 \simeq y_2 \simeq y$. If the selection function is a slowly varying function, so that the thickness of the shell being observed is also of order the depth, the integration range on x may be taken as being infinite. For small angles, we then obtain **Limber's equation**:

$$w(\theta) = \int_0^\infty y^4 \phi^2\, dy \int_{-\infty}^{\infty} \xi\left(\sqrt{x^2 + y^2\theta^2} \right)\, dx \tag{16.100}$$

(see sections 51 and 56 of Peebles 1980). Theory usually supplies a prediction about the linear density field in the form of the power spectrum, and so it is convenient to recast Limber's equation:

$$w(\theta) = \int_0^\infty y^4 \phi^2\, dy \int_0^\infty \pi\, \Delta^2(k)\, J_0(ky\theta)\, dk/k^2. \tag{16.101}$$

The form $\phi \propto y^{-1/2} \exp[-(y/y^*)^2]$ is often taken as a reasonable approximation to the Schechter function, and this gives

$$w(\theta) = \frac{\pi}{2\Gamma^2\left(\frac{5}{4}\right)} \int_0^\infty \Delta^2(k)\, e^{-(k\theta y^*)^2/8} \left[1 - \tfrac{1}{8}(k\theta y^*)^2 \right] \frac{dk}{k^2 y^*}. \tag{16.102}$$

The power-spectrum version of Limber's equation is already in the form required for the relation to the angular power spectrum ($w = \int \Delta_\theta^2 J_0(K\theta)\, dK/K$), and so we obtain the direct small-angle relation between spatial and angular power spectra:

$$\Delta_\theta^2 = \frac{\pi}{K} \int \Delta^2\left(\frac{K}{y}\right)\, y^5 \phi^2(y)\, dy. \tag{16.103}$$

This is just a convolution in log space, and is considerably simpler to evaluate and interpret than the $w - \xi$ version of Limber's equation.

Finally, note that it is not difficult to make allowance for spatial curvature in the above discussion. Write the Robertson–Walker metric in the form

$$c^2 d\tau^2 = c^2 dt^2 - R^2 \left(\frac{dr^2}{1 - kr^2} + r^2\theta^2 \right); \tag{16.104}$$

for $k = 0$, the notation $y = R_0 r$ was used for comoving distance, where $R_0 = (c/H_0)|1 - \Omega|^{-1/2}$. The radial increment of comoving distance was $dx = R_0\, dr$, and the comoving distance between two objects was $(dx^2 + y^2\theta^2)^{1/2}$. To maintain this version of Pythagoras's theorem, we clearly need to keep the definition of y and redefine radial distance: $dx = R_0\, dr\, C(y)$, where $C(y) = [1 - k(y/R_0)^2]^{-1/2}$. The factor $C(y)$ appears in the non-Euclidean comoving volume element, $dV \propto y^2 C(y)\, dy$, so that we now require the normalization equation for ϕ to be

$$\int_0^\infty y^2 \phi(y)\, C(y)\, dy = 1. \tag{16.105}$$

The full version of Limber's equation therefore gains two powers of $C(y)$, but one of these is lost in converting between $R_0\, dr$ and dx:

$$w(\theta) = \int_0^\infty [C(y)]^2 y^4 \phi^2\, dy \int_{-\infty}^\infty \xi\left(\sqrt{x^2 + y^2\theta^2}\right) \frac{dx}{C(y)}. \tag{16.106}$$

The net effect is therefore to replace $\phi^2(y)$ by $C(y)\phi^2(y)$, so that the full power-spectrum equation is

$$\Delta_\theta^2 = \frac{\pi}{K} \int \Delta^2\left(\frac{K}{y}\right) C(y)\, y^5 \phi^2(y)\, dy. \tag{16.107}$$

It is also straightforward to allow for evolution. The power version of Limber's equation is really just telling us that the angular power from a number of different radial shells adds incoherently, so we just need to use the actual evolved power at that redshift. These integral equations can be inverted numerically to obtain the real-space 3D clustering results from observations of 2D clustering; see Baugh & Efstathiou (1993; 1994).

SCALING One important consequence follows from the above formula for the angular power spectrum. Suppose we increase the depth of the survey by some factor D, but keep the functional form of $\phi(y)$ the same. The angular clustering therefore scales as

$$\Delta'^2_\theta(K) = \frac{1}{D} \Delta_\theta^2\left(\frac{K}{D}\right); \tag{16.108}$$

i.e. the clustering pattern moves to smaller angles, reducing the observed clustering, and is also diluted by a further factor $1/D$. This test is an excellent way of discriminating true spatial clustering from spurious effects caused by foreground extinction or magnitude errors. Prior to redshift surveys, it was the only means of establishing galaxy clustering as a real effect.

16.7 Measuring the clustering spectrum

CORRELATION-FUNCTION ESTIMATION The correlation function is usually estimated by starting from its definition as the excess probability of finding a pair,

$$dP = \rho_0^2\, [1 + \xi(r)]\, dV_1 dV_2. \tag{16.109}$$

To implement this definition, we need to calculate the expected numbers of pairs in the absence of clustering, taking into account the sample limits. In practice, this number is usually estimated by creating a random catalogue much larger in number than the sample under study, and by counting pairs either within the two catalogues or between catalogues, giving a variety of possible estimators for $1+\xi$ as a ratio of pair counts, usually symbolized as follows:

$$\begin{aligned} 1+\xi_1 &= \langle DD\rangle/\langle RR\rangle \\ 1+\xi_2 &= \langle DD\rangle/\langle DR\rangle \\ 1+\xi_3 &= \langle DD\rangle\langle RR\rangle/\langle DR\rangle^2 \\ 1+\xi_4 &= 1+\langle (D-R)^2\rangle/\langle RR\rangle. \end{aligned} \tag{16.110}$$

In all cases one is effectively just measuring the expected number of pairs by monte-carlo integration; within a bin covering a specified range of radii, the observed number of pairs is counted, and $1+\xi$ is the ratio of this number to the number expected. To be explicit, suppose there are respectively n and m objects in the real and mock catalogues. There are $n(n-1)/2$ distinct pairs in the data, but nm cross-pairs between the data and the random catalogue. So, considering ξ_2 as an example, if DD stands for the number of distinct data pairs in a given range of r, DR stands for the number of cross pairs in the same range, multiplied by $n/2m$. For a large volume, all estimators should be equivalent; however, for a small sample some will be less robust, since the expected pair count may be sensitive to whether there is a rich cluster close to the sample boundary. Many authors have preferred to use ξ_2, but Hamilton (1993b) has shown that ξ_3 and ξ_4 are superior. Probably a sensible approach in practice is to try several estimators: if they differ greatly, then the dataset being used may simply not be up to the job.

Having made an estimate of $\xi(r)$, we now need to consider error bars. In the absence of clustering, $\langle\xi\rangle=0$ and $\langle\xi^2\rangle=1/N_p$, where N_p is the number of independent pairs in a given bin of radius (Peebles 1980). For non-zero ξ, this suggests the commonly used 'Poisson error bar'

$$\frac{\Delta\xi}{1+\xi}=\frac{1}{\sqrt{N_p}}. \tag{16.111}$$

This will usually be a lower limit to the uncertainty in ξ; Peebles (1973) showed that the rhs should be increased by a factor of approximately $1+4\pi nJ_3$, $4\pi J_3$ being the volume integral of ξ out to the radius of interest and n being the number density (see also Kaiser 1986a). The problem with this expression is that J_3 may be hard to estimate. It has been suggested (Ling, Frenk & Barrow 1986) that the correct errors can be estimated via the **bootstrap resampling** method. This involves making repeated random drawings (with replacement) of new sets of galaxies from the catalogue under analysis, in each case estimating ξ by treating the bootstrap sample as real data. The resulting estimates will be biased in general (clearly, since several galaxies may be selected more than once), but the variance in the estimates gives a measure of the uncertainty in ξ (although see Mo, Jing & Börner 1992). The philosophy is clear: use the data themselves to estimate the sampling fluctuations. However, one is always concerned with sampling errors in the sense that large-scale variations in properties of the density field cause the area under study not to be a totally fair sample; such problems can never be treated by the bootstrap method.

POWER-SPECTRUM ESTIMATION In an ideal sample, the correlation function can be transformed to give the power spectrum directly. However, in the presence of the noise that is inevitable with limited data, this is not always the best way of proceeding: it can be better to measure the power spectrum directly. Consider the Fourier relation between ξ and Δ^2, $\xi = \int \Delta^2(k)[\sin(kr)/(kr)]dk/k$. The result for ξ at large separations is a mixture of modes, with a large 'leakage' of high-k power because the window function $\sin(kr)/(kr)$ declines rather slowly. Since the small-scale power is often large, this can swamp any signal from low k. Worse, the *uncertainties* in ξ are dominated by this small-scale term, so it is hard to subtract the unwanted small-scale contribution with any accuracy. A further problem on large scales is that $\xi(r)$ is very sensitive to assumptions about the mean density, because $1+\xi \propto \bar{n}^{-1}$ and $\xi \ll 1$; it is impossible to measure ξ once it falls below the uncertainty in $\bar{n}$. In contrast, uncertainty in the mean level of the density field just scales all the power-spectrum coefficients by the same factor. So, suppose we do the simplest thing and take the discrete transform of the data. In the continuum case, we would like to evaluate the Fourier coefficients

$$\delta_k = \frac{1}{V}\int \delta(\mathbf{x})\exp(i\mathbf{k}\cdot\mathbf{x})\, d^3x; \tag{16.112}$$

in practice, with a discrete set of N galaxies, we take the sum

$$\delta_k = \sum N^{-1}\exp(i\mathbf{k}\cdot\mathbf{x}). \tag{16.113}$$

In the absence of clustering, these coefficients execute a random walk on the Argand plane. The expectation value of the power is evaluated by splitting 3D space into a large number of **microcells**, which are so small that the probability of multiple occupation is negligible. It is then only necessary to consider occupation numbers $n_i = 0$ or 1 (cf. Peebles 1980), so that $n_i^2 = n_i$, leading to

$$\langle|\delta_k|^2\rangle = \sum n_i^2/N^2 = 1/N. \tag{16.114}$$

For $N \gg 1$ the central limit theorem yields an exponential distribution for a single mode:

$$P(|\delta_k|^2 > X) = \exp(-NX). \tag{16.115}$$

This is the distribution of the **Poisson noise** or **shot noise** that will overlay the true clustering signal that we are trying to measure.

The analysis so far has assumed a uniform background density, but any real survey will be subject to a varying probability of selection of objects; some areas of sky will not be surveyed, and distant objects will be lost, as described above through the **selection function** $\phi(y)$. The product of these factors means that the number density of objects is multiplied by a **window function** or **mask** $f(\mathbf{x})$: $n(\mathbf{x}) \to f(\mathbf{x})n(\mathbf{x})$. The corrections this introduces are as follows.

(1) The Fourier coefficients that are obtained are proportional to the transform of $f(\mathbf{x})[1+\delta(\mathbf{x})]$, so the transform of f needs to be subtracted.

(2) The selection function has a smaller effective volume than any cube in which it is embedded; this affects the amplitude of the power spectrum.

We now expand on these points a little (for details, see Peebles 1973; Peacock & Nicholson 1991). We first need to subtract the transform of the selection function, $f(\mathbf{x})$, which vanishes in the usual case of a uniform distribution:

$$\delta_k = \sum \frac{1}{N}\exp(i\mathbf{k}\cdot\mathbf{x}_i) - \frac{1}{N}\int n_b(\mathbf{x})\exp(i\mathbf{k}\cdot\mathbf{x})\, d^3x. \tag{16.116}$$

We can now subtract the Poisson contribution from the power; the quantity

$$P(k) \equiv \frac{1}{m} \sum |\delta_k|^2 - \frac{1}{N} \tag{16.117}$$

may therefore seem at first sight to be a good estimator for the true power spectrum, averaged over a region of k-space containing m modes.

However, we have not yet properly accounted for the window function. After subtracting the transform of the mean density, what remains is a density field that is the ideal infinite field $\delta(\mathbf{x})$ multiplied by the window function. This has the effect of convolving the Fourier coefficients of the density field with the transform of f:

$$\delta_k^{\text{obs}} = \delta_k^{\text{true}} * f_k. \tag{16.118}$$

In the absence of phase correlations between clustering pattern and mask (the **fair sample hypothesis**), is is straightforward to show that this also gives a convolution of power spectra:

$$\boxed{P_{\text{obs}} = P_{\text{true}} * |f_k|^2} \tag{16.119}$$

(see Peacock & Nicholson 1991; Feldman *et al.* 1994). Apart from smoothing the power spectrum, this also changes the normalization, as may be seen by considering the case of a uniform cubical survey of volume V embedded in a larger cube of volume V'. Using the density of states for the larger volume causes a multiple counting of modes, and therefore causes $P(k)$ to be too high by a factor V'/V. In general, we must therefore apply the correction

$$\boxed{P(k) \to P(k) \frac{\left(\int f\, d^3x\right)^2}{\int f^2\, d^3x \int d^3x}.} \tag{16.120}$$

For a uniform survey in which Poisson noise dominates, Webster (1976) proved that different δ_k are uncorrelated and that the result of summing over a band in k-space containing m modes should be an approximately Gaussian variable with variance $2m/N^2$ (the factor 2 arises because only half the modes are independent owing to the reality of the density field). This procedure may be used to yield minimal **Poisson error bars** on the power, once the effective number of independent modes is scaled as above to allow for the selection function.

The above procedure weights each galaxy equally, which is valid enough for samples of low galaxy density where shot noise dominates the weak clustering signal. In the opposite extreme, where there are very many galaxies in some sub-volumes, it may be preferable to give galaxies reduced weight, since the uncertainty on the power spectrum will then be determined by cosmic variance in the density field, independent of how many galaxies are sampled. Feldman *et al.* (1994) discuss this issue, and show that the optimum weight to give each galaxy is

$$w = \frac{1}{1 + \bar{n}\, P}, \tag{16.121}$$

where $\bar{n}$ is the mean number density at a given position and P is the non-dimensionless power spectrum (as in Peebles' convention, with $V = 1$). This expression does the intuitively correct thing of weighting galaxies equally at low density and volumes equally at high density. The power-spectrum analysis is straightforward to extend to the case of weights; the Fourier coefficients become

$$\delta_k = \left(\sum w_i\right)^{-1} \sum w_i \exp(i\mathbf{k}\cdot\mathbf{x}) \tag{16.122}$$

and the shot noise is modified to

$$\langle|\delta_k|^2\rangle = \left(\sum w_i\right)^{-2} \sum w_i^2. \tag{16.123}$$

The effect of the weight function is to boost the contribution from poorly sampled regions of space, where $\bar{n}$ is low. This means that the weight function combines with the window function to create a new effective window function for the density field. We should therefore replace f by wf in the above procedure for converting the Fourier coefficients δ_k to an estimate of the power spectrum P.

INTEGRAL CONSTRAINTS We should now say a little more about the normalization effects that arise because the mean density is unknown and must be estimated from the sample to hand. This can clearly cause a systematic underestimate of power in modes whose wavelengths approach the size of the sample volume. Similar problems exist in estimating the correlation function within a finite volume; since we use the observed mean number density $\bar{n}$ instead of the global $\langle n\rangle$, we have

$$(1+\xi)_{\rm true} = (1+\xi)_{\rm obs}(\bar{n}/\langle n\rangle)^2, \tag{16.124}$$

and hence

$$\boxed{\langle(1+\xi)_{\rm true}\rangle = (1+\xi)_{\rm obs}(1+\sigma^2),} \tag{16.125}$$

where σ^2 is the fractional rms density variation over a set of volumes that are the size of the sample. We have here ignored any correlation between $\xi_{\rm obs}$ and fluctuations in $\bar{n}$; this will be correct to the extent that they are much smaller than correlations between $\bar{n}$ and $\xi_{\rm true}$ for a given field. The final effect is very nearly to subtract a constant from ξ. To apply this correction, however, we need to know the contribution to σ^2 from modes of wavelength larger than the sample, which is a serious drawback of the method.

With the power spectrum, the main effect of any offset in mean density is just to scale the amplitudes of all modes by some factor. However, the relative amplitude of large-wavelength modes is affected in a more subtle way, as follows. We shift the true average level of δ by subtracting the mean value over our sample:

$$\delta' = \delta - \int \delta(\mathbf{x}) f(\mathbf{x})\, d^3x. \tag{16.126}$$

In what follows, we will need to distinguish carefully between the power spectrum of the true density field $P(k)$ and the power spectrum of the renormalized field P', and between their convolved counterparts P_* and P'_*. By subtracting a constant from the density, the renormalized power spectrum P' differs from the true P by a delta-function spike of power at $k=0$. Since we have established that the power spectrum we estimate is the true one convolved with $|f_k|^2$, this spike is also convolved, so that P'_* differs from P_* by a term proportional to $|f_k|^2$. Now, using the observed mean density forces $P'_*(0)$ to be zero; this determines the constant of proportionality, giving $P'_*(k) = P_*(k) - P_*(0)|f_k|^2$, or

$$P'_*(k) = P_*(k)\left[1 - |f_k|^2 \frac{P_*(0)}{P_*(k)}\right]. \tag{16.127}$$

The effect that this has depends on the power spectrum; consider power-law spectra with $P \propto k^n$. Clearly, for an $n = 0$ spectrum, the 'normalization damping' factor is just $(1 - |f_k|^2)$; only modes with $|f_k|^2 \gtrsim 0.1$ will be affected at all significantly. This is the precise expression of the obvious constraint that a sample of depth D cannot be used to measure the clustering spectrum for wavelengths greater than D.

16.8 The observed clustering spectrum

The history of attempts to quantify galaxy clustering goes back to Hubble's demonstration that the distribution of galaxies on the sky was non-uniform. The major post-war landmarks were the angular analysis of the Lick catalogue, described in Peebles (1980), and the analysis of the CfA redshift survey (Davis & Peebles 1983). It has taken some time to obtain data on samples that greatly exceed these in depth, but several pieces of work appeared around the start of the 1990s which clarified many of the discrepancies between different surveys and which paint a relatively consistent picture of large-scale structure. Perhaps the most significant of these surveys have been the **APM survey** (automatic plate-measuring machine survey; see Maddox *et al.* 1990b and Maddox *et al.* 1996), the **CfA survey** (Center for Astrophysics survey; see Huchra *et al.* 1990) and the **LCRS** (Las Campanas redshift survey; see Shectman *et al.* 1996), together with a variety of surveys based on galaxies selected at 60 μm by the infrared astronomy satellite, **IRAS** (Saunders *et al.* 1991; Fisher *et al.* 1993). These surveys are sensitive to rather different galaxy populations: the APM, CfA and LCRS surveys select in blue light and are sensitive to stellar populations of intermediate age; the IRAS emission originates from hot dust, associated with bursts of active star formation. A compilation of clustering results for a variety of tracers was given by Peacock & Dodds (1994); the results are shown in figure 16.4.

There is a wide range of power measured, ranging over perhaps a factor 20 between the real-space APM galaxies and the rich Abell clusters. Are these measurements all consistent with one Gaussian power spectrum for mass fluctuations? Corrections for redshift-space distortions and nonlinearities can be applied to these data to reconstruct the linear mass fluctuations, subject to an unknown degree of bias. The simplest assumption for this is that the bias is a linear response of the galaxy-formation process, and may be taken as independent of scale:

$$\Delta^2_{\rm tracer} = b^2 \Delta^2_{\rm mass}. \tag{16.128}$$

There thus exist five free parameters that can be adjusted to optimize the agreement between the various estimates of the linear power spectrum; these are Ω and the four bias parameters for Abell clusters, radio galaxies, optical galaxies and IRAS galaxies ($b_{\rm A}$, $b_{\rm R}$, $b_{\rm O}$, $b_{\rm I}$). However, only two of these really matter: Ω and some measure of the overall level of fluctuations. For now, we take the IRAS bias parameter to play this latter role. Once these two are specified, the other bias parameters are well determined, principally

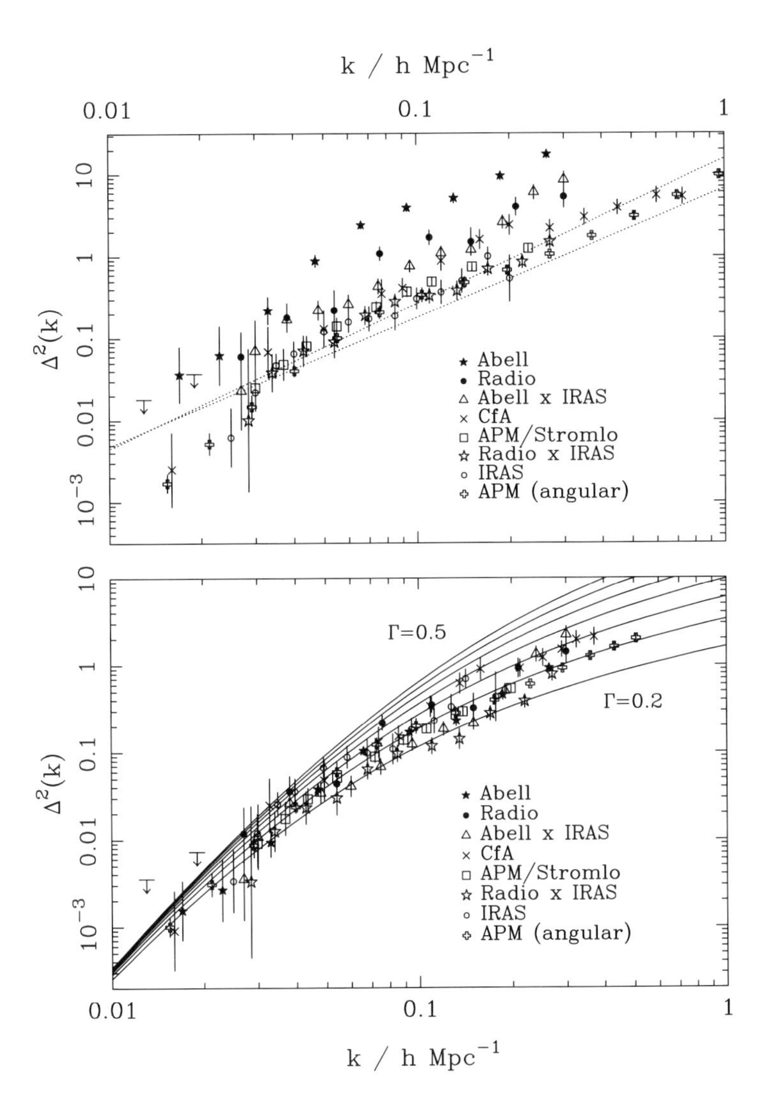

Figure 16.4. A compilation of power-spectrum data, adapted from Peacock & Dodds (1994). The upper panel shows raw power-spectrum data in the form $\Delta^2 \equiv d\sigma^2/d\ln k$; all data with the exception of the APM power spectrum are in redshift space. The two dotted lines shown for reference are the transforms of the canonical real-space correlation functions for optical and IRAS galaxies ($r_0 = 5\,h^{-1}$ Mpc and $3.78\,h^{-1}$ Mpc and slopes of 1.8 and 1.57 respectively). The lower panel shows the results of correcting these datasets for different degrees of bias and for nonlinear evolution. There is an excellent degree of agreement, particularly in the detection of a break around $k = 0.03h$. The data are compared to various CDM models, assuming scale-invariant initial conditions, with the same large-wavelength normalization. Values of the fitting parameter $\Gamma = 0.5$, 0.45, …, 0.25, 0.2 are shown. The best-fit model has $\Gamma = 0.25$.

from the linear data at small k, and have the approximate ratios

$$b_{\rm A} : b_{\rm R} : b_{\rm O} : b_{\rm I} = 4.5 : 1.9 : 1.3 : 1 \tag{16.129}$$

(Peacock & Dodds 1994). The reasons why different galaxy tracers may show different strengths of clustering are discussed above and in chapter 17. Rich clusters are inevitably biased with respect to the mass, simply through the statistics of rare high-density regions. Massive ellipticals such as radio galaxies share some of this bias through the effect of morphological segregation, which says that the E/S0 fraction rises in clusters to almost 100%, by comparison with a mean of 20% (see chapter 13). At high overdensities, the fraction of optical galaxies that are IRAS galaxies declines by a factor $\simeq 3$–5 (Strauss *et al.* 1992), reflecting the fact that IRAS galaxies are mainly spirals. Generally, the analysis of galaxies of a given type assumes that the luminosity function is independent of environment so that bias is independent of luminosity. This is not precisely true, and the amplitude of ξ does appear to rise slightly for galaxies of luminosity above a few times L^*. (Valls-Gabaud *et al.* 1989; Loveday *et al.* 1995; Benoist *et al.* 1996). However, for the bulk of the galaxies in a given population, it is a good approximation to say that **luminosity segregation** can be neglected.

The various reconstructions of the linear power spectrum for the case $\Omega = b_{\rm I} = 1$ are shown superimposed in figure 16.4, and display an impressive degree of agreement. This argues very strongly that what we measure from large-scale galaxy clustering has a direct relation to mass fluctuations, rather than being an optical illusion caused by non-uniform galaxy-formation efficiency (Bower *et al.* 1993). If effects other than gravity were dominant, the shape of spectrum inferred from clusters would be very different at large scales, contrary to observation.

LARGE-SCALE POWER-SPECTRUM DATA AND MODELS It is interesting to ask whether the power spectrum contains any features or whether it is consistent with a single smooth curve. A convenient description is in terms of the CDM power spectrum, which is $\Delta^2(k) \propto k^{n+3} T_k^2$. We shall use the BBKS approximation for the transfer function given in chapter 15:

$$T_k = \frac{\ln(1+2.34q)}{2.34q}\,[1 + 3.89q + (16.1q)^2 + (5.46q)^3 + (6.71q)^4]^{-1/4}, \tag{16.130}$$

where $q \equiv (k/\,h\,{\rm Mpc}^{-1})\Gamma^{-1}$. For models with zero baryon content, the shape parameter $\Gamma = \Omega h$, but the general scaling depends on the baryonic density (Sugiyama 1995):

$$\Gamma = \Omega h \exp\left[-\Omega_{\rm B}(1 + \sqrt{2h}/\Omega)\right]. \tag{16.131}$$

The CDM spectrum is very commonly used as a basis for comparison with cosmological observations, and it is essential to realize that this can be done in two ways. The CDM physics can be accepted, in which case the best-fitting value of Γ constrains Ω, h and $\Omega_{\rm B}$. Alternatively, the CDM spectrum can be used as a completely empirical fitting formula, which is assumed to approximate some different set of physics. For example, the MDM model includes CDM and an admixture of massive neutrinos; over a limited range of k, this will appear similar to a CDM spectrum, but with an effective value of Γ that is very much less than Ωh, because of the way in which neutrinos remove small-scale power in this case (see chapter 15).

The normalization of the spectrum is specified by the rms density variation averaged over $8\,h^{-1}\,{\rm Mpc}$ spheres; for CDM-like spectra, this measures power at an effective wavenumber well approximated by

$$\sigma_8^2 = \Delta^2(k_{\rm eff}), \quad k_{\rm eff} / h\,{\rm Mpc}^{-1} = 0.172 + 0.011\,[\ln(\Gamma/0.34)]^2. \tag{16.132}$$

Fitting this spectrum to the large-scale linearized data of figure 16.4 requires the parameters

$$\Gamma \simeq 0.25 + 0.3(1/n - 1), \tag{16.133}$$

in agreement with many previous arguments suggesting that an apparently low-density model is needed; the linear transfer function does not bend sharply enough at the break wavenumber if a 'standard' high-density $\Gamma = 0.5$ model is adopted. For any reasonable values of h and baryon density, a high-density CDM model is not viable. Even a high degree of 'tilt' in the primordial spectrum (Cen *et al.* 1992) does not help change this conclusion unless n is set so low that major difficulties result when attempting to account for microwave-background anisotropies (see chapter 18).

An important general lesson can also be drawn from the lack of large-amplitude features in the power spectrum. This is a strong indication that collisionless matter is deeply implicated in forming large-scale structure. Purely baryonic models contain large bumps in the power spectrum around the Jeans' length prior to recombination ($k \sim 0.03\Omega h^2$ Mpc^{-1}), whether the initial conditions are isocurvature or adiabatic; see chapter 15. It is hard to see how such features can be reconciled with the data, beyond a 'visibility' in the region of 20%.

SMALL-SCALE CLUSTERING AND BIAS It should clearly be possible to reach stronger conclusions by using the data at larger k. However, here the assumption of linear bias is clearly very weak, and we need to understand the scale dependence of the bias. Mann, Peacock & Heavens (1997) argue that this can be constrained empirically: we know from cluster M/L ratios that the density of light in high-density regions must be boosted by about a factor 5 if $\Omega = 1$. There is also the possibility that galaxy formation may be suppressed in voids. These features can all be accounted for in **local bias models**, in which the galaxy density ρ_g is some nonlinear function of the mass density ρ_m. It has been argued on the basis of numerical simulations that it is physically reasonable to expect a direct relation of this sort between mass and light (Cen & Ostriker 1992). General arguments (Coles 1993) require any local bias model to produce a systematic steepening of the correlation function, so that the effective bias is larger on small scales.

We can illustrate the sort of effect that might be contemplated by using a lognormal process as an analytic model for the nonlinear density field (see below). A simple illustrative example of the possible nonlinear modifications of density involved in local bias is a power law: $\rho_g \propto \rho_m^B$. The correlation function for the lognormal process is obtained below, and the power-law case is a simple extension, which is left as an exercise:

$$1 + \xi_g = (1 + \xi_m)^{B^2}. \tag{16.134}$$

The number B is the linear bias for $\xi \ll 1$, but the boost to the correlation function is increased when the mass correlations are nonlinear. Note that these effects do not need to be extreme; if $\Omega = 1$, we need to boost cluster M/L values by a factor 5 at overdensities of a few hundred (see chapter 12), suggesting that $B \simeq \ln(5 \times 500)/\ln 500 = 1.26$. In this model, a linear bias of 1.26 would give an effective bias of 3.9 at $\xi = 100$, so the distortion of shape of the correlation function is only moderate. Inspired by this result, a flexible fitting formula for the effects of bias on power spectra is

$$1 + \Delta^2(k) \to (1 + b_1 \Delta^2)^{b_2}, \tag{16.135}$$

where $(b_1 b_2)^{1/2}$ would be the bias parameter in the linear regime. Although we cannot test this by observing the mass directly, we can consider the relative clustering of optically selected galaxies and IRAS galaxies. At high overdensities, the fraction of optical galaxies that are also IRAS galaxies declines by a factor $\simeq 4$ from the mean (Strauss *et al.* 1992), reflecting the fact that IRAS galaxies are mainly spirals, whereas morphological segregation means that clusters are dominated by ellipticals. This behaviour is similar to the relation between optical light and mass: if $\Omega = 1$, we know that the optical light in clusters must be enhanced with respect to the mass by a factor of about 5, as above. Figure 16.5 shows that the real-space clustering of APM and IRAS galaxies does indeed display a relative bias that is a monotonic function of scale, fitted extremely well by relative bias parameters of $b_1 = 1.2$, $b_2 = 1.1$ or $b_{\rm APM}/b_{\rm IRAS} \simeq 1.15$. Of particular interest is the inflection in the spectrum around $k \simeq 0.1\, h\,{\rm Mpc}^{-1}$, which seems likely to be real, since it is seen in two rather different datasets, which probe different regions of space.

Figure 16.5 indicates that there is now excellent knowledge of the power spectrum of optically selected galaxies. Because this is a real-space measurement, obtained by deprojection of angular clustering, the results range from the large-scale linear regime to the very nonlinear regime. Although we do not know exactly how the galaxy spectrum relates to that of the light, the IRAS results reassure us that bias will be a slowly varying function of scale. With these points in mind, we should now carry out a more demanding test on the $\Gamma = 0.25$ spectrum, which was shown above to describe the large-scale spectrum well. Figure 16.6 compares three different models with the data: the Einstein–de Sitter model and open and flat $\Omega = 0.3$ models. These are given a normalization fixed from the abundance of rich clusters, as discussed in chapter 17:

$$\boxed{\sigma_8 = 0.57\,\Omega^{-0.56}} \tag{16.136}$$

(White, Efstathiou & Frenk 1993). There is a tension between the data and all these models, in that it seems impossible to fit both large scales and small scales simultaneously. Without bias, the correct amplitude of small-scale clustering requires $\sigma_8 \simeq 0.7$; this greatly under-predicts the clustering at $k \simeq 0.1\, h\,{\rm Mpc}^{-1}$ unless $\Gamma \lesssim 0.1$ is adopted, in which case the power at $k \simeq 0.02\, h\,{\rm Mpc}^{-1}$ is greatly exceeded. The $\Omega = 1$ model needs a lower normalization, and so does not exceed the small-scale data, but it suffers from related problems of shape. Any hypothetical bias that would scale the mass spectrum to that of light would need to be non-monotonic, with a smaller effect at $k \simeq 2\, h\,{\rm Mpc}^{-1}$ than on larger scales. The conclusion seems to be clear: the CDM spectrum is of the wrong shape, and does not give a correct description of the observed galaxy clustering.

These ideas can be illustrated with a simple empirical model. Consider a spectrum in the form of a break between two power laws:

$$\Delta^2(k) = \frac{(k/k_0)^\alpha}{1 + (k/k_1)^{\alpha - \beta}}. \tag{16.137}$$

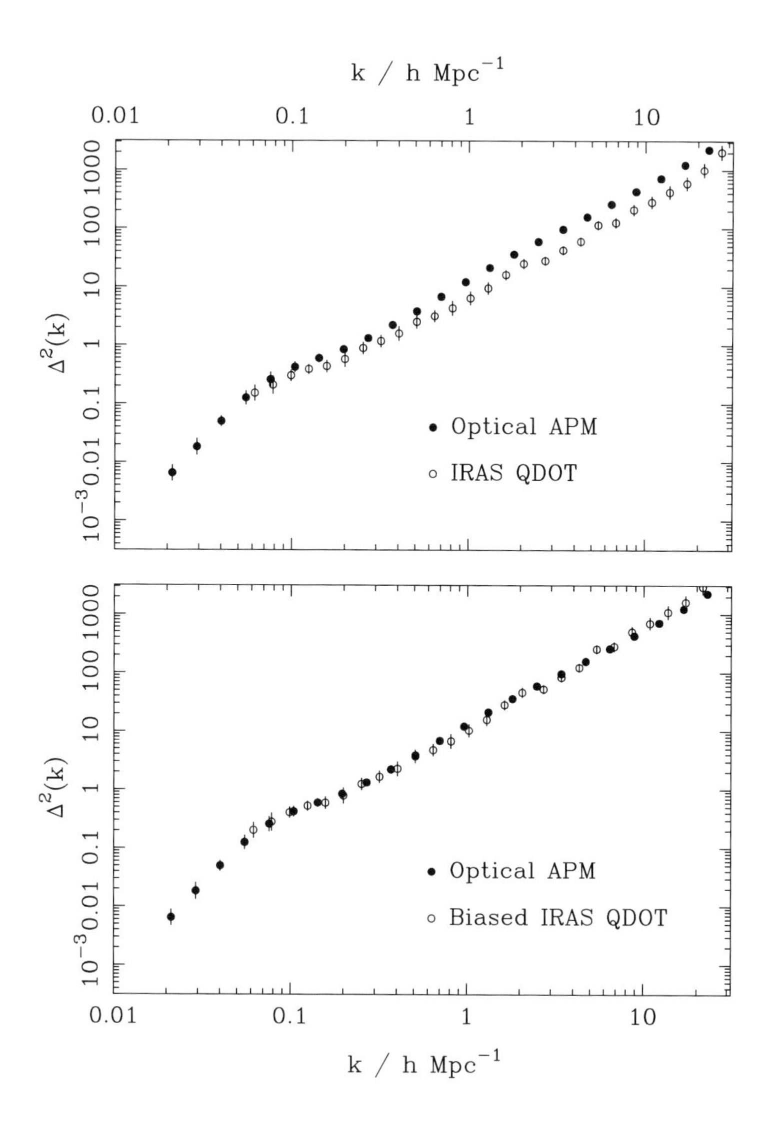

Figure 16.5. The upper panel shows the real-space power spectra of APM and IRAS galaxies, as deduced by Baugh & Efstathiou (1993; 1994) and Saunders *et al.* 1992. The lower panel shows the same data with a two-parameter scale-dependent boost to the IRAS results. The agreement is outstanding, indicating that the real-space clustering properties of optical galaxies are known to very high precision.

As shown in figure 16.7, the nonlinear power that results from this linear spectrum matches the data very nicely, if we choose the parameters

$$\begin{aligned} k_0 &= 0.3\, h\,\mathrm{Mpc}^{-1} \\ k_1 &= 0.05\, h\,\mathrm{Mpc}^{-1} \\ \alpha &= 0.8 \\ \beta &= 4.0. \end{aligned} \tag{16.138}$$

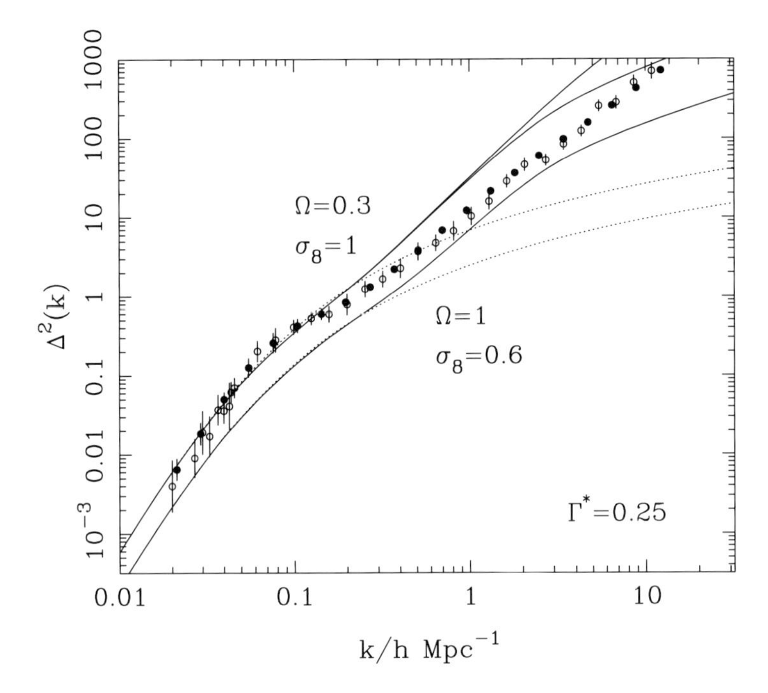

Figure 16.6. The clustering data for optical galaxies, compared to three models with $\Gamma = 0.25$: $\Omega = 1$, $\sigma_8 = 0.6$; $\Omega_m = 0.3$, $\Omega_v = 0$, $\sigma_8 = 1$; $\Omega_m = 0.3$, $\Omega_v = 0.7$, $\sigma_8 = 1$. The linear spectra are shown dotted; solid lines denote evolved nonlinear spectra. All these models are chosen with a normalization that is approximately correct for the rich-cluster abundance and large-scale peculiar velocities. In all cases, the shape of the predicted spectrum fails to match observation. The high-density model would require a bias that is not a monotonic function of scale, whereas the low-density models exceed the observed small-scale clustering (figure adapted from Peacock 1997).

A value of $\beta = 4$ corresponds to a scale-invariant spectrum at large wavelengths, whereas the effective small-scale index is $n = -2.2$. We consider below the physical ways in which a spectrum of this shape might arise.

A further general point that emerges from the comparison in figure 16.6 is a longstanding puzzle: the clustering data continue as an unbroken $n \simeq -1$ power law up to $\Delta^2 \sim 10^3$. As the regime of virialized clustering is reached, a break to a flatter slope would be expected; only the open models fail to show this feature, which is a robust prediction for any smooth linear spectrum. Of course, this change occurs where the clustering is particularly nonlinear and it is difficult to have a firm idea what the effects of bias would be in this regime. Nevertheless, it would be something of a coincidence if bias managed to steepen the spectrum just where the gravitational dynamics are causing it to flatten.

Finally, it is important to note that bias of either sign may have to be considered. A high-density universe with the cluster-normalized value of σ_8 predicts clustering well

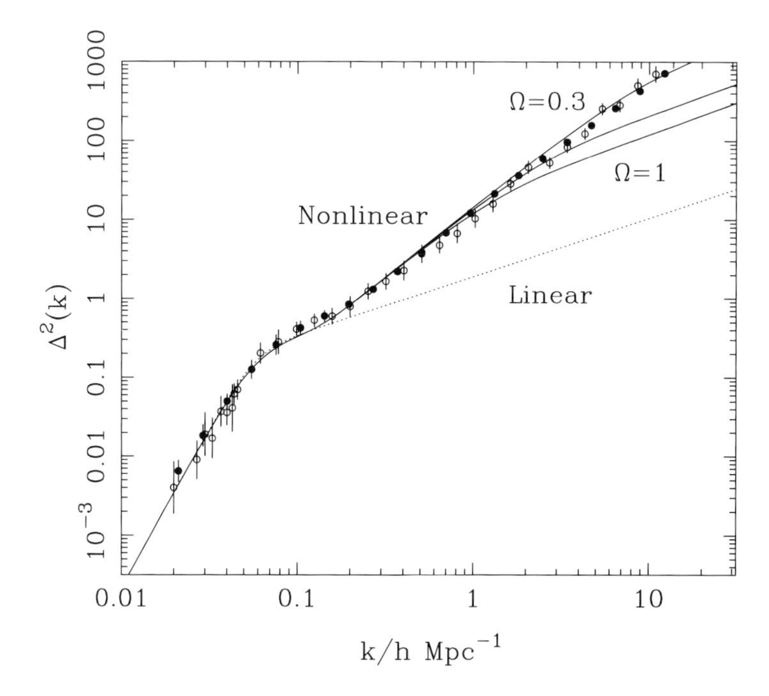

Figure 16.7. An empirical double power-law model for the power spectrum provides an extremely good fit to the optical-galaxy power spectrum but requires an open universe and a sharp break in the spectrum, with a rather flat ($n < -2$) high-k behaviour.

below that observed. However, the opposite is true for low Ω. If we simply take the APM power spectrum and ignore nonlinear corrections, the apparent value of σ_8 is about 0.9. Contrast this with the prediction of the cluster-normalization formula, which requires $\sigma_8 = 1.4$ for $\Omega = 0.2$, or 2.1 for $\Omega = 0.1$. Thus, low-density models inevitably require significant **antibias**, and we would have to consider the possibility that galaxy formation was suppressed in high-density regions. Bias in this sense has one advantage over positive bias, since it will tend to make the predicted small-scale spectrum less steep, which figure 16.6 suggests may be required in order to match the data. However, as discussed above, it is implausible that the scale dependence of the bias will be very extreme; a model that matches the data at $k \simeq 1\,h\,\mathrm{Mpc}^{-1}$ will probably significantly undershoot the 'bump' at $k \simeq 0.1\,h\,\mathrm{Mpc}^{-1}$. This large-scale feature therefore has a critical importance in the interpretation of large-scale structure. If it is correct, then the simplest CDM models fail and must be replaced by something more complicated. However, if future observations should yield lower power values at this point, then a low-density CDM model with antibias would provide a model for large-scale structure that is attractive in many ways (Jing *et al.* 1997).

CLUSTERING EVOLUTION So much for present-day clustering. If large-scale structure really did originate through gravitational instability, we should see lower clustering in

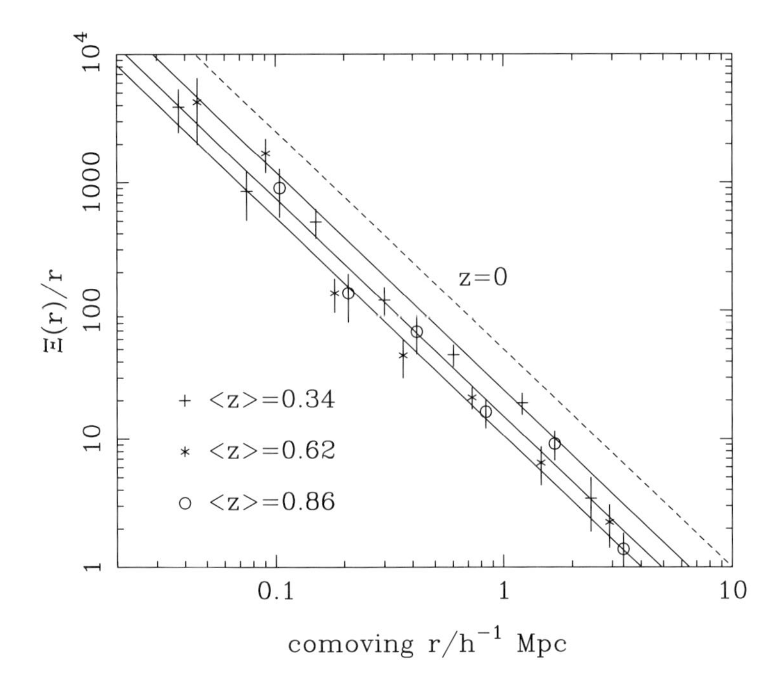

Figure 16.8. The results of the Canada-France redshift survey (Le Fèvre *et al.* 1996) on the projected correlation function as a function of redshift, assuming $\Omega = 1$. The solid lines show the model $r_0 = 4.5\, h^{-1}$ Mpc, $\gamma = 1.7$, $\epsilon = 1.2$ for the three redshifts at which points are plotted. The broken line shows the corresponding clustering at $z = 0$. It appears that the correlation function maintains its power-law shape, but changes its amplitude at about the linear-theory rate.

the past. Recall that the predicted evolution of the correlation function may be written as

$$\boxed{\xi(r, z) = \left(r/r_0\right)^{-\gamma}\, (1 + z)^{-(3-\gamma+\epsilon)},} \tag{16.139}$$

where ϵ parameterizes the departures from stable clustering (see chapter 15). For many years, this relation could only be tested against the angular clustering of faint galaxies, where the uncertain redshift distribution muddied the interpretation. However, the Canada–France redshift survey (CFRS; see Le Fèvre *et al.* 1996), has made a reasonably firm set of measurements out to redshift $z \simeq 1$ (figure 16.8). Le Fèvre *et al.* (1996) analysed their data in terms of the projected correlation function $\Xi(r)$,

$$\Xi(r) = 2 \int_r^\infty \xi(x)\, \frac{x\, dx}{(x^2 - r^2)^{1/2}}, \tag{16.140}$$

which has the advantage that it is independent of peculiar velocities and yields a direct real-space measure of clustering. The results indicate a redshift-independent power-law slope of about $\gamma = 1.7$, similar to that measured locally, and a high rate of evolution $\epsilon \simeq 1$.

This is something of a paradox, since $\epsilon = 1$ is roughly the linear-theory rate of evolution for $\Omega = 1$: how can clustering evolve at the linear rate when it is in the nonlinear regime? One possible solution is to note that the quasilinear evolution is much more rapid, as discussed above. If we ignore bias or its evolution, these results would require a linear growth function for density perturbations of $D(z) \simeq (1+z)^{-0.5}$, which would be satisfied by $\Omega \simeq 0.3$ for an open universe, or by $\Omega \simeq 0.1$ for a flat Λ model (Peacock 1997). Alternatively, if we insist on $\Omega = 1$, the slow apparent rate of clustering evolution says that galaxies at $z = 1$ must be much more strongly biased than they are today.

CMB ANISOTROPIES Moving to higher redshifts still, a consistent model must match the normalization of the mass fluctuations on large scales inferred from fluctuations in the cosmic microwave background (CMB). This topic is discussed at length in chapter 18, but it is worth anticipating a few issues here. CMB anisotropies are simple on angular scales above a few degrees, because they are dominated by just gravitational redshift; on smaller scales, a variety of processes involving the hydrodynamic response of the photon–baryon fluid come into play, and the interpretation is more complicated. With regard to large-scale anisotropies, note that we have no evidence for any clustering power for $k \lesssim 0.02\, h\,\mathrm{Mpc}^{-1}$, or wavelengths $\lambda \gtrsim 300\, h^{-1}$ Mpc. The present-day comoving horizon distance is $r_{\mathrm{H}} = 2c/H_0$ if $\Omega = 1$, or larger for low densities. The scales probed by large-scale structure therefore probe angles no larger than 3 degrees on the sky. So far, the best detections of CMB anisotropies have come at around 20-degree scales, so consistency with the CMB is always possible, depending on how we choose to extrapolate the large-scale spectrum. In fact, the conclusion in chapter 18 is that the observed CMB is very close to what would be expected if the clustering spectrum continued to larger scales with a scale-invariant shape; it appears that gravitational instability can account for the growth of structure in the universe since a redshift of 1000.

SUMMARY The empirical clustering spectrum now appears to be known with reasonable precision; what is the physical interpretation of the data? On large scales, it is interesting that a CDM spectrum with $\Gamma \simeq 0.25$ is a reasonable description of the data. Since $\Gamma \simeq 0.9\Omega h$ for reasonable baryon content, this argues strongly against an $\Omega = 1$ universe; however, low-density universes would be consistent, especially if experimental error allowed Γ closer to 0.2. The big question is then whether the CDM transfer function has the correct shape. If it does, then a low-Ω CDM model is certainly a simple enough universe to inhabit. As discussed above, the critical area of the clustering spectrum that will answer this question is the 'APM bump' around $k \simeq 0.1\, h\,\mathrm{Mpc}^{-1}$.

The simplest way of saving the Einstein–de Sitter model would be to increase the energy density in relativistic species beyond the standard contribution from CMB photons plus three species of massless neutrino. This would delay the epoch of matter–radiation equality and thus increase the wavelength at which the spectrum breaks, lowering the apparent value of Γ. The break wavenumber in the CDM spectrum scales as $k \propto \Omega h^2 \theta^{-1/2}$, where θ is the ratio of the relativistic energy density to the standard value; an approximate doubling of the number of relativistic degrees of freedom would therefore allow a model with $\Omega = 1$ to yield a consistent spectral shape. As discussed in chapter 15, this can be achieved if there exists a massive neutrino that decays to weakly interacting particles, boosting the ultrarelativistic background density. The apparent value of Ωh depends on the mass and lifetime of the particle roughly as

$$(\Omega h)_{\rm apparent} \simeq \Omega h \left\{ 1 + 0.15 \left[\left(\frac{m}{\rm keV} \right)^2 \frac{\tau}{\rm yr} \right]^{2/3} \right\}^{-1/2} \tag{16.141}$$

(Bardeen, Bond & Efstathiou 1987; Bond & Efstathiou 1991), so a range of masses is possible.

However, if the smoothly varying CDM spectrum is not a correct description of reality, we have to find ways to reduce the small-scale power. An immediate possibility is damping, which results if the dark-matter particle has non-zero rest mass. The general case is warm dark matter (**WDM**): $\Omega = 1$ is supplied by a particle somewhat less abundant than neutrinos. According to BBKS, the transfer function is modified in the warm dark matter case to $T_{\rm WDM} = T_{\rm CDM} \exp[-(kR_D)/2-(kR_D)^2/2]$, where the damping length is $R_D = 0.2\,(\Omega h^2)^{1/3}\,(m/{\rm keV})^{-4/3}$ Mpc. However, this produces virtually negligible damping at the point in the power spectrum where it is required, $k \simeq 0.1\,h\,{\rm Mpc}^{-1}$, unless the mass is low, of the order of 50 eV, for all h values. We thus end up with all the well-known problems of the hot dark matter (**HDM**) picture for galaxy formation (e.g. Hut & White 1984).

Interesting alternatives with high density are either mixed dark matter (MDM; see Holtzman 1989, Klypin *et al.* 1993), or non-Gaussian pictures such as cosmic strings plus HDM, where the lack of a detailed prediction for the power spectrum helps ensure that the model has not yet been excluded (Albrecht & Stebbins 1992). Isocurvature CDM is also attractive in that it gives a rather sharper bend at the break scale, as the data seem to require. However, this model conflicts strongly with limits on CMB anisotropies, and cannot be correct (chapter 18). Mixed dark matter seems rather *ad hoc*, but may be less so if it is possible to produce both hot and cold components from a single particle, with a Bose condensate playing the role of the cold component (Madsen 1992; Kaiser, Malaney & Starkman 1993). The main problems with an MDM model are those generic to any model with a very flat high-k spectrum in a high-Ω universe: difficulty in forming high-redshift objects and difficulty in achieving a steep correlation function on small scales. Nevertheless, the MDM spectrum is the only existing model in the literature that does seem to have about the desired shape. The parameters needed to produce a good fit are $\Omega h \simeq 0.4$ and about 30% of the dark matter in the form of light massive neutrinos. The value of Ωh is high enough to be reasonable for $\Omega = 1$ and a sensible Hubble constant, given possible baryonic effects on the transfer function, but it makes no sense at all if $\Omega \simeq 0.3$. We are faced with an unpleasant choice. If we live in an $\Omega = 1$ MDM universe, then all the difficulties with high-z clustering and the lack of virialization signatures in present small-scale clustering will have to be accounted for by bias. Alternatively, we might inhabit a low-density universe with little bias, and a fluctuation spectrum of the MDM shape. However, the physical origin of such a spectrum is presently unknown.

16.9 Non-Gaussian density fields

Although the power spectrum is of unquestionable importance, it says nothing about the phases of perturbations, or about whether the primordial fluctuations constituted a Gaussian random field. Although many inflation models would yield Gaussian fluctuations, we do not wish to assume that the real density field must have had this character; for example, all topological-defect models produce non-Gaussian perturbations. The key requirement is to introduce phase correlations between modes, and this will clearly happen when a variety of modes all respond to the gravity of a single defect. When the effects of many defects combine, as with overlapping string wakes, the central limit theorem will tend to give Gaussian answers, and so the real density field could be Gaussian on large scales, but progressively more non-Gaussian on small scales.

It is a complicated numerical problem to calculate the density perturbations that result from topological defects, so much attention has been devoted to heuristic models that are simple modifications of Gaussian fields $\delta_{\rm G}$ (see e.g. Coles & Barrow 1987). Many of these cases can be given motivation through more complicated models of inflation, in which the existence of multiple scalar fields can yield products of Gaussian fluctuations (e.g. Yi & Vishniac 1993). As a very simple case, consider the χ^2 field,

$$\delta_{\chi^2} = \delta_{\rm G}^2 - \sigma^2; \tag{16.142}$$

σ^2 is the variance of the Gaussian field, and subtracting this enforces $\langle\delta\rangle = 0$, as required by the meaning of the average density. This distribution is always skewed towards positive δ, however small we make σ. An interesting variant on this model is to assume that the whole density of the universe is contributed at early times by the potential term of a Gaussian scalar field with a simple mass term, $\rho = m\phi_{\rm G}^2/2$. Dividing by the mean density gives $1 + \delta = \phi_{\rm G}^2/\langle\phi_{\rm G}^2\rangle$; this is therefore the special case of a χ^2 field with $\sigma = 1$, and the universe would be born mildly nonlinear (Peebles 1997). These simple models are markedly non-Gaussian: a simple inspection of the one-point histogram of field values serves to demonstrate this. A more subtle case would be to apply a monotonic transformation $\delta \to f(\delta)$, so that the one-point distribution became Gaussian. If the non-Gaussian field were generated via a non-monotonic transformation, then the resulting field would be Gaussian distributed, but still non-Gaussian (i.e. the higher-order correlations would have a more complicated behaviour). The universe is unlikely to play such a dirty trick on us, but more subtle non-Gaussian signatures are possible from a product of two independent fields. This allows us to construct non-Gaussian models in which the local character of the linear density fluctuations is Gaussian, but with a spatial modulation or **intermittency**, so that there are 'quiet' and 'noisy' parts of space:

$$\delta(\mathbf{r}) \to \delta(\mathbf{r})\,[1 + M(\mathbf{r})]. \tag{16.143}$$

Such a model was proposed on phenomenological grounds by Peebles (1983).

Even if we leave aside primordial non-Gaussianity, the real universe must be non-Gaussian because it is nonlinear. A Gaussian field of rms σ has a non-zero probability of producing the unphysical $\delta < -1$, however small we make σ. Once we reach $\sigma \simeq 1$, it is clear that very substantial alterations to the density distribution are required: gravitational collapse will yield very high positive densities, and the distribution must be strongly skewed in order to allow this while preserving $\langle\delta\rangle = 0$. An interesting model that shows this feature but which is Gaussian in the limit of small σ is the **lognormal density field** (Coles & Jones 1991):

$$1 + \delta_{\rm LN} = \exp(\delta_{\rm G} - \sigma^2/2). \tag{16.144}$$

These authors argue that this analytical form is a reasonable approximation to the exact nonlinear evolution of the mass density distribution function. As a bonus, the model can also handle the effects of a biased galaxy distribution, since it is similar in form to the high-peak behaviour used in some models for biased galaxy formation (see chapter 17).

One of the nice features of these exact non-Gaussian models is that the underlying Gaussian field is still there, and so the joint distribution of the density at n points is still known. This means that the statistical properties for these model density fields are especially simple to calculate. For example, consider the two-point correlation function for the χ^2 field:

$$\begin{aligned}\xi(r) &= \langle \delta(x)\delta(x+r) \rangle \\ &= \left\langle [\delta_{\rm G}^2(x) - \sigma^2][\delta_{\rm G}^2(x+r) - \sigma^2] \right\rangle \\ &= \left\langle \delta_{\rm G}^2(x)\delta_{\rm G}^2(x+r) \right\rangle - \sigma^4. \end{aligned} \tag{16.145}$$

The first term looks nasty, but is easily treated: denote $\delta_{\rm G}(x)$ by y_1 and $\delta_{\rm G}(x+r)$ by y_2, so that the required integral is

$$\begin{aligned}\langle y_1^2 y_2^2 \rangle &= \iint y_1^2 y_2^2 \, P(y_1, y_2) \, dy_1 \, dy_2 \\ P(y_1, y_2) &= \frac{1}{2\pi\sigma^2(1-\psi^2)^{1/2}} \exp\left[-\frac{y_1^2 + y_2^2 - 2\psi y_1 y_2}{2\sigma^2(1-\psi^2)} \right], \end{aligned} \tag{16.146}$$

where $\psi(r) \equiv \xi_{\rm G}(r)/\sigma^2$. Performing the integral gives $\langle y_1^2 y_2^2 \rangle = \sigma^4(1+2\psi^2)$, so that

$$\xi_{\chi^2}(r) = 2\xi_{\rm G}^2(r). \tag{16.147}$$

By the same method, the correlation for the lognormal model is

$$\xi_{\rm LN} = \exp \xi_{\rm G} - 1, \tag{16.148}$$

which says that ξ on large scales is unaltered by nonlinearities in this model; they only add extra small-scale correlations.

VORONOI TESSELLATIONS Another approximate model for the non-Gaussian universe that results from gravitational instability was inspired by the observation that visual examination of redshift survey data has often suggested a **cellular structure** for the universe (Einasto, Joeveer & Saar 1980). This need not be inconsistent with the idea of Gaussian initial conditions, as was first shown by Zeldovich (1970), who emphasized the tendency for gravitational collapse to yield flattened structures (pancakes). Subsequent extension of his work in the **adhesion model** (Gurbatov, Saichev & Shandarin 1989) confirms this conclusion. These workers show that, at late times, matter will tend to congregate in a geometrical structure consisting of a network of pancakes, which meet in filaments, which in turn intersect at clusters. This amounts to saying that the structure of the universe tends to be dominated by the voids; because they are underdense, they expand faster than average, sweeping all material into a small volume where the expanding void walls meet (see Hoffman, Salpeter & Wasserman 1983; Bertschinger 1985a). Such a process is generic to gravitational evolution, and will certainly result from Gaussian initial conditions.

A simple model of this process is provided by the **Voronoi tessellation**. Start with a number of randomly-distributed seeds; for each seed, construct the plane that bisects the line joining that seed to its neighbours. These planes define a cell around each seed such that all points in the cell are closer to that seed than to any other – the spatial analogue

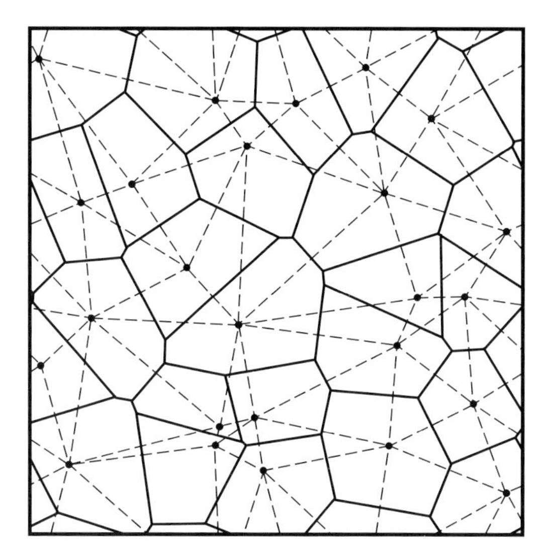

Figure 16.9. Illustrating the construction of the Voronoi tessellation in 2D. The points mark the seeds – notionally peaks in the gravitational potential about which voids develop. If all voids overexpand at the same rate, matter accumulates in the cell walls, which are equidistant from their nearest seeds. Each seed is thus surrounded by a 'zone of influence' that encloses those points closer to that seed than any other. The cell walls form perpendicular bisectors of the broken lines connecting seeds.

of the Wigner-Seitz k-space cells encountered in solid-state physics. A 2D example of this process is illustrated in figure 16.9. In the adhesion model, the seeds would be identified with large-scale pits in the gravitational potential. The Voronoi model is inexact in that these pits are neither unclustered nor all of the same depth. In practice, the influence of this imprecision is small, and the Voronoi model is useful both as a quantitative and heuristic tool for studying a cellular distribution of galaxies (van de Weygaert & Icke 1989; van de Weygaert 1991).

The predictions of the model are fixed by the density of seeds n_s or by the inter-seed separation $r_s \equiv (n_s)^{-1/3}$. There are three different sites for galaxies in the Voronoi network, nodes, filaments and walls, and they all have different clustering properties. Nodes and filaments have approximately $\xi \propto r^{-2}$, whereas walls have $\xi \propto r^{-1}$. In all cases, ξ reaches unity for $r/r \simeq 0.1$–0.3, so that a sensible choice for the inter-seed separation is about $40\, h^{-1}$ Mpc (comparable to the observed size of typical voids). The slopes for walls and filaments are easy to understand, as they arise through geometry. The number of neighbours within r of a point on a sheet or in a filament grows respectively as r^2 or r, whereas the expected number for uniform density grows as r^3. This gives $1+\xi \propto r^{-1}$ for walls and $1+\xi \propto r^{-2}$ for filaments. In other words, clustering is induced simply by the fact that galaxies are distributed within a fixed **geometrical skeleton**, which is a radically different point of view from the usual assumption that clustering reflects nonlinear dynamics. Since the observed correlation function is close to $\xi \propto r^{-2}$, this suggests that in practice galaxy clustering is dominated by filaments; this is an important insight.

SKEWNESS AND KURTOSIS In order to determine observationally whether the density field of the universe is Gaussian, we need to look at more subtle statistics than just the power spectrum. In principle, one might use the higher-order n-point correlation functions, since these are directly related to the power spectrum for a Gaussian field. On small scales, nonlinear evolution must produce non-Gaussian behaviour; for any σ, a Gaussian density distribution will produce a tail to unphysical $\delta < -1$ values, and so the real distribution must be skew, with a tail towards large values of δ. The question is whether the density field is non-Gaussian to a greater degree than that induced by gravitational evolution.

The lowest-order non-Gaussian signatures are probability distributions for density that display asymmetry (**skewness**) or a non-Gaussian degree of 'peakiness' (**kurtosis**). These deviations depend on the third and fourth moments of the one-point distribution, and may be measured through the following skewness and kurtosis parameters (scaled versions of skewness and kurtosis):

$$\begin{aligned} S &\equiv \frac{\langle \delta^3 \rangle}{[\langle \delta^2 \rangle]^2} \\ K &\equiv \frac{\langle \delta^4 \rangle - 3\langle \delta^2 \rangle^2}{[\langle \delta^2 \rangle]^3}; \end{aligned} \tag{16.149}$$

these can be calculated through second-order gravitational perturbation theory (see Peebles 1980) and should be constants of order unity. For density fields smoothed with a uniform sphere filter, the expected result is

$$S = \frac{34}{7} - (3 + n), \tag{16.150}$$

where n is the index of the density power spectrum (see e.g. Bernardeau 1994b). This result will not be derived here, but it is interesting to compare with a simple analytical model, which is the lognormal field discussed above. This writes the density fluctuation in terms of a generating Gaussian field, g, with rms σ: $\delta = \exp(g - \sigma^2/2) - 1$. For small σ the field is effectively Gaussian, but larger σ causes the distribution to become more skewed – as is necessary in order to avoid generating unphysical negative densities. The skewness parameter is easily calculated for this model, and is

$$S = 3 + \langle \delta^2 \rangle. \tag{16.151}$$

In the linear regime, this goes to $S = 3$. Coles & Frenk (1991) argue that most approximate nonlinear models give results close to this value. Since the observed skewness is deduced by filtering highly nonlinear data, it is not clear that the result from second-order theory should be perfect, so it is reassuring that the result seems robust for any physically sensible distribution function.

Gaztañaga (1992) showed that the observed skewness parameter for the APM galaxy survey was approximately constant with scale, at the value expected for nonlinear gravitational evolution of a Gaussian field. Does this mean (i) that conditions are Gaussian; (ii) that $b = 1$? This is possible, but the effects of bias need to be understood first (see chapter 17). As a simple example, consider a power-law modification of a lognormal field: $\rho' \propto \rho^b$. It is again easy to find the skewness parameter, which now becomes

$$S = 2 + (1 + \langle \delta^2 \rangle)^{b^2}. \tag{16.152}$$

In the linear regime, the skewness is $S = 3$, independent of b, and so this sort of model would not violate the observation that the moments $\langle\delta^3\rangle$ and $\langle\delta^2\rangle$ are in the correct ratio for straightforward gravitational evolution without bias. This is a symptom of a general problem: a biased mass distribution can closely resemble one which is unbiased, but which has a stronger degree of dynamical evolution.

TOPOLOGY An interesting alternative probe of Gaussianity was suggested by Gott, Melott & Dickinson (1986): the topology of the density field. To visualize the main principles, it will help to think initially about a 2D field. Two extreme non-Gaussian fields would consist either of discrete 'hotspots' surrounded by uniform density or the opposite case of discrete 'coldspots'; the picture in either case is a set of polka dots. Both of these cases are clearly non-Gaussian just by symmetry: the contours of average density will be simply-connected circles containing regions that are all either above or below the mean density, whereas in a Gaussian field (or in any symmetric case), the numbers of hotspots and coldspots must balance.

One might think that things would be much the same in 3D: the obvious alternatives are 'meatball' or 'Swiss cheese' models. However, there is a third topological possibility: the **sponge topology**. In our two previous examples, high- and low-density regions were distinguished by their **connectivity** (whether it is possible to move continuously between all points in a given set). In contrast, in a sponge, both classes of region are connected: it is possible to swim to any point through the holes, or to burrow to any point within the body of the sponge. Again, just by the symmetry between overdensity and underdensity, a Gaussian field in 3D must have a sponge-like topology.

The above discussion has focused on the properties of contour surfaces. These properties can be studied quantitatively via the **genus**: the number of 'holes' in a surface (zero for a sphere, one for a doughnut etc.). This is related to the Gaussian curvature of the surface, $K = 1/(r_1 r_2)$, where r_1 and r_2 are the two principal radii of curvature, via the **Gauss–Bonnet theorem** (see e.g. Dodson & Poston 1977)

$$C \equiv \int K\, dA = 4\pi(1 - G), \tag{16.153}$$

where G is the genus. Alternatively, topological results are sometimes quoted in terms of the **Euler–Poincaré characteristic**, χ, which equals $-2G$ (see Bardeen *et al.* 1986). The genus can also be viewed as one of a number of general measures of isodensity surfaces in 3D space. The others are the volume contained within the surface, the surface area and mean curvature; this is similar to the genus, but is evaluated by integrating $(1/r_1 + 1/r_2)/2$ over the surface, rather than $1/(r_1 r_2)$. Collectively, these are known as the **Minkowski functionals** (see Mecke, Buchert & Wagner 1994).

For Gaussian fields, the expectation value of the genus per unit volume, denoted by g, is (see Hamilton, Gott & Weinberg 1986)

$$\boxed{g = \frac{1}{4\pi^2}\left[\frac{-\xi''(0)}{\xi(0)}\right]^{3/2}(1 - \nu^2)e^{-\nu^2/2}.} \tag{16.154}$$

For the median density contour ($\nu = 0$), the curvature is negative, implying that the surface has genus greater than unity. For $|\nu| > 1$, however, the curvature is positive, as expected if there are no holes. The contours become simply-connected balls around either isolated peaks or voids.

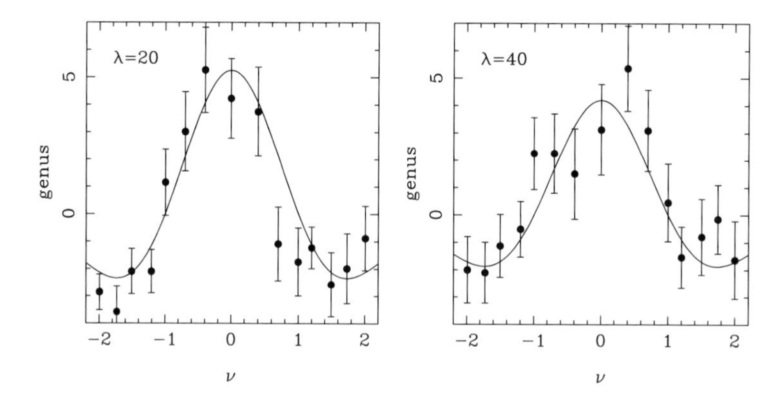

Figure 16.10. Results from the genus analysis applied to 3D data from the QDOT IRAS redshift survey (Moore *et al.* 1992). For large filtering lengths ($20\,h^{-1}$ Mpc and $40\,h^{-1}$ Mpc are shown), the results are perfectly consistent with the Gaussian prediction. Such plots constitute the strongest evidence we have that cosmic structure did indeed form via gravitational instability from Gaussian primordial fluctuations.

It is interesting to note that the genus carries some information about the shape of the power spectrum, not in the behaviour with ν but in the overall scaling. For Gaussian filtering,

$$g = \frac{1}{4\pi^2 R_{\rm G}^3} \left(\frac{3+n}{3} \right)^{3/2} (1-\nu^2) e^{-\nu^2/2}, \tag{16.155}$$

and so the effective spectral index can be determined in this way.

A similar procedure can be carried out in 2D (see Melott *et al.* 1989; Coles & Plionis 1991). The Gauss–Bonnet theorem is now

$$C \equiv \int K\, dA = 2\pi(1-G), \tag{16.156}$$

where $1-G$ is the number of isolated contours minus the number of contour loops within other loops (sometimes the 2D genus is defined with the opposite sign; our convention follows that in 3D and the signs below are consistent). The result for the 2D genus per unit area is [problem 16.6]

$$\boxed{g = -\frac{1}{(2\pi)^{3/2}} \left[\frac{-\xi''(0)}{\xi(0)} \right] \nu\, e^{-\nu^2/2}.} \tag{16.157}$$

The 3D case is analogous but, as always, more messy; see BBKS.

The application of this method to real data is illustrated in figure 16.10. Once the galaxy density field is filtered on scales large enough to reach the regime of linear density fluctuations, the results are consistent with the Gaussian prediction. This of course raises the worry that, by smoothing over many small regions, the central limit theorem will inevitably yield a Gaussian-like result in all cases. However, detailed simulations have shown that the genus statistic is capable of detecting non-Gaussian initial conditions, so the results of figure 16.10 have to be considered as important evidence that the primordial density field was at least close to being Gaussian.

16.10 Peculiar velocity fields

A research topic that has assumed increasing importance since about 1986 is the subject of deviations from the Hubble flow. Although the relation $v = Hr$ is a good approximation, it has long been known that individual galaxies have random velocities of a few hundred $\mathrm{km\,s^{-1}}$ superimposed on the general expansion. An exciting development has been the realization that these peculiar velocities display large-scale coherence in the form of **bulk flows** or **streaming flows**, which gives us the chance to probe very large-scale density fluctuations in the universe, and perhaps even to measure its mean density. Detailed reviews of these developments are given by Dekel (1994) and Strauss & Willick (1995). The peculiar-velocity field was studied in chapter 15, and this section builds on the main results derived there, which may be summarized as follows.

In linear perturbation theory, the peculiar velocity is parallel to the peculiar gravitational acceleration $\mathbf{g} = \boldsymbol{\nabla}\delta\Phi/a$:

$$\delta\mathbf{v} = \frac{2f(\Omega)}{3H\Omega}\mathbf{g}, \tag{16.158}$$

where

$$f(\Omega) \equiv \left(\frac{a}{\delta}\right)\frac{d\delta}{da} \simeq \Omega^{0.6}. \tag{16.159}$$

Alternatively, we can work in Fourier terms, where

$$\delta\mathbf{v}_{\mathbf{k}} = -\frac{iHf(\Omega)a}{k}\delta_k\hat{\mathbf{k}}. \tag{16.160}$$

In either case, the linear peculiar-velocity field satisfies the continuity relation

$$\boldsymbol{\nabla}\cdot\mathbf{v} = -Hf(\Omega)\delta \tag{16.161}$$

(where the divergence is in terms of proper coordinates). Since the fractional density perturbation (denoted by δ) is unobservable directly, one makes a connection with the galaxy distribution via the linear bias parameter

$$\delta_{\rm light} = b\,\delta_{\rm mass}. \tag{16.162}$$

The combination

$$\beta \equiv \Omega^{0.6}/b \tag{16.163}$$

can therefore be measured in principle, given the observed velocity field plus a large deep redshift survey from which the density perturbation field can be estimated.

REFERENCE FRAMES Discussions of data in this field can be confusing because there are several possible reference frames in which the velocity field can be discussed.

(1) **Heliocentric frame**. This uses redshifts as observed from the Earth (give or take the few tens of $\mathrm{km\,s^{-1}}$ induced by our orbit around the sun – usually an insignificant correction).

(2) **Local-group frame**. This frame makes corrections for the motion of the sun with respect to the nearest few galaxies (mainly due to the rotation of the Milky Way). A variety of slightly different numbers are possible according to how the local group is defined; normally the IAU convention of a velocity of $300\,\mathrm{km\,s^{-1}}$ towards galactic coordinates $\ell = 90°$, $b = 0$ is adopted (i.e. the Milky Way rotates in a left-handed sense). Galactocentric redshifts are therefore *increased* relative to heliocentric redshifts by a velocity of $300 \sin\ell \cos b$.

(3) **CMB frame**. The local group frame was originally defined on the expectation that it would be close to a Machian rest frame. However, from the 'absolute' point of view given by motion relative to the microwave background, the heliocentric frame is more nearly at rest. The motion of the Earth with respect to the background is about $371\,\mathrm{km\,s^{-1}}$ towards $\ell = 264°$, $b = 48°$ (Fixsen *et al.* 1996).

A further point to note is that redshifts are often quoted in velocity units. This does *not* assume the special-relativistic form $1 + z = [(1 + v/c)/(1 - v/c)]^{1/2}$; the 'velocity' quoted is just $v \equiv cz$. It is often the case that distances will be quoted in velocity units also, and care is needed when converting to e.g. comoving distance ($r = (2c/H_0)[1-(1+z)^{-1/2}]$ for $\Omega = 1$). At small distances, v and r will be in a ratio of 100 ($1000\,\mathrm{km\,s^{-1}} \simeq 10\,h^{-1}\,\mathrm{Mpc}$), but nonlinearities become significant at higher redshift ($10000\,\mathrm{km\,s^{-1}} \simeq 97.6\,h^{-1}\,\mathrm{Mpc}$).

THE COSMIC VIRIAL THEOREM Viewed with a resolution of ~ 1 Mpc, the galaxy clustering pattern is highly nonlinear, consisting of systems from clusters of galaxies down to small groups that are reasonably well mixed. This suggests that the assumption of stable gravitational equilibrium that is used to analyse individual clusters might be applied in a statistical sense to the whole universe. We first give a rough analysis, which illustrates the main idea, before giving the exact answer.

If we view the density field as being constructed from the superposition of a number of clusters, then the correlation function $\xi(r)$ is roughly the density profile of a cluster in the limit where clusters are well separated. The mass interior to r is therefore $M = (4\pi/3)\rho_0 r^3 \bar{\xi}$, where $\bar{\xi}(r)$ is the volume-averaged correlation function. The orbital speed is just $v^2 = GM/r$, which would correspond to a one-dimensional velocity dispersion of $\sigma^2 = GM/2r$. For a pair of galaxies at separation r in the same cluster, the pairwise dispersion will typically be $\sqrt{2}$ times the one-galaxy dispersion at $r/2$:

$$v_p^2 \simeq 2GM/r \simeq (8\pi G/3)\rho_0 r^2 \bar{\xi}(r). \qquad (16.164)$$

Eliminating ρ_0 in terms of H_0 and Ω, and taking $\xi(r) = (r/5\,h^{-1}\,\mathrm{Mpc})^{-1.8}$ for optically-selected galaxies, which implies $\bar{\xi} = (r/8.3\,h^{-1}\,\mathrm{Mpc})^{-1.8}$, we get

$$\frac{v_p}{\mathrm{km\,s^{-1}}} \simeq 670\,\Omega^{1/2} \left(\frac{r}{h^{-1}\,\mathrm{Mpc}}\right)^{0.1}. \qquad (16.165)$$

A more exact treatment (section 75 of Peebles 1980) gives a higher coefficient ($1600\,\mathrm{km\,s^{-1}}$ instead of $670\,\mathrm{km\,s^{-1}}$). A pairwise dispersion at $1\,h^{-1}\,\mathrm{Mpc}$ of 300–$400\,\mathrm{km\,s^{-1}}$ thus suggests that the universe is open ($\Omega \simeq 0.1$) if optically-selected galaxies trace mass.

Other methods of estimating Ω operate on larger scales, where the assumption of virialization is clearly false. Instead, one operates close to the linear regime of velocity perturbations.

INFALL AND OUTFLOW The classical topic in velocity fields beyond our local group has long been the perturbation to the Hubble flow caused by the local supercluster, centred on the Virgo cluster at a heliocentric $cz \simeq 1140\,\mathrm{km\,s^{-1}}$. This density concentration should retard the expansion at the radius of our galaxy (which lies in the periphery of the local supercluster). The redshift of Virgo is therefore smaller than expected for its distance; this is the phenomenon of **Virgocentric infall**. Consider the application of linear theory to this situation:

$$\delta \mathbf{v} = \frac{2\Omega^{0.6}}{3H\Omega}\,\mathbf{g}. \tag{16.166}$$

If we assume spherical symmetry, it is easy to calculate the gravitational acceleration $\mathbf{g}$ in terms of the mean density contrast $\bar{\delta}$ interior to our galaxy. We then end up with the following relation between the fractional velocity and density perturbations:

$$\frac{\delta v}{v} = \bar{\delta}\,\frac{\Omega^{0.6}}{3}(1+\bar{\delta})^{-1/4}. \tag{16.167}$$

The last factor in the above expression is an empirical nonlinear correction taken from Yahil (1985), which closely approximates the exact solution of the spherical collapse problem.

The relevant density contrast for Virgo is $\bar{\delta} \simeq 3$ in optical galaxies. The infall velocity has historically generated considerable controversy, largely because of a failure to distinguish clearly between the component of the local group's motion in the direction of Virgo ($\simeq 400\,\mathrm{km\,s^{-1}}$) and the relative velocity. The observed redshift is $1055\,\mathrm{km\,s^{-1}}$ (local group frame), whereas the distance in velocity units appears to be $1186\,\mathrm{km\,s^{-1}}$, or an infall of around $230\,\mathrm{km\,s^{-1}}$. Using the above equation thus yields $\Omega = 0.12$ if optical light traces mass.

While we are discussing the velocity perturbations caused by spherically symmetric density concentrations, we should also consider *underdense* regions. The history here is quite interesting: when apparently empty regions of space up to $30\,h^{-1}\,\mathrm{Mpc}$ in radius were discovered, it seemed initially that very large peculiar velocities were implied: to evacuate such a space over the age of the universe, particles from the centre of the void would appear to require a mean speed of order $3000\,\mathrm{km\,s^{-1}}$. However, it turns out that realistic numbers are a good deal less spectacular (see Bertschinger 1985a). Both this calculation and Virgocentric infall are interesting, but difficult to apply to the real universe because of the restriction to spherical symmetry. However, the increasingly accurate mapping of the galaxy distribution via redshift surveys has opened the way to more general techniques.

COSMOLOGICAL DIPOLES The well-determined absolute motion of the Earth with respect to the microwave background provides one of the possible general methods of estimating the cosmological density parameter Ω. Given a galaxy redshift survey, an estimate can be made of the gravitational acceleration of the local group produced by large-scale galaxy clustering. In perturbation theory, the relation between $\mathbf{v}$ and $\mathbf{g}$ can then be used to derive the peculiar velocity at a point in terms of the surrounding density field:

$$\mathbf{v} = \frac{H_0}{4\pi}\,\Omega^{0.6} \int \frac{\delta}{r^2}\,\hat{\mathbf{r}}\,d^3r \tag{16.168}$$

(Peebles 1980).

In fact, a reasonable answer for the predicted dipole can be obtained without a redshift survey, but just using angular data. Any flux-limited survey will have a radial distribution peaked around some characteristic distance that is a function of the flux limit. This means that the anisotropy of the galaxy distribution in a given flux bin provides a convolved measure of the density dipole averaged over a range of distance bins; by suitable combinations of the data as a function of flux density, one can estimate the peculiar gravity. In terms of the luminosity function $\phi(L) \equiv d^2N/(dV\, d\ln L)$, the flux distribution is

$$\frac{dN}{d\ln S} = \int (1+\delta)\, \phi(Sr^2)\, d^3r. \tag{16.169}$$

The flux-weighted dipole $\int (dN/d\ln S)\, S\, d\ln S$ is then very close to being proportional to the desired gravity integral; it differs only by the radial weighting function $\int Sr^2\phi(Sr^2)\, d\ln S$. If $\phi(L) \propto 1/L$, this is exactly constant, and so the flux and gravity dipoles align. The luminosity function is of course not exactly of this idealized form, but in practice the radial weight varies slowly. This method was used by Yahil, Walker & Rowan-Robinson (1986) and by Meiksin & Davis (1986) to derive the value $\Omega^{0.6}/b \simeq 1$ for IRAS galaxies, well before detailed redshift surveys were available. For optical galaxies, a lower value $\Omega^{0.6}/b \simeq 0.3$ was found using this method (Lynden-Bell, Lahav & Burstein 1989; Kaiser & Lahav 1989). This is a larger difference than one would expect simply from the known 20% difference in bias factors, so it does raise a question about whether the flux-based method gives a robust answer. In principle, it should be possible to obtain better answers from a redshift survey; it is then possible to construct the density field $\delta(\mathbf{r})$ and integrate to obtain a direct estimate of the predicted peculiar velocity at $r = 0$. This exercise was performed by Rowan-Robinson *et al.* (1990), who obtained $\Omega^{0.6}/b = 0.85 \pm 0.15$ as a result of analysing the QDOT IRAS redshift survey.

The main difficulty with the direct method of calculating the peculiar acceleration acting on the local group is that the integral must be cut off at large radii (a limit of $R = 150\, h^{-1}$ Mpc in the case of Rowan-Robinson *et al.* 1990). In order to see whether this truncation matters, we can ask how the dipole velocity signal is distributed over components at different distances. Consider the usual Fourier decomposition of the density field, $\delta = \sum_k \delta_k \exp(-i\mathbf{k}\cdot\mathbf{x})$. For growing-mode perturbations, $\mathbf{v}$ is parallel to $\mathbf{k}$, so that the peculiar velocity is

$$\mathbf{v} = \frac{Hf}{4\pi} \sum_k \left[-i\hat{\mathbf{k}}\, \delta_k \int \sin(\mathbf{k}\cdot\mathbf{x})\, \hat{\mathbf{k}}\cdot\hat{\mathbf{x}}\, \frac{d^3x}{x^2} \right], \tag{16.170}$$

choosing $\mathbf{x} = 0$ as the origin. The integral may be performed by using $\mathbf{k}$ as a polar axis. For the case where the d^3x integral is over a spherical region of radius R, we obtain

$$\mathbf{v} = -iHf \sum_k \frac{\delta_k\, W_k}{k}\, \hat{\mathbf{k}}, \tag{16.171}$$

where $W_k = [1 - (kR)^{-1} \sin(kR)]$. This is identical to the result of convolving the linear velocity field with a 'high-pass' window function, whose transform is W_k. Does this filtering matter? Consider the mean square value of this velocity:

$$\langle v^2 \rangle = H^2 f^2 \int \frac{\Delta^2(k)}{k^2}\, W_k^2\, \frac{dk}{k}; \tag{16.172}$$

W_k cuts off the contributions from power at $k \lesssim 1/R$. This truncation will be unimportant provided the cutoff corresponds to scales beyond the peak in $k^{-2}\Delta^2(k)$, which occurs empirically at $k \simeq 0.05\, h\, \mathrm{Mpc}^{-1}$. It is therefore clear that maximum radii of $R \simeq 100\,h^{-1}$ Mpc are barely large enough for the purpose of obtaining the full dipole signal.

We can quantify this point by writing down the correlation between estimates of the dipole velocity made with two different filterings, which will be written as respectively $\mathbf{v}^a$ and $\mathbf{v}^b$. Because $\mathbf{v}_k \propto \mathbf{k}$, it is easy to see by symmetry that orthogonal components of these velocities are uncorrelated; the covariance matrix for parallel components may therefore be written as

$$C^{ab} \equiv \langle v_i^a v_i^b \rangle = \frac{H^2 f^2}{3} \int \Delta^2(k)\, W_k^a W_k^b \,\frac{dk}{k^3}, \tag{16.173}$$

where there is no summation over i and $\Delta^2(k)$ is the spatial power spectrum in dimensionless form. In the case of Gaussian statistics, the covariance matrix suffices to give the joint distribution for the two velocities, which may be reduced to

$$dP = \frac{v_a^2 v_b^2}{\pi \Gamma^{3/2}} \exp\left[-\frac{(C_{bb} v_a^2 + C_{aa} v_b^2 - 2C_{ab} v_a v_b \mu)}{2\Gamma}\right] dv_a\, dv_b\, d\mu, \tag{16.174}$$

where $\mu = \hat{\mathbf{v}}^a \cdot \hat{\mathbf{v}}^b$ is the cosine of the misalignment angle and $\Gamma \equiv C_{aa}C_{bb} - C_{ab}^2$ (see Vittorio, Juszkiewicz & Davis 1986; Lahav, Kaiser & Hoffman 1990).

The obvious application of this formula is to the joint distribution of the true dipole velocity and the contribution from radii $< R$. We shall be interested not so much in the magnitudes of the two velocities as in their ratio. Things can be presented most simply in terms of the dimensionless variables

$$\begin{aligned} A &\equiv \mu\, C_{ab}/\sqrt{C_{aa}C_{bb}} \equiv \mu\, r \\ B &\equiv \frac{v_a}{v_b}\sqrt{\frac{C_{bb}}{C_{aa}}} \equiv \frac{v_a}{v_b}\, F, \end{aligned} \tag{16.175}$$

and integrating over one of the velocities gives

$$dP = \frac{8(1-r^2)^{3/2}}{\pi r} \frac{B^2\, dA\, dB}{(1 + B^2 - 2AB)^3}, \tag{16.176}$$

where A & B cover the ranges $-r < A < r$, $B > 0$. The problem is thus characterized by the two numbers r (the velocity correlation coefficient) and F (the ratio of the expected velocity dispersions on the two scales). Using this expression, Peacock (1992) showed that, even at large correlations r, the exact dipole will tend to align very well with that estimated from the gravity inside R, but with a magnitude that can be very different unless $r > 0.95$. Surveys with depths approaching $300\,h^{-1}$ Mpc are required in order for the peculiar acceleration to converge; estimates of Ω made using shallower surveys need to be treated with caution.

WEIGHING THE UNIVERSE The power of velocity fields is that they sample scales large enough that density perturbations are fully in the linear regime. In combination with large redshift surveys to define the spatial distribution of light, this has allowed not only a test of the assumption that large-scale clustering reflects gravitational instability but also a much more powerful extension of velocity-based methods for estimating the global density. We showed above how a knowledge of the density field surrounding the local group could be used to estimate the motion of the local group and hence predict the CMB dipole. Given a deep enough redshift survey, it is possible to use the same

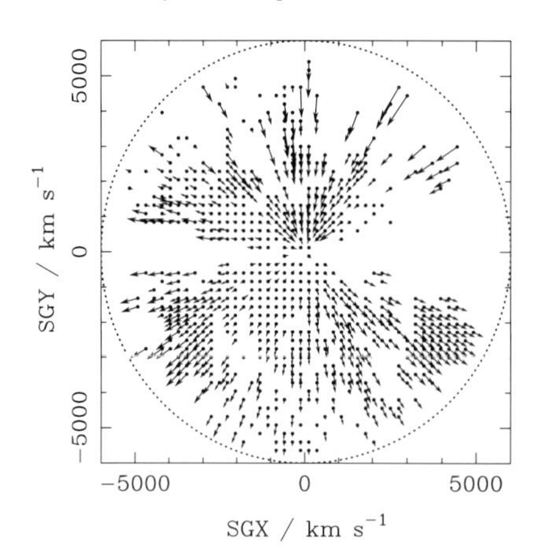

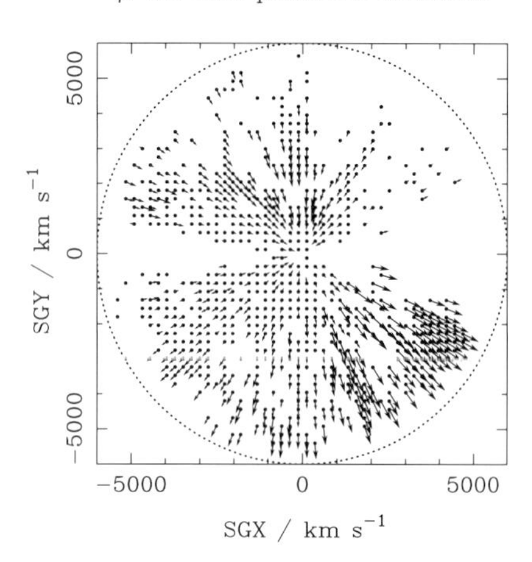

Figure 16.11. A comparison of peculiar velocities inferred from the infrared Tully–Fisher method with predictions from the IRAS galaxy density field, for galaxies within 30° of the supergalactic plane and within 6000 km s^{-1}, assuming $\beta\Omega^{0.6}/b = 0.5$. The data of Davis, Nusser & Willick (1996) have been averaged onto a grid for clarity in regions of high galaxy density.

method to predict the peculiar velocity for any point in our local region. If we know the galaxy density perturbation field δ_g, then the peculiar velocity of a point is given by

$$\mathbf{v} = \beta\,\frac{H_0}{4\pi}\int \frac{\delta_g}{r^2}\,\hat{\mathbf{r}}\,d^3r, \tag{16.177}$$

where the $\mathbf{r}$-coordinate system is centred on the point of interest. The only subtlety is that δ_g is not observed directly in real space but is deduced in redshift space. In practice, this can be corrected in an iterative way: δ_g is used to predict $\mathbf{v}$, the galaxy redshifts are corrected for the peculiar velocities, and the exercise is repeated until a stable real-space estimate of δ_g is obtained. Both this correction and the final prediction for $\mathbf{v}$ depend on β, and so it should be possible to estimate β from a comparison between the predicted velocity field and the peculiar velocities derived using distance indicators such as the Tully–Fisher method (see chapter 5). Figure 16.11 shows a recent result from one of these studies (Davis, Nusser & Willick 1996). Since most galaxies are concentrated towards the supergalactic plane defined by the local supercluster, it makes sense to plot the velocity vectors projected onto this plane. Since thousands of galaxies are involved, the velocity field is shown averaged onto a grid. The right-hand panel shows the field predicted from the gravity field due to the local density distribution.

This figure shows that, within 5000 km s^{-1}, the velocity field appears to be dominated by a few distinct regions in which the flow is nearly coherent. Of these features, the one that has received the most attention is the outward flow seen near SGX $= -4000$ km s^{-1}, SGY $= -1000$ km s^{-1}. This suggests the existence of a single large mass concentration at somewhat larger radii, which has been dubbed the **great attractor** (Dressler *et al.* 1987). Popular discussions of this object have sometimes given the impression of some mysterious concentration of mass that is detected only through its gravitational attraction. However, it should be clear from figure 16.11 that this is implausible; the overwhelming impression is that the observed and predicted velocity fields follow each other with reasonable fidelity, strongly suggesting that it should be possible to see the great attractor. This is not a totally straightforward process, since the long-range nature of gravity leaves some ambiguity over the distance at which the mass responsible for the peculiar velocities may lie. However, it is clear that the region of sky towards the great attractor contains many particularly rich superclusters, so there is no shortage of candidates (see e.g. Hudson 1993). The general agreement between the observed flows and the predictions of gravitational instability is extremely encouraging; furthermore, the amplitude of the prediction scales with β, and $\beta = 0.5$ seems to give a reasonable overall match (Davis, Nusser & Willick 1996; Willick *et al.* 1996; see figure 16.11). This is very satisfying, as it agrees with the determinations from clustering anisotropy discussed earlier.

Could we make the comparison the other way around, and predict the density from the velocities? At first sight, this seems impossible, since observations only reveal *radial* components of velocity. However, in linear theory, vorticity perturbations become negligible relative to the growing mode for times sufficiently long after the perturbations are created. It is therefore very tempting to make the assumption that the linear velocity field should be completely irrotational at the present epoch. Furthermore, **Kelvin's circulation theorem** (see section 8 of Landau & Lifshitz 1959) guarantees that the flow will remain irrotational even in the presence of nonlinearities, provided these are not so large as to cause dissipative processes. Dissipation does of course operate on the smallest scales (galaxies rotate, after all), but this should not affect the large-scale motions. We are therefore driven to write

$$\mathbf{v} = -\boldsymbol{\nabla}\psi. \tag{16.178}$$

The problem is now solved in principle: the **velocity potential** ψ can be estimated by integrating the peculiar velocities along radial lines of sight. The unobservable transverse components can then be recovered by differentiation of the potential. In practice, this is a nontrivial problem, given that we are dealing with a limited number of galaxies, each of which has a rather noisy velocity estimate, of 20% precision at best. Nevertheless, by averaging over large numbers of galaxies to produce a smoothed representation of the radial velocity field on a grid, it is possible to use this method. The practical application goes by the name of **POTENT** (Bertschinger & Dekel 1989; Dekel *et al.* 1990; Bertschinger *et al.* 1990).

Before moving on, we might ask how we could *test* the assumption of potential flow. There is no general technique, but certain consistency tests are possible. For example, suppose the velocity field consisted of a set of rotating bodies. The potential-flow analysis would give a potential that had discontinuities between lines of sight that pass either side of a rotation axis; differentiation would then lead to very large (infinite for solid-body rotation) inferred velocities in the plane of the sky. The fact that the recovered velocity fields do not show such gross anisotropies between radial and transverse components is reassuring, but some smaller degree of vorticity is harder to rule out. A better test comes from the comparison between the recovered velocity field and estimates of the density field. The continuity relation says that the divergence of the velocity field should scale directly with δ, with the constant of proportionality being β. Armed with the POTENT velocity field, and an estimate of δ from a redshift survey, one can test this relation point-by-point throughout the sample volume. There are two ways this can be done: either predict velocities from densities and compare velocity fields, or predict densities from velocities and compare densities. Neither of these is ideal. A density comparison involves differentiation of a noisy velocity field, but is local. The prediction of velocities is less noisy, but involves non-local integration over the density field. At present, results from this method favour $\beta_{\rm IRAS} \simeq 1$ (Dekel 1994); this about a factor 2 higher than the results quoted earlier, and the reason for the discrepancy is not understood.

Possible caveats concerning these important developments are twofold. At present, the density–velocity studies do little more than establish that there is a gradient in density increasing towards the great attractor, and that galaxies stream in that direction. We need to have a volume of space containing many independent structures, to confirm that the predicted and observed velocity fields really do correlate in detail. The other point is that although the continuity equation is a prediction of gravitational instability theory it applies to many other mechanisms for generating velocities (e.g. explosions). An apparent $\Omega^{0.6}/b$ of order unity simply says that the velocities have been transporting particles for a substantial fraction of the Hubble time. Nevertheless, the velocities are a prediction of the instability theory, and they are seen much as expected; it is hard to ask more of a theory than that.

Problems

(16.1) Show that the two-point function in k space for a homogeneous random statistical process is always a delta function:

$$\langle \delta_k(\mathbf{k})\delta_k^*(\mathbf{k}')\rangle = \frac{(2\pi)^3}{V} P(k)\, \delta_{\rm D}(\mathbf{k}-\mathbf{k}'). \tag{16.179}$$

Note that this applies for both Gaussian and non-Gaussian processes.

(16.2) Consider a power spectrum with small-scale truncation:

$$\Delta^2 = (k/k_0)^{n+3} \exp(-k/k_c). \tag{16.180}$$

Show that the correlation function is

$$\xi(r) = \frac{(k_c/k_0)^{n+3}}{y\,(1+y^2)^{1+n/2}}\, \Gamma(2+n)\, \sin[(2+n)\arctan y], \tag{16.181}$$

where $y = k_c r$. For this model, ξ is negative at large r for $n > 0$; for what values of n does $\bar{\xi}$ stay positive at large r?

(16.3) Suppose we begin with a totally uniform matter distribution and then group it into discrete chunks as uniformly as possible. Show that conservation of mass corresponds to a spectrum of $n = 2$, whereas if the grouping is done in terms of local dipoles that conserve momentum, the minimal spectrum induced by discreteness is $n = 4$.

(16.4) By writing down the joint distribution for a Gaussian density field sampled at two points, show that the mean density profile around a given point has a radial dependence that is proportional to the correlation function, $\xi(r)$. Is this behaviour likely to hold for non-Gaussian fields?

(16.5) Derive the relation between the 3D power spectrum of a random density field, and the 1D spectrum of the fluctuations seen along a 'skewer': a randomly chosen line through the field (hint: the 1D and 3D correlation functions are identical, through isotropy).

(16.6) Show how to obtain an expression for the genus per unit area of a 2D Gaussian random field.

17 Galaxy formation

17.1 The sequence of galaxy formation

The discussion of galaxy evolution in chapter 13 raised many basic questions about the process of galaxy formation: did bulges form first, and did they accrete disks later? What is the importance of galaxy mergers? What sets the form of the galaxy luminosity function? In addition, we have seen in chapters 12 and 16 that an $\Omega = 1$ universe requires galaxy formation to be biased in favour of high-density environments; how could such a bias have arisen? The purpose of this chapter is to present some of the theoretical tools with which these questions may be tackled. We start with a simple overview of two contrasting ways in which collapsed objects like galaxies could form.

DISSIPATIONLESS COLLAPSE What will be the final state of an object that breaks away from the background and undergoes gravitational collapse? If the matter of the object is collisionless (either purely dark matter, or stars), this is a relatively well-posed problem, which should be capable of a clear solution.

The analytical approach has concentrated on **gravitational thermodynamics**, and sought an equilibrium solution. This has turned out to be a subtle and paradoxical problem, whose main analysis goes back to a classic paper by Lynden-Bell (1967). Imagine initially that the self-gravitating body consists of gas, so that it is reasonable to look for an equilibrium solution in the form of a configuration of constant temperature. In this case, the gas particles would have a Maxwellian distribution of velocities, with a velocity dispersion σ_v that was independent of position. Such self-gravitating isothermal distributions were studied in chapter 12, where it was shown that the density at large radii tended to

$$\rho = \frac{\sigma_v^2}{2\pi G r}, \tag{17.1}$$

with the possibility of a constant-density core at small radii. Lynden-Bell showed that such a distribution might still be expected for a system of collisionless particles, even though these lack the microscopic energy exchange that produces thermal equilibrium in a gas. Collisionless particles change their total energy E only if they experience a fluctuating gravitational potential, $\dot{E} = m\dot{\Phi}$; if the potential changes are sufficiently chaotic, this can scramble particle energies as effectively as microscopic interactions. In the initial phases of collapse, the potential is indeed highly time dependent, and Lynden-Bell coined the term **violent relaxation** for this mechanism, by which particles exchange energy via collective gravitational effects. Notice that the energy transfer is proportional to mass: this means that equilibrium would lead to equipartition of velocity, rather than equipartition of energy. In a cluster of galaxies, we would therefore expect high-mass and low-mass galaxies to have the same spatial distribution. For the more massive galaxies to sink to the bottom of the potential well requires individual gravitational encounters [problem 17.1].

The isothermal sphere has several puzzling features. The density is proportional to σ_v^2, whereas one might have thought that heating the system would cause the particles to disperse, lowering the density. The reason that this does not happen is the **virial theorem** [problem 17.2]: the kinetic and potential energies for a equilibrium gravitating body satisfy $2K + V = 0$, so that the total energy is $E = K + V = -K$. Such systems thus have a negative specific heat: increasing the 'temperature' (raising K) reduces the total energy and makes the system more strongly bound. Of course, these arguments are not easy to apply to the isothermal sphere, because its mass diverges. This happens because the Maxwellian distribution contains particles of arbitrarily large velocities, whose orbits can reach any given radius. In practice, there will be a limit to the existence of such stars, if only because tidal effects at large radii will remove them from the system. If a maximum total energy for particles is imposed, then it is possible to construct more sensible configurations, which fall below the isothermal $1/r^2$ profile at large radii (King 1966). Empirically, such models can bear a strong resemblance to de Vaucouleurs' $r^{1/4}$ law; the significance of this in terms of the energy distribution in systems with $r^{1/4}$ density profiles was explored by Binney (1982).

These models are still very much idealizations, since it is not to be expected that violent relaxation could go to completion. Consider the spherical infall model studied in chapter 15, where infalling shells of matter are expected to reach virial equilibrium at around 200 times the critical density. The shells that fall in early will very probably relax, but the last shells probably will not. The structure of the collapsed object will therefore retain a memory of the conditions under which it formed: it will have a characteristic density and size dictated by the density of the universe at the time of formation. This argument sets a limit to the redshift at which the central parts of galaxies could have formed. For a spherically symmetric body, the circular orbital velocity is

$$v_c^2 = \frac{GM}{r} = \frac{4\pi G}{3}\,\bar{\rho}\,r^2, \tag{17.2}$$

and the mean density must exceed $f_c \simeq 200$ times the background at formation (dissipation could increase the density further after virialization):

$$\bar{\rho} \gtrsim f_c\,(1+z)^3\,\frac{3\Omega H^2}{8\pi G} \quad \Rightarrow \quad (1+z) \lesssim \left(\frac{2}{\Omega f_c}\right)^{1/3} \frac{(v_c/\,\mathrm{km\,s^{-1}})^{2/3}}{(r/\,h^{-1}\,\mathrm{Mpc})^{2/3}}. \tag{17.3}$$

Since the central parts of luminous galaxies have rotation speeds of a few hundred $\mathrm{km\,s^{-1}}$ at radii of a few kpc, this suggests a maximum redshift of about 20, unless Ω is very low.

Given the observational reality of dark matter, these calculations for collisionless collapse have established themselves as a standard part of the current picture for galaxy formation. White & Rees (1978) were the first to suggest that the formation process must be in two stages, in which baryons condense within the potential wells defined by the collisionless collapse of **dark-matter haloes**. This simplifies the problem in many ways, since the complex fluid-mechanical and radiative behaviour of the gas can be ignored initially. The next section will develop these pure dark-matter methods in some detail.

DISSIPATIONAL COLLAPSE An aspect of this problem that we will not consider in detail here is whether the White–Rees picture gives a reasonable account of the appearance of individual galaxies. This depends partly on the detailed structure of dark-matter haloes, as revealed by numerical simulations, but also on how the baryonic material settles in the dark-matter halo. The simplest possibility would be that the gas turns into stars at an early stage of the collapse. There would then be no separation between stars and dark matter: the stars would behave as collisionless particles, and the radial mass profile would then have the same shape as that of the dark matter. An early and influential argument for something along these lines was given by Eggen, Lynden-Bell & Sandage (1962) – the **ELS argument**. They noted that metal-poor stars in the Solar neighbourhood have highly eccentric orbits and therefore argued that the spheroid stars must have formed in a time short compared to the collapse time ($\sim 10^8$ years), so that their orbits mixed collisionlessly.

The opposite is true for spiral disks, which are supported by rotation, and whose stars have mainly circular orbits. From this, it is possible to argue that disks formed much more slowly and appeared relatively late on the cosmological scene, as follows. It is believed (see below) that galaxies acquire angular momentum as a result of tidal torques during protogalactic growth. This is quantified via the dimensionless spin parameter that relates angular momentum, binding energy and mass:

$$\boxed{\lambda \equiv \frac{J\,|E|^{1/2}}{GM^{5/2}},} \tag{17.4}$$

where it is shown below that $\lambda \simeq 0.05$ is expected at collapse. If we now allow the gas to dissipate energy (by heating due to shocks when the orbits of different parcels of gas attempt to cross, followed by radiation), then it will settle in the halo, conserving J and M. If the gas dominated the density, its binding energy $|E|$ would scale with radius as $1/r$, and in order to reach rotational support ($\lambda \simeq 1$), the gas would have to collapse by a factor ~ 400 in radius. If the dark matter dominates, however, the gas only has to collapse by the smaller factor $1/\lambda$ (the halo has a flat rotation curve, so that the angular momentum per unit mass is $\propto r$; the gas starts collapse with angular momentum too small for rotational support by a factor λ, so it needs to collapse by the reciprocal of this factor). In either case, this is a substantial degree of collapse, and it is likely to happen slowly since the radiative cooling will be less efficient at first while the gas is still diffuse. This also gives some insight into morphological segregation: in a dense cluster environment, extended gas will not easily settle on an individual galaxy.

The redshift z_f at which disk formation was completed can be estimated if there is no exchange of angular momentum between disk and halo as the disk forms, so that J/M for the gas remains at its initial value. If the gas settles into an exponential disk with a constant rotational velocity, one can then deduce [problem 17.3] that the scalelength of the disk satisfies

$$r_{\rm disk} \simeq 3\,h^{-1}{\rm kpc}\left(\frac{v_c}{100\,{\rm km\,s^{-1}}}\right)\frac{H_0}{H(z_f)} \tag{17.5}$$

(see Fall & Efstathiou 1980; Mao, Mo & White 1997). The last factor equals $(1+z)^{-1}(1+\Omega z)^{-1/2}$ for a matter-only model, and disks forming at redshifts significantly above unity would be much smaller than those observed today. This argument for recent disk formation contrasts with the above suggestion that dissipationless collapse in spheroidal components may have occurred at much higher redshift. This separation in characteristic redshifts has for many years been the basis for a 'standard' view in which galaxy formation is a bimodal process (see e.g. Larson 1976; Gunn 1982).

Finally, it is interesting to ask what effect dissipative infall has on the overall mass profile of a galaxy: it appears that stars and gas dominate the gravity in the central parts of galaxies, and this must also modify the dark-matter density. A useful approximation here is to assume that the angular momentum of dark-matter particles is unaffected as baryons accumulate at the centre. For circular motions, $J = mvr$ and $v^2 = GM/r$, so that $J \propto (Mr)^{1/2}$, suggesting that $rM(< r)$ would be an adiabatic invariant during disk growth. More explicitly,

$$r[M_{\rm B}(< r) + M_{\rm DM}(< r)] = r_i M_i(< r_i), \tag{17.6}$$

which relates the initial mass distribution to the final distributions of baryons and dark matter (Flores *et al.* 1993). Since the final baryon distribution can be deduced directly from the light profile of the galaxy and the final dark distribution from the rotation curve, the initial halo profile can be deduced (or a hypothetical halo profile can be used to predict the rotation curve). As an illustration of this technique, consider what happens to an isothermal-sphere dark halo that grows a baryonic concentration in its centre. Initially, the mass profile is $M(< r_i) = Ar_i$, of which a fraction α is baryonic and a fraction $1 - \alpha$ is dark. Let the final baryonic mass profile be $M_{\rm B}(< r)$, so that the adiabatic relation is

$$r[M_{\rm B}(< r) + M_{\rm DM}(< r)] = Ar_i^2. \tag{17.7}$$

Conserving dark matter implies

$$M_{\rm DM}(< r) = (1 - \alpha)M_i(< r_i) = (1 - \alpha)Ar_i, \tag{17.8}$$

which allows r_i to be eliminated in terms of $M_{\rm DM}$, giving a quadratic equation for $M_{\rm DM}$ whose solution is

$$M_{\rm DM}(< r) = Ar\,\frac{(1-\alpha)^2}{2}\left[1 + \frac{4M_{\rm B}(< r)}{Ar(1-\alpha)^2}\right]^{1/2}. \tag{17.9}$$

At large r, this corresponds to the original halo profile, compressed by a factor $(1 - \alpha)$; at smaller r, where the baryon mass is significant, there is a more complex deviation from a power law. Realistic cases, which take into account the ways in which real haloes deviate from the simple isothermal sphere, can be solved numerically along the same lines (Navarro, Frenk & White 1996). For most galaxies, these comparisons work out quite well, implying that the observed dark-matter distributions are much as expected from the gravitational collapse of cold dark matter. The exception is in dwarf galaxies, where the rotation curves indicate that dark matter dominates and yet the observed rotation velocity profile is not as predicted (Moore 1994). It is not yet clear how serious a problem this is.

17.2 Hierarchies and the Press–Schechter approach

We have seen in chapter 15 that the perturbations that survive recombination are of two distinct classes. In some cases, such as cold dark matter, primordial fluctuations survive on very small scales (**small-scale damping**); in other cases, such as hot dark matter or adiabatic baryons, the perturbation field is dominated by fluctuations on scales of order the horizon at $z_{\rm eq}$ (**large-scale damping**). The consequences for galaxy formation are radically different. In the former case, nonlinear collapse of sub-galactic mass units can be the first event that occurs after recombination. These will then cluster together in a **hierarchy**, forming successively more massive systems as time progresses. Hierarchies are also known as **bottom-up** pictures for galaxy formation. Conversely, in large-scale damping, pancakes of cluster or supercluster size are the first structures to form. Galaxies must then be presumed to form through dissipative processes occurring in the shocked gas that forms at pancake collapse (the **top-down** picture).

As described, it is clear that top-down pictures are unappealing in that the physics of galaxy formation is likely to be very complex and messy. Reality *is* often like this: the formation of stars is a good example of a fundamental process where is is very hard to understand what is going on. In contrast, the computational simplicity of hierarchies has led to much more detailed work being performed on them. Theoretical prejudice aside, however, the universe *looks* like a hierarchy, displaying many small groups of galaxies (e.g. the Milky Way's own Local Group), which exist within superclusters that are only mildly nonlinear. If the linear power spectrum did contain a cutoff at large wavenumbers, we can be confident that this must lie at relatively small scales. Nonlinear clustering under gravity will tend to erase the evidence for any initial coherence scale once the universe has evolved to the point where its rms linear density contrast is of order unity, since a power-law correlation function to $\xi \gtrsim 1000$ will then become established (see chapter 16). Today, the rms contrast is unity on a scale of around $8\,h^{-1}$ Mpc, and we also observe clustering at $z \simeq 1$, where this scale would have been closer to $1\,h^{-1}$ Mpc. This means that a cutoff at a few tenths of a Mpc cannot be ruled out, but we have no evidence against the idea that the density field contains a fluctuation amplitude that diverges as we look at ever smaller scales.

It may seem obvious that an infinitely nonlinear density field cannot be analysed within the bounds of linear theory, but a way forward was identified by Press & Schechter (1974; PS). The critical assumption in the PS analysis is that, even if the field is nonlinear, the amplitude of large-wavelength modes in the final field will be close to that predicted from linear theory. For this to hold requires the 'true' large-scale power to exceed that generated via nonlinear coupling of small-scale modes, which in turn requires a spectral index $n < 1$ (e.g. Williams *et al.* 1991). We now proceed by recognizing that, for a massive clump to undergo gravitational collapse, the average overdensity in a volume containing that mass should (as usual) exceed some threshold of order unity, δ_c, independent of substructure. The location and properties of these bound objects can thus be estimated by an artificial smoothing (or filtering) of the initial linear density field. If the filter function has some characteristic length R_f, then the typical size of filtered fluctuations will be $\sim R_f$ and they can be assigned a mass $M \sim \rho_0 R_f^3$. The exact analytic form of the filter function is arbitrary and is often taken to be a Gaussian for analytic convenience.

Now, if the density field is Gaussian, the probability that a given point lies in a region with $\delta > \delta_c$, the **critical overdensity** for collapse, is

$$p(\delta > \delta_c \mid R_f) = \frac{1}{2}\left[1 - \mathrm{erf}\left(\frac{\delta_c}{\sqrt{2}\,\sigma(R_f)}\right)\right], \tag{17.10}$$

where $\sigma(R_f)$ is the linear rms in the filtered version of δ. The PS argument now takes this probability to be proportional to the probability that a given point has ever been part of a collapsed object of scale $> R_f$. This is really to assume that the only objects that exist at a given epoch are those that have only just reached the $\delta = \delta_c$ collapse threshold; if a point has $\delta > \delta_c$ for a given R_f, then it will have $\delta = \delta_c$ when filtered on some larger scale and will be counted as an object of the larger scale. The problem with this argument is that half the mass remains unaccounted for: PS therefore simply multiplied the probability by a factor 2. Note that this procedure need not be confined to Gaussian fields; all we need is the functional form of $p(\delta > \delta_c \mid R_f)$. The analogue of the factor 2 problem remains: what should be done with the points having $\delta < 0$?

Accepting the PS factor 2 (which will be justified properly below), the fraction of the universe condensed into objects with mass $> M$ can be written in the universal form

$$F(> M) = 1 - \mathrm{erf}(\nu/\sqrt{2}), \tag{17.11}$$

where $\nu = \delta_c/\sigma(M)$ is the threshold in units of the rms density fluctuation. This integral probability is related to the mass function $f(M)$ (defined such that $f(M)\, dM$ is the comoving number density of objects in the range dM) via

$$Mf(M)/\rho_0 = |dF/dM|, \tag{17.12}$$

where ρ_0 is the total comoving density. Thus,

$$\boxed{\frac{M^2 f(M)}{\rho_0} = \frac{dF}{d\ln M} = \left|\frac{d\ln\sigma}{d\ln M}\right| \sqrt{\frac{2}{\pi}}\, \nu \, \exp\left(-\frac{\nu^2}{2}\right).} \tag{17.13}$$

We have expressed the result in terms of the **multiplicity function**, $M^2 f(M)/\rho_0$, which is the fraction of the mass carried by objects in a unit range of $\ln M$. For power-law fluctuation spectra, the multiplicity function thus always has the same shape (a skew-negative hump around $\nu \simeq 1$); changing the spectral index only alters the mass scale via $\nu = (M/M_*)^{(n+3)/6}$. Notice that the low-mass slope is generally rather steep:

$$f(M) \propto M^{(n-9)/6}. \tag{17.14}$$

This is quite a serious problem for applying this analysis to galaxies, where the faint end of the luminosity function scales roughly as $\phi \propto L^{-1}$. Such a slope would require $n = 3$; the practical value of the fluctuation-spectrum index on galaxy scales is $n \lesssim -2$. Consequently, most models of galaxy formation based on dark-matter hierarchies are faced with an embarrassing surplus of low-mass dark-matter haloes.

APPLICATION TO GALAXY CLUSTERS The most important practical application of the PS formalism is to rich clusters. As already discussed, these are the most massive nonlinear systems in the current universe, so a study of their properties should set constraints on the shape and normalization of the power spectrum on large scales. These issues are discussed by e.g. Henry & Arnaud (1991), who sidestep the issue of what mass to assign to a cluster by using the observed distribution of temperatures (see below for the relation between mass and virial temperature). White, Efstathiou & Frenk (1993) work directly from the abundance of rich Abell clusters, whose masses they assume to be known from the sort of detailed modelling described in chapter 12. Typical rich-cluster masses are $10^{14.5} h^{-1} M_\odot$, and this mass is contained within a Lagrangian radius of $6.5\Omega^{-1/3}\, h\,\mathrm{Mpc}^{-1}$; it is therefore normally only a small extrapolation to say that the cluster results probe the density field averaged in spheres of radius $8\, h\,\mathrm{Mpc}^{-1}$. Fitting their data with the PS form for top-hat filtering and $\delta_c = 1.69$, Henry and Arnaud deduce a *linear-theory* rms in $8h^{-1}$ Mpc spheres of $\sigma_8 = 0.59 \pm 0.02$ for $\Omega = 1$. Similarly, White, Efstathiou & Frenk (1993) deduce $\sigma_8 = 0.57 \pm 0.06$. These calculations have been checked against numerical simulations, so the result therefore seems rather robust and well determined. For models with low density, the required normalization rises: for a lower mean density, a greater degree of collapse is needed to yield the observed number of objects of a given mass. The scaling with Ω is approximately the same as is required in the cosmic virial theorem to keep velocity dispersions constant (reasonably enough, given that these determine the cluster masses):

$$\boxed{\sigma_8 \simeq (0.5 - 0.6)\, \Omega^{-0.56}.} \tag{17.15}$$

This knowledge of σ_8 as a measure of the true mass inhomogeneity (the **cluster normalization**) is a fundamental cosmological datum; most methods return statistics that are governed by an unknown degree of bias, and it is nice to have a measure that refers directly to the mass. Of course, this result applies only for a Gaussian density field, but the evidence is that the degree of non-Gaussianity on these large scales is small (see chapter 16). Furthermore, this number is consistent with what one would infer by combining other measures of clustering. Leaving aside nonlinear corrections, we know from the observed inhomogeneities on these scales that $b\sigma_8 \simeq 1$ for optically selected galaxies. Studies of redshift-space distortions and peculiar velocities imply $\Omega^{0.6}/b \simeq 0.5$; combining these two results gives $\sigma_8 \Omega^{0.6} \simeq 0.5$, independently of the degree of bias. Alternatively, we learn that $b \simeq 2\Omega^{0.6}$; mass roughly traces light if $\Omega \simeq 0.3$.

APPLICATION TO HIGH-REDSHIFT SYSTEMS At high redshifts, the scale of nonlinearity must be smaller; exactly how it evolves gives information on Ω and the small-scale power spectrum. If we have observations on the number density of some type of object of mass M, then the collapsed fraction must be at least $F_c \sim Mn/\rho_b = n\,(4\pi R^3/3)$, where ρ_b is the background density and R is the radius of the Lagrangian sphere that contains mass M. Knowing F_c, the number $\sigma(R)$ can then be inferred. For rare objects, this is a pleasingly robust process: a large error in F_c will give only a small error in $\sigma(R)$, because the abundance is exponentially sensitive to σ.

Now, total masses are ill defined observationally, and a more robust quantity to use is the velocity dispersion. Virial equilibrium for a halo of mass M and proper radius r demands a circular orbital velocity of

$$v_c^2 = \frac{GM}{r}. \tag{17.16}$$

For a spherically collapsed object this velocity can be converted directly into a Lagrangian comoving radius containing the mass of the object (e.g. White, Efstathiou & Frenk 1993)

$$\frac{R}{h^{-1}\,\mathrm{Mpc}} = \frac{2^{1/2}(v_c/100\,\mathrm{km\,s^{-1}})}{\Omega_m^{1/2}(1+z_c)^{1/2} f_c^{1/6}} \tag{17.17}$$

(recall that $H_0 \equiv 100h\,\mathrm{km\,s^{-1}Mpc^{-1}}$). Here, z_c is the redshift of virialization, Ω_m is the *present* value of the matter density parameter and f_c is the density contrast at virialization relative to the background, which is well approximated by $f_c = 178/\Omega_m^{0.7}(z_c)$ with only a slight sensitivity to whether Λ is non-zero (chapter 15).

What this shows is that the higher characteristic density of objects that collapse at high redshifts means that a system of given velocity dispersion can be achieved with a lower mass at high z than at present. This trend makes it harder to detect evolution in the mass distribution through looking at the abundance of massive systems. Rich clusters are rare objects today, and since σ_8 will change with redshift in a way that depends on Ω, then in principle the abundance of high-redshift clusters provides a test for Ω that is very sensitive: for high Ω, the abundance of massive clusters should be exponentially suppressed at high z. However, given that what we observe is the abundance of clusters as a function of velocity dispersion (or virial temperature, through X-ray observations), the majority of this sensitivity is lost (Kaiser 1986b). It is still possible to infer something about Ω, but more careful modelling is required.

A more fruitful approach is to use the abundances of high-z objects to probe the power spectrum on much smaller scales than σ_8, where it is hard to make direct measurements today. The best application of this argument is to damped Lyα absorption systems, which show that substantial amounts of neutral hydrogen (HI) existed at high redshifts (see chapter 12). If the fraction of baryons in the virialized dark-matter halos equals the global value $\Omega_{\rm B}$, then data on these systems can be used to infer a lower limit to the total fraction of matter that has collapsed into bound structures at high redshifts. The highest abundance of HI is found at $\langle z \rangle \simeq 3.2$, corresponding to $\Omega_{\rm HI} \simeq 0.0025h^{-1}$ (Lanzetta *et al.* 1991, Storrie-Lombardi *et al.* 1996). If we take a nucleosynthetic measure of $\Omega_{\rm B}h^2 \simeq 0.02$ (chapter 9), then the collapsed fraction corresponding to these systems is quite high:

$$F_c = \frac{\Omega_{\rm HI}}{\Omega_{\rm B}} = 0.12h \qquad (17.18)$$

(e.g. Mo & Miralda-Escudé 1994). As discussed below, the photoionizing background prevents virialized gaseous systems with circular velocities of less than about $50\,{\rm km\,s^{-1}}$ from cooling efficiently, so R cannot be much less than $0.2\,h^{-1}$ Mpc, but cannot exceed the $\simeq 1\,h^{-1}$ Mpc appropriate for present-day giant galaxies. We conclude that, at $z = 3.4$, $\sigma(R) \gtrsim 1$ for $R \simeq 0.5\,h\,{\rm Mpc}^{-1}$. This is an invaluable measure of power on scales that are today erased by nonlinear evolution. If $\Omega = 1$, it implies that $\sigma \gtrsim 4$ today, and this is quite a large number for models in which the power spectrum is very flat on small scales, as discussed in chapter 16. This observation is more consistent with a low-density universe, since there is less evolution in σ between $z = 3.2$ and the present.

RANDOM WALKS AND CONDITIONAL MASS FUNCTIONS The factor 2 'fudge' has long been recognized as the crucial weakness of the PS analysis. What one has in mind is that the mass from lower-density regions accretes onto collapsed objects, but it does not seem correct for this to cause a doubling of the total number of objects. To see where the error arises, consider figure 17.1, which illustrates the **random trajectory** taken by the filtered field δ at some fixed point, as a function of filtering radius R. This starts at $\delta = 0$ at $R = \infty$, and develops fluctuations of increasing amplitude as we move to smaller R. Thus, if $\delta < \delta_c$ at a given point, it is quite possible that it will exceed the threshold at some other point; indeed, if the field variance diverges as $R \to 0$, it is inevitable that the threshold will be exceeded. So, instead of ignoring all points below threshold at a given R, we should find the **first upcrossing** of the random trajectory, the largest value of R for which $\delta = \delta_c$.

This analysis is most easily performed for one particular choice of filter: sharp truncation in k-space. Decreasing R then corresponds to adding in new k-space shells,

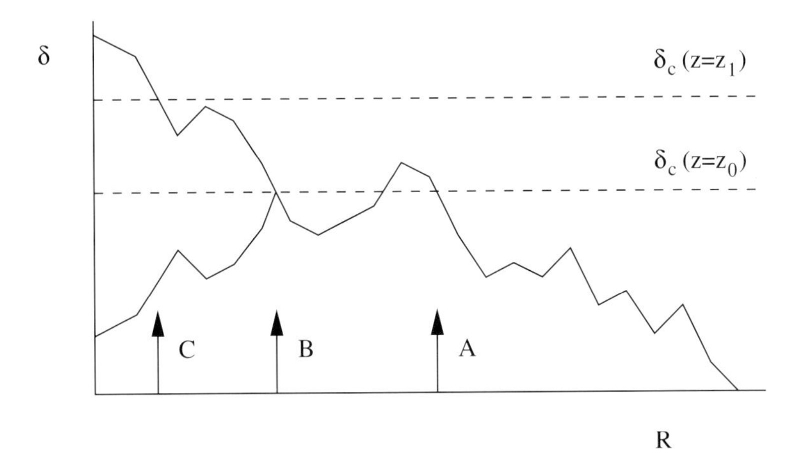

Figure 17.1. This plot shows the value of the density perturbation field δ, filtered on a scale R and sampled at a single point of space for different R values. If the filter function corresponds to abrupt truncation in k-space, the $\delta(R)$ trajectory is a random walk. As we move to smaller R, the rms δ increases, and at point A first crosses the collapse threshold. The trajectory can subsequently fall below threshold and recross at B. However, the point is associated with a collapsed system of size R_A, not R_B. The random walk is symmetric, and both rising and falling trajectories from B are equally likely. The conditional mass function may be obtained by considering two thresholds, one for the present and one for high redshifts. In this case, the point is in a system of scale R_A at $z = z_0$, but inhabited the smaller system of scale R_C at $z = z_1$.

all of which are independent for a Gaussian field. The $\delta(R)$ trajectory is then just a random walk (in R-space, not real space; we are always considering just a single position in space), and the solution is very easy. Consider a point on the walk that has reached the threshold; its subsequent motion will be symmetric, so that it is equally likely to be found above the threshold as below at some smaller R. The points above threshold at a given value of R all correspond to trajectories that crossed at some higher value of R; we can get the corresponding distribution of points below threshold that once crossed the threshold simply by taking the part of the δ-distribution lying above threshold and reflecting it in the threshold. The probability that the threshold has never been crossed (the **survival probability**) is then obtained by subtracting this reflected distribution from the overall Gaussian distribution:

$$\frac{dP_S}{d\delta} = \frac{1}{\sqrt{2\pi}\,\sigma}\left[\exp\left(-\frac{\delta^2}{2\sigma^2}\right) - \exp\left(-\frac{(\delta - 2\delta_c)^2}{2\sigma^2}\right)\right]. \tag{17.19}$$

Integrating this and subtracting it from unity gives the probability that the threshold has been crossed at least once:

$$1 - P_S = 1 - \operatorname{erf}\left(\frac{\delta_c}{\sqrt{2}\,\sigma}\right); \tag{17.20}$$

this is just twice the unconstrained probability of lying above the threshold, thus supplying the missing factor 2. Unfortunately, the above analysis is only valid for this special choice of filter: using k-space filters that are differentiable leads to just the original PS form (without the factor 2) at high mass. The surplus probability must then shift to low masses, so that the shape of the function changes. Which filter is the best choice cannot be said in advance; we are left with the empirical fact that the PS formula does fit N-body results quite well (see Lacey & Cole 1994).

One useful extension of the random-walk model is that it allows a calculation of the **conditional multiplicity function**: given a particle in a system of mass M_0 at some epoch a_0, what was the distribution of masses where that particle resided at some earlier epoch a_1? This is now just the random walk with two absorbing barriers, at δ_c/a_0 and δ_c/a_1; we want the probability that the second has not been crossed, subject to the first having been crossed at M_0 (see figure 17.1). The solution for the integral mass distribution is obtained by the same reflection argument as above, giving

$$F(>\mu) = 1 - \mathrm{erf}\left[\frac{\nu(t^{-1}-1)}{\sqrt{2(\mu^{-1}-1)}}\right], \tag{17.21}$$

where $\nu \equiv \delta_c/(a_0\sigma_0)$, $t \equiv a_1/a_0$ and $\mu \equiv \sigma^2(M_0)/\sigma^2(M_1)$. The function $F(>\mu)$ tells us that the early histories of particles that end up in different mass systems are markedly different over a large range of expansion factors. As $t \to 0$, the distribution loses the 'memory' of its destination mass M_0, and the distribution of M_1 becomes identical to that in the unconstrained case. As $t \to 1$, however, the distribution of M_1 peaks around M_0, but it is clearly more sharply peaked for large ν. In other words, the large-ν systems have atypical precursor masses over a greater part of their past. This may provide a way of understanding many of the systematics of galaxy systems in terms of their merger histories; see Bower (1991) and Lacey & Cole (1993).

These analytical descriptions of how mass gathers into clumps in hierarchical clustering have been shown to describe the results of numerical simulations extremely well (Bond *et al.* 1991) and they provide a powerful tool for understanding the growth of cosmological structure. For example, it is now possible to answer an apparent paradox of the simple Press–Schechter theory, in which all collapsed objects considered are those that formed only at the instant considered (i.e. they sit exactly at the threshold $\delta = \delta_c$), even though commonsense says that some objects must have formed at high redshift and survived unchanged to the present day. The conditional mass function indeed allows us to define something close to the typical formation epoch for a clump, which we might take as the time when the probability is 0.5 that the precursor mass is at least half the final mass. From the above expression for the conditional mass function, this gives a formation redshift

$$z_f = \frac{O(1)}{\nu}, \tag{17.22}$$

where the exact coefficient depends on the spectrum. This expression says that the formation redshift depends on the final mass of the clump; if it is $\gg M^*$, the formation epoch will be in the recent past, whereas low-mass clumps survive for longer. As a specific example, this calculation implies that massive clusters in an Einstein–de Sitter model form at very low redshifts. About 30% of Abell clusters will have doubled their mass since as recently as $z = 0.2$ (as against only 5% if $\Omega = 0.2$). The observed frequency of substructure in clusters is used on this basis by Lacey & Cole (1993) to argue that

$\Omega \gtrsim 0.5$ (although lower Ω is allowed in models with vacuum energy, since what matters for this analysis is the linear growth suppression factor).

GALAXY MERGERS The PS method can also yield an approximate expression for the merger rate in a gravitational hierarchy. A simple argument is given by considering the time derivative of the mass function:

$$M^2 f(M) \propto \nu e^{-\nu^2/2} \quad \Rightarrow \quad M^2 \dot{f}(M) \propto |\dot{\nu}| \left(\nu^2 - 1\right) e^{-\nu^2/2}. \tag{17.23}$$

This expression must correspond to the difference between the creation and destruction rates for bound structures. In the limit $\nu \to \infty$, we expect to have only creation (since pre-existing structures are very rare), whereas for $\nu \to 0$ there should be only destruction (since the universe fragmented into low-mass objects long ago). It is therefore tempting to identify these two terms directly, deducing

$$\begin{aligned} \text{creation rate} &\propto \nu^2 e^{-\nu^2/2} \\ \text{destruction rate} &\propto e^{-\nu^2/2}, \end{aligned} \tag{17.24}$$

although this is clearly just a guess for $\nu \simeq 1$ (see Sasaki 1994; Kitayama & Suto 1996).

For a more sophisticated treatment, we need to consider the limit of the conditional mass function at two nearly equal epochs, since the time derivative of the conditional function must be related to the merger rate. First, we need to invert the above reasoning, which gives the differential conditional probability

$$P(M_1, z \mid M_0, z = 0) = dF/dM_1. \tag{17.25}$$

Conditional probability definitions then imply

$$P(M_0 \mid M_1) = \frac{P(M_1 \mid M_0)\, P(M_0)}{P(M_1)}. \tag{17.26}$$

We now have the probability that an object of mass M_1 at redshift z becomes incorporated into one of mass M_0 at redshift $z = 0$. The **merger rate** $\mathscr{M}$, at which an object of mass M_0 accretes objects of mass ΔM, is then

$$\boxed{\mathscr{M}(M_0, \Delta M) = \left(\frac{dP(M_0 \mid M_0 - \Delta M)}{dt_1} \right)_{t_1 \to t_0}} \tag{17.27}$$

(Lacey & Cole 1993).

Although this expression for the merger rate looks somewhat messy when written out in full, its main content is that objects tend to merge with systems whose size is similar to their own. For masses $\ll M^*$ (which is today an Abell cluster mass), the overall merger rate is not a strong function of redshift. A simple argument for how things scale is to say that, in this low-mass limit, the predicted number density of objects scales as $n \propto 1/\sigma(M)$. This density declines with time as a result of mergers, so the merger rate is $\propto \dot{n}$. In an Einstein–de Sitter universe, $\sigma \propto (1+z)^{-1}$ and $t \propto (1+z)^{-2/3}$, so the merger rate scales as

$$dn/dt \propto (1+z)^{5/2}. \tag{17.28}$$

However, all of this applies only to dark haloes; for galaxies themselves, it is easy to argue that mergers must be suppressed today. Even if dark-matter haloes merge, there is no guarantee that the stellar cores will do so as well (merger of two clusters just gives a larger cluster). The key parameter is the relative velocity of two galaxies: only if this is below the relative escape velocity ($\simeq 200\,\mathrm{km\,s^{-1}}$) will a merger be probable. For collapsed clusters, the velocity dispersion is $\sigma_v \simeq GM/r \propto M^{2/3}\rho^{1/3} \propto M^{2/3}(1+z)$; since theories such as cold dark matter predict that M^* will change quite rapidly with redshift, σ_v was usually smaller in the past. This epoch dependence of merging may be important in understanding some of the observations of galaxy evolution; for example, if we assume that AGN are short-lived events, with a constant lifetime, that are triggered by mergers, then the comoving AGN density should track the rate of mergers. Hence, the sort of epoch dependence of galaxy merging required to account for AGN density evolution, $(1+z)^5$, say, may in principle be produced (Carlberg 1990).

GALAXY MERGERS AND CANNIBALISM Since the stellar parts of galaxies are compact in comparison with their dark haloes, it is a reasonable first approximation to treat them as point particles that remain collisionless even though their dark haloes may have merged. However, the baryonic cores do have a finite extent, and the merger of the baryonic parts of galaxies is not ruled out. To see how common this event might be, let us calculate the cross-section for the process.

The criterion that determines whether a merger will occur is the quantity of energy that is transferred to the stellar system by the tidal forces of a close encounter. If this is a significant fraction of the orbital energy of the perturbing body, a merger is likely. Consider a mass M, travelling at velocity V, that passes within a distance b of a star. In the **impulsive approximation** b is sufficiently large that the small velocity impulse given to the star can be ignored during the encounter, and the star can be treated as being stationary. It is then easy to show that the impulse imparted to the star is perpendicular to the track of the perturbing mass:

$$\delta v_\perp = \frac{2GM}{b\,V}. \tag{17.29}$$

Now consider such effects integrated over the body of the galaxy. We are only interested in the differential effect, i.e. the heating of the stellar system, so we should replace $1/b$ by $(\mathbf{r}\cdot\hat{\mathbf{b}})/b^2$, where b is now the impact parameter with respect to the centre of mass of the galaxy. The total energy transferred is an integral over the stellar mass of the galaxy: $\Delta E = \int \frac{1}{2}(\delta v_\perp)^2\, dm$, which is easily evaluated for a spherically symmetric galaxy, in terms of the mass-weighted mean-square radius $\langle r^2\rangle$. Now consider two identical galaxies with relative velocity V. Each has kinetic energy $E = MV^2/8$, and they merge if $\Delta E > E$, giving the critical impact parameter

$$\boxed{b_{\rm crit}^{-1} = \frac{V}{(32G^2M^2\langle r^2\rangle/3)^{1/4}}.} \tag{17.30}$$

The content of this is that $b_{\rm crit} \simeq (v_c/V)\,r$, where r is the galaxy size, v_c its internal circular velocity and n the number density. The merger rate $\Gamma = \pi b^2 n V$ thus declines as $1/V$. For typical numbers, this is unimportant ($\Gamma/H_0 \ll 1$) in clusters today, but would have mattered in the past when typical galaxy groups were denser and had lower velocity dispersions.

The impulsive deflection calculation can be used in a different way to look at mergers. Even if galaxies avoid a close collision, they can still lose orbital energy through the phenomenon of **dynamical friction**. Consider a body of mass M moving through a sea of low-mass particles m, which we take to have a velocity dispersion $\ll V$, so that they can be considered to be at rest. In the rest frame of the main body, each particle suffers a velocity impulse perpendicular to its path as before of $\delta v_\perp = 2GM/Vb$, where b is the impact parameter. The particle therefore undergoes deflection through an angle

$$\alpha = \frac{\delta v_\perp}{V} = \frac{2GM}{V^2 b}. \tag{17.31}$$

In return, the main body suffers an impulse both parallel and perpendicular to its path. However, when we integrate over collisions with all test particles, the perpendicular components will cancel, leaving a net loss of forward momentum, i.e. a gravitational drag. To see how this happens from a macroscopic point of view, think of the body as experiencing a stream of fluid in its rest frame; gravitational deflection of the fluid focuses streamlines and yields a density enhancement behind the body. The gravitational attraction of this mass concentration is in the opposite direction to the main body's velocity, as required. Returning to the microscopic viewpoint, the fractional change in parallel momentum per collision is $\Delta p = mV\alpha^2/2$ for small deflection angles, whereas the rate of collisions as a function of b is $\Gamma = 2\pi b\, nV\, db$, where n is the number density of test particles. Integrating this to get the total rate of change of parallel momentum gives the total drag force for dynamical friction:

$$\boxed{F = \frac{4\pi G^2 M^2}{V^2}\,\rho\,\ln\left(\frac{b_{\rm max}}{b_{\rm min}}\right).} \tag{17.32}$$

This depends only on the mass density of the background fluid, ρ, not on the mass of the test particles, as expected from the fluid argument. The result also depends on the ratio of maximum to minimum impact parameters, but only logarithmically, as with the **Gaunt factor** in bremsstrahlung radiation. The lower limit may be taken as $b_{\rm min} = GM/V^2$, where the deflection is through 90°, and the impulsive analysis breaks down. For typical galaxy numbers of $10^{11} M_\odot$ and $300\,{\rm km\,s^{-1}}$, this is roughly 10 kpc. The maximum is the size of the system of test particles, although whether $b_{\rm max}$ is taken as a cluster scale of 1 Mpc or as the horizon size of order Gpc makes satisfyingly little difference to the logarithmic term: $\Lambda \equiv \ln(b_{\rm max}/b_{\rm min}) \simeq 5$ in practice. Finally, what happens if we relax the assumption that the test particles are at rest? The drag force will cut off when the velocity of the main body becomes comparable to the velocity dispersion σ, and a rough approximation would be to replace V^{-2} by $(V^2+\sigma^2)^{-1}$. At low velocities, it is probably best to think thermodynamically: the collisions in the stellar system will tend to relax the velocity distribution towards a Maxwellian; particles nearly at rest will tend to be 'heated up', just as those moving faster than the average will be decelerated. To finish the calculation, we should work out the timescale for orbital decay via dynamical friction:

$$\tau_{\rm D} = \frac{p}{\dot{p}} = \frac{V^3}{4\pi G^2 M\rho\Lambda}. \tag{17.33}$$

To put some numbers in this, we can eliminate the density by assuming that the galaxy moves in an isothermal halo with $\rho = \sigma_v^2/(2\pi G r^2)$:

$$\frac{\tau_{\rm D}}{\rm Gyr} \simeq 2\left(\frac{V}{1000\,{\rm km\,s^{-1}}}\right)^3\left(\frac{\sigma_v}{300\,{\rm km\,s^{-1}}}\right)^{-2}\left(\frac{r}{100\,{\rm kpc}}\right)^2\left(\frac{M}{10^{12}M_\odot}\right)^{-1}, \tag{17.34}$$

so it is not implausible that this process could cause orbital decay within a Hubble time in the centres of clusters. The result would be the process of **galactic cannibalism**, as named by Hausman & Ostriker (1978): one galaxy in a cluster is able to grow by consuming some of its neighbours.

What would be the observational signatures that would help us to decide whether a galaxy had undergone merging in its past history? This is clearly a problem where numerical simulation can play a major part, but it is possible to gain a reasonable insight into the process by scaling arguments. We can make progress if we are prepared to assume **homologous mergers** – i.e. that the light distribution before and after merging obeys some universal functional form:

$$\frac{\Sigma(r)}{\Sigma_0} = f\left(\frac{r}{r_0}\right), \tag{17.35}$$

where Σ is the projected luminosity density, or surface brightness. The universal function might in practice be something close to de Vaucouleurs' law, but this is irrelevant for what follows. Assuming constant mass-to-light ratios, and that the central parts of the galaxy are dominated by stellar mass, the scalings for the galaxy mass and binding energy are

$$\begin{aligned} M &\sim \Sigma_0\, r_0^2 \\ E &\sim \frac{GM^2}{r_0} \sim G\Sigma_0^2\, r_0^3. \end{aligned} \tag{17.36}$$

We now want to add galaxies together in a way that conserves both mass and energy, and there are two interesting extreme behaviour modes for the cannibal:

(1) Eat small galaxies. The mass of the cannibal increases, but low-mass galaxies have very small binding energies, so we effectively increase M at constant E, which implies the scaling $\Sigma_0 \propto M^{-3}$.

(2) Eat someone your own size. From the formula for the merging rate, this is the preferred option. Now the binding energy increases in proportion to the mass, and so we have a different scaling $\Sigma_0 \propto M^{-1}$.

The observed scaling requires an appropriate projection of the fundamental plane to get a relation between Σ_0 and M that does not depend on velocity dispersion. This is approximately $\Sigma_0 \propto M^{-2}$, midway between the two extreme models. This is hardly solid proof that elliptical galaxies have participated in merging, but it is certainly quite consistent with the idea (see e.g. Capaccioli, Caon & D'Onofrio 1992).

What does the cannibalistic growth of a galaxy do to its total luminosity within a given aperture? If r_0 is small, then $L \propto \Sigma_0 r_0^2 \propto M$ in both cases; at late times r_0 is large and so $L \propto \Sigma_0 r^2$, which scales as M^{-1} or M^{-3} in cases (1) and (2) respectively. This means that the aperture magnitude of the cannibal can in practice be very nearly constant for large parts of its history as it passes through a broad maximum. It has been suggested that this could provide an explanation for the tendency of **brightest cluster galaxies** (BCGs) to act as very good standard candles. It might be thought that no explanation is required beyond accounting for the existence of a high-luminosity cutoff in the galaxy luminosity function, but it can be shown that BCGs are more unusual than simply being the largest of a few hundred random samplings of the Schechter function (Bhavasar & Barrow 1985). The gap in luminosity between the brightest and second brightest galaxy is larger than can be accounted for just from statistics: some extra process must have been in action. Further evidence of the unusual status of BCGs is that many of them are classified as **cD galaxies**. This is a marvellously obscure piece of terminology, which uses the symbol for a supergiant star to denote a supergiant galaxy. This is certainly what cD galaxies are; apart from their status as BCGs, they are also distinguished by departures from the homologous assumption made above. At large radii, cD galaxies show a low surface-brightness halo that lies above the de Vaucouleurs' law that fits the inner parts. This halo has no clear edge, and in spectacular cases can extend up to 1 Mpc throughout the core of the cluster (Schombert 1988). In many cases, it is difficult to know whether a distinction should be drawn between a cD halo and general **intracluster light**. In both cases, it seems likely that the origin of the starlight is material either tidally stripped from the outer parts of galaxies or resulting from tidal disruption.

Apart from the case of BCGs, opinions among astronomers divide over the importance of merging in the history of galaxies. It is particularly difficult to know if the process played a significant role in spiral galaxies, because the spiral disks are cold in the stellar-dynamical sense: the velocity dispersion of orbits perpendicular to the disk is $\lesssim 10\%$ of the circular velocity. Mergers that added more than perhaps 10% to the mass of a spiral galaxy since $z \simeq 1$ would probably 'heat' these orbits and thicken the disks unacceptably (Tóth & Ostriker 1992).

TIDAL TORQUES AND THE HUBBLE SEQUENCE Apart from the clear differences in their stellar populations, the other striking distinction between spiral disks and bulges is that the former are supported against gravity by rotation, whereas for the latter it is primarily the 'pressure' of the anisotropic stellar orbits that is responsible. This may be quantified by introducing a dimensionless **spin parameter** that relates angular momentum, binding energy and mass:

$$\boxed{\lambda \equiv \frac{J\,|E|^{1/2}}{G\,M^{5/2}}.} \qquad (17.37)$$

Rotational support corresponds to $\lambda \simeq 1$, and observed values for spiral disks are indeed of this order. Conversely, ellipticals have $\lambda \simeq 0.05$ and are pressure supported. This is a luminosity-dependent statement, however, since low-luminosity ellipticals have a significant component of rotation. It was thought at one time that ellipticals and bulges differed in their spin properties, with the latter having a higher degree of rotation. However, the general higher luminosity of ellipticals coupled with the spin–luminosity correlation appears to account for this (Kormendy 1982). Since ellipticals and bulges also occupy the same fundamental plane in luminosity–size–velocity-dispersion space, their formation histories seem likely to have been very similar.

It was first suggested by Hoyle (in 1949, in an obscure US government report; see the discussion in section 22 of Peebles 1993) that this angular momentum arises from tidal torques between protogalaxies. We can make a rough estimate of the value of the spin parameter expected via this process, as follows. Convolve the density field with a filter of scale R, to pick out objects of the mass scale of interest. The density field at early times then consists of blobs of comoving size and separation $\sim R$. Since the blobs are not spherically symmetric, each will be spun up by the tidal force of other nearby blobs. The total torque produced, T, must be of order

$$T \sim \text{tidal acceleration} \times \text{mass} \times \text{lever arm}$$
$$\Rightarrow \; T \sim \frac{\delta GM}{a^2R^2} \times \frac{M}{2} \times \frac{aR}{2} = \frac{\delta GM^2}{4Ra}, \tag{17.38}$$

regarding some central blob as roughly a dipole, so that half of the mass of the blob feels the force from a blob a distance R away. The effective mass causing the tidal force is, however, only the excess mass above uniform density, i.e. $M\delta$, where δ is the fractional density perturbation of the attracting blobs. Lastly, the various powers of the scale factor a allow for the fact that R is a comoving length.

Since the density perturbations grow as $\delta \propto a$, we see that the torque is constant and angular momentum grows linearly with time:

$$J \propto t \tag{17.39}$$

(for $\Omega = 1$). This linear growth terminates once the object being spun up goes nonlinear and collapses, the time of collapse dictating the final angular momentum acquired. This nonlinear process is what accounts for the generation of angular momentum, even though the growing-mode linear velocity perturbation field is irrotational, and Kelvin's circulation theorem shows that vorticity will not subsequently be generated if the flow is dissipationless. We can model these final stages by the spherical model, for which the density contrast at turnround is $\rho/\bar{\rho} = 9\pi^2/16$. Since the mass of the object is $M \simeq (R/2)^3 4\pi\rho_0/3$, this gives the proper radius of the object at turnround, in terms of which

$$E = -\frac{3GM^2}{5r_{\rm prop}} \tag{17.40}$$

(for a uniform sphere). It is now a straightforward process to obtain an estimate of the spin parameter by combining E and M with the angular momentum acquired by the time of turnround $J = Tt$. The answer depends on the degree of overdensity of the object under consideration: more extreme overdensities will collapse earlier and acquire less angular momentum. If we say that the object has overdensity ν times the rms fractional density contrast, then the parameter δ_c introduced earlier is $\simeq 1/\nu$ (since turnround occurs at approximately unit density contrast). Assembling all these arguments gives the approximate answer

$$\lambda \simeq \frac{1}{40\nu}. \tag{17.41}$$

This is independent of the mass scale and epoch of collapse, as would be expected in a flat universe that lacks a natural scale.

This argument has neglected several factors of order unity, as well as more subtle complications such as the tendency of higher density peaks to cluster, so that a higher torque than average applies for these. However, this seems to be a small effect by comparison with the gross fact that higher peaks collapse earlier. Detailed calculations and numerical simulations give values for λ roughly three times the above figure (Barnes & Efstathiou 1987; Heavens & Peacock 1988). It is interesting to ask what the interpretation of the ν^{-1} scaling might be. Early bias models suggested that elliptical galaxies could be associated with higher-ν peaks than spirals, so that the difference in spin properties between the early and late types might be understood in this way. However, there is in fact too broad a spread in λ at a given ν for this hypothesis to be tenable. More plausibly, there could be a relation to the observed mass dependence of λ, since low-mass objects would be associated with lower-ν peaks in Press–Schechter type analyses of the growth of structure in hierarchical density fields.

17.3 Cooling and the intergalactic medium

So much for the hierarchical assembly of collisionless objects from the dark-matter field. We now have to face up in more detail to the complex behaviour of the baryonic component, which must be able to **dissipate** and turn into stars. There are many reasons why galaxies might not form, even if there exist dark-matter potential wells of galactic mass: (i) the gas might be prevented from falling in by drag forces from the cosmic background radiation; (ii) the gas might be too hot to fall in, with a temperature that exceeds the virial temperature of the dark halo; (iii) the gas might heat as it falls in, and will only continue to collapse if this energy can be radiated away.

RADIATION DRAG Electrons moving with respect to the isotropic CMB experience a net drag force as they Compton-scatter photons (see chapter 14). This converts to a bulk acceleration, which depends on the ionization, x:

$$-\frac{\dot{v}}{v} = \frac{4}{3}\left(\frac{\sigma_{\rm T}\, U_{\rm rad}\, x}{m\, c}\right) \equiv t_{\rm drag}^{-1}. \tag{17.42}$$

Comparing $t_{\rm drag}$, which is $\propto (1+z)^{-4}$, with the expansion timescale

$$t = \frac{6.7h^{-1}\,{\rm Gyr}}{\sqrt{\Omega}\,(1+z)^{3/2}}, \tag{17.43}$$

gives the redshift at which radiation drag ceases to be important, allowing baryonic material to separate from radiation and fall into dark-matter potential wells: $1+z \lesssim 210\,(\Omega h^2)^{1/5} x^{-2/5}$. This is the redshift obtained from using the proton mass in the equation for the drag force. If the electron mass had been used instead, this would have yielded the redshift at which radiation drag is able to cool random electron motions, i.e. the redshift down to which the matter temperature equals the radiation temperature. This is a factor $\sqrt{1836}$ lower: $1+z \lesssim 10\,(\Omega h^2)^{1/5} x^{-2/5}$.

This calculation suggests that the temperature of the intergalactic medium (IGM) obeyed the universal $T \propto a^{-1}$ law until quite recently, rather than the more rapid adiabatic $T \propto a^{-3(\Gamma-1)}$ that would be expected in the absence of radiation. The gas may therefore be so hot that it cannot fall into dark-matter haloes (i.e. the gas temperature exceeds the virial temperature defined below). Although we earlier derived an observed upper limit of 20–30 for the redshift of galaxy formation, many fluctuation spectra may not allow objects to form at this redshift. For example, in the CDM model, Blanchard *et al.* (1992) show that the virial velocity of typical haloes at these redshifts is $\sim$ 10 K, too small to accrete gas at $\sim$ 100 K. Galaxy formation would commence only at lower redshifts in this model.

THE COOLING FUNCTION Once gas begins to settle into dark-matter haloes, mergers will heat it up to the virial temperature via shocks; in order for the gas to form stars, it must be able to undergo **radiative cooling**, to dispose of this thermal energy. Clearly, the redshift of collapse clearly needs to be sufficiently large that there is time for an object to cool between its formation at redshift $z_{\rm cool}$ (when $\delta\rho/\rho \simeq \delta_c$) and the present epoch. We shall show below that $z_{\rm cool}$ is a function of mass; it is therefore possible to put cooling into the Press–Schechter machinery simply by using the *mass-dependent* threshold, $\nu(M) = \delta_c[1 + z_{\rm cool}(M)]/\sigma_0(M)$, in the mass function.

The cooling function for a plasma in thermal equilibrium has been calculated by Raymond, Cox & Smith (1976) and by Sutherland & Dopita (1993). For an H + He plasma with an He abundance by mass of $Y = 0.25$ and some admixture of metals, their results for the cooling time $t_{\rm cool} \equiv 3kT/[2\Lambda(T)n]$ may be approximated as roughly

$$\frac{t_{\rm cool}}{\rm yr} = 1.8 \times 10^{24} \left(\frac{\rho_{\rm B}}{M_\odot\,{\rm Mpc}^{-3}}\right)^{-1} \left(T_8^{-1/2} + 0.5 f_m T_8^{-3/2}\right)^{-1}, \tag{17.44}$$

where $T_8 \equiv T/10^8K$. The $T^{-1/2}$ term represents bremsstrahlung cooling and the $T^{-3/2}$ term approximates the effects of atomic line emission. The parameter f_m governs the metal content: $f_m = 1$ for Solar abundances; $f_m \simeq 0.03$ for no metals. In this model, where so far dissipation has not been considered, the baryon density is proportional to the total density, the collapse of both resulting from purely gravitational processes. $\rho_{\rm B}$ is then a fraction $\Omega_{\rm B}/\Omega$ of the virialized total density. This is itself some multiple f_c of the background density at virialization (which we refer to as 'collapse'):

$$\rho_c = f_c\, \rho_0\, (1 + z_c)^3. \tag{17.45}$$

The virialized potential energy for constant density is $3GM^2/(5r)$, where the radius satisfies $4\pi\rho_c r^3/3 = M$. This energy must equal $3MkT/(\mu m_p)$, where $\mu = 0.59$ for a plasma with 75% hydrogen by mass. Hence, using $\rho_0 = 2.78 \times 10^{11} \Omega h^2\, M_\odot {\rm Mpc}^{-3}$, we obtain the **virial temperature**:

$$\boxed{T_{\rm virial}/{\rm K} = 10^{5.1} (M/10^{12} M_\odot)^{2/3}\, (f_c \Omega h^2)^{1/3}\, (1 + z_c).} \tag{17.46}$$

Knowing T, we can now ask whether there is time for the object to have cooled by the present; for $\Omega = 1$, this means we must solve $t_{\rm cool} = \frac{2}{3} H_0^{-1}[1 - (1 + z_c)^{-3/2}]$. If only atomic cooling was important, the solution to this would be

$$1 + z_c = (1 + M/M_{\rm cool})^{2/3}, \quad M_{\rm cool}/M_\odot = 10^{13.1}\, f_m f_c^{1/2} \Omega_{\rm B} \Omega^{-1/2}. \tag{17.47}$$

For high metallicity, where bremsstrahlung only dominates at $T \gtrsim 10^8 K$, this equation for z_c will be a reasonable approximation up to $z_c \simeq 10$, at which point Compton cooling will start to operate. Given that we expect at least some enrichment rather early in the progress of the hierarchy, we shall keep things simple by using just the above expression for z_c.

We see that cooling is rapid for low masses, where the luminous and dark-mass functions are expected to coincide. Given that cooling of massive objects is ineffective, probability in the mass function must therefore accumulate at intermediate masses: the numbers of faint galaxies relative to bright are decreased. If $M_{\rm cool} \ll M_c$, then there is a power-law region between these two masses that differs from the PS slope: $M^2 f(M) \propto M^{2/3 + (n+3)/6}$; i.e. there is an effective change in n to $n + 4$. This may be relevant for the galaxy luminosity function, where the faint-end slope is close to constant numbers per magnitude. For constant mass-to light ratio, this implies $M^2 f(M) \propto M$, and apparently requires $n = 3$. Alternatively, a spectral index more in accord with large-scale structure observations of $n \lesssim -1$ gives a Press–Schechter slope much steeper than the

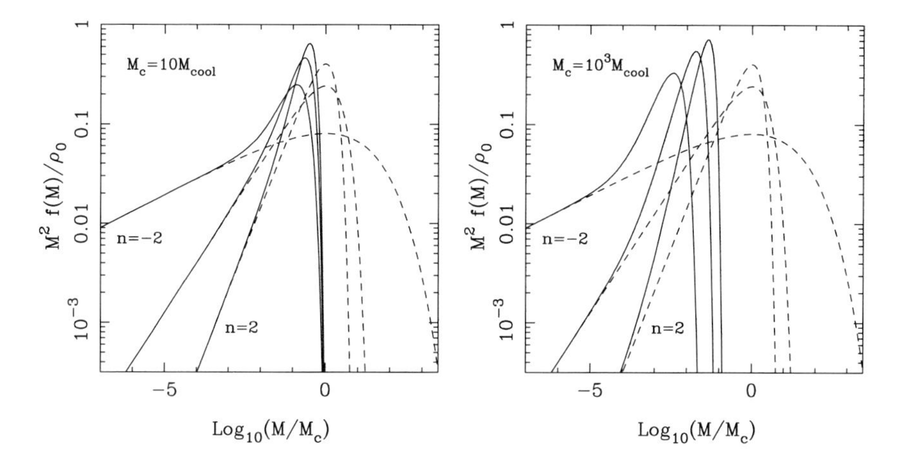

Figure 17.2. The multiplicity functions resulting from the PS formalism with the adoption of the cooling threshold criterion, $\delta_c \rightarrow \delta_c(1 + M/M_{\rm cool})^{2/3}$. The solid lines show the mass functions with cooling, and the broken lines show the mass functions without cooling; the two sets of curves coincide only at very low masses. M_c is the mass scale with $\sigma(M_c) = \delta_c$; the two panels show results for two different ratios of M_c to $M_{\rm cool}$. Note the relative insensitivity to $M_{\rm cool}$ and that over several orders of magnitude in mass, the relative numbers of low-mass objects are greatly reduced by cooling.

observed galaxy luminosity function. There have been many full and complicated studies of galaxy formation that show in detail why this is not really as much of a paradox as it initially seems, but the above simplified discussion of cooling illustrates the main way in which the naive analysis goes wrong. This is illustrated for power-law spectra in figure 17.2.

GALAXY MASSES AND THE COOLING CATASTROPHE There is a danger with this analysis that star formation in the first generation of the hierarchy may be so efficient that all gas immediately becomes locked up in low-mass objects. This seems implausible; the low binding energy should allow supernovæ to unbind the matter (Dekel & Silk 1986), leading to a very low overall efficiency of star formation. Indeed, the high fraction of baryonic material in galaxy clusters that is in the form of gas may indicate that such a process did take place. However, the criterion of equating cooling time with look-back time will work only if an object is able to cool undisturbed over this time; if subsequent generations of the hierarchy collapse while the object is still cooling quasistatically, then the gas will be reheated and collapse may never occur (see White & Rees 1978). Objects are immune to this effect if the cooling time is shorter than the free-fall time, which turns out to be simply a criterion on mass (Silk 1977; Rees & Ostriker 1977). If we relate the collapse time, $t_{\rm coll} \equiv [3\pi/(32G\rho_c)]^{1/2}$, to the cooling time via $t_{\rm cool} = t_{\rm coll}$, then the above figures for atomic cooling yield a mass

$$M_{\rm coll}/M_\odot = 10^{13.5} f_m \Omega_{\rm B} \Omega^{-1}. \tag{17.48}$$

This is of the same order as $M_{\rm cool}$, and so very massive structures may be truncated still

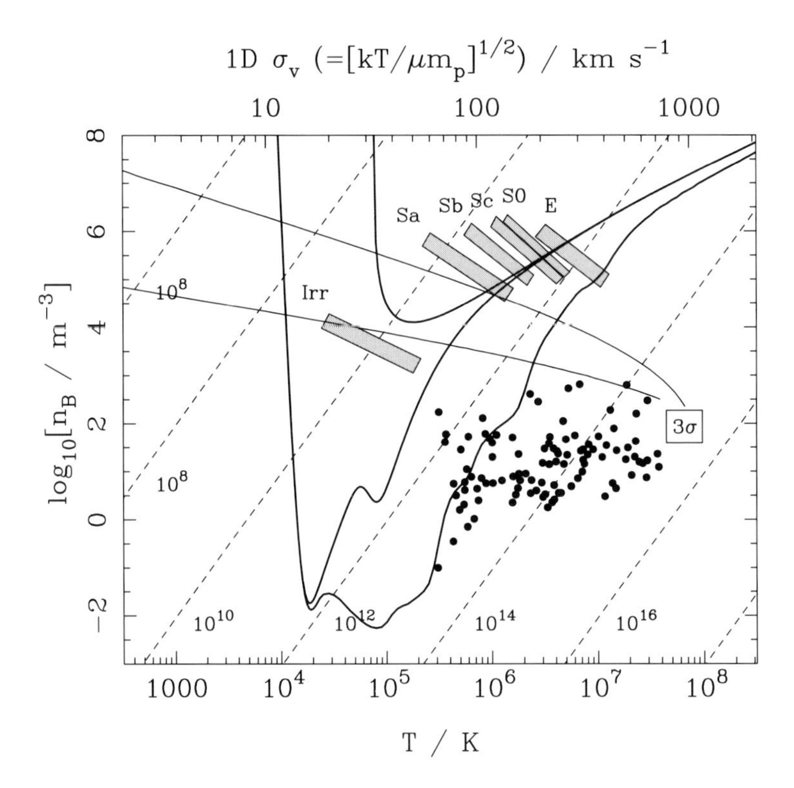

Figure 17.3. The cosmological cooling diagram, adapted from Blumenthal *et al.* (1984). This is a plot of baryonic density against temperature, but the conversion to velocity dispersion is shown above the plot. The diagonal broken lines are lines of constant mass (in Solar units), assuming 10% of the total mass to be baryonic. The solid curves with minima show the locus along which the cooling time equals the collapse time; objects above these curves can cool efficiently. The three curves show (bottom to top): Solar metals; zero metals; zero metals photoionized with an $\alpha = 1$ spectrum. The two nearly horizontal solid lines indicate the results for virialized 3σ fluctuations according to the empirical fluctuation spectrum for $\Omega = 1$ (lower curve) and 0.3 (upper curve). The locations of clusters (solid points) and galaxies (shaded areas) in this diagram (plotted assuming $h = 0.65$) provide convincing evidence that it is dissipation that governs whether a collapsed object is described as a galaxy or as a system of galaxies.

more abruptly than we have assumed above. Detailed application of these arguments to the CDM model was made by Blumenthal *et al.* (1984), who concluded that cooling allows a very good understanding of the delineation between galaxies and clusters (see figure 17.3).

The remarkable conclusion of this section is that simple considerations of microphysics have allowed us to understand the characteristic masses of galaxies. Collapsed haloes that exceed the critical mass will contain gas that cannot cool efficiently:

when we look at a $10^{15}\, M_\odot$ cluster of galaxies, we see a system with hot X-ray gas between the galaxies, rather than a single huge stellar system. The order of magnitude of the critical galaxy mass may be expressed in terms of fundamental constants by writing the coefficient for atomic cooling in these terms, rather than just numerically. This exercise gives the characteristic galaxy mass as

$$\boxed{M_{\rm G} \sim \alpha^5 \left(\frac{\hbar c}{G m_p^2}\right)^2 \left(\frac{m_p}{m_e}\right)^{1/2} m_p,} \tag{17.49}$$

where $\alpha \simeq 1/137$ is the fine-structure constant (Silk 1977; Rees & Ostriker 1977). By contrast, the typical mass of a star is $[\hbar c/(G m_p^2)]^{3/2} m_p$ (see chapter 5).

PHOTOIONIZATION In the years since 1977, it has been appreciated that there is a potentially serious missing ingredient in this analysis of cooling processes. The limit on galaxy masses arises because atomic cooling is strong for temperatures between roughly 10^4 and 10^5 K, and inefficient for higher temperatures. However, this cooling rate assumes that the gas is ionized collisionally, so that there is a substantial neutral component at these temperatures, and the gas is effectively completely neutral for $T \lesssim 10^4$ K. The dominant contributor to the cooling is then collisional excitation due to collisions between atoms and electrons; recombination radiation, from collisions between ions and electrons, is an order of magnitude lower (e.g. Katz, Weinberg & Hernquist 1996).

This means that the low-temperature peak in the cooling curve is completely removed if the gas is exposed to ionizing radiation, since the recombination rate is very little affected but the critical atomic emission vanishes. The critical ionizing flux can be estimated by the same arguments as used in chapter 14 in the context of quasar emission lines: the ionization is set by balancing the rate of photoionization with the recombination rate. In practical units, this gives

$$\frac{1-x}{x^2} = 10^{-7.0}\,(3+\alpha)\,(J_{-21})^{-1}\,\Omega_{\rm B} h^2\,(1+z)^3 \tag{17.50}$$

(Efstathiou 1992). Here, x is the fractional ionization, and J_{-21} measures the level of the ionizing continuum at the Lyman limit, $\lambda = 912$ Å:

$$J_\nu = J_{-21}\,(\nu/\nu_{\rm L})^{-\alpha} \times 10^{-21}\,{\rm erg\,s^{-1}\,Hz^{-1}\,cm^{-2}\,sr^{-1}}. \tag{17.51}$$

This particular reference level is chosen because it is approximately the background at high redshifts inferred from the proximity effect in quasar absorbers. Unless the ionizing flux drops very rapidly beyond the redshifts where quasars are known, it therefore seems that the neutral fraction will be very small at all redshifts of interest, making it difficult for galaxies to be formed at all.

This cure for the cooling catastrophe is worse than the disease, but there is room for a sensible compromise: massive galaxies contain sufficient gas that they are self-shielding. The central parts can then remain neutral and will cool efficiently as in the old picture. The optical depth through a collapsed object of rotational velocity V can be shown to be [problem 17.4]

$$\tau = 10^{-4.2}\,\alpha\,(J_{-21})^{-1}\,\left(V/100\,{\rm km\,s^{-1}}\right)\,[f_c \Omega_{\rm B} h^2 (1+z)^3]^{3/2}. \tag{17.52}$$

For the usual collapse factor $f_c = 178$ and assuming $\Omega_{\rm B} h^2 = 0.02$, the critical rotation velocity at $z \simeq 3$ will be several hundred km s^{-1}. Small galaxies at high redshift, with velocities of only tens of $\mathrm{km\,s^{-1}}$, will therefore have trouble cooling. Things do not improve much by the present, either: J_ν evolves roughly as $(1+z)^4$, which is largely cancelled by the redshift dependence of the mean density. Photoionization is therefore probably at least a good part of the reason why the high Press–Schechter abundance of low-mass galaxies is not seen.

COOLING FLOWS The concept of the cooling time is critical in allowing us to understand why clusters of galaxies exist as systems of distinct galaxies, rather than being a single mega-galaxy of luminosity $\gtrsim 100 L_\odot$. The latter systems may be produced in the far future, but the intergalactic gas is too hot to have cooled by the present. The gas in clusters of galaxies is therefore usually treated as being in hydrostatic equilibrium. However, in the densest central parts of rich clusters, the bremsstrahlung cooling time is in fact slightly shorter than a Hubble time. This opens the way to the phenomenon of the **cooling flow**, in which the gas undergoes a slow subsonic inflow as the loss of energy from the centre reduces the pressure that holds up the outer parts of the cluster. The flow rate comes from energy conservation: the energy being radiated from the central gas must balance the thermal energy of the inflowing gas, plus the work done in compressing this gas in order to match the higher central pressure, i.e. the enthalpy flow rate equals the radiated luminosity:

$$L(<r) = \frac{5kT}{2\mu m}\,\dot{M}, \tag{17.53}$$

where μm is the mean mass per particle in the plasma. If the cluster were isothermal, and the mass flow rate were independent of r, this would imply a central delta function in the X-ray emission. This is not seen in practice, and so the conclusion is that $\dot{M}(r)$ cannot be constant: mass must drop out of the cooling flow over a range of radii. In practice, the required radial dependence is very roughly $\dot{M} \propto r$; most of the mass is deposited at large radii. The effects are, however, more noticeable at small r, because the deposition rate per unit volume is

$$\dot{\rho} = \frac{1}{4\pi r^2}\,\frac{d\dot{M}}{dr}. \tag{17.54}$$

The majority of clusters would be expected to have such flows; indeed, it is possible that they are universal but sometimes go unrecognized owing to inadequate X-ray data. The mass deposition rates inferred are high: typically 50–100 $M_\odot\,\mathrm{yr^{-1}}$, with some reaching ten times this figure (see e.g. Fabian, Nulsen & Canizares 1991; Allen *et al.* 1993). This is enough to assemble a large galaxy over a Hubble time, so one would expect spectacular associated optical phenomena if the cooled gas were turning into stars. In fact, the limits to the star-formation rate in the outer parts of cluster galaxies are at the level of a few $M_\odot\,\mathrm{yr^{-1}}$, provided the IMF is normal.

This paradox has caused some controversy over the reality of cooling flows. Is it possible that some source of heat exists that balances the energy radiated in X-rays? The total energetics are large, but a huge reservoir of energy exists in the thermal energy of the gas in the outer parts of the cluster. It is possible that heat conduction rather than advection may be the mechanism by which some of this energy is transported inwards (Bregman 1992).

MAGNETIC FIELDS As a last illustration of the complications gas can bring, what about magnetic fields? These are of central importance in stars and star formation, and yet are almost never mentioned in polite cosmological society. The reason for this should be that gravity is the dominant force on large scales, and so everything else can be neglected. This is certainly true for superclusters, but we cannot be so sure for individual galaxies. It is easy to estimate how large B would have to be in order to wreck gravity-only approaches, as follows (see e.g. Coles 1992). Let there exist a magnetic field tangled on some comoving scale λ: this will produce a pressure gradient $|\nabla P|/a \sim B^2/(\mu_0 a\lambda)$, which in turn yields a peculiar acceleration $\dot{\mathbf{v}} = \boldsymbol{\nabla} P/(a\rho)$. The continuity equation says $\dot{\delta} = -\boldsymbol{\nabla}\cdot\mathbf{v}/a \sim v/(a\lambda)$, so that

$$\ddot{\delta} \sim \frac{B^2}{\mu_0 (a\lambda)^2 \rho}. \tag{17.55}$$

To apply this, multiply by t^2 to get the order of magnitude of the density disturbance created by the field. In order to have an important influence on structure formation by the present, we need $\delta \gtrsim 10^{-3}$ at recombination ($a \simeq 10^{-3}$). This criterion gives the critical magnetic field at recombination as

$$B_{\rm crit} \sim 10^{-6}\,(\lambda/\,h^{-1}\,\mathrm{Mpc})\,(\Omega_{\rm B} h^2)^{1/2}\ \mathrm{T}. \tag{17.56}$$

Since frozen magnetic flux in a plasma scales as $B \propto a^{-2}$, the intergalactic field today would have to be of order 10^{-12} T. Compressing by a factor of 10^5 in density would yield fields of order a few nT, comparable with those seen in the interstellar medium.

It is therefore possible that magnetic fields did play a significant role in galaxy formation. It is thought more likely that they did did not: the interstellar magnetic field may well have been generated by dynamo effects in which turbulence amplifies a tiny seed field by a huge factor. Such seed fields can be produced by nonlinear effects in the first generation of stars and supernovae, so a primordial field is not necessarily required. However, the above argument should make us uneasy, in that primordial fields at an important level are not so easy to rule out observationally.

17.4 Chemical evolution of galaxies

Important clues to the early life histories of galaxies are provided by the systematics of the distribution of elements within and between galaxies. Our assumption is that star formation began with primordial material essentially lacking in ‘metals’ (elements more massive than helium); these were then generated during stellar evolution, and deposited in the interstellar medium by the agency of supernova explosions. This means that subsequent generations of stars will have increasing **metallicities**, and we should be able to use this as an age estimator for stellar populations. A complicating factor is that galaxies do not form closed boxes of gas: new gas may be added later through accretion, and enriched gas may be lost from galaxies altogether if the potential well cannot contain the supernova-heated gas. These effects can and do lead to patterns in the element distributions that give us hints about the conditions in early galaxies. We now give some of the basic formulae of this subject; for more details, consult e.g. Lynden-Bell (1992) or Pagel (1994; 1997).

First, some notation: denote the mass in a galaxy in respectively stars and interstellar gas by s and g, and let Z be the fractional abundance of metals in the gas. The idea will be that a given generation of stars form and take a ‘reading’ of the gas abundance at that time; subsequent evolution changes the abundances in these stars very little for the majority (which are low mass and long lived). Some fraction β of the mass turned into stars in each generation will be returned essentially instantaneously without being altered, by the stellar winds from short-lived unstable massive stars. Some fraction of the remaining mass is processed into metals and again returned almost immediately, because the massive stars that make supernovae have lifetimes of only $\sim 10^7$ years. Define the **yield** y as the fraction of the gas turned into stars that is returned as metals. Finally, we can allow for the possibility of **galactic winds** in which some of the gas is driven completely out of the galaxy by supernova energy; let f denote the fraction of the gas returned by stars that is retained in the galaxy. We can then write down immediately the equation of chemical evolution:

$$\boxed{\frac{d(Zg)}{ds} = -\frac{Z}{1-\beta} + yf + \frac{\beta}{1-\beta} Zf.} \tag{17.57}$$

The equation just expresses conservation of metals, the three terms on the right having the following interpretations: loss of metals from gas in making stars; metals gained from supernova debris; metals returned unprocessed. The factors $(1-\beta)^{-1}$ arise because changing a mass δM of gas into stars only produces a lasting increase in stellar mass of $\delta s = (1-\beta)\delta M$. Notice that this equation does not require conservation of gas; if primordial gas is accreted onto the galaxy, this alters the total amount of gas g at a given s, but the chemical evolution equation describes the evolution of metals only, so there is no explicit term corresponding to accretion.

The goal of chemical evolution studies is to solve the basic equation above and so find $Z(s)$. This gives the evolutionary history of a galaxy using the total mass of stars as a time variable. The metallicity $Z(s)$ is that of the interstellar gas at 'time' s, and also that of the stars forming at that time. Inverting the relation to get $s(Z)$ gives the integral mass distribution of stars as a function of metallicity. The nice aspect of this approach is that it is completely independent of the details of the star-formation history: it does not matter whether stars form in isolated bursts, or continuously.

THE G-DWARF PROBLEM Consider the simplest possible solution: the galaxy is a closed box of fixed total mass. This says that $f = 1$ and moreover that the gas mass is just $g = g_0 - s$. We then have to solve $d[Z(g_0 - s)]/ds = y - Z$, which is easily integrated to give the differential metallicity distribution

$$\frac{d(s/g_0)}{d(Z/y)} = \exp\left[\frac{-(Z - Z_0)}{y}\right], \tag{17.58}$$

where the initial metallicity is Z_0. We see that the majority of stars inevitably lie mainly at the lowest metallicities, with only a rare tail to high metallicities.

This prediction fares not so badly in the metal-poor parts of the Galaxy (the bulge and halo), but is not at all in accord with observations in the Solar neighbourhood: stars small enough to have lifetimes long enough to have survived from the start of this process (the 'G dwarfs' – stars like the Sun) have metallicities that are roughly Gaussianly distributed. The predicted large numbers of low-metal stars are not present in the disk population: this has become known as the **G-dwarf problem**. This conclusion is quite robust: if all the gas is present initially, then the metals arising from the first generations of stars are diluted heavily by the large amounts of primordial gas.

Substantial enrichment only occurs in the late stages of the process, once most of the gas has been used up. The obvious way out of this predicament is to drop the assumption that the gas is all present at the start, and instead to let it build up gradually.

The incorporation of gas growth by accretion requires in principle some detailed physical model for how this arises. To obtain some insight into the effects on galactic chemical evolution, however, it makes sense to keep to the simplest model possible (following Lynden-Bell 1992). First note that the fundamental equation can be solved by multiplying through by an integrating factor W to rewrite the equation as

$$(ZgW)' = yfW \quad \Rightarrow \quad Z(s) = \frac{1}{gW} \int yfW \, ds. \tag{17.59}$$

The required factor is the solution of $W'/W = (Dg)^{-1}$, where $D(s) = (1-\beta)/[1-\beta f(s)]$ is a number close to unity that allows for mass loss. Lynden-Bell pointed out that it is possible to choose a convenient form for W, and hence $g(s)$, that gives a neat closed-form solution for the chemical evolution. First, however, we need to deal with the issue of gas lost by the galaxy.

SUPERNOVA-DRIVEN WINDS One of the most striking systematic trends in galaxy metallicities is that small galaxies are less metal-rich than large galaxies (see chapter 13). Moreover, the central parts of galaxies are systematically more metal-rich than are their outskirts. The main mechanism that suggests itself as an explanation for this is that gas that is returned by supernovae can be ejected with an energy that is too large for it to be bound to the galaxy. The fraction of gas retained (f in the above) will therefore be an increasing fraction of the mass of the galaxy. Dwarf galaxies, with shallow potential wells, can lose all their gas in this way very soon after the onset of star formation. Indeed, 'self-destructing' dwarfs are one of the suggested explanations for the fact that the comoving density of low-luminosity galaxies was apparently higher in the past (see Babul & Rees 1992). How are we to incorporate this into our model? If we view star formation as occurring at the bottom of a dark-matter potential well into which gas is periodically poured, then f is just some number that is independent of time. However, the visible regions of galaxies consist of self-gravitating baryonic material, so the energy needed to remove the gas from the region of star formation will increase as the stellar mass increases. According to what is now known as the Faber–Jackson law, but which was first enunciated (and directly in the form we need) by Fish (1964), the potential depth of elliptical galaxies is proportional to the square root of their mass. This suggests

$$f(s) = \left(1 + \sqrt{s_c/s}\right)^{-1}, \tag{17.60}$$

which interpolates between a retained fraction proportional to the potential well depth for small potentials and f tending to unity for very large masses. If we now assume $W(s) = s/(s_\infty - s)$, where s_∞ is the final mass of the galaxy, it is easy to verify that the solution for the gas evolution is

$$g(s) = D(s)^{-1} s_\infty \, \frac{s}{s_\infty} \left(1 - \frac{s}{s_\infty}\right), \tag{17.61}$$

i.e. a physically sensible gas mass that rises from zero initially as accretion operates, but which declines to zero at large times once accretion has ceased and the remaining gas is processed into stars. The corresponding solution for the metallicity distribution is

$$\frac{Z(s)}{y} = 2D(s)\,\frac{s_\infty^2}{s^2}\int_0^{\sqrt{s/s_\infty}} \frac{x^4}{(1-x^2)(x+\sqrt{s_c/s_\infty})}\,dx. \tag{17.62}$$

This illustrative model successfully cures the problems of the closed box. Consider small metallicities, and set $D(s) = 1$ for simplicity: for $s \ll s_c$, the metallicity distribution rises as $ds/dZ \propto Z^3$, rather than falling exponentially as in the closed box; for $s \gg s_c$, mass-loss tames the rising metallicity curve, and ds/dZ goes to a constant before falling as s becomes comparable with s_∞. The model thus produces differential metallicity distributions that are peaked in the required way, low metallicities being suppressed even more strongly in the case of shallow potential wells ($s_c/s_\infty \lesssim 1$); more massive galaxies are predicted to have higher metallicities. Similarly, metallicity gradients are predicted *within* galaxies because the outer parts are less strongly bound.

However, the analysis given here ignores the fact that galaxies may form via mergers, and this can alter the situation significantly. Suppose that star formation were triggered only by mergers: the fraction of gas processed (and hence the mean metallicity) would be highest for the systems that had merged the largest number of times – the most massive galaxies. Similarly, the central metallicity would be higher if merger products preferentially reached the centre, or if the merger starburst were triggered there (Larson 1976; Tinsley & Larson 1979). It is plausible that both mergers and supernova winds are important, and it is not easy to distinguish which is the dominant process.

17.5 Biased galaxy formation

HISTORY One of the major advances of cosmology in the 1980s was the realization that the distribution of galaxies need not trace the underlying density field. The main motivation for such a view may in fact be traced back to 1933 with Zwicky's measurement of the dark matter in the Coma cluster. A series of ever more detailed studies of cluster masses have confirmed his original numbers: if the Coma mass-to-light ratio is universal, then the density parameter of the universe is $\Omega \simeq 0.2$. Those who argued that the value $\Omega = 1$ was more natural (a greatly increased camp after the advent of inflation) were therefore forced to postulate that the efficiency of galaxy formation was enhanced in dense environments: **biased galaxy formation**. This desire to reconcile cluster M/L ratios with $\Omega = 1$ probably remains the strongest motivation for considering a high degree of bias.

We can note immediately that a consequence of this bias in density will be to affect the velocity statistics of galaxies relative to dark matter. Both galaxies and dark-matter particles follow orbits in the overall gravitational potential well of a cluster; if the galaxies are to be more strongly concentrated towards the centre, they must clearly have smaller velocities than the dark matter. This is the phenomenon known as **velocity bias** (Carlberg, Couchman & Thomas 1990). The effect can be understood quantitatively through the example of the equation of hydrostatic equilibrium (see chapter 12). On the assumption of a constant velocity dispersion σ_v, the equation is

$$\frac{GM(<r)}{r} = -\sigma_v^2\left(\frac{d\ln\rho}{d\ln r}\right), \tag{17.63}$$

where ρ is the density of the tracer population. For fixed M and a population of tracer particles obeying $\rho \propto r^{-\gamma}$, the velocity dispersion scales as $\sigma_v \propto \gamma^{-1/2}$; the more concentrated the particles are towards $r = 0$, the cooler they are. A similar effect was discussed in the context of galaxy haloes in chapter 4.

An argument for bias at the opposite extreme of density arose through the discovery of large **voids** in the galaxy distribution (Kirshner *et al.* 1981). There was a reluctance to believe that such vast regions could be truly devoid of matter – although this was at a time before the discovery of large-scale velocity fields. This tendency was given further stimulus through the work of Davis, Efstathiou, Frenk & White (1985), who were the first to calculate N-body models of the detailed nonlinear structure arising in CDM-dominated universes. Since the CDM spectrum curves slowly between effective indices of $n = -3$ and $n = 1$, the correlation function steepens with time. There is therefore a unique epoch when ξ will have the observed slope of -1.8. Davis *et al.* identified this epoch as the present and then noted that, for $\Omega = 1$, it implied a rather low *amplitude* of fluctuations: $r_0 = 1.3h^{-2}$ Mpc. An independent argument for this low amplitude came from the size of the peculiar velocities in CDM models. If the spectrum were normalized so that $\sigma_8 \simeq 1$, as seen in the galaxy distribution, the pairwise dispersion σ_p would then be $\simeq 1000$–$1500\,\mathrm{km\,s^{-1}}$, more than three times the observed value. What seemed to be required was a galaxy correlation function that was an amplified version of that for mass. This was exactly the phenomenon analysed for Abell clusters by Kaiser (1984), and thus was born the idea of **high-peak bias**: bright galaxies form only at the sites of high peaks in the initial density field (figure 17.4). This was developed in some analytical detail by Bardeen *et al.* (1986), and was implemented in the simulations of Davis *et al.*, leading to the conclusion that bias allowed the $\Omega = 1$ $h = 1/2$ CDM model to give a good match to observation. Since the mid-1980s, fashion has moved in the direction of low-Ω universes, which removes many of the original arguments for bias. However, the lesson of the attempts to save the $\Omega = 1$ universe is that cosmologists have learned to be wary of assuming that light traces mass. The assumption is now that the galaxy density field is guilty of bias, until it is shown to be innocent.

As was shown by Kaiser (1984), the high-peak model produces a linear amplification of large-wavelength modes. This is likely to be a general feature of other models for bias, so it is useful to introduce the **linear bias parameter**:

$$\boxed{\left(\frac{\delta\rho}{\rho}\right)_{\mathrm{galaxies}} = b \left(\frac{\delta\rho}{\rho}\right)_{\mathrm{mass}}.} \tag{17.64}$$

This seems a reasonable assumption when $\delta\rho/\rho \ll 1$. Galaxy clustering on large scales therefore allows us to determine mass fluctuations only if we know the value of b. When we observe large-scale galaxy clustering, we are only measuring $b^2\xi_{\mathrm{mass}}(r)$ or $b^2\Delta^2_{\mathrm{mass}}(k)$. A more difficult question is how the effective value of b would be expected to increase on nonlinear scales; this is discussed in Chapter 16. Even the large-scale linear bias relation cannot be taken for granted, however; if galaxy formation is not an understood process, then in principle studies of galaxy clustering may tell us nothing useful about the statistics of the underlying potential fluctuations against which we would like to test inflationary theories. It is possible to construct models (e.g. Bower *et al.* 1993) in which the large-scale modulation of the galaxy density is entirely non-gravitational in nature.

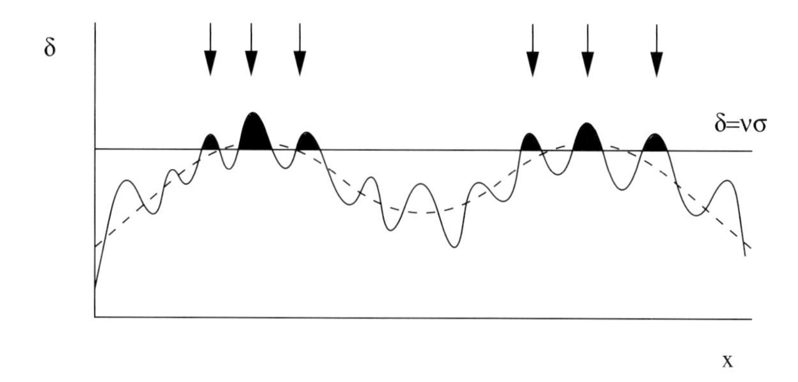

Figure 17.4. The high-peak bias model. If we decompose a density field into a fluctuating component on galaxy scales, together with a long-wavelength 'swell' (shown broken), then those regions of density that lie above a threshold in density of ν times the rms will be strongly clustered. If proto-objects are presumed to form at the sites of these high peaks (shaded, and indicated by arrows), then this is a population with Lagrangian bias – i.e. a non-uniform spatial distribution even prior to dynamical evolution of the density field. The key question is the physical origin of the threshold; for massive objects such as clusters, the requirement of collapse by the present imposes a threshold of $\nu \gtrsim 2$. For galaxies, there will be no bias without additional mechanisms to cause star formation to favour those objects that collapse first.

MECHANISMS FOR BIAS Why should the galaxy distribution be biased at all? In the context of the high-peak model, attempts were made to argue that the first generation of objects could propagate disruptive signals, causing neighbours in low-density regions to be 'still-born'. However, it turned out to be hard to make such mechanisms operate: the energetics and required scale of the phenomenon are very large (Rees 1985; Dekel & Rees 1987). A more promising idea, known as **natural bias**, was introduced by White *et al.* (1987). This relied on the idea that an object of a given mass will collapse sooner if it lies in a region of large-scale overdensity, leading to an enhanced abundance of such objects with respect to the mean. This picture can be analysed quite easily within the Press–Schechter picture, as shown by Cole & Kaiser (1989). The Press–Schechter expression for the number density of haloes depends on the parameter $\nu = \delta_c/\sigma(M)$:

$$\frac{dn}{d\ln M} \propto \nu \exp\left(-\frac{\nu^2}{2}\right). \tag{17.65}$$

This equation applies to the population as a whole, but it will vary from place to place depending on the large-scale density field. In the **peak–background split**, we imagine that the true large-scale density components are added to a pre-existing small-scale density field. The effect of this is to perturb the threshold: where the large-scale component is $\delta = \epsilon$, the small-scale component only needs to reach $\delta = \delta_c - \epsilon$ in order to achieve collapse. The number density is therefore modulated:

$$n \rightarrow n + \frac{dn}{d\nu}\frac{d\nu}{d\epsilon}\,\epsilon = n\left[1 + \epsilon\frac{(\nu^2 - 1)}{\sigma\nu}\right] \tag{17.66}$$

(because $\nu = (\delta_c - \epsilon)/\sigma$). This gives a bias in the number density of haloes in Lagrangian space: $\delta n/n = b_{\rm L}\epsilon$, where the Lagrangian bias is $b_{\rm L} = (\nu^2 - 1)/\nu\sigma$. In addition to this modulation of the halo properties, the large-scale disturbance will move haloes closer together where ϵ is large, giving a density contrast of $1+\epsilon$. If $\epsilon \ll 1$, the overall fractional density contrast of haloes is therefore the sum of the dynamical and statistical effects: $\delta_{\rm halo} = \epsilon + b_{\rm L}\epsilon$. The overall bias in Eulerian space ($b = \delta_{\rm halo}/\epsilon$) is therefore

$$\boxed{b(\nu) = 1 + \frac{\nu^2 - 1}{\nu\sigma} = 1 + \frac{\nu^2 - 1}{\delta_c}.} \tag{17.67}$$

This says that M^* haloes are unbiased, low-mass haloes are antibiased and high-mass haloes are positively biased, eventually reaching the $b = \nu/\sigma$ value expected for high peaks.

The high-peak calculation is undoubtedly relevant to present-day haloes, and to the correlations of Abell clusters; the problem is to know how to apply it to galaxies. At high redshift, when M^* is of order a galaxy mass, the relevant ϵ is $\epsilon_0/(1 + z_f)$, and galaxies could be strongly biased relative to the mass at that time. Indeed, there is good evidence that this is the case. Steidel *et al.* (1997) have used the Lyman-limit technique to select galaxies around redshifts $2.5 \lesssim z \lesssim 3.5$ and found their distribution to be highly inhomogeneous. The apparent value of σ_8 for these objects is of order unity, whereas the present value of $\sigma_8 \simeq 0.6$ should have evolved to about 0.13 at these redshifts (for $\Omega = 1$). This suggests a bias parameter of $b \simeq 8$, or $\nu \simeq 3.6$. For low-density models, the bias values are somewhat lower, but still substantially in excess of unity. Although the masses of these high-redshift objects are presently uncertain, it is clearly reasonable that galaxy-scale objects should be rare peaks at these redshifts. If $\nu \simeq 3.6$ at $z = 3$, then this translates to a present-day value of $\sigma = 1.9$ on the scale of these objects. If we adopt a Lagrangian radius of $1\,h^{-1}$ Mpc (which contains roughly the total mass of an L^* galaxy), then the spectral index that connects this value with $\sigma_8 = 0.6$ is $n = -1.9$. This is consistent with the discussion in chapter 16, where it was shown that the linear fluctuation spectrum appears to be rather flat on small scales. If this picture is correct, then it would be reasonable to claim that the Lyman-limit galaxies are the long-sought **primaeval galaxies**, since they are objects of galaxy scale seen very close to the time of their first collapse.

So much for high redshifts; can this sort of mechanism produce substantial bias at $z = 0$? Galaxies can clearly be born biased, but as the mass fluctuations grow, gravitational dynamics will tend to dilute the bias. If the Lagrangian component to the biased density field is kept unaltered, then the present-day bias will be

$$b(\nu) = 1 + \frac{\nu^2 - 1}{(1 + z_f)\delta_c}. \tag{17.68}$$

If $b = 8$ at $z = 3$, this suggests a current level of bias of $b = 2.8$, so the $z = 3$ Lyman-limit galaxies would certainly be strongly biased today if they maintained their identity to the present. However, the mechanism of high-peak bias only works by selecting rare objects that collapse early: if we wait a little, until all fluctuations of a given mass scale have collapsed, then there will be no bias. It is probably more reasonable to apply the high-peak bias formula to the present galaxies, using the appropriate value of ν. If $\nu = 3.6$ at $z = 3$, it will have fallen to below unity by the present, and the overall galaxy distribution will show little or no bias. The only way in which this conclusion can be changed is if it is assumed that galaxies never form below a threshold of $\nu \simeq 2$–3 (Bardeen *et al.* 1986). However, as discussed above, such a threshold would have to involve large-scale non-gravitational feedback processes, which do not seem plausible.

A further problem with trying to produce present-day bias just from the dynamics of halo formation is that the resulting objects will not have the right dynamical properties to correspond to real galaxies. The virial velocity dispersion resulting from gravitational collapse was derived above: $V \propto M^{1/3}(1+z_c)^{1/2}$. In practice, Gaussian statistics will give an order-unity dispersion in $1+z_c$, and so a large dispersion in V at fixed M. However, real galaxies display a tight Tully–Fisher correlation between luminosity and circular velocity. Cole & Kaiser (1989) showed that this problem could only be cured if star formation is biased (perhaps by epoch-dependent efficiency) in order to produce more stars in the galaxies that collapse earlier. Suppose an object collapsing at redshift z generates a stellar luminosity

$$L \propto M^{\alpha}(1+z)^{\beta}; \tag{17.69}$$

a perfect Tully–Fisher relation then requires $\beta = 3\alpha/2$. This is easily proved from $V \propto M^{1/3}(1+z_c)^{1/2}$: the above condition removes any redshift dependence and leaves $V \propto L^{1/3\alpha}$. The conventional Tully–Fisher slope of 1/4 then implies $\alpha = 4/3$, $\beta = 2$. If we identify haloes with galaxies, then a strong epoch dependence of star-forming efficiency seems to be needed. This is in principle a further way of understanding how 'natural bias' effects could enhance galaxies in clusters: galaxies form earlier in clusters and will therefore be brighter if $\beta > 1$. In fact, Cole and Kaiser argue that a somewhat stronger epoch dependence ($\beta \gtrsim 3$) is required to achieve sufficient bias to understand cluster mass-to-light ratios. In any case, such systematic behaviour would be inconsistent with the overall homogeneity of statistical properties such as the luminosity function or colour–luminosity relations, which show little dependence on local density.

In order to go beyond this illustrative analysis, we need some physical ideas for why there should be a dependence of luminosity on collapse epoch and mass of halo. The way this issue has been addressed in recent research is to accept that galaxy formation is a complicated process, in which many mechanisms combine to dictate the final stellar mass of a galaxy. This chapter has surveyed the main ingredients of the recipe: (i) the cooling rate in virialized haloes dictates the rate at which stars can form; (ii) energy returned from supernovae can slow the conversion of gas to stars, or even cause gas to be lost from the galaxy altogether; (iii) mergers will occur to combine stellar units, at a rate that depends on environment and on epoch; (iv) following mergers, galaxies may settle towards the centres of cluster-scale haloes, through the operation of **dynamical friction** (West & Richstone 1988; Couchman & Carlberg 1992). These mechanisms clearly allow early-collapsing haloes (which will tend to end up in rich clusters, as with high-peak bias) to form more stars: for a given mass, these objects will be denser at a given redshift. Cooling is more efficient, and the potential well is deeper, allowing the gas to be contained more effectively. Counteracting this, stellar populations that formed at high redshifts will have faded more and become redder by the present.

To put all these effects together, it is necessary to have a large computer code that determines the star-formation history of a given halo, and predicts its appearance using the spectral synthesis codes discussed in chapter 13. Two approaches are being followed in this regard. The brute-force method is to perform N-body simulations in which the evolution of both collisionless dark matter and dissipative gas is tracked and the physical state of the gas (i.e. its ability to cool) is used as a cue to insert star formation. The stars in turn are allowed to feed energy back into the gas, simulating the effects of mass loss and supernovae. This approach removes the need for any approximations such as the spherical model, which features heavily in any theoretical discussion. It is, however, limited in the range of densities and mass scales that can be probed, despite the rapid increase in the computing power that can be devoted to this problem. In particular, the controlling processes of star formation occur on such a small scale that they have to be inserted entirely by hand. This limitation has motivated models for **semi-analytic galaxy formation**, in which a Press–Schechter approach is used to deal with the dark-matter hierarchy. Combining cooling criteria for gas and a recipe for star formation with spectral synthesis models allows observational properties of such universes to be predicted (Kauffmann, White & Guiderdoni 1993; Cole *et al.* 1994; Baugh *et al.* 1997; White 1997). Since the statistical clustering properties of dark haloes can be predicted (see chapter 16; also Mo & White 1996), the bias of various classes of galaxy can also be found. Kauffmann, Nusser & Steinmetz (1997) discuss these issues in detail. They find that minimal bias is expected for low-density models where the normalization σ_8 is high; conversely, for high-density models with $\sigma_8 \simeq 0.5$, bias of up to $b = 1.5$ is naturally expected, with even larger values for $> L^*$ galaxies.

OUTSTANDING ISSUES It can fairly be stated that our understanding of biased galaxy formation is now enormously more sophisticated than when the idea was first suggested in the early 1980s. However, these advances have brought us only a little nearer to answering the one question that really counts: is the galaxy distribution biased or not? Let us summarize some of the key arguments.

(1) If $\Omega = 1$, then bias exists. Both the M/L ratios in clusters and the value of σ_8 inferred from the cluster abundance agree on this. Conversely, if $\Omega = 0.2$–0.3, then the effects must be relatively weak. Lower densities still would require antibias.

(2) Different classes of galaxy are biased relative to one another. Blue-selected galaxies have bias parameters about a factor 1.2 larger than far-infrared-selected IRAS galaxies. Elliptical galaxies and $> L^*$ galaxies have bias parameters that are a further factor of about 1.5 larger.

(3) The relative proportions of disk and elliptical galaxies and highly luminous galaxies alter greatly in the centres of rich clusters (luminosity and morphology segregation). However, outside these regions, the galaxy population appears to be virtually uniform in its properties, with all types inhabiting the same network of voids, walls and filaments.

Unfortunately, these observations can be used to argue with reasonable conviction for both high- and low-density universes. Point (1) merely emphasizes that the issues of bias and Ω are inextricably linked. The second point is encouraging for believers in bias, since it proves that different galaxies do have a wide range of clustering properties. For example, it is well understood how and why the spatial correlations of rich clusters will be an amplified version of the underlying density correlations (at least for Gaussian statistics). Given this, it is a safe bet that the clustering properties of elliptical galaxies also overestimate the density correlations; the known phenomenon of morphological segregation means that ellipticals are closely associated with clusters. We know in both the cases of clusters and ellipticals that we are excluding the low-density universe simply through our observational selection, so an unrepresentative answer would be expected. However, even the most weakly clustered population (IRAS galaxies) would need to be clustered with about $b = 2$ relative to the mass in order for an Einstein–de Sitter universe to be tenable. If we ask whether IRAS galaxies could favour high-density environments, this sounds implausible: the observations show them to be abruptly suppressed in clusters relative to optical galaxies (Strauss *et al.* 1992), and it is easy to believe that this is due to the fragility of disks. The intracluster gas almost certainly represents in part material that might have formed disks in a less stressful environment.

The arguments now seem to turn against the idea of strong bias, and the large-scale galaxy distribution is consistent with this. Away from clusters, all galaxies follow the same overall 'skeleton' of large-scale structure, independent of Hubble type (Thuan *et al.* 1987; Babul & Postman 1990; Mo, McGaugh & Bothun 1994). Countering this, we have to acknowledge that all populations of galaxies share common voids, and it is not easy to estimate the amount of mass that may reside in these regions. A substantial fraction of the entire volume of the universe is empty of galaxies; if these zones have a density that is close to the mean, then the total mass could be substantially underestimated. Even in simulations of low-density universes, the density in the voids is non-zero, and so a cutoff in the formation of galaxies at very low densities is always required. It is reasonable that such a cutoff should exist for disk galaxies, in which star formation is associated with spiral structure, which is triggered by differential rotation. Kennicutt (1989) showed that star formation is greatly suppressed in disks with surface densities below a critical threshold of about $4\, M_\odot \mathrm{pc}^{-2}$. Such low-density disks might be expected in late-forming galaxies and in voids. Certainly, examples are known of substantial disks that consist largely of neutral hydrogen that has not been processed into stars (Bothun *et al.* 1987; Impey & Bothun 1997).

The problem is that the mechanisms that produce bias seem to work more effectively in exactly the high-density universes that empirically require bias. If Ω is high, then the cluster abundance says that the true mass distribution must have a lower degree of dynamical evolution. This means that the density in the voids is closer to the mean, making it easier to hide large amounts of mass there. Conversely, if Ω is low, the voids have very nearly been emptied by gravity, so it makes very little difference whether or not galaxies form there in proportion to the mass. Also, the case of high Ω and low normalization causes the biases of galaxy haloes to increase in two ways. For a low normalization, the nonlinear mass scale today becomes comparable to that of the most massive galaxies, bringing in high-peak bias. Since the fluctuation spectrum is convex, the spectral index at the nonlinear point is more negative than would be the case for a higher normalization. This means that the statistical mechanisms that bias halo correlations work more effectively (the relative amplitude of the large-scale modulation in the peak-background split is larger). In short, bias works in such a way as to conspire rather effectively against our determining how strongly it is operating.

There are two ways to evade this stubborn degeneracy. The first is to look at the accounting of the total baryonic material in clusters. Within the central $\simeq 1$ Mpc, the masses in stars, X-ray emitting gas and total dark matter can be determined with reasonable accuracy (perhaps 20% rms), and this allows a minimum baryon fraction to be determined:

$$\boxed{\frac{M_{\rm baryons}}{M_{\rm total}} \gtrsim 0.009 + 0.050\, h^{-3/2}} \qquad (17.70)$$

(White *et al.* 1993). This equation is often referred to as the **baryon catastrophe**, for the following reasons. Assume for now that the baryon fraction in clusters is representative of the whole universe, and adopt the primordial nucleosynthesis prediction of $\Omega_B h^2 = 0.0125$. This gives an equation for Ω:

$$\Omega \lesssim 0.25 \left(h^{1/2} + 0.18h^2\right)^{-1}, \qquad (17.71)$$

which is a limit varying between 0.33 and 0.21 for h between 0.5 and 1. This is a catastrophe for the Einstein–de Sitter universe, in that clusters have to be biased not only in the light they emit, but also in the sense of containing a larger than average baryon fraction. However, producing a large-scale separation of dark matter and baryons on scales that are little past the turn-round phase is very difficult. If the density parameter is really unity, it appears that the nucleosynthesis density must be too low by at least a factor 3.

The other way forward is to look at the clustering pattern at different redshifts, exploiting the fact that the rate of evolution depends on density. Clustering evolution was discussed in chapter 16, where it was shown that the rate of evolution is slower than would be expected for an $\Omega = 1$ universe. This is consistent with the idea that the density is a few tenths of critical, and that bias is a small effect at $z = 0$ and at $z = 1$. However, if $\Omega = 1$, then this observation proves that the degree of bias at $z = 1$ is higher than at present. If so, this observation can be tested quite easily by looking at the M/L ratios of high-redshift clusters. At the present era, these indicate Ω values of several tenths, and increasing bias at high redshift would suppress this to an apparent $\Omega < 0.1$. Conversely, for an unbiased low-density universe, we would expect to measure the true Ω at high redshift, which evolves towards unity. This is a clear difference, and the indications are already that gravitational-lens signatures can be seen in some high-z clusters, proving them to be true large-mass concentrations (Luppino & Kaiser 1997).

Whatever the eventual resolution of this debate, the concept of bias in galaxy formation has been a vital element of cosmology. Even if it should turn out that the degree of bias is not large in practice, it is abundantly clear that we cannot rely in advance on this being true. Early work assumed without question that galaxies number density varied in direct proportion to mass density, but we now know that this is something that cannot be true in detail. A straightforward reading of the evidence favours a simple low-density model, but we always need to be vigilant in case the subtleties of bias have fooled us. Certainly, the Einstein–de Sitter universe occupies a position of huge importance in cosmology, and a high level of proof will rightly be required before it can be given up.

Problems

(17.1) Show that two-body **gravitational relaxation** due to collisions will be important in a system of N particles when the system is old enough that a typical particle has crossed the system a number of times equal to

$$n_{\rm relax} = \frac{N}{12 \ln N}. \tag{17.72}$$

Show that collisions are unimportant in galaxies, but marginally significant in clusters.

(17.2) Prove the virial theorem, $2K + V = 0$, for an equilibrium self-gravitating system.

(17.3) By conserving angular momentum for the gas in a $\lambda = 0.05$ halo, derive the characteristic disk scale length as a function of formation redshift,

$$r_{\rm disk} \simeq 3\, h^{-1}{\rm kpc}\, \left(v_c/100\,{\rm km\,s^{-1}}\right)\, H_0/H(z_f). \tag{17.73}$$

(17.4) Derive the optical depth of a collapsed halo to ionizing radiation:

$$\tau = 10^{-4.2}\, \alpha\, (J_{-21})^{-1}\, (V/100\,{\rm km\,s^{-1}})\, [f_c \Omega_{\rm B} h^2 (1+z)^3]^{3/2}. \tag{17.74}$$

18 Cosmic background fluctuations

18.1 Mechanisms for primary fluctuations

At the last-scattering redshift ($z \simeq 1000$), gravitational instability theory says that fractional density perturbations $\delta \gtrsim 10^{-3}$ must have existed in order for galaxies and clusters to have formed by the present. A long-standing challenge in cosmology has been to detect the corresponding fluctuations in brightness temperature of the cosmic microwave background (CMB) radiation, and it took over 25 years of ever more stringent upper limits before the first detections were obtained, in 1992. The study of CMB fluctuations has subsequently blossomed into a critical tool for pinning down cosmological models.

This can be a difficult subject; the treatment given here is intended to be the simplest possible. For technical details see e.g. Bond (1997), Efstathiou (1990), Hu & Sugiyama (1995), Seljak & Zaldarriaga (1996); for a more general overview, see White, Scott & Silk (1994) or Partridge (1995). The exact calculation of CMB anisotropies is complicated because of the increasing photon mean free path at recombination: a fluid treatment is no longer fully adequate. For full accuracy, the Boltzmann equation must be solved to follow the evolution of the photon distribution function. A convenient means for achieving this is provided by the public domain **CMBFAST** code (Seljak & Zaldarriaga 1996). Fortunately, these exact results can usually be understood via a more intuitive treatment, which is quantitatively correct on large and intermediate scales. This is effectively what would be called local thermodynamic equilibrium in stellar structure: imagine that the photons we see each originated in a region of space in which the radiation field was a Planck function of a given characteristic temperature. The observed brightness temperature field can then be thought of as arising from a superposition of these fluctuations in thermodynamic temperature.

We distinguish **primary anisotropies** (those that arise due to effects at the time of recombination) from **secondary anisotropies**, which are generated by scattering along the line of sight. There are three basic primary effects, illustrated in figure 18.1, which are important on respectively large, intermediate and small angular scales:

(1) Gravitational (Sachs–Wolfe) perturbations. Photons from high-density regions at last scattering have to climb out of potential wells, and are thus redshifted.

(2) Intrinsic (adiabatic) perturbations. In high-density regions, the coupling of matter and radiation can compress the radiation also, giving a higher temperature.

(3) Velocity (Doppler) perturbations. The plasma has a non-zero velocity at

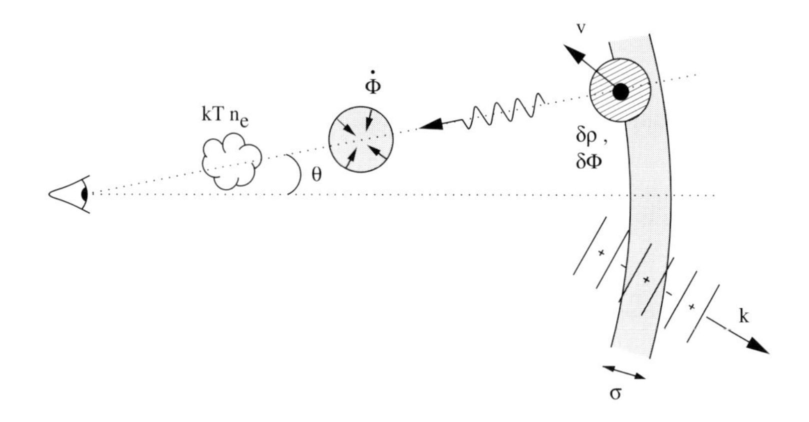

Figure 18.1. Illustrating the physical mechanisms that cause CMB anisotropies. The shaded arc on the right represents the last-scattering shell; an inhomogeneity on this shell affects the CMB through its potential, adiabatic and Doppler perturbations. Further perturbations are added along the line of sight by time-varying potentials (the Rees–Sciama effect) and by electron scattering from hot gas (the Sunyaev–Zeldovich effect). The density field at last scattering can be Fourier-analysed into modes of wavevector **k**. These spatial perturbation modes have a contribution that is in general damped by averaging over the shell of last scattering. Short-wavelength modes are more heavily affected (i) because more of them fit inside the scattering shell, and (ii) because their wavevectors point more nearly radially for a given projected wavelength.

recombination, which leads to Doppler shifts in frequency and hence brightness temperature.

To make quantitative progress, the next step is to see how to predict the size of these effects in terms of the spectrum of mass fluctuations.

THE TEMPERATURE POWER SPECTRUM The statistical treatment of CMB fluctuations is very similar to that of spatial density fluctuations. We have a 2D field of random fluctuations in brightness temperature, and this can be analysed by the same tools that are used in the case of 2D galaxy clustering (see chapter 16).

Suppose that the fractional temperature perturbations on a patch of sky of side L are Fourier expanded:

$$\begin{aligned}\frac{\delta T}{T}(\mathbf{X}) &= \frac{L^2}{(2\pi)^2}\int T_K \exp(-i\mathbf{K}\cdot\mathbf{X})\, d^2K \\ T_K(\mathbf{K}) &= \frac{1}{L^2}\int \frac{\delta T}{T}(\mathbf{X}) \exp(i\mathbf{K}\cdot\mathbf{X})\, d^2X,\end{aligned} \tag{18.1}$$

where $\mathbf{X}$ is a 2D position vector on the sky, and $\mathbf{K}$ is a 2D wavevector. This is only a valid procedure if the patch of sky under consideration is small enough to be considered flat; we give the full machinery below. We will normally take the units of length to be units of angle on the sky, although they could also in principle be h^{-1} Mpc at a given redshift. The relation between angle and comoving distance on the last-scattering sphere requires the comoving angular-diameter distance to the last-scattering sphere; because of its high redshift, this is effectively identical to the horizon size at the present epoch, $R_{\rm H}$:

$$\begin{aligned} R_{\rm H} &= \frac{2c}{\Omega_m H_0} \quad \text{(open)} \\ R_{\rm H} &\simeq \frac{2c}{\Omega_m^{0.4} H_0} \quad \text{(flat);} \end{aligned} \tag{18.2}$$

the latter approximation for models with $\Omega_m + \Omega_v = 1$ is due to Vittorio & Silk (1991).

As with the density field, it is convenient to define a dimensionless power spectrum of fractional temperature fluctuations,

$$\boxed{\mathcal{T}^2 \equiv \frac{L^2}{(2\pi)^2}\, 2\pi K^2 \, |T_K|^2,} \tag{18.3}$$

so that $\mathcal{T}^2$ is the fractional variance in temperature from modes in unit range of $\ln K$. The corresponding dimensionless spatial statistic is the two-point correlation function

$$\boxed{C(\theta) = \left\langle \frac{\delta T}{T}(\psi) \frac{\delta T}{T}(\psi + \theta) \right\rangle,} \tag{18.4}$$

which is the Fourier transform of the power spectrum, as usual:

$$C(\theta) = \int \mathcal{T}^2(K)\, J_0(K\theta)\, \frac{dK}{K}. \tag{18.5}$$

Here, the Bessel function comes from the angular part of the Fourier transform:

$$\int \exp(ix\cos\phi)\, d\phi = 2\pi J_0(x). \tag{18.6}$$

Now, in order to predict the observed anisotropy of the microwave background, the problem we must solve is to integrate the temperature perturbation field through the **last-scattering shell**. In order to do this, we assume that the sky is flat; we also neglect curvature of the 3-space, although this is only strictly valid for flat models with $k = 0$. Both these restrictions mean that the results are not valid for very large angles. Now, introducing the Fourier expansion of the 3D temperature perturbation field (with coefficients $T_k^{\rm 3D}$) we can construct the observed 2D temperature perturbation field by integrating over k-space and optical depth:

$$\frac{\delta T}{T} = \frac{V}{(2\pi)^3} \iint T_k^{\rm 3D}\, e^{-i\mathbf{k}\cdot\mathbf{r}}\, d^3k\, e^{-\tau}\, d\tau. \tag{18.7}$$

A further simplification is possible if we approximate $e^{-\tau} d\tau$ by a Gaussian in comoving radius:

$$\exp(-\tau)\, d\tau \propto \exp[-(r - r_{\rm LS})^2/2\sigma_r^2]\, dr. \tag{18.8}$$

This says that we observe radiation from a last-scattering shell centred at comoving distance $r_{\rm LS}$ (which is very nearly identical to $r_{\rm H}$, since the redshift is so high), with a thickness σ_r. The section on recombination in chapter 9 showed that the appropriate value of σ_r is approximately

$$\sigma_r = 7\,(\Omega h^2)^{-1/2}\ \text{Mpc}. \tag{18.9}$$

An intuitively useful way of thinking about the integral for the observed temperature perturbation is as a two-stage process: we produce a temperature field that is convolved in the radial direction, and then say that we observe a single shell that slices through this convolved field at the radius of last scattering. If the observed CMB is a slice in the xy-plane, the effect of the last-scattering convolution in the z-direction is $T_k^{3D} \to T_k^{3D} \exp(-k_z^2 \sigma_r^2/2)$ (z will briefly denote the Cartesian coordinate in the redshift direction, not redshift itself). As a result of this radial convolution and the angular dependence of the Doppler scattering term, the temperature spatial power spectrum is anisotropic. Nevertheless, we can still write down 2D and 3D Fourier-transform expressions for the correlation function in the plane $z = 0$ (taking the origin to be in the centre of the last-scattering shell):

$$\begin{aligned} C_{3D} &= \int \frac{\mathcal{T}_{3D}^2}{4\pi k^3} e^{-i\mathbf{k}\cdot\mathbf{x}} \, dk_x \, dk_y \, dk_z \\ C_{2D} &= \int \frac{\mathcal{T}_{2D}^2}{2\pi K^2} e^{-i\mathbf{K}\cdot\mathbf{x}} \, dK_x \, dK_y. \end{aligned} \tag{18.10}$$

Note the distinction between k and K – wavenumbers in 3D and 2D respectively. The definition of $\mathcal{T}_{3D}^2$ as the dimensionless power spectrum of spatial variations in temperature is analogous to the 3D spatial power spectrum:

$$\mathcal{T}_{3D}^2 = \frac{V}{(2\pi)^3} \, 4\pi k^3 \, |T_k^{3D}|^2 \tag{18.11}$$

(the 2D equivalent was written above just as $\mathcal{T}^2$, but sometimes it will be convenient for clarity to add an explicit subscript 2D). Equating the two expressions for $C(\mathbf{x})$ gives the usual expression relating 2D and 3D power spectra, derived in chapter 16, which we shall write in the slightly different form

$$\boxed{\mathcal{T}_{2D}^2(K) = K^2 \int_0^\infty \mathcal{T}_{3D}^2 \left(\sqrt{K^2 + w^2}\right) e^{-w^2\sigma_r^2} \, \frac{dw}{(w^2 + K^2)^{3/2}}.} \tag{18.12}$$

This simple expression gives the 2D spectrum as a projection, to which all modes with wavelength shorter than the projected wavelength of interest contribute; short-wavelength modes that run nearly towards the observer have a much longer apparent wavelength on the sky; see figure 18.1. The integral will generally be dominated by the contribution around $w = 0$, unless $\mathcal{T}_{3D}^2$ is a very rapidly increasing function, in which case what matters will be the small-scale cutoff governed by the width of the last-scattering shell.

The 2D power spectrum is thus a smeared version of the 3D one: any feature that appears at a particular wavenumber in 3D will cause a corresponding feature at the same wavenumber in 2D. A particularly simple converse to this rule arises when there are *no* features: the 3D power spectrum is scale invariant ($\mathcal{T}_{3D}^2 =$ constant). In this case, for scales large enough that we can neglect the radial smearing from the last-scattering shell,

$$\mathcal{T}_{2D}^2 = \mathcal{T}_{3D}^2 \tag{18.13}$$

so that the pattern on the CMB sky is scale invariant also. To apply the above machinery for a general spectrum, we now need quantitative expressions for the spatial temperature anisotropies.

SACHS–WOLFE EFFECT This is the dominant large-scale effect, and arises from potential perturbations at last scattering. These have two effects: (i) they redshift the photons we see, so that an overdensity *cools* the background as the photons climb out, $\delta T/T = \delta\Phi/c^2$; (ii) they cause time dilation at the last-scattering surface, so that we seem to be looking at a younger (and hence *hotter*) universe where there is an overdensity. The time dilation is $\delta t/t = \delta\Phi/c^2$; since the time dependence of the scale factor is $a \propto t^{2/3}$ and $T \propto 1/a$, this produces the counterterm $\delta T/T = -(2/3)\delta\Phi/c^2$. The net effect is thus one-third of the gravitational redshift:

$$\boxed{\frac{\delta T}{T} = \frac{\delta\Phi}{3c^2}.} \tag{18.14}$$

This effect was originally derived by Sachs & Wolfe (1967) and bears their name (SW effect). It is common to see the first argument alone, with the factor 1/3 attributed to some additional complicated effect of general relativity. However, in weak fields, general relativistic effects should already be incorporated within the concept of gravitational time dilation; the above argument shows that this is indeed all that is required to explain the full result.

To relate to density perturbations, use Poisson's equation $\nabla^2\delta\Phi_k = 4\pi G\rho\delta_k$. The effect of ∇^2 is to pull down a factor of $-k^2/a^2$ (a^2 because k is a comoving wavenumber). Eliminating ρ in terms of Ω and $z_{\rm LS}$ gives

$$T_k = -\frac{\Omega(1+z_{\rm LS})}{2}\left(\frac{H_0}{c}\right)^2\frac{\delta_k(z_{\rm LS})}{k^2}. \tag{18.15}$$

DOPPLER SOURCE TERM The effect here is just the Doppler effect from the scattering of photons by moving plasma:

$$\boxed{\frac{\delta T}{T} = \frac{\delta\mathbf{v}\cdot\hat{\mathbf{r}}}{c}.} \tag{18.16}$$

Using the standard expression for the linear peculiar velocity from chapter 16, the corresponding k-space result is

$$T_k = -i\sqrt{\Omega(1+z_{\rm LS})}\left(\frac{H_0}{c}\right)\frac{\delta_k(z_{\rm LS})}{k}\,\hat{\mathbf{k}}\cdot\hat{\mathbf{r}}. \tag{18.17}$$

ADIABATIC SOURCE TERM This is the simplest of the three effects mentioned at the start of this chapter:

$$\boxed{T_k = \frac{\delta_k(z_{\rm LS})}{3},} \tag{18.18}$$

because $\delta n_\gamma/n_\gamma = \delta\rho/\rho$ and $n_\gamma \propto T^3$. However, this simplicity conceals a paradox. Last scattering occurs only when the universe recombines, which occurs at roughly a fixed temperature such that $kT \sim \chi$, the ionization potential of hydrogen. Surely, then, we should just be looking back to a surface of constant temperature? Hot and cold spots should normalize themselves away, so that the last-scattering sphere appears uniform. The solution is that a denser spot recombines *later*: it is therefore less redshifted and appears hotter. In algebraic terms, the observed temperature perturbation is

$$\left(\frac{\delta T}{T}\right)_{\rm obs} = -\frac{\delta z}{1+z} = \frac{\delta\rho}{\rho}, \tag{18.19}$$

where the last expression assumes linear growth, $\delta \propto (1+z)^{-1}$. Thus, even though a more correct picture for the temperature anisotropies seen on the sky is of a crinkled surface at constant temperature, thinking of hot and cold spots gives the right answer. Any observable cross-talk between density perturbations and delayed recombination is confined to effects of order higher than linear.

We now draw the above results together to form the spatial power spectrum of CMB fluctuations in terms of the power spectrum of mass fluctuations at last scattering:

$$\boxed{\mathcal{T}^2_{\rm 3D} = \left[(f_{\rm A} + f_{\rm SW})^2(k) + f_{\rm V}^2(k)\mu^2\right] \Delta_k^2(z_{\rm LS}).} \tag{18.20}$$

There is no cross term between the adiabatic and Sachs–Wolfe terms proportional to δ and the Doppler term proportional to $i\delta$: $|a\delta + ib\delta|^2 = (a\delta + ib\delta)(a\delta^* - ib\delta^*)$. The dimensionless factors can be written most simply as

$$\begin{aligned} f_{\rm SW} &= -\frac{2}{(kD_{\rm LS})^2} \\ f_{\rm V} &= \frac{2}{kD_{\rm LS}} \\ f_{\rm A} &= 1/3, \end{aligned} \tag{18.21}$$

where

$$D_{\rm LS} = \frac{2c}{\Omega_m^{1/2} H_0}(1+z_{\rm LS})^{-1/2} = 184(\Omega h^2)^{-1/2}\ {\rm Mpc} \tag{18.22}$$

is the comoving horizon size at last scattering (a result that is independent of whether there is a cosmological constant).

We can see immediately from these expressions the relative importance of the various effects on different scales. The Sachs–Wolfe effect dominates for wavelengths $\gtrsim 1h^{-1}$ Gpc; Doppler effects then take over but are almost immediately dominated by adiabatic effects on the smallest scales.

SMALL-SCALE FLUCTUATIONS The above expressions apply to perturbations for which only gravity has been important up till last scattering, i.e. those larger than the horizon at $z_{\rm eq}$. For smaller wavelengths, a variety of additional physical processes act on the radiation perturbations, generally reducing the predicted anisotropies. An accurate treatment of these effects is not really possible without a more complicated analysis, as is easily seen by considering the thickness of the last-scattering shell, $\sigma_r = 7(\Omega h^2)^{-1/2}$ Mpc. This clearly has to be of the same order of magnitude as the photon mean free path at this time; on any smaller scales, a fluid approximation for the radiation is inadequate and a proper solution of the Boltzmann equation is needed. Nevertheless, some qualitative insight into the small-scale processes is possible. The radiation fluctuations will be damped relative to the baryon fluid by photon diffusion, characterized by the Silk-damping scale, $\lambda_{\rm S} = 2.7(\Omega\,\Omega_{\rm B} h^6)^{-1/4}$ Mpc (see chapter 15). Below the horizon scale at

$z_{\rm eq}$, $16(\Omega h^2)^{-1}$ Mpc, there is also the possibility that dark-matter perturbations can grow while the baryon fluid is still held back by radiation pressure, which results in adiabatic radiation fluctuations that are less than would be predicted from the dark-matter spectrum alone. In principle, this suggests a suppression factor of $(1+z_{\rm eq})/(1+z_{\rm LS})$, or roughly a factor 10. In detail, the effect is an oscillating function of scale, since we have seen in chapter 15 that baryonic perturbations oscillate as sound waves when they come inside the horizon:

$$\delta_b \propto (3c_{\rm S})^{1/4} \exp\left(\pm i \int k c_{\rm S}\, d\tau\right); \tag{18.23}$$

here, τ stands for conformal time. There is thus an oscillating signal in the CMB, depending on the exact phase of these waves at the time of last scattering. These oscillations in the fluid of baryons plus radiation cause a set of **acoustic peaks** in the small-scale power spectrum of the CMB fluctuations (see below).

It is clear that small-scale CMB anisotropies are a complex area, because of the near-coincidence between $z_{\rm eq}$ and $z_{\rm LS}$, and between σ_r, $r_{\rm H}(z_{\rm eq})$ and $\lambda_{\rm S}$. To some extent these complications can be ignored, because the finite thickness of the last-scattering shell smears out small-scale perturbations in any case. However, the damping is exponential in $(k\mu)^2$ and so modes with low μ receive little damping; averaging over all directions gives a reduction in power that goes only $\propto k^{-1}$. In the absence of the other effects listed above, small-scale adiabatic fluctuations would still dominate the anisotropy pattern.

LARGE-SCALE FLUCTUATIONS The flat-space formalism becomes inadequate for very large angles; the proper basis functions to use are the spherical harmonics:

$$\frac{\delta T}{T}(\hat{\mathbf{q}}) = \sum a_\ell^m Y_{\ell m}(\hat{\mathbf{q}}), \tag{18.24}$$

where $\hat{\mathbf{q}}$ is a unit vector that specifies direction on the sky. Since the spherical harmonics satisfy the orthonormality relation $\int Y_{\ell m} Y^*_{\ell' m'}\, d^2q = \delta_{\ell\ell'}\delta_{mm'}$, the inverse relation is

$$a_\ell^m = \int \frac{\delta T}{T}\, Y^*_{\ell m}\, d^2q. \tag{18.25}$$

The analogues of the Fourier relations for the correlation function and power spectrum are (see also chapter 16).

$$\boxed{\begin{aligned} C(\theta) &= \frac{1}{4\pi} \sum_\ell \sum_{m=-\ell}^{m=+\ell} |a_\ell^m|^2\, P_\ell(\cos\theta) \\ |a_\ell^m|^2 &= 2\pi \int_{-1}^{1} C(\theta)\, P_\ell(\cos\theta)\, d\cos\theta. \end{aligned}} \tag{18.26}$$

These are exact relations, governing the actual correlation structure of the observed sky. However, the sky we see is only one of infinitely many possible realizations of the statistical process that yields the temperature perturbations; as with the density field, we are more interested in the **ensemble average power**. A common notation is to define C_ℓ as the expectation value of $|a_\ell^m|^2$:

$$C(\theta) = \frac{1}{4\pi} \sum_\ell (2\ell+1)\, C_\ell\, P_\ell(\cos\theta), \qquad C_\ell \equiv \left\langle |a_\ell^m|^2 \right\rangle, \tag{18.27}$$

where now $C(\theta)$ is the ensemble-averaged correlation. For small θ and large ℓ, we saw in chapter 16 how the exact form reduces to a Fourier expansion:

$$C(\theta) = \int_0^\infty \mathcal{T}^2(K)\, J_0(K\theta)\, \frac{dK}{K}, \qquad \mathcal{T}^2(K = \ell + \tfrac{1}{2}) = \frac{(\ell + \frac{1}{2})(2\ell + 1)}{4\pi}\, C_\ell. \tag{18.28}$$

The effect of filtering the microwave sky with the beam of a telescope may be expressed as a multiplication of the C_ℓ, as with convolution in Fourier space:

$$C_{\rm S}(\theta) = \frac{1}{4\pi} \sum_\ell (2\ell + 1)\, W_\ell^2\, C_\ell\, P_\ell(\cos\theta). \tag{18.29}$$

When the telescope beam is narrow in angular terms, the Fourier limit can be used to deduce the appropriate ℓ-dependent filter function. For example, for a Gaussian beam of **FWHM** (full-width to half maximum) 2.35σ, the filter function is $W_\ell = \exp(-\ell^2\sigma^2/2)$.

For the large-scale temperature anisotropy, we have already seen that what matters is the Sachs–Wolfe effect, for which we have derived the spatial anisotropy power spectrum. The spherical harmonic coefficients for a spherical slice through such a field can be deduced using the results for large-angle galaxy clustering in chapter 16, in the limit of a selection function that goes to a delta function in radius:

$$\boxed{C_\ell^{\rm SW} = 16\pi \int (kD_{\rm LS})^{-4} \Delta_k^2(z_{\rm LS})\, j_\ell^2(kR_{\rm H})\, \frac{dk}{k},} \tag{18.30}$$

where the j_ℓ are **spherical Bessel functions** (see chapter 10 of Abramowitz & Stegun 1965). This formula, derived by Peebles (1982), strictly applies only to spatially flat models, since the Fourier expansion of the density field is invalid in an open model. Nevertheless, since the curvature radius R_0 subtends an angle of $\Omega/[2(1-\Omega)^{1/2}]$, even the lowest few multipoles are not seriously affected by this point, provided $\Omega \gtrsim 0.1$.

For simple mass spectra, the integral for the C_ℓ can be performed analytically. The case of most practical interest is a scale-invariant spectrum ($\Delta_k^2 \propto k^4$), for which the integral scales as

$$C_\ell = \frac{6}{\ell(\ell + 1)}\, C_2 \tag{18.31}$$

(see equation 6.574.2 of Gradshteyn & Ryzhik 1980). The direct relation between the mass fluctuation spectrum and the multipole coefficients of CMB fluctuations mean that either can be used as a measure of the normalization of the spectrum. One measure that has become common is to work in terms of the amplitude of the quadrupole ($\ell = 2$), by means of the rms temperature fluctuation $Q_{\rm rms}$ produced just by the $\ell = 2$ term(s) in the spherical harmonic expansion:

$$Q_{\rm rms}^2 = \frac{1}{4\pi} \sum_{m=-2}^{m=+2} |a_2^m|^2. \tag{18.32}$$

Unfortunately, although the quadrupole is the largest-scale intrinsic anisotropy signal (the intrinsic dipole is unobservable, owing to the Earth's motion), it is not a good choice as a reference point, for several reasons. First, the large-scale temperature pattern is subject to corruption by emission from the Milky Way, and it is better to work at galactic latitudes $|b| \gtrsim 20^\circ$; second, the intrinsic quadrupole is badly affected by **cosmic**

variance. The C_ℓ coefficients are the average of $|a_\ell^m|^2$ over an ensemble, and so the $Q^2_{\rm rms}$ value seen by a given observer is distributed like χ^2 with five degrees of freedom [problem 18.3]. A more useful quantity is the ensemble-averaged quadrupole, since this relates directly to the power spectrum:

$$Q^2_{\rm rms-ps} \equiv \frac{5}{4\pi}\, C_2. \tag{18.33}$$

However, in practice power is measured over a range of multipoles, centred at $\ell > 2$, so that the value at $\ell = 2$ is really an extrapolation that assumes a specific index for the spectrum. The most common choice is a scale-invariant spectrum, and so the clumsy quantity $Q_{{\rm rms-ps}|n=1}$ is used as a way of expressing the normalization of a scale-invariant spectrum. More generally, following our discussion of small-scale anisotropies, it makes sense to define a broad-band measure of the 'power per log ℓ':

$$\boxed{\mathcal{T}^2(\ell) = \frac{\ell(\ell+1)}{2\pi}\, C_\ell,} \tag{18.34}$$

so that $\mathcal{T}^2$ is constant for a scale-invariant spectrum. This is a close relation of another measure that is sometimes encountered, $Q^2(\ell) = (5/12)\mathcal{T}^2(\ell)$, which is an obvious generalization of the Q notation for the quadrupole amplitude. Finally, whichever measure is adopted, there is still the choice of units. The temperature fluctuation $\Delta T/T$ is dimensionless, but anisotropy experiments generally measure ΔT directly, independently of the mean temperature. It is therefore common practice to quote numbers like Q in units of μK.

ISOCURVATURE AND OTHER EFFECTS The above effects are the most important anisotropy mechanisms, but there are other sources for anisotropies that must sometimes be considered. The most obvious alteration to the above exposition comes when the gravitational potential changes with time along the photon trajectory. The photon then accumulates a redshift as it travels, which translates to a temperature perturbation:

$$\frac{\delta T}{T} = \int \frac{\dot{\Phi}}{c^2}\,\frac{d\ell}{c}, \tag{18.35}$$

where $d\ell$ is the element of proper distance. This effect does not operate in an Einstein–de Sitter universe, for which $\Phi_k = -4\pi G\rho a^2 \delta_k$, so that $\Phi_k \propto \delta_k/a$. The linear growth law for $\Omega = 1$ is $\delta_k \propto a$, so potential perturbations only change with time either when things become nonlinear or when Ω diverges significantly from unity. In the former case, one speaks of the **Rees–Sciama effect** (1968); in the latter case, it is the **integrated Sachs–Wolfe effect**. Nonlinear effects can be important on small scales, but of course there are also other sources of anisotropy to consider, principally Comptonization from moving gas (Vishniac 1987; Hu *et al.* 1994). Because the integrated Sachs–Wolfe effect in $\Omega \neq 1$ models only becomes important at low redshifts, the main effect tends to be on large angular scales. However, since there are problems with fundamentals such as the applicability of Fourier analysis on large scales for open models, this is not such a handicap. At $\ell \gtrsim 10$, the error made by ignoring the time dependence of the potential is small (Kofman *et al.* 1993). There is nevertheless a form of the integrated SW effect at last scattering itself, which arises because the universe is still not entirely matter dominated at this time, and this must be included for high-accuracy work (Hu & Sugiyama 1995).

A more general difficulty is that the analysis so far has implicitly assumed adiabatic initial conditions, where radiation, dark matter and baryons undergo a common density perturbation. This is not true of isocurvature initial conditions, where the perturbation is to the equation of state only, so that the photon-to-baryon ratio can vary spatially; at $t = 0$, this produces a perturbation to the matter density only. However, the subsequent evolution of the perturbations is adiabatic: outside the horizon, matter and photons respond to gravity in the same way. We have seen in chapter 15 that the matter perturbation decays at late times, as the universe becomes more matter dominated:

$$\delta_m \rightarrow \frac{4\delta_i}{y}, \tag{18.36}$$

where $y = \rho_m/\rho_r = a/a_{\rm eq}$. Since the matter perturbation tends to zero, the conserved entropy per baryon must manifest itself in a perturbation to the radiation:

$$\frac{\delta T}{T} \rightarrow -\frac{\delta_i}{3}. \tag{18.37}$$

(the factor $1/3$ arises because photon number density and hence entropy density scale as T^3). This is the **isocurvature effect**: the radiation is left cooler where the matter was originally denser. It is not really a new mechanism, but a case where the adiabatic perturbation caused by the perturbation to the photon number density is not proportional to the matter density.

The isocurvature effect is most conveniently compared with the Sachs–Wolfe effect, since it has the same scale dependence as the perturbations that would be predicted by the normal route of calculating the potential perturbations from the large-wavelength part of the present-day spectrum and applying $\delta T/T = \delta\Phi/(3c^2)$ (Efstathiou & Bond 1986). However, the amplitude is larger: the overall effect is that a given observed present-day power spectrum, if it originated from isocurvature perturbations, will produce large-angle CMB anisotropies six times larger than if the initial fluctuations were adiabatic [problem 18.4]. Since we shall argue below that data on galaxy clustering matches moderately well onto CMB data, this is some evidence that the initial conditions were not isocurvature.

REIONIZATION In a variety of models for galaxy formation, it is likely that the intergalactic medium will become reionized at high redshift. Certainly, we know empirically from the Gunn–Peterson test in quasars that such **reheating** did occur, and at a redshift in excess of 5 (see chapter 12). What produced the required energy input into the IGM is still not understood; the most common suggestion is that the first generation of star formation was responsible, but many more exotic possibilities have been discussed, amongst them decaying elementary particles and explosive energy output from an early generation of protoquasars. The consequences for the microwave background of this reionization depend on the Thomson-scattering optical depth:

$$\tau = \int \sigma_{\rm T}\, N_e\, \frac{c}{H_0}\, \frac{dz}{(1+z)^2\sqrt{1+\Omega z}}. \tag{18.38}$$

If we re-express the electron number density in terms of the gas density parameter as

$$N_e = \Omega_{\rm g}\, \frac{3H_0^2}{8\pi G \mu m_p}\, (1+z)^3, \tag{18.39}$$

where the parameter μ is approximately 1.143 for a gas of 25% helium by mass, and we then perform the integral over redshift, we get

$$\tau = 0.04h \, \frac{\Omega_{\rm G}}{\Omega} \left[(3\Omega + \Omega z - 2)\sqrt{1+\Omega z} - (3\Omega - 2)\right]. \tag{18.40}$$

The critical redshift at which the optical depth is unity is then (if $z \gg 1$)

$$z_{\rm crit} = \left(\frac{25}{\sqrt{\Omega}\, h \, \Omega_{\rm g}} \right)^{2/3}. \tag{18.41}$$

This can be substantially less than the normal last-scattering redshift at $z \simeq 1065$. The thickness of this new last-scattering shell will be $\Delta z \sim z_{\rm crit}$, and CMB perturbations are erased up to the comoving scale corresponding to this redshift interval. This is easily seen to be equal within a factor of order unity to the comoving horizon size at that time, $d_{\rm H} = (2c/H_0)(\Omega z)^{-1/2}$, which subtends an angle of $\sqrt{\Omega/z}$ radians. As discussed in chapter 17, the onset of structure formation might correspond to a redshift of as high as 30; the optical depth back to this era will exceed unity if $\Omega_{\rm g} h \gtrsim 0.15$. This is an implausibly high IGM density, and the effects of reionization would be more likely to produce $\sim 10\%$ alterations in the small-scale or intermediate-scale CMB fluctuations in a realistic case. However, some non-Gaussian models could start structure formation at $z \gtrsim 100$, by which point an optical depth of unity is easily achievable. Perturbations on scales of up to the order of a few degrees might be erasable in this way.

Even if reionization cannot yield sufficient optical depth to damp the primordial fluctuations, ionized gas can add fluctuations to the microwave sky through Comptonization. This process was studied in chapter 12, where it was shown that the important parameter for CMB perturbations is not so much the optical depth as the 'y-parameter':

$$y \equiv \int \sigma_{\rm T} n_e \frac{kT}{m_e c^2} \, d\ell. \tag{18.42}$$

Significant CMB distortions require an effect that exceeds the intrinsic anisotropies: $y \gtrsim 10^{-5}$. This will be possible even for optical depths as low as 10^{-4}, if the gas is hot enough. As was shown in chapter 12, the existing measurements of the CMB spectrum limit the global amounts of hot gas to a very low level. These effects are mainly important in the direction of rich clusters of galaxies (the **Sunyaev–Zeldovich effect**), where temperature fluctuations at the 10^{-5} level are now routinely detected (e.g. Jones *et al.* 1993). Such systems are rare, and so their overall contribution to the rms sky fluctuations have not been significant to date. However, in future high-precision CMB studies the SZ signal will be another contaminant that must be subtracted, as with galactic foreground emission. Fortunately, the characteristic frequency dependence of the SZ effect (which changes sign at the peak of the CMB) allows it to be clearly distinguished from intrinsic anisotropies.

18.2 Characteristics of CMB anisotropies

CRITICAL ANGLES We are now in a position to understand the characteristic angular structure of CMB fluctuations. The change-over from scale-invariant Sachs–Wolfe fluctuations to fluctuations dominated by Doppler scattering has been shown to occur at $k \simeq D_{\rm LS}$. This is one critical angle (call it θ_1); its definition is $\theta_1 = D_{\rm LS}/R_{\rm H}$, and for a matter-only model it takes the value

$$\theta_1 = 1.8\,\Omega^{1/2}\ \text{degrees.} \tag{18.43}$$

For flat low-density models with significant vacuum density, $R_{\rm H}$ is smaller; θ_1 and all subsequent angles would then be larger by about a factor $\Omega^{-0.6}$ (i.e. θ_1 is roughly independent of Ω in flat Λ-dominated models).

The second dominant scale is the scale of last-scattering smearing set by $\sigma_r = 7(\Omega h^2)^{-1/2}$ Mpc. This subtends an angle

$$\theta_2 = 4\,\Omega^{1/2}\ \text{arcmin.} \tag{18.44}$$

Finally, a characteristic scale in many density power spectra is set by the horizon at $z_{\rm eq}$. This is $16(\Omega h^2)^{-1}$ Mpc and subtends

$$\theta_3 = 9h^{-1}\ \text{arcmin,} \tag{18.45}$$

independently of the value of Ω. This is quite close to θ_2, so that alterations in the transfer function are an effect of secondary importance in most models.

We therefore expect that all scale-invariant models will have similar CMB power spectra: a flat Sachs–Wolfe portion down to $K \simeq 1\,\text{degree}^{-1}$, followed by a bump where Doppler and adiabatic effects come in, which turns over on arcminute scales through damping and smearing. This is illustrated well in figure 18.2, which shows some detailed calculations of 2D power spectra, generated with the CMBFAST package. From these plots, the key feature of the anisotropy spectrum is clearly the peak at $\ell \sim 100$. This is often referred to as the **Doppler peak**, but it is not so clear that this name is accurate. Our simplified analysis suggests that Sachs–Wolfe anisotropy should dominate for $\theta > \theta_1$, with Doppler and adiabatic terms becoming of comparable importance at θ_1, and adiabatic effects dominating at smaller scales. There are various effects that cause the simple estimate of adiabatic effects to be too large, but they clearly cannot be neglected for $\theta < \theta_1$. A better name, which is starting to gain currency, is the **acoustic peak**. In any case, it is clear that the peak is the key diagnostic feature of the CMB anisotropy spectrum: its height above the SW 'plateau' is sensitive to $\Omega_{\rm B}$ and its angular location depends on Ω and Λ. It is therefore no surprise that many experiments are currently attempting accurate measurements of this feature. Furthermore, it is apparent that sufficiently accurate experiments will be able to detect higher 'harmonics' of the peak, in the form of smaller oscillations of amplitude perhaps 20% in power, around $\ell \simeq 500$–1000. These features arise because the matter–radiation fluid undergoes small-scale oscillations, the phase of which at last scattering depends on wavelength, since the density oscillation varies roughly as $\delta \propto \exp(ic_{\rm S}k\tau)$; see above, and problem 15.4. Accurate measurement of these oscillations would pin down the sound speed at last scattering, and help give an independent measurement of the baryon density.

Ω-DEPENDENCE AND NORMALIZATION It is not uncommon to encounter the claim that the level of CMB fluctuations is inconsistent with a low-density universe, and in particular that a high density in collisionless dark matter is required. In fact, this statement is something of a fallacy, and it is worth examining the issue of density dependence in some detail.

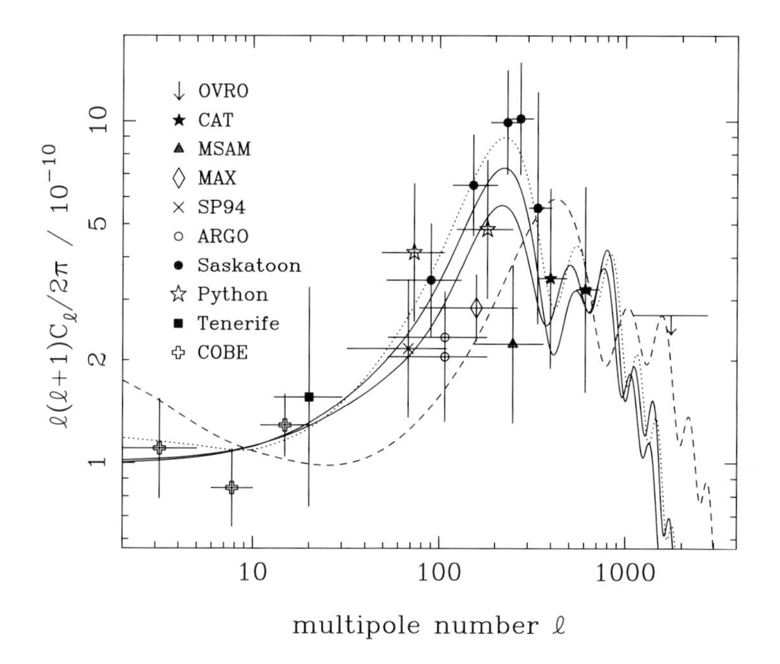

Figure 18.2. Angular power spectra $\mathcal{T}^2(\ell) = \ell(\ell+1)C_\ell/2\pi$ for the CMB, plotted against angular wavenumber ℓ in radians^{-1}. The experimental data are an updated version of the compilation described in White, Scott & Silk (1994), communicated by M. White; see also Hancock *et al.* (1997). Various model predictions for adiabatic scale-invariant CDM fluctuations are shown. The two solid lines correspond to $(\Omega, \Omega_B, h) = (1, 0.05, 0.5)$ and (1,0.1,0.5), with the higher Ω_B increasing power by about 20% at the peak. The dotted line shows a flat Λ-dominated model with $(\Omega, \Omega_B, h) = (0.3, 0.05, 0.65)$; the broken line shows an open model with the same parameters. Note the very similar shapes of all the curves. The normalization has been set to the large-scale amplitude, and so any dependence on Ω is quite modest. The main effects are that open models shift the peak to the right, and that the height of the peak increases with Ω_B and h.

Suppose we perform calculations assuming some mass power spectrum and $\Omega = 1$. If we now change Ω while keeping the shape and normalization of the power spectrum fixed, there are two effects: the power spectrum is translated both horizontally and vertically. The horizontal translation is quite simple: the main angles of importance scale as $\Omega^{1/2}$ so the CMB pattern shifts to smaller scales as Ω is reduced (unless Λ is important, in which case the shift is almost negligible; see above). To predict the vertical shift, it will suffice to consider the Sachs–Wolfe portion of the spectrum. This is

$$\mathcal{T}^2_{\rm SW} = \frac{4}{(kD_{\rm LS})^4}\,\Delta^2(z_{\rm LS}). \tag{18.46}$$

To relate δ at last scattering to its present value, we need the Ω-dependent growth-suppression factor for density perturbations:

$$\delta(z_{\rm LS}) \simeq \frac{\delta_0}{1+z_{\rm LS}} \, [g(\Omega)]^{-1} \tag{18.47}$$

i.e. there is less growth in low-density universes. Including the Ω-dependence of $D_{\rm LS}$ gives

$$\mathcal{T}^2_{\rm SW} = \frac{1}{4}\left(\frac{ck}{H_0}\right)^{-4} \Delta_0^2 \, \Omega^2 \, [g(\Omega)]^{-2}. \tag{18.48}$$

The approximate power-law dependence of g is $\Omega^{0.65}$ for open models or $\Omega^{0.23}$ for flat models, so it appears that low-density universes predict lower fluctuations. This is clearly contrary to the common idea that low-density universes are ruled out owing to the freeze-out of density-perturbation growth requiring higher fluctuations at last scattering. The fractional density fluctuations are indeed higher, but the potential fluctuations that are observable depend on the *total* density fluctuation, which is lower.

PREDICTIONS FROM GALAXY CLUSTERING However, Δ_0^2 here is the *mass* power spectrum, and the normalization we deduce from the light will generally depend on Ω. The effect this has depends on the scale at which we normalize. One common approach is to use σ_8, the rms density contrast in spheres of radius $8\,h^{-1}$ Mpc, since this is closely related to the abundance of rich clusters of galaxies. Alternatively, the amplitude of large-scale peculiar velocities measures the fractional density fluctuation on a somewhat larger scale – in both cases with a dependence of approximately $\delta_0 \propto \Omega^{-0.6}$. Note that these determinations make no use of galaxy clustering data at all: the present-day potential fluctuations are measurable directly, and yield $\delta_0\Omega^{0.6}$. What we deduce from galaxy clustering, conversely, is the biased quantity $b\delta_0$, and so galaxy clustering observations allow the parameter $\beta \equiv \Omega^{0.6}/b$ to be measured (as well as the shape of the spectrum, of course). The requirement for larger matter fluctuations in the case of lower Ω now makes the density dependence of the Sachs–Wolfe effect very weak: roughly $(\Delta^2_{\rm SW})^{1/2} \propto \Omega^{-0.25}$ (open) or $\Omega^{0.17}$ (flat).

How then can we constrain Ω from CMB observations? One possible route arises because the transfer function on small scales is rather sensitive to the total density, since modes with wavelengths below $r_{\rm H}(z_{\rm eq}) = 16(\Omega h^2)^{-1}$ Mpc have their amplitudes reduced. For a given primordial amplitude, this reduction clearly increases as Ω decreases, and is particularly severe for pure baryon universes where the small-scale power is removed by Silk damping. In the extreme case, low-density universes can have very little power on 8-Mpc wavelengths, and so normalizing on this scale gives silly answers. The number σ_8 measures the total rms density fluctuation after filtering with a sphere, and the only way this number can be large in models with a damping cutoff is to set the amplitude of 100-Mpc modes high. This is the real reason why low-density universes have often been associated with very large CMB fluctuations. However, it was argued in chapter 16 that the real universe does not seem to display such a large-scale coherence length. In any case, now that we have clustering data on 100-Mpc scales, it makes much more sense to fix the normalization there, in which case the predicted amplitude of large-scale temperature fluctuations loses almost all dependence on Ω, as discussed above.

Following this discussion, it should be clear that it was possible to make relatively clear predictions of the likely level of CMB anisotropies, even in advance of the first detections. What was required was a measurement of the typical depth of large-scale potential wells in the universe, and many lines of argument pointed inevitably to numbers of order 10^{-5}. This was already clear from the existence of massive clusters of galaxies with velocity dispersions of up to $1000\,\mathrm{km\,s^{-1}}$:

$$v^2 \sim \frac{GM}{r} \quad \Rightarrow \quad \frac{\Phi}{c^2} \sim \frac{v^2}{c^2}, \tag{18.49}$$

so the potential well of a cluster is of order 10^{-5} deep. More exactly, the abundance of rich clusters is determined by the amplitude σ_8, which measures $[\Delta^2(k)]^{1/2}$ at an effective wavenumber of very nearly $0.17\,h\,\mathrm{Mpc}^{-1}$ (see chapter 16). If we assume that this is a large enough scale that what we are measuring is the amplitude of any scale-invariant spectrum, then the earlier expression for the temperature power spectrum gives

$$\sqrt{\mathcal{T}^2_{\mathrm{SW}}} \simeq 10^{-5.7}\,\Omega\,\sigma_8\,[g(\Omega)]^{-1}. \tag{18.50}$$

There were thus strong grounds to expect that large-scale fluctuations would be present at about the 10^{-5} level, and it was a significant boost to the credibility of the gravitational-instability model that such fluctuations were eventually seen.

18.3 Observations of CMB anisotropies

Prior to April 1992, no CMB fluctuations had been detected, but existing limits were close to the interesting 10^{-5} level. Continued non-detection of fluctuations at a much lower level would have forced a fundamental overhaul of our ideas about cosmological structure formation, so this period was a critical time for cosmology.

COBE DMR DATA April 23 1992 saw the announcement by the COBE DMR team of the first detection of CMB fluctuations (Smoot *et al.* 1992). COBE is an acronym for NASA's **cosmic background explorer** satellite; launched in November 1989, this carried several experiments to probe the large-scale radiation field over the wavelength range 1 μm to 1 cm. The one concerned with the CMB at $\lambda \gtrsim 1$ mm was the **differential microwave radiometer** (DMR). The DMR experiment made a map of the sky with an angular resolution set by its 7° FWHM beam. This resolution means that only the low-order multipoles are accessible; DMR thus probes pure Sachs–Wolfe, and so it is easy to relate the DMR detection of sky fluctuations to a limit on the power spectrum.

In the case of the COBE measurements, the simplest and most robust datum in the initial detection reports was just the sky variance (convolved to 10° FWHM resolution, in order to suppress the noise a little), i.e. $C_{\mathrm{S}}(0)$ with $\sigma = 4.25°$. The expected result for this for pure Sachs–Wolfe anisotropies can be predicted to almost perfect accuracy by using a small-angle approximation:

$$\begin{aligned} C_{\mathrm{S}}(0) &= \frac{\Omega^2}{g^2(\Omega)} \int 4\left[k\left(\frac{2c}{H_0}\right)\right]^{-4} \Delta^2(k)\, W^2(kR_{\mathrm{H}})\,\frac{dk}{k} \\ W^2(y) &= [1 - j_0^2(y) - 3j_1^2(y)]\,F(y\sigma)/(y\sigma), \end{aligned} \tag{18.51}$$

where a Gaussian beam of FWHM 2.35σ is assumed, and

$$F(x) \equiv \exp(-x^2)\int_0^x \exp t^2\,dt \tag{18.52}$$

is Dawson's integral. The terms involving the spherical Bessel functions j_ℓ correspond to the subtraction of the unobservable monopole and dipole terms. The window function is relatively sharply peaked and so the COBE variance essentially picks out the power at a given characteristic scale, which is well approximated as follows (σ in radians):

$$\begin{aligned} C_{\rm S}(0) &= 1.665\,\frac{\Omega^2}{g^2(\Omega)}\,4[k_{\rm S}(2c/H_0)]^{-4}\Delta^2(k_{\rm S}) \\ k_{\rm S}R_{\rm H} &= \frac{0.54}{\sigma} + 2.19(n-1) \end{aligned} \tag{18.53}$$

(Peacock & Dodds 1994). The original reported value was $C^{1/2}(0) = 1.10 \pm 0.18 \times 10^{-5}$ (Smoot *et al.* 1992). For scale-invariant spectra, this corresponds to an rms quadrupole $Q_{\rm rms-ps} = 15.0 \pm 2.5$ μK. The final results from four years of data are consistent with the initial announcement: a scale-invariant fitted $Q_{\rm rms-ps} = 18.0 \pm 1.8$ μK. Only weak constraints on the spectrum index can be set, but the results are certainly consistent with the scale-invariant prejudice: $n = 1.2 \pm 0.3$ (Bennett *et al.* 1996).

This detection required a large number of systematic effects to be eliminated in order to be certain that the reported signal was indeed cosmological. The COBE experiment had the advantage of being a satellite in a stable environment, with no atmospheric fluctuations to contend with. By observing the same piece of sky many times from different parts of its orbit, any signals that related to low-level interference from the Earth or Sun could be eliminated. The most serious remaining problem was astronomical foregrounds – principally emission from the Milky Way. The DMR experiment observed with three different frequency channels, 31.5, 53 and 90 GHz, each of which had two independent receivers. At these wavelengths, the main contaminants are galactic synchrotron and bremsstrahlung emission (brightness temperatures varying roughly as $\nu^{-2.8}$ and ν^{-2}, respectively). A constant-temperature cosmological signal can thus be picked up by averaging the channels so as to make the galactic signal vanish (and analysing only high galactic latitudes for good measure). Even after the systematics have been dealt with, the individual DMR receivers had significant thermal noise; in the four-year data set, the full-resolution combined map suffers a noise of around 30 μK at any given position. Since the sky map has 1844 pixels, the excess cosmological variance can be detected with huge significance, but some have criticized COBE on the grounds that it failed to detect individual structures in the CMB. Such comments are unwarranted, since all that any CMB experiment can do is to measure multipole components of the temperature perturbation field down to some limit. COBE was unable to measure individual multipole coefficients to its full resolution ($\ell \simeq 20$), but did perfectly well to $\ell \simeq 10$: the hot and cold spots on the CMB sky at 20° resolution were correctly identified even in the first-year data.

SMALL-SCALE EXPERIMENTS Eventually, the CMB sky will be revisited by satellite experiments with resolutions well below 1 degree. In the meantime, experiments that seek to improve on the COBE map have to work either from the ground or from balloons. In either case, they are forced to work with restricted patches of the sky, and have to contend with variable atmospheric emission. As a result, multiple-beam experiments designed to remove atmospheric emission are the norm. The simplest strategy is to switch rapidly between either two beams separated by an angle θ (chopping) or between three beam positions in a line, each separated by θ (chopping plus nodding). It is then possible to form combinations of these signals that reduce the atmospheric contribution: $T_2 - T_1$ in the two-beam case or $T_2-(T_1+T_3)/2$ in the three-beam case. The first case gives a signal insensitive to the mean atmosphere, whereas the second also cancels any contribution from a constant gradient in atmospheric emission. Squaring these expressions and taking expectation values shows that the rms fluctuations measured in such experiments are

$$\Delta T/T = \begin{cases} (2[C(0) - C(\theta)])^{1/2} & \text{(two-beam)} \\ \left(2[C(0) - C(\theta)] - \frac{1}{2}[C(0) - C(2\theta)]\right)^{1/2} & \text{(three-beam)} \end{cases} \tag{18.54}$$

Putting in the relation between power spectrum and $C(\theta)$, we see that three-beam experiments produce an effective 2D window function,

$$\begin{aligned} \left\langle (\delta T/T)^2 \right\rangle &= \int \mathcal{T}^2 \, W_K^2 \, dK/K \\ W_K^2 &= \left[\tfrac{3}{2} - 2J_0(K\theta) + \tfrac{1}{2}J_0(2K\theta)\right] e^{-K^2\theta_s^2}, \end{aligned} \tag{18.55}$$

where $2.35\theta_s$ is the beam FWHM and θ is the beam throw. The filter function peaks at some effective wavelength that is very close to 2θ. In general, all such experiments can be viewed in this way as observing the CMB sky with some effective window function,

$$\boxed{\left\langle (\delta T/T)^2 \right\rangle = \frac{1}{4\pi} \sum_\ell (2\ell + 1) \, W_\ell^2 \, C_\ell} \tag{18.56}$$

(see Partridge 1995 for more complex observing strategies).

A common means for analysing such experiments, especially in the case of null detections, has been to assume that the sky possesses a correlation function and a power spectrum which are Gaussian in form:

$$\begin{aligned} C(\theta) &= C(0) \exp(-\theta^2/2\theta_c^2) \\ \mathcal{T}^2(K) &= C(0)(K\theta_c)^2 \exp[-(K\theta_c)^2/2], \end{aligned} \tag{18.57}$$

and to quote $C(0)$ or limits thereto as a function of assumed θ_c. A better procedure is to say that the overall filter function is rather narrow, peaking at some effective wavenumber, so that one effectively determines a limit to the power at that wavenumber (obtained by using the filter function to predict the rms for a constant power spectrum). These two methods are closely related, since the most sensitive θ_c is that where the power spectrum peaks at the effective wavenumber of the experiment. The Gaussian power spectrum peaks at $K = \sqrt{2}/\theta_c$, at which point $\mathcal{T}^2(K) = 0.736C(0)$; these formulae can be used to convert between the two conventions for quoting results. The power-spectrum convention is certainly to be preferred as small-scale experiments move from limits to actually *measuring* the CMB power spectrum.

Fortunately, many experiments now quote their results in this direct form. A selection of the best existing power measurements are collected in figure 18.2, and make a fascinating picture. It appears that there is now clear evidence for the existence of a peak in $\mathcal{T}^2$ at multipole numbers of a few hundred, of a plausible location and height to correspond to the acoustic oscillations expected in gravitational instability.

18.4 Conclusions and outlook

Having reviewed the physical mechanisms that cause anisotropies in the microwave background, and summarized the observational situation, it is time to ask what conclusions can be drawn. In order to narrow the field of possibilities, the discussion will concentrate on models with primordial fluctuations that are adiabatic and Gaussian. As well as being the simplest models, they will also turn out to be in reasonably good agreement with observation. Isocurvature models suffer from the high amplitude of the large-scale perturbations, and do not become any more attractive when modelled in detail (Hu, Bunn & Sugiyama 1995). Topological defects were for a long time hard to assess, since accurate predictions of their CMB properties were difficult to make. Recent progress does, however, indicate that these theories may have difficulty matching the main details of CMB anisotropies, even as they are presently known (Pen, Seljak & Turok 1997).

INFLATIONARY PREDICTIONS Matching the CMB sky with what we know of mass inhomogeneities today is important for physical cosmology, since we have seen that the anisotropies depend on the cosmological parameters Ω, Λ, Ω_B and h. As if this were not already potentially a rich enough prize, CMB anisotropies also offer the chance to probe the very earliest phases of the big bang, and to test whether the expanding universe really did begin with an inflationary phase. Let us recall from chapter 11 what predictions inflation makes for the fluctuation spectrum. Inflation is driven by a scalar field ϕ, with a potential $V(\phi)$. As well as the characteristic energy density of inflation, V, this can be characterized by two dimensionless parameters

$$\begin{aligned} \epsilon &\equiv \frac{m_P^2}{16\pi}\,(V'/V)^2 \\ \eta &\equiv \frac{m_P^2}{8\pi}\,(V''/V), \end{aligned} \tag{18.58}$$

where m_P is the Planck mass, $V' = dV/d\phi$, and all quantities are evaluated towards the end of inflation, when the present large-scale structure modes were comparable in size to the inflationary horizon. Prior to transfer-function effects, the primordial fluctuation spectrum is specified by a horizon-scale amplitude (extrapolated to the present) δ_H and a slope n:

$$\Delta^2(k) = \delta_H^2 \left(\frac{ck}{H_0}\right)^{3+n}. \tag{18.59}$$

The inflationary predictions for these numbers are

$$\begin{aligned} \delta_H &\sim \frac{V^{1/2}}{m_P^2\,\epsilon^{1/2}} \\ n &= 1 - 6\epsilon + 2\eta, \end{aligned} \tag{18.60}$$

which leaves us in the unsatisfactory position of having two observables and three parameters.

The critical ingredient for testing inflation by making further predictions is the possibility that, in addition to scalar modes, the CMB could also be affected by gravitational waves (following the original insight of Starobinsky 1985). We therefore distinguish explicitly between scalar and tensor contributions to the CMB fluctuations by using appropriate subscripts. The former category are those described by the Sachs–Wolfe effect, and are gravitational potential fluctuations that relate directly to mass fluctuations. Chapter 11 showed that the relative amplitude of tensor and scalar contributions depended on the inflationary parameter ϵ alone:

$$\frac{C_\ell^{\rm T}}{C_\ell^{\rm S}} \simeq 12.4\epsilon \simeq 6(1-n). \tag{18.61}$$

The second relation to the **tilt** (which is defined to be $1-n$) is less general, as it assumes a polynomial-like potential, so that η is related to ϵ. If we make this assumption, inflation can be tested by measuring the tilt and the tensor contribution. For simple models, this test should be feasible: $V = \lambda\phi^4$ implies $n \simeq 0.95$ and $C_\ell^{\rm T}/C_\ell^{\rm S} \simeq 0.3$. To be safe, we need one further observation, and this is potentially provided by the spectrum of $C_\ell^{\rm T}$. Suppose we write separate power-law index definitions for the scalar and tensor anisotropies:

$$C_\ell^{\rm S} \propto \ell^{n_{\rm S}-3}, \qquad C_\ell^{\rm T} \propto \ell^{n_{\rm T}-3}. \tag{18.62}$$

From the discussion of the Sachs–Wolfe effect, we know that, on large scales, the scalar index is the same as index in the matter power spectrum: $n_{\rm S} = n = 1 - 6\epsilon + 2\eta$. By the same method, it is easily shown that $n_{\rm T} = 1 - 2\epsilon$ (although different definitions of $n_{\rm T}$ are in use in the literature; the convention here is that $n = 1$ always corresponds to a constant $\mathcal{T}^2(\ell)$). Finally, then, we can write the **inflationary consistency equation**:

$$\boxed{\frac{C_\ell^{\rm T}}{C_\ell^{\rm S}} = 6.2(1-n_{\rm T}).} \tag{18.63}$$

The slope of the scalar perturbation spectrum is the only quantity that contains η, and so $n_{\rm S}$ is not involved in a consistency equation, since there is no independent measure of η with which to compare it.

From the point of view of an inflationary purist, the scalar spectrum is therefore an annoying distraction from the important business of measuring the tensor contribution to the CMB anisotropies. A certain degree of degeneracy exists here (see Bond *et al.* 1994), since the tensor contribution has no acoustic peak; $C_\ell^{\rm T}$ is roughly constant up to the horizon scale and then falls. A spectrum with a large tensor contribution therefore closely resembles a scalar-only spectrum with smaller $\Omega_{\rm B}$ (and hence a relatively lower peak). One way in which this degeneracy may be lifted is through polarization of the CMB fluctuations. A non-zero polarization is inevitable because the electrons at last scattering experience an anisotropic radiation field. Thomson scattering from an anisotropic source will yield polarization, and the practical size of the fractional polarization P is of the order of the quadrupole radiation anisotropy at last scattering: $P \gtrsim 1\%$. Furthermore, the polarization signature of tensor perturbations differs from that of scalar perturbations (e.g. Seljak 1997; Hu & White 1997); the different contributions to the total unpolarized C_ℓ can in principle be disentangled, allowing the inflationary test to be carried out.

IMPLICATIONS OF LARGE-SCALE ANISOTROPIES Despite the above discussion, it will be convenient to compare the present-day mass distribution with the CMB data by considering only scalar perturbations at first. Possible complications due to tensor contributions can be brought in a little later. The best and cleanest anisotropy measurements are those due to COBE, and we have seen above how the large-scale Sachs–Wolfe anisotropy can be calculated. It is possible to argue in both directions, either using the mass spectrum to predict the CMB, or vice versa; we shall start with the latter route. Górski *et al.* (1995), Bunn, Scott & White (1995), and White & Bunn

(1995) discuss the large-scale normalization from the two-year COBE data in the context of CDM-like models. The final four-year COBE data favour very slightly lower results, and we scale to these in what follows. For scale-invariant spectra and $\Omega = 1$, the best normalization is

$$\text{COBE} \quad \Rightarrow \quad \Delta^2(k) = \left(\frac{k}{0.0737\, h\, \text{Mpc}^{-1}} \right)^4, \qquad (18.64)$$

which is equivalent to $Q_{\text{rms-ps}} = 18.0\, \mu\text{K}$, or $\delta_{\text{H}} = 2.05 \times 10^{-5}$.

For low-density models, the earlier discussion suggests that the power spectrum should depend on Ω and the growth factor g as $P \propto g^2/\Omega^2$. Because of the time dependence of the gravitational potential (the integrated Sachs–Wolfe effect) and because of spatial curvature, this expression is not exact, although it captures the main effect. From the data of White & Bunn (1995), a better approximation is

$$\Delta^2(k) \propto \frac{g^2}{\Omega^2}\, g^{0.7}. \qquad (18.65)$$

This applies for low-Ω models both with and without vacuum energy, with a maximum error of 2% in density fluctuation provided $\Omega > 0.2$. Since the rough power-law dependence of g is $g(\Omega) \simeq \Omega^{0.65}$ and $\Omega^{0.23}$ for open and flat models respectively, we see that the implied density fluctuation amplitude scales approximately as $\Omega^{-0.12}$ and $\Omega^{-0.69}$ respectively for these two cases. The dependence is weak for open models, but vacuum energy implies much larger fluctuations for low Ω.

What if we consider a **tilted spectrum**? To see the effect of $n \neq 1$, we need to know the effective k at which COBE determines the spectrum. We saw earlier that the Sachs–Wolfe contribution to the rms sky fluctuations filtered with a Gaussian beam of FWHM 2.35σ effectively measured the power at $kR_{\text{H}} = (0.54/\sigma) + 2.19(n-1)$. For 10° resolution, $\sigma = 0.0742$, and so the effective wavenumber is approximately

$$k_{\text{eff}}^{\text{COBE}} = 0.0012\, h\, \text{Mpc}^{-1} \times \begin{cases} \Omega^{1.0} & \text{(open)} \\ \Omega^{0.4} & \text{(flat),} \end{cases} \qquad (18.66)$$

ignoring the small n-dependent correction. This is a scale at least 20 times beyond the largest wavelength on which large-scale structure is reliably measured, and so the effects of tilt will be substantial. We will adopt from chapter 16 the following measure of power on the largest reliable scales:

$$\Delta^2_{\text{opt}}(k = 0.02\, h\, \text{Mpc}^{-1}) \simeq 0.005 \pm 0.0015. \qquad (18.67)$$

Furthermore, we know from redshift-space distortions and the value of σ_8 inferred from the cluster abundance (chapter 17) that the corresponding number for the mass must be lower if $\Omega = 1$. As before, $b \simeq 1.6$ seems the best guess for $\Omega = 1$. Since σ_8 from clusters scales as $\Omega^{-0.56}$, this suggests that $\Delta^2_{\text{mass}}(k = 0.02\, h\, \text{Mpc}^{-1}) \simeq 0.002\Omega^{-1.1}$. We can compare this number with the COBE prediction, scaling the COBE-predicted amplitude with Ω as above and pivoting the power-law spectrum about k_{eff}; clearly there will be a unique value of n that matches prediction and observation. The only problem is that, although $k = 0.02\, h\, \text{Mpc}^{-1}$ is a very large scale, the power measured there is not quite the primordial value. Two physical models that fit the shape of the large-scale clustering spectrum were discussed in chapter 16: (i) $\Gamma = 0.25$ CDM; (ii) $\Gamma = 0.4$, $f_\nu = 0.3$ MDM. At $k = 0.02\, h\, \text{Mpc}^{-1}$, the transfer functions for these models are 0.69 and 0.81 respectively. We therefore adopt a mean of 0.75, and scale the inferred primordial mass fluctuations upwards by $1/(0.75)^2$.

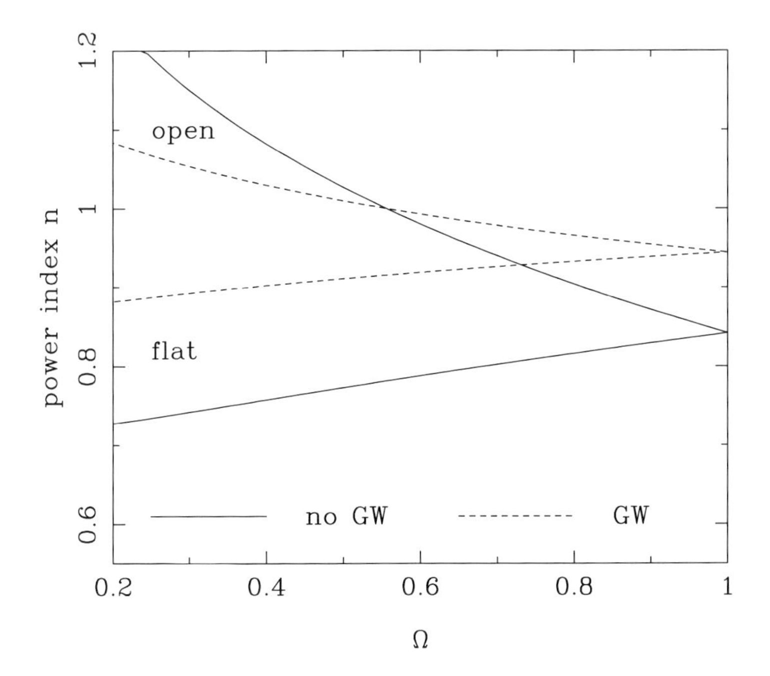

Figure 18.3. A plot of the power-law index of the spectrum needed to reconcile the COBE level of CMB fluctuations with the inferred present-day mass fluctuations on the largest scales, $k = 0.02\,h\,\mathrm{Mpc}^{-1}$, allowing for the effects of the transfer function as described in the text. Both open models and flat Λ-dominated models are shown. Solid lines denote models without gravity waves; broken lines show the effect of adding gravity waves with the usual coupling to tilt. A scale-invariant spectrum ($n = 1$) requires an open universe with $\Omega \simeq 0.6$. For lower densities, Λ-dominated flat models with gravity waves come closest to simple inflationary predictions ($n \simeq 0.9$).

Figure 18.3 shows the values of n that are required to reconcile this measure of primordial large-scale structure with COBE. Gravity waves are treated in two distinct ways: in the first case they are ignored; in the second they are added in with the above inflationary coupling to tilt, $C_\ell^{\mathrm{T}}/C_\ell^{\mathrm{S}} = 6(1-n)$. Figure 18.3 has several interesting features:

(1) For $\Omega = 1$, a significant tilt is needed: $n \simeq 0.84$ without tensors, rising to $n = 0.95$ if they are included.

(2) Going to low-density Λ-dominated models requires a greater degree of tilt. Even though the inferred mass fluctuations today increase for low Ω, the Ω-dependence of the CMB fluctuations in Λ models is even stronger.

(3) Conversely, the weak CMB dependence on Ω for open models means that the required tilt changes rapidly with Ω. A scale-invariant spectrum with no tensor contribution is consistent with the data if $\Omega \simeq 0.6$.

Some of these results look more attractive than others. High degrees of tilt are not expected in simple models of inflation (e.g. $n = 0.95$ for $V = \lambda\phi^4$). Moreover, large tilt causes problems with the CMB on smaller scales. COBE normalization corresponds to $\ell \simeq 20$, whereas we have seen that there is mounting evidence for a peak at $\ell \simeq 200$. The existence of tilt thus reduces the amplitude of this peak relative to COBE by a factor 10^{1-n}, which is equal to 1.4 for $\Omega = 1$ without gravity waves. Things are just as bad if gravity waves are included: the tilt is less, but the gravity-wave component has no peak, so that the reduction of the relative height of the peak is again a factor 1.4. Looking at figure 18.2, we see that $n = 1$ models with a reasonable baryon content in fact get the height of the peak about right; to allow the tilted variants, the baryon content would have to bc boosted above what is allowed from nucleosynthesis.

Although the preference for $n < 1$ is thus a negative feature of flat models, it may not be fatal, since the inferred tilt depends on the accuracy of the large-scale clustering measurements. Changing the assumed large-scale power by a factor 1.5 changes n by 0.14 without gravity waves, or 0.06 with gravity waves. If flat models are to survive, the true power at $k = 0.02\,h\,\mathrm{Mpc}^{-1}$ would thus need to be larger by a factor of order 1.5, and current data do not exclude this possibility.

IMPLICATIONS OF SMALL-SCALE ANISOTROPIES Stronger diagnostics for Ω and Λ come from the intermediate-scale and small-scale CMB anisotropies. The location of the peak at $\ell \simeq 200$ is sensitive to Ω, since it measures directly the angular size of the horizon at last scattering, which scales as $\ell \propto \Omega^{-1/2}$ for open models. The cutoff at $\ell \simeq 1000$ caused by last-scattering smearing also moves to higher ℓ for low Ω; if Ω were small enough, the smearing cutoff would be carried to large ℓ, where it would be inconsistent with the upper limits to anisotropies on 10-arcminute scales. For flat models with $\Lambda \neq 0$, the Ω-dependence is much weaker, which is one possible way of detecting Λ, should other arguments favour $\Omega < 1$.

It is straightforward to explore the details of this argument, in order to see how the predicted level of temperature fluctuations depends on spatial curvature. At last scattering, the dynamical effects of curvature and any cosmological constant are negligible. For a given Ω_m, Ω_B and h, the temperature fluctuations in flat and open models therefore differ only through the curvature dependence of the present-day horizon size and the different growth factors in flat and open models. There is therefore a simple recipe for taking $\mathcal{T}^2(\ell)$ for a flat model and converting the predictions to an open universe of the same present-day mass fluctuation spectrum: (i) move the results to the right (higher ℓ) by a factor $\Omega^{-0.6}$ to allow for the different horizon size; (ii) scale the results up to allow for the difference in density growth factors. It is then easy to see that, for example, the open equivalent of the vacuum-dominated universe shown in figure 18.2 will greatly exceed the small-scale CMB limits. This unpleasant conclusion can be softened by lowering the value of σ_8 for open models, but usually extreme values $\sigma_8 \lesssim 0.5$ are required.

This tendency for open models to violate the upper limits to arcminute-scale anisotropies is is a long-standing problem, which allowed Bond & Efstathiou (1984) to deduce the following limit on CDM universes:

$$\Omega \gtrsim 0.3h^{-4/3} \tag{18.68}$$

(in the absence of reionization, with a spectrum normalization that is independent of Ω, thus not allowing for the possibility of bias).

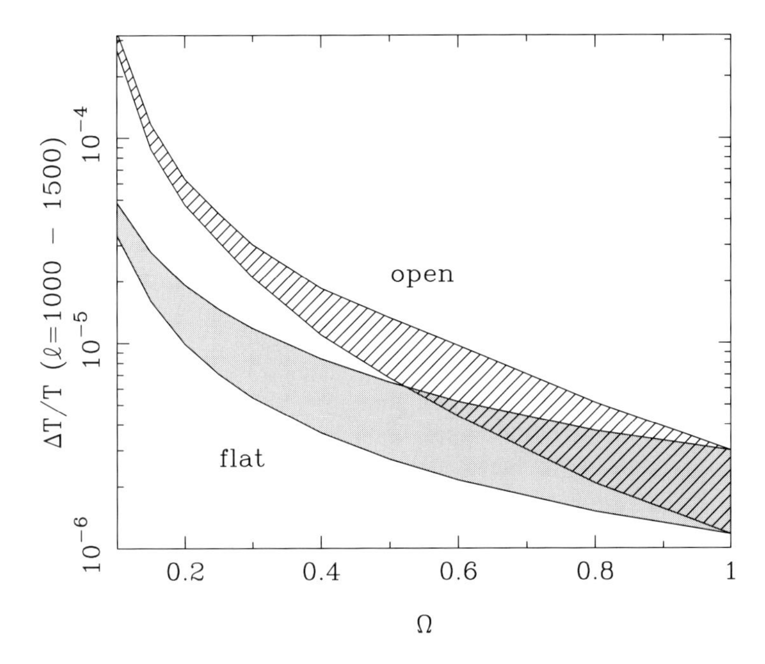

Figure 18.4. The power in CMB temperature fluctuations around $\ell \sim 1000$ can be predicted from present-day large-scale structure, where the normalization is known from the cluster abundance. This plot shows the fractional rms temperature fluctuation from modes with $1000 < \ell < 1500$. The result depends on the baryon content: the lower and upper edges of each shaded band correspond to $\Omega_{\rm B} = 0.01$ and $0.2\,\Omega$, respectively. A Hubble parameter of $h = 0.65$ is assumed, and the results scale roughly inversely with this choice. From figure 18.2, the observational upper limit to $\Delta T/T$ on this scale is between 1 and 2×10^{-5}, and this is much more easily satisfied by flat models than by open models.

An appealing feature of these arguments is that they can be made rather robust, since the high-ℓ fluctuations correspond to the projection on the sky of modes that can be investigated today. In particular, the cluster-scale normalization σ_8 was shown in chapter 16 to depend on wavenumbers $k \simeq 0.2\,h\,{\rm Mpc}^{-1}$. Since the horizon size is $R_{\rm H} = 6000\,\Omega^{-1}\,h^{-1}\,{\rm Mpc}$ for an open model, these scales appear at $\ell \simeq 1200\,\Omega^{-1}$, and so it should be possible to predict $\Delta T/T$ directly. Figure 18.4 shows an attempt at such a prediction. What has been done here is to use the CMBFAST code of Seljak & Zaldarriaga (1996) to calculate the small-scale anisotropies in a variety of $n = 1$ CDM models, normalized to satisfy $\sigma_8 = 0.57\,\Omega^{-0.56}$. Although the results may therefore seem tied to a particular model, the above arguments suggest they should be general. Because we are looking directly at the scales where σ_8 normalization is carried out, the fact that the fluctuation spectrum changes shape with Ω is of no interest – this will affect other issues such as the agreement with COBE. If the dark matter is not cold, the results will be lower limits, since collisionless damping could reduce the present-day power on the σ_8 scale.

In fact, the main uncertainty in converting from the estimate of the gravitational potential fluctuations measured via large-scale structure to $\Delta T/T$ is the baryon content of the universe. As was seen in figure 18.2, the 'bump' in the power spectrum relative to the Sachs–Wolfe portion is more pronounced for models with a high baryon fraction. This reflects the different amounts of growth that occur in the dark matter and in the baryonic fluids between $z_{\rm eq}$ and $z_{\rm LS}$. After the universe becomes matter dominated, dark-matter fluctuations on all scales can grow, whereas baryonic fluctuations below the Jeans length are still behaving as acoustic waves whose amplitude changes very slowly. Thus, in a DM-dominated model, the baryonic fluctuations are reduced relative to the overall mass fluctuations. This effect does not occur in pure baryon models, and so the fluctuations are higher, albeit only by a factor 2–3 relative to the Sachs–Wolfe portion of the spectrum. It is again a common misconception to claim that this growth difference is the reason why baryon-only models predict much larger fluctuations, so that non-baryonic dark matter is required. As mentioned earlier, the main difference is usually that one compares baryon-only models of low density with non-baryonic models of high density, but normalizing to a small-scale measure such as σ_8.

In figure 18.4, the shaded band of data therefore shows a conservative range of $\Omega_{\rm B}$, from 0.01 to $0.2\,\Omega$. Despite this uncertainty, the verdict is very clearly against open models. Figure 18.2 showed that the observational upper limit to fluctuations in this regime is $\mathcal{T}^2 \lesssim 3 \times 10^{-10}$, suggesting that the upper limit to $\Delta T/T$ for the multipole range $\ell = 1000$–1500 is between 10^{-5} and 2×10^{-5}. Flat models survive this restriction provided Ω is not lower than 0.2, whereas open models require much higher densities: $\Omega \gtrsim 0.5$. This argument is direct and very nearly model independent; it appears that the only way in which low-density open universes can be salvaged is by overturning some of the assumptions that have been made. Either the cluster-scale normalization must be lowered, which would probably require non-Gaussian statistics, or very substantial reionization effects must be invoked to damp away the $\ell \sim 1000$ power. By contrast, $k = 0$ models survive the confrontation with the small-scale CMB data with ease.

CODA The study of anisotropies in the CMB is presently one of the most exciting observational areas in cosmology, as a plethora of experiments map out the anisotropy spectrum over a wide range of scales. The fact that these detections are at the $\lesssim 10^{-5}$ level makes an amusing contrast with the early days of the subject, when fluctuations of order 10^{-3} were expected, based on the simplistic formula '$\delta T/T = (1/3)\delta\rho/\rho$ and galaxies must be made by $z = 3$'. Today, we have a much more sophisticated appreciation of the scales that are accessible to observation, plus much improved data on the inhomogeneity of the local universe.

The COBE detection and smaller-scale measurements are enormously encouraging indications of the overall correctness of the picture of structure formation via gravitational instability, but they leave open many possibilities. These will be constrained by the information that resides in the $\ell \gtrsim 100$ peaks in the anisotropy spectrum. At present, all we can say is that there are hints of a acoustic peak in the spectrum at $\ell \simeq 200$ and a sharp fall by $\ell \simeq 1000$. If confirmed, these facts would make it very difficult to sustain the idea of an open universe. Nevertheless, both the standard $k = 0$ alternatives presently face serious difficulties in the form of timescale problems for $\Omega = 1$ and negative q_0 for vacuum-dominated models. It is too early to say how this debate will be resolved, but it is fascinating to speculate. The most radical possibility would be to accept that the CMB requires $k = 0$ but that this cannot be provided by the constant-density vacuum equation of state. Something like a scalar field that continues to roll down the potential today would work (Peebles & Ratra 1988). It would certainly be wonderful if cosmology were able to prove the existence of a component to the universe with an equation of state that differs from all the common alternatives. However, it would be foolish to bet against the view that improved data will increasingly come to resemble one of the more conventional models.

Definitive measurements of the CMB fluctuation spectrum will require a new generation of experiments, which are expected to yield results in the first decade of the twenty-first century. As well as accurate large-scale mapping, these probes will measure the fine-scale anisotropy down to $\ell \sim 1000$. Measurements of the higher harmonics of the acoustic oscillations will be sensitive to the fine details of the physical conditions at last scattering: these will either rule out all the standard range of models or determine the nature of dark matter and measure very accurately the main cosmological parameters, if the present framework is correct (Bond, Efstathiou & Tegmark 1997; Zaldarriaga, Spergel & Seljak 1997).

Finally, the deepest attraction offered by CMB studies is the possibility of testing inflation. We have seen that one characteristic prediction of inflation is the existence of tensor anisotropies, and have discussed how these may in principle be detected via their contribution to the CMB anisotropy spectrum, and also through the polarization of CMB fluctuations. These will be challenging observations, but ones whose importance it would be difficult to overstate. The detection of the inflationary background of gravitational waves would give us experimental evidence on the nature of physics at almost the Planck energy. It is astonishing to realize that this might be accomplished within a mere 100 years from the first discovery of the expansion of the universe. The present is undoubtedly a golden age for cosmology, but perhaps the best is yet to come.

Problems

(18.1) In terms of the 2D power spectrum of the temperature fluctuations on the sky, $\Delta^2(K)$, we are often interested in the moments

$$\sigma_n^2 = \int_0^\infty \mathcal{T}^2(K)\, K^{2n}\, \frac{dK}{K}. \tag{18.69}$$

From this definition, show that the behaviour of the temperature autocorrelation function near the origin is

$$\frac{C(\theta)}{\sigma_0^2} \simeq 1 - \frac{\gamma^2}{2}\left(\frac{\theta}{\theta^*}\right)^2 + \frac{\gamma^2}{16}\left(\frac{\theta}{\theta^*}\right)^4 + \cdots, \tag{18.70}$$

and give the combination of moments that define the parameters γ and θ^*.

(18.2) Using the projection equation for the 2D temperature power spectrum as an integral over the 3D spectrum, obtain the corresponding integral that gives the correlation function $C(\theta)$ in terms of $\mathcal{T}^2_{\rm 3D}$.

(18.3) Consider the spherical harmonic coefficients a_ℓ^m for a 3D density field sampled on the surface of a sphere. Show that, for a Gaussian density field, these coefficients have either a one- or two-dimensional Gaussian distribution, depending on whether $m = 0$ or $m \neq 0$. What is the probability distribution of the quadrupole signal? Is the CMB sky a Gaussian random field?

(18.4) Show that a CDM universe with a given matter power spectrum will generate large-scale temperature perturbations that are six times higher for isocurvature initial conditions than for adiabatic initial conditions (i.e. show that the isocurvature effect of residual entropy perturbations is five times the Sachs–Wolfe effect from gravitational potential perturbations).

(18.5) A microwave background experiment consists of three telescope beams with equal Gaussian profiles of FWHM 2.35σ, arranged in an equilateral triangle of side θ. If the experiment measures the rms temperature fluctuation with respect to the mean of the three beams, how is this related to $C(0)$ and $C(\theta)$, where $C(\theta)$ is the correlation function of the microwave sky after smoothing with the Gaussian beam?

Hints for solution of the problems

Problem 1.1. At cylindrical radius r, a star appears to move at $v = r\omega$, implying time dilation by a factor $\Gamma \simeq 1 + r^2\omega^2/2c^2$; this is not observed. However, in order to maintain the stars in circular orbits, a centripetal acceleration $a = v^2/r$ is needed. The potential is $\Phi = r^2\omega^2/2$, so gravitational blueshift of the radiation cancels the kinematic redshift (at least to order r^2).

Problem 1.2. The equation of motion is $dU^\mu/d\tau = F^\mu/m$, where $F^\mu/m = (0, g)$ in the rest frame, so that $F^\mu/m = (\gamma\beta g, \gamma g)$ in the lab frame. Since $U^\mu = (\gamma, \gamma\beta)$, both parts of the equation imply $\gamma' = \gamma\beta g = (\gamma^2 - 1)^{1/2} g$, which integrates to $\gamma = \cosh g\tau$, implying $\beta = \tanh g\tau$. These latter equations integrate to give the required $x(\tau)$, $t(\tau)$. A quick way of verifying the solution is to remember that $A^\mu A_\mu = -a_0^2$, where a_0 is the proper acceleration.

Expanding to get the lowest-order correction to Newtonian behaviour gives

$$x \simeq \frac{g\tau^2}{2} + \frac{g^3\tau^4}{24c^2} \simeq \frac{gt^2}{2} - \frac{g^3t^4}{8c^2}.$$

In six months' acceleration, the distance travelled is 0.132 light years, making a total of 0.264 light years.

Problem 1.3. A set of observers all with constant proper acceleration g will obey $gt = \sinh g\tau$, $gx = (\cosh g\tau - 1) + x'$, where their positions at $t = 0$ are at $x = x'$; x' is just the constant of integration in the equation of motion. The observers all move in synchrony and maintain the same Δx at constant t. Observers at $x' = 0$ and $x' = L$ thus do not define the same frame as observers at the tail and nose of a rocket. Let the tail observer have a proper acceleration g; for proper time τ, $\gamma = \cosh g\tau$, and the lab length of the rocket should be L/γ. This means that an observer at the nose of the rocket will be left behind by a separate observer who undergoes proper acceleration g, and who starts at $x' = L$ at $t = 0$. This is reasonable enough: the set of independent observers all start accelerating at $t = 0$ exactly, whereas the nose of the rocket cannot learn until at least $t = L/c$ that it needs to start accelerating, by which time the accelerating neighbour has a relative velocity of $v = gL/c$, and will start to pull ahead.

If the nose of the rocket is not a trajectory of proper acceleration g, it might still be approximated by a locus of some constant proper acceleration. The given relations define a proper acceleration $a_0 = g/(1 + gx')$, and this does the job. Consider two events at the same t at $x' = 0$ and $x' = L$. The spatial separation is $\Delta x = L/\gamma + O([gL]^2)$, where $\gamma = \cosh gt'$ as before. The x' spatial coordinates thus length-contract as required.

They also undergo 'gravitational' time dilation, since $d\tau = \sqrt{dt'^2 - dx'^2} = (1 + gx')dt'$ at constant x'. The time coordinate t' therefore looks like time in the accelerated frame, measured by the observer at the base of the rocket; they see clocks ahead of them running fast. Overall, the (Minkowski) metric can be rewritten as

$$d\tau^2 = (1 + gx')^2 dt'^2 - dx'^2,$$

with no cross-term. See chapter 6 of Misner, Thorne & Wheeler (1973) for more details.

Problem 1.4. The element of length on a sphere is $ds = (r^2 d\theta^2 + r^2 \sin^2\theta\, d\phi^2)^{1/2}$. The total length is thus $s = r\int [1 + \sin^2\theta\, (\phi')^2]^{1/2}\, d\theta$. To make this stationary, apply Euler's equation, which gives the nonlinear 'equation of motion'

$$[1 + \sin^2\theta\, (\phi')^2]\, (\phi'' \sin^2\theta + \phi' \sin 2\theta) = (\phi')^2 \sin\theta \cos\theta + \phi''\phi' \sin^2\theta.$$

Clearly a great circle with constant ϕ is a solution, provided the start and end points have the same ϕ. Otherwise, the sensible method of solution is to choose rotated coordinates so that this is so. Trying to solve the equation without such geometrical insights is not attractive.

Problem 1.5. The derivative $\partial A/\partial x^\mu$ is a general 4-vector, so we can write it as $A_{;\mu}$. The derivative of a vector is $B_{\mu;\nu} = B_{\mu,\nu} - \Gamma^\alpha_{\mu\nu} B_\alpha$. Because of the symmetry of Γ, they vanish in the antisymmetric curl combination, but not in div. To re-express the divergence in terms of $\sqrt{-g}$, we need $\delta\sqrt{-g} = \frac{1}{2}\sqrt{-g}\, g^{\mu\nu}\delta g_{\mu\nu}$ [problem 7.2], so that $\Gamma^\mu_{\mu\nu} = (\ln\sqrt{-g})_{,\nu}$. The general d'Alembertian must be $\Box A \equiv (g^{\mu\nu} A_{;\nu})_{;\mu}$, and this is similarly re-expressable as

$$\Box A = g^{\mu\nu}\, A_{,\mu\nu} + \frac{1}{\sqrt{-g}}\, \left(\sqrt{-g}\, g^{\mu\nu}\right)_{,\mu} A_{,\nu}.$$

Problem 1.6. Any identity of the form 'tensor = 0' can be proved in an inertial frame, and then transformed out. In an inertial frame, all Christoffel symbols Γ vanish and covariant derivatives become simple partial derivatives. The Riemann tensor is then simplified to $R_{\alpha\beta\gamma\delta} = g_{\alpha\gamma,\beta\delta} - g_{\alpha\delta,\beta\gamma} + g_{\beta\delta,\alpha\gamma} - g_{\beta\gamma,\alpha\delta}$. From this, it is a matter of index manipulation to prove the identities obeyed by the Riemann tensor and its contracted forms.

Problem 1.7. Write the vector as $\mathbf{u} = (dx^i/d\lambda)\, \mathbf{e}_i$. Now, as for a normal derivative, the covariant derivative should obey the **Leibniz property**: $\boldsymbol{\nabla}_\mathbf{u}(f\mathbf{v}) = f\, \boldsymbol{\nabla}_\mathbf{u}\mathbf{v} + (\boldsymbol{\nabla}_\mathbf{u} f)\, \mathbf{v}$, remembering that $\boldsymbol{\nabla}_\mathbf{u} f = df/d\lambda$. Applying this to the component form of $\mathbf{u}$ gives

$$\boldsymbol{\nabla}_\mathbf{u}\mathbf{u} = (dx^i/d\lambda)(dx^j/d\lambda)\Gamma^k_{ij}\mathbf{e}_k + (d^2x^k/d\lambda^2)\mathbf{e}_k,$$

as required. Similar reasoning works for components of the covariant derivative of $\mathbf{g} = g^{ij}\mathbf{e}_i \otimes \mathbf{e}_j$, using $\boldsymbol{\nabla}_\mathbf{u}(\mathbf{a}\otimes\mathbf{b}) = (\boldsymbol{\nabla}_\mathbf{u}\mathbf{a})\otimes\mathbf{b} + \mathbf{a}\otimes(\boldsymbol{\nabla}_\mathbf{u}\mathbf{b})$.

Problem 2.1. If enthalpy is $w = \rho + p$, and $\gamma = \cosh\theta$, then the jump conditions are $w_1 s_1 c_1 = w_2 s_2 c_2$ and $w_1 s_1^2 + p_1 = w_2 s_2^2 + p_2$, where s and c stand for $\sinh\theta$ and $\cosh\theta$. Eliminating the downstream velocity gives

$$\beta_1^2 = \frac{(p_2 - p_1)(\rho_2 + p_1)}{(\rho_2 - \rho_1)(\rho_1 + p_2)},$$

and the same for downstream by exchanging subscripts. The equation of state is $p = (\Gamma - 1)(\rho - n)$, where n is rest-mass density (proportional to number density). This has the conservation equation $n_1 s_1 = n_2 s_2$; together with the equations for β_1 and β_2, this gives the **shock adiabat** as

$$(\rho_2 + p_1)w_2/n_2^2 = (\rho_1 + p_2)w_1/n_1^2.$$

Solving these in general is hard, especially because the upstream and downstream equations of state will differ. It is possible to treat the strong-shock case (a) as follows. The Lorentz factor of the shock velocity difference (downstream as viewed by upstream) is, from above,

$$\gamma_\Delta^2 = \frac{(\rho_1 + p_2)(\rho_2 + p_1)}{w_1 w_2},$$

which is obviously symmetric. Taking this plus the shock adiabat and letting $p_1 \to 0$ gives

$$\frac{n_2}{n_1} = \frac{\Gamma_2 \gamma_\Delta + 1}{\Gamma_2 - 1},$$

or a stronger compression than nonrelativistic shocks. The ultrarelativistic case (b) is easier, since $\Gamma = 4/3$ upstream and downstream, leading to $\gamma_1^2 = (9/8)(1 + \rho_2/3\rho_1)$; the shock speed is $c/\sqrt{3}$ for compressions close to unity, becoming ultrarelativistic as the compression ratio rises (see Blandford & McKee 1976 for further details).

Problem 2.2. The only non-zero parts of the energy–momentum tensor are the density $T^{00} = (M/4\pi R^2)\delta(r - R)$ and the azimuthal mass flux density $T^{0\phi} = T^{00} R\Omega \sin\theta$. Thus, the only perturbations to the metric are h^{00} and $h^{0\phi}$. It is the latter term that looks like a rotation: the metric will be of the form $d\tau^2 = g^{00}dt^2 - dr^2 - r^2 d\theta^2 - r^2 \sin^2\theta(d\phi - \omega dt)^2$. To first order, the angular velocity at which inertial frames are 'dragged' is $\omega = g^{0\phi}/r\sin\theta$. In this case, $g^{0\phi} = h^{0\phi}$, so it only remains to solve for this component of the perturbed potential.

Rather than integrating over the source, it is more straightforward to use the gravitational Poisson equation $\nabla^2 h^{\mu\nu} = 16\pi G T^{\mu\nu}$. For potentials ψ with the angular dependence of a spherical harmonic of order ℓ, the Laplacian is $\nabla^2\psi = \psi'' + (2/r)\psi' - \ell(\ell+1)\psi/r^2$. Here, we seek a term that has the same $\ell = 1$ dependence as the source, so the equation becomes $\psi'' + (2/r)\psi' - 2\psi/r^2 = (4GM/R^2)\delta(r - R)$. Interior and exterior solutions go as r and r^{-2}; matching ψ at $r = R$ and integrating gives the required solution: $\omega = 4GM\Omega/(3c^2 R)$.

Problem 2.3. According to e.g. section 9.3 of Jackson (1975), the power radiated by an oscillating electric quadrupole is

$$P = \frac{\omega^6}{360c^5} Q_{ij} Q_{ij}$$

in cgs units; in SI, a factor $1/4\pi\epsilon_0$ should be added. The definition of Q_{ij} is as for the moment of inertia tensor, but using charge density rather than matter density. There is also a factor 3 difference in the definition. Allowing for this, and making the usual $1/4\pi\epsilon_0 \to G$ substitution, we get

$$P = \frac{G\omega^6}{40c^5} I_{ij} I_{ij},$$

which is smaller than the actual result by a factor 4, allowing for a factor 1/2 in averaging the quadrupole formula over one cycle.

Problem 2.4. We need to use the quadrupole formula and evaluate $I_{ij} = \sum_k m_k [x_i x_j - (r^2/3)\delta_{ij}]$, summing over the masses of the bodies. The second part is constant and so can be neglected, since we need $\dddot{I}$. The two-body nature of the system is accommodated by using the total separation of the bodies instead of coordinates relative to the centre of mass, and using the reduced mass μ. If the orbit is in the xy-plane, the only four non-zero time-varying components are thus

$$I_{11} = \mu a^2 \cos^2 \omega t$$
$$I_{22} = \mu a^2 \sin^2 \omega t$$
$$I_{21} = I_{12} = \mu a^2 \sin \omega t \cos \omega t.$$

Take the third time derivative, square, add and time average to find $\langle \dddot{I}_{ij} \dddot{I}_{ij} \rangle = 32\mu^2 a^4 \omega^6$ (remember the semimajor axis a of an orbit is defined to be the full binary separation, not the distance from the centre of mass). Adding in Kepler's law $\omega^2 a^3 = G(m_1 + m_2) \equiv GM$ gives the required answer. The orbital energy is $|E| = Gm_1 m_2/2a^2 = G\mu M/2a^2$; the expression for $\dot{E}$ is therefore a differential equation for $a(t)$, which is easily solved to give $a/a_0 = (1 - t/\tau)^{1/4}$, where

$$\tau = \frac{5c^5}{256G^3} \frac{a_0^4}{\mu M^2}.$$

Applying this classical approach to an atomic orbit, we now have $\omega^2 = Ze^2/(4\pi\epsilon_0 a^3 m_e)$ and $|E| = Ze^2/(8\pi\epsilon_0 a)$. If we take a typical radius as the Bohr radius $a = \hbar/(\alpha m_e c)$, where α is the fine-structure constant, and consider $Z = 1$, the typical lifetime becomes

$$\tau = \frac{5\hbar^2}{256Gc\alpha^6 m_e^3} \simeq 10^{27.5} \text{ yr}.$$

Problem 2.5. In outline, the proof is as follows. Use spherical symmetry to argue that the metric must involve only two unknown functions: g_{tt} and g_{rr}, written as

$$d\tau^2 = e^{\Phi} dt^2 - e^{\Lambda} dr^2 - r^2 d\psi^2.$$

The use of the Euclidean form for the transverse part is just a definition of r, not an extra assumption. Sadly, there is now no alternative but a moderate amount of tedious work in order to evaluate the components of the Einstein tensor. The non-zero Christoffel symbols are

$$\Gamma^0_{00} = \dot{\Phi}/2, \quad \Gamma^0_{01} = \Phi'/2, \quad \Gamma^0_{11} = \dot{\Lambda} e^{\Lambda - \Phi}/2$$
$$\Gamma^1_{00} = \Phi' e^{\Phi - \Lambda}/2, \quad \Gamma^1_{11} = \Lambda'/2, \quad \Gamma^1_{10} = \dot{\Lambda}/2, \quad \Gamma^1_{22} = -re^{-\Lambda}, \quad \Gamma^1_{33} = -r \sin^2 \theta e^{-\Lambda}$$
$$\Gamma^2_{12} = 1/r, \quad \Gamma^2_{33} = -\sin\theta \cos\theta$$
$$\Gamma^3_{13} = 1/r, \quad \Gamma^2_{23} = \cos\theta / \sin\theta.$$

From these, the Einstein equations come out as

$$-8\pi G T^0_0 = e^{-\Lambda}(r^{-2} - \Lambda'/r) - r^{-2}$$
$$-8\pi G T^1_1 = e^{-\Lambda}(\Phi'/r^2 - r^{-2}) - r^{-2}$$
$$-8\pi G T^1_0 = e^{-\Lambda} \dot{\Lambda}/r.$$

Outside the matter (which could mean inside a shell), the third of these says that Λ is independent of time, and the first integrates to $\exp(-\Lambda) = 1 - 2GM/r$, where $M = \int_0^r 4\pi r^2\, T^0_0\, dr$. The first two equations give $\Phi' + \Lambda' = 0 \Rightarrow \Phi = f(t) - \Lambda$. However, time can always be redefined so that $\exp[f(t)/2]dt \rightarrow dt'$, so we can always choose $\Phi = -\Lambda$, establishing the Schwarzschild form.

Problem 2.6. First use the conservation of angular momentum to recast the energy conservation equation as one relating radius and angle:

$$\frac{L^2}{2r^4}\left(\frac{dr}{d\phi}\right)^2 - \frac{GM}{r} + \frac{L^2}{2r^2} - \frac{L^2 GM}{c^2 r^3} = E.$$

This is a differential equation that can be integrated to give the orbit $r(\phi)$ as a function of the two parameters L and E. It helps to make the substitution $u \equiv 1/r$, which turns the equation into

$$\left(\frac{du}{d\phi}\right)^2 = \frac{2E}{L^2} + \frac{2GM}{L^2}\,u - u^2 + \frac{2GM}{c^2}\,u^3.$$

In the Newtonian case (where the last term is absent), this can be integrated to give a $\cos^{-1}$ integral, proving that the orbit is an ellipse. Substituting the equation for an ellipse and equating coefficients for each of the constant and oscillating terms gives the orbital elements in terms of the energy and angular momentum:

$$\frac{1}{r} = \frac{1}{\ell}(1 + e\cos\phi), \qquad \ell = \frac{L^2}{GM}, \qquad e = \sqrt{1 + \frac{2EL^2}{(GM)^2}}.$$

Now treat the extra term as a perturbation about this Newtonian solution, $u_0(\phi)$, writing $u = u_0 + u_1/\ell$. To first order in u_1, this gives

$$-\frac{du_1}{d\phi}e\sin\phi + u_1 e\cos\phi = \frac{GM}{c^2\ell}(1 + e\cos\phi)^3.$$

This is a messy equation, but the result we are after can be extracted by writing $u = \{1 + e\cos[(1+\alpha)\phi]\}/\ell$, i.e. a precessing ellipse with $u_1 = -e\alpha\phi\sin\phi$. The equation for the perturbation is now

$$e^2\alpha\sin^2\phi = \frac{GM}{c^2\ell}(1 + e\cos\phi)^3,$$

so there must other distortions to the orbit than a simple precession. However, these are oscillating or constant, and there is no need to solve for them. We can therefore pick out from the rhs the only term with the correct $\sin^2\phi$ dependence, and conclude that the perihelion advance is described by

$$\alpha = -3\frac{GM}{c^2\ell}.$$

Problem 2.7. Adding in the centrifugal part, the total effective potential is $\Phi = -GM/|r-a| + L^2/2r^2$. The condition for a circular orbit is $\Phi' \equiv \partial\Phi/\partial r = 0$, and the marginally stable orbit requires $\Phi'' = 0$. Solving these yields $r_{\rm ms} = 3a$; comparing the Schwarzschild solution shows that this corresponds to the choice $a = r_{\rm S}$.

Problem 2.8. Work in units of r_S for radius and r_S/c for time, so that the equations for infall from infinity and for null radial motion become respectively

$$t = \ln \frac{\sqrt{r}+1}{\sqrt{r}-1} - 2\sqrt{r} - \tfrac{2}{3} r^{3/2}$$

$$t = t_0 + r_0 - r - \ln \frac{r-1}{r_0 - 1}.$$

Now assume that the start of the motion is at large r_0, where the infall time $\simeq -2r_0^{3/2}/3$. However, this assumes infall from infinity, whereas it is easy to integrate the Newtonian equation of motion to show that collapse from rest at r_0 takes a time $\pi r_0^{3/2}/2$. After the particle has fallen to radii much smaller than r_0, the motion is indistinguishable from infall from infinity, apart from this time offset. We can therefore use the infall equation, and take $t_0 = -\pi r_0^{3/2}/2 + \Delta t$. Close to the horizon, with $r = 1 + \epsilon$, the infall solution is $t = -\ln \epsilon + O(1)$, whereas the null solution is $t = -\ln \epsilon + t_0 + O(r_0)$. Since t_0 is of order $-r_0^{3/2}$, the infall time at very small ϵ will be later than the null time, allowing contact with the infalling body, provided t_0 does not change sign. The maximum allowed delay before sending a signal is thus

$$\Delta t < \pi r_0^{3/2}/2.$$

Problem 3.1. As with the Schwarzschild metric, there is no way of avoiding a moderate amount of work in this calculation. To minimize the effort, it is best to write the metric in the form

$$d\tau^2 = dt^2 - R^2(t)\left[\frac{dr^2}{1-kr^2} + r^2(d\theta^2 + \sin^2\theta\, d\phi^2)\right],$$

so that the Schwarzschild results from problem 2.5 can be used. Now write the energy–momentum tensor as

$$T^\mu_\nu = (\rho + p)U^\mu U_\nu - p g^\mu_\nu.$$

In the fluid rest frame, this is just $\mathrm{diag}(\rho, -p, -p, -p)$, as in special relativity, because $g^\mu_\nu = \mathrm{diag}(1,1,1,1)$; this is independent of whether the metric is curved. This is why it is simpler to work with T^μ_ν than $T_{\mu\nu}$. The two independent Einstein equations are

$$G^0_0 = 3(\dot{R}^2 + k)/R^2 = 8\pi G\rho$$

$$G^1_1 = (2R\ddot{R} + \dot{R}^2 + k)/R^2 = 8\pi G p.$$

These can be reassembled into conservation of energy and the Friedmann equation.

Problem 3.2. The redshift is $1 + z = R_{\rm obs}/R_{\rm em}$. We want

$$\begin{aligned}\frac{dz}{dt_{\rm obs}} &= \frac{\dot{R}_{\rm obs}}{R_{\rm em}} - \frac{\dot{R}_{\rm em} R_{\rm obs}}{R_{\rm em}^2}\frac{dt_{\rm em}}{dt_{\rm obs}}\\ &= H_0(R_{\rm obs}/R_{\rm em}) - (1+z)^{-1}(R_{\rm obs}/R_{\rm em})\, H(z)\\ &= (1+z)H_0 - H(z) = -H_0(1+z)[\sqrt{1+\Omega z} - 1],\end{aligned}$$

where the last expression is for a matter-dominated model. For $z = \Omega = 1$, this implies $\delta v/v = 10^{-9.4}h$ in a decade. In practice, any such change would be masked by random accelerations from neighbouring galaxies (Lake 1981; Phillipps 1982).

Problem 3.3. Here we need the geometry for Scheuer's derivation of the RW metric. The parallax angle is $\beta = d\xi/dr$, and $\xi = S_k(r)\,d\psi$. The comoving parallax distance is thus $\xi/\beta = S_k(r)/C_k(r)$. For a matter-dominated model, we can use Mattig's equation to obtain

$$D_{\rm P} = \frac{R_0 S_k(r)}{C_k(r)} \simeq \frac{c}{H_0}\left(z - \frac{2+\Omega}{4}z^2\right).$$

Problem 3.4. Use a time coordinate t that is the look-back time to a given epoch. It is then easy to obtain the approximate time dependence of the scale factor via a Taylor expansion:

$$R(t) \simeq R_0(1 - H_0 t - q_0 H_0^2 t^2/2).$$

Now assume $1 + z = R_0/R$ and eliminate t in terms of comoving distance $r = \int dt/R$. Since $S_k(r) \simeq r$ to second order, this recovers the second-order distance–redshift relation, whose physical content is that objects are brighter in a high-q_0 universe because the deceleration has left them closer.

An alternative argument that does not use the RW metric directly is also illuminating. It ought to be possible at small distances to think of the redshift as being a combination of a global Doppler shift and a gravitational blueshift: from the point of view of distant objects, we are at the centre of a sphere of mass whose radius is our mutual separation. As their photons fall towards us, the potential shift should yield a blueshift (ignoring all more distant matter, as usual). We then write

$$1 + z = \sqrt{\frac{1+\beta}{1-\beta}}\left(1 + \frac{\Delta\phi}{c^2}\right).$$

To evaluate the gravitational blueshift, we need the gravitational potential difference between the centre and edge of a sphere of radius R and mass M. The radial acceleration at radius r is $a = GMr/R^3$, and the potential is $\Phi = \int a\,dr = GM/2R$; this says that $z_{\rm grav} \simeq -\Omega D^2/4$ to second order. Expanding the special relativity redshift formula to get the velocity parameter β, we find that the 'effective' recession velocity needed in this interpretation is

$$\beta \simeq D + (1 + \Omega/2)\,D^2.$$

This expression makes perfect Newtonian sense. Suppose $\Omega = 0$, so that we have undecelerated expansion. The distance to a given object is just $d = vt$, but the relevant time is the time of emission, so this becomes $d = v(t - d/c)$ if we allow for light-travel time. Since $t = 1/H$ and $d = (c/H)D$, this gives $\beta = D + D^2$ to second order. Finally, the $\Omega/2$ term expresses the effect of deceleration: β is observed to increase at large D because the universe expanded faster then. Quantitatively, this may be derived in the same way as for $R(t)$ in the first part of this problem. The overall conclusion is that, to second order, it is correct to think of the cosmological redshift as a combination of Doppler and gravitational redshifts (see Bondi 1947).

Problem 3.5. The redshift–time relation is $1 + z = \exp[H(t_{\rm obs} - t_{\rm em})]$. A null geodesic has $R\,dr = c\,dt = c\,dz/\dot{z} = c\,dz/[(1+z)H]$. Since $R = R_0/(1+z)$, this integrates to $R_0 r = cz/H$. For this model, $k = 0$, hence giving the volume. The flux-redshift relation is $S = L/[R_0^2 r^2 (1+z)^{1+\alpha}] \propto (1+z)^{-(3+\alpha)}$ for $z \gg 1$. The count slope is

$\beta = d\ln N/d\ln S$, where $dN \propto (1+z)^{-3}dV$, which integrates to $d\ln N/d\ln z = 1/\ln z$. Dividing by $d\ln S/d\ln z$ gives the result.

Problem 4.1. Consider a source of area A_1 and a detector of area A_2; let the source subtend a solid angle of Ω_1 at the detector and let the detector subtend Ω_2 at the source. The total flux of energy through the detector is $BA_1\Omega_2$ (surface brightness times area times solid angle into which radiation is emitted). The specific intensity I is this flux divided by $A_2\Omega_1$; I is equal to the source brightness if $A_1/A_2 = \Omega_1/\Omega_2$, which is true because $\Omega = A/r^2$.

For non-static observers, we have to transform the photon phase space. Elements dy, dp_y are unaffected; dx length-contracts by γ, whereas dp is boosted by γ. The volume $d^3x\, d^3p$ is thus invariant, so the phase-space density is invariant. Now, the number density of photons is the energy flux density divided by c and by the photon energy $\hbar\omega$, so the number of photons in dV is

$$N = \frac{I_\omega\, d\Omega\, d\omega\, dV}{\hbar\omega\, c} = \frac{I_\omega\, d\Omega\, \omega^2 d\omega\, dV}{\hbar\omega^3\, c}.$$

The volume element in momentum space is $d^3p = (\hbar^3/c^3)d\Omega\, d\omega$, so the invariant phase-space density is $c^2 I_\omega/(\hbar^4\omega^3)$. The form of black body radiation is $I_\omega \propto \omega^3\bar{n}$, where $\bar{n} = [\exp(\hbar\omega/kT)+1]^{-1}$ is the occupation number, which is clearly invariant. A scaling of frequency under Lorentz transformation thus still looks like thermal radiation, but with a scaled temperature $T'/T = \omega'/\omega$. The CMB dipole thus does not prove absolute motion, and can be interpreted as a velocity only subject to the assumption that the intrinsic background is isotropic.

Problem 4.2. The bend equation is $\alpha = 2\int \nabla_2\Phi\, d\ell$, and $\alpha \propto \nabla_2\psi$ by definition, so that $\nabla_2^2\psi$ involves $2\int \nabla_2^2\Phi\, d\ell$. Now, $\nabla_2^2\Phi = \nabla^2\Phi - \partial^2\Phi/\partial\ell^2$; the latter term integrates to zero if the gravitational force vanishes at large distances from the lens, leaving a Poisson equation for ψ. This leads in principle to difficulties of justification if the lens extends along the whole line of sight. However, the 2D potential formalism is then not valid: it has been derived only for the case of a single lens plane, and effectively assumes that the deflections arising at different distances can be linearly superimposed. This can only be true if the lens is geometrically thin.

Problem 4.3. The previous problem showed that there is a 2D Poisson equation $\nabla^2\psi = 2\Sigma/\Sigma_c$. In 2D, $\nabla^2\psi = (r\psi')'/r + r^{-2}\partial^2\psi/\partial\theta^2$, where primes denotes $\partial/\partial r$. A quick way to derive this is to write $\boldsymbol{\nabla} = \hat{\mathbf{r}}(\partial/\partial r) + \hat{\boldsymbol{\theta}} r^{-1}(\partial/\partial\theta)$, and use $\partial\hat{\mathbf{r}}/\partial\theta = \hat{\boldsymbol{\theta}}$ and $\partial\hat{\boldsymbol{\theta}}/\partial\theta = -\hat{\mathbf{r}}$. Now we have $(r\psi') = \int 2r\Sigma/\Sigma_c\, dr = M/(\pi\Sigma_c)$, where $\alpha \propto \psi'$. Inserting the definition of Σ_c proves the result.

The surface density is $\Sigma(R) = 2\int_R^\infty \rho(r)\, dx$, where $x = \sqrt{r^2 - R^2}$. For an isothermal sphere, $\rho = \sigma^2/(2\pi G r^2)$, so that $\Sigma = [\sigma^2/(\pi G R)]\int_1^\infty [y(y^2-1)^{1/2}]^{-1} dy$, where $y = r/R$. Substituting $y = 1/x$ and $x = \sin\theta$ solves it.

Problem 4.4. The lens equation for a point source is $\theta = \theta_{\rm S} + \theta_{\rm E}^2/\theta$, implying $\theta = [\theta_{\rm S} \pm \sqrt{\theta_{\rm S}^2 + 4\theta_{\rm E}^2}]/2$. The amplification is $A = |\theta/\theta_{\rm S}|\ |d\theta/d\theta_{\rm S}|$ and it is easily shown that this gives

$$A = \frac{1}{2\sqrt{1+4/x^2}}\left(1 + 2/x^2 \pm \sqrt{1+4/x^2}\right).$$

Problem 4.5. In the limit of low optical depth, the probability of exceeding amplification A is

$$P(>A) = \frac{2\tau}{(A^2-1) + A\sqrt{A^2-1}}.$$

The effect of statistical lensing amplification is to convolve the distributions of $\ln S$ and $\ln A$. Let $F(\ln A)$ be the differential distribution of $\ln A$, and $G(\ln S)$ be the differential distribution of $\ln S$:

$$G'(\ln S) = \int G(\ln S - \ln A)\, F(\ln A)\, d\ln A = \int G(\ln S - \ln A)\, f(A)\, dA.$$

For a power law of integral slope β, $G(x) = \exp(-\beta x)$, so that

$$G'/G = \int A^{\beta} f(A)\, dA.$$

From respectively probability and flux conservation, there can be no change for $\beta = 0$ or 1. For $\beta > 2$, the $f \propto A^{-3}$ tail causes the effect to diverge (or to be limited by some maximum amplification).

The $G(\ln S - \ln A)$ factor clearly gives the A-dependent weight to be applied to the intrinsic probability of obtaining amplification A (to be normalized by the overall G'/G factor). This **amplification bias** allows the distribution to be weighted towards high A, even where the luminosity function is unchanged by lensing. A reasonable first approximation to $f(A)$ would be to take the amplification of the closest lens to the line of sight: $f(A)\, dA = dP\, \exp(-P)$, where P is given above. The integrals then need to be done numerically. A crude approximation that includes the high-A tail is $f(A) = \delta(1) + 2\tau/A^3$, which predicts $G'/G = 1 + \tau/(1-\beta/2)$.

Problem 5.1. Solve the problem assuming a star of uniform density. Introducing a density profile only changes things by some dimensionless constants. In virial equilibrium the potential and kinetic energies are of the same order: $GM^2/R \sim nkTR^3$, so that $T \propto M/R$. Now, if photons are scattered inside the star with mean free path λ $(= (n\sigma_{\rm T})^{-1} \propto R^3/M)$, the typical time for escape is that taken to diffuse a distance R, $t \sim R^2/c\lambda$. The luminosity corresponds to radiating away the energy content $E \propto R^3T^4$ over this time, so $L \propto RT^4\lambda \propto R^4T^4/M$. Using the virial temperature above, this is just $L \propto M^3$. Putting in the dimensional constants gives

$$L \simeq \frac{G^4 m_p^5}{\hbar^3 c^2 \sigma_{\rm T}}\, M^3.$$

Since another way of writing L is to say that the surface radiates like a black body, $L \propto R^2T^4$, we additionally deduce $T \propto M^{1/2}$, $R \propto M^{1/2}$. These relations in fact apply reasonably well to stars several times more massive than the sun. They fail for the sun (i) because some energy is transported via convection; (ii) because cooler stars have their opacity dominated by ionic processes: free–free and bound–free scattering, where **Kramer's opacity** applies: $\lambda^{-1} \propto nT^{-3.5}$.

Problem 5.2. Assume that galaxies are **homologous systems**, i.e. that they have radial density profiles of the same functional form, differing only by a scaling in size and in overall density: $\rho(r) = \rho_0 f[r/r_0]$. By dimensions, or directly, the maximum circular velocity must therefore scale as $V^2 \propto GM/r_0$, where $M \propto \rho_0 r_0^3$. This gives

$$V^4 \propto M^2/r_0^2 \propto (M/L)^2\, L^2/r_0^2 \propto (M/L)^2\, L\, \Sigma_0,$$

where $\Sigma_0 = L/r_0^2$ is a characteristic surface brightness. The Tully–Fisher relation requires $M/L \propto \Sigma_0^{-1/2}$.

Problem 5.3. Consider a given solid angle of the surface of the galaxy. Let this be divided into a large number M of microscopic pixels. The total flux density I of this area is $I = \sum_i n_i S_i$, where the ith cell contains n stars with flux density S. We can easily consider microcells, so that usually $n = 0$, and rare cells have $n = 1$. The mean flux is

$$\langle I \rangle = \left\langle \sum_i n_i S_i \right\rangle = M \langle n \rangle \langle S \rangle,$$

where $\langle S \rangle$ is the mean stellar flux density, averaged over the stellar luminosity function: $\langle S \rangle = \int S\, dN(S) / \int dN(S)$. If the expectation number of stars in the region is m, then $\langle n \rangle = m/M$ and $\langle I \rangle = m \langle S \rangle$. For $\langle I^2 \rangle$ the reasoning is similar, except that $\langle S_i S_j \rangle = \langle S^2 \rangle$ if $i = j$, otherwise $\langle S \rangle^2$. Similarly, $\langle n_i n_j \rangle = m/M$ if $i = j$, otherwise m^2/M^2 (because $n = 0$ or 1, so that $n^2 = n$, and the numbers are independent):

$$\langle I^2 \rangle = M(M-1)(m^2/M^2)\langle S \rangle^2 + M(m/M)\langle S^2 \rangle.$$

The fluctuation in I is thus $\sigma_I^2 = \langle I^2 \rangle - \langle I \rangle^2 = m\langle S^2 \rangle - (m^2/M)\langle S \rangle^2$. Now, we assumed $m \ll M$, so the second term can be dropped:

$$\frac{\sigma_I}{\langle I \rangle} = \frac{1}{\sqrt{m}} \frac{\langle S^2 \rangle^{1/2}}{\langle S \rangle} = \frac{1}{\sqrt{m}} \left(1 + \frac{\sigma_S^2}{\langle S \rangle^2} \right)^{1/2}.$$

The fluctuations are thus the expected $1/\sqrt{m}$ Poisson noise, weighted by a term that depends on the width of the luminosity function. The number of stars, m, is unknown, but is eliminated via the total flux:

$$\left(\frac{\sigma_I}{\langle I \rangle} \right)^2 \langle I \rangle = \frac{\langle S^2 \rangle}{\langle S \rangle} = D^{-1} \frac{\langle L^2 \rangle}{\langle L \rangle}.$$

The luminosity $\langle L^2 \rangle / \langle L \rangle$ therefore acts as a standard candle and allows the distance to be measured in terms of observables.

The practical means of observing this signature involve taking a galaxy image and dividing by the square root of the smooth image profile (via a model fit). The result should be a field of uniform-amplitude white noise, made coherent on small scales by convolution with the point-spread function of the image. A Fourier transform is the easiest way of estimating the overall level of the Poisson noise. The main difficulties are that the discreteness noise will be amplified at large r by detector and sky noise. Also, foreground stars and globular clusters must be identified and subtracted, and allowance made for objects of this kind that are too faint to be seen individually.

Problem 5.4. Let the vector normal to the plane of the ring be at θ to the line of sight. Define a system of coordinates with the xy plane in the ring; coordinates of points on the ring are $R(\cos\phi, \sin\phi, 0)$, and the vector towards the observer is $(\sin\theta, 0, \cos\theta)$ (so $\phi = 0$ is the point on the ring closest to the observer). If the supernova explodes at

$t = 0$, the ring lights up at $t = R/c$, but different parts are seen to light up at different times. What matters for calculating the time delay is the projection of the ring vector onto the line of sight, which is $R \cos\phi \sin\theta$. The part of the ring at angle ϕ is therefore seen to light up later than the supernova by $(R/c)(1 - \cos\phi \sin\theta)$. Since $\cos\theta = b/a$ is observable in terms of the (angular) ellipse axes and $a = R/D$, this lag determines D. For the subsequent light curve, let the supernova flash emit energy E, which is scattered by the ring. The observed flux density is

$$S = \frac{dE}{dt} = \frac{E\, d\phi/\pi}{(R/c)\sin\phi \sin\theta\, d\phi}.$$

Now define a new time origin relative to the time at which the ring at $\phi = \pi/2$ is seen: $t = \tau\cos\phi$, where $\tau = R\sin\theta/c$. This gives $S = (E/\pi)/\sqrt{\tau^2 - t^2}$, so the flux has a cusp at $t = \pm\tau$. This response assumes that the ring simply reflects the brief supernova flash. In practice the excited gas has a subsequent emissivity that declines more slowly with time, and this must be convolved with the (observed) supernova light curve to predict the light curve of the ring. Independently of the time lag, this determines τ, which gives another estimate of the distance D.

Problem 6.1. The Schrödinger equation (S) is $i\hbar\dot{\psi} = [V - (\hbar^2/2m)\nabla^2]\psi$. Now take $\psi S^* - \psi^* S$, which gives $-i\hbar(\psi\dot{\psi}^* + \psi^*\dot{\psi}) = -(\hbar^2/2m)(\psi\nabla^2\psi^* - \psi^*\nabla^2\psi)$. The lhs is $-i\hbar\partial|\psi^2|/\partial t$, and the rhs is $-(\hbar^2/2m)\boldsymbol{\nabla}\cdot(\psi\boldsymbol{\nabla}\psi^* - \psi^*\boldsymbol{\nabla}\psi)$. To see this last step, operate $\boldsymbol{\nabla}$ on the product terms inside the brackets; this gives $\psi\nabla^2\psi^* - \psi^*\nabla^2\psi$ from differentiating the second parts of the product, plus the vanishing term $\boldsymbol{\nabla}\psi \cdot \boldsymbol{\nabla}\psi^* - \boldsymbol{\nabla}\psi^* \cdot \boldsymbol{\nabla}\psi$ from the second. This give the conservation equation with $\rho = |\psi|^2$ as a density, and $\mathbf{j} = (i\hbar/2m)(\psi\boldsymbol{\nabla}\psi^* - \psi^*\boldsymbol{\nabla}\psi)$ as a current. If this is not expressing conservation of probability, then what else could it be? Note that this does not prove that $|\psi|^2$ is the only conserved density that can be obtained from the Schrödinger equation.

Problem 6.2. The Klein–Gordon equation is $(\Box + m^2)\phi = 0$. Use the same trick as for the Schrödinger equation: $\phi^*(KG) - \phi(KG)^*$ gives the conservation equation. Plane wave solutions should be $\phi \propto \exp(i\mathbf{k}\cdot\mathbf{x} - i\omega t)$, where $E = \hbar\omega$; the Klein–Gordon equation converts this to $(-E^2 + k^2 + m^2)\phi = 0$, so $E = \pm\sqrt{k^2 + m^2}$. Now, $\partial^\mu\phi = (\dot{\phi}, -\boldsymbol{\nabla}\phi) = (-iE, -i\mathbf{k})\phi$, so that $J^0 = 2E|\phi|^2$. This is positive only if $E > 0$, so it cannot be a probability density; the only obvious physical alternative for a conserved quantity is a charge density, suggesting that the negative-energy solutions are to be interpreted as applying to antiparticles.

Problem 6.3. First consider what the expectation of an operator looks like for a single state $|\psi\rangle$: $\langle A\rangle \equiv \langle\psi|A|\psi\rangle$. If the state is expanded in some basis $|\psi\rangle = \sum_i a_i|i\rangle$, then

$$\langle A\rangle = \sum_{i,j} a_i a_j^*\, \langle j|A|i\rangle = \sum_{i,j} \rho_{ij}A_{ji} = \mathrm{Tr}(\rho A).$$

The name 'density matrix' is clearly applicable here, since $\rho = |\psi\rangle\langle\psi|$, and $\rho_{ij} = \langle i|\rho|j\rangle = a_i a_j^*$. In the general case,

$$\begin{aligned}\mathrm{Tr}(A\rho) &= \sum_n \langle n|A\rho|n\rangle = \sum_{n,i} p_i\langle n|A|i\rangle\, \langle i|n\rangle \\ &= \sum_{n,i} p_i\langle i|n\rangle\, \langle n|A|i\rangle = \sum_i p_i\langle i|A|i\rangle,\end{aligned}$$

where the last step uses completeness, $\sum_n |n\rangle\langle n| = 1$; thus $\langle A\rangle$ is just the weighted sum of $\langle A\rangle$ over each of the possible pure states.

For a thermal ensemble, the occupation probabilities are Boltzmann factors, so

$$\rho = \sum_i \exp(-E_i/kT)\, |i\rangle\langle i|.$$

Now, $\exp(-E_i/kT)\,|i\rangle = \exp(-H/kT)\,|i\rangle$ (expand the exponential term by term) and completeness says $\sum_i |i\rangle\langle i| = 1$, so $\rho = \exp(-H/kT)$.

Problem 6.4. This is a sensible definition of entropy: it is additive, because the microstates of two combined independent systems have new microstates whose occupation probabilities are $p = p_i p_j$. Using $\ln p = \ln p_i + \ln p_j$, with $\sum p_i = 1$ by definition, shows that entropy adds. Now, the golden rule says that to first order the transition rate $\Gamma_{i\to j} = (2\pi/\hbar)\rho|\langle j|\Delta H|i\rangle|^2$, in terms of the density of states and the matrix element. However, this assumes $p_i = 1$, $p_j = 0$, so we should multiply by $p_i - p_j$ to get the rate allowing for reverse transitions. In equilibrium, the p's are equal (detailed balance), leading to $S = k \ln W$, where W is the number of microstates. Out of equilibrium, transitions between i and j cause $dS = k \ln(p_j/p_i)\, dp_i$, where we have shown that $dp_i = -dp_j \propto (p_j - p_i)$. Thus $dS > 0$, irrespective of whether p_i or p_j is the larger. The time asymmetry arises by assuming that the scattering rates tend to equalize probabilities. This asymmetry is not imposed via the Schrödinger equation (in which ψ^* plays the role of the time-reversed wave function), but via causality. The normal derivation of the golden rule switches on a perturbation at $t = 0$ and allows it to persist indefinitely. We therefore assume that the states to be perturbed have no foreknowledge of the perturbations to be applied to them, which is not true for the time-reversed problem. Waldram (1985) gives an illuminating discussion of this approach.

Problem 6.5. The force law is $\dot{\mathbf{p}} = e(\mathbf{E}+\mathbf{v}\wedge\mathbf{B})$. Momentum is a polar vector that changes direction under spatial inversion (being proportional to velocity). The electric field is also polar, since it satisfies $\boldsymbol{\nabla}\cdot\mathbf{E} = \rho/\epsilon_0$ and $\boldsymbol{\nabla}$ changes sign under inversion. The magnetic field is an axial vector, since it satisfies $\boldsymbol{\nabla}\wedge\mathbf{B} = \mu_0(\mathbf{j} + \epsilon_0\dot{\mathbf{E}})$. The Lorentz force therefore changes sign under spatial inversions, as required for electromagnetism to be invariant.

Problem 6.6. In the Dirac representation, all the γ^μ are real except γ^2, which is imaginary; these matrices anticommute: $\gamma^\mu\gamma^\nu = -\gamma^\nu\gamma^\mu$ if $\mu \neq \nu$. Consider the Dirac equation $[\gamma^\mu(P_\mu - eA_\mu) - mc]\psi = 0$ and multiply on the left by γ^0. Moving γ^0 through the equation to premultiply ψ gives

$$\left[\gamma^0(P_0 - eA_0) - \gamma^i(P_i - eA_i) - mc\right]\gamma^0\psi = 0.$$

If we invert spatial axes and change the sign of P_i and A_i, the Dirac equation is restored, so γ^0 is the parity operator (actually any multiple of it; different definitions can be found in the literature). Similarly, $\gamma^2\gamma^{\mu *} = \gamma^2\gamma^\mu = -\gamma^\mu\gamma^2$ if $\mu \neq 2$, and $\gamma^2\gamma^{2*} = -\gamma^2\gamma^2$ because γ^2 is imaginary. Hence $\gamma^2\gamma^{\mu *} = -\gamma^\mu\gamma^2$. Taking the complex conjugate of the Dirac equation changes the sign of $P_\mu = i\hbar\partial_\mu$, which is restored by commuting through γ^2. This recoves the Dirac equation if the sign of e is flipped, so that $\gamma^2\psi^*$ is the charge-conjugate wave function. Finally, the T operator is similar, except that each value of μ must be considered separately to show that $\gamma^1\gamma^3\gamma^\mu = -\gamma^\mu\gamma^1\gamma^3$ unless $\mu = 0$, where the sign is opposite. This combination thus effectively flips the time part of P_μ, as required for the time-reversal operator.

The overall combination $CPT\psi$ is $CP\gamma^1\gamma^3\psi^* = C\gamma^0\gamma^1\gamma^3\psi^* = \gamma^2\gamma^0\gamma^1\gamma^3\psi$. The matrix representing CPT is then $\gamma^2\gamma^0\gamma^1\gamma^3 = \gamma^0\gamma^1\gamma^2\gamma^3 = -i\gamma^5$ (anticommuting γ^2 twice), so that γ^5 would do as a representation. Since $\gamma^5\gamma^\mu = -\gamma^\mu\gamma^5$, we recover the Dirac equation when t, x, e all have their signs reversed.

Problem 6.7. The angular momentum operators are $L_x = -i\hbar(y\,\partial/\partial z - z\,\partial/\partial y)$ etc., from which the Hermitian conjugate ladder operators are defined: $u = L_x + iL_y$; $d = L_x - iL_y$. These commute with total angular momentum L^2, but not with L_z: if $L_z\psi = b\psi$, $L_z(u\psi) = (b+\hbar)u\psi$. Similarly, d lowers the eigenvalue of L_z by $\hbar$. Now suppose there is a highest eigenstate of L_z, with eigenvalue c, such that $u\psi = 0$. Not only is this intuitively reasonable, but $du = L^2 - L_z^2 - \hbar L_z$ and $\langle u\psi \mid u\psi\rangle = \langle\psi|du|\psi\rangle$, so the wave function cannot be normalized if L_z is too big. Now, use the equation for du plus $du\psi = 0$, to show that the eigenvalue of L^2 is $L^2 = c(c+\hbar)$. Now do the same trick with the lowest eigenstate: the eigenvalue here must be $c - n\hbar$, where n is an integral number of lowerings from the top state. In this case, we have $ud\psi = 0$, where $ud = L^2 - L_z^2 + \hbar L_z$, which gives $L^2 = (c - n\hbar)^2 - (c - n\hbar)$; together with the previous equation, this gives $c = n\hbar/2$. The fundamental quantum of angular momentum is therefore $\hbar/2$, even though successive L_z states are separated by twice this amount.

Problem 7.1. The perturbation to the Hamiltonian is $[(\mathbf{p} - e\mathbf{A})^2 - p^2]/2m = -e\mathbf{p}\cdot\mathbf{A}/2m - e\mathbf{A}\cdot\mathbf{p}/2m + e^2A^2/2m$. In the first term, $\mathbf{p}$ operates on both $\mathbf{A}$ and ψ, but can be permuted with $\mathbf{A}$ because $\mathbf{p}\cdot\mathbf{A} \propto \boldsymbol{\nabla}\cdot\mathbf{A}$, which vanishes if we choose the Coulomb gauge. The first-order perturbation is thus $-e\mathbf{A}\cdot\mathbf{p}/m$. To see the effect of this in a matrix element $\langle i|\delta H|j\rangle$, note that $[H, x] = -i\hbar p_x/m$, so the effective perturbation is $ei\mathbf{A}\cdot\mathbf{r}\Delta E/\hbar$. Now, $\Delta E/\hbar = \omega_{ij}$, with $\dot{A} = -i\omega A$; furthermore $\mathbf{E} = \dot{\mathbf{A}}$ in the Coulomb gauge, so the perturbation is just the electrostatic-like term $e\mathbf{E}\cdot\mathbf{r}$, as might have been guessed. The A^2 term, viewed from a field-theory point of view, contains the ability to create two photons and therefore allows transitions to proceed with **two-photon emission**. This ability to degrade high-energy photons to low-energy ones is of great importance in cosmological recombination.

Problem 7.2. The determinant is $g = \epsilon_{\alpha\beta\ldots} g_{1\alpha} g_{2\beta}\ldots$. The relation we require is $\delta g = g\, g^{\mu\nu}\,\delta g_{\mu\nu}$. In matrix terms, this is $\delta|G| = |G|\,\mathrm{Tr}(G^{-1}\delta G)$. Since the metric is symmetric, a simple way of proving this relation is to use a set of coordinates in which the metric 'matrix' G is diagonal. The determinant is independent of the coordinates, and the cyclic property of traces may be used to show that $\mathrm{Tr}(G^{-1}\delta G)$ is also coordinate independent (put $G \to \tilde{R}GR$ etc.). Even though the perturbation δG is non-diagonal, the trace just picks out the diagonal components: $\mathrm{Tr}(G^{-1}\delta G) = \sum \delta\lambda_i/\lambda_i$, where λ_i are the diagonal components. This is clearly equal to $\delta \ln|G|$, since the determinant has no term that is linear in the off-diagonal elements, thus completing the proof. Alternatively, a more elegant route is to note that $|G + \delta G| = |G|\,|G^{-1}(G + \delta G)| = 1 + \mathrm{Tr}(G^{-1}\delta G)$, since $|AB| = |A|\,|B|$.

Now consider the Hilbert action: $S \propto \int R\sqrt{-g}\, d^4x^\mu$. The required variation is

$$\delta\left(R\sqrt{-g}\right) = \delta\left(R^{\mu\nu} g_{\mu\nu}\sqrt{-g}\right) = \left(R^{\mu\nu} - \tfrac{1}{2}Rg^{\mu\nu}\right)\sqrt{-g}\,\delta g_{\mu\nu} + \sqrt{-g}\, g_{\mu\nu}\,\delta R^{\mu\nu}.$$

It is possible to show that the integral over the last term vanishes, by a rather long argument that reduces $R^{\mu\nu}$ to its component parts (see e.g. chapter 10 of Peebles 1993). If we define the variation in the matter action to be $\delta S = -\frac{1}{2}\int T^{\mu\nu}\,\delta g_{\mu\nu}\,\sqrt{-g}\,d^4x^\mu$, then the variation in the total action vanishes if $G^{\mu\nu} \propto T^{\mu\nu}$.

Problem 7.3. Suppose there exist n Grassmann numbers x_i, all of which anticommute with each other:

$$\{x_i, x_j\} = 0.$$

This means that $x_i^2 = 0$, and functions made up as power series in the x_i (with ordinary complex numbers as coefficients) can have at most 2^n terms (e.g. $f = a + bx_1 + cx_2 + dx_1x_2$ in the $n = 2$ case). Calculus also takes an unfamiliar form when Grassmann variables are used. Differentiation of a normal complex number by a Grassmann number should obviously be defined to vanish; derivatives among Grassmann variables must include a negative sign because of the anticommutation of these variables:

$$\frac{d}{dx_i}\,x_j = \delta_{ij} - x_j\,\frac{d}{dx_i}.$$

The lack of quadratic terms in any function defined as a series expansion means that second derivatives always vanish and that derivatives also anticommute:

$$\left\{\frac{d}{dx_i}, \frac{d}{dx_j}\right\} = 0.$$

Lastly, the operation of Grassmann integration cannot be defined as the simple inverse of differentiation; since $d^2/dx_i^2 = 0$, the normal rules of integration will not recover d/dx_i. We can at least think about definite integration from $-\infty$ to $+\infty$, in which case the integral is translationally invariant: $\int f(x)\,dx = \int f(x+y)\,dx$. Consider the case of a single Grassmann variable: $f(x) = a + bx$. The translated integral should be $\int f(x+y)\,dx = (a+by)\int dx + b\int x\,dx$, which is only independent of y if $\int dx = 0$. If the second integral is normalized by definition so that $\int x\,dx = 1$, then this shows that integration and differentiation are identical for Grassmann variables:

$$\int f(x)\,dx = b = \frac{df(x)}{dx}.$$

Problem 7.4. Write $\int_x^\infty e^{-t^2}\,dt = \int_x^\infty (2te^{-t^2})\,(2t)^{-1}\,dt$ and integrate by parts repeatedly. The series clearly diverges as the factorial-like nature of the numerator will overwhelm the power-law behaviour of the denominator for large orders. Consider summing the series to orders x^0, x^{-2} and x^{-4}. Relative to the true answer, the sums at $x = 1$ are respectively 1.319, 0.660, 1.155; the corresponding figures at $x = 3$ are 1.051, 0.992, 0.997. In both cases, the fractional error is of order the last term omitted, as is characteristic in such series. Although the extra terms do not help much at small x, they are very effective at larger x.

Problem 7.5. When a virtual pair is produced, the uncertainty principle says that it can live for up to $t \sim \hbar/(mc^2)$, implying a maximum separation of $x \sim ct \sim \hbar/mc$. Estimating a characteristic density as $\rho \sim m/x^3$ gives $\rho \sim c^3m^4/\hbar^3$. Using the Planck mass for m (2.177×10^{-8} kg) gives $\rho = 10^{96.71}$ kg, which exceeds the present cosmological density by a factor $10^{122.44}(\Omega h^2)^{-1}$.

Problem 7.6. First clarify what is meant by the wave function for a particle at position x. We mean the eigenstates $|q\rangle$ of the position operator: $x|q\rangle = q|q\rangle$, which clearly means that the spatial state $|q\rangle$ must be a delta function. Thus, the usual Schrödinger wavefunction is $\psi(q,t) = \langle q|\psi t\rangle$, or (more conveniently) $\psi(q,t) = \langle q,t|\psi\rangle$ in the Heisenberg representation.

Now write the (1D) propagator equation for ψ:

$$\psi(q',t') = \int G(q',t'\,;\,q,t)\,\psi(q,t)\,dq.$$

In Dirac notation, the propagator is $G = \langle q',t'|q,t\rangle$, as may be seen by using completeness to write $\langle q',t'|\psi\rangle = \int \langle q',t'|q,t\rangle\,\langle q,t|\psi\rangle\,dq$ and comparing with the Schrödinger version. This idea of propagation can be extended by splitting the time interval into a number of time slices separated by τ:

$$\langle q',t'|q,t\rangle = \int dq_1\,dq_2,\cdots\ \langle q_1,t_1|q,t\rangle\,\langle q_2,t_2|q_1,t_1\rangle\cdots$$

Now, $\langle q_{i+1},t_{i+1}|q_i,t_i\rangle = \langle q_{i+1}|e^{-iH\tau/\hbar}|q_i\rangle \simeq \langle q_{i+1}|(1 - iH\tau/\hbar)|q_i\rangle$. To evaluate this, note that the first part gives a delta function, which can be written as an integral over momentum:

$$\langle q_{i+1}|q_i\rangle = \delta(q_{i+1} - q_i) = \frac{1}{2\pi\hbar}\int \exp\left(\frac{ip(q_{i+1} - q_i)}{\hbar}\right)\,dp.$$

The second part can be written similarly:

$$\langle q_{i+1}|H|q_i\rangle = \frac{1}{2\pi\hbar}\int \exp\left(\frac{ip(q_{i+1} - q_i)}{\hbar}\right)\,H(p,q)\,dp,$$

which is obvious enough for the parts of H that are independent of p. The only tricky part is $p^2/2m$, for which

$$\langle q_{i+1}|p^2|q_i\rangle = \int dp'dp\,\langle q_{i+1}|p^2|p'\rangle\,\langle p'|p^2|p\rangle\,\langle p|q_i\rangle$$

(using completeness twice). Finally, use $\langle q|p'\rangle = (2\pi\hbar)^{-1/2}\exp(-p'q/\hbar)$ (momentum-space representation of the delta function again), plus $\langle p'|p^2|p\rangle = p^2\langle p'|p\rangle = p^2\delta(p'-p)$ to give the required expression. It only remains to take the limit $\tau \to 0$ so that $\prod_i \exp(X_i\tau) \to \exp(\int X\,dt)$, and Feynman's path integral is obtained.

Problem 8.1. Consider $x^\mu \to x^\mu + \epsilon^\mu$, which produces

$$\delta\mathcal{L} = \frac{\partial\mathcal{L}}{\partial\phi}\,\delta\phi + \frac{\partial\mathcal{L}}{\partial\phi_{,\mu}}\,\delta\phi_{,\mu},$$

since there is no explicit spatial dependence. Now use $\delta\phi = \epsilon^\nu\phi_{,\nu}$ and $\delta\phi_{,\mu} = \epsilon^\nu\phi_{,\mu\nu}$ to obtain

$$\frac{d\mathcal{L}}{dx^\mu}\,\epsilon^\mu = \frac{\partial\mathcal{L}}{\partial\phi}\,\phi_{,\nu}\epsilon^\nu + \frac{\partial\mathcal{L}}{\partial\phi_{,\mu}}\,\phi_{,\mu\nu}\epsilon^\nu.$$

Next, use the Euler–Lagrange equation to replace $\partial\mathcal{L}/\partial\phi$ by $(\partial\mathcal{L}/\partial\phi_{,\mu})_{,\mu}$, giving

$$\left(\frac{\partial\mathcal{L}}{\partial\phi_{,\mu}}\,\phi_{,\nu}\right)_{,\mu}\,\epsilon^\nu = \frac{d\mathcal{L}}{dx^\mu}\epsilon^\mu = g^{\mu\nu}\frac{d\mathcal{L}}{dx^\mu}\epsilon^\nu$$

(note the distinction between total and partial derivatives: $\partial\mathscr{L}/\partial x^\mu = 0$ but $d\mathscr{L}/dx^\mu \neq 0$). Finally, arguing that ϵ^ν is arbitrary gives the required conservation equation. The reasoning for an internal symmetry is similar.

Problem 8.2. Denote the identity by e and the inverse of x by x^{-1}. If there is another element a, for which $x \circ a = x$, then operating on the left by x^{-1} gives $e \circ a = e \Rightarrow a = e$. Similarly, if $x \circ b = e$ then again operating by x^{-1} gives $b = x^{-1}$.

Problem 8.3. Consider the group consisting of four elements: the identity (I), reflections in two perpendicular mirror planes (M_1 and M_2), and a rotation about the axis where the mirrors join (R). The group table is easily constructed, and in particular says that $M_1 \circ R = M_2$. This says that, when standing in front of a mirror, we cannot distinguish the result of a front–back inversion from a left–right inversion followed by a rotation. A reasonable interpretation of the reflected image is thus a left–right flipped person facing us, but it is not the only interpretation. If we allow a third (horizontal) mirror that passes through our waist, the same reasoning applies: the single inversion in the facing mirror will be equivalent to a reflection through the waist mirror, plus a rotation. The first step would swap head and feet, but the rotation undoes this; our reflection does not appear to be standing on its head.

Problem 8.4. The potential clearly has a minimum at $\phi^\dagger\phi = -\mu^2/2\lambda = v^2/2$. Now, ϕ has two complex components: $\phi = (\phi_a, \phi_b)$, which may be written in terms of four real fields via $\phi_a = (\phi_1 + i\phi_2)/\sqrt{2}$ and $\phi_b = (\phi_3 + i\phi_4)/\sqrt{2}$. The minimum criterion is then $\phi_1^2 + \phi_2^2 + \phi_3^2 + \phi_4^2 = v^2$, which is an $O(4)$ symmetry. Choosing the particular direction for the ground state of $\phi_1 = \phi_2 = \phi_4 = 0$ is allowed through this symmetry, but therefore corresponds to fixing three position angles. This is the right number of degrees of freedom for the longitudinal parts of $W^\pm$ and Z^0, which must disappear when the bosons gain mass.

Problem 8.5. The gravitational equation of motion for a particle including forces is

$$\partial U^\mu/\partial\tau + \Gamma^\mu_{\alpha\beta} U^\alpha U^\beta = F^\mu/m.$$

In an inertial frame instantaneously at rest with the particle, the measured 4-acceleration would be F^μ/m, and so the rest-frame acceleration would be $a_0^2 = -F^\mu F_\mu/m^2$. Rather than evaluate the Γ terms directly, we can go to the known equations for free fall (chapter 2) and note that the required terms are the components of $\partial A^\mu/\partial\tau$ in these equations when r is set constant. These are $t'' = 0$ and $r'' = -GM/r^2$. Hence $a_0 = \sqrt{g_{rr}}GM/r^2$. The Unruh–Davies temperature is proportional to this acceleration, but gravitational redshift removes the $\sqrt{g_{rr}}$ term when the radiation reaches infinity.

Problem 8.6. For commutators, the relation of the number operator to a and $a^\dagger$ was given in chapter 7. For anticommutators, keep $N = a^\dagger a$ and impose $a^\dagger a + aa^\dagger = 1$. These imply $Na = a(1-N)$ and $Na^\dagger = a^\dagger(1-N)$, so that both $a^\dagger|n\rangle \propto |n-1\rangle$ and $a|n\rangle \propto |n-1\rangle$. The normalization comes from $\langle an|an\rangle = 1 = \langle n|a^\dagger a|n\rangle = n$ and $\langle a^\dagger n|a^\dagger n\rangle = 1 = \langle n|aa^\dagger|n\rangle = 1 - n$. thus $a^\dagger|n\rangle = \sqrt{1-n}|n-1\rangle$ and $a|n\rangle = \sqrt{n}|n-1\rangle$. The only possible values of n are thus 0 and 1; attempts to create particles in the $|0\rangle$ vacuum more than once result in a null outcome: this is the exclusion principle. The result for the Hamiltonian then follows as in the bosonic case. For the unbroken supersymmetric vacuum, the zero-point energy from each mode is identically zero, since fermion and boson contributions are equal and opposite.

Problem 9.1. The idea of ignoring the interaction between particles in a plasma becomes progressively more exact as the temperature of the plasma increases. Consider a single two-body collision between an electron and an ion of charge Ze. The deflection angle in the centre-of-momentum frame is

$$\tan\left(\frac{\theta}{2}\right) = \frac{Ze^2}{4\pi\epsilon_0 \mu V^2 b},$$

for nonrelativistic collisions, where b is the impact parameter and μ is the reduced mass of the electron. There is thus an effective size for the ion to act as a large-angle scatterer, which equals

$$b_c \sim \frac{Ze^2}{4\pi\epsilon_0\, kT},$$

using the thermal equilibrium condition $\langle \mu V^2/2\rangle = 3kT/2$. This same relation also holds in the ultrarelativistic limit: the deflection angle decreases by a factor γ, but $\gamma\mu V^2$ is the electron energy E, and $\langle E\rangle = 3kT$ for an ultrarelativistic gas. The interaction zone around the ion thus shrinks to zero at the temperature increases. However, since the number density of particles in an equilibrium background scales as $n \sim (kT/\hbar c)^3$, a fraction $\sim (Z/137)^3$ of particles will always be interacting at a given time. This allows the plasma to remain in equilibrium, while ensuring that interactions can be ignored for almost all particles.

Problem 9.2. Consider a system in contact with a heat bath:

$$dS_{\rm tot} = dS + dS_{\rm bath} = dS + \frac{dE_{\rm bath}}{T} + \frac{P}{T}\, dV_{\rm bath} - \frac{\mu}{T}\, dN_{\rm bath},$$

applying the second law to the bath. If we define the **availability** as $dA = dE + P dV - \mu dN - T dS$, then $dS_{\rm tot} > 0 \Rightarrow dA < 0$, so the availability will be minimized; at constant N, V and T, this corresponds to minimizing $F = E - TS$. For fluctuations, argue that the number of ways of arranging the total system is dominated by the entropy of the bath, since it has infinitely more degrees of freedom. Thus,

$$p_i \propto \exp(dS_{\rm bath}/k) \propto \exp[-(E_i - \mu N_i)/kT],$$

which is the **Gibbs' factor**. If the particle number is fixed, this becomes just the Boltzmann factor. Since the entropy is $S = k\sum p_i \ln p_i$, then for a Boltzmann system where $p_i = \exp(-E_i/kT)/Z$, the free energy is just $F = -kT \ln Z$ in terms of the **partition function** Z.

Problem 9.3. The initial proper energy density is $u = nm/a^3$. If this energy were instantaneously converted to radiation at scale factor a_τ, we would get the thermal relation for the comoving entropy density: $s/k = 4u/(3kT)$, where T is determined by inverting the thermal energy density $u_\gamma(T)$. This gives an entropy density at decay of

$$a_\tau^3 s_\tau/k = a_\tau^{3/4}\, \frac{4nm}{3} \left(\frac{30\, nm\, \hbar^3 c^3}{\pi^2 g_*}\right)^{-1/4}.$$

A better argument is to say that the present energy density is the sum of the redshifted contributions from all decays:

$$u = \int a(t) mn\, e^{-t/\tau}\, dt/\tau = \sqrt{\pi} a_\tau\, nm$$

(the initial density is $\propto a^{-3}$, but the radiation density redshifts as a^4). Apart from the $\sqrt{\pi}$ factor, this would give us just the same current entropy density as the first argument. The crude argument therefore underestimates the entropy density by a factor $\pi^{3/8} \simeq 1.54$.

Problem 9.4. The covariant form of the Boltzmann equation must be given in terms of the change in f with some variable λ that labels the spacetime trajectory under consideration:

$$\frac{df}{d\lambda} = C,$$

where C represents the evolution of phase-space density under collisions. By the chain rule, this equation may be written in the expanded form

$$\frac{df}{d\lambda} = \frac{\partial f}{\partial x^\mu}\frac{dx^\mu}{d\lambda} + \frac{\partial f}{\partial p^\mu}\frac{dp^\mu}{d\lambda} = C.$$

If we choose proper time as λ, the first term is $p^\mu \partial_\mu f$ and is covariant. For the second term, we need the geodesic equation of motion for momentum, which gives the final form:

$$\left(p^\mu \frac{\partial}{\partial x^\mu} - \Gamma^\mu_{\alpha\beta} p^\alpha p^\beta \frac{\partial}{\partial p^\mu}\right) f = C.$$

Problem 9.5. Write down the t and x components of the relevant 4-vectors for photons travelling towards the observer: the 4-frequency is $k^\mu = (\omega, -\omega\cos\theta)$; the 4-current is $J^\mu = (n, -n\cos\theta)$. Transforming these gives $\omega' = \omega\gamma(1+\beta\cos\theta)$ and $\omega'\cos\theta' = \omega\gamma(\cos\theta + \beta)$. Dividing gives

$$\cos\theta' = \frac{\beta + \cos\theta}{1 + \beta\cos\theta} \quad \Rightarrow \quad d\cos\theta' = \frac{d\cos\theta}{\gamma^2(1+\beta\cos\theta)^2} = d\cos\theta \left(\frac{\omega}{\omega'}\right)^2.$$

Similarly, $n' = n\gamma(1+\beta\cos\theta)$. Now, $I\, d\omega\, d\cos\theta$ is a rate of transport of energy per unit area, and so equals $nc\hbar\omega$. If the Doppler factor is $\omega' = \mathcal{D}\omega$, then

$$I'\, d\omega'\, d\cos\theta' = \mathcal{D}^2 I\, d\omega\, d\cos\theta;$$

putting in the transformations of ω and $d\cos\theta$ gives $I' = \mathcal{D}^3 I$, so that I/ω^3 is invariant.

Problem 9.6. The volume element $d^3p_\nu = 4\pi p_\nu^2\, dp_\nu$, so integration over the delta function gives $4\pi(Q - \epsilon_e)^2$. Now convert the integral over electron momentum to one over energy: for $c = 1$, $\epsilon_e^2 = p_e^2 + m_e^2 \Rightarrow p_e\, dp_e = \epsilon_e\, d\epsilon_e$. The allowed range of energy is m_e to Q (in the decay to the three bodies e, p, ν, the electron can carry between all and none of the kinetic energy). Since $Q = 2.531 m_e$, this gives an integral proportional to m_e^5, as can be seen by dimensions.

Problem 10.1. Set up a coordinate system in which the string lies along the $\hat{z}$ axis, the deficit wedge lies along the $\hat{x}$ axis, and two photons travel on either side of the string with initial momentum vectors $\mathbf{p} = (0, p\cos\theta, p\sin\theta)$. To first order, the string does not affect p_y or p_z, but p_x for one of the photons changes to $-8\pi G\mu p_y$. If we now form the cross product of the two photon momenta $\mathbf{p}_1 \wedge \mathbf{p}_2$, this vector is easily shown to be perpendicular to the initial momentum and of magnitude $8\pi G\cos\theta$, so the string acts as a simple gravitational lens with this deflection angle.

Problem 10.2. The quadrupole formula for gravitational radiation reads

$$-\dot{E} = \frac{G}{5c^5}\left\langle \dddot{I}_{ij}\dddot{I}^{ij}\right\rangle,$$

where I^{ij} is the reduced quadrupole tensor; $I \sim ML^2$, where $M \sim 2\pi L\mu$ for a loop. Since the speed of 'sound' is c for a string, the period of oscillation is $\sim 1/L$ and hence $-\dot{E} \sim G\mu^2$. Equating this to $-\dot{M}c^2$ shows that L declines linearly with time: $L - L_i \sim G\mu(t - t_i)$, which causes a cutoff at $L \sim G\mu t$.

Problem 10.3. If there is no loop formation, we can consider the strings independently. The string network will then have much the same appearance when plotted in comoving units at all times. However, the physical mass per unit length does not change, and so the mass per unit comoving length scales as $R(t)$. If the same strings cross a given comoving volume at all times, this means that the comoving density scales as $R(t)$ also, so that the proper density scales as $R/R^3 = R^{-2}$. Both radiation and matter decay faster, so the strings will eventually dominate.

If the strings are oriented isotropically, they can be treated as a fluid with $p = -\rho/3$ and $\rho \propto R^{-2}$. The first property is required thermodynamically: $p\,dV = d(\rho V) = \rho\,dV + V\,d\rho$, but $d\rho/\rho = -(2/3)dV/V$. However, it is not needed, because Friedmann's equation is independent of the pressure; all we have to do is add a new fluid contribution to the total density. Since the active density $\rho + 3p$ vanishes, it is not so surprising that the scale factor obeys the undecelerated law $R(t) \propto t$. Another way of seeing this is from the Friedmann equation; $\dot{R}^2 - (8\pi G/3)\rho R^2 = -kc^2$ with $\rho \propto 1/R^2$ implies constant $\dot{R}$. We can then immediately add strings to the normal distance–redshift relation:

$$R_0\,dr = \frac{c}{H_0}\,dz\ \left[(1 - \Omega_m - \Omega_s)(1+z)^2 + \Omega_m(1+z)^3 + \Omega_s(1+z)^2\right]^{-1/2}.$$

This is easy to solve when Ω_m is negligible, giving $R_0\,dr = (c/H_0)\ \ln(1+z)$, where $R_0 = (c/H_0)|\Omega_{\rm tot} - 1|^{-1/2}$, as usual. Dąbrowski & Stelmach (1989) give the general case, when matter, radiation and vacuum are all included.

Problem 10.4. Work in the string rest frame, where the deficit wedge has a semi-angle of $\theta = 4\pi G\mu$. The momentum of a particle that hits this wedge is $p = \gamma mv$, and this is redirected through an angle θ by the wedge, giving a transverse momentum $p\theta$. This is unchanged on transforming back from the string frame, but the particle is nonrelativistic, so $\delta v = p\theta/m = 4\pi G\mu\gamma v$. This velocity will give the particle a comoving displacement $x(t)$ towards the sheet-like wake. According to the Zeldovich approximation dealt with in chapter 15, the *proper* displacement r obeys the equation $\ddot{r}/r = 8\pi G\rho/3$. If we put $r(t) = R(t)x(t)$, then we get the differential equation (for $\Omega = 1$)

$$\ddot{x} + (4/3t)\dot{x} - (2/3t^2)x = 0.$$

This clearly has power-law solutions $x \propto t^{2/3}$ and t^{-1}, or $r \propto t^{4/3}$ and $t^{-1/3}$. The solution with $r = 0$ at $t = t_i$ must be

$$r(t) = A[(t/t_i)^{4/3} - (t/t_i)^{-1/3}] \quad \Rightarrow \quad \dot{r}(t) = (A/3t_i)\,[4(t/t_i)^{2/3} + (t/t_i)^{-1}].$$

Imposing the condition $\dot{r} = \delta v$ at $t = t_i$ gives the constant A. The perturbation extends infinitely far above the wake, but the critical height h is where the velocity cancels the Hubble expansion, $\dot{r} = (\dot{R}/R)h$. Again for $\Omega = 1$, this gives

$$h = \frac{6}{5}\,\delta v\, t_i \left(\frac{t}{t_i}\right)^{4/3}.$$

Multiplying by 2 gives the total wake thickness; the surface density is $\Sigma = 2h\rho$.

Problem 11.1. The Friedmann equation is $m_{\rm P}^2 H^2 = (8\pi/3)(V + \dot{\phi}^2/2) - k m_{\rm P}^2/R^2$. Start by assuming that the curvature term is negligible; we have to check that it stays so, i.e. that V does not decline too quickly. If $V \gg \dot{\phi}^2/2$, then we have the usual exponential expansion, so it is interesting to look for a solution where the kinetic term is always some constant fraction of the potential term: $\dot{\phi}^2/2 = \alpha V$. Now consider the evolution equation for ϕ, $\ddot{\phi} + 3H\dot{\phi} = -V'$; the $\ddot{\phi}$ term can be dealt with via $\ddot{\phi} = \dot{\phi}\, d(\dot{\phi})/d\phi = (\dot{\phi}^2/2)' = \alpha V'$. Similarly, $\dot{\phi} = -\sqrt{2\alpha V}$ (choose the negative square root, so that the field rolls down the potential towards the origin; the opposite sign gives a declining potential, so that ϕ rolls away from $\phi = 0$). The evolution equation then becomes

$$(1+\alpha)V' = \sqrt{48\pi\alpha(1+\alpha)/m_{\rm P}^2}\; V \quad \Rightarrow \quad V \propto \exp\left(\sqrt{\frac{48\pi\alpha}{m_{\rm P}^2\,(1+\alpha)}}\;\phi\right).$$

To get the time dependence of $R(t)$, we need the equation of state, $p = \dot{\phi}^2/2 - V = (\alpha - 1)V$. Now, conservation of energy says $d\ln\rho/d\ln(R^3) = -1 - p/\rho = -2\alpha/(1+\alpha)$; in the Friedmann equation, we therefore have $\rho R^2 \propto R^{(2-4\alpha)/(1+\alpha)}$. Looking for a power-law solution gives $R \propto t^p$ with $p = (1+\alpha)/3\alpha$. Since $V \propto H^2$, this gives $V(t) \propto t^{-2} \propto R^{-2/p}$. We therefore get a negligible curvature term if $p > 1$, which also solves the horizon problem. Note that this is an exact solution of the evolution equation, rather than one that assumes slow rolling. The slow-rolling parameter is $\epsilon = p/2$, so power-law inflation with large p illustrates nicely the approach to the slow-rolling limit.

Problem 11.2. Rewrite the potential scaled so that effectively $b = \lambda = 1$:

$$(\lambda/b)^4 V/\lambda = (aT^2\lambda/b^2)y^2 - y^3 + y^4,$$

where $y \equiv \lambda\phi/b$. If we define $A \equiv (aT^2\lambda/b^2)$, then solving for $V' = 0$ gives $y = 0$ and $y = (3 \pm \sqrt{9 - 32A})/8$, the positive root being a minimum. The first critical temperature is thus $A_1 = 9/32$. Solving for $V = 0$ gives $y = [1 \pm \sqrt{1 - 4A}]/2$; at the degenerate point, the second minimum is clearly just at $V = 0$, so that gives $A_2 = 1/4$. Putting the expression for y at the local maximum into V gives (to lowest order) $y_{\rm max} \simeq 2A/3$, $(\lambda/b)^4 V_{\rm max}/\lambda \simeq 4A^3/27$.

Problem 11.3. The classical amplitude from quantum fluctuations is $\delta\phi = H/2\pi$, and a new disturbance of the same rms will be added for every $\Delta t = 1/H$. The slow-rolling equation says that the trajectory is $\dot{\phi} = -V'/3H$; we also have $H^2 = 8\pi V/3m_{\rm P}^2$, so that the classical change in ϕ is $\Delta\phi = -m_{\rm P}^2 V'/8\pi V$ in a time $\Delta t = 1/H$. Consider $V = \lambda|\phi|^n/(n m_{\rm P}^{n-4})$, for which these two changes in ϕ will be equal at $\phi \sim \phi^* = m_{\rm P}/\lambda^{1/(n+2)}$. For smaller ϕ, the quantum fluctuations will have a negligible effect on the classical trajectory; for larger ϕ, the equation of motion will become stochastic. Notice that the behaviour is still below the quantum gravity regime, because $V(\phi^*) \sim \lambda^{2/(n+2)} m_{\rm P}^4 \ll m_{\rm P}^4$, if the field is weakly coupled. The field undergoes a random walk, with $\Delta\phi = \sqrt{N}\, H/2\pi$, where $N = H\Delta t$ is the number of e-foldings. The probability distribution of ϕ, $P(\phi)$, should thus obey the diffusion equation:

$$\frac{\partial P(\phi)}{\partial t} = \frac{\partial^2}{\partial \phi^2}\,[D P(\phi)]\,, \qquad D = \frac{H^3}{8\pi^2},$$

remembering $x_{\rm rms} = \sqrt{2Dt}$ for a spatial diffusion process. Half of space is thus driven to infinite values of ϕ.

Problem 11.4. The slow-rolling condition says $\dot\phi = -V'/3H$ and near-exponential expansion says $H \propto \sqrt{V}$. For $V \propto \phi^n$, this gives $\dot\phi \propto -\phi^\beta$, where $\beta = (n/2) - 1$. The solution is

$$\frac{\phi}{\phi_{\rm initial}} = (1 - t/t_{\rm final})^{1/(1-\beta)},$$

saying that $\phi = 0$ is reached at $t = t_{\rm final}$, which will be close to the time at which slow rolling breaks down. The horizon amplitude is $\delta_{\rm H} \propto V^{3/2}/V' \propto \phi^{\beta+2}$. We now have to relate time to wavenumber. Modes leave the de Sitter horizon at $k/a = H$, and near-exponential expansion says that the scale factor relative to the end of inflation is $a = \exp[H(t - t_{\rm final})]$. Thus, k determines time, which determines ϕ, giving $\delta_{\rm H} \propto [-\ln(k r_{\rm H})]^\alpha$, where $\alpha = (\beta + 2)/(1 - \beta)$ and $r_{\rm H}$ is the horizon size at the end of inflation. Large-scale modes that leave early, at large ϕ, get a larger amplitude: polynomial potentials thus predict $n < 1$.

Problem 11.5. Consider the wave equation for weak gravitational fields, $\Box \bar h^{\mu\nu} = -16\pi G T^{\mu\nu}$. We want to obtain this from a Lagrangian, so that we have the Lagrangian density for gravitational waves, to compare with $\mathcal{L} = \partial^\mu\phi\partial_\mu\phi/2$ for a scalar field. The direct route is to take the gravitational Lagrangian $\mathcal{L} = R/8\pi G$, and obtain the weak-field expansion $R \simeq \bar h^{\mu\nu,\alpha}\bar h_{\mu\nu,\alpha}/2$. This suggests that we treat $\bar h^{\mu\nu}/\sqrt{16\pi G}$ as an effective scalar field. Similar reasoning will apply to any field with a quadratic term in the Lagrangian, so that the corresponding equation of motion is the wave equation.

Problem 12.1. The equation of motion for a bound electron is $\ddot x + \Gamma\dot x + \omega_0^2 x = eE/m$. The second term on the lhs corresponds to radiation damping. If the electric field oscillates at frequency ω, then

$$\ddot x = \frac{eE}{m}\left(\frac{-\omega^2}{\omega_0^2 - \omega^2 - i\omega\Gamma}\right).$$

The rate of radiation is proportional to $(\ddot x)^2$, and dividing by the Poynting vector gives the cross-section

$$\sigma = \sigma_{\rm T}\,\frac{\omega^4}{(\omega^2 - \omega_0^2)^2 + (\Gamma\omega)^2}$$

(the normalization can be deduced immediately; we must get the Thomson cross section when ω_0 and Γ are zero). Near to resonance, this looks like a Lorentzian. To convert it to the required form, remember that the standard rate of energy loss is $-\dot{E} = e^2(\ddot{x})^2/(6\pi\epsilon_0 c^3)$, so that $\Gamma = e^2\omega_0^2/(6\pi\epsilon_0 c^3 m)$.

Problem 12.2. Confusion is possible when writing something like $\sigma([1+z]\omega_0)$. Remember that $\sigma(x)$ is a function of x such that $\int \sigma(x)\,dx = 10^{-5.16}$; we need to apply this where $x = \omega_0(1+z)$. The actual integral is one over z, but we can use $dz = d([1+z]\omega_0)/\omega_0$. If we finally argue that the integrand is sharply peaked around $\omega_0(1+z) = \omega(\mathrm{Ly}\alpha)$ (remember that ω_0 is the observed frequency), then all functions of z can be taken out of the integral. One power of $(1+z)$ is lost by rewriting dz again in terms of the fixed frequency of Lyα: $dz/(1+z) = d([1+z]\omega_0)/\omega(\mathrm{Ly}\alpha)$. This gives the Gunn–Peterson result. If the comoving density of clouds scales as $(1+z)^\gamma$, then the proper density of neutral hydrogen scales as $(1+z)^{3+\gamma}$ and for $z \gg 1$ $\tau \propto (1+z)^{3/2+\gamma}$, or roughly $(1+z)^{3.3}$. If $\tau = 1$ at $z = 3$, continuing evolution of the cloud population would produce a suppression of the continuum by a factor of 45 at $z = 5$.

Problem 12.3. This is an exercise that needs a little care, as it combines a number of elements. Consider first scattering by an electron with velocity β. Let the incoming photon make angle θ with the electron's velocity (choose $\theta = 0$ to be antiparallel to the velocity for convenience). In the frame of the electron, this angle is θ_1, and the scattered photon makes an angle θ_2 with the velocity (now choosing $\theta_2 = 0$ to be parallel to the velocity). Double Lorentz transformation of 4-frequency gives the lab frequency after scattering: $\omega'' = \gamma^2\omega(1+\beta\cos\theta)(1+\beta\cos\theta_2)$, and the angle relation $\cos\theta = (\cos\theta_1 - \beta)/(1-\beta\cos\theta_1)$. We need to expand this and average over scatterings, remembering that (i) head-on scatterings (small θ) are more frequent owing to aberration; (ii) the scattering cross-section depends on angle.

Point (i) requires the number density of photons seen by the electron, which is boosted from the lab density $n \propto d\cos\theta$ by a factor $\gamma(1+\beta\cos\theta)$. Point (ii) requires the differential cross-section for Thomson scattering; this is $d\sigma/d\Omega = (3\sigma_{\mathrm{T}}/16\pi)(1+\cos^2\psi)$, where ψ is the scattering angle in the rest frame. If we use polar coordinates such that the incoming photon is at $\phi = 0$, then writing down the vector components for incoming and scattered photon momenta gives a scalar product $\cos\psi = \sin\theta_1\sin\theta_2\cos\psi - \cos\theta_1\cos\theta_2$, and averaging over ϕ gives $\langle(1+\cos^2\psi)\rangle = \cos^2\theta_1\cos^2\theta_2 + \sin^2\theta_1\sin^2\theta_2/2$. Having these ingredients, it remains to expand $\delta\omega/\omega$ times the boost factor to second order in β, multiply by the cross-section, and integrate over $d\cos\theta_1$ and $d\cos\theta_2$. The result is (eventually) $\langle\delta\omega/\omega\rangle = 4\beta^2/3$ and $\langle(\delta\omega/\omega)^2\rangle = 2\beta^2/3$. Now we can integrate over the Maxwellian, for which $\langle\beta^2\rangle = 3kT/mc^2$.

The only thing now missing is the Compton recoil effect: in the rest frame, this gives

$$\frac{\omega'}{\omega} = \left[1 + \frac{\hbar\omega}{mc^2}(1-\cos\psi)\right]^{-1}.$$

To first order, averaging over angle gives $\langle\delta\omega/\omega\rangle = -\hbar\omega/mc^2$. This has the same power of m as the frequency change from transforming in and out of the moving frame, and so must be included; however, it is a negligible higher-order correction to $\langle(\delta\omega/\omega)^2\rangle$. In summary, therefore, the first Fokker–Planck coefficient contains a positive term from various second-order corrections to frequency and scattering rate (but not from the head-on effect alone), and a negative one from recoil. To order β^2, these cancel at $\hbar\omega = 4kT$.

Problem 12.4. Classically, we need $e^2/4\pi\epsilon_0 r \sim kT$, implying $T \simeq 10^{10.2}$ K for $r = 10^{-15}$ m. In practice, in the Sun fusion occurs at temperatures 100 times smaller. The classical barrier is penetrated by quantum tunnelling. For two ions of charge $Z_1 e$ and $Z_2 e$, the wavenumber of the wave function is found from

$$E = \frac{\hbar^2 k^2}{2m} + \frac{Z_1 Z_2 e^2}{4\pi\epsilon_0 r},$$

so that k is imaginary for r less than the classical limit. The tunnelling probability is given approximately by assuming $\psi \propto \exp(\int ik\, dx)$ irrespective of slow changes in k (cruder than WKB), implying a decaying probability density in the classically forbidden region:

$$P \simeq \left| \exp\left(-\int_0^{r_c} |k|\, dr \right) \right|^2,$$

where $k(r_c) = 0$. Doing the integral gives $P = \exp[-(E_G/E)^{1/2}]$, where the **Gamow energy** is

$$E_G = (\pi\alpha Z_1 Z_2)^2\, 2mc^2,$$

which is 493 keV for *pp* fusion. The actual reaction rate will depend on the balance between this suppression probability and the thermal energy distribution, for which the probability of exceeding an energy E is mainly governed by the Boltzmann factor $\exp(-E/kT)$. The product of these factors has a **Gamow peak** at $E = [(kT)^2 E_G/4]^{1/3}$, giving a peak probability of

$$P_{\rm max} = \exp\left[-\frac{3}{2^{2/3}} \left(\frac{E_G}{kT} \right)^{1/3} \right],$$

and the overall fusion rate scales with this factor times powers of E and kT. For *pp* fusion in the Sun, typically $kT \simeq 2\,$keV, so that $d\ln P/d\ln T \simeq 3.9$, and the fusion rate scales roughly as T^4.

Problem 13.1. The luminosity of a giant-dominated galaxy is determined by the rate at which stars evolve off the main sequence, which depends on the stellar mass function:

$$L_G \propto \frac{dN}{dm} \frac{dm_{\rm turnoff}}{dt}.$$

If we take $L_{\rm star} \propto m^\gamma$ with $\gamma \simeq 3$–5, then the main-sequence lifetime is $\tau \propto m/L_{\rm star} \propto m^{1-\gamma}$. Alternatively, the turnoff mass scales with time as $m_{\rm turnoff} \propto t^{-1/(\gamma-1)}$. Differentiating to get $\dot{m}_{\rm turnoff}$, and assuming a power-law IMF with $dN/dm \propto m^{-(1+x)}$ and $x \simeq 1$–2, then it is easy to show that $L_G \propto m_{\rm turnoff}^{\gamma-1-x}$ and hence $d\ln L_G/d\ln t = (\gamma - 1 - x)/(1 - \gamma)$. This is roughly -0.5 for $x \simeq 1.5$, $\gamma \simeq 4$. Note that the giant luminosities are assumed constant for simplicity; they certainly scale much more slowly with mass than main-sequence luminosities.

Problem 13.2. It turns out that large values of Q are not strongly excluded by the counts. Consider the limit of very large Q, and think about objects all with the same (evolving) luminosity L and density ρ. The number of objects above a given flux density, S, is just $\int \rho r^2\, dr$. Since $S \propto L/r^2$, $L \propto 1/\rho$ and the limiting radius changes very slowly with S, this yields an integral count with $N \propto S^{-1}$, rather close to the observed sub-Euclidean slope. The way to constrain such models is then through the redshift distributions; high-Q models predict lower mean redshifts.

Problem 13.3. The total metal density from stellar nucleosynthesis is

$$\rho Z = \frac{1}{\epsilon c^2} \iint \rho(L,t)\, L\, \frac{dL}{L}\, dt,$$

where the relation between flux density and bolometric luminosity is $L = 4\pi BSD^2\,(1+z)$; D is comoving distance and B is a 'bolometric correction'. Note the factor 4π, which is needed to get a total power in W, rather than the more usual W sr^{-1}. Now, the observed joint redshift–flux distribution is given by

$$\rho(L,z)\, \frac{dL}{L}\, D^2\, dr = N(S,z)\, \frac{dS}{S}\, dz,$$

where dr is the radial increment of comoving distance, which is simply related to cosmological time via $c\, dt = dr/(1+z)$. Because of the flat-spectrum assumption, we get $cL\, dt = 4\pi SD^2\, dr$; our integral therefore reduces to an expression for the metal density which is independent of redshift and proportional just to the optical background:

$$\rho Z = \frac{4\pi B}{\epsilon c^3} \iint N(S,z)\, S\, \frac{dS}{S}\, dz = \frac{4\pi B}{\epsilon c^3} I_\nu.$$

Problem 13.4. Assume that there exist at $z \simeq 10$–20 objects with comoving density ρ similar to that of massive galaxies today, emitting in CO lines during this period ($\sim 10^8$ years) at a healthy luminosity in any one line. Scale to $L_9 \equiv L/(10^9\ h^{-2}\ L_\odot)$. Do the sums for $\Omega = 1$. Anything else will make it harder by roughly a factor Ω^{-2} in flux (open model; $\Omega^{-0.8}$ for $k = 0$ via vacuum energy). The characteristic surface density is

$$\begin{aligned}\frac{N}{\text{sr}^{-1}} &= \rho\, D^2 \left(\frac{c}{H_0}\right) \frac{dz}{(1+z)^{3/2}} \\ &= 1.59 \times 10^{10}\ h^{-3} \rho\, \frac{\delta\nu}{\nu} \quad (z = 10) \\ &= 1.42 \times 10^{10}\ h^{-3} \rho\, \frac{\delta\nu}{\nu} \quad (z = 20).\end{aligned}$$

where $D(z)$ is comoving distance $R_0 S_k(r)$. If $\rho \simeq 0.01 h^3$ and a practical bandwidth is $\simeq 0.003$, that implies something like one object per 30 arcmin^2.

The bolometric flux density will be

$$S_{\text{bol}} = \frac{L}{4\pi D^2 (1+z)^2} \simeq 4.2 \times 10^{-18}\ L_9\ (1+z)^{2.35}\ \text{W m}^{-2},$$

where the latter expression is a power-law fit to the numbers at $z = 10$ and 20. Now suppose the object emits with a velocity width of $\sigma_{100} \times 100\ \text{km s}^{-1}$, giving a fractional rms line width ϵ:

$$S_\nu^{\text{obs}} = \frac{S_{\text{bol}}}{\sqrt{2\pi}\ \nu_{\text{obs}}\ \epsilon}.$$

In practical units, this is

$$S_\nu/\mathrm{mJy} = 504\,(1+z)^{-2.35}\,L_9\,\sigma_{100}^{-1}\,\nu_{\mathrm{GHz}}^{-1},$$

so in practice sub-mJy sensitivity would be required, which is potentially achievable.

The main problem for this method is level populations. Relative to $J = 0$, the $J = 1$ level corresponds to a temperature of 5.5 K, $J = 2$ to 16.5 K and so on. At low redshifts, the CMB temperature is low enough that collisional effects mainly govern the populations of the first few upper levels. However, since the CMB temperature at $z = 10$ is 30 K, it is clear that the first few levels will be almost completely depopulated, leaving only the weaker higher-order transitions. This is a general problem at all redshifts: we want to observe transitions at low enough frequency that high-sensitivity heterodyne methods can be used ($\nu \lesssim 15\,\mathrm{GHz}$, say), whereas the only transitions whose level populations are not destroyed by the CMB must satisfy $\hbar\omega/k \gtrsim 2.7(1+z)\mathrm{K} \Rightarrow \nu_{\mathrm{obs}} \gtrsim 50\,\mathrm{GHz}$, independent of redshift. At $z = 15$, this requires transitions of rest frequency at least 1 THz – more exotic than the common molecular coolants at low z.

Problem 14.1. The 4-frequency for photons emitted from the accelerated electron is $k^\mu = (\omega, \omega\cos\phi)$, where ϕ is the angle between the velocity vector and the photon momentum. Transforming this and eliminating the lab-frame frequency $\omega' = \omega\gamma(1+\beta\cos\phi)$ gives the lab-frame angle $\cos\phi' = (\cos\phi+\beta)/(1+\beta\cos\phi)$. Differentiating this gives $d\cos\phi' = d\cos\phi[\gamma(1+\beta\cos\phi)]^{-2}$, so that solid-angle elements change by a factor $(\omega/\omega')^2$. In the rest frame, half the photons go into the hemisphere $\phi = \pi/2$, which aberrates to $\cos\phi' = \beta$ or

$$\sin\phi' = \sqrt{1-\beta^2} = \gamma^{-1}.$$

For highly relativistic speeds, this goes to $\phi' = 1/\gamma$, so the bulk of the emission is confined to a forward-pointing cone of this semi-angle. Now, for an electron in a circular orbit with angular frequency ω_g, how long does this cone take to sweep over an observer? It might seem that the electron would need to be at two positions in its orbit differing in angle by $1/\gamma$ in order for the cone to sweep over an observer, so the time should be $\Delta t = \gamma^{-1}/\omega_g$. However, the *observed* time taken to travel this segment of the orbit is shorter because of light-travel-time effects. If the radius of the orbit is R, then the time taken by the electron to cover $R\Delta\phi$ is $R\Delta\phi/v$, but the light travel time across this segment is $R\Delta\phi/c$, and it is the difference that is observed:

$$\Delta t_{\mathrm{obs}} = \frac{R\Delta\phi}{c}\left(\beta^{-1}-1\right).$$

Now, $R\omega_g/c = \beta$, so that $\Delta t_{\mathrm{obs}} = \omega_g^{-1}\gamma^{-1}\left(1-\beta^{-1}\right) \simeq \omega_g^{-1}\gamma^{-3}/2$.

Problem 14.2. Consider monoenergetic photons in the lab frame travelling at some angle θ to the electron's velocity (choosing $\theta = 0$ to correspond to photons travelling in the opposite direction to the electron). The lab number density of photons is $nc = [I(\theta)/\hbar\omega]\;d\Omega = [2\pi I(\theta)/\hbar\omega]\;d\cos\theta$. Transforming the number density to the electron frame by writing down the 4-current gives $n' = n\gamma(1+\beta\cos\theta)$. In the electron frame, the rate of scattering is $\Gamma' = \sigma_{\mathrm{T}} n' c$, so that the total rate at which photons are scattered is

$$\Gamma' = \frac{2\pi\sigma_{\mathrm{T}}}{\hbar\omega}\int I(\theta)\,\gamma(1+\beta\cos\theta)\,d\cos\theta.$$

In the case where the radiation field is isotropic in the lab frame, $I = Uc/4\pi$ and we get $\Gamma' = \gamma\sigma_T cU/\hbar\omega$. Now allowing for time dilation, the rate at which a lab observer perceives the electron to scatter photons is $\Gamma = \sigma_T cU/\hbar\omega$, exactly as if the electron were stationary.

Problem 14.3. It is inelegant to use quantum concepts to derive what is essentially a classical phenomenon. At least in the ultrarelativistic case, a simple alternative exists. Here, the electron perceives all the radiation to be 'head-on', so that we just need to find the Poynting vector in the electron's frame. For isotropic radiation, this is easily done by writing the energy–momentum tensor in covariant form: $T^{\mu\nu} = (U_0/3c^2)\left(4U^\mu U^\nu - c^2 g^{\mu\nu}\right)$. In the ultrarelativistic case, we therefore obtain $T'^{10} = 4\gamma^2 U_0/3$ immediately.

Problem 14.4. It is easiest to retreat from the conservation equation and look directly at the 'equation of motion' for the energy of a single electron, $\dot{E} = -bE^2$. Integrating this from an energy of E_0 at $t = 0$ gives

$$E = \frac{E_0}{1 + E_0 bt} \quad \Rightarrow \quad E_0 = \frac{E}{1 - Ebt}.$$

If the initial *integral* energy distribution is $N_0(> E) \propto E_0^{-(x-1)}$, then the monotonic nature of the evolution in energy means that $N(> E)$ at a later time is the same in terms of E_0, so can be immediately written down in terms of E: $N(> E) \propto [E/(1 - Ebt)]^{-(x-1)}$. Differentiating gives the corresponding expression for $N(E)$.

This solution is the Green function for the problem. Suppose now that the source Q is constant at E^{-x} for times between $t_0 = 0$ and $t_0 = T$. To get the solution at time t, we just integrate the source times the Green function:

$$N(E, t) = \int E^{-x}[1 - Eb(t - t_0)]^{x-2}\, dt_0.$$

To set the limits of integration, remember that the source acts from 0 to T, but the solution only applies for $t - t_0 < 1/Eb \Rightarrow t_0 > t - 1/Eb$. There are then two regimes:

$$E < 1/bt \qquad N = E^{-x} \int_0^T [1 - Eb(t - t_0)]^{x-2}\, dt_0$$

$$E > 1/bt \qquad N = E^{-x} \int_{t-1/Eb}^T [1 - Eb(t - t_0)]^{x-2}\, dt_0.$$

Taking the simplest case of a continuously active source, $T = t$, and we get

$$E < 1/bt \qquad N = \frac{E^{-x}}{(x-1)Eb}\,[1 - (1 - Ebt)^{x-1}]$$

$$E > 1/bt \qquad N = \frac{E^{-x}}{(x-1)Eb},$$

which gives a continuously curving transition between slopes of x and $x + 1$.

Problem 14.5. Measuring luminosity in units of L^*, the weighted probability distribution is $dp/d\ln L \propto L^{x-\alpha}\exp(-L)$. The mean value of L is thus

$$\langle L \rangle = \frac{\int L^{x-\alpha}\exp(-L)\, dL}{\int L^{x-1-\alpha}\exp(-L)\, dL} = x - \alpha,$$

and similarly $\langle L^2 \rangle = (1 + x - \alpha)(x - \alpha)$. The fractional scatter is

$$\frac{\sigma(L)}{\langle L \rangle} = \left(\frac{\langle L^2 \rangle}{\langle L \rangle^2} - 1 \right)^{1/2} = (x - \alpha)^{-1/2}.$$

A scatter of $\sigma(L)/\langle L \rangle = 0.4$ thus requires $x = 6.25 + \alpha$, a very rapid onset of activity.

Problem 14.6. The comoving density of mass in black holes is

$$\rho_\bullet c^2 = \int n(L, z)\,(gL/\epsilon)\, dt\, d\ln L,$$

where g is the bolometric correction, ϵ is the efficiency and $n\, d\ln L$ is a comoving number density of sources. Now, $dt = R_0 dr/(1 + z)$ and $L = S[R_0 S_k(r)]^2 (1 + z)^{1+\alpha}$, where S is the flux density. $[R_0 S_k(r)]^2 R_0 dr$ is the comoving volume element for unit solid angle, and the observed flux–redshift distribution is $n(S, z)\, d\ln S\, dz = n(L, z)\, d\ln L\, dV$, because the Jacobian between $(\ln S, z)$ space and $(\ln L, z)$ space is unity. Putting this together gives

$$\rho_\bullet c^2 = \frac{g}{\epsilon} \int n(S, z)\, S\, (1 + z)^\alpha\, d\ln S\, dz,$$

which is observable.

Problem 15.1. Consider initially just $\Omega = 1$. For this, we have the exact density contrast of a sphere

$$\frac{\rho}{\rho_b} = \frac{[6(\theta - \sin\theta)]^2}{8(1 - \cos\theta)^3} \simeq 1 + \frac{3\theta^2}{20} + \frac{37\theta^4}{2800} + \cdots,$$

and the approximate linear density contrast

$$1 + \delta_{\rm lin} = 1 + \frac{3}{20}\,[6(\theta - \sin\theta)]^{2/3} \simeq 1 + \frac{3\theta^2}{20} - \frac{\theta^4}{200} + \cdots.$$

This clearly reproduces the higher-order deviations very poorly. Bernardeau's approximation, however, gets the θ^4 term correct within 4%:

$$\left(1 - \frac{2\delta_{\rm lin}}{3}\right)^{-3/2} \simeq 1 + \frac{3\theta^2}{20} + \frac{11\theta^4}{800} + \cdots.$$

Better still, it predicts collapse at $\delta_{\rm lin} = 1.5$ as against the exact 1.686. At turnround ($\theta = \pi$), the approximate density contrast is only 14% higher than the exact 5.552. This makes an impressive contrast with the comparative failure of the Zeldovich approximation in this case: $\rho/\rho_b = (1 - \delta_{\rm lin}/3)^{-3}$, which only collapses at $\delta_{\rm lin} = 3$. Given the expressions in the main text for the spherical model and linear growth when $\Omega \neq 1$, it is possible to show that Bernardeau's model does even better for low Ω, and is exact in the $\Omega = 0$ limit.

Problem 15.2. The basic perturbation equation is $\ddot{\delta} + 2H\dot{\delta} = 4\pi G \rho_0 \delta$. In the case of small-scale perturbations, neutrino free-streaming will have erased structure. The neutrinos contribute to the total density of the universe, but not to the driving term on the rhs, which is multiplied by a factor $\Omega_m/(\Omega_m + \Omega_\nu)$. In the case of critical total density, $a \propto t^{2/3}$, so that $H = 2/3t$ and the rhs is $2(1 - \Omega_\nu)/3t^2$. Looking for power-law behaviour $\delta \propto t^\alpha$ implies $\alpha(\alpha - 1) + 4\alpha/3 = 2(1 - \Omega_\nu)/3$, giving the required solution.

Problem 15.3. For a particle with comoving displacement $\mathbf{x}$, the proper peculiar velocity is $\mathbf{v} = a\dot{\mathbf{x}}$. The normal Lagrangian for a particle moving in a potential would be $L = mv^2/2 - m\Phi$, which suggests $L = a^2 m\dot{x}^2 - m\Phi$ in cosmology. It is actually easiest to consider an action integral in terms of conformal time: $\delta \int L\, d\eta = 0$, so that the candidate Lagrangian is $L = a(mx'^2 - m\Phi)$, where primes denote conformal time derivatives. It is now easy to apply Euler's equation, since $\partial L/\partial(x') = amx'$ and $\partial L/\partial x = -am\nabla\Phi$. The equation of motion is then $(ax')' = -a\nabla\Phi$, as required.

Problem 15.4. It will be convenient to rewrite the perturbation equation in terms of conformal time $d\eta = dt/a$: $(a\delta_b')' + k^2 c_S^2 a\delta_b = (a\delta_d')'$, where primes denote conformal time derivatives. Suppose we put δ_d in the growing mode, $\delta_d = \eta^2$; also, $a = \eta^2$ (choosing a suitable constant of proportionality), so that $(a\delta_d')' = 6a$. We have the equation of a damped oscillator, so the solution is a sum of a particular integral and the homogeneous solution with the rhs set to zero. The particular integral is clearly $\delta_b = 6/k^2c_S^2$ if c_S is constant. For the homogeneous part, we need to solve $\delta_b'' + 2\delta_b'/\eta + k^2c_S^2\delta_b = 0$. It will help to eliminate the linear term, and this can always be done. For an equation $y'' + fy' + gy = 0$, define $v = y/p$, where $p(x) = \exp[-\int(f/2)\, dx]$; the equation is then $v'' + (p''/p + fp'/p + g)v = 0$. This is the equation of motion for a harmonic oscillator, and can be solved approximately by the **WKB method**: $v'' + A^2 v = 0 \Rightarrow v \simeq A^{-1/2}\exp(\pm\int A\, dx)$. In our case, the integrating factor is just $p = 1/\eta$, and the homogeneous solution is $\eta\delta_b \propto (kc)^{-1/2}\exp[\pm\int kc_S\, d\eta]$. At late times, $c_S \propto a^{-1/2}$, so the wave amplitude declines adiabatically as $a^{-1/4}$. For low baryon density, where $c_S \simeq c/\sqrt{3}$, the solution is exact. Treating the sound speed as constant, we can add the particular and homogeneous solutions to satisfy $\delta_b \propto \eta^2$ at early times:

$$\frac{\delta_b}{\delta_d} = \frac{6}{k^2c_S^2\eta^2}\left[1 - \frac{\sin(kc_S\eta)}{kc_S\eta}\right] \simeq 1 - \frac{k^2c_S^2\eta^2}{20}.$$

The oscillatory part never changes the overall sign, reasonably enough: the 'pressure bounce' will not completely expel the baryons from the potential well they are falling into.

Problem 15.5. Let $y = a/a_{\rm eq}$. The growing mode is $\delta \propto y^2$ outside the horizon; inside, at $y \ll 1$, the 'growing' mode is a constant and the decaying mode is $\propto \ln y$. Matching δ and $d\delta/dy$ at $y_{\rm entry}$ gives the growing-mode amplitude as $\delta_{\rm grow}/\delta_{\rm entry} \simeq -2\ln y_{\rm entry}$. Following horizon entry, the perturbations inside the horizon do not undergo the linear y^2 growth between entry and matter–radiation equality, so the transfer function is approximately $T \simeq 2y^2\ln(1/y)$. Clearly, the factor 2 cannot be relied on, since the assumption of sudden horizon entry is invalid. In radiation domination the horizon grows $\propto y$, and perturbations enter the horizon when the horizon size is $1/k$: thus $y_{\rm entry}^{-1} \simeq kr_{\rm H}$, where $r_{\rm H}$ is the horizon size at matter–radiation equality.

Problem 15.6. We need to compare the motion of mass shells in a perturbation with those in the unperturbed universe, which we can treat as the motion of some other sphere:

$$\begin{aligned} r &= Ak[1 - C_k(\eta)] \\ t &= Bk[\eta - S_k(\eta)], \end{aligned}$$

It is easily shown that the development angle η can be expressed in terms of the density parameter at a given time:

$$\Omega(z) = \frac{2}{1 + C_k(\eta)}$$

(use $\Omega = -2q$). The development angles for the universe and for the perturbation can be related because the time coordinate must be the same in both cases. The density contrast of the perturbation is the ratio of M/r^3 for the two spheres; since $B^2 = GM/A^3$, this just involves the ratio of the two B-parameters, which is known through the agreement of the time coordinates. The density contrast is then

$$1 + \delta = \frac{(\sin\theta - \theta)^2(1 - \cos\theta)^{-3}}{[S_k(\eta) - \eta]^2|1 - C_k(\eta)|^{-3}}.$$

If virialization is assumed to occur at $\theta = 3\pi/2$ (i.e. after collapse by a factor 2 from maximum expansion), then the above expression can be evaluated readily by eliminating η in terms of Ω. The answer can be approximated by

$$1 + \delta_{\rm vir} \simeq 147\Omega^{-0.7}.$$

It is common to replace 147 by 178, on the assumption that virialization would only be complete once the spherical model dictated collapse to a point. This allows the background a little extra time to expand, raising the contrast.

Problem 16.1. Write down the definition of δ_k twice and multiply for different wavenumbers, using the reality of δ:

$$\delta_k \delta_{k'}^* = \frac{1}{V^2} \int \delta(\mathbf{r})\delta(\mathbf{r} + \mathbf{x})\, e^{-i\mathbf{k}'\cdot\mathbf{x}}\, d^3x \int e^{i\mathbf{r}\cdot(\mathbf{k}-\mathbf{k}')}\, d^3r.$$

Performing the ensemble average for a stationary statistical process gives $\langle\delta(\mathbf{r})\delta(\mathbf{r} + \mathbf{x})\rangle = \xi(x)$, independent of r. The integral over r can now be performed, showing that $\langle\delta_k(\mathbf{k})\delta_k^*(\mathbf{k}')\rangle$ vanishes unless $\mathbf{k} = \mathbf{k}'$ in the discrete case, or that in the continuum limit there is a delta function in k-space.

Problem 16.2. The integral for ξ is a standard result. It cannot be integrated to obtain $\bar{\xi}$ unless n is an integer. Explicit results leaving out the $(k_c/k_0)^{n+3}$ factor are $\bar{\xi} = 3(y - \arctan y)/y^3$ ($n = -1$); $\bar{\xi} = 3[y^{-3}\arctan y - (y^2 + y^4)^{-1}]$ ($n = 0$); $\bar{\xi} = 6(1 + y^2)^{-2}$ ($n = 1$); $\bar{\xi} = 24(1 + y^2)^{-3}$ ($n = 2$); $\bar{\xi} = (120 - 24y^2)(1 + y^2)^{-4}$ ($n = 3$). The integrated correlation function thus remains positive everywhere unless $n > 2$, in which case it oscillates at large radii.

Problem 16.3. Suppose we begin with a uniform density and create a perturbation field δ by moving parcels of matter. The Fourier transform of the resulting perturbations looks like

$$\delta_k = \frac{1}{V} \int \delta(\mathbf{x}) \exp(i\mathbf{k} \cdot \mathbf{x})\, d^3x.$$

Now, the Taylor expansion of $\exp(i\mathbf{k} \cdot \mathbf{x})$ is $1 + i\mathbf{k} \cdot \mathbf{x} - (\mathbf{k} \cdot \mathbf{x})^2/2 + \cdots$; the integral over the first two terms vanishes by conservation of mass and momentum, but the third survives. This tells us that if we try to create a power spectrum that goes to zero at small wavelengths more rapidly than $\delta_k \propto k^2$, we will fail: discreteness of matter produces the $n = 4$ **minimal spectrum**.

Problem 16.4. The two-point distribution is governed by the covariance matrix. Put the density perturbations (x & y) at the two points in a vector $\mathbf{V}$:

$$C_{ij} = \langle V_i V_j \rangle = \begin{pmatrix} \sigma^2 & \xi \\ \xi & \sigma^2 \end{pmatrix}.$$

Inverting this and putting into the joint normal distribution gives

$$p(x,y) = [2\pi\sigma^2(1-r^2)^{1/2}]^{-1} \exp\left\{-(x^2+y^2-2rxy)/[2\sigma^2(1-r^2)]\right\},$$

where $r \equiv \xi/\sigma^2$. To obtain the mean field at a second point (we choose the y-field), we need

$$\langle y \rangle = \int_{-\infty}^{\infty} y\, p(x,y)\, dy = rx \left[\frac{\exp(-x^2/2\sigma^2)}{\sqrt{2\pi}\sigma}\right].$$

The second part of this expression is the unconditional distribution of x. The first factor gives the conditional mean field: $\langle y \rangle = rx = x\xi/\sigma^2$.

For non-Gaussian fields, a similar result might be expected, at least around high peaks, since one can argue that the small-scale correlations reflect the density profiles of the high peaks in the field. An example that illustrates this is to consider one of the simplest possible non-Gaussian fields that can be generated from a Gaussian field: $\delta' = (\delta^2 - \sigma^2)$. This has zero mean by construction, and its variance is $\langle \delta'^2 \rangle = \langle \delta^4 \rangle - \sigma^4 = 2\sigma^4$. The correlation function is

$$\xi_{\delta'} = \iint (x^2-\sigma^2)(y^2-\sigma^2)\, p(x,y)\, dx\, dy = 2\sigma^4 r^2 = 2\xi_\delta^2.$$

Carrying out a similar integration to above gives $\langle (y^2-\sigma^2) \rangle$ at given x: $\langle (y^2-\sigma^2) \rangle = r^2(x^2-\sigma^2)$, or

$$\langle \delta'_y \rangle = \frac{\xi_{\delta'}}{\xi_{\delta'}(0)}\, \delta'_x,$$

which is identical to the relation obeyed by the Gaussian field.

Problem 16.5. Consider a skewer consisting of data along the 3D x_1-axis; the 1D field is

$$\delta(x_1) = \frac{L^3}{(2\pi)^3} \int \delta_k \exp(ik_1 x_1)\, d^3k,$$

and the correlation function (common to both 3D field and 1D skewer because of statistical isotropy) is

$$\xi(r) = \frac{L^3}{(2\pi)^3} \int |\delta_k|^2 \exp(ik_1 r)\, d^3k.$$

Taking the 1D transform of this gives

$$P_{\rm 1D}(k) = \frac{L^2}{(2\pi)^2} \int P_{\rm 3D}\left(\sqrt{k^2+k_2^2+k_3^2}\right) dk_2\, dk_3 = \frac{L^2}{2\pi} \int_{|k|}^{\infty} P_{\rm 3D}(y)\, y\, dy.$$

Problem 16.6. First, it helps to choose a special xy coordinate system, in which a contour line is horizontal at the origin, i.e. $\delta_1' = 0$. Let the equation of the contour line be $y(x)$; since $y' \equiv dy/dx = 0$ at the origin, the curvature is just y'' if $\delta_2' > 0$, or $-y''$ if $\delta_2' < 0$ (distinguishing high-density regions that are 'inside' or 'outside' the contour). Making a Taylor expansion of δ about the origin and differentiating twice, we find for the curvature

$$K = -\delta_{11}''/|\delta_2'|$$

in all cases. The contribution to the curvature integral from the element $dx\,dy$ is therefore $K\,dx$; we just need the probability of crossing a contour in dy and we are done. As for peaks, this is achieved by saying we wish to consider the contour at $\delta = \delta_c$ and changing variables in the Gaussian probability distribution from $d\delta$ to dy, bringing in a factor $|\delta_2'|$ as the Jacobian. The contribution to the genus per unit area is then

$$\frac{dg}{dA} = -\frac{1}{2\pi}(-\delta_{11}'')\, P_{\rm G}(\delta = \delta_c)\, d^3\delta''\, d^2\delta',$$

where $P_{\rm G}$ is shorthand for the Gaussian probability density. Finally, we must allow for having assumed that the coordinate system is always oriented such that $\delta_1' = 0$. This is achieved by replacing $d^2\delta'$ by $|\delta_2'|\, d\phi\, d|\delta_2'|$, where ϕ is a polar angle. Integrating ϕ over 0 to π gives the final answer, with no sign constraints on δ' or δ'':

$$\frac{dg}{dA} = -\frac{1}{2}(-\delta_{11}'')|\delta_2'|\, P_{\rm G}(\delta = \delta_c, \delta_1' = 0)\, d^3\delta''\, d\delta_2'.$$

Integration yields the formula for g.

Problem 17.1. First work out the velocity kick imparted to a stationary star of mass m by a similar star that passes with velocity v. In the impulsive approximation, where the moving star is assumed to follow a straight line, it is easy to integrate the transverse force to show that $\delta v_\perp = 2Gm/bv$, where b is the impact parameter. This must also be the transverse perturbation to the velocity of the moving star. The orbit of this star will be affected by such collisions when the number of independent collisions satisfies $n|\delta v_\perp|^2 \simeq v^2$. Treat the galaxy as a uniform sphere of radius R, so that in crossing once, a star undergoes

$$dn = 2R\left(\frac{N}{4\pi R^3/3}\right) 2\pi b\, db$$

collisions. The integrated squared perturbation to the transverse velocity is

$$|\Delta v_\perp|^2 = \int |\delta v_\perp|^2\, dn = 12N\left(\frac{Gm}{Rv}\right)^2 \ln\Lambda,$$

where $\Lambda = b_{\rm max}/b_{\rm min}$. From the virial theorem, we know that $v^2 \simeq GNm/R$, so that

$$|\Delta v_\perp|^2 = v^2\,\frac{12\ln\Lambda}{N}$$

on a single crossing. The number of crossings for relaxation is thus $N/(12\ln\Lambda)$. The log term can be rewritten, since $b_{\rm min}$ is the point at which $|\delta v_\perp| = v$ for a single collision, $b_{\rm min} = 2Gm/v^2$. Since $b_{\rm max} \simeq R$, the virial theorem says $\ln\Lambda \simeq \ln N$.

For galaxies and clusters, $N = 10^{11}$ and 10^3 respectively, so that the critical numbers of crossings are $10^{8.5}$ and 12. The sizes and velocities of galaxies and clusters are such that they are 10–100 crossing times old, so that galaxies are very firmly collisionless systems, but clusters are not (although any dark-matter particles making up the cluster dark matter certainly are collisionless).

Problem 17.2. Define the **virial** of a system of mutually gravitating point masses as $v = \sum_i \mathbf{f}_i \cdot \mathbf{r}_i$, i.e. the summed dot product of the force and position vectors for each particle. This can be rewritten as

$$v = \sum_i m_i \ddot{\mathbf{r}}_i \cdot \mathbf{r}_i = \frac{d^2}{dt^2} \sum_i \frac{m_i r_i^2}{2} - 2 \sum_i \frac{m_i \dot{r}_i^2}{2}.$$

In time average, the first term of the rhs will tend to zero (since the integral gives a boundary term, so the average over time T scales as T^{-1}). This applies for single orbiting particles or more complex systems, so that $\langle v \rangle = -2K$, where K is the kinetic energy. Now write the pair forces explicitly:

$$v = \sum_{i \neq j} \frac{-G m_i m_j}{|\mathbf{r}_i - \mathbf{r}_j|^3} (\mathbf{r}_i - \mathbf{r}_j) \cdot \mathbf{r}_i.$$

Finally write the same thing again, swapping i and j, and add the two expressions:

$$2v = \sum_{i \neq j} \frac{-G m_i m_j}{|\mathbf{r}_i - \mathbf{r}_j|} = 2V.$$

The virial is thus equal to the gravitational potential, so proving $2K + V = 0$ in time average.

Problem 17.3. For an exponential disk with surface density $\propto \exp(-r/h)$ and a constant rotational velocity v_g, $J/M = 2 v_g h$. From the virial theorem, $|E| = K = (3/2) M \sigma_h^2$, where σ_h is the one-dimensional velocity dispersion in the halo. These relations eliminate J and $|E|$ from λ in terms of M. The mass is eliminated in terms of the halo rotation velocity, $GM/r = v_h^2$, assuming that the density is f_c times the background density at collapse, $\rho_b(z) = 3\Omega(z) H^2(z)/(8\pi G)$. This gives

$$h = \frac{\lambda}{2 v_g} \left(\frac{3\sigma_h^2}{2} \right)^{-1/2} v_h^3 \left[\frac{f_c}{2} \Omega(z) H^2(z) \right]^{-1/2}.$$

Finally, $\sigma_h = v_h/\sqrt{2}$ for an isothermal sphere, $f_c \simeq 178/\Omega^{0.7}$, and we assume that the disk settles to approximately the halo velocity, $v_g \simeq v_h$.

Problem 17.4. The cross-section for ionization of hydrogen is $\sigma = (2.8 \times 10^{25}/\nu^3)\,\mathrm{m}^2$, or $\sigma = \sigma_0 (\nu/\nu_\mathrm{L})^{-3}$ (Rybicki & Lightman 1979). The optical depth to ionizing photons is

$$\tau = N_{\mathrm{HI}} \frac{\int (\sigma J_\nu / h\nu)\, d\nu}{\int (J_\nu / h\nu)\, d\nu}.$$

This gives $\tau = N_{\mathrm{HI}} \sigma_0 \alpha/(3 + \alpha)$, where N_H is the neutral hydrogen column density, which is $(1 - x)$ times the total hydrogen column. For x close to unity, $1 - x = 10^{-7.9}(3 + \alpha)(n_\mathrm{H}^{\mathrm{tot}}/\mathrm{m}^{-3})(\mathrm{J}_{-21})^{-1}$. Now, the cosmological total hydrogen density is $n_\mathrm{H}^{\mathrm{tot}}/\mathrm{m}^{-3} = 8.4(1+\mathrm{z})^3 \Omega_\mathrm{B} \mathrm{h}^2$, and a collapsed object has a density f_c times this number. For uniform density, the column is $N_\mathrm{H} = 2 r n_\mathrm{H}$, where r may be eliminated in terms of density and circular velocity via $V^2 = GM/r$ as usual. This gives τ in terms of J, V, f_c, $\Omega_\mathrm{B} h^2$ and z, as required.

Problem 18.1. The moments can be cast in terms of derivatives of the correlation function, $C(\theta) = \int_0^\infty \Delta^2(K)\, J_0(K\theta)\, dK/K$. Differentiating J_0 gives

$$\sigma_n^2 = (-1)^n \frac{2^{2n}(n!)^2}{(2n)!}\, C^{2n}(0).$$

We can thus construct the BBKS parameters for the Gaussian temperature field

$$\theta^* \equiv \sqrt{2}\,\frac{\sigma_1}{\sigma_2}, \qquad \gamma \equiv \frac{\sigma_1^2}{\sigma_0\sigma_2},$$

and the expansion of $C(\theta)$ follows.

Problem 18.2. It is simple enough to Fourier-transform the expression for the projected power to deduce the temperature correlation function (convert from 2D wavenumbers on the sky in $h\,\mathrm{Mpc}^{-1}$ to angular wavenumbers and use $\mu^2 = w^2/(w^2 + K^2)$, where μ is the cosine of the angle between $\mathbf{k}$ and the line of sight to the observer). However, the temperature correlation function can also be found directly from the equation for $\delta T/T$ by introducing spherical polars to perform the k-space integral:

$$C(\Delta\psi) = \frac{1}{2}\int_0^\infty \int_{-1}^1 \mathcal{T}^2(k)\, J_0\left(k\Delta r\sqrt{1-\mu^2}\right)\, e^{-k^2\sigma_r^2\mu^2}\,\frac{dk}{k}\, d\mu,$$

where $\Delta r = R_{\rm H}\Delta\psi$ is the comoving separation at last scattering of two points an angle $\Delta\psi$ apart.

Problem 18.3. The spherical harmonic coefficients are $a_\ell^m = \int (\delta T/T)\, Y_{\ell m}^*\, d\Omega$. The temperature perturbation field $\delta T/T$ is a Gaussian process; the harmonic coefficients are sums over a Gaussian field, so they themselves are also Gaussian-distributed variables. The only complication is the form of the spherical harmonic functions:

$$Y_{\ell m}(\theta,\phi) = \sqrt{\frac{2\ell+1}{4\pi}\frac{(\ell-m)!}{(\ell+m)!}}\, e^{im\phi}\, P_\ell(\cos\theta).$$

For $m = 0$, the functions are real; for $m \neq 0$ they are complex and have real and imaginary parts, that will be uncorrelated (because $\int \sin\phi\, \cos\phi\, d\phi = 0$). The $m \neq 0$ coefficients therefore have a distribution that is a 2D Gaussian on the complex plane. In the main text, it is shown that the expectation value of $|a_\ell^m|^2$ is independent of m; the rms of the $m = 0$ coefficient is thus $\sqrt{2}$ larger than that of the real or imaginary part of the $m \neq 0$ coefficients.

For the quadrupole, $\ell = 2$, and the mean-squared sky temperature fluctuations are

$$\left\langle [\delta T/T]^2 \right\rangle = C(0) = \sum_{m=-2}^{m=+2} |a_\ell^m|^2.$$

Now, the various coefficients of different m are uncorrelated, as may be proved by taking the definition of a_ℓ^m:

$$\langle a^{*m'}_{\ell'}\, a_\ell^m\rangle = \iint \langle \Delta_1\Delta_2\rangle\, Y_{\ell' m'}\, Y_{\ell m}^*\, d\Omega_1\, d\Omega_2.$$

Replacing $\langle\Delta_1\Delta_2\rangle$ by $C(\theta)$ and using orthonormality of the $Y_{\ell m}$'s proves the result, which is analogous to the fact that the two-point function for the density field in k-space is a delta function. The quadrupole amplitude squared thus has a distribution like χ^2 with five degrees of freedom.

Although the temperature field is a Gaussian field in the sense that fluctuations between realizations have a normal distribution, this is not so for any given realization. Suppose all the spherical-harmonic coefficients were zero except those for $\ell = 1$; the temperature field would vary as $\cos\theta$ in terms of the angle away from some apex, and this does not give a Gaussian distribution when we make a histogram of field values at various places on the celestial sphere. Similarly, the rms amplitude of the quadrupole temperature pattern need not be the same as its rms expectation value; we could easily live in a universe where the quadrupole seen by us is on the low side of the ensemble's distribution (and probably this is just what has happened, given the low quadrupole signal seen by COBE). However, if many $Y_{\ell m}$'s are important, the central limit theorem means that any small portion of the sky will resemble a sample of a 2D Gaussian field.

Problem 18.4. The transfer function for isocurvature perturbations (not to be confused with a temperature) has the following large-wavelength behaviour for an $\Omega = 1$ CDM universe:

$$T_{\rm iso}(k) = \frac{2}{15}\left(\frac{ck}{H_0}\right)^2 .$$

Now, the Sachs–Wolfe effect is

$$\Phi_k/3c^2 = -\tfrac{1}{2}\,(ck/H_0)^{-2}\;T_k\delta_i = -\delta_i/15.$$

This has the same sign as the isocurvature effect, but is five times smaller. The scale dependence is the same even though the SW effect depends on potential and the isocurvature effect depends on initial density perturbation. The reason is that the transfer function tilts the output density spectrum by two powers of k on large scales (the opposite of the adiabatic case, where the slope changes by two powers of k on *small* scales). The overall effect is thus that a given observed present-day power spectrum will produce large-angle CMB anisotropies six times larger if it originated from isocurvature perturbations than if the initial fluctuations were adiabatic.

Problem 18.5. The signal measured for a single sample is $x^2 = \sum_i [\Delta_i - \bar{\Delta}]^2$, where $\bar{\Delta} = (\Delta_1 + \Delta_2 + \Delta_3)/3$. Here, Δ denotes a sample of the CMB $\delta T/T$ anisotropy field (or rather, the field after convolution with the telescope beam). Expanding the signal and taking expectation values over many independent samples gives

$$\langle x^2\rangle = \sum_i \langle\Delta_i^2\rangle + \langle\bar{\Delta}^2\rangle - 2\langle\Delta_i\bar{\Delta}\rangle.$$

From the definition of $\bar{\Delta}$,

$$\langle\bar{\Delta}^2\rangle = \langle\Delta_i\bar{\Delta}\rangle = C(0)/3 + 2C(\theta)/3,$$

since $\langle\Delta_i\Delta_j\rangle = C(\theta) + [C(0) - C(\theta)]\delta_{ij}$. This leaves $\langle x^2\rangle = 2[C(0) - C(\theta)]$, the same as for a two-beam experiment.

Bibliography and references

Bibliography

Here is a list of textbooks that are especially recommended for further reading. The levels vary from the mathematical to the more intuitive, but all have invaluable insights to offer. Particularly relevant general textbooks, which give more technical details on many of the topics treated here, are

Börner G. (1988) *The Early Universe*, Springer

Coles P., Lucchin F. (1995) *Cosmology: The Formation and Evolution of Cosmic Structure*, Wiley

Kolb E.W., Turner M.S. (1990) *The Early Universe*, Addison-Wesley

Narlikar J.V., Padmanabhan T. (1986) *Gravity, Gauge Theories and Quantum Cosmology*, Reidel

Peacock J.A., Heavens A.F., Davies A., eds. (1990) *Physics of the Early Universe*, Adam Hilger

Peebles P.J.E. (1993) *Principles of Physical Cosmology*, Princeton University Press

Schaeffer R., Silk J., Spiro M., Zinn-Justin J., eds. (1997) *Cosmology and Large-scale Structure, Proc. 60th Les Houches School*, Elsevier

Useful general references for details of the astronomical aspects of the subject are

Allen C.W. (1973) *Astrophysical Quantities*, third edition, Athlone Press, University of London

Lang K.R. (1980) *Astrophysical Formulae*, second edition, Springer

More specific recommended references include:

Part 1: Gravitation and relativity

Weinberg S. (1972) *Gravitation and Cosmology*, Wiley

Rindler W. (1977) *Essential Relativity*, second edition, Springer

Misner C.W., Thorne K.S., Wheeler J.A. (1973) *Gravitation*, Freeman

Berry M.V. (1976) *Principles of Cosmology and Gravitation*, Cambridge University Press

Part 2: Classical cosmology

Gunn J.E., Longair M.S., Rees M.J. (1978) *Observational Cosmology, Proc. Eighth Saas-Fee Course*, eds. A. Maeder, L. Martinet, G. Tammann, Geneva Observatory Press

Raine D.J. (1981) *The Isotropic Universe*, Adam Hilger
Schneider P., Ehlers J., Falco E. (1992) *Gravitational Lenses*, Springer
Rowan-Robinson M. (1985) *The Cosmological Distance Scale*, Freeman

Part 3: Basics of quantum fields
Sakurai J.J. (1967) *Advanced Quantum Mechanics*, Addison-Wesley
Itzykson C., Zuber J.-B. (1980) *Quantum Field Theory*, McGraw Hill
Collins P.D.B., Martin A.D., Squires E.J. (1989) *Particle Physics and Cosmology*, Wiley
Kane G. (1987) *Modern Elementary Particle Physics*, Addison-Wesley

Part 4: The early universe
Weinberg S. (1977) *The First Three Minutes*, Basic Books
Vilenkin A., Shellard E.P.S. (1994) *Cosmic Strings and Other Topological Defects*, Cambridge University Press
Ellis J. (1997) in *Cosmology and Large-scale Structure, Proc. 60th Les Houches School*, eds. R. Schaeffer, J. Silk, M. Spiro, J. Zinn-Justin, Elsevier, p715

Part 5: Observational cosmology
Binney J., Tremaine S. (1987) *Galactic Dynamics*, Princeton University Press
Rybicki G.B., Lightman A.P. (1979) *Radiation Processes in Astrophysics*, Wiley
Weedman D.W. (1986) *Quasar Astronomy*, Cambridge University Press

Part 6: Galaxy formation and clustering
Efstathiou G., Silk J. (1983) *Fund. Cosm. Phys.*, **9**, 1
Peebles, P.J.E. (1980) *The Large-scale Structure of the Universe*, Princeton University Press
Padmanabhan T. (1993) *Structure Formation in the Universe*, Cambridge University Press

Electronic information

Increasingly, it is possible to access professional scientific information efficiently through the internet. Things are evolving quickly here, so some of the following addresses will not stay accurate indefinitely, but they should on the whole remain useful as a starting point.

• Abstracts and (sometimes) full text of past scientific papers:

http://adswww.harvard.edu/ads_abstracts.html

• Archive of preprints. The best place to look for the latest research results (mirrored in several places):

http://xxx.lanl.gov/
http://babbage.sissa.it/
http://xxx.soton.ac.uk/

• Physical constants and particle properties:

http://www-pdg.lbl.gov/pdg.html

- Major astronomical data archives:

http://stdatu.stsci.edu/dss/dss_form.html
http://simbad.u-strasbg.fr/
telnet://ned@ned.ipac.caltech.edu/

- Sources of major cosmology computer codes, for N-body and CMB work:

http://zeus.ncsa.uiuc.edu:8080/GC3_software_archive.html
http://star-www.maps.susx.ac.uk/~pat/consort.html
http://arcturus.mit.edu/cosmics/
http://arcturus.mit.edu:80/~matiasz/CMBFAST/cmbfast.html

- National science centres with a variety of useful links:

http://www.stsci.edu/
http://www.roe.ac.uk/
http://www.ast.cam.ac.uk/

References

This section contains the detailed references given in the text of each chapter. These are mainly to professional research journals, which can often be accessed electronically, as described above. The journal names are given in the usual abbreviated notation, which should not be too hard to decode. For example: *Ann. Rev. Astr. Astrophys.* = Annual Reviews of Astronomy and Astrophysics; *Astr. J.* = Astronomical Journal; *Astr. Astrophys.* = Astronomy and Astrophysics; *Astrophys. J.* = Astrophysical Journal; *Astrophys. J. Suppl.* = Astrophysical Journal Supplement Series; *Mon. Not. R. Astr. Soc.* = Monthly Notices of the Royal Astronomical Society; *Q.J.R. Astr. Soc.* = Quarterly Journal of the Royal Astronomical Society; *Phys. Rev.* = Physical Review; *Phys. Rev. Lett.* = Physical Review Letters; *Proc. Astr. Soc. Pacif.* = Proceedings of the Astronomical Society of the Pacific; *Rev. Mod. Phys.* = Reviews of Modern Physics. The *Annual Reviews* series in particular is a mine of invaluable reading, with several relevant articles each year, in addition to those cited here. Especially recent work is cited only by its reference number in the electronic preprint archives mentioned above: astro-ph/yymmnnn. Those without access to journals will often be able to use astro-ph to locate the preprint form of articles published from 1992 onwards.

Abell G.O. (1958) *Astrophys. J. Suppl.*, **3**, 211
Abell G.O., Corwin H.G., Olowin R.P. (1989) *Astrophys. J. Suppl.*, **70**, 1
Abramowicz M.A., Ellis G.F.R., Lanza A. (1990) *Astrophys. J.*, **361**, 470
Abramowitz M., Stegun I.A. (1965) *Handbook of Mathematical Functions*, Dover
Adler R.J. (1981) *The Geometry of Random Fields*, Wiley
Albrecht A., Stebbins A. (1992) *Phys. Rev. Lett.*, **69**, 2615
Alcock C. *et al.* (1993) *Nature*, **365**, 621
Alcock C. *et al.* (1996) *Astrophys. J.*, **461**, 84
Allen C.W. (1973) *Astrophysical Quantities*, third edition, Athlone Press, University of London
Allen S.W., Fabian A.C., Johnstone R.M., White D.A., Daines S.J., Edge A.C., Stewart G.C. (1993) *Mon. Not. R. Astr. Soc.*, **262**, 901

Allen S.W., Fabian A.C., Kneib J.P. (1996) *Mon. Not. R. Astr. Soc.*, **279**, 615
Aller L.H. (1984) *Physical Processes in Gaseous Nebulae*, D. Reidel
Allington-Smith J.R., Ellis R.S., Zirbel E., Oemler A. (1993) *Astrophys. J.*, **404**, 521
Alpher R.A., Bethe H., Gamow G. (1948) *Phys. Rev.*, **73**, 803
Amaldi U. *et al.* (1991) *Phys. Lett.*, **B260**, 447
Angel J.R.P., Stockman H.S. (1980) *Ann. Rev. Astr. Astrophys.*, **8**, 321
Antonucci R.R.J., Miller J.S. (1985) *Astr. J.*, **297**, 621
Arnould M., Takahashi K. (1990) in *Astrophysical Ages and Dating Methods*, eds. E. Vangioni-Flam *et al.*, Editions Frontières, Paris, p325
Arp H.C. (1981) *Astrophys. J.*, **250**, 31
Atwater H.A. (1974) *Introduction to General Relativity*, Pergamon
Avni Y., Bahcall J.N. (1980) *Astrophys. J.*, **235**, 694
Babul A., Postman M. (1990) *Astrophys. J.*, **359**, 280
Babul A., Rees M.J. (1992) *Mon. Not. R. Astr. Soc.*, **255**, 346
Bahcall J.N. (1984) *Astrophys. J.*, **287**, 926
Bahcall J.N. (1989) *Neutrino Astrophysics*, Cambridge University Press
Bahcall J.N., Casertano S. (1985) *Astrophys. J.*, **293**, L7
Bahcall N.A. (1979) *Astrophys. J.*, **232**, 689
Bahcall N.A., Lubin L.M. (1994) *Astrophys. J.*, **426**, 515
Bahcall N.A., Soneira R.M. (1983) *Astrophys. J.*, **270**, 20
Bailey M.E., MacDonald J. (1981) *Nature*, **289**, 659
Bailin D., Love J. (1994) *Supersymmetric Gauge Field Theory and String Theory*, IOP Publishing
Baker J.C. (1997) *Mon. Not. R. Astr. Soc.*, **286**, 23
Balian R., Schaeffer R. (1989) *Astr. Astrophys.*, **220**, 1
Bardeen J.M. (1980) *Phys. Rev. D*, **22**, 1882
Bardeen J.M., Bond J.R., Efstathiou G. (1987) *Astrophys. J.*, **321**, 28
Bardeen J.M., Bond J.R., Kaiser N., Szalay A.S. (1986) *Astrophys. J.*, **304**, 15
Barnes J., Efstathiou G. (1987) *Astrophys. J.*, **319**, 575
Baron E., White S.D.M. (1987) *Astrophys. J.*, **322**, 585
Barrow J.D., Tipler F.J. (1986) *The Anthropic Cosmological Principle*, Oxford University Press
Barthel P.D. (1989) *Astrophys. J.*, **336**, 606
Baugh C.M., Cole S., Frenk C.S., Lacey C.G. (1997) astro-ph/9703111
Baugh C.M., Efstathiou G. (1993) *Mon. Not. R. Astr. Soc.*, **265**, 145
Baugh C.M., Efstathiou G. (1994) *Mon. Not. R. Astr. Soc.*, **267**, 323
Baym G. (1968) *Lectures on Quantum Mechanics*, Benjamin-Cummings
Bell J.S. (1964) *Physics*, **1**, 195
Bell J.S., Leinaas J.M. (1987) *Nucl. Phys.*, **B284**, 488
Bennett C.L. *et al.* (1996) *Astrophys. J.*, **464**, L1
Benoist C., Maurogordato S., da Costa L.N., Cappi A., Schaeffer R. (1996) *Astrophys. J.*, **472**, 452
Bernardeau F. (1994a) *Astrophys. J.*, **427**, 51
Bernardeau F. (1994b) *Astrophys. J.*, **433**, 1
Berry M.V. (1976) *Principles of Cosmology and Gravitation*, Cambridge University Press
Bertschinger E. (1985a) *Astrophys. J. Suppl.*, **58**, 1
Bertschinger E. (1985b) *Astrophys. J. Suppl.*, **58**, 39
Bertschinger E. (1997) in *Cosmology and Large-scale Structure, Proc. 60th Les Houches School*, eds. R. Schaeffer, J. Silk, M. Spiro, J. Zinn-Justin, Elsevier, p273
Bertschinger E., Dekel A. (1989) *Astrophys. J.*, **336**, L5
Bertschinger E., Dekel A., Faber S.M., Dressler A. Burstein D. (1990) *Astrophys. J.*, **364**, 370
Bessell M.S., Brett J.M. (1988) *Proc. Astr. Soc. Pacif.*, **100**, 1134
Bessell M.S., Stringfellow G.S. (1993) *Ann. Rev. Astr. Astrophys.*, **21**, 433
Bhavasar S.P., Barrow J.D. (1985) *Mon. Not. R. Astr. Soc.*, **213**, 857
Binggeli B., Sandage A.R., Tammann G.A. (1988) *Ann. Rev. Astr. Astrophys.*, **26**, 509
Binney J.J. (1982) *Mon. Not. R. Astr. Soc.*, **200**, 951
Binney J.J., Mamon G. A. (1982) *Mon. Not. R. Astr. Soc.*, **200**, 361
Binney J.J., Tremaine S. (1987) *Galactic Dynamics*, Princeton University Press
Biretta J.A., Zhou F., Owen F.N. (1995) *Astrophys. J.*, **447**, 582
Birkinshaw M., Hughes J.P. (1994) *Astrophys. J.*, **420**, 33
Birrell N.D., Davies P.C.W. (1982) *Quantum Fields in Curved Space*, Cambridge University Press
Blades J.C., Turnshek D., Norman C.A. (1988) *QSO Absorption Lines: Probing the Universe, Proc. 2nd STScI Symposium Series*, Cambridge University Press

Blair D.G. (1991) *The Detection of Gravitational Waves*, Cambridge University Press
Blanchard A., Valls-Gabaud D., Mamon G.A. (1992) *Astr. Astrophys.*, **264**, 365
Blandford R.D., Kochanek C.S. (1988) *Astrophys. J.*, **321**, 658
Blandford R.D., McKee C.F. (1976) *Phys. Fluids*, **19**, 1130
Blandford R.D., Narayan R. (1986) *Astrophys. J.*, **310**, 568
Blandford R.D., Narayan R. (1992) *Ann. Rev. Astr. Astrophys.*, **30**, 311
Blandford R.D., Netzer H., Woltjer L. (1990) *Active Galactic Nuclei, 20th Saas-Fee Course*, Springer
Blandford R.D., Rees M.J. (1974) *Mon. Not. R. Astr. Soc.*, **169**, 395
Blandford R.D., Rees M.J. (1978) in *Pittsburgh Conference on BL Lac Objects*, ed. A.M. Wolfe, University of Pittsburgh Press, p328
Blandford R.D., Znajek R.L. (1977) *Mon. Not. R. Astr. Soc.*, **179**, 433
Blumenthal G.R., Faber S.M., Primack J.R., Rees M.J. (1984) *Nature*, **311**, 517
Boesgaard A.M., Steigman G. (1985) *Ann. Rev. Astr. Astrophys.*, **23**, 319
Bond J.R. (1997) in *Cosmology and Large-scale Structure, Proc. 60th Les Houches School*, eds. R. Schaeffer, J. Silk, M. Spiro, J. Zinn-Justin, Elsevier, p469
Bond J.R., Cole S., Efstathiou G., Kaiser N. (1991) *Astrophys. J.*, **379**, 440
Bond J.R., Crittenden R., Davis R.L., Efstathiou G., Steinhardt P.J. (1994) *Phys. Rev. Lett.*, **72**, 13
Bond J.R., Efstathiou G. (1984) *Astrophys. J.*, **285**, L45
Bond J.R., Efstathiou G. (1987) *Mon. Not. R. Astr. Soc.*, **226**, 655
Bond J.R., Efstathiou G. (1991) *Phys. Lett. B*, **265**, 245
Bond J.R., Efstathiou G., Tegmark M. (1997) astro-ph/9702100
Bondi H. (1947) *Mon. Not. R. Astr. Soc.*, **107**, 411
Börner G. (1988) *The Early Universe*, Springer
Bothun G.D., Impey C.D., Malin D.F., Mould J.R. (1987) *Astr. J.*, **94**, 23
Bouchet F.R., Bennett D.P., Stebbins A. (1988) *Nature*, **335**, 410
Bouchet F.R., Colombi S., Hivon E., Juszkiewicz R. (1995) *Astr. Astrophys.*, **296**, 575
Bower R.G. (1991) *Mon. Not. R. Astr. Soc.*, **248**, 332
Bower R.G., Coles P., Frenk C.S., White S.D.M. (1993) *Astrophys. J.*, **405**, 403
Bower R.G., Lucey J.R., Ellis R.S. (1992) *Mon. Not. R. Astr. Soc.*, **254**, 601
Box G.E.P., Tiao G.C. (1992) *Bayesian Inference in Statistical Analysis*, Wiley
Boyle B.J., Couch W.J. (1993) *Mon. Not. R. Astr. Soc.*, **264**, 604
Boyle B.J., Fong R., Shanks T., Peterson B.A. (1987) *Mon. Not. R. Astr. Soc.*, **227**, 717
Boyle B.J., Griffiths R.E., Shanks T., Stewart G.C., Georgantopoulos I. (1993) *Mon. Not. R. Astr. Soc.*, **260**, 49
Boyle B.J., Shanks T., Peterson B.A. (1988) *Mon. Not. R. Astr. Soc.*, **235**, 935
Branch D. (1992) *Astrophys. J.*, **392**, 35
Brandenberger R.H. (1990) in *Physics of the Early Universe, Proc. 36th Scottish Universities Summer School in Physics*, eds. J.A. Peacock, A.F. Heavens, A.T. Davies, Adam Hilger, p281
Bransden B.H., Joachain C.J. (1989) *Introduction to Quantum Mechanics*, Longman
Bregman J.N. (1992) in *Clusters and Superclusters of Galaxies*, NATO ASI C366, ed. A.C. Fabian, Kluwer, p119
Broadhurst T.J., Ellis R. S., Glazebrook K. (1992) *Nature*, **355**, 55 (BEG)
Broadhurst T.J., Ellis R.S., Koo D.C., Szalay A.S. (1990) *Nature*, **343**, 726
Broadhurst T.J., Ellis R. S., Shanks T. (1988) *Mon. Not. R. Astr. Soc.*, **235**, 827
Broadhurst T.J., Taylor A.N., Peacock J.A. (1995) *Astrophys. J.*, **438**, 49
Bruzual A. G., Charlot S. (1993) *Astrophys. J.*, **405**, 538
Bucher M., Goldhaber A.S., Turok N. (1995) *Phys. Rev. D*, **52**, 3314
Bunn E.F., Scott D., White M. (1995) *Astrophys. J.*, **441**, 9
Burbidge E.M., Burbidge G.R., Fowler W.A., Hoyle F. (1957) *Rev. Mod. Phys.*, **29**, 547
Burke W.L. (1981) *Astrophys. J.*, **244**, L1
Burrows A., Liebert J. (1993) *Rev. Mod. Phys.*, **65**, 301
Butcher H.R. (1987) *Nature*, **328**, 127
Butcher H.R., Oemler G. (1978) *Astrophys. J.*, **219**, 18
Cahn R.N., Goldhaber G. (1989) *The Experimental Foundations of Particle Physics*, Cambridge University Press
Calzetti D., Livio M., Madau P. (1995) *Extragalactic Background Radiation*, Cambridge University Press
Canizares C.R. (1981) *Nature*, **291**, 620
Canizares C.R. (1982) *Astrophys. J.*, **263**, 508
Capaccioli M., Caon N., D'Onofrio M. (1992) *Mon. Not. R. Astr. Soc.*, **259**, 323
Carlberg R.G. (1990) *Astrophys. J.*, **350**, 505

Carlberg R.G., Couchman H.M.P., Thomas P.A. (1990) *Astrophys. J.*, **352**, L29
Carlberg R.G., Yee H.K.C., Ellingson E. (1997) *Astrophys. J.*, **478**, 462
Carr B. (1994) *Ann. Rev. Astr. Astrophys.*, **32**, 531
Carroll S.M., Press W.H., Turner E.L. (1992) *Ann. Rev. Astr. Astrophys.*, **30**, 499
Carter B. (1983) *Phil. Trans. R. Soc.*, **A370**, 347
Cartwright D.E., Longuet-Higgins M.S. (1956) *Proc. R. Soc. Lond.*, **A237**, 212
Celnikier L.M. (1989) *Basics of Cosmic Structures*, Editions Frontières
Cen R., Gnedin N.Y., Kofman L.A., Ostriker J.P. (1992) *Astrophys. J.*, **399**, L11
Cen R., Ostriker J.P. (1992) *Astrophys. J.*, **399**, L113
Cheng T.-P., Li L.-F. (1984) *Gauge Theory of Elementary Particle Physics*, Oxford University Press
Chiba M., Yoshii Y. (1997) *Astrophys. J.*, **490**, L73
Christenson J.H. *et al.* (1964) *Phys. Rev. Lett.*, **13**, 138
Chuang I., Durrer R., Turok N., Yurke B. (1991) *Science*, **251**, 1336
Cimatti A., di Serego Alighieri S., Fosbury R.A.E., Salvati M., Taylor D. (1993) *Mon. Not. R. Astr. Soc.*, **264**, 421
Clayton D.D. (1983) *Principles of Stellar Evolution and Nucleosynthesis*, University of Chicago Press
Colberg J.M. *et al.* (1997) astro-ph/970208
Cole S., Aragón-Salamanca A., Frenk C.S., Navarro J.F., Zepf S.E. (1994) *Mon. Not. R. Astr. Soc.*, **271**, 781
Cole S., Fisher K.B., Weinberg D.H. (1994) *Mon. Not. R. Astr. Soc.*, **267**, 785
Cole S., Fisher K.B., Weinberg D.H. (1995) *Mon. Not. R. Astr. Soc.*, **275**, 515
Cole S., Kaiser N. (1989) *Mon. Not. R. Astr. Soc.*, **237**, 1127
Coles P. (1992) *Comments on Astrophys.*, **16**, 45
Coles P. (1993) *Mon. Not. R. Astr. Soc.*, **262**, 1065
Coles P., Barrow J.D. (1987) *Mon. Not. R. Astr. Soc.*, **228**, 407
Coles P., Frenk C.S. (1991) *Mon. Not. R. Astr. Soc.*, **253**, 727
Coles P., Jones B.J.T. (1991) *Mon. Not. R. Astr. Soc.*, **248**, 1
Coles P., Lucchin F. (1995) *Cosmology: The Formation and Evolution of Cosmic Structure*, Wiley
Coles P., Plionis M (1991) *Mon. Not. R. Astr. Soc.*, **250**, 75
Colless M. (1989) *Mon. Not. R. Astr. Soc.*, **237**, 799
Collins G.W. III (1989) *Fundamentals of Stellar Astrophysics*, Freeman
Collins P.D.B., Martin A.D., Squires E.J. (1989) *Particle Physics and Cosmology*, Wiley
Cooke J.H., Kantowski R. (1975) *Astrophys. J.*, **195**, L11
Copson E.T. (1965) *Asymptotic Expansions*, Cambridge University Press
Corson E.M. (1953) *Introduction to Tensors, Spinors and Relativistic Wave Equations*, Blackie
Couch W.J., Sharples R.M. (1987) *Mon. Not. R. Astr. Soc.*, **229**, 423
Couchman H.M.P. (1991) *Astrophys. J.*, **368**, L23
Couchman H.M.P., Thomas P.A., Pearce F.R. (1995) *Astrophys. J.*, **452**, 797
Couchman H.M.P., Carlberg R.G. (1992) *Astrophys. J.*, **389**, 453
Cowan J.J., Thielemann F.-K., Truran J. (1991) *Ann. Rev. Astr. Astrophys.*, **29**, 447
Cowie L.L., Hu E.M., Songaila A. (1995a) *Nature*, **377**, 603
Cowie L.L., Songaila A., Hu E.M., Cohen J.G. (1996) *Astr. J.*, **112**, 839
Cowie L.L., Songaila A., Kim T.-S., Hu E.M. (1995b) *Astr. J.*, **109**, 1522
Cowley A.P. (1992) *Ann. Rev. Astr. Astrophys.*, **30**, 287
Dąbrowski M.P., Stelmach J. (1987) *Astr. J.*, **94**, 1373
Dąbrowski M.P., Stelmach J. (1989) *Astr. J.*, **97**, 978
Dabrowski Y., Fabian A.C., Iwasawa K., Lasenby A.N., Reynolds C.S. (1997) *Mon. Not. R. Astr. Soc.*, **288**, L11
Dalton G.B., Efstathiou G., Maddox S.J., Sutherland, W. (1992) *Astrophys. J.*, **390**, L1
Davidson K., Netzer H. (1979) *Rev. Mod. Phys.*, **51**, 715
Davies P.C.W. (1975) *J. Phys. A.: Gen. Phys.*, **8**, 365
Davis M., Efstathiou G., Frenk C.S., White S.D.M. (1985) *Astrophys. J.*, **292**, 371
Davis M., Nusser A., Willick J.A. (1996) *Astrophys. J.*, **473**, 22
Davis M., Peebles P.J.E. (1983) *Astrophys. J.*, **267**, 465
de Laix A., Krauss L.M., Vachaspati T. (1997) *Phys. Rev. Lett.*, **79**,
Dekel A. (1994) *Ann. Rev. Astr. Astrophys.*, **32**, 371
Dekel A., Bertschinger E., Faber S.M. (1990) *Astrophys. J.*, **364**, 349
Dekel A., Rees M.J. (1987) *Nature*, **326**, 455
Dekel A., Silk J. (1986) *Astrophys. J.*, **303**, 39
Dicke R.H. (1961) *Nature*, **192**, 440

Disney M.J. (1976) *Nature*, **263**, 573

Djorgovski S., Davis M. (1987) *Astrophys. J.*, **313**, 59

Dodson C.T.J., Poston T. (1977) *Tensor Geometry*, Pitman

Dressler A. (1980) *Astrophys. J.*, **236**, 351

Dressler A. (1984) *Astrophys. J.*, **281**, 512

Dressler A. (1987) *Astrophys. J.*, **317**, 1

Dressler A., Lynden-Bell D., Burstein D., Davies R.L., Faber S.M., Terlevich R.J., Wegner G. (1987) *Astrophys. J.*, **313**, 42

Dressler A., Oemler A., Couch W.J., Smail I., Ellis R.S., Barger A., Butcher H., Poggianti B.M., Sharples R.M. (1997) *Astrophys. J.*, **490**, 577

Dressler A., Oemler A., Sparks W.B., Lucas R.A. (1994) *Astrophys. J.*, **435**, L23

Driver S. *et al.* (1994) *Mon. Not. R. Astr. Soc.*, **268**, 393

Duncan R.C., Vishniac E.T., Ostriker J.P. (1991) *Astrophys. J.*, **368**, L1

Dunlop J.S., Peacock J.A. (1990) *Mon. Not. R. Astr. Soc.*, **247**, 19

Dunlop J.S., Peacock J.A., Jimenez R., Dey A., Spinrad H., Stevens D., Windhorst R. (1996) *Nature*, **381**, 581

Dwek E., Felten J.E. (1992) *Astrophys. J.*, **387**, 551

Dyer C.C., Roeder R.C. (1973) *Astrophys. J.*, **180**, L31

Efstathiou G. (1990) in *Physics of the Early Universe, Proc. 36th Scottish Universities Summer School in Physics of the Early Universe*, eds. J.A. Peacock, A.F. Heavens, A.T. Davies, Adam Hilger, p361

Efstathiou G. (1992) *Mon. Not. R. Astr. Soc.*, **256**, 43

Efstathiou G. (1995) *Mon. Not. R. Astr. Soc.*, **274**, L73

Efstathiou G., Bernstein G., Katz N., Tyson T., Guhathakurta P. (1991) *Astrophys. J.*, **380**, 47

Efstathiou G., Bond J.R. (1986) *Mon. Not. R. Astr. Soc.*, **218**, 103

Efstathiou G., Bond J.R., White S.D.M. (1992) *Mon. Not. R. Astr. Soc.*, **258**, 1P

Efstathiou G., Davis M., White S.D.M., Frenk C.S. (1985) *Astrophys. J. Suppl.*, **57**, 241

Efstathiou G., Ellis R.S., Peterson B.A. (1988a) *Mon. Not. R. Astr. Soc.*, **232**, 431

Efstathiou G., Frenk C.S., White S.D.M., Davis M. (1988b) *Mon. Not. R. Astr. Soc.*, **235**, 715

Efstathiou G., Kaiser N., Saunders W., Lawrence A., Rowan-Robinson M., Ellis R.S., Frenk C.S. (1990) *Mon. Not. R. Astr. Soc.*, **247**, 10

Efstathiou G., Rees M.J. (1988) *Mon. Not. R. Astr. Soc.*, **230**, 5

Efstathiou G., Silk J. (1983) *Fund. Cosm. Phys.*, **9**, 1

Eggen O.J., Lynden-Bell D., Sandage A.R. (1962) *Astrophys. J.*, **136**, 748

Einasto J., Joeveer M., Saar E. (1980) *Mon. Not. R. Astr. Soc.*, **193**, 353

Einstein A., Podolsky B., Rosen N. (1935) *Phys. Rev.*, **47**, 777

Eisenstein D.J., Hu W. (1997) astro-ph/9709112

Eke V.R., Cole S., Frenk C.S. (1996) *Mon. Not. R. Astr. Soc.*, **282**, 263

Ellis J. (1997) in *Cosmology and Large-scale Structure, Proc. 60th Les Houches School*, eds. R. Schaeffer, J. Silk, M. Spiro, J. Zinn-Justin, Elsevier, p715

Ellis R.S., Colless M., Broadhurst T., Heyl J., Glazebrook K. (1996) *Mon. Not. R. Astr. Soc.*, **280**, 235

Englert F., Brout R. (1964) *Phys. Rev. Lett.*, **13**, 321

Faber S.M. (1982) in *Astrophysical Cosmology, Proc. Vatican study week*, eds. H.A. Brück, G.V. Coyne, M.S. Longair, p191

Faber S.M., Jackson R.E. (1976) *Astrophys. J.*, **204**, 668

Faber S.M. *et al.* (1989) *Astrophys. J. Suppl.*, **69**, 763

Fabian A.C., Nulsen P.E.J., Canizares C.R. (1991) *Astr. Astrophys. Review*, **2**, 191

Falco E.E., Gorenstein M.V., Shapiro I.I. (1991) *Astrophys. J.*, **372**, 364

Fall S.M., Efstathiou G. (1980) *Mon. Not. R. Astr. Soc.*, **193**, 189

Fall S.M., Pei Y.C. (1993) *Astrophys. J.*, **402**, 479

Fanaroff B.L., Riley J.M. (1974) *Mon. Not. R. Astr. Soc.*, **167**, 31

Feast M.W., Catchpole R.M. (1997) *Mon. Not. R. Astr. Soc.*, **286**, L1

Feigelson E.D., Babu G.J. (1992) *Astrophys. J.*, **397**, 55

Feldman H.A., Kaiser N., Peacock J.A. (1994) *Astrophys. J.*, **426**, 23

Felten J.E. (1976) *Astrophys. J.*, **207**, 700

Felten J.E., Isaacman R. (1986) *Rev. Mod. Phys.*, **58**, 689

Fernández-Soto A., Lanzetta K.M., Barcons X., Carswell R.F., Webb J.K., Yahil A. (1996) *Astrophys. J.*, **460**, L85

Feynman R.P. (1972) *Lectures on Statistical Mechanics*, Addison-Wesley

Feynman R.P., Morinigo F.B., Wagner W.G. (1995) *Feynman Lectures on Gravitation*, Addison-Wesley

Fish R. (1964) *Astrophys. J.*, **139**, 284

Fisher K.B. (1995) *Astrophys. J.*, **448**, 494
Fisher K.B., Davis M., Strauss M.A., Yahil A., Huchra J.P. (1993) *Astrophys. J.*, **402**, 42
Fisher K.B., Davis M., Strauss M.A., Yahil A., Huchra J.P. (1994) *Mon. Not. R. Astr. Soc.*, **267**, 927
Fixsen D.J., Cheng E.S., Gales J.M., Mather J.C., Shafer R.A., Wright E.L. (1996) *Astrophys. J.*, **473**, 576
Flores R., Primack J.P., Blumenthal G.R., Faber S.M. (1993) *Astrophys. J.*, **412**, 443
Fowler W.A. (1988) *Q.J.R. Astr. Soc.*, **28**, 87
Francis P.J., Hewett P.C., Foltz C.B., Chaffee F.H., Weymann R.J., Morris S.L. (1991) *Astrophys. J.*, **373**, 465
Freedman W.L., Madore B.F. (1990) *Astrophys. J.*, **365**, 186
Freedman W.L. *et al.* (1994) *Nature*, **371**, 757
Freeman K. (1987) *Ann. Rev. Astr. Astrophys.*, **25**, 603
Frei Z., Gunn J.E. (1994) *Astr. J.*, **108**, 1476
Frogel J.A., Persson S.E., Aaronson M., Matthews K. (1978) *Astrophys. J.*, **220**, 75
Gallego J., Zamorano J., Aragon-Salamanca A., Rego M. (1995) *Astrophys. J.*, **455**, L1
Gardner J.P., Sharples R.M., Frenk C.S., Carrasco B.E. (1997) *Astrophys. J.*, **480**, L99
Garnavich P.M. *et al.* (1997) astro-ph/9710123
Gamow G. (1946) *Phys. Rev.*, **70**, 527
Gaztañaga E. (1992) *Astrophys. J.*, **398**, L17
Ghisellini G., Celotti A., George I.M., Fabian A.C. (1992) *Mon. Not. R. Astr. Soc.*, **258**, 776
Glashow S.L. (1961) *Nucl. Phys.*, **22**, 579
Glazebrook K., Peacock J.A., Collins C.A., Miller L. (1994) *Mon. Not. R. Astr. Soc.*, **266**, 65
Goldschmidt P., Miller L., La Franca F., Cristiani S. (1992) *Mon. Not. R. Astr. Soc.*, **256**, 65P
Górski K.M., Ratra B., Sugiyama N., Banday A.J. (1995) *Astrophys. J.*, **444**, L65
Gott J.R. III, Melott A.L., Dickinson M. (1986) *Astrophys. J.*, **306**, 341
Gould A. (1994) *Astrophys. J.*, **426**, 542
Gradshteyn I.S., Ryzhik I.M. (1980) *Table of Integrals, Series and Products*, Academic Press
Green M.B., Schwarz J.H., Witten E. (1987) *Superstring Theory*, vols. 1 and 2, Cambridge University Press
Grevesse N., Anders E. (1991) in *Solar Interior and Atmosphere*, eds. A.N. Cox, W.C. Livingston, M.S. Matthews, University of Arizona Press
Griest K. (1991) *Astrophys. J.*, **366**, 412
Grigoriev D., Shaposhnikov M., Turok N. (1992) *Phys. Lett.*, **275B**, 395
Grinstein B., Wise M.B. (1986) *Astrophys. J.*, **310**,
Grogin N.A., Narayan R. (1996) *Astrophys. J.*, **464**, 92
Groth E.J., Peebles P.J.E. (1977) *Astrophys. J.*, **217**, 385
Guilbert P.W., Fabian A.C. (1986) *Mon. Not. R. Astr. Soc.*, **220**, 439
Gull S.F. (1991) in *Maximum Entropy in Action*, eds. B. Buck, V.A. Macaulay, Oxford University Press, p171
Gunn J.E. (1982) in *Astrophysical Cosmology, Proc. Vatican Study Week*, eds. H.A. Brück, G.V. Coyne, M.S. Longair, Pontifical Academy Press, p233
Gunn J.E., Gott J.R. III (1972) *Astrophys. J.*, **176**, 1
Gunn J.E., Longair M.S., Rees M.J. (1978) in *Observational Cosmology, Proc. 8th Saas-Fee Course*, eds. A. Maeder, L. Martinet, G. Tammann, Geneva
Gunn J.E., Oke J. B. (1975) *Astrophys. J.*, **195**, 255
Gurbatov S.N., Saichev A.I., Shandarin S.F. (1989) *Mon. Not. R. Astr. Soc.*, **236**, 385
Gush H.P., Halpern M., Wishnow E. (1990) *Phys. Rev. Lett.*, **65**, 537
Guth A.H. (1981) *Phys. Rev. D*, **23**, 347
Halliwell J.J. (1991) *Scientific American*, **265**, 76
Hamilton A.J.S. (1993a) *Astrophys. J.*, **406**, L47
Hamilton A.J.S. (1993b) *Astrophys. J.*, **417**,
Hamilton A.J.S., Gott J.R. III, Weinberg D.H. (1986) *Astrophys. J.*, **309**, 1
Hamilton A.J.S., Kumar P., Lu E., Matthews A. (1991) *Astrophys. J.*, **374**, L1
Hamilton D. (1985) *Astrophys. J.*, **297**, 371
Hancock S. *et al.* (1997) *Mon. Not. R. Astr. Soc.*, **289**, 505
Hausman M.A., Ostriker J.P. (1978) *Astrophys. J.*, **195**, 255
Hawking S.W. (1974) *Nature*, **248**, 30
Hawking S.W. (1976) *Phys. Rev. D*, **13**, 191
Hayes D.S., Latham D.W (1975) *Astrophys. J.*, **197**, 593
Heath D. (1977) *Mon. Not. R. Astr. Soc.*, **179**, 351
Heavens A.F., Peacock J.A. (1988) *Mon. Not. R. Astr. Soc.*, **232**, 339
Heavens A.F., Taylor A.N. (1995) *Mon. Not. R. Astr. Soc.*, **275**, 483

Hehl F. W., Von Der Heyde P., Kerlick G.D. (1976) *Zeitschrift für Naturforschung*, **31a**, 524
Henry J.P., Arnaud K.A. (1991) *Astrophys. J.*, **372**, 410
Hernquist L., Bouchet F.R., Suto Y. (1991) *Astrophys. J. Suppl.*, **75**, 231
Hestenes D. (1985) *New Foundations for Classical Mechanics*, D. Reidel
Hestenes D., Sobczyk G. (1984) *Clifford Algebra to Geometric Calculus*, D. Reidel
Hewett P.C, Foltz C.B., Chaffee F.H. (1993) *Astrophys. J.*, **406**, L43
Higgs P.W. (1964) *Phys. Lett.*, **12**, 132
Hindmarsh M. (1990) in *The Formation and Evolution of Cosmic Strings*, eds. G.W. Gibbons, S.W. Hawking, T. Vachaspati, Cambridge University Press, p527
Hockney R.W., Eastwood J.W. (1988) *Computer Simulations Using Particles*, IOP Publishing
Hoffman G.L., Salpeter E.E., Wasserman I. (1983) *Astrophys. J.*, **268**, 527
Hoffman Y. (1988) *Astrophys. J.*, **328**, 489
Holtzman J.A. (1989) *Astrophys. J. Suppl.*, **71**, 1
Hoyle F., Dunbar D.N.F., Wemsel W.A., Whaling W. (1953) *Phys. Rev.*, **92**, 1095
Hu W., Bunn E.F., Sugiyama N. (1995) *Astrophys. J.*, **447**, L59
Hu W., Scott D., Silk J. (1994) *Phys. Rev. D*, **49**, 648
Hu W., Sugiyama N. (1995) *Astrophys. J.*, **444**, 489
Hu W., White M. (1997) *New Astronomy*, **2**, 323
Huchra J.P., Geller M.J., de Lapparant V., Corwin H.G. (1990) *Astrophys. J. Suppl.*, **72**, 433
Hudson M.J. (1993) *Mon. Not. R. Astr. Soc.*, **265**, 43
Hughes J.P. (1989) *Astrophys. J.*, **337**, 21
Hughes P., ed. (1991) *Beams and Jets in Astrophysics*, Cambridge University Press
Hut P., White S.D.M. (1984) *Nature*, **310**, 637
Iliopoulos J. (1990) in *Physics of the Early Universe, Proc. 36th Scottish Universities Summer School in Physics*, eds. J.A. Peacock, A.F. Heavens, A.T. Davies, Adam Hilger, p311
Impey C., Bothun G. (1997) *Ann. Rev. Astr. Astrophys.*, **35**, 267
Irwin M.J., Webster R.L., Hewett P.C., Corrigan R.T., Jedrzejewski R.I. (1989) *Astr. J.*, **98**,
Itzykson C., Zuber J.-B. (1980) *Quantum Field Theory*, McGraw-Hill
Iwasawa K. *et al.* (1996) *Mon. Not. R. Astr. Soc.*, **282**, 1038
Jackson J.D. (1975) *Classical Electrodynamics*, Wiley
Jacoby G.H., Ciardullo R., Ford H.C. (1990) *Astrophys. J.*, **356**, 332
Jacoby G.H., Ciardullo R., Walker A.R. (1990) *Astrophys. J.*, **365**, 471
Jacoby G.H. *et al.* (1992) *Proc. Astr. Soc. Pacif.*, **104**, 599
Jain B., Mo H.J., White S.D.M. (1995) *Mon. Not. R. Astr. Soc.*, **276**, L25
Jeffries H. (1962) *Asymptotic Approximations*, Oxford University Press
Jensen L.G., Szalay A.S. (1986) *Astrophys. J.*, **305**, L5
Jimenez R., Thejll P., Jørgensen U., MacDonald J., Pagel B. (1996) *Mon. Not. R. Astr. Soc.*, **282**, 926
Jing Y.P., Mo H.J., Börner G. (1997) astro-ph/9707106
Jones B.J.T., Wyse R.F.G. (1985) *Astr. Astrophys.*, **149**, 144
Jones M. *et al.* (1993) *Nature*, **365**, 320
Joshi A.W. (1982) *Elements of Group Theory for Physicists*, Wiley
Jungman G., Kamionkowski M., Griest K. (1996) *Phys. Reports*, **267**, 198
Kaiser N. (1984) *Astrophys. J.*, **284**, L9
Kaiser N. (1986a) *Mon. Not. R. Astr. Soc.*, **219**, 785
Kaiser N. (1986b) *Mon. Not. R. Astr. Soc.*, **222**, 323
Kaiser N. (1987) *Mon. Not. R. Astr. Soc.*, **227**, 1
Kaiser N., Lahav O. (1989) *Mon. Not. R. Astr. Soc.*, **237**, 129
Kaiser N., Malaney R.A., Starkman G.D. (1993) *Phys. Rev. Lett.*, **71**, 1128
Kaiser N., Peacock J.A. (1991) *Astrophys. J.*, **379**, 482
Kaiser N., Squires G. (1993) *Astrophys. J.*, **404**, 441
Kaiser N., Stebbins A. (1984) *Nature*, **310**, 391
Kaluza T. (1921) *Preus. Acad. Wiss.*, **K1**, 966
Kane G. (1987) *Modern Elementary Particle Physics*, Addison-Wesley
Kaneda H. *et al.* (1995) *Astrophys. J.*, **453**, L13
Kang H. *et al.* (1994) *Astrophys. J.*, **430**, 83
Kashlinsky A. (1991) *Astrophys. J.*, **376**, L5
Katz N., Weinberg D.H., Hernquist L. (1996a) *Astrophys. J. Suppl.*, **105**, 19
Katz N., Weinberg D.H., Hernquist L., Miralda-Escudé J. (1996b) *Astrophys. J.*, **457**, 57
Kauffmann G., Nusser A., Steinmetz M. (1997) *Mon. Not. R. Astr. Soc.*, **286**, 795
Kauffmann G., White S.D.M., Guiderdoni B. (1993) *Mon. Not. R. Astr. Soc.*, **264**, 201

Kendall M.G., Stuart A. (1958) *The Advanced Theory of Statistics*, vol. 1, Griffin
Kennicutt R.C. (1983) *Astrophys. J.*, **272**, 54
Kennicutt R.C. (1989) *Astrophys. J.*, **344**, 685
Kerr R.P. (1963) *Phys. Rev. Lett.*, **11**, 237
Kibble T.W.B. (1967) *Phys. Rev.*, **155**, 1554
Kibble T.W.B. (1976) *J. Phys.*, **A9**, 1387
Kim C.W., Pevsner A. (1993) *Neutrinos in Physics and Astrophysics*, Harwood Academic Publishers
King C.R., Ellis R.S. (1985) *Astrophys. J.*, **288**, 456
King I.R. (1966) *Astr. J.*, **71**, 64
Kippenhahn R., Weigert R. (1990) *Stellar Structure and Evolution*, Springer
Kirshner R.P., Oemler A., Schechter P.L., Shectman S.A. (1981) *Astrophys. J.*, **248**, L57
Kitayama T., Suto Y. (1996) *Astrophys. J.*, **469**, 480
Klein O. (1926) *Nature*, **118**, 516
Klein O., Nishina Y. (1929) *Z. Phys.*, **52**, 853
Klypin A., Holtzman J., Primack J., Regős E. (1993) *Astrophys. J.*, **416**, 1
Kneib J.-P., Ellis R.S., Smail I., Couch W.J., Sharples R.M. (1996) *Astrophys. J.*, **471**, 643
Kochanek C.S. (1996) *Astrophys. J.*, **466**, 638
Kochanek C.S., Blandford R. (1988) *Astrophys. J.*, **321**, 676
Kodama H., Sasaki M. (1984) *Prog. Theor. Phys. Suppl.*, **78**, 1
Kofman L., Gnedin N., Bahcall N. (1993) *Astrophys. J.*, **413**, 1
Kofman L., Linde A. (1987) *Nucl. Phys.*, **B282**, 555
Kofman L., Linde A., Starobinsky A.A. (1997) hep-ph/9704452
Kolb E.W. (1989) *Astrophys. J.*, **344**, 543
Kolb E.W., Turner M.S. (1990) *The Early Universe*, Addison-Wesley
Kormendy J. (1982) in *Proc. 12th Saas-Fee Course, Morphology and Dynamics of Galaxies*, eds. L. Martinet, M. Mayor, Geneva Observatory Press
Kormendy J., Richstone D. (1995) *Ann. Rev. Astr. Astrophys.*, **33**, 581
Kristian J., Sachs R.K. (1966) *Astrophys. J.*, **143**, 379
Kuijken K., Gilmore G. (1991) *Astrophys. J.*, **367**, L9
Kundić T. *et al.* (1997) *Astrophys. J.*, **482**, 68
Lacey C., Cole S. (1993) *Mon. Not. R. Astr. Soc.*, **262**, 627
Lacey C., Cole S. (1994) *Mon. Not. R. Astr. Soc.*, **271**, 676
Lacey C., Silk, J. (1991) *Astrophys. J.*, **381**, 14
Lahav O., Kaiser N., Hoffman Y. (1990) *Astrophys. J.*, **352**, 448
Lahav O., Lilje P.B., Primack J.R., Rees M.J. (1991) *Mon. Not. R. Astr. Soc.*, **251**, 128
Lake K. (1981) *Astrophys. J.*, **247**, 17
Landau L.D., Lifshitz E.M. (1959) *Fluid Mechanics*, Pergamon
Landolt A.U. (1983) *Astr. J.*, **88**, 439
Lang K.R. (1980) *Astrophysical Formulae*, second edition, Springer
Langston G.I., Conner S.R., Lehar J., Burke B.F., Weiler K.W. (1990) *Nature*, **344**, 43
Lanzetta K., Wolfe A.M., Turnshek D.A., Lu L., McMahon R.G., Hazard C. (1991) *Astrophys. J. Suppl.*, **77**, 1
Larson R.B. (1976) *Mon. Not. R. Astr. Soc.*, **176**, 31
Le Fèvre O. *et al.* (1996) *Astrophys. J.*, **461**, 534
Lee B.W., Weinberg S. (1977) *Phys. Rev. Lett.*, **39**, 165
Lee T.D., Yang C.N. (1956) *Phys. Rev.*, **104**, 254
Liddle A.R., Lyth D. (1993) *Phys. Reports*, **231**, 1
Liebert J.W. (1980) *Ann. Rev. Astr. Astrophys.*, **18**, 363
Lilly S.J. (1988) *Astrophys. J.*, **333**, 161
Lilly S.J., Cowie L.L., Gardner J.P. (1991) *Astrophys. J.*, **369**, 79
Lilly S.J., Le Fèvre O., Hammer F., Crampton D. (1996) *Astrophys. J.*, **460**, L1
Lilly S.J., Longair M.S. (1984) *Mon. Not. R. Astr. Soc.*, **211**, 833
Lilly S.J., Tresse L., Hammer F., Crampton D., Le Fèvre O. (1995) *Astrophys. J.*, **455**, 108
Lind K.R., Blandford R.D. (1985) *Astrophys. J.*, **295**, 358
Linde A. (1986) *Phys. Lett.*, **175B**, 295
Linde A. (1989) *Inflation and Quantum Cosmology*, Academic Press
Lindgren L. *et al.* (1992) *Astr. Astrophys.*, **258**, L134
Ling E.N., Frenk C.S., Barrow J.D. (1986) *Mon. Not. R. Astr. Soc.*, **223**, 21P
Liu R., Pooley G.G., Riley J.M. (1992) *Mon. Not. R. Astr. Soc.*, **257**, 545

Llewellyn Smith C.H. (1990) in *Physics of the Early Universe, Proc. 36th Scottish Universities Summer School in Physics*, eds. J.A. Peacock, A.F. Heavens, A.T. Davies, Adam Hilger, p145
Longair M.S. (1978) in *Observational Cosmology, Proc. 8th Saas-Fee Course*, eds. A. Maeder, L. Martinet, G. Tammann, Geneva Observatory Press
Longair M.S. (1986) *Theoretical Concepts in Physics*, Cambridge University Press
Longair M.S., Ryle M., Scheuer P.A.G. (1973) *Mon. Not. R. Astr. Soc.*, **164**, 243
Longair M.S., Seldner M. (1979) *Mon. Not. R. Astr. Soc.*, **189**, 433
Loveday J. (1997) *Astrophys. J.*, **489**, 29
Loveday J., Efstathiou G., Maddox S.J., Peterson B.A. (1996) *Astrophys. J.*, **468**, 1
Loveday J., Maddox S.J., Efstathiou G., Peterson B.A. (1995) *Astrophys. J.*, **442**, 457
Loveday J., Peterson B. A., Efstathiou G., Maddox, S. J. (1992) *Astrophys. J.*, **390**, 338
Lumsden S.L., Heavens A.F., Peacock J.A. (1989) *Mon. Not. R. Astr. Soc.*, **238**, 293
Lumsden S.L., Nichol R.C., Collins C.A., Guzzo L. (1992) *Mon. Not. R. Astr. Soc.*, **258**, L1
Luppino G.A., Kaiser N. (1997) *Astrophys. J.*, **475**, L20
Lynden-Bell D. (1967) *Mon. Not. R. Astr. Soc.*, **136**, 101
Lynden-Bell D. (1977) *Nature*, **270**, 396
Lynden-Bell D. (1978) *Physica Scripta*, **17**, 185
Lynden-Bell D. (1992) in *Elements and the Cosmos, Proc. 31st Herstmonceux Conference*, eds. M.G. Edmunds, R.J. Terlevich, Cambridge University Press, p270
Lynden-Bell D., Lahav O., Burstein D. (1989) *Mon. Not. R. Astr. Soc.*, **241**, 325
Ma C. (1996) *Astrophys. J.*, **471**, 13
Madau P. *et al.* (1996) *Mon. Not. R. Astr. Soc.*, **283**, 1388
Maddox S.J., Sutherland W.J., Efstathiou G., Loveday J., Peterson B.A. (1990a) *Mon. Not. R. Astr. Soc.*, **242**, 43P
Maddox S.J., Efstathiou G., Sutherland W.J., Loveday J. (1990b) *Mon. Not. R. Astr. Soc.*, **247**, 1P
Maddox S.J., Efstathiou G., Sutherland W.J. (1996) *Mon. Not. R. Astr. Soc.*, **283**, 1227
Madsen J. (1992) *Phys. Rev. Lett.*, **69**, 571
Mandl F., Shaw G. (1984) *Quantum Field Theory*, Wiley
Mann R.G., Peacock J.A., Heavens A.F. (1997) astro-ph/9708031
Mao S., Mo H.J., White S.D.M. (1997) astro-ph/9712167
Maoz D., Rix H.-W. (1993) *Astrophys. J.*, **416**, 425
Marley M.S., Saumon D., Guillot T., Freedman R.S., Hubbard W.B., Burrows A., Lunine J.I. (1996) *Science*, **272**, 1919
Marzke R., Huchra J.P., Geller M. (1994) *Astrophys. J.*, **428**, 43
Mather J.C. *et al.* (1990) *Astrophys. J.*, **354**, L37
Mathews G.J., Kajino T., Orito M. (1996) *Astrophys. J.*, **456**, 98
Matsubara T. (1995) *Astrophys. J. Suppl.*, **101**, 1
Mattig W. (1958) *Astron. Nachr.*, **284**, 109
McCarthy P.J. (1993) *Ann. Rev. Astr. Astrophys.*, **31**, 639
McClelland J., Silk J. (1977) *Astrophys. J.*, **217**, 331
McGaugh S.S. (1994) *Nature*, **367**, 538
Mecke K.R., Buchert T., Wagner H. (1994) *Astr. Astrophys.*, **288**, 697
Meiksin A., Davis M. (1986) *Astr. J.*, **91**, 191
Meiksin A., Madau P. (1993) *Astrophys. J.*, **412**, 34
Meisenheimer K. *et al.* (1989) *Astr. Astrophys.*, **219**, 63
Melott A.L., Cohen A.P., Hamilton A.J.S., Gott J.R. III, Weinberg D.H. (1989) *Astrophys. J.*, **345**, 618
Merritt D. (1987) *Astrophys. J.*, **313**, 121
Mestel L. (1952) *Mon. Not. R. Astr. Soc.*, **112**, 583
Mészáros P. (1974) *Astr. Astrophys.*, **37**, 225
Miller G.E., Scalo J.M. (1979) *Astrophys. J. Suppl.*, **41**, 513
Miller L., Peacock J.A., Mead A.R.G. (1990) *Mon. Not. R. Astr. Soc.*, **244**, 207
Milne E.A. (1948) *Kinematical Relativity*, Clarendon Press
Miralda-Escudé J., Babul A. (1995) *Astrophys. J.*, **449**, 18
Misner C.W., Thorne K.S., Wheeler J.A. (1973) *Gravitation*, Freeman
Miyoshi M. *et al.* (1995) *Nature*, **373**, 127
Mo H.J., Jing Y.P., Börner G. (1992) *Astrophys. J.*, **392**, 452
Mo H.J., Jing Y.P., Börner G. (1993) *Mon. Not. R. Astr. Soc.*, **264**, 825
Mo H.J., McGaugh S.S., Bothun G.D. (1994) *Mon. Not. R. Astr. Soc.*, **267**, 129
Mo H.J., Miralda-Escudé J. (1994) *Astrophys. J.*, **430**, L25
Mo H.J., White S.D.M. (1996) *Mon. Not. R. Astr. Soc.*, **282**, 347

Moore B. (1994) *Nature*, **370**, 629
Moore B. *et al.* (1992) *Mon. Not. R. Astr. Soc.*, **256**, 477
Moore G. (1996) *Nucl. Phys.*, **B480**, 689
Moore W.J. (1992) *Schrödinger*, Cambridge University Press
Mukhanov V.F., Feldman H.A., Brandenberger R.H. (1992) *Phys. Reports*, **215**, 203
Murdoch H.S., Hunstead R.W., Pettini M., Blades J.C. (1986) *Astrophys. J.*, **309**, 19
Narayan R. (1989) *Astrophys. J.*, **339**, L53
Narayan R. (1996) *Astrophys. J.*, **462**, 136
Narlikar J.V., Padmanabhan T. (1986) *Gravity, Gauge Theories and Quantum Cosmology*, Reidel
Navarro J.F., Frenk C.S., White S.D.M. (1996) *Astrophys. J.*, **462**, 563
Oke J.B. (1970) *Astrophys. J.*, **170**, 193
Omnès R. (1994) *The Interpretation of Quantum Mechanics*, Princeton University Press
Opal Consortium (1990) *Phys. Lett.*, **B240**, 497
Orr M.J.L., Browne I.W.A. (1982) *Mon. Not. R. Astr. Soc.*, **200**, 1067
Osmer P.S. (1981) *Astrophys. J.*, **247**, 762
Osmer P.S. (1982) *Astrophys. J.*, **253**, 28
Osterbrock D.E. (1974) *Astrophysics of Gaseous Nebulae*, Freeman
Osterbrock D.E. (1989) *Astrophysics of Gaseous Nebulae and Active Galactic Nuclei*, University Science Books
Ostriker J.P. (1965) *Astrophys. J.*, **140**, 1056
Ostriker J.P., Heisler J. (1984) *Astrophys. J.*, **278**, 1
Ostriker J.P., Peebles P.J.E., Yahil A. (1974) *Astrophys. J.*, **193**, L1
Paczynski B. (1986) *Astrophys. J.*, **310**, 503
Padmanabhan T. (1993) *Structure Formation in the Universe*, Cambridge University Press
Padoan P., Jimenez R. (1997) *Astrophys. J.*, **475**, 580
Pagel B.E.J. (1994) in *The Formation and Evolution of Galaxies*, eds. C. Muñoz-Tuñón, F. Sánchez, Cambridge University Press, p151
Pagel B.E.J. (1997) *Nucleosynthesis and Chemical Evolution of Galaxies*, Cambridge University Press
Pais A. (1982) *Subtle is the Lord... The Science and Life of Albert Einstein*, Oxford University Press
Panagia N., Gilmozzi R., Macchetto F., Adorf H.-M., Kirshner R. (1991) *Astrophys. J.*, **380**, L23
Partridge R.B. (1995) *3K: The Cosmic Microwave Background*, Cambridge University Press
Peacock J.A. (1982) *Mon. Not. R. Astr. Soc.*, **199**, 987
Peacock J.A. (1985) *Mon. Not. R. Astr. Soc.*, **217**, 601
Peacock J.A. (1986) *Mon. Not. R. Astr. Soc.*, **223**, 113
Peacock J.A. (1987) in *Astrophysical Jets and Their Engines*, ed. W. Kundt, D. Reidel, p185
Peacock J.A. (1992) *Mon. Not. R. Astr. Soc.*, **258**, 581
Peacock J.A. (1997) *Mon. Not. R. Astr. Soc.*, **284**, 885
Peacock J.A., Dodds S.J. (1994) *Mon. Not. R. Astr. Soc.*, **267**, 1020
Peacock J.A., Dodds S.J. (1996) *Mon. Not. R. Astr. Soc.*, **280**, L19
Peacock J.A., Heavens A.F. (1985) *Mon. Not. R. Astr. Soc.*, **217**, 805
Peacock J.A., Heavens A.F., Davies A., eds. (1990) *Physics of the Early Universe*, Adam Hilger
Peacock J.A., Nicholson D. (1991) *Mon. Not. R. Astr. Soc.*, **253**, 307
Peacock J.A., West M.J. (1992) *Mon. Not. R. Astr. Soc.*, **259**, 494
Peccei R.D., Quinn H.R. (1977) *Phys. Rev. D*, **16**, 1791
Peebles P.J.E. (1973) *Astrophys. J.*, **185**, 413
Peebles P.J.E. (1974) *Astrophys. J.*, **32**, 197
Peebles P.J.E. (1980) *The Large-scale Structure of the Universe*, Princeton University Press
Peebles P.J.E. (1982) *Astrophys. J.*, **263**, L1
Peebles P.J.E. (1983) *Astrophys. J.*, **274**, 1
Peebles P.J.E. (1987) *Nature*, **327**, 210
Peebles P.J.E. (1993) *Principles of Physical Cosmology*, Princeton University Press
Peebles P.J.E. (1997) *Astrophys. J.*, **483**, L1
Peebles P.J.E., Ratra B. (1988) *Astrophys. J.*, **325**, L17
Pei Y.C. (1993) *Astrophys. J.*, **404**, 436
Pei Y.C., Fall S.M. (1995) *Astrophys. J.*, **454**, 69
Pelt J., Kayser R., Refsdal S., Schramm T. (1996) *Astr. Astrophys.*, **305**, 97
Pen U.-L., Seljak U., Turok N. (1997) astro-ph/9704165
Penrose R. (1969) *Nuovo Cimento*, **1**, 252
Penrose R., Rindler W., (1984) *Spinors and Space-time*, vol. 1, Cambridge University Press
Penston M.V. (1969) *Mon. Not. R. Astr. Soc.*, **144**, 425

Penzias A.A., Wilson R.W. (1965) *Astrophys. J.*, **142**, 419
Perlmutter S. *et al.* (1997a) *Astrophys. J.*, **476**, L63
Perlmutter S. *et al.* (1997b) *Astrophys. J.*, **483**, 565
Perlmutter S. *et al.* (1997c) astro-ph/9712212
Persic M., Salucci P. (1995) *Astrophys. J. Suppl.*, **99**, 501
Peterson B.M. (1993) *Proc. Astr. Soc. Pacif.*, **105**, 247
Pettini M., Smith L.J., Hunstead R.W., King D.L. (1994) *Astrophys. J.*, **426**, 79
Phillipps S. (1982) *Astrophys. Lett.*, **22**, 123
Phillips M.M. (1993) *Astrophys. J.*, **413**, L105
Pierce M.J., Welch D.L., McClure R.D., VandenBergh S., Racine R., Stetson P.B. (1994) *Nature*, **371**, 385
Pippard A.B. (1957) *Elements of Classical Thermodynamics*, Cambridge University Press
Pogosyan D., Starobinsky A.A. (1995) *Astrophys. J.*, **447**, 465
Pound R.V., Rebka G.G. (1960) *Phys. Rev. Lett.*, **4**, 337
Press W.H., Gunn J.E. (1973) *Astrophys. J.*, **185**, 397
Press W.H., Rybicki G.B., Hewitt J.N. (1992) *Astrophys. J.*, **385**, 416
Press W.H., Schechter P. (1974) *Astrophys. J.*, **187**, 425
Press W.H., Teukolsky S.A., Vetterling W.T., Flannery B.P. (1992) *Numerical Recipes*, second edition, Cambridge University Press
Press W.H., Vishniac E.T. (1980) *Astrophys. J.*, **239**, 1
Pringle J.A. (1981) *Ann. Rev. Astr. Astrophys.*, **19**, 137
Puget J.-L. *et al.* (1996) *Astr. Astrophys.*, **308**, L5
Raghavan R.S. (1995) *Science*, **267**, 45
Raine D.J. (1981) *The Isotropic Universe*, Adam Hilger
Raine D.J., Heller M. (1981) *The Science of Space-time*, Pergamon
Rajagopal K., Turner M.S., Wilczek F. (1991) *Nucl. Phys. B*, **358**, 447
Rarita W., Schwinger J. (1941) *Phys. Rev.*, **70**, 61
Ratcliffe A., Shanks T., Broadbent A., Parker Q.A., Watson F.G., Oates A.P., Fong R., Collins C.A. (1996) *Mon. Not. R. Astr. Soc.*, **281**, L47
Raymond J.C., Cox D.P., Smith B.W. (1976) *Astrophys. J.*, **204**, 290
Rees M.J. (1967) *Mon. Not. R. Astr. Soc.*, **135**, 345
Rees M.J. (1984) *Astrophys. J.*, **22**, 471
Rees M.J. (1985) *Mon. Not. R. Astr. Soc.*, **213**, 75
Rees M.J., Ostriker J.P. (1977) *Mon. Not. R. Astr. Soc.*, **179**, 541
Rees M.J., Sciama D.W. (1968) *Nature*, **217**, 511
Refsdal S. (1966) *Mon. Not. R. Astr. Soc.*, **132**, 101
Regős E., Geller M.J. (1989) *Astr. J.*, **98**, 755
Reines F., Sobel H., Pasierb E. (1980) *Phys. Rev. Lett.*, **45**, 1307
Renn J., Sauer T., Stachel J. (1997) *Science*, **275**, 184
Renton P. (1990) *Electroweak Interactions*, Cambridge University Press
Ressell M.T., Turner M.S. (1990) *Comments on Astrophys.*, **14**, 323
Rhee G. (1991) *Nature*, **350**, 211
Rice S.O. (1954) in *Selected Papers on Noise and Stochastic Processes*, ed. Wax N., Dover, p133
Riess A., Press W., Kirshner R. (1995) *Astrophys. J.*, **438**, L17
Riess A., Press W., Kirshner R. (1996) *Astrophys. J.*, **473**, 88
Rindler W. (1977) *Essential Relativity*, second edition, Springer
Rindler W. (1982) *Introduction to Special Relativity*, Clarendon Press
Robinson A., Terlevich R.J. eds. (1994) *The Nature of Compact Objects in Active Galactic Nuclei*, Cambridge University Press
Robson I. (1996) *Active Galactic Nuclei*, Wiley
Rocca-Volmerange B., Guiderdoni B. (1988) *Astr. Astrophys. Suppl.*, **75**, 93
Rogerson J.B., York D.G. (1973) *Astrophys. J.*, **186**, L95
Ross G.G. (1985) *Grand Unified Theories*, Addison-Wesley
Rowan-Robinson M. (1968) *Mon. Not. R. Astr. Soc.*, **138**, 445
Rowan-Robinson M. (1985) *The Cosmological Distance Scale*, Freeman
Rowan-Robinson M. *et al.* (1990) *Mon. Not. R. Astr. Soc.*, **247**, 1
Rybicki G.B., Lightman A.P. (1979) *Radiation Processes in Astrophysics*, Wiley
Ryder L.H. (1985) *Quantum Field Theory*, Cambridge University Press
Sachs R.K. (1961) *Proc. R. Soc.*, **264**, 309
Sachs R.K. (1973) in *Relativity, Astrophysics and Cosmology*, ed. W. Israel, Reidel
Sachs R.K., Wolfe A.M. (1967) *Astrophys. J.*, **147**, 73

Sadler E.M., Jenkins C.R., Kotanyi C.G. (1989) *Mon. Not. R. Astr. Soc.*, **240**, 591
Sakharov A.D. (1967) *JETP Lett.*, **5**, 24
Sakurai J.J. (1964) *Invariance Principles and Elementary Particles*, Princeton University Press
Sakurai J.J. (1967) *Advanced Quantum Mechanics*, Addison-Wesley
Salam A. (1968) in *Elementary Particle Theory*, eds. W. Svartholm *et al.*, Stockholm
Salpeter E.E. (1955) *Astrophys. J.*, **121**, 161
Sandage A. (1965) *Astrophys. J.*, **151**, 1560
Sandage A. (1972) *Astrophys. J.*, **178**, 25
Sandage A., Tammann G.A. (1993) *Astrophys. J.*, **415**, 1
Sandage A., Tammann G.A., Yahil A. (1979) *Astrophys. J.*, **232**, 352
Sandage A., Visvanathan N. (1978) *Astrophys. J.*, **225**, 742
Sanders R.H. (1983) *Astrophys. J.*, **266**, 73
Sarazin C.L. (1988) *X-ray Emissions from Clusters of Galaxies*, Cambridge University Press
Sasaki S. (1994) *Proc. Astr. Soc. Japan*, **46**, 427
Sasselov D.D. *et al.* (1996) astro-ph/9612216
Saunders W., Frenk C., Rowan-Robinson M., Efstathiou G., Lawrence A., Kaiser N., Ellis R., Crawford J., Xia X.-Y., Parry I. (1991) *Nature*, **349**, 32
Saunders W., Rowan-Robinson M., Lawrence A. (1992) *Mon. Not. R. Astr. Soc.*, **258**, 134
Scalo J.M. (1986) *Fund. Cosm. Phys.*, **11**, 1
Scheuer P.A.G. (1974) *Mon. Not. R. Astr. Soc.*, **166**, 513
Scheuer P.A.G. (1987) in *Superluminal Radio Sources*, eds. J.A. Zensus, T.J. Pearson, Cambridge University Press, p104
Schiff L.I. (1955) *Quantum Mechanics*, McGraw-Hill
Schild R., Thomson D.J. (1997) *Astr. J.*, **113**, 130
Schmidt B.P., Kirshner R.P., Eastman R.G. (1992) *Astrophys. J.*, **395**, 366
Schmidt M. (1963) *Nature*, **197**, 1040
Schmidt M. (1968) *Astrophys. J.*, **151**, 393
Schmidt M. (1970) *Astrophys. J.*, **162**, 371
Schmidt M., Green R.F. (1983) *Astrophys. J.*, **269**, 352
Schneider D.P., Gunn J.E., Hoessel J.G. (1983) *Astrophys. J.*, **264**, 337
Schneider P. (1984) *Astr. Astrophys.*, **140**, 119
Schneider P. (1985) *Astr. Astrophys.*, **143**, 413
Schneider P., Ehlers J., Falco E. (1992) *Gravitational Lenses*, Springer
Schombert J.M. (1988) *Astrophys. J.*, **328**, 475
Schutz B.F. (1980) *Geometrical Methods of Mathematical Physics*, Cambridge University Press
Schweizer F., Van Gorkom J.H., Seitzer P. (1989) *Astrophys. J.*, **338**, 770
Seljak U. (1997) *Astrophys. J.*, **482**, 6
Seljak U., Zaldarriaga M. (1996) *Astrophys. J.*, **469**, 437
Seyfert C.K. (1943) *Astrophys. J.*, **97**, 28
Shanks T., Boyle B.J. (1994) *Mon. Not. R. Astr. Soc.*, **271**, 753
Shanks T., Fong R., Boyle B.J., Peterson B.A. (1987) *Mon. Not. R. Astr. Soc.*, **227**, 739
Shanks T., Stevenson P.R.F., Fong R., MacGillivray H.T. (1984) *Mon. Not. R. Astr. Soc.*, **206**, 767
Shanks T. *et al.* eds. (1991) *Observational Tests of Cosmological Inflation*, NATO ASI C264, Kluwer
Shapiro P., Martel H., Villumsen J.V., Owen J.M. (1996) *Astrophys. J. Suppl.*, **103**, 269
Shapiro S.L., Teukolsky S.A. (1983) *Black Holes, White Dwarfs and Neutron Stars: The Physics of Compact Objects*, Wiley
Shaw M., Tadhunter C., Dickson R., Morganti R. (1995) *Mon. Not. R. Astr. Soc.*, **275**, 703
Shectman S.A., Landy S.D., Oemler A., Tucker D.L., Lin H., Kirshner R.P., Schechter P.L. (1996) *Astrophys. J.*, **470**, 172
Shull J.M., Thronson H.A., eds. (1993) *The Environments and Evolution of Galaxies*, Kluwer
Silk J.I. (1977) *Astrophys. J.*, **211**, 638
Slipher V.M. (1918) *Lowell Observatory Bulletin*, **3**, 59
Smith E.P., Heckman T.M., Bothun G.D., Romanishin W., Balick B. (1986) *Astrophys. J.*, **306**, 64
Smith M.S., Kawano L.H., Malaney R.A. (1993) *Astrophys. J. Suppl.*, **85**, 219
Smith P.F., Lewin J.D. (1990) *Phys. Reports*, **187**, 1
Smoot G.F., Gorenstein M.V., Muller R.A. (1977) *Phys. Rev. Lett.*, **39**, 898
Smoot G.F. *et al.* (1992) *Astrophys. J.*, **396**, L1
Soltan A. (1982) *Mon. Not. R. Astr. Soc.*, **200**, 115
Songaila A., Cowie L.L., Lilly S.J. (1990) *Astrophys. J.*, **348**, 371
Spooner N.J.C., ed. (1997) *The Identification of Dark Matter*, World Scientific

Squires G., Kaiser N., Babul A., Fahlman G., Woods D., Neumann D.M., Böringer H. (1996) *Astrophys. J.*, **461**, 572
Starobinsky A.A. (1985) *Sov. Astr. Lett.*, **11**, 133
Steidel C.C., Adelberger K.L., Dickinson M., Giavalisco M., Pettini M., Kellogg M. (1997) astro-ph/9708125
Steidel C.C., Giavalisco M., Pettini M., Dickinson M., Adelberger K.L. (1996) *Astrophys. J.*, L17
Storrie-Lombardi L.J., Irwin M.J., McMahon R.G. (1996) *Mon. Not. R. Astr. Soc.*, **282**, 1330
Strauss M.A., Davis M., Yahil Y., Huchra J.P. (1992) *Astrophys. J.*, **385**, 421
Strauss M.A., Willick J.A. (1995) *Phys. Reports*, **261**, 271
Subrahmanian K., Cowling S.A. (1986) *Mon. Not. R. Astr. Soc.*, **219**, 333
Sugiyama N. (1995) *Astrophys. J. Suppl.*, **100**, 281
Sutherland R.S., Dopita M.A. (1993) *Astrophys. J. Suppl.*, **88**, 253
Sutherland W.J. (1988) *Mon. Not. R. Astr. Soc.*, **234**, 159
Tóth G., Ostriker J.P. (1992) *Astrophys. J.*, **389**, 5
Tanvir N., Shanks T., Ferguson H.C., Robinson D.R.T. (1995) *Nature*, **377**, 27
Tayler R.J. (1994) *The Stars: Their Structure and Evolution*, Cambridge University Press
Taylor G.L., Dunlop J.S., Hughes D.H., Robson E.I. (1996) *Mon. Not. R. Astr. Soc.*, **283**, 930
Terlevich R. *et al.* (1992) *Mon. Not. R. Astr. Soc.*, **255**, 713
Thorne K.S. (1987) in *300 Years of Gravitation*, eds. S. Hawking, W. Israel, Cambridge University Press, p330
Thorne K.S., Price R.H., Macdonald D.A. (1986) *Black Holes: The Membrane Paradigm*, Yale University Press
Thuan T.X., Gott J.R. III, Schneider S.E. (1987) *Astrophys. J.*, **315**, L93
Tinsley B.M., Larson R.B. (1979) *Mon. Not. R. Astr. Soc.*, **186**, 503
Tonry J.L. (1991) *Astrophys. J.*, **373**, L1
Tonry J.L., Schneider D.P. (1988) *Astr. J.*, **96**, 807
Treille D. (1994) *Rep. Prog. Phys.*, **57**, 1137
Tremaine S., Gunn J.E. (1979) *Phys. Rev. Lett.*, **42**, 407
Tully R.B., Fisher J.R. (1977) *Astrophys. J.*, **54**, 661
Turner E.L., Ostriker J.P., Gott J.R. (1984) *Astrophys. J.*, **284**, 1
Turner M.S., Wilczek F., Zee A. (1983) *Phys. Lett.*, **125B**, 35
Turok N. (1991) in *Proc. Nobel Symp. no. 79.*, eds. J.S. Nilsson, B. Gustafsson, B.-S. Skagerstam, *Physica Scripta,*, **T36**, 135
Tyson J.A. (1988) *Astr. J.*, **96**, 1
Tyson J.A., Wenk R.A., Valdes F. (1990) *Astrophys. J.*, **349**, L1
Tytler D., Fan X.-M., Burles S. (1996) *Nature*, **381**, 207
Udalski A. *et al.* (1994) *Astrophys. J.*, **426**, L69
Ulrich M.H., Boksenberg A., Bromage G.E., Clavel J., Elvius A., Penston M.V., Perola G.C., Pettini M., Snijders M.A.J., Tanzi E.G., Tarenghi M. (1984) *Mon. Not. R. Astr. Soc.*, **209**, 479
Unruh W.G. (1976) *Phys. Rev. D*, **14**, 870
Unwin S.C., Cohen M.H., Biretta J.A., Pearson T.J., Seilstad G.A., Walker R.C., Simon R.S., Linfield R.P. (1985) *Astrophys. J.*, **289**, 109
Urry C.M., Padovani P. (1995) *Proc. Astr. Soc. Pacif.*, **107**, 803
Valls-Gabaud D., Alimi J.-M., Blanchard A. (1989) *Nature*, **341**, 215
van Albada T.S., Bahcall J.N., Begeman K., Sancisi R. (1985) *Astrophys. J.*, **295**, 305
VandenBergh S. (1991) in *The Formation and Evolution of Star Clusters,*, ed. K. Janes, P.A.S.P. Conf. Ser., **13**, 183
VandenBergh S. (1996) *Proc. Astr. Soc. Pacif.*, **108**, 1091
van de Weygaert R. (1991) *Mon. Not. R. Astr. Soc.*, **249**, 159
van de Weygaert R., Icke V. (1989) *Astrophys. J.*, **213**, 1
Vilenkin A. (1984) *Astrophys. J.*, **282**, L51
Vilenkin A. (1986) *Nature*, **322**, 613
Vilenkin A., Shellard E.P.S. (1994) *Cosmic Strings and Other Topological Defects*, Cambridge University Press
Vishniac E.T. (1987) *Astrophys. J.*, **322**, 597
Vittorio N., Juszkiewicz R., Davis M. (1986) *Nature*, **323**, 132
Vittorio N., Silk J. (1991) *Astrophys. J.*, **385**, L9
Waldram J.R. (1985) *The Theory of Thermodynamics*, Cambridge University Press
Walker T.P., Steigman G., Kang H.-S., Schramm D.M., Olive K.A. (1991) *Astrophys. J.*, **376**, 51
Wall J.V. (1994) *Aust. J. Phys.*, **47**, 625

Walsh D., Carswell R.F., Weymann R.J. (1979) *Nature*, **279**, 381
Wandel A., Mushotzky R.F. (1986) *Astrophys. J.*, **306**, L61
Wardle J.F.C. (1977) *Nature*, **269**, 563
Warren S.J., Hewett P.C., Osmer P.S. (1994) *Astrophys. J.*, **421**, 412
Webster A.S. (1976) *Mon. Not. R. Astr. Soc.*, **175**, 61
Webster R.L., Hewett P.C., Harding M.E., Wegner G.A. (1988) *Nature*, **336**, 358
Weedman D.W. (1986) *Quasar Astronomy*, Cambridge University Press
Weinberg S. (1967) *Phys. Rev. Lett.*, **19**, 1264
Weinberg S. (1972) *Gravitation and Cosmology*, Wiley
Weinberg S. (1976) *Astrophys. J.*, **208**, L1
Weinberg S. (1977) *The First Three Minutes*, Basic Books
Weinberg S. (1989) *Rev. Mod. Phys.*, **61**, 1
Weinberg S. (1996) *The Quantum Theory of Fields*, vol. 1, Cambridge University Press
Wesson P.S., Valle K., Stabell R. (1987) *Astrophys. J.*, **317**, 601
West M.J., Richstone D.O. (1988) *Astrophys. J.*, **335**, 532
Wheeler J.A., Zurek W.H. (1983) *Quantum Theory and Measurement*, Princeton University Press
White M., Bunn E.F. (1995) *Astrophys. J.*, **450**, 477
White M., Scott D., Silk J. (1994) *Ann. Rev. Astr. Astrophys.*, **32**, 319
White S.D.M. (1979) *Mon. Not. R. Astr. Soc.*, **186**, 145
White S.D.M. (1990) in *Physics of the Early Universe, Proc. 36th Scottish Universities Summer School in Physics*, eds. J.A. Peacock, A.F. Heavens, A.T. Davies, Adam Hilger, p1
White S.D.M. (1997) in *Cosmology and Large-scale Structure, Proc. 60th Les Houches School*, eds. R. Schaeffer, J. Silk, M. Spiro, J. Zinn-Justin, Elsevier, p349
White S.D.M., Davis M., Efstathiou G., Frenk C.S. (1987) *Nature*, **330**, 451
White S.D.M., Efstathiou G., Frenk C.S. (1993) *Mon. Not. R. Astr. Soc.*, **262**, 1023
White S.D.M., Navarro J.F., Evrard A.E., Frenk C.S. (1993) *Nature*, **366**, 429
White S.D.M., Rees M. (1978) *Mon. Not. R. Astr. Soc.*, **183**, 341
Will C. (1993) *Theory and Experiment in Gravitational Physics*, second edition, Cambridge University Press
Williams B.G., Heavens A.F., Peacock J.A., Shandarin S.F. (1991) *Mon. Not. R. Astr. Soc.*, **250**, 458
Williams R.E. *et al.* (1996) *Astr. J.*, **112**, 1335
Willick J.A., Strauss M.A., Dekel A., Kollatt T. (1996) astro-ph/9612240
Wills B.J., Browne I.W.A. (1986) *Astrophys. J.*, **302**, 56
Wilson T.L., Rood R.T. (1994) *Ann. Rev. Astr. Astrophys.*, **32**, 191
Winget D. *et al.* (1987) *Astrophys. J.*, **315**, L77
Winter A.J.B., Wilson D.M.A., Warner P.J., Waldram E.M., Routledge D., Nicol A.T., Boysen R.C., Bly D.W.J., Baldwin J.E. (1980) *Mon. Not. R. Astr. Soc.*, **192**, 931
Worthey G. (1994) *Astrophys. J. Suppl.*, **95**, 107
Worthey G., Faber S.M., Gonzalez J.J. (1992) *Astrophys. J.*, **398**, 69
Wu C.S., Ambler E., Hayward R., Hoppes D., Hudson R. (1957) *Phys. Rev.*, **105**, 1413
Wu Z.C. (1993) *No-Boundary Universe*, Human Science and Technology Press
Yahil A. (1985) in *The Virgo Cluster, Proc. ESO workshop*, eds. O.-G. Richter, B. Binggeli, p359
Yahil A., Walker D., Rowan-Robinson M. (1986) *Astrophys. J.*, **301**, L1
Yang C.N., Mills R. (1954) *Phys. Rev.*, **96**, 191
Yates M.G., Miller L., Peacock J.A. (1989) *Mon. Not. R. Astr. Soc.*, **240**, 129
Yi I., Vishniac E.T. (1993) *Astrophys. J. Suppl.*, **86**, 333
Young P. (1981) *Astrophys. J.*, **244**, 756
Young P., Gunn J.E., Kristian J., Oke J.B., Westphal J.A. (1980) *Astrophys. J.*, **241**, 507
Zaldarriaga M., Spergel D.N., Seljak U. (1997) astro-ph/9702157
Zeldovich Y.B. (1970) *Astr. Astrophys.*, **5**, 84
Ziman J.M. (1972) *Principles of the Theory of Solids*, second edition, Cambridge University Press

Useful numbers and formulae

The purpose of this appendix is to collect together some of the physical constants, unit conversions, and formulae that are most useful in practice for cosmological calculations. More comprehensive listings and discussions of numerical constants are given in a variety of sources. For physical constants, see e.g. the Particle Data Group as cited in the bibliography; for astronomical ones see Allen (1973) or Lang (1974). Uncertainties are usually not given; except in cases of definitions (such as c), only sufficient figures to be useful in practice are given. It may be assumed that the uncertainties are no worse than about one digit in the smallest decimal place given.

Physical constants

speed of light (exact)	$c = 2.99792458 \times 10^{8}\ \mathrm{m\,s^{-1}}$
gravitational constant	$G = 6.6726 \pm 0.0009 \times 10^{-11}\ \mathrm{m^3kg^{-1}s^{-2}}$
Planck's constant	$\hbar = 1.054572 \times 10^{-34}\ \mathrm{J\,s}$
Boltzmann's constant	$k = 1.3806 \times 10^{-23}\ \mathrm{J\,K^{-1}}$
permeability of vacuum (exact)	$\mu_0 = 4\pi \times 10^{-7}\ \mathrm{N\,A^{-2}}$
permittivity of vacuum (exact)	$\epsilon_0 = (c^2\mu_0)^{-1}\ \mathrm{F\,m^{-1}}$
electron mass	$m_e = 9.109389 \times 10^{-31}\ \mathrm{kg}$
proton mass	$m_p = 1.672623 \times 10^{-27}\ \mathrm{kg}$
neutron mass	$m_n = 1.674929 \times 10^{-27}\ \mathrm{kg}$
electron charge	$e = -1.602177 \times 10^{-19}\ \mathrm{C}$
fine-structure constant	$\alpha = e^2/(4\pi\epsilon_0\hbar c) = 1/137.03599$
Thomson cross-section	$\sigma_{\mathrm{T}} = e^4/(6\pi\epsilon_0^2 m_e^2 c^4) = 6.65246 \times 10^{-29}\ \mathrm{m^2}$

cgs unit conversions

energy	$1\ \mathrm{erg} = 10^{-7}\ \mathrm{W}$
magnetic flux density	$1\ \mathrm{gauss} = 10^{-4}\ \mathrm{T}$
charge	$1\ \mathrm{esu} = 0.1(c/\mathrm{m\,s^{-1}})^{-1}\ \mathrm{C}$

Astronomical units

parsec	$1\ \mathrm{pc} = 3.0856 \times 10^{16}\ \mathrm{m}$
year (sidereal)	$1\ \mathrm{yr} = 3.155815 \times 10^{7}\ \mathrm{s}$

Solar mass	$1\ M_\odot = 1.989 \times 10^{30}$ kg
Solar luminosity	$1\ L_\odot = 3.90 \times 10^{26}$ W
angström	$1\ \text{Å} = 1 \times 10^{-10}$ m
jansky	$1\ \text{Jy} = 10^{-26}\ \text{W m}^{-2}\ \text{Hz}^{-1}$
keV as a temperature unit	$1\ \text{keV} = 1.1605 \times 10^7$ K

Cosmological quantities

dimensionless Hubble parameter	$h = H_0/100\ \text{km s}^{-1}\text{Mpc}^{-1}$
Hubble time	$H_0^{-1} = 9.7776 \times 10^9\ h^{-1}$ yr
Hubble radius	$c/H_0 = 2997.9\ h^{-1}$ Mpc
density of universe	$\rho = 1.8791 \times 10^{-26}\ \Omega h^2\ \text{kg m}^{-3}$
	$= 2.7755 \times 10^{11}\ \Omega h^2\ M_\odot \text{Mpc}^{-3}$
relativistic density	$\Omega_r = 1.681\Omega_\gamma = 4.183 \times 10^{-5} h^{-2}$
hydrogen density	$n_{\text{H}} = 8.42\,\Omega_{\text{B}} h^2\ \text{m}^{-3}$
electron density	$n_e = 9.83\,\Omega_{\text{B}} h^2\ \text{m}^{-3}$
mass per particle	$\mu = 0.593\, m_p$
mass per electron	$\mu_e = 1.143\, m_p$
(assuming 25% He by mass)	

Natural and Planck units

energy	$1\ \text{GeV} = 1.6022 \times 10^{-10}$ J
temperature	$1\ \text{GeV} = 1.1605 \times 10^{13}$ K
mass	$1\ \text{GeV} = 1.7827 \times 10^{-27}$ kg
length	$1\ \text{GeV}^{-1} = 1.9733 \times 10^{-16}$ m
time	$1\ \text{GeV}^{-1} = 6.6522 \times 10^{-25}$ J
Planck mass	$m_{\text{P}} = (\hbar c/G)^{1/2} = 2.1767 \times 10^{-8}$ kg
Planck length	$\ell_{\text{P}} = (\hbar G/c^3)^{1/2} = 1.6161 \times 10^{-35}$ m
Planck time	$t_{\text{P}} = (\hbar G/c^5)^{1/2} = 5.3906 \times 10^{-44}$ s

Formulae for Friedmann models

Robertson–Walker metric; recommended form ($d\psi$ is angular separation on sky):

$$c^2 d\tau^2 = c^2 dt^2 - R^2(t)\left[dr^2 + S_k^2(r)\, d\psi^2\right].$$

Alternative form:

$$c^2 d\tau^2 = c^2 dt^2 - R^2(t)\left(\frac{dr'^2}{1 - kr'^2} + r'^2 d\psi^2\right).$$

Curvature-dependent function for comoving radial distance:

$$S_k(r) = \begin{cases} \sin r & (k = 1) \\ \sinh r & (k = -1) \\ r & (k = 0). \end{cases}$$

The cosine-like analogue of $S_k(r)$:

$$C_k(r) \equiv \sqrt{1 - kS_k^2(r)} = \begin{cases} \cos r & (k=1) \\ \cosh r & (k=-1) \\ 1 & (k=0). \end{cases}$$

Comoving separation between two events as an analogue of the spherical cosine rule:

$$C_k(r_{12}) = C_k(r_1)C_k(r_2) + kS_k(r_1)S_k(r_2)\cos\psi.$$

Friedmann equation in integral form:

$$\dot{R}^2 - \frac{8\pi G}{3}\rho R^2 = -kc^2.$$

Friedmann equation in differential form:

$$\ddot{R} = -\frac{4\pi G R}{3}\left(\rho + \frac{3p}{c^2}\right).$$

Conservation equation:

$$\dot{\rho} = -3\left(\rho + \frac{p}{c^2}\right)\frac{\dot{R}}{R}.$$

Dust + radiation + vacuum equation of state:

$$\frac{8\pi G\rho}{3} = H_0^2(\Omega_v + \Omega_m a^{-3} + \Omega_r a^{-4}).$$

Density parameter:

$$\Omega \equiv \frac{\rho}{\rho_c} = \frac{8\pi G\rho}{3H^2}.$$

Friedmann equation in terms of density parameters:

$$\frac{kc^2}{H^2R^2} = \Omega_m + \Omega_r + \Omega_v - 1.$$

Curvature radius and dimensionless scale factor:

$$R_0 = \frac{c}{H_0}\,|1-\Omega|^{-1/2}, \qquad a(t) = R(t)/R_0.$$

Definition of deceleration parameter and relation to density parameters:

$$q \equiv -\frac{\ddot{R}R}{\dot{R}^2} = \frac{\Omega_m}{2} + \Omega_r - \Omega_v.$$

Epoch dependence of H:

$$H^2(a) = H_0^2\left[\Omega_v + \Omega_m a^{-3} + \Omega_r a^{-4} - (\Omega - 1)a^{-2}\right].$$

Epoch dependence of total Ω:

$$\Omega(a) - 1 = \Omega_m(a) + \Omega_v(a) - 1 = \frac{\Omega - 1}{1 - \Omega + \Omega_v a^2 + \Omega_m a^{-1} + \Omega_r a^{-2}}.$$

Comoving distance in differential form:

$$\begin{aligned} R_0\,dr &= \frac{c}{H(z)}\,dz \\ &= \frac{c}{H_0}\frac{dz}{\sqrt{(1-\Omega)(1+z)^2 + \Omega_v + \Omega_m(1+z)^3 + \Omega_r(1+z)^4}}. \end{aligned}$$

Comoving distance in integral form (for zero vacuum density):

$$R_0 S_k(r) = \left(\frac{2c}{H_0}\right) \frac{\Omega_m z + (\Omega_m + 2\Omega_r - 2)[\sqrt{1 + \Omega_m z + \Omega_r(z^2 + 2z)} - 1]}{[\Omega_m^2 + 4\Omega_r(\Omega_r + \Omega_m - 1)]\,(1+z)}.$$

Density–time relations for $k = 0$:

$$t = \sqrt{\frac{1}{6\pi G\rho}} \qquad \text{(matter domination)}$$

$$t = \sqrt{\frac{3}{32\pi G\rho}} \qquad \text{(radiation domination).}$$

Age of the universe for $k = 0$ and zero vacuum energy:

$$H_0 t = \frac{2}{3\Omega_m^2}\left[(\Omega_m a - 2\Omega_r)\sqrt{\Omega_r + \Omega_m a} + 2\Omega_r^{3/2}\right].$$

Simple approximation for the age of the universe with matter and Λ:

$$H_0 t_0 \simeq \tfrac{2}{3}\,(0.7\Omega_m + 0.3 - 0.3\Omega_v)^{-0.3}.$$

Comoving horizon length for matter-dominated models:

$$R_0 r_{\rm H} = R_0 \int_z^\infty dr \simeq \frac{2}{\sqrt{\Omega z}}\frac{c}{H_0}.$$

Comoving horizon length at matter–radiation equality:

$$R_0 r_{\rm H}(z_{\rm eq}) = 2(\sqrt{2} - 1)\,(\Omega z_{\rm eq})^{-1/2}\frac{c}{H_0} = \frac{16.0}{\Omega h^2}\,\text{Mpc}.$$

Present comoving horizon length:

$$R_{\rm H} = \frac{2c}{\Omega_m H_0} \qquad \text{(open)}$$

$$R_{\rm H} \simeq \frac{2c}{\Omega_m^{0.4} H_0} \qquad \text{(flat).}$$

Formulae for matter-dominated models

Epoch dependence of H and Ω:

$$H(z) = H_0(1+z)\sqrt{1+\Omega z}$$

$$\Omega(z) = \frac{\Omega(1+z)}{1+\Omega z}.$$

Age of the universe:

$$t = H^{-1}\left[\frac{1}{1-\Omega} + \frac{k\Omega}{2[k(\Omega-1)]^{3/2}}\,C_k^{-1}\left(\frac{2-\Omega}{\Omega}\right)\right].$$

Useful approximation for the age:

$$t \simeq H(z)^{-1}[1 + \Omega(z)^{0.6}/2]^{-1}.$$

Comoving distance in differential form:

$$R_0\,dr = \left(\frac{c}{H_0}\right)\frac{dz}{(1+z)\sqrt{1+\Omega z}}.$$

Comoving distance in integral form:

$$R_0 S_k(r) = \left(\frac{2c}{H_0}\right) \frac{\Omega z + (\Omega - 2)\left(\sqrt{1+\Omega z} - 1\right)}{\Omega^2(1+z)}.$$

Cosine-like function:

$$C_k(r) = \frac{(2-\Omega)(2-\Omega+\Omega z) + 4(\Omega-1)\sqrt{1+\Omega z}}{\Omega^2(1+z)}.$$

Cycloid solution for the scale factor:

$$R = k\frac{GM}{c^2}[1 - C_k(\eta)]$$
$$ct = k\frac{GM}{c^2}[\eta - S_k(\eta)].$$

Characteristic mass in cycloid solution:

$$\frac{GM}{c^2} = \frac{4\pi}{3}\rho R^3 = \frac{c}{H_0}\frac{\Omega[k(\Omega-1)]^{3/2}}{2}.$$

Observations in cosmology

Flux density:

$$S_\lambda = \frac{d|\nu|}{d|\lambda|} S_\nu = \frac{c}{\lambda^2} S_\nu.$$

Brightness temperature:

$$kT_{\rm B} = \frac{c^2}{2\nu^2} I_\nu.$$

Comoving volume element:

$$dV = 4\pi[R_0 S_k(r)]^2 R_0\, dr.$$

Proper length and angular-diameter distance:

$$d\ell = d\psi\, R_0 S_k(r)/(1+z) = D_{\rm A}\, d\psi.$$

Flux density and luminosity distance:

$$S_\nu(\nu_0) = \frac{L_\nu([1+z]\nu_0)}{4\pi R_0^2 S_k^2(r)(1+z)} = \frac{(1+z)\, L_\nu([1+z]\nu_0)}{4\pi\, D_{\rm L}^2}.$$

Absolute magnitude and K-correction:

$$m = M + 5\log_{10}\left(\frac{D_{\rm L}}{10\ \rm pc}\right) + K(z).$$

K-correction for power-law spectrum $S_\nu \propto \nu^{-\alpha}$:

$$K(z) = 2.5(\alpha - 1)\log_{10}(1+z).$$

Surface brightness:

$$I_\nu(\nu_0) = \frac{B_\nu([1+z]\nu_0)}{(1+z)^3} \quad \Rightarrow \quad \frac{I_\nu}{\nu^3} = \text{invariant}.$$

Large-scale structure

Fourier transform convention:

$$F(x) = \left(\frac{L}{2\pi}\right)^n \int F_k(k) \exp(-i\mathbf{k}\cdot\mathbf{x})\, d^n k$$
$$F_k(k) = \left(\frac{1}{L}\right)^n \int F(x) \exp(i\mathbf{k}\cdot\mathbf{x})\, d^n x.$$

Jeans length:

$$\lambda_{\rm J} = c_{\rm S}\sqrt{\frac{\pi}{G\rho}}.$$

Jeans mass; definition:

$$M_{\rm J} \equiv \rho\lambda_{\rm J}^3 = \left(\frac{\gamma\pi k T}{G\mu m_p}\right)^{3/2} \rho^{-1/2}$$

Jeans mass; practical units:

$$M_{\rm J} = 10^{5.3}(T/{\rm K})^{3/2}(n/{\rm m}^{-3})^{-1/2}\, M_\odot.$$

First-order perturbations; comoving units:

$$\dot{\delta} = -\boldsymbol{\nabla}\cdot\mathbf{u}$$
$$\ddot{\delta} + 2\frac{\dot{a}}{a}\dot{\delta} = \delta\left(4\pi G\rho_0 - \frac{c_s^2 k^2}{a^2}\right).$$

Pressure-free perturbation growth ($k = 0$ growing mode):

$$\delta \propto \eta^2.$$

Spherical model:

$$R = k\frac{GM}{c^2}[1 - C_k(\eta)]$$
$$ct = k\frac{GM}{c^2}[\eta - S_k(\eta)].$$

Matter-dominated pressure-free growing mode:

$$\delta \propto \frac{2k[1 - C_k(\eta)]^2 - 3S_k(\eta)[\eta - S_k(\eta)]}{[1 - C_k(\eta)]^2}.$$

Matter-dominated conformal time–redshift relation:

$$\eta = C_k^{-1}\left[1 - \frac{2(\Omega - 1)}{\Omega(1+z)}\right].$$

Growth-suppression factor, for zero radiation density:

$$g(\Omega) \equiv \frac{\delta(z=0,\Omega)}{\delta(z=0,\Omega=1)} \simeq \tfrac{5}{2}\Omega_m\left[\Omega_m^{4/7} - \Omega_v + (1 + \tfrac{1}{2}\Omega_m)(1 + \tfrac{1}{70}\Omega_v)\right]^{-1}.$$

Peculiar velocity and peculiar acceleration:

$$\delta\mathbf{v} = \frac{2g(\Omega)}{3H\Omega}\mathbf{g}.$$

Peculiar velocity and density perturbation in Fourier space:

$$\delta \mathbf{v}_{\mathbf{k}} = -\frac{iHg(\Omega)a}{k}\delta_k \hat{\mathbf{k}}.$$

3D dimensionless power spectrum:

$$\Delta^2(k) \equiv \frac{V}{(2\pi)^3} 4\pi k^3 P(k) = \frac{2}{\pi}k^3 \int_0^\infty \xi(r) \frac{\sin kr}{kr} r^2 \, dr.$$

3D power-law spectrum:

$$\xi(r) = \left(\frac{r}{r_0}\right)^{-\gamma} \quad \Rightarrow \quad \Delta^2(k) = \frac{2}{\pi}(kr_0)^\gamma \, \Gamma(2-\gamma) \sin \frac{(2-\gamma)\pi}{2}.$$

Miscellaneous

The de Vaucouleurs' surface brightness profile:

$$I(r) = I_e \exp[-(7.67(r/r_e)^{1/4} - 1)].$$

Its integrated luminosity:

$$L_{\text{tot}} \simeq 22.7 \, I_e \, r_e^2.$$

Density–radius relation for the singular isothermal sphere:

$$\rho = \frac{\sigma_v^2}{2\pi G r^2}.$$

Equation of hydrostatic equilibrium for collisional gas:

$$M(r) = -\frac{kT(r)r}{\mu m_p G}\left(\frac{d\ln T}{d\ln r} + \frac{d\ln \rho}{d\ln r}\right).$$

The Comptonization parameter:

$$y \equiv \int \sigma_{\text{T}} \, n_e \frac{kT}{m_e c^2} \, d\ell.$$

Sunyaev–Zeldovich effect ($x = h\nu/kT$):

$$\frac{\delta I_\nu}{I_\nu} = \frac{\delta n}{n} = -y \frac{xe^x}{e^x - 1}\left[4 - x \coth\left(\frac{x}{2}\right)\right].$$

The Einstein-ring radius:

$$\theta_{\text{E}} = \left(\frac{4GM}{c^2} \frac{D_{\text{LS}}}{D_{\text{L}} D_{\text{S}}}\right)^{1/2} = \left(\frac{M}{10^{11.09} M_\odot}\right)^{1/2} \left(\frac{D_{\text{L}} D_{\text{S}}/D_{\text{LS}}}{\text{Gpc}}\right)^{-1/2} \quad \text{arcsec}$$

The redshift of matter–radiation equality, assuming three species of massless neutrinos:

$$1 + z_{\text{eq}} = 23\,900 \, \Omega h^2 \, (T/2.73\,\text{K})^{-4}.$$

Synchrotron emissivity:

$$\epsilon_\nu = 2.14 \times 10^{-30}(x-1) \, N_0 \, a(\alpha) \left(\frac{B}{100\,\text{nT}}\right)^{1+\alpha} \left(\frac{\nu}{4\,\text{kHz}}\right)^{-\alpha} \quad \text{W Hz}^{-1}.$$

Bremsstrahlung emissivity:

$$\epsilon_\nu/\text{Wm}^{-3}\text{Hz}^{-1} = 6.8 \times 10^{-32} \, T^{-1/2} \, n_e^2 \, e^{-h\nu/kT} \, G(\nu, T).$$

Gaunt-factor approximation:

$$G \simeq 1 + \log_{10}\left(kT/h\nu\right).$$

Eddington luminosity:

$$L_{\rm Edd} = 4\pi GM \left(\frac{m_e c}{\sigma_{\rm T}}\right) = 10^{39.1}\left(\frac{M}{10^8 M_\odot}\right)\ {\rm W}.$$

Press–Schechter mass function:

$$M^2 \frac{dn}{dM} = \frac{2\delta_c}{\sqrt{2\pi}\,\sigma}\,\rho_0 \left|\frac{d\ln\sigma}{d\ln M}\right| \exp\left(-\frac{\delta_c^2}{2\sigma^2}\right).$$

Virial temperature for objects that collapse at redshift z_c, with collapse factor f_c:

$$\frac{T_{\rm virial}}{\rm K} = 10^{5.1}\left(\frac{M}{10^{12}M_\odot}\right)^{2/3} (f_c\Omega h^2)^{1/3}\,(1+z_c).$$

CMB correlation function, where $W_\ell = \exp[-\ell^2\sigma^2/2]$ for a Gaussian beam of FWHM 2.35σ:

$$C(\theta) = \frac{1}{4\pi}\sum_\ell (2\ell+1)\,W_\ell^2\,C_\ell\,P_\ell(\cos\theta).$$

Conversion between B1950 celestial coordinates and galactic coordinates:

$$\begin{aligned}
\cos b\cos(\ell - 33^\circ) &= \cos\delta\cos(\alpha - 282.25^\circ)\\
\cos b\sin(\ell - 33^\circ) &= \cos\delta\sin(\alpha - 282.25^\circ)\cos 62.6^\circ + \sin\delta\sin 62.6^\circ\\
\sin b &= \sin\delta\cos 62.6^\circ - \cos\delta\sin(\alpha - 282.25^\circ)\sin 62.6^\circ\\
\cos\delta\sin(\alpha - 282.25^\circ) &= \cos b\sin(\ell - 33^\circ)\cos 62.6^\circ - \sin b\sin 62.6^\circ\\
\sin\delta &= \cos b\sin(\ell - 33^\circ)\sin 62.6^\circ + \sin b\cos 62.6^\circ.
\end{aligned}$$

Index